MATHEMATICAL ROUTES TO THE PHYSICAL WORLD

MATHEMATICAL ROUTES TO THE PHYSICAL WORLD

Shijie Yang

清華大學出版社
Tsinghua University Press

World Scientific

Published by

World Scientific Publishing Co. Pte. Ltd.
5 Toh Tuck Link, Singapore 596224
USA office: 27 Warren Street, Suite 401-402, Hackensack, NJ 07601
UK office: 57 Shelton Street, Covent Garden, London WC2H 9HE

Library of Congress Control Number: 2025015485

British Library Cataloguing-in-Publication Data
A catalogue record for this book is available from the British Library.

MATHEMATICAL ROUTES TO THE PHYSICAL WORLD

ISBN 978-981-98-1355-1 (hardcover)
ISBN 978-981-98-1356-8 (ebook for institutions)
ISBN 978-981-98-1357-5 (ebook for individuals)

For any available supplementary material, please visit
https://www.worldscientific.com/worldscibooks/10.1142/14325t=suppl

Desk Editor: Nambirajan Karuppiah

Typeset by Stallion Press
Email: enquiries@stallionpress.com

About the Author

Yang Shijie is currently a professor of physics at Beijing Normal University in China and works in the field of theoretical condensed matter physics. From 1983 to 1987, he studied at Peking University. He acquired the master's degree at Fudan University in 1990 and the PhD at the Institute of Theoretical Physics, Chinese Academy of Sciences in 2001.

Contents

Chapter 1

Complex Functions

1.1 Complex Numbers

When solving quadratic equations with real coefficients, it is often encountered that the square root of a negative real number arises. Initially, people simply regarded this as meaningless and concluded that the equation had no solution. However, it was later discovered through practice that the square root of a negative real number is a common and unavoidable issue. This led to the introduction of the imaginary unit, denoted as $\mathrm{i} \equiv \sqrt{-1}$, which satisfies $\mathrm{i}^2 = -1$. Consequently, real numbers are extended to complex numbers which are composed of a pair of real numbers:

$$z = (x, y) \overset{\text{def}}{=} x + \mathrm{i}y(x, y) \in \mathbb{R}$$

Here, x and y are respectively called the real part and the imaginary part of the complex number, denoted as $x = \operatorname{Re} z$, $y = \operatorname{Im} z$. If two complex numbers are equal, i.e., $z_1 = z_2$, it requires

$$\operatorname{Re} z_1 = \operatorname{Re} z_2, \quad \operatorname{Im} z_1 = \operatorname{Im} z_2$$

The set of all ordered pairs of real numbers forms the complex number set, which is denoted by $\mathbb{C}$.

1.1.1 *Operational rules*

The complex conjugate of a complex number z is defined as $\bar{z} \overset{\text{def}}{=} x - \mathrm{i}y$, and sometimes, z^* is also used to represent the complex conjugate.

The operational rules for complex numbers are defined as follows.

- *Addition and subtraction*

$$z_1 \pm z_2 = (x_1 \pm x_2) + \mathrm{i}(y_1 \pm y_2)$$

That is, the real parts and the imaginary parts are added and subtracted separately. The addition operation of complex numbers satisfies commutativity and associativity.

- *Multiplication*

$$z_1 \cdot z_2 = (x_1 x_2 - y_1 y_2) + \mathrm{i}(x_1 y_2 + x_2 y_1)$$

Multiplication is performed according to the distributive law of real numbers, and the result is still a complex number. It can be easily verified that the multiplication of complex numbers satisfies commutativity, associativity, and distributivity. For complex conjugates, it holds that $\overline{z_1 \cdot z_2} = \overline{z_1} \cdot \overline{z_2}$.

- *Division*

$$\frac{z_1}{z_2} = \frac{x_1 x_2 + y_1 y_2}{x_2^2 + y_2^2} + \mathrm{i}\frac{x_2 y_1 - x_1 y_2}{x_2^2 + y_2^2}$$

where $z_2 \neq 0$. The result of complex division is still a complex number.

For a set consisting of a certain type of numbers, if the arithmetic operations of addition, subtraction, multiplication, and division among its elements yield numbers that still belong to the same set, then the set is called a field. For example, the set of all rational numbers $\mathbb{Q}$ forms the rational number field, and the set of all real numbers $\mathbb{R}$ forms the real number field.

By performing addition, subtraction, multiplication, and division operations on complex numbers, we still obtain a complex number. Therefore, the set of complex numbers $\mathbb{C}$ forms the complex number field. Unlike real numbers, however, complex numbers cannot be compared in terms of size, so the complex number field is not an ordered field.

Reflection: Why is it improper if the following multiplication definitions of the complex number are adopted?

(a) $z_1 \cdot z_2 = x_1x_2 + \mathrm{i}y_1y_2$; (b) $z_1 \cdot z_2 = (x_1x_2 + y_1y_2) + \mathrm{i}(x_1y_2 + x_2y_1)$.

1.1.2 *Complex plane*

A two-dimensional Cartesian coordinate system represents an ordered pair of real numbers (x, y). Any complex number can be corresponded to a unique point on the plane, with the horizontal and vertical axes respectively denoting the real and the imaginary parts of the complex number. This plane is called the complex plane, as shown in Figure 1.1. By adopting polar coordinates, we can also write

$$z = x + \mathrm{i}y = r(\cos\theta + \mathrm{i}\sin\theta)$$

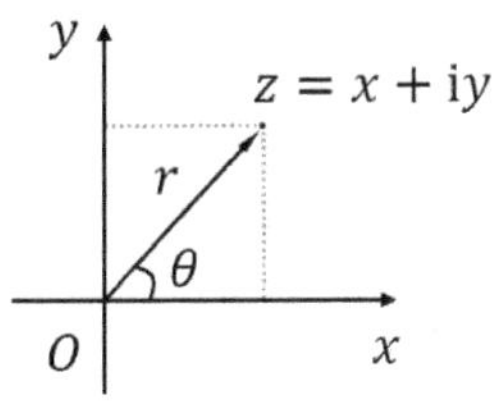

Figure 1.1.

The modulus of a complex number is defined as

$$|z| \overset{\text{def}}{=} (z \cdot \bar{z})^{1/2} = r = \sqrt{x^2 + y^2}$$

and the argument of a complex number is

$$\arg z \overset{\text{def}}{=} \theta + 2k\pi \ (k \in \mathbb{Z})$$

where $\mathbb{Z}$ represents the set of integers. The argument with $k = 0$ is usually referred to as the principal argument.

As shown in Figure 1.2, the addition and subtraction of two complex numbers satisfy the parallelogram rule and the triangle rule, respectively. Using the product-to-sum identities of trigonometric functions, complex multiplication can be expressed as

$$z_1 \cdot z_2 = r_1r_2[\cos(\theta_1 + \theta_2) + \mathrm{i}\sin(\theta_1 + \theta_2)] \tag{1.1.1}$$

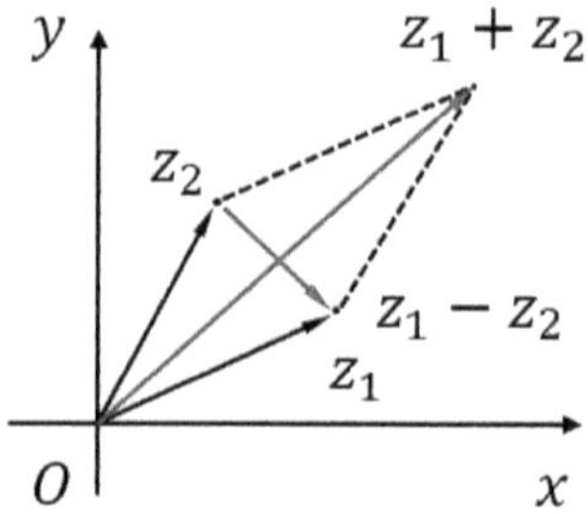

Figure 1.2.

Equation (1.1.1) shows that complex multiplication exhibits a mapping relationship:

$$f(\theta_1) \cdot f(\theta_2) \longmapsto f(\theta_1 + \theta_2)$$

This implies an exponential function relationship between complex numbers and arguments. In fact, it will be shown that a complex number z can be represented as

$$z = r(\cos\theta + \mathrm{i}\sin\theta) = r\mathrm{e}^{\mathrm{i}\theta} \tag{1.1.2}$$

This important relationship is known as Euler's formula, which connects exponential functions with trigonometric functions in the complex number field. From this, it follows that the product (division) of two complex numbers is equal to the product (division) of their moduli and the sum (difference) of their arguments:

$$z_1 \cdot z_2 = r_1 r_2 \mathrm{e}^{\mathrm{i}(\theta_1+\theta_2)}, \quad \frac{z_1}{z_2} = \frac{r_1}{r_2}\mathrm{e}^{\mathrm{i}(\theta_1-\theta_2)}$$

Taking the modulus $r = 1$, Euler's formula leads to De Moivre's formula

$$(\cos\theta + \mathrm{i}\sin\theta)^n = \cos n\theta + \mathrm{i}\sin n\theta \tag{1.1.3}$$

and the legendary most perfect mathematical formula follows immediately.

$$\mathrm{e}^{\mathrm{i}\pi} + 1 = 0$$

Example 1.1. Express the complex numbers in Cartesian form:

$$\text{(a) } (1+\mathrm{i})^{100}; \quad \text{(b) } \sqrt[4]{1-\mathrm{i}}$$

Solution:

$$\text{(a) } (1+\mathrm{i})^{100} = (\sqrt{2}\mathrm{e}^{\frac{\pi \mathrm{i}}{4}})^{100} = 2^{50}\mathrm{e}^{25\pi \mathrm{i}} = -2^{50}$$

$$\text{(b) } \sqrt[4]{1-\mathrm{i}} = (\sqrt{2}\mathrm{e}^{-\frac{\pi \mathrm{i}}{4}+2k\pi \mathrm{i}})^{\frac{1}{4}} = \sqrt[8]{2}\mathrm{e}^{-\frac{\pi \mathrm{i}}{16}+\frac{k\pi \mathrm{i}}{2}}$$

$$= \sqrt[8]{2}\mathrm{e}^{-\frac{\pi \mathrm{i}}{16}}, \sqrt[8]{2}\mathrm{e}^{\frac{7\pi \mathrm{i}}{16}}, \sqrt[8]{2}\mathrm{e}^{\frac{15\pi \mathrm{i}}{16}}, \sqrt[8]{2}\mathrm{e}^{\frac{23\pi \mathrm{i}}{16}}$$

It should be emphasized that taking roots may result in multiple distinct complex values. In Example 1.1, the four complex values form the vertices of a regular quadrilateral inscribed in a circle with a radius of $\sqrt[8]{2}$, and all of them raised to the fourth power equal $1-\mathrm{i}$.

Example 1.2. Express the complex numbers in Cartesian form:

$$\text{(a) } \mathrm{i}^{\mathrm{i}}; \quad \text{(b) } \mathrm{i}^{-\mathrm{i}}$$

Solution:

$$\text{(a) } \mathrm{i}^{\mathrm{i}} = \mathrm{e}^{(\frac{\pi \mathrm{i}}{2}+2k\pi \mathrm{i})\mathrm{i}} = \mathrm{e}^{-\frac{\pi}{2}}\mathrm{e}^{-2k\pi} \quad (k \in \mathbb{Z})$$

$$\text{(b) } \mathrm{i}^{-\mathrm{i}} = \mathrm{e}^{-(\frac{\pi \mathrm{i}}{2}+2k\pi \mathrm{i})\mathrm{i}} = \mathrm{e}^{\frac{\pi}{2}}\mathrm{e}^{2k\pi} \quad (k \in \mathbb{Z})$$

Note: When multiplying two multivalued complex numbers, such as

$$(\mathrm{i}^{\mathrm{i}}) \cdot (\mathrm{i}^{-\mathrm{i}}) = \mathrm{e}^{2n\pi} \quad (n \in Z)$$

it can be observed that

$$(\mathrm{i}^{\mathrm{i}}) \cdot (\mathrm{i}^{-\mathrm{i}}) \neq \mathrm{i}^0 = 1$$

This means that the usual multiplication operation does not apply to complex number operations with multiple values. The multivalued nature of complex numbers constitutes the rich content of complex analysis, which will be further discussed in the subsequent text.

Example 1.3. Represent an ellipse equation with foci at z_1 and z_2 using complex numbers.

Solution: An ellipse is a set of points in a plane where the sum of the distances to two fixed points is constant. It is convenient to represent it using complex numbers:

$$|z-z_1| + |z-z_2| = 2a$$

where $2a$ is the major axis of the ellipse.

1.1.3 *Stereographic projection*

Stereographic projection is a mapping that establishes a one-to-one correspondence between each point in the extended complex plane $\bar{\mathbb{C}}$ (the closure of the complex plane $\mathbb{C}$, which includes the point at infinity) and a point on the unit sphere $\mathbb{S}$, known as the Riemann sphere. The point at infinity corresponds to the north pole of the sphere. This mapping is illustrated in Figure 1.3.

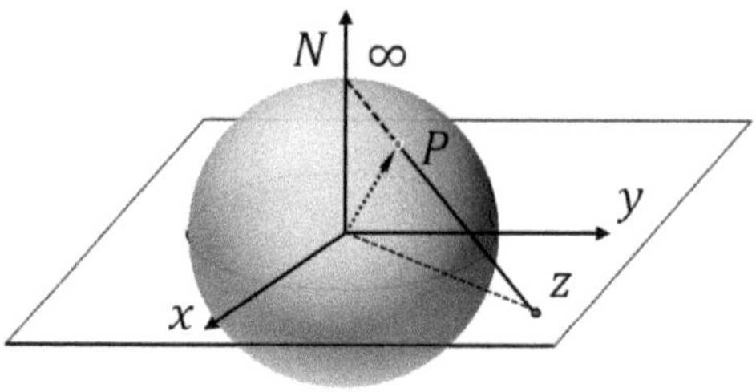

Figure 1.3.

Let P be a point on the Riemann sphere $\mathbb{S}$ with Cartesian coordinates (X, Y, Z), satisfying the equation $X^2 + Y^2 + Z^2 = 1$. The projection point $z(x, y)$ on the closed complex plane $\bar{\mathbb{C}}$ has the corresponding relationship:

$$X = \frac{2x}{1 + |z|^2}, \quad Y = \frac{2y}{1 + |z|^2}, \quad Z = \frac{|z|^2 - 1}{1 + |z|^2} \tag{1.1.4}$$

1.1.4 *Fundamental theorem of algebra*

The Algebraic Fundamental Theorem states that any polynomial equation has at least one complex root.

If a polynomial equation $p(z) = 0$ has a root z_1, then it can be factored as $(z - z_1)$ multiplied by other factors. If $p(z)$ is a polynomial of degree n, then it can be subsequently decomposed into n such factors. Therefore, we conclude that any nth-degree polynomial equation has exactly n complex roots.

Several proofs of the Algebraic Fundamental Theorem will be presented in subsequent chapters.

ENDNOTE

Italian mathematician Cardano, in the 16th century, was perhaps one of the earliest proponents of imaginary numbers as square roots of negative numbers. However, he did not attribute any practical significance to imaginary numbers and considered them merely as "imaginary quantities". For quadratic algebraic equations, there is no requirement for the existence of solutions, and therefore, imaginary numbers were not necessary. However, for cubic equations, the appearance of imaginary numbers became inevitable. For example, in the equation $x^3 + px + q = 0$, the solution is known as Cardano's formula:

$$x = \sqrt[3]{-\frac{q}{2} + \sqrt{\left(\frac{q}{2}\right)^2 + \left(\frac{p}{3}\right)^3}} + \sqrt[3]{-\frac{q}{2} - \sqrt{\left(\frac{q}{2}\right)^2 + \left(\frac{p}{3}\right)^3}}$$

When $(q/2)^2 + (p/3)^3 < 0$, the formula involves taking the square root of a negative number. However, it is not valid to disregard it simply because the equation has no real solutions. In fact, the expression does represent a real root of the equation, even though its representation involves imaginary numbers. For example, according to Cardano's formula, one of the roots of the equation $x^3 - 15x - 4 = 0$ is

$$x = \sqrt[3]{2 + 11\sqrt{-1}} + \sqrt[3]{2 - 11\sqrt{-1}}$$

It may appear to involve imaginary numbers. In 1572, Bombelli speculated that the two parts of the formula might have the form of $2 + n\sqrt{-1}$ and $2 - n\sqrt{-1}$, respectively. Taking their cube roots and utilizing $(\sqrt{-1})^2 = -1$, it was indeed found that

$$\sqrt[3]{2 + 11\sqrt{-1}} = 2 + \sqrt{-1}, \quad \sqrt[3]{2 - 11\sqrt{-1}} = 2 - \sqrt{-1}$$

Therefore, Cardano's formula represents the real solution $x=4$, even though both parts involve "imaginary quantities". This fact compelled people to seriously consider the necessity and objectivity of imaginary numbers.

Cardano also posed an intellectual challenge: can 10 be divided into two parts such that their product is equal to 40? His answer was that if 10 is written as $5+\sqrt{-15}$ and $5-\sqrt{-15}$, this goal can be achieved.

The extensive application of complex numbers began mainly with Euler. However, in his algebraic work in 1770, he still wrote the following awkward remark: "All mathematical expressions of the form $\sqrt{-1}$ are impossible, imaginary quantities; they are neither something nor more than something, nor less than something". The geometric representation of complex numbers first appeared in a paper by Danish surveyor Wessel in 1797, where he provided an intuitive graphical interpretation of complex number operations. Gauss was the first to systematically expound complex numbers and their rules of operation. He also provided the first complete proof of the Fundamental Theorem of Algebra.

Given the tremendous success and profound meaning of complex numbers, people began to seek more general types of numbers. As early as the 3rd century AD, Diophantus pointed out in his book *Arithmetica*: The number 65 can be expressed as the sum of two squares in two different ways, namely $65 = 7^2 + 4^2 = 8^2 + 1^2$. This is because $65 = 13 \times 5$, and both $13 = 2^2 + 3^2$ and $5 = 1^2 + 2^2$. It is more commonly expressed as a theorem, which states that the product of the sum of two square numbers can be expressed as the sum of two square numbers:

$$(a^2 + b^2)(c^2 + d^2) = (ac - bd)^2 + (bc + ad)^2$$

Here $a, b, c, d \in \mathbb{N}$. If we represent the number pair (a, b) as $a+ib$, then the above equation is precisely the multiplication rule for complex numbers. In other words, by taking the modulus on both sides of the following equation,

$$(a+\mathrm{i}b)(c+\mathrm{i}d)=(ac-bd)+\mathrm{i}(ac+bd)$$

Hamilton recognized the importance of ordered real number pairs in multiplication, leading him to explore larger arrays known as ternions and quaternions. However, he faced significant obstacles when dealing with the multiplication of ternions. Regardless of how he defined the multiplication, he couldn't perform division operations. A more profound reason for this difficulty was the absence of a "product theorem for the sum of three square numbers". For example,

$$3=1^2+1^2+1^2, \quad 21=1^2+2^2+4^2$$

but 3×21 cannot be expressed as the sum of three square numbers.

On the other hand, Bachet and Lagrange proved that the product of the sum of four square numbers can be expressed as the sum of four square numbers:

$$\begin{aligned}&(a^2+b^2+c^2+d^2)(\alpha^2+\beta^2+\gamma^2+\delta^2)\\&\quad=(a\alpha-b\beta-c\gamma-d\delta)^2+(a\beta+b\alpha+c\delta-d\gamma)^2\\&\qquad+(a\gamma-b\delta+c\alpha+d\beta)^2+(a\delta+b\gamma-c\beta+d\alpha)^2\end{aligned}$$

However, the significance of this result was not fully appreciated. When Hamilton shifted his focus from ternion arrays to quaternion arrays, everything became clear. He introduced three imaginary symbols (i,j,k) that possess the following properties:

$$\mathrm{i}^2=\mathrm{j}^2=\mathrm{k}^2=-1$$

and the non-commutative cyclic relationships:

$$\begin{aligned}\mathrm{ij}&=-\mathrm{ji}=\mathrm{k}\\\mathrm{jk}&=-\mathrm{kj}=\mathrm{i}\\\mathrm{ki}&=-\mathrm{ik}=\mathrm{j}\end{aligned}$$

By these symbols, the quaternion is expressed as

$$(a, b, c, d) = a + b\mathrm{i} + c\mathrm{j} + d\mathrm{k}$$

Its general multiplication is defined as

$$\begin{aligned}
&(a + b\mathrm{i} + c\mathrm{j} + d\mathrm{k})(\alpha + \beta\mathrm{i} + \gamma\mathrm{j} + \delta\mathrm{k}) \\
&\quad = (a\alpha - b\beta - c\gamma - d\delta) + (a\beta + b\alpha + c\delta - d\gamma)\mathrm{i} \\
&\qquad + (a\gamma - b\delta + c\alpha + d\beta)\mathrm{j} + (a\delta + b\gamma - c\beta + d\alpha)\mathrm{k}
\end{aligned}$$

The theorem of the sum of four square numbers implies the existence of a multiplication rule for the modulus of quaternions, which allows for the definition of quaternion multiplication. Soon, John Graves and later Arthur Cayley further discovered octonions, sometimes referred to as Cayley numbers or Cayley–Graves numbers:

$$\alpha + \beta\mathrm{i} + \gamma\mathrm{j} + \delta\mathrm{k} + \varepsilon\mathrm{l} + \zeta\mathrm{m} + \eta\mathrm{n} + \theta\mathrm{o}$$

The base vectors for octonions are respectively called $(1, \mathrm{i}, \mathrm{j}, \mathrm{k}, \mathrm{l}, \mathrm{m}, \mathrm{n}, \mathrm{o})$ corresponding to the "theorem of the sum of eight square numbers". Octonions appear quite peculiar as they not only fail to satisfy the commutative property of multiplication but also lack the associative property. For example,

$$(\mathrm{m} \times \mathrm{j}) \times \mathrm{i} \neq \mathrm{m} \times (\mathrm{j} \times \mathrm{i})$$

The specific relationships can be represented by Figure 1.4, where the arrow direction determines the order of multiplication: $\mathrm{j} \times \mathrm{l} = \mathrm{n}, \mathrm{i} \times \mathrm{j} = \mathrm{k}$, and so on. This multiplication related to the order of combination is similar to the ploy of braiding. In effect, it was only at this point that people realized that associativity is a fundamental property of multiplication, a term coined by Hamilton himself. Quaternions and octonions, as well as multidimensional arrays constructed from them, are collectively referred to as hypercomplex numbers.

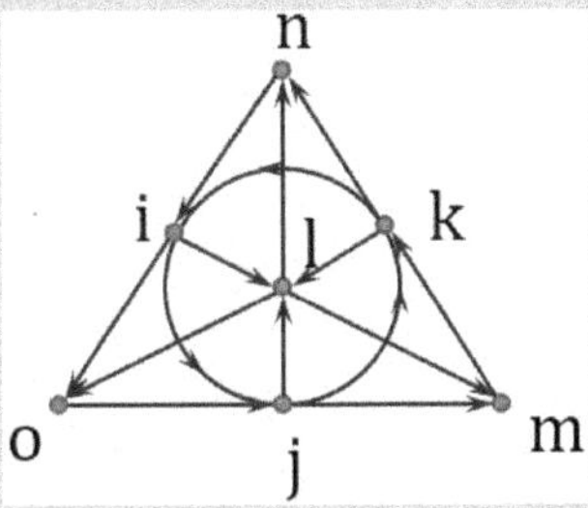

Figure 1.4.

Although Hamilton confidently proclaimed that quaternions were the most perfect numbers in nature, quaternions did not achieve the same level of success as complex numbers. It was initially expected that quaternions could be used to describe common three-dimensional vector fields in physics, but the attempt was not very successful, despite their significant contribution to Maxwell's theory of electromagnetic fields. Vector analysis, established by Gibbs and Heaviside, essentially disassembled the complete and unified quaternions into several independent components. Although vector analysis became widely popular in physics and yielded fruitful results (with Heaviside being the first to present the modern form of Maxwell's equations), it left many mathematicians frustrated in their pursuit of perfection. In fact, quaternion multiplication represents a certain type of spatial rotation. When Cartan identified quaternions with spinors, they precisely described the quantization of spin for microscopic particles.

So, is there no end to the extension of number systems? Not quite! Hurwitz proved that apart from $n = 2, 4, 8$, there are no other perfect square numbers for n, which means that there can be no further extension of fundamental number systems beyond octonions in higher dimensions.

Each time the dimension doubles, the extended number system sacrifices some of its properties: From one dimension to two dimensions, it sacrifices orderliness. From two dimensions to four dimensions, it sacrifices commutativity. From four dimensions to eight dimensions, it sacrifices associativity. If we forcefully extend the system from eight dimensions to 16 dimensions, we would have to sacrifice division. However, since the dam of ideas has already burst, mathematicians have continued to expand their horizons. In a more general sense, they have constructed closed algebraic structures with only addition and multiplication, known as "rings". By discarding addition, they form "groups", or by preserving only addition, they form "monoids"...

Complex numbers possess functions and properties that cannot be replaced by real numbers. Although they are algebraic masterpieces, there is no reason to expect them to be directly related to the operation of the real world. However, as the 20th century unfolded, astonishing news arrived from the field of new physics: the microscopic world is entirely governed by the complex number system! For mathematicians who had been working in isolation, this was not only highly encouraging but also secretly gratifying. By the way, in recent years, there have been some active signs of octonions in the exploration of the fundamental physics of interactions. Could they be the key to unlocking a new world?

Exercises

[1] Write multiple values of complex numbers:

$$\text{(a) } (-1)^{\mathrm{i}};\quad \text{(b) } \sqrt[\mathrm{i}]{2+\mathrm{i}};\quad \text{(c) } (1+\mathrm{i}\sqrt{3})^{1-\mathrm{i}};\quad \text{(d) } \sqrt{\sqrt[\mathrm{i}]{3}}$$

[2] Let $\omega_0, \omega_1, \omega_2, \ldots, \omega_{n-1} (n > 1)$ be the n roots of the equation $z^n - 1 = 0$. Prove that

$$\text{(a) } \omega_0 + \omega_1 + \omega_2 + \cdots + \omega_{n-1} = 0;$$

$$\text{(b) } \omega_0^k + \omega_1^k + \omega_2^k + \cdots + \omega_{n-1}^k = \begin{cases} 0 & (1 \le k \le n-1); \\ n & (k = n); \end{cases}$$

$$\text{(c) } \omega_0\omega_1\omega_2\cdots\omega_{n-1} = (-1)^{n-1}$$

[3] Prove that

(a) $\cos\varphi + \cos 2\varphi + \cos 3\varphi + \cdots + \cos n\varphi = \dfrac{\sin\frac{n\varphi}{2}\cos\frac{(n+1)\varphi}{2}}{\sin\frac{\varphi}{2}}$;

(b) $\sin\dfrac{\pi}{n}\sin\dfrac{2\pi}{n}\sin\dfrac{3\pi}{n}\cdots\sin\dfrac{(n-1)\pi}{n} = \dfrac{n}{2^{n-1}}$

Hint: Express the roots of the equation $z^n - 1 = 0$ in the form of product factors.

[4] Solve the equation:

$$(z+1)^5 = z^5$$

Answer: $z = -\frac{1}{2}(1 + \mathrm{i}\cot\frac{k\pi}{5})\ (k = 0, 1, 2, 3, 4)$

[5] Prove that the sum of the interior angles of a triangle is equal to π.

[6] Let $A, B \in \mathbb{R}$ and $a \in \mathbb{C}$. We need to prove that when $|a|^2 > AB$, the equation $Az\bar{z} + \bar{a}z + a\bar{z} + B = 0$ represents either a circle or a straight line, and find the position of the center and the radius.

Hint: The equation is rewritten as

$$\left|z + \frac{a}{A}\right|^2 = \frac{|a|^2}{A^2} - \frac{B}{A}$$

[7] Prove the spherical projection relation:

$$X = \frac{2x}{1+|z|^2}, \quad Y = \frac{2y}{1+|z|^2}, \quad Z = \frac{|z|^2 - 1}{1+|z|^2}$$

Hint: The parametric equation of the line passing through points N and P can be expressed as $Q(t) = N + t(P - N)$.

[8] Explain the reason of

$$6 = \sqrt[3]{18 + 26\sqrt{-1}} + \sqrt[3]{18 - 26\sqrt{-1}}$$

[9] Given the arithmetic equation

$$m^2 + n^2 = (7^2 + 9^2)(13^2 + 20^2) \quad (m, n \in \mathbb{N})$$

we are asked to find the integer solutions (m, n)?

Answer: $(89, 257)$

[10] Given the arithmetic equation

$$m^2 + n^2 + k^2 + l^2 = 3 \times (1^2 + 2^2 + 3^2 + 4^2) \quad (m, n, k, l \in \mathbb{N})$$

we are asked to find the integer solutions (m, n, k, l)?
Answer: $(2, 9, 1, 2)$

[11] Prove that the three solutions of the cubic equation $x^3 + px + q = 0$ are

$$x_1 = A + B, \quad x_2 = \omega A + \omega^2 B, \quad x_3 = \omega^2 A + \omega B$$

where

$$A = \sqrt[3]{-\frac{q}{2} + \sqrt{\left(\frac{q}{2}\right)^2 + \left(\frac{p}{3}\right)^3}}, \quad B = \sqrt[3]{-\frac{q}{2} - \sqrt{\left(\frac{q}{2}\right)^2 + \left(\frac{p}{3}\right)^3}}$$

$$\omega = \frac{-1 + \mathrm{i}\sqrt{3}}{2}, \quad \omega^3 = 1$$

Hint: Set $x = z - \frac{p}{3z}$.

1.2 Functions of One Complex Variable

1.2.1 *Mappings and regions*

Let E be a set of complex numbers $z = x + \mathrm{i}y$. If every element z in E is mapped to one or more elements w in another set of complex numbers B, then w is called a function of the complex variable z, denoted as $w = f(z)$, $z \in E$. If each z corresponds to a unique w in B, then the function $f(z)$ is called single-valued. If a z corresponds to multiple w values, then $f(z)$ is called a multivalued function, as shown in Figure 1.5, where $B = B_1 \bigcup B_2$.

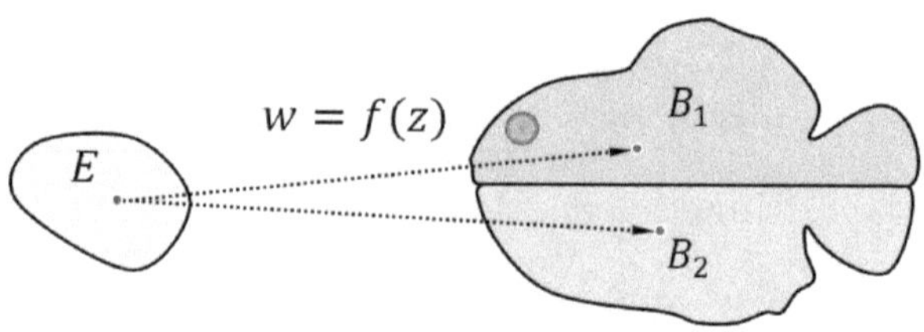

Figure 1.5.

Let us introduce some basic concepts as follows:

(1) **Neighborhood:** A set of points in the complex plane, consisting of all points within a circle with center z_0 and arbitrarily small positive radius ε, is called a neighborhood of z_0.
(2) **Interior point:** If both z_0 and the points within its neighborhood belong to the set E, then z_0 is called an interior point of E.
(3) **Exterior point:** If neither z_0 nor the points within its neighborhood belong to the set E, then z_0 is called an exterior point of E.
(4) **Open set:** If every point within E is an interior point, then E is called an open set.
(5) **Boundary point:** If within the neighborhood of z_0, there are points that belong to E as well as points that do not belong to E, then z_0 is called a boundary point of the set E. The collection of all boundary points forms the boundary of the set E.
(6) **Connectedness:** If any two points in the set E can be connected by a continuous curve that lies entirely within E, then E is called connected. If such a curve cannot be found, then E is called disconnected (Figure 1.6(a)).

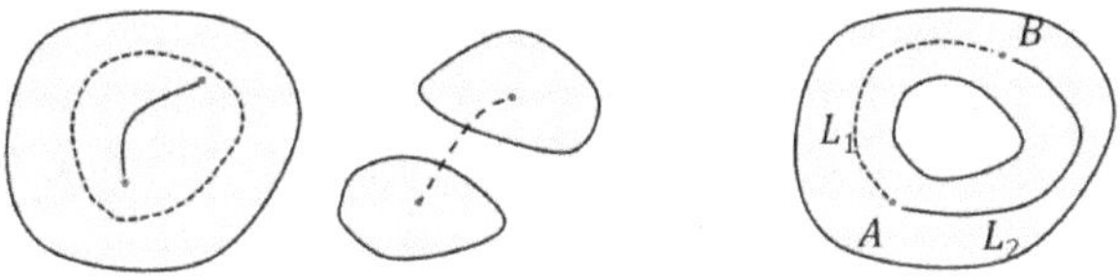

(a) (b)

Figure 1.6.

(7) **Region:** If the set B consists entirely of interior points and is mutually connected, then it is called a region.
(8) **Closed region:** The set formed by the region B and its boundary is called a closed region, denoted as $\bar{B}$.
(9) **Simply connected region:** If any closed curve within the region can be continuously shrunk to a point, as shown in Figure 1.6(a), then the region is called simply connected. In a simply connected region, all paths connecting any two points can be continuously deformed into each other, and they are topologically equivalent, known as path homotopy.

(10) **Multiply connected region:** Two points within the region can be connected by two or more topologically distinct continuous curves belonging to the same set, as shown in Figure 1.6(b). L_1 cannot be continuously deformed into L_2, or in other words, the closed loop formed by L_1 and L_2 in the figure cannot be continuously shrunk to a point.

Using Chinese characters as examples, Figure 1.7(a) is simply connected, Figure 1.7(b) is disconnected, and Figures 1.7(c)–1.7(e) are multiply connected. Both Figures 1.7(c) and 1.7(d) have two "holes" or genus, and they are topologically equivalent, implying that one graph can be continuously transformed (without cutting) into the other. In mathematics, they are called homeomorphic to each other.

(a) (b) (c) (d) (e)

Figure 1.7.

1.2.2 *Elementary functions of complex variables*

Using the complex number $z = x + \mathrm{i}y$ as the independent variable, the following common elementary functions of complex variables are defined:

(1) Exponential function

$$\mathrm{e}^z \stackrel{\text{def}}{=} \lim_{n\to\infty}\left(1+\frac{z}{n}\right)^n \tag{1.2.1}$$

For arbitrary $z \in \mathbb{C}$, we have

$$\left|\left(1+\frac{z}{n}\right)^n\right| = \left|1+\frac{z}{n}\right|^n = \left(1+\frac{2x}{n}+\frac{x^2+y^2}{n^2}\right)^{n/2}$$

and

$$\arg\left(1+\frac{z}{n}\right)^n = n\arg\left(1+\frac{z}{n}\right) = n\arctan\frac{y/n}{1+x/n}$$

Hence,

$$\lim_{n\to\infty}\left|\left(1+\frac{z}{n}\right)^n\right| = \mathrm{e}^x$$

$$\lim_{n\to\infty}\arg\left(1+\frac{z}{n}\right)^n = y$$

The magnitude and argument of the exponential function are $|\mathrm{e}^z| = \mathrm{e}^x$ and $\arg\mathrm{e}^z = y$, respectively. This leads to Euler's formula:

$$\mathrm{e}^z = \mathrm{e}^{x+\mathrm{i}y} = \mathrm{e}^x(\cos y + \mathrm{i}\sin y)$$

From Euler's formula, it can be deduced that the exponential function is periodic along the imaginary axis: $\mathrm{e}^{z+2\pi\mathrm{i}} = \mathrm{e}^z$. It can also be easily proven that the exponential function satisfies the relationship:

$$\mathrm{e}^{z_1+z_2} = \mathrm{e}^{z_1}\cdot\mathrm{e}^{z_2}$$

(2) Trigonometric functions

$$\sin z \overset{\text{def}}{=} \frac{\mathrm{e}^{\mathrm{i}z}-\mathrm{e}^{-\mathrm{i}z}}{2\mathrm{i}}, \qquad \cos z \overset{\text{def}}{=} \frac{\mathrm{e}^{\mathrm{i}z}+\mathrm{e}^{-\mathrm{i}z}}{2} \tag{1.2.2}$$

It can be verified that complex trigonometric functions have properties similar to real trigonometric functions, satisfying

$$\sin^2 z + \cos^2 z = 1$$

$$\sin(z_1 \pm z_2) = \sin z_1 \cos z_2 \pm \cos z_1 \sin z_2$$

$$\cos(z_1 \pm z_2) = \cos z_1 \cos z_2 \mp \sin z_1 \sin z_2$$

Similarly, complex tangent and cotangent functions can also be defined.

We mention that only in the complex domain, trigonometric functions are related to exponential functions.

Example 1.4. Set $z = x + \mathrm{i}y$, we need to prove

(a) $|\sin z|^2 = \sin^2 x + \sinh^2 y$; (b) $|\cos z|^2 = \cos^2 x + \sinh^2 y$

Proof. (a) According to the definition, we have

$$|\sin z|^2 = \frac{1}{4}|\mathrm{e}^{\mathrm{i}z} - \mathrm{e}^{-\mathrm{i}z}|^2 = \frac{1}{4}(\mathrm{e}^{\mathrm{i}x-y} - \mathrm{e}^{-\mathrm{i}x+y})(\mathrm{e}^{-\mathrm{i}x-y} - \mathrm{e}^{\mathrm{i}x+y})$$
$$= \frac{1}{4}(\mathrm{e}^{-2y} + \mathrm{e}^{2y} - \mathrm{e}^{-2\mathrm{i}x} - \mathrm{e}^{2\mathrm{i}x}) = \sin^2 x + \sinh^2 y$$

The proof for (b) is similar. It can be seen that $|\sin z|$ and $|\cos z|$ can be greater than 1, which is different from real functions. □

(3) Hyperbolic functions

$$\sinh z \overset{\text{def}}{=} \frac{\mathrm{e}^z - \mathrm{e}^{-z}}{2}, \quad \cosh z \overset{\text{def}}{=} \frac{\mathrm{e}^z + \mathrm{e}^{-z}}{2}$$

Hyperbolic functions are also periodic along the imaginary axis and have the following relationships:

$$\sinh z = -\mathrm{i}\sin(\mathrm{i}z), \quad \cosh z = \cos(\mathrm{i}z)$$

(4) Logarithmic function

The logarithmic function is the inverse of the exponential function. Since the exponential function is periodic, the logarithmic function is a multivalued function:

$$\ln z \overset{\text{def}}{=} \ln|z| + \mathrm{i}(\arg z + 2k\pi) \quad (k \in \mathbb{Z}) \tag{1.2.3}$$

The case $k = 0$ is called the principal value of the logarithm denoted as

$$\mathrm{Ln}\, z = \ln|z| + \mathrm{i}\arg z \quad (0 \leq \arg z < 2\pi)$$

Example 1.5. Compute the logarithmic values:

$$\text{(a) } \ln(-1); \quad \text{(b) } \ln(1+\mathrm{i})$$

Solution: According to the definition, we have

(a) $\ln(-1) = \ln \mathrm{e}^{\pi\mathrm{i}+2k\pi\mathrm{i}} = \pi\mathrm{i} + 2k\pi\mathrm{i} \quad (k \in \mathbb{Z})$;

(b) $\ln(1+\mathrm{i}) = \ln(\sqrt{2}\mathrm{e}^{\frac{\pi\mathrm{i}}{4}} + 2k\pi\mathrm{i}) = \ln\sqrt{2} + \dfrac{\pi\mathrm{i}}{4} + 2k\pi\mathrm{i} \quad (k \in \mathbb{Z})$

(5) Inverse trigonometric/hyperbolic functions

Suppose that the inverse function of the sine function is $w = \sin^{-1} z$, i.e.,

$$\sin w = \frac{e^{iw} - e^{-iw}}{2i} = z$$

The solution is given by

$$e^{iw} = iz \pm \sqrt{1 - z^2}$$

Taking the logarithms, we obtain

$$\sin^{-1} z = -i \ln(iz \pm \sqrt{1 - z^2})$$

To obtain a genuine function, we must restrict the domain. The inverse sine function is defined as

$$\arcsin z \overset{\text{def}}{=} -i \ln(iz + \sqrt{1 - z^2}) = i \operatorname{Ln}(\sqrt{1 - z^2} - iz)$$

Similarly, the inverse cosine function is defined as

$$\arccos z \overset{\text{def}}{=} -i \ln(z + \sqrt{z^2 - 1}) = \frac{\pi}{2} - \arcsin z$$

and the inverse tangent function is

$$\operatorname{atan} z \overset{\text{def}}{=} -\frac{i}{2} \operatorname{Ln}\left(\frac{1 + iz}{1 - iz}\right)$$

Since trigonometric functions are periodic, the inverse trigonometric functions are multivalued functions and can be expressed using logarithmic functions.

In the same way, the inverse hyperbolic functions are defined as

$$\operatorname{arcsinh} z \overset{\text{def}}{=} \operatorname{Ln}(z + \sqrt{1 + z^2}), \quad \operatorname{arccosh} z \overset{\text{def}}{=} \operatorname{Ln}(z + \sqrt{z^2 - 1})$$

(6) Power function

$$z^\alpha \overset{\text{def}}{=} e^{\alpha \ln z}$$

When α is an integer, the power function is a single-valued function. When α is a rational number, the fractional power function becomes a multivalued function. For example, when $\alpha = 1/2$, it is called a radical function:

$$\sqrt{z} = \sqrt{r}\left(\cos\frac{\theta + 2k\pi}{2} + \mathrm{i}\sin\frac{\theta + 2k\pi}{2}\right) \quad (k = 0, 1)$$

For each z, there are two different function values: $\sqrt{z} = \pm\sqrt{r}\mathrm{e}^{\mathrm{i}\theta/2}$. When α is an irrational number or a complex number, the power function becomes a multivalued function with infinitely many values.

(7) General exponential function

$$a^z \stackrel{\text{def}}{=} \mathrm{e}^{z \ln a}$$

where $a \neq 0, \infty$. This function is not a function in the usual sense because

$$\ln a = \ln|a| + \mathrm{i}\arg a + 2k\pi\mathrm{i} \quad (k \in \mathbb{Z})$$

has infinitely many values. Therefore, the function has multiple values. However, it is not a multivalued function, and its multivalued nature does not come from the increase in the argument of the independent variable z. It should be regarded as a collection of different functions:

$$\{\mathrm{e}^{z(\ln|a| + \mathrm{i}\arg a)}\mathrm{e}^{2\mathrm{i}k\pi z} \ (k \in \mathbb{Z})\}$$

Usually, only the principal value of the logarithm, $k = 0$, is taken to represent the general exponential function.

Example 1.6. Finding Roots of Equations:

$$\text{(a)}\sin z = 0; \quad \text{(b)}\sin z = 2$$

Solution: (a) Using the definition

$$\sin z = \frac{1}{2\mathrm{i}}(\mathrm{e}^{\mathrm{i}z} - \mathrm{e}^{-\mathrm{i}z}) = 0$$

then

$$\mathrm{e}^{2\mathrm{i}z} = 1 \longrightarrow z = k\pi \quad (k \in \mathbb{Z})$$

(b) Using

$$\sin z = \frac{1}{2\mathrm{i}}(\mathrm{e}^{\mathrm{i}z} - \mathrm{e}^{-\mathrm{i}z}) = 2$$

we have

$$\mathrm{e}^{2\mathrm{i}z} - 4\mathrm{i}\mathrm{e}^{\mathrm{i}z} - 1 = 0 \longrightarrow \mathrm{e}^{\mathrm{i}z} = (2 \pm \sqrt{3})\mathrm{i}$$

Hence,

$$z = -\mathrm{i}\ln(2 \pm \sqrt{3}) + \frac{\pi}{2} + 2k\pi \quad (k \in \mathbb{Z})$$

Discussion: Trigonometric functions are periodic along the real axis, and exponential functions are periodic along the imaginary axis. Is it possible that a function is periodic along both the real and imaginary axes?

ENDNOTE

The complex logarithm function has caused significant confusion in the history of mathematics. Based on the following observations,

$$\frac{\mathrm{d}z}{1+z^2} = \frac{\mathrm{d}z}{2(1+z\sqrt{-1})} + \frac{\mathrm{d}z}{2(1-z\sqrt{-1})}$$

Johann Bernoulli concluded the result that the imaginary logarithm represents a circular sector, namely,

$$\arctan z = \frac{1}{2\sqrt{-1}} \ln \frac{\sqrt{-1} - z}{\sqrt{-1} + z}$$

However, Bernoulli had limited understanding of complex logarithms. In his debate with Gottfried Leibniz regarding the logarithms of negative numbers, Bernoulli asserted that they were real numbers and insisted that $\ln x = \ln(-x)$. His reasoning was as follows:

$$(-x)^2 = x^2 \longrightarrow 2\ln x = 2\ln(-x)$$

and

$$\frac{\mathrm{d}}{\mathrm{d}x}\ln(-x) = \frac{1}{x} = \frac{\mathrm{d}}{\mathrm{d}x}\ln x$$

On the other hand, Leibniz argued that the logarithm of a negative number is imaginary, and $\ln(-x) \neq \ln x$. His reasoning was based on the series expansion:

$$\ln(1+x) = x - \frac{x^2}{2} + \frac{x^3}{3} - \frac{x^4}{4} + \cdots$$

Substituting $x = -2$, we have

$$\ln(-1) = -2 - \frac{4}{2} - \frac{8}{3} - \cdots$$

All terms on the right side are negative, indicating that $\ln(-1)$ must be related to some imaginary number. In this debate, Jean le Rond d'Alembert sided with Bernoulli. Euler, on the other hand, believed that by following Bernoulli's approach, we would necessarily obtain

$$(x\sqrt{-1})^4 = x^4 \longrightarrow \ln x + \ln\sqrt{-1} = \ln x$$

which would lead to $\ln\sqrt{-1} = 0$. However, Bernoulli himself had discovered

$$\frac{\ln\sqrt{-1}}{\sqrt{-1}} = \frac{\pi}{2}$$

Euler provided the correct answer to the debate: the logarithm of a negative number is a set of infinitely many values. According to Euler's formula, $\mathrm{i}x = \ln(\cos x + \mathrm{i}\sin x)$, any given complex number has infinitely many logarithmic values:

$$\ln 1 = 2k\pi\mathrm{i}, \quad \ln(-1) = (2k+1)\pi\mathrm{i} \quad (k \in \mathbb{Z})$$

Nevertheless, Euler's reasoning was rather peculiar. Gauss eventually explained the multivalued nature of the logarithm function by recognizing it as an integral function.

Common elementary functions such as exponential functions and trigonometric functions are single periodic functions. We can also define a type of spiral functions:

$$\sin p(z) \stackrel{\text{def}}{=} \frac{\varphi^z - \psi^z}{2}, \quad \cos p(z) \stackrel{\text{def}}{=} \frac{\varphi^z + \psi^z}{2}$$

where

$$\varphi = \frac{\alpha + \beta}{2}, \quad \psi = \frac{\alpha - \beta}{2} \ (\alpha, \beta \in \mathbb{R})$$

When $\alpha > \beta > 0$, it becomes the ordinary exponential function. By taking

$$\alpha^2 - \beta^2 = 4 \longrightarrow \varphi \cdot \psi = 1$$

the spiral function degenerates into hyperbolic functions. When $\alpha < \beta < 0$, it becomes the trigonometric functions. Furthermore, when $0 < \alpha < \beta$, the function maps a straight line in the z-plane into a spiral curve, with

$$\beta^2 - \alpha^2 = 4 \longrightarrow \varphi \cdot \psi = -1$$

The spiral function has the following properties:

$$\sin p(z+1) = \alpha \sin p(z) + \sin p(z-1)$$
$$\sin p(2z) = 2 \sin p(z) \cdot \cos p(z)$$

Taking $\alpha = 1$ and $\beta = \sqrt{5}$, it reflects the Fibonacci sequence, with the golden ratio $\varphi = \frac{1+\sqrt{5}}{2}$, as mentioned in the endnote of Section 8.4. Taking $\alpha = 2$ and $\beta = 2\sqrt{2}$, it corresponds to the Pell numbers, with the silver ratio $\delta_s = 1+\sqrt{2}$. Taking $\alpha = n$ and $\beta = \sqrt{n^2+4}$, we obtain the general metallic ratio:

$$\delta_n = \frac{n + \sqrt{n^2+4}}{2} \quad (n \in \mathbb{N})$$

which is a root of the equation $x^2 - nx - 1 = 0$. In this sense, the spiral function can also be called the metallic function.

Some people refer to the values of $\alpha = 2\sqrt{3}$, $\beta = 4$, and $\varphi = \sqrt{3} + 2$ as the platinum ratio. It is a root of the equation $x^2 - 4x + 1 = 0$.

Exercises

[1] Find the value of

$$\text{(a)}\ \ln{(1+\mathrm{i})^{2\mathrm{i}}}; \qquad \text{(b)}\ \sin(a + \mathrm{i}b) \quad (a, b \in \mathbb{R})$$

[2] Prove that $\mathrm{i}^{\mathrm{i}^{\mathrm{i}}}$ is not equal to $\mathrm{i}^{\mathrm{i}\cdot\mathrm{i}} = \mathrm{i}^{-1}$.

[3] Prove the following formulas:

$$\text{(a)}\ \sin(x + \mathrm{i}y) = \sin x \cosh y + \mathrm{i} \cos x \sinh y;$$

$$\text{(b)}\ \cos(x + \mathrm{i}y) = \cos x \cosh y - \mathrm{i} \sin x \sinh y;$$

$$\text{(c)}\ \tanh\left(\frac{z}{2}\right) = \frac{\sinh x + \mathrm{i} \sin y}{\cosh x + \cos y};$$

$$\text{(d)}\ \coth\left(\frac{z}{2}\right) = \frac{\sinh x - \mathrm{i} \sin y}{\cosh x - \cos y}$$

[4] Prove the expression for inverse trigonometric functions:

$$\text{(a)}\ \arcsin z = -\mathrm{i}\mathrm{Ln}(\mathrm{i}z + \sqrt{1 - z^2});$$

$$\text{(b)}\ \arctan z = -\frac{\mathrm{i}}{2}\mathrm{Ln}\frac{1 + \mathrm{i}z}{1 - \mathrm{i}z}$$

[5] Find the roots of the following equations:

$$\text{(a)}\ \tan z = \mathrm{i}; \quad \text{(b)}\ z^2 + 2z\cos\lambda + 1 = 0 \ (0 < \lambda < \pi)$$

Answer: (a) No solution; (b) $-\mathrm{e}^{\pm\mathrm{i}\lambda}$

[6] Prove that the spiral function satisfying

$$\sin p(z + 1) = \alpha \sin p(z) + \sin p(z - 1)$$

1.3 Derivatives of Complex Functions

1.3.1 *Limits and derivatives*

(1) *Function limits*

Suppose that the complex function $f(z)$ is defined in a punctured neighborhood of z_0. If for any small positive ε, there exists a positive δ such that for $0 < |z - z_0| < \delta$, $|f(z) - A| < \varepsilon$, then A is called the limit of $f(z)$ as z approaches z_0, denoted as $\lim_{z \to z_0} f(z) = A$.

Since z is a complex number, it can approach z_0 from different directions. The existence of a limit for a function implies that the function values $f(z)$ must approach the same limit value A. Conversely, if the limit values from different approaching directions are distinct, then the function does not have a limit at that point. Looking back at the limit definition of real-valued functions, the limits from the left and right sides of a point must be the same, indicating the logical consistency.

Example 1.7. Find the limit of the function:

$$\lim_{z \to 1} \frac{z\overline{z} + 2z - \overline{z} - 2}{z^2 - 1}$$

Solution: First, consider the approaching route $z \to 1$ along the horizontal direction. Let $z = x$, one has

$$\lim_{z \to 1} \frac{z\overline{z} + 2z - \overline{z} - 2}{z^2 - 1} = \lim_{x \to 1} \frac{x^2 + 2x - x - 2}{x^2 - 1} = \frac{3}{2}$$

Next, consider the approaching route $z \to 1$ along the vertical direction. In this case, let $z = 1 + \mathrm{i}y$ with $y \to 0$, so

$$\lim_{z \to 1} \frac{z\overline{z} + 2z - \overline{z} - 2}{z^2 - 1} = \lim_{y \to 0} \frac{(1 + y^2) + 2(1 + \mathrm{i}y) - (1 - \mathrm{i}y) - 2}{(1 + \mathrm{i}y)^2 - 1} = \frac{3}{2}$$

Therefore, the limit of the function at $z = 1$ is

$$\lim_{z \to 1} \frac{z\overline{z} + 2z - \overline{z} - 2}{z^2 - 1} = \frac{3}{2}$$

Practice: Choose the approaching route $z \to 1$ along any direction by taking the parameter equation $z = 1 + \delta e^{\mathrm{i}\theta}$ with $\delta \to 0$. Find the limit of the function.

(2) *Function Continuity*

A complex function $f(z)$ is said to be continuous at $z = z_0$ if it satisfies the following conditions: the function value $f(z_0)$ as well as the function limit $\lim_{z\to z_0} f(z)$ exists, and the function value equals the limit:

$$\lim_{z\to z_0} f(z) = f(z_0)$$

The sufficient and necessary condition for the complex function $f(z) = u(x, y) + \mathrm{i}v(x, y)$ to be continuous at $z_0 = x_0 + \mathrm{i}y_0$ is that both the real-valued functions $u(x, y)$ and $v(x, y)$ are continuous at the point (x_0, y_0).

(3) *Derivatives of complex functions*

Let $w = f(z)$ be a single-valued function defined in a region B. At a point z_0 within B, if the limit

$$\lim_{\Delta z\to 0} \frac{\Delta w}{\Delta z} = \lim_{z\to z_0} \frac{f(z) - f(z_0)}{z - z_0}$$

exists, then the function $f(z)$ is said to be differentiable at the point z_0, and the limit value is called the derivative of the function $f(z)$ at z_0, denoted as

$$f'(z_0) \stackrel{\text{def}}{=} \left.\frac{\mathrm{d}f(z)}{\mathrm{d}z}\right|_{z=z_0}$$

The existence of the function derivative implies that the limit is independent of the approach towards z_0. In fact, in one-dimensional real variable functions, the derivative must be the same from both the left and right sides of the point for the function to be differentiable. Thus, the derivative of complex functions is a natural extension of the derivative of real-valued functions. However, the derivative rules for complex functions impose stricter restrictions, as we will see later. These restrictions endow complex functions a more delicate local structure compared to real-valued functions.

Example 1.8. Proving that function $f(z) = \bar{z}$ is continuous everywhere on the complex plane but not differentiable everywhere.

Proof. Since $f(z) = x - \mathrm{i}y$, both its real and imaginary parts are continuous functions, so $f(z)$ is continuous everywhere on the z-plane. Take an arbitrary point $z_0 = x_0 + \mathrm{i}y_0$ on the complex plane. According to the definition of the derivative, the limit when approaching z_0 from the horizontal direction is

$$\lim_{\Delta z \to 0} \frac{\Delta f(z)}{\Delta z} = \lim_{x \to x_0} \frac{x - x_0}{x - x_0} = 1$$

The limit when approaching z_0 from the vertical direction is

$$\lim_{\Delta z \to 0} \frac{\Delta f(z)}{\Delta z} = \lim_{y \to y_0} \frac{-\mathrm{i}y + \mathrm{i}y_0}{\mathrm{i}y - \mathrm{i}y_0} = -1$$

Since the two limits are not equal, the function is not differentiable at z_0. Since z_0 is an arbitrary point, the function $f(z) = \bar{z}$ is not differentiable at any point on the entire complex plane. □

1.3.2 *Cauchy–Riemann Conditions*

Let $f(z) = u(x, y) + \mathrm{i}v(x, y)$, the function $f(z)$ is differentiable if and only if the real functions $u(x, y)$ and $v(x, y)$ are both continuous and have partial derivatives that satisfy the Cauchy–Riemann conditions:

$$\frac{\partial u}{\partial x} = \frac{\partial v}{\partial y}, \quad \frac{\partial v}{\partial x} = -\frac{\partial u}{\partial y} \tag{1.3.1}$$

Proof. The derivative of $f(z)$ along the direction parallel to the real axis is

$$\frac{\partial f(z)}{\partial x} = \frac{\partial u}{\partial x} + \mathrm{i}\frac{\partial v}{\partial x}$$

and along the direction parallel to the imaginary axis is

$$\frac{\partial f(z)}{\mathrm{i}\partial y} = \frac{\partial u}{\mathrm{i}\partial y} + \frac{\partial v}{\partial y}$$

According to the definition of the derivative for complex functions, the derivative value is independent of the approach to a point. Hence, the two derivatives must be equal. Consequently, the real and imaginary parts of these two equations should be equal separately:

$$\frac{\partial u}{\partial x} = \frac{\partial v}{\partial y}, \quad \frac{\partial v}{\partial x} = -\frac{\partial u}{\partial y}$$

The Cauchy–Riemann conditions imply that for a differentiable complex function, its real and imaginary parts are closely related. In other words, not every complex function is differentiable. The fact that the real and imaginary parts of a complex function are individually differentiable does not guarantee that the complex function itself is differentiable.

The Cauchy–Riemann conditions can also be expressed in polar coordinates as

$$\frac{\partial u}{\partial \rho} = \frac{1}{\rho}\frac{\partial v}{\partial \phi}, \quad \frac{1}{\rho}\frac{\partial u}{\partial \phi} = -\frac{\partial v}{\partial \rho} \tag{1.3.2}$$

Proof is left as an exercise. □

1.3.3 *Derivative rules*

Based on the definition of derivatives, it can be easily shown that complex functions follow the same derivative rules as real functions. The rules are listed as follows:

$$\frac{\mathrm{d}}{\mathrm{d}z}(w_1 \pm w_2) = \frac{\mathrm{d}w_1}{\mathrm{d}z} \pm \frac{\mathrm{d}w_2}{\mathrm{d}z}$$

$$\frac{\mathrm{d}}{\mathrm{d}z}(w_1 \cdot w_2) = w_1\frac{\mathrm{d}w_2}{\mathrm{d}z} + w_2\frac{\mathrm{d}w_1}{\mathrm{d}z}$$

$$\frac{\mathrm{d}}{\mathrm{d}z}\left(\frac{w_1}{w_2}\right) = \frac{w_1' w_2 - w_1 w_2'}{w_2^2}$$

$$\frac{\mathrm{d}w}{\mathrm{d}z} = 1\Big/\frac{\mathrm{d}z}{\mathrm{d}w}, \quad \frac{\mathrm{d}F(w)}{\mathrm{d}z} = \frac{\mathrm{d}F}{\mathrm{d}w}\cdot\frac{\mathrm{d}w}{\mathrm{d}z}$$

Let $f(z) = u(x, y) + \mathrm{i}v(x, y)$ be differentiable at the point $z = x + \mathrm{i}y$. Then

$$\frac{\mathrm{d}f(z)}{\mathrm{d}z} = \frac{\partial u}{\partial x} + \mathrm{i}\frac{\partial v}{\partial x} = \frac{\partial v}{\partial y} - \mathrm{i}\frac{\partial u}{\partial y}$$

Starting from the definition, the derivative formulas for the following elementary complex functions can be directly proven. They have the same form as the derivatives of the corresponding real functions:

$$\frac{\mathrm{d}e^z}{\mathrm{d}z} = \mathrm{e}^z, \quad \frac{\mathrm{d}\ln z}{\mathrm{d}z} = \frac{1}{z}$$

$$\frac{\mathrm{d}\sin z}{\mathrm{d}z} = \cos z, \quad \frac{\mathrm{d}\cos z}{\mathrm{d}z} = -\sin z$$

$$\frac{\mathrm{d}\sinh z}{\mathrm{d}z} = \cosh z, \quad \frac{\mathrm{d}\cosh z}{\mathrm{d}z} = \sinh z$$

Exercises

[1] Let α be a real number. Prove that for $z \neq 0$,

$$\frac{\mathrm{d}z^{\alpha}}{\mathrm{d}z} = \alpha z^{\alpha-1}$$

Hint: Use the definition of derivatives and differentiate along the x-direction.

[2] Study the differentiability of the complex function,

$$f(z) = (3x^2y - x^3) + \mathrm{i}xy$$

[3] Prove the derivative formula:

$$\frac{\mathrm{d}\sin z}{\mathrm{d}z} = \cos z$$

[4] Based on the definition of derivatives, derive the Cauchy–Riemann conditions in polar coordinates:

$$\frac{\partial u}{\partial \rho} = \frac{1}{\rho}\frac{\partial v}{\partial \phi}, \quad \frac{1}{\rho}\frac{\partial u}{\partial \phi} = -\frac{\partial v}{\partial \rho}$$

[5] Prove the following derivative formulas:

$$\text{(a)}\ \frac{\mathrm{d}}{\mathrm{d}z}\left(\frac{w_1}{w_2}\right) = \frac{w_1' w_2 - w_1 w_2'}{w_2^2}; \qquad \text{(b)}\ \frac{\mathrm{d}F(w)}{\mathrm{d}z} = \frac{\mathrm{d}F}{\mathrm{d}w}\cdot\frac{\mathrm{d}w}{\mathrm{d}z}$$

1.4 Analytic Functions

1.4.1 *Analyticity and singularity*

Let $f(z)$ be a function that is differentiable at every point in the neighborhood of z_0. Then the function $f(z)$ is said to be analytic at z_0. If $f(z)$ is analytic at every point in a region B, it is called an analytic function in the region B.

Points where a complex function is not differentiable are called singular points of the function. If a complex function is analytic at a certain point, it must be differentiable at that point. Conversely, if it is not differentiable at a certain point, such as the function $f(z) = |z|^2$, which is only differentiable at $z = 0$ and not differentiable at any other point in its neighborhood, then the function is not analytic at $z = 0$.

Sufficient and necessary conditions for analyticity: A function $f(z)$ is analytic in a region B if and only if its real and imaginary parts are differentiable in B and they satisfy the Cauchy–Riemann conditions at every point in B.

1.4.2 *Fundamental properties*

(1) *Family of orthogonal curves*

Let $f(z) = u(x, y) + \mathrm{i}v(x, y)$ be an analytic function in a region B. Then, $u(x, y) = C_1$ and $v(x, y) = C_2$ are two families of orthogonal curves in B.

Proof. Using the Cauchy–Riemann conditions, we have

$$\nabla u \cdot \nabla v = \frac{\partial u}{\partial x}\frac{\partial v}{\partial x} + \frac{\partial u}{\partial y}\frac{\partial v}{\partial y} = -\frac{\partial u}{\partial x}\frac{\partial u}{\partial y} + \frac{\partial u}{\partial y}\frac{\partial u}{\partial x} = 0$$

Since the gradient direction represents the normal direction of a plane curve, the two curves are mutually perpendicular at their intersection point.

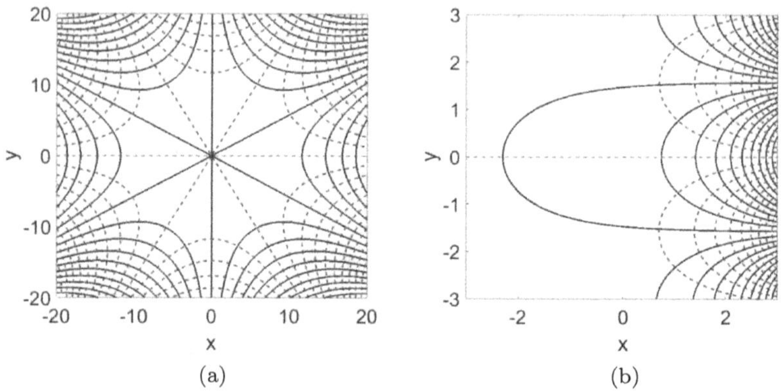

Figure 1.8.

Figure 1.8 illustrates the contour plots of the real part (solid lines) and the imaginary part (dashed lines) of the analytic functions at given values: (a) $f(z) = z^3$; (b) $f(z) = \mathrm{e}^z$. These curves form an orthogonal grid. □

(2) *Harmonic functions*

If $f(z) = u(x, y) + \mathrm{i}v(x, y)$ is an analytic function in a region B, then $u(x, y)$ and $v(x, y)$ are both harmonic functions in B, satisfying the Laplace equations respectively,

$$\Delta u \equiv \nabla^2 u = 0, \quad \Delta v \equiv \nabla^2 v = 0$$

Proof. Using the Cauchy–Riemann conditions, differentiate both sides with respect to x or y separately and then add them together, then

$$\nabla^2 u = 0, \quad \nabla^2 v = 0$$

The functions $u(x, y)$ and $v(x, y)$ are called conjugate harmonic functions.

The real and imaginary parts of an analytic function are related. Figure 1.11 visually shows the texture maps of the real part (a) and the imaginary part (b) of the analytic function $f(z) = \sin z$. Given either the real or imaginary part, the analytic function can be completely determined.

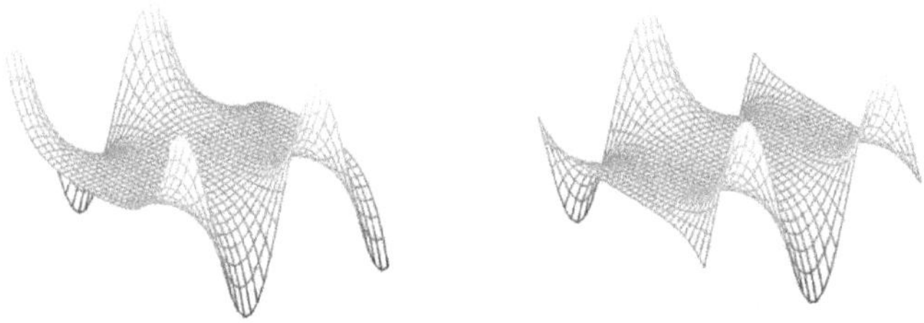

(a) (b)

Figure 1.9.

□

Reflection: Consider Figure 1.9 carefully and summarize the basic characteristics of analytic functions.

Example 1.9. Given that the real part of a certain analytic function is $u(x, y) = x^2 - y^2$, we need to determine the analytic function.

Solution: First, we verify that $u(x, y)$ indeed satisfies the Laplace equation:

$$\Delta u(x, y) = 0$$

Making use of the Cauchy–Riemann conditions, we have

$$\frac{\partial u}{\partial x}=\frac{\partial v}{\partial y}=2x,\qquad \frac{\partial u}{\partial y}=-\frac{\partial v}{\partial x}=-2y$$

$$\mathrm{d}v=\frac{\partial v}{\partial x}\mathrm{d}x+\frac{\partial v}{\partial y}\mathrm{d}y=2y\mathrm{d}x+2x\mathrm{d}y=\mathrm{d}(2xy)\longrightarrow v=2xy+c$$

Hence, the analytic function is given by

$$f(z)=(x^2-y^2)+2\mathrm{i}xy+c=z^2+c$$

Example 1.10. To determine the analytic function, we are given its imaginary part as

$$v(x,y)=\sqrt{-x+\sqrt{x^2+y^2}}$$

Solution: First, we verify that $v(x,y)$ indeed satisfies the Laplace equation. It is convenient to use polar coordinates for this problem, i.e.,

$$v(\rho,\varphi)=\sqrt{\rho(1-\cos\phi)}=\sqrt{2\rho}\sin\frac{\phi}{2}$$

According to the Cauchy–Riemann conditions,

$$\frac{\partial u}{\partial\rho}=\frac{\partial v}{\rho\partial\phi}=\frac{1}{\sqrt{2\rho}}\cos\frac{\phi}{2}$$

and

$$\frac{\partial u}{\rho\partial\phi}=-\frac{\partial v}{\partial\rho}=-\frac{1}{\sqrt{2\rho}}\sin\frac{\phi}{2}$$

we get

$$u=\sqrt{2\rho}\cos\frac{\phi}{2}+c$$

Therefore, the analytical function is

$$f(z)=\sqrt{2\rho}\cos\frac{\phi}{2}+\mathrm{i}\sqrt{2\rho}\sin\frac{\phi}{2}+c=\sqrt{2z}+c$$

(3) *Region mappings*

Analytic functions map a certain region on the z-plane to the corresponding region on the w-plane. Figure 1.10(a) shows that the function $f(z) = z^3$ maps a sector with an angle of $\pi/3$ on the z-plane to the upper half-plane on the w-plane. Figure 1.10(b) illustrates the fractional linear function

$$f(z) = \frac{z - \mathrm{i}a}{z + \mathrm{i}a}$$

which maps the upper half z-plane to the interior of the unit circle.

As for the topic of region mapping, we will pause here. Chapter 6 will specifically study the properties and applications of various region mappings.

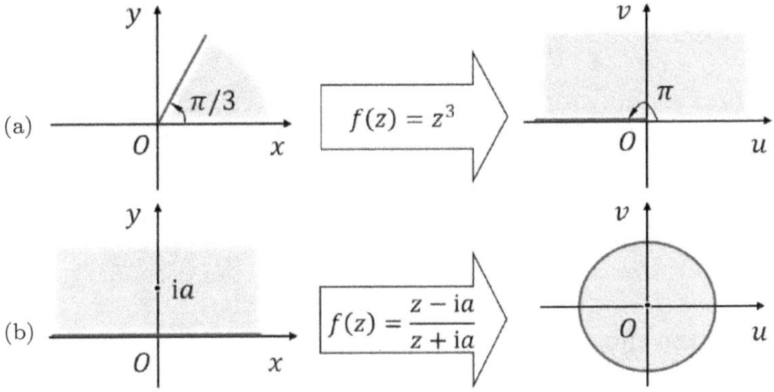

Figure 1.10.

Exercises

[1] Determine the analytic functions, given the real part by

(a) $u(x, y) = \mathrm{e}^x \sin y$; (b) $u(x, y) = x^2 - y^2 + xy$, $f(0) = 0$;

(c) $u(x, y) = \dfrac{2 \sin 2x}{\mathrm{e}^{2y} + \mathrm{e}^{-2y} - 2 \cos 2x}$, $f\left(\dfrac{\pi}{2}\right) = 0$;

(d) $x^4 - 6x^2y^2 + y^4$, $f(0) = 0$

Answers: (a) $-\mathrm{i}\mathrm{e}^z + c$; (b) $z^2\left(1 - \frac{\mathrm{i}}{2}\right)$; (c) $\cot z$; (d) z^4

[2] Can we construct an analytic function with the following imaginary part?

$$v(x,y) = x^3 - 3xy$$

[3] If the function $f(z) = u + iv$ is analytic in a certain region, prove that the Jacobian determinant is given by

$$\det \mathrm{J} = \frac{\partial u}{\partial x}\frac{\partial v}{\partial y} - \frac{\partial u}{\partial y}\frac{\partial v}{\partial x} = |f'(z)|^2$$

[4] Plot the images of the following curves under the mapping $f(z) = z^2$:

$$\text{(a) } |z-1| = 1; \qquad \text{(b) } y = \frac{1}{x}$$

Hint: (a) Use polar coordinates; (b) Choose $z = x + \frac{i}{x}$
Answer:

(a) 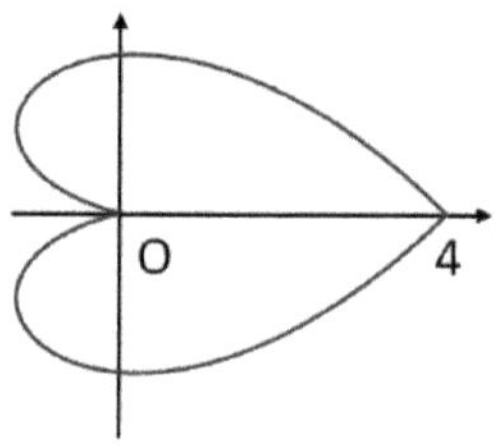

; (b) 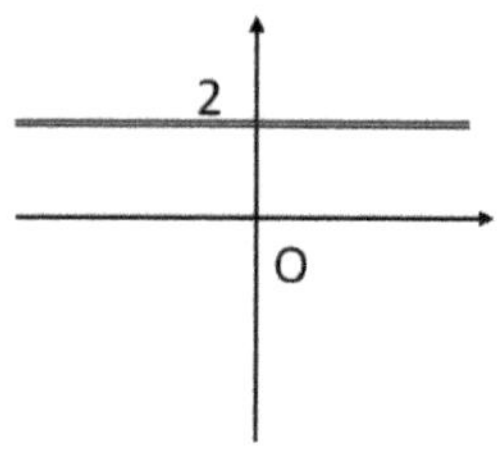

[5] Determine the region to which the function $f(z) = \sin z$ maps the given region:

$$-\frac{\pi}{2} \leq \operatorname{Re} z \leq \frac{\pi}{2}, \quad \operatorname{Im} z \geq 0$$

Answer: The upper half-plane $\mathbb{H}$.

[6] Find an analytic function that maps the unit circle to the right half-plane.
Answer: $\zeta = \frac{1-z}{1+z}$

1.5 Multivalued Functions

1.5.1 *Branch point and branch cut*

Consider the square root function $w = \sqrt{z-1}$. When z travels along the contour L_1 shown in Figure 1.11, encircling the point $z = 1$ once in the complex plane, the value w of the function does not return to its initial value. Only when z completes another loop around that point does the function w return to its initial value. Therefore, for the same independent variable z, there exist two distinct function values:

$$\begin{cases} w_1 = \sqrt{|z-1|}\mathrm{e}^{\frac{\mathrm{i}}{2}\arg(z-1)} \\ w_2 = \sqrt{|z-1|}\mathrm{e}^{\frac{\mathrm{i}}{2}\arg(z-1)} + \mathrm{i}\pi \end{cases}$$

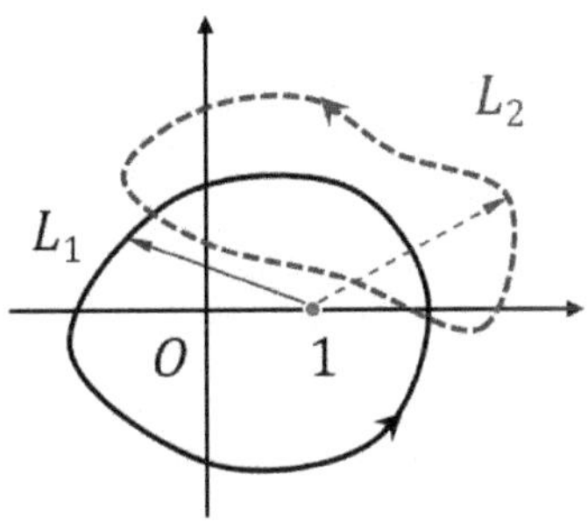

Figure 1.11.

The functions w_1 and w_2 are referred to as the two branches of the multivalued function. However, when z travels around any closed loop that does not contain $z = 1$, such as the loop L_2 shown in Figure 1.11, the function returns to its initial value. So, the point $z = 1$ is called a branch point of the multivalued function. If z needs to complete $n + 1$ loops for the function value w to return to its initial value, then that point is referred to as an nth order branch point.

In fact, besides the point $z = 1$, the infinity $z = \infty$ is also a branch point of the function $w = \sqrt{z-1}$, which can be identified by letting $z - 1 = 1/\zeta$, where $\zeta = 0$ is a first-order branch point of the function.

In order to visually represent the multivalued function on the complex plane, a branch cut is introduced along the positive real axis between the two branch points $z = 1$ and $z = \infty$, as shown in Figure 1.12. It is defined that the argument along the upper edge of the branch cut $\arg(z-1) = 0$ and the argument along the lower edge is $\arg(z - 1) = 2\pi$. The corresponding function branch is w_1, with $0 \leq \arg w_1 \leq \pi$, representing the upper half of the w-plane. Only when z crosses the branch cut and re-enters the upper half-plane, the argument along the branch cut becomes $\arg(z-1) = 2\pi$, and the argument along the lower edge becomes $\arg(z-1) = 4\pi$. The corresponding function branch becomes w_2, with $\pi \leq \arg w_2 \leq 2\pi$, forming the lower half of the w-plane.

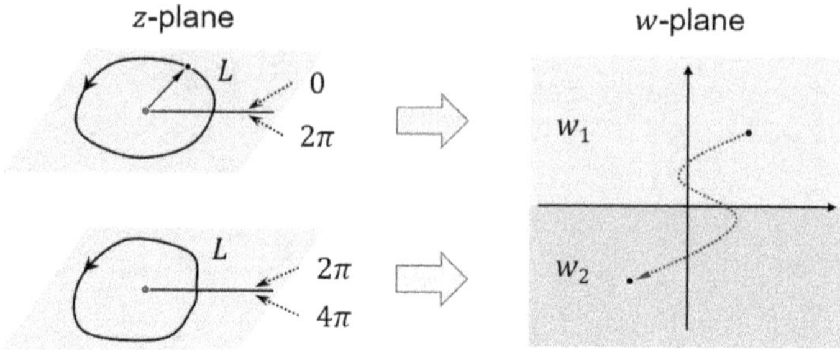

Figure 1.12.

1.5.2 *Riemann surface*

(1) *Square root function*

Whenever z crosses a branch cut, the function value enters another branch. By using the branch cut as a connecting link, different z-planes are intertwined and glued together, forming a Riemann surface. Figure 1.13 represents the double-layered Riemann surface of the multivalued function $w = \sqrt{z-1}$. If z moves on the complex plane without crossing the branch cut, the function will remain in

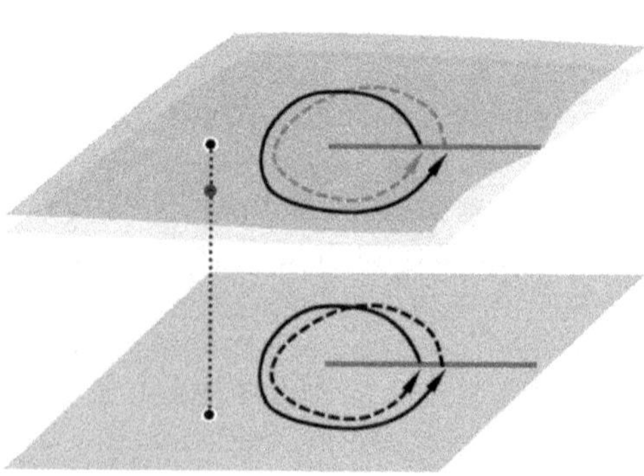

Figure 1.13.

the same branch, thus having single-valuedness. Only when z crosses the branch cut does the function enter another branch.

(2) *Logarithmic function*

$$w = \ln z = \ln |z| + \mathrm{i}(\arg z + 2k\pi) \quad (k \in \mathbb{Z})$$

The logarithmic function is an infinite-valued function which has two branch points 0 and ∞, through which a branch cut is made. The z-plane with an argument in the range $0 \leq \arg z \leq 2\pi$ corresponds to a strip-shaped region in the w-plane parallel to the real axis with a width of 2π. Whenever z completes a loop around the branch point $z = 0$ and crosses the branch cut, the function value enters another branch, which corresponds to an adjacent strip-shaped region in the w-plane. Figure 1.14 illustrates the periodic mapping relationship of the logarithmic function. Figure 1.15 represents the infinitely layered Riemann surface of the logarithmic function, where different branch

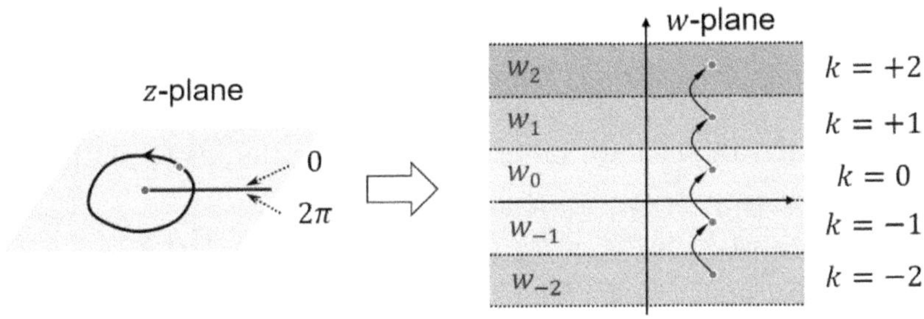

Figure 1.14.

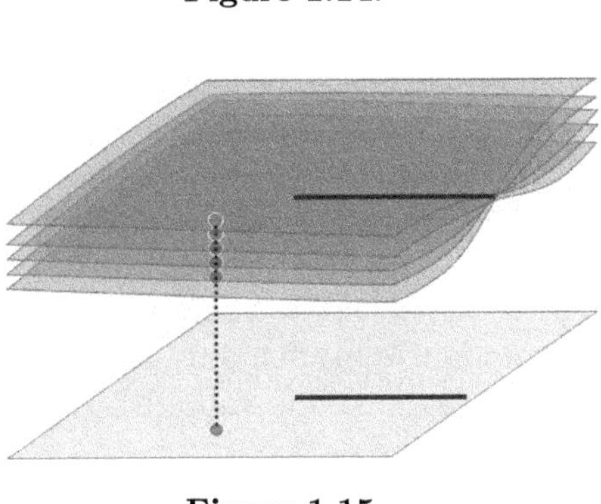

Figure 1.15.

sheets are intertwined and glued together at the branch cut. Each time z crosses the branch cut, the function value enters the next branch.

Example 1.11. Draw the Riemann surface of the multivalued function:

$$w = \sqrt{z(z-1)}$$

Solution: The function has two branch points at $z = 0$ and $z = 1$. In this case, $z = \infty$ is not a branch point. A branch cut is made between the two branch points, and the two branch sheets are intertwined and glued together. The resulting Riemann surface is shown in Figure 1.16.

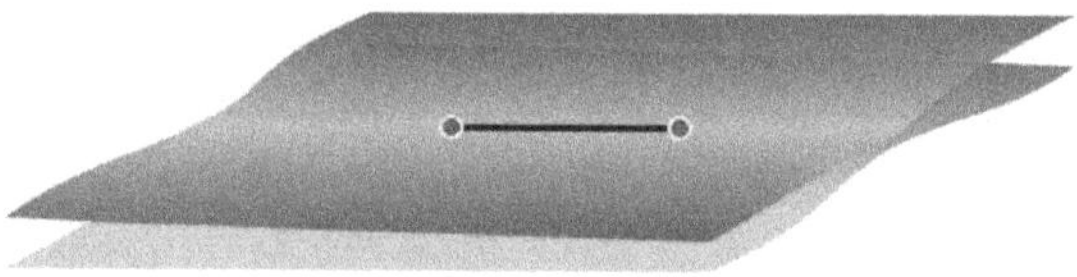

Figure 1.16.

Reflection: How can we determine if $z = \infty$ is a branch point of the function?

1.5.3 *Complex projective curves*

In topology, a complex curve corresponds to a surface, known as a Riemann surface, and the key lies in the branch points of the function. Consider a multivalued function: $w = \sqrt{z}$. A branch cut is made between its two branch points, $z = 0$ and $z = \infty$. From the perspective of the Riemann sphere, a unit sphere is covered by two layers of Riemann surfaces, and the two layers intersect and glue together at the branch cut, as shown in Figure 1.17(b). Topologically, this is equivalent to the closed sphere shown in Figure 1.17(c), where the dashed line represents the branch cut. Therefore, the square root function can be viewed as a single-valued function defined on the sphere.

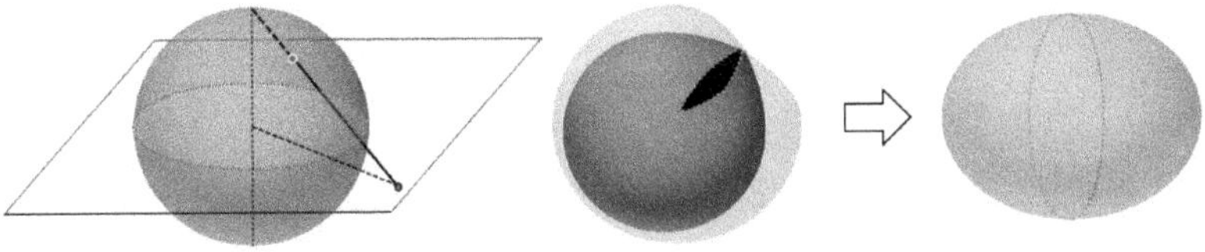

(a) (b) (c)

Figure 1.17.

This method can be extended to other complex algebraic curves, such as the cubic curve equation:

$$w^2 = z(z - \alpha)(z - \beta)$$

It defines a two-layer covering surface of the unit sphere, with two branch cuts $[0, \alpha]$ and $[\beta, \infty]$. After appropriate continuous deformations, it can be observed that the topological surface is a torus-shaped annulus, as shown in Figure 1.18. So, the cubic curve can be viewed as a single-valued function defined on a torus.

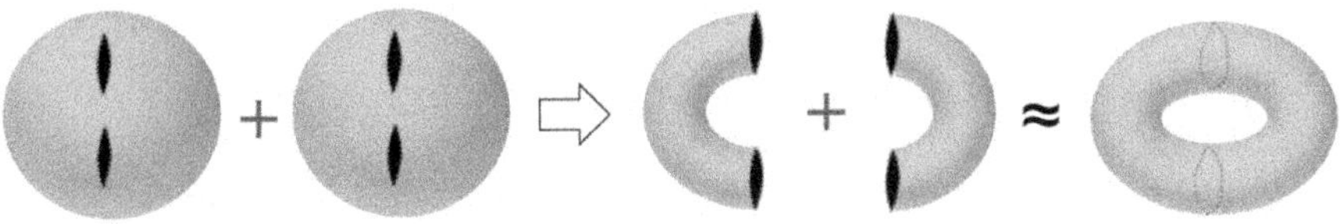

Figure 1.18.

Möbius proved that any closed surface in space is topologically equivalent to a genus surface shown in Figure 1.19. Therefore, closed surfaces can be topologically classified based on the number of holes, and the topological invariant is the genus number. For example, the genus number of the torus in Figure 1.18 is 1, while the genus number of the sphere in Figure 1.17 is 0. Generally, for algebraic curves of the form

$$w^2 = z(z - \alpha_1)(z - \alpha_2) \cdots (z - \alpha_{2n})$$

the genus number is n.

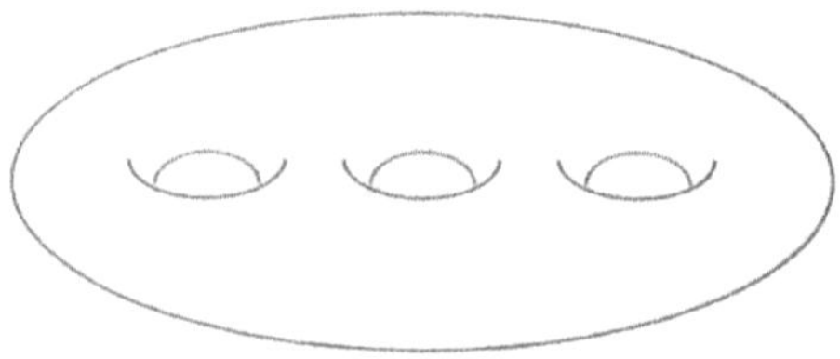

Figure 1.19.

ENDNOTE

From the mid-19th century onwards, geometry underwent three revolutionary evolutions, two of which were related to Riemann, and they had a profound impact on 20th-century algebra. Riemann's first idea was presented in his doctoral thesis in 1851, where he introduced Riemann's surface and expounded the Riemann mapping theorem, bridging the gap between function theory and geometry, marking the true beginning of topology. Riemann's second idea was presented in his habilitation thesis "On the Hypotheses that Underlie Geometry" in 1854, where he proposed studying the structure of surfaces from within the surfaces themselves. This intrinsic geometry could be conveniently extended to higher-dimensional spaces, leading to the establishment of modern differential geometry and tensor analysis and becoming the mathematical foundation of general relativity later on.

The branch cut doesn't necessarily have to be a straight line; any smooth curve connecting the two branch points will suffice. Although constructing Riemann surfaces can be done by drawing a branch cut between the branch points, it should be noted that the branch cut is a virtual concept, and there is no the complex plane. From a topological perspective, the branch cut is not necessary for constructing Riemann surfaces. One can envision the Riemann surface as a spiral distinct line in surface centered around the branch points and smoothly connecting the last layer with the first layer. In principle, the relative positions of the branch points are not important. What matters is determining the orders of the branch points and the relationship of the function values between different branches to represent the Riemann surface.

Let's now examine the branch point properties of the multivalued function:

$$f(z) = \sqrt{1+\sqrt{z}}$$

The branch points of the function are $z = 0, 1, \infty$. Among them, $z = 0$ is a first-order branch point with double multiplicity, implying that it is a double first-order branch point. $z = 1$ is a simple first-order branch point, which is the branch point for taking $\sqrt{z} = -1$ when $z = 1$. Finally, the infinity $z = \infty$ is a third-order branch point. The characteristics of each branch point can be represented by the Riemann surface shown in Figure 1.20.

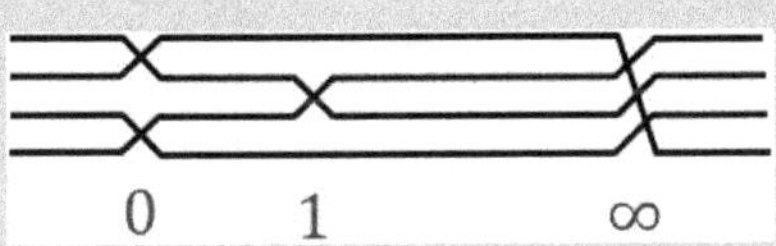

Figure 1.20.

Exercises

[1] Determine whether the following functions are single-valued or multivalued:

$$\text{(a) } \sqrt{z}\sin\sqrt{z}; \qquad \text{(b) } \sqrt{z}\cos\sqrt{z}$$

[2] Find the branch points of the functions:

$$\text{(a) } \sqrt[3]{z^2+1}; \qquad \text{(b) } \ln(\cos z)$$

[3] Plot the mapping relationship of the function: $f(z) = \sqrt[3]{1-z}$.

[4] Draw the branch cuts of the multivalued functions:

$$\text{(a) } \sqrt{z(z^2+1)}; \quad \text{(b) } \sqrt{z^2 - \frac{1}{z}}$$

[5] Assuming that the argument along the upper part of the branch cut is $\arg(z-2) = 0$, find the value of the function $f(z) =$

$z\sqrt[3]{z-2}$ along a path that does not cross the branch cut and reaches $z = 3$.
Answer: $f(3) = 3e^{\frac{2\pi i}{3}}$

[6] For the function $f(z) = z + \sqrt{z-1}$, draw a branch cut $[1, -\infty]$ and assume $f(2) = 1$. Find the values of $f(-3)$ along the upper and lower parts of the branch cut.
Answer: $f(-3) = -3 - 2i$; $f(-3) = -3 + 2i$

[7] The three branch points of multivalued function $f(z) = \sqrt[3]{z^2 + 2z + 2}$ are $z = -1 \pm I$ and $z = \infty$. Taking $f(0) = \sqrt[3]{2}$, we need to find the values of $f(-3)$ when z travels along the three routes L_1, L_2 and L_3 to point $z = -3$, respectively, as shown in Figure 1.21.

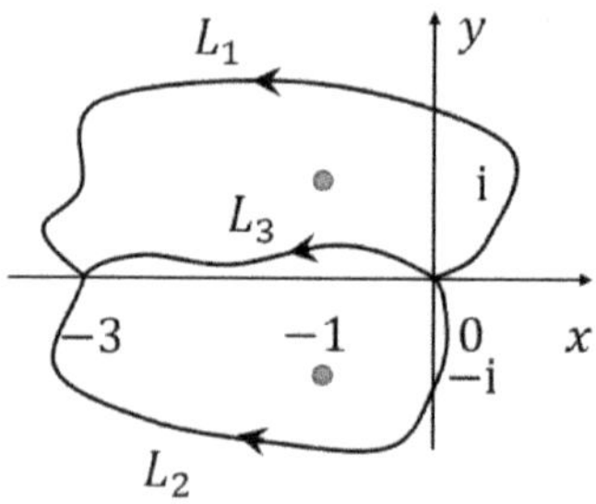

Figure 1.21.

Answer: $f(-3) = \sqrt[3]{5}e^{\frac{2\pi}{3}i}$; $\sqrt[3]{5}e^{-\frac{2\pi}{3}i}$; $\sqrt[3]{5}$

[8] Inverse sine function is defined as

$$f(z) = \arcsin z = -i\,\mathrm{Ln}(iz + \sqrt{1 - z^2})$$

Taking $f(0) = 0$, we need to find the values of $f(0)$ after z travels clockwise around $z = -1$ once and n times, as well as simultaneously around $z = -1$ and $z = +1$ twice.
Answer: $f(0) = \pi$; $n\pi$; 4π

[9] Plot the Riemann surface of the multivalued function:

$$f(z) = \sqrt{z^3 - 1}$$

[10] Plot the Riemann surface of the multivalued functions:

$$\text{(a) } f(z) = \sqrt[3]{z^2(z-1)}; \quad \text{(b) } f(z) = \sqrt{1 + \sqrt{z^2 - 1}}$$

Answer:

(a) 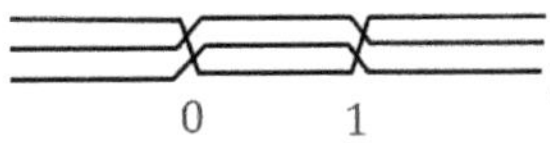

; (b) 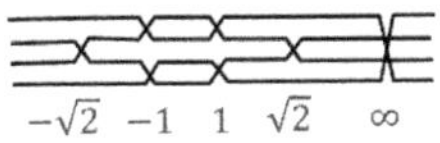

1.6 Complex Potential Functions

In physics and engineering, various fields are often studied. If the field being investigated is uniform in a certain direction in space, it is only necessary to study the two-dimensional plane distribution perpendicular to that direction, which is called a two-dimensional planar field. For a two-dimensional vector field, taking the plane perpendicular to a certain direction as the XOY plane, the points on it can be represented by $z = x + \mathrm{i}y$. At first glance, a planar vector field with components A_x and A_y can be expressed as

$$\boldsymbol{A} = A(z) = A_x(x, y) + \mathrm{i}A_y(x, y)$$

However, this approach is incorrect! In order for $A(z)$ to have the properties of an analytic function, it implies that there are Cauchy–Riemann relationships between the real and imaginary parts, which means that A_x and A_y are not independent.

Can a planar vector field be represented by an analytic function? We know that if the vector field is irrotational, such as the electrostatic field, a scalar field called the electric potential, also known as a planar scalar field, can be introduced to describe it. If the electric potential function is taken as the real part of a complex function, since the real and imaginary parts of an analytic function are closely related, what should be the imaginary part?

1.6.1 *Planar electrostatic field*

Let's consider a two-dimensional electrostatic field:

$$\boldsymbol{E} = E_x(x, y)\mathbf{i} + E_y(x, y)\mathbf{j}$$

where the field components $E_x(x, y)$ and $E_y(x, y)$ have continuous partial derivatives. If this electrostatic field is irrotational, $\nabla \times \boldsymbol{E} = 0$, there exists an electrostatic potential function $u(x, y)$ such that $\mathbf{E} =$

$-\nabla u$. It gives

$$E_x = -\frac{\partial u}{\partial x}, \quad E_y = -\frac{\partial u}{\partial y} \tag{1.6.1}$$

If this electrostatic field is also source-free, $\nabla \cdot \boldsymbol{E} = 0$, implying that

$$\frac{\partial E_x}{\partial x} + \frac{\partial E_y}{\partial y} = 0$$

then there exists an electric flux function $v(x, y)$ such that

$$E_x = -\frac{\partial v}{\partial y}, \quad E_y = \frac{\partial v}{\partial x}$$

Thus, a complex potential function can be constructed as

$$f(z) = u(x, y) + \mathrm{i}v(x, y) \tag{1.6.2}$$

The complex potential function $f(z) = u(x, y) + \mathrm{i}v(x, y)$ is an analytic function. This is because, on one hand, it satisfies the Cauchy–Riemann conditions:

$$\frac{\partial u}{\partial x} = \frac{\partial v}{\partial y}, \quad \frac{\partial v}{\partial x} = -\frac{\partial u}{\partial y}$$

On the other hand, we have both $\nabla^2 u = 0$ and $\nabla^2 v = 0$. Therefore, $u(x, y)$ and $v(x, y)$ are conjugate harmonic functions.

The field strength of a planar electrostatic field can be represented by the complex potential as

$$E = E_x + \mathrm{i}E_y = -\frac{\partial u}{\partial x} - \mathrm{i}\frac{\partial u}{\partial y} = -\frac{\partial u}{\partial x} + \mathrm{i}\frac{\partial v}{\partial x} = -\overline{f'(z)} \tag{1.6.3}$$

Thus, it is proven that a two-dimensional electrostatic field without charge can be described by an analytic function — the complex potential, with the real and imaginary parts representing the electrostatic potential and electric flux function, respectively.

The contour lines $u(x, y) = D$ represent equipotential lines, while the contour lines $v(x, y) = C$ represent electric field lines. That is, for a given constant, one can draw an electric field line in the plane. $v(x, y)$ is called the flux function, and its physical significance is that the difference between two electric field lines' constants represents the electric flux passing through the region between those two lines, as shown in Figure 1.22.

$$\Phi_{AB} = \int_A^B \boldsymbol{E} \cdot \boldsymbol{n} \mathrm{d}l = C_1 - C_4$$

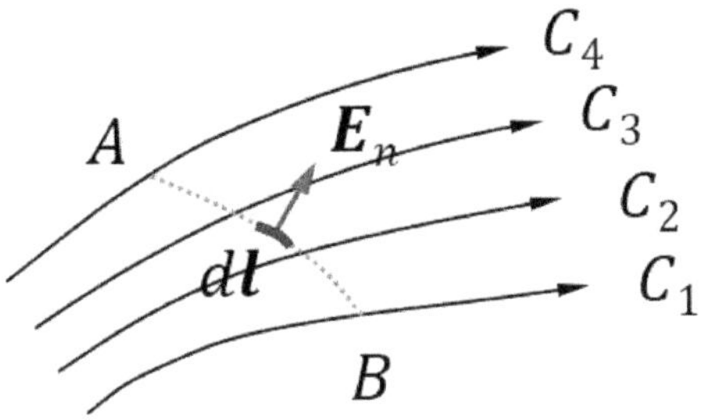

Figure 1.22.

Example 1.12. Given that the electric field lines of the planar electrostatic field are parametric parabolas of the form $y^2 = c^2 + 2cx (c > 0)$, we need to find the equations of the equipotential lines.

Solution: Solving the electric field line equation for the parameter c, we have

$$c = -x \pm \sqrt{x^2 + y^2}$$

Since $c > 0$, we choose the positive sign:

$$-x + \sqrt{x^2 + y^2} = c$$

The equation for the electric field lines is $v(x, y) = c$. Does this mean that we can take the imaginary part of the complex potential as $v(x, y) = -x + \sqrt{x^2 + y^2}$? However, upon verification, this $v(x, y)$ does not satisfy the Laplace equation, i.e., it is not a harmonic function. Therefore, one needs to find another function. Let's assume

$$t = -x + \sqrt{x^2 + y^2}$$

and let $v = F(t)$. As this $v(x, y)$ can also be constant, it gives us the equation for the electric field lines. According to the requirement of harmonic functions, we have

$$\frac{\partial v}{\partial x} = F'(t)\left[\frac{x}{\sqrt{x^2+y^2}} - 1\right]$$

$$\frac{\partial^2 v}{\partial x^2} = F''(t)\left[\frac{x}{\sqrt{x^2+y^2}} - 1\right]^2 + F'(t)\left[\frac{1}{\sqrt{x^2+y^2}} - \frac{x^2}{(x^2+y^2)^{3/2}}\right]$$

$$= F''(t)\left[\frac{x}{\sqrt{x^2+y^2}} - 1\right]^2 + F'(t)\frac{y^2}{(x^2+y^2)^{3/2}}$$

Similarly, we have

$$\frac{\partial^2 v}{\partial y^2} = F''(t)\left[\frac{y}{\sqrt{x^2+y^2}}\right]^2 + F'(t)\frac{x^2}{(x^2+y^2)^{3/2}}$$

Substituting these into the Laplace equation, it yields

$$2F''(t)\left[1 - \frac{x}{\sqrt{x^2+y^2}}\right] + F'(t)\frac{1}{\sqrt{x^2+y^2}} = 0$$

Solving this differential equation, we get

$$F(t) = C_1\sqrt{t} + C_2$$

Therefore, the equation for the electric field lines is

$$v = F(t) = C_1\sqrt{t} + C_2 = C_1\sqrt{-x + \sqrt{x^2+y^2}} + C_2$$

Referring to the result from Example 1.10 in Section 1.4, the equation of the potential lines is

$$u = C_1\sqrt{2\rho}\cos\frac{\phi}{2} + C_3$$

Converting back to the Cartesian coordinate system, we obtain the equation satisfied by the equipotential lines:

$$y^2 = c^2 - 2cx$$

Hence, the orthogonal curves to the parabolas are also parabolas.

1.6.2 *Planar velocity field*

Consider a two-dimensional vector field $\boldsymbol{v}$ representing the steady flow velocity field of an incompressible ideal fluid. If the velocity field is irrotational, $\nabla \times \boldsymbol{v} = 0$, we can introduce a velocity potential function $\phi(x, y)$ such that

$$v_x = \frac{\partial \phi}{\partial x}, \quad v_y = \frac{\partial \phi}{\partial y}$$

If the velocity field has no sources or leaks, i.e., the divergence is zero,

$$\nabla \cdot \mathbf{v} = \frac{\partial v_x}{\partial x} + \frac{\partial v_y}{\partial y} = 0$$

then we can introduce a stream function $\psi(x, y)$ such that

$$v_x = \frac{\partial \psi}{\partial y}, \quad v_y = -\frac{\partial \psi}{\partial x}$$

Hence, we have the equations

$$\frac{\partial \phi}{\partial x} = \frac{\partial \psi}{\partial y}, \quad \frac{\partial \phi}{\partial y} = -\frac{\partial \psi}{\partial x}$$

which are the Cauchy–Riemann conditions. It can also be shown that $\phi(x, y)$ and $\psi(x, y)$ are conjugate harmonic functions. Therefore, we define the complex potential as

$$f(z) = \phi(x, y) + \mathrm{i}\psi(x, y)$$

which is also an analytic function. The two-dimensional velocity field is represented by the complex potential as

$$\boldsymbol{v} = \nabla\phi = \frac{\partial \phi}{\partial x} + \mathrm{i}\frac{\partial \phi}{\partial y} = \frac{\partial \phi}{\partial x} - \mathrm{i}\frac{\partial \psi}{\partial x} = \overline{f'(z)} \tag{1.6.4}$$

1.6.3 *Planar heat flux field*

Similarly, we can use a complex potential to describe the temperature field without heat sources by using $u(x, y)$ represent the temperature distribution (real part) and $v(x, y)$ represent the heat flux function within the plane (imaginary part). Further details are omitted here.

Exercises

[1] Given the equation of equipotential lines as $x^2 + y^2 = c$, find the complex potential.
Answer: $a \ln z + b$

[2] Given that the electric field lines are a family of circles tangent to the real axis at the origin, find the complex potential.
Answer: $\frac{a}{z} + b$

[3] Prove that the electric flux passing through points A and B, as shown in Figure 1.22, is given by

$$\Phi_{AB} = \int_A^B \boldsymbol{E} \cdot \boldsymbol{n} \mathrm{d}l = C_1 - C_4$$

Hint: $\int_A^B \boldsymbol{E} \cdot \boldsymbol{n} \mathrm{d}l = - \int_A^B \mathrm{d}v$

Chapter 2

Path Integrals

2.1 Complex Line Integrals

2.1.1 *Definition of integral*

Let L be a piecewise smooth curve on the complex plane. Along the curve, we take a series of points $z_1, z_2, \ldots, z_{n-1}$ from the starting point to the endpoint, dividing the curve into n small segments, as shown in Figure 2.1. On each segment $[z_{k-1}, z_k]$, we choose a point ζ_k and take the sum $\sum_{k=1}^{n} f(\zeta_j)(z_k - z_{k-1})$. By taking the limit of $n \to \infty$, with $\Delta z_k = z_k - z_{k-1} \to 0$, the integral of a complex function along the curve L is defined as follows:

$$\int_L f(z)\mathrm{d}z \stackrel{\text{def}}{=} \lim_{n\to\infty} \sum_{k=1}^{n} f(\zeta_j)(z_k - z_{k-1}) \tag{2.1.1}$$

The integral of a complex function, known as the complex line integral, depends on the path L. When computing the integral, it is generally necessary to specify the path taken on the complex plane. The reverse path of L is denoted by L^-. If the integral path of a single-valued function forms a closed loop, it is called a closed contour integral. Conventionally, the integral is defined as positive when traversed counterclockwise.

Note that $z_k - z_{k-1}$ is not the length $|z_k - z_{k-1}|$ of the line segment. It is a "directed" line segment, carrying an angular factor.

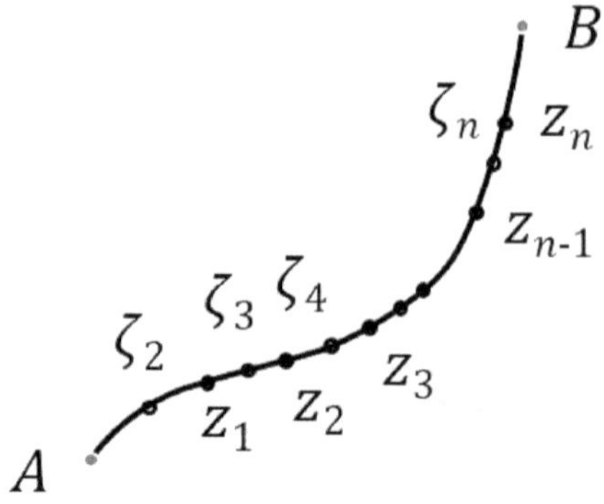

Figure 2.1.

2.1.2 *Basic properties*

The path integral of a complex function has the following properties:

[1] $\int_L [Af(z) + Bg(z)]\mathrm{d}z = A\int_L f(z)\mathrm{d}z + B\int_L g(z)\mathrm{d}z$

[2] $\int_{L_1+L_2} f(z)\mathrm{d}z = \int_{L_1} f(z)\mathrm{d}z + \int_{L_2} f(z)\mathrm{d}z$

[3] $\int_{L^-} f(z)\mathrm{d}z = -\int_L f(z)\mathrm{d}z$

[4] $|\int_L f(z)\mathrm{d}z| \leq \int_L |f(z)||\mathrm{d}z|$

[5] $|\int_L f(z)\mathrm{d}z| \leq Ml$

where A and B are constants. In the last two properties, $M = \max|f(z)|$ represents the maximum value of $|f(z)|$ along the integration path, and l is the length of path L. These properties, which are analogous to the properties of integrals in real variable functions, can be directly proven from the definition of the line integral.

2.1.3 *Computing path integrals*

(1) Conversion to integrals of two real variables

$$\int_L f(z)\mathrm{d}z = \int_L [u(x,y)\mathrm{d}x - v(x,y)\mathrm{d}y] + \mathrm{i}\int_L [v(x,y)\mathrm{d}x + u(x,y)\mathrm{d}y] \tag{2.1.2}$$

Example 2.1. Calculate the integrals

$$\text{(a)} \int_{L_1} \operatorname{Re} z \, \mathrm{d}z; \qquad \text{(b)} \int_{L_2} \operatorname{Re} z \, \mathrm{d}z$$

where the integration paths L_1 and L_2 are shown in Figure 2.2.

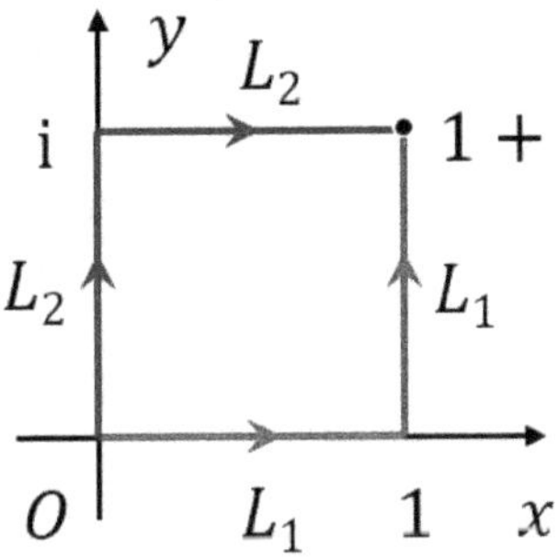

Figure 2.2.

Solution: (a) The integrals along path L_1 is

$$\int_{L_1} \operatorname{Re} z \, \mathrm{d}z = \int_0^1 x \, \mathrm{d}x + \int_0^1 1 \cdot \mathrm{i} \mathrm{d}y = \frac{1}{2} + \mathrm{i}$$

(b) While the integral along path L_2 is

$$\int_{L_2} \operatorname{Re} z \, \mathrm{d}z = \int_0^1 0 \cdot \mathrm{i} \mathrm{d}y + \int_0^1 x \, \mathrm{d}x = \frac{1}{2}$$

We see that the two results are different, indicating that the integral outcome depends on the chosen path.

(2) Parameterization of paths

$$\int_L f(z) \mathrm{d}z = \int_A^B f[z(t)] z'(t) \mathrm{d}t \tag{2.1.3}$$

Reflection: The integral of a complex function generally depends on the path. Under what conditions does the integral only depend on the starting and ending positions and not on the path chosen?

Exercises

[1] Prove the inequality:

$$\left|\int_L f(z)\mathrm{d}z\right| \leq \int_L |f(z)||\mathrm{d}z|$$

[2] Let L be a closed curve in the complex plane, with its four vertices at $z_1 = 0$, $z_2 = 1$, $z_3 = 1+\mathrm{i}$, and $z_4 = \mathrm{i}$. We calculate the following integrals:

(a) $\oint_L (z^2+1)\mathrm{d}z$; (b) $\oint_L (|z|^2+1)\mathrm{d}z$

(c) $\oint_L \mathrm{e}^{\pi\bar{z}}\mathrm{d}z$; (d) $\oint_L \dfrac{1}{z^2-\frac{1}{2}(1+\mathrm{i})}\mathrm{d}z$

2.2 Cauchy's Theorem

2.2.1 *Simply connected domain*

Cauchy's theorem states that if a function $f(z)$ is analytic within a simply connected region B, then along any smooth closed path C within B, we have

$$\oint_C f(z)\mathrm{d}z = 0 \tag{2.2.1}$$

Proof. Let $f(z) = u(x,y)+\mathrm{i}v(x,y)$, and consider the contour integral along the path C:

$$\oint_C f(z)\mathrm{d}z = \oint_C [u(x,y)\mathrm{d}x - v(x,y)\mathrm{d}y] + \mathrm{i}\oint_C [v(x,y)\mathrm{d}x + u(x,y)\mathrm{d}y]$$

Since $f(z)$ is analytic within the simply connected region B, both the real and imaginary parts are differentiable. We can apply Green's theorem

$$\oint_C [P\mathrm{d}x + Q\mathrm{d}y] = \iint_\sigma \left(\frac{\partial Q}{\partial x} - \frac{\partial P}{\partial y}\right)\mathrm{d}x\mathrm{d}y$$

to convert the contour integral to a double integral over the region enclosed by the contour, which yields

$$\oint_C f(z)\mathrm{d}z = \iint_\sigma \left(-\frac{\partial v}{\partial x} - \frac{\partial u}{\partial y}\right)\mathrm{d}x\mathrm{d}y + \mathrm{i}\iint_\sigma \left(\frac{\partial u}{\partial x} - \frac{\partial v}{\partial y}\right)\mathrm{d}x\mathrm{d}y$$

The two parts are equal to zero from the Cauchy–Riemann conditions. □

The fact that the contour integral of a complex function is zero for any closed path is equivalent to the closed path being continuously deformable to a point.

Example 2.2. Prove the integral

$$\oint_{|z|=1} \frac{1}{z^2+2z+2}\mathrm{d}z = 0$$

Proof. The singular points of the function to be integrated are given by $z^2+2z+2=0$, which implies $z_{1,2} = -1 \pm \mathrm{i}$. Since both of these singular points are located outside the integration path, as shown in Figure 2.3, the function being integrated is analytic within the unit circle $|z| = 1$. Therefore, according to Cauchy's theorem, the contour integral along this path must be zero. □

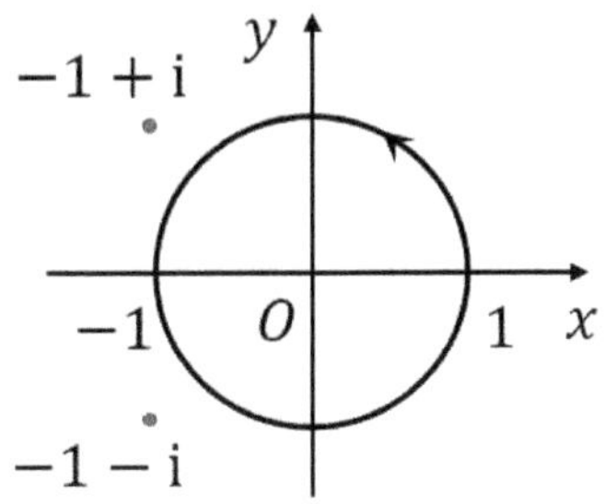

Figure 2.3.

From Cauchy's theorem, we know that for an analytic function over a simply connected region, as long as the starting and ending points remain fixed, the integral value of the function remains constant regardless of the path taken. Thus,

$$\oint_C f(z)\mathrm{d}z = 0 \longrightarrow \int_{L_1} f(z)\mathrm{d}z + \int_{L_2} f(z)\mathrm{d}z = 0$$

It reduces to

$$\int_{L_1} f(z)\mathrm{d}z = \int_{L_2^-} f(z)\mathrm{d}z$$

As shown in Figure 2.4, the path L_1 can be continuously deformed into L_2. It is referred as homotopy between paths L_1 and L_2.

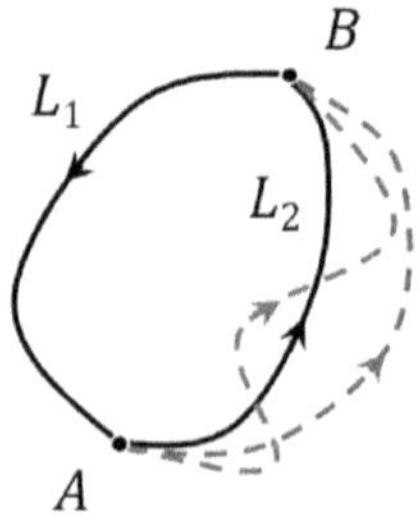

Figure 2.4.

2.2.2 *Multiply connected domain*

In the case of a multiply connected domain, let B be a region surrounded by $C_0, C_1, C_2, \ldots, C_n$. Cauchy's theorem states that if a function $f(z)$ is analytic within B, then we have

$$\oint_{C_0} f(z)\mathrm{d}z = \sum_{k=1}^{n} \oint_{C_k} f(z)\mathrm{d}z \tag{2.2.2}$$

Proof. As shown in Figure 2.5(a), establish bridge paths between the outer contour C_0 and all the inner contours $C_1, C_2, \ldots, C_n$. These paths form the boundary of a simply connected region, as shown in Figure 2.5(b). Since the total effect of integrating along these bridge paths is zero and does not contribute to the contour integral, according to Cauchy's theorem for simply connected domains, we have

$$\begin{aligned}\oint_{C_0+C_1^-+C_2^-+\cdots} f(z)\mathrm{d}z &= \oint_{C_0} f(z)\mathrm{d}z + \oint_{C_1^-} f(z)\mathrm{d}z \\ &\quad + \oint_{C_2^-} f(z)\mathrm{d}z + \cdots + \oint_{C_n^-} f(z)\mathrm{d}z = 0\end{aligned}$$

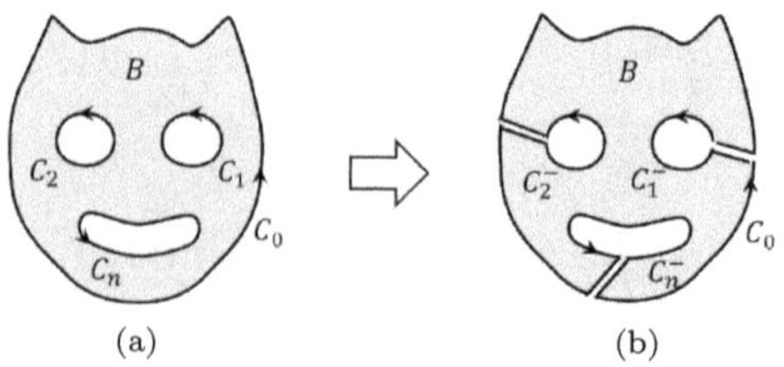

Figure 2.5.

Therefore,

$$\oint_{C_0} f(z)\mathrm{d}z = \sum_{k=1}^{n} \oint_{C_k} f(z)\mathrm{d}z$$

□

Example 2.3. Compute the integral

$$\oint_{|z|=2} \frac{3z-1}{z(z-1)}\mathrm{d}z$$

Solution: The function to be integrated has two singular points, $z_1 = 0$ and $z_2 = 1$, both of which lie inside the circular integration path $|z| = 2$. Let's consider two small circular paths C_1 and C_2 centered at the singular points, as shown in Figure 2.6. According to Cauchy's theorem for multiply connected domains, the integration along the path $|z| = 2$ is equal to the sum of the integrations along the two small circular paths:

$$\oint_{|z|=2} \frac{3z-1}{z(z-1)}\mathrm{d}z = \oint_{C_1} \frac{3z-1}{z(z-1)}\mathrm{d}z + \oint_{C_2} \frac{3z-1}{z(z-1)}\mathrm{d}z$$

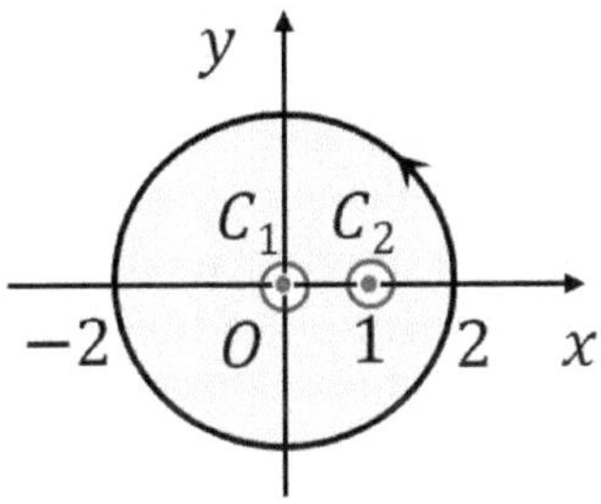

Figure 2.6.

For the path C_1, taking the parameterization of the circle as $z = \varepsilon e^{i\theta}$, where $\varepsilon \to 0$, we have

$$\oint_{C_1} \frac{3z-1}{z(z-1)}\mathrm{d}z = \int_0^{2\pi} \frac{3\varepsilon \mathrm{e}^{\mathrm{i}\theta} - 1}{\varepsilon \mathrm{e}^{\mathrm{i}\theta}(\varepsilon \mathrm{e}^{\mathrm{i}\theta} - 1)} \mathrm{i}\varepsilon \mathrm{e}^{\mathrm{i}\theta}\mathrm{d}\theta \xrightarrow{\varepsilon \to 0} \int_0^{2\pi} \mathrm{i}\mathrm{d}\theta = 2\pi\mathrm{i}$$

For the path C_2, taking the parameterization as $z - 1 = \varepsilon \mathrm{e}^{\mathrm{i}\theta}$ and letting $\varepsilon \to 0$, we obtain

$$\oint_{C_2} \frac{3z-1}{z(z-1)}\mathrm{d}z = \int_0^{2\pi} \frac{3\varepsilon \mathrm{e}^{\mathrm{i}\theta} + 2}{(1 + \varepsilon \mathrm{e}^{\mathrm{i}\theta})\varepsilon \mathrm{e}^{\mathrm{i}\theta}} \mathrm{i}\varepsilon \mathrm{e}^{\mathrm{i}\theta}\mathrm{d}\theta \xrightarrow{\varepsilon \to 0} \int_0^{2\pi} 2\mathrm{i}\mathrm{d}\theta = 4\pi\mathrm{i}$$

The final result is

$$\oint_{|z|=2} \frac{3z-1}{z(z-1)}\mathrm{d}z = 6\pi\mathrm{i}$$

2.2.3 *Primitive functions*

If $f(z) = \frac{d}{dz}F(z)$, then $F(z)$ is called a primitive function of $f(z)$. According to Cauchy's theorem, if $f(z)$ is analytic within a simply connected region B, then the integral along any path within B, $\int f(z)dz$, depends only on the initial and final points and is independent of the integration path. When the initial point $z_0 \in B$ is fixed, this integral defines a single-valued function of the endpoint z, known as a potential function in physics, denoted as follows:

$$F(z) = \int_{z_0}^{z} f(\zeta)\mathrm{d}\zeta$$

The primitive function of a $f(z)$ is generally expressed as $F(z) + c$, where c is an arbitrary constant. It is called the indefinite integral of the function $f(z)$,

$$\int f(z)\mathrm{d}z = F(z) + c$$

For a non-simply connected region, such as the region G shown in Figure 2.7, which contains a singularity, the integral of the function cannot be expressed in the form of an indefinite integral and requires specifying the integration path.

Example 2.4. Compute the integral:

$$\int_0^1 z\cos z\,\mathrm{d}z$$

Solution: In this example, the integration path is not specified, which means the integrand is analytic in the entire complex plane. Therefore, we can choose any smooth integration path, such as integrating along the real axis segment $[0, 1]$. Thus,

$$\int_0^1 z\cos z\,\mathrm{d}z = \int_0^1 x\cos x\,\mathrm{d}x = \sin 1 + \cos 1 - 1$$

If a different smooth path is chosen, it may not be technically feasible to compute the integral value, but the result will always be the same.

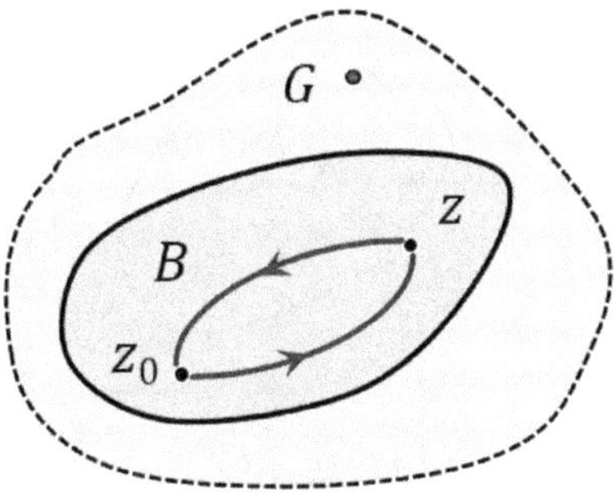

Figure 2.7.

Example 2.5. Prove the integral:

$$\oint_L \frac{\mathrm{d}z}{(z-a)^n} = \begin{cases} 2\pi\mathrm{i} & (n=1) \\ 0 & (n \neq 1) \end{cases} \tag{2.2.3}$$

where L is an arbitrary closed contour enclosing the point a and $n \in \mathbb{Z}$.

Proof. The integrand has only one singularity at $z = a$, as shown in Figure 2.8. Let's draw a circular contour C with center at a and radius r. According to Cauchy's theorem for multiply connected domains, the integration along the contour L is equal to the integration along the circular contour C:

$$\oint_L \frac{\mathrm{d}z}{(z-a)^n} = \oint_C \frac{\mathrm{d}z}{(z-a)^n} = \int_0^{2\pi} \frac{\mathrm{i}r\mathrm{e}^{\mathrm{i}\theta}}{r^n\mathrm{e}^{in\theta}}\mathrm{d}\theta$$

When $n = 1$, we have

$$\int_0^{2\pi} \frac{\mathrm{i}r\mathrm{e}^{i\theta}}{r^n\mathrm{e}^{in\theta}}\mathrm{d}\theta = \mathrm{i}\int_0^{2\pi} 1\mathrm{d}\theta = 2\pi\mathrm{i}$$

When $n \neq 1$, we have

$$\int_0^{2\pi} \frac{\mathrm{i}r\mathrm{e}^{\mathrm{i}\theta}}{r^n\mathrm{e}^{\mathrm{i}n\theta}}\mathrm{d}\theta = \frac{\mathrm{i}}{r^{n-1}}\int_0^{2\pi} \mathrm{e}^{\mathrm{i}(1-n)\theta}\mathrm{d}\theta = 0$$

□

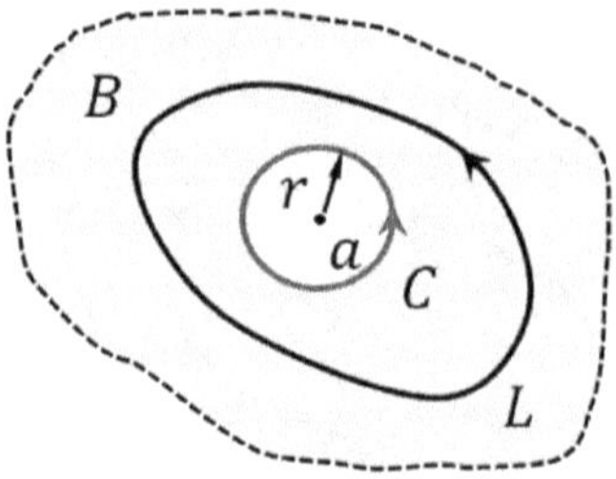

Figure 2.8.

Discussion: Why is the $n = 1$ case special. It is because the primitive function of the integrand is a multivalued logarithmic function. When the complex variable z completes one full loop around the point a, the argument of the logarithmic function increases by exactly 2π. For other integer values of $n \neq 1$, the primitive function of the integrand is a single-valued power function. As a result, the contour integral evaluates to zero.

ENDNOTE

If a vector field is irrotational or curl-free, a scalar potential function can be introduced, as observed in electrostatic or gravitational fields. For magnetic fields, although a vector potential can be introduced due to the absence of magnetic charges, Ampere's loop theorem states that $\oint_\Gamma \boldsymbol{B} \cdot \mathrm{d}\boldsymbol{l} = \mu_0 I$, and the magnetic field lines are closed. Therefore, a scalar potential function cannot be introduced. However, nothing is absolute. If a region is defined, such as the region B shown in Figure 2.9, where no current passes through, the integral of magnetic field intensity along any closed loop is zero. In this case, a magnetic scalar potential φ_{m} can be introduced, $\boldsymbol{H} = \nabla\varphi_{\mathrm{m}}$, which leads to the introduction of scalar fields and complex potentials. It will be mentioned later that complex potential is a more general concept, even applicable to vortical fields with non-zero circulation.

By the way, the magnetic scalar potential φ_{m} is a very useful concept in electrical engineering.

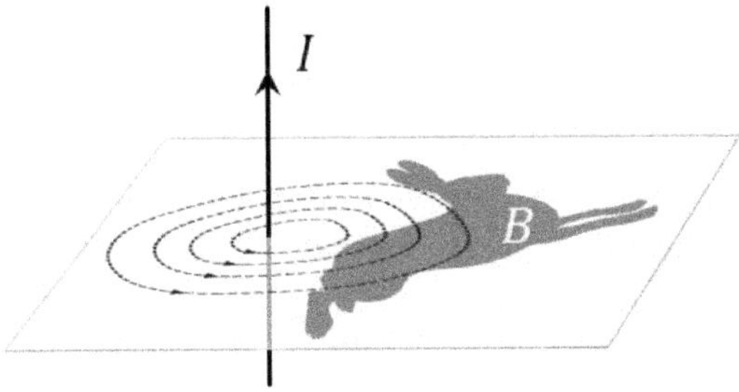

Figure 2.9.

Exercise

[1] Compute the integral:

$$\oint_{|z|=2} \frac{z^2}{(z^2+1)(z-3)} \mathrm{d}z$$

[2] Compute the integral along the contour L shown in Figure 2.10(a):

$$\oint_L \frac{\sin z}{z^2 - z} \mathrm{d}z$$

[3] Compute the integral along the contour L shown in Figure 2.10(b):

$$\oint_L \frac{\mathrm{e}^z - \mathrm{e}^{-z}}{z^4} \mathrm{d}z$$

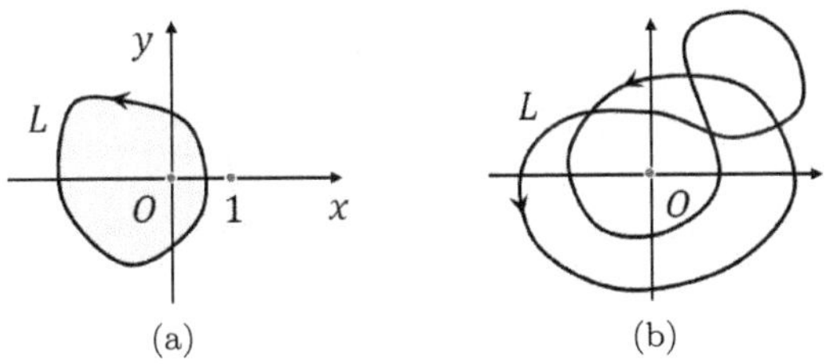

Figure 2.10.

2.3 Cauchy's Integral Formula

2.3.1 *Simply connected domain*

Let $f(z)$ be analytic in a simply connected region B and continuous on $\bar{B}$. For any point ζ in B, consider any closed contour L that includes the point ζ. Then, we have Cauchy's integral formula:

$$f(\zeta) = \frac{1}{2\pi \mathrm{i}} \oint_L \frac{f(z)}{z-\zeta} \mathrm{d}z \tag{2.3.1}$$

Proof. Taking a circular path C_ε centered at ζ with radius ε, as shown in Figure 2.11, Cauchy's theorem tells

$$\frac{1}{2\pi \mathrm{i}} \oint_L \frac{f(z)}{z-\zeta} \mathrm{d}z = \frac{1}{2\pi \mathrm{i}} \oint_{C_\varepsilon} \frac{f(z)}{z-\zeta} \mathrm{d}z$$

Considering the continuity of the analytic function $f(z)$ at the point ζ, it has $f(z) \to f(\zeta)$ as $\varepsilon \to 0$. Hence,

$$\frac{1}{2\pi \mathrm{i}} \oint_{C_\varepsilon} \frac{f(z)}{z-\zeta} \mathrm{d}z = \frac{1}{2\pi \mathrm{i}} \oint_{C_\varepsilon} \frac{f(\zeta)}{z-\zeta} \mathrm{d}z = \frac{f(\zeta)}{2\pi \mathrm{i}} \oint_{C_\varepsilon} \frac{1}{z-\zeta} \mathrm{d}z = f(\zeta)$$

The formula can also be written as

$$f(z) = \frac{1}{2\pi \mathrm{i}} \oint_L \frac{f(\zeta)}{\zeta - z} \mathrm{d}\zeta \tag{2.3.2}$$

It implies that an artificial singularity is introduced for the analytic function within the closed contour L, and the integral value around this singularity is 2πi. □

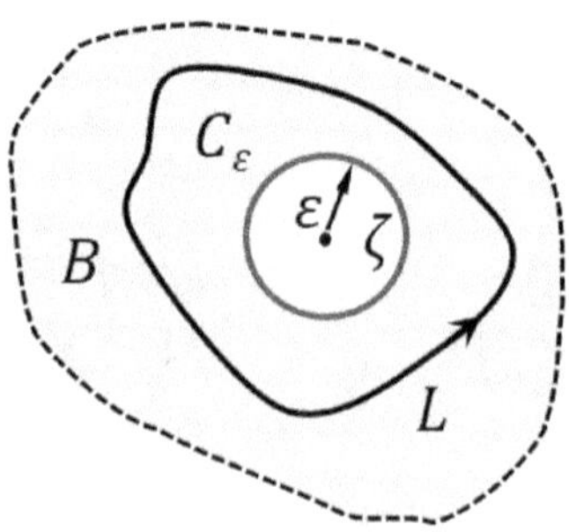

Figure 2.11.

Example 2.6. Calculate the integral:

$$\oint_{|z|=3} \frac{e^z}{z(z^2+1)} dz$$

Solution: The singular points of the integrand are $z_{1,2,3} = 0, \pm i$. Taking small circular contours C_1, C_2, C_3 centered at each singularity, according to Cauchy's theorem for multiply connected domains, the contour integral can be transformed into the sum of integrals along each small contour:

$$\oint_{|z|=3} \frac{e^z}{z(z^2+1)} dz = \oint_{C_1} \frac{\left[\frac{e^z}{z^2+1}\right]}{z} dz + \oint_{C_2} \frac{\left[\frac{e^z}{z(z+i)}\right]}{z-i} dz + \oint_{C_3} \frac{\left[\frac{e^z}{z(z-i)}\right]}{z+i} dz$$

By separately using Cauchy's integral formula, we have

$$\begin{aligned}\oint_{|z|=3} \frac{e^z}{z(z^2+1)} dz &= 2\pi i \left[\left.\frac{e^z}{z^2+1}\right|_{z=0} + \left.\frac{e^z}{z(z+i)}\right|_{z=i} + \left.\frac{e^z}{z(z-i)}\right|_{z=-i} \right] \\ &= 2\pi i(1 - \cos 1)\end{aligned}$$

Cauchy's integral formula states that the value of an analytic function at any point within a simply connected domain is completely determined by its values on the boundary. Physically, an analytic function corresponds to a source-free and circulation-free plane vector field, such as an electrostatic field, whose potential satisfies Laplace's equation. Its distribution within the region is fully determined by its values on the boundary. Cauchy's integral formula can be used to derive two types of Poisson's formulas for a half-plane and a circular domain, respectively, which are applications for specific boundary conditions. In the following, we illustrate the example of a half-plane.

Example 2.7. Assume $f(z)$ is analytic in the upper half-plane and as $z \to \infty$, $|f(z)| \to 0$. Consider the semicircular contour shown in Figure 2.12. We want to prove Poisson's formula:

$$f(z) = \frac{y}{\pi} \int_{-\infty}^{\infty} \frac{f(\xi, 0)}{(\xi - x)^2 + y^2} d\xi$$

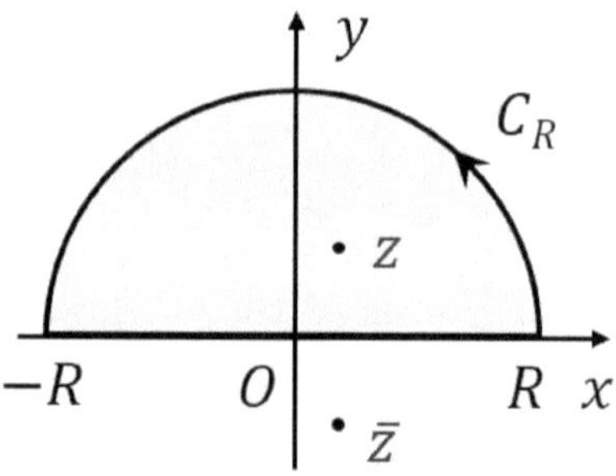

Figure 2.12.

Proof. According to Cauchy's integral formula,

$$f(z) = \frac{1}{2\pi \mathrm{i}} \oint_L \frac{f(\zeta)}{\zeta - z} \mathrm{d}\zeta = \frac{1}{2\pi \mathrm{i}} \int_{-R}^{R} \frac{f(\xi)}{\xi - z} \mathrm{d}\xi + \frac{1}{2\pi \mathrm{i}} \int_{C_R} \frac{f(\zeta)}{\zeta - z} \mathrm{d}\zeta$$

Since $|\zeta - z| \sim R$ in the second term, its modulus satisfies

$$\left| \frac{1}{2\pi \mathrm{i}} \int_{C_R} \frac{f(\zeta)}{\zeta - z} \mathrm{d}\zeta \right| \leq \frac{1}{2\pi} \int_{C_R} \left| \frac{f(\zeta)}{\zeta - z} \right| |\mathrm{d}\zeta| \leq \max_{\zeta \in C_R} |f(\zeta)| \xrightarrow{R \to \infty} 0$$

Therefore,

$$f(z) = \frac{1}{2\pi \mathrm{i}} \int_{-\infty}^{\infty} \frac{f(\xi)}{\xi - z} \mathrm{d}\xi \ (\xi \in \mathbb{R})$$

This means that the value of the function at any point in the upper half-plane is fully determined by the values of $f(z)$ on the real axis. Additionally, since the complex conjugate $\bar{z}$ of z lies outside the integration contour, according to Cauchy's theorem, we know that

$$\frac{1}{2\pi \mathrm{i}} \oint_C \frac{f(\zeta)}{\zeta - \bar{z}} \mathrm{d}\zeta = \frac{1}{2\pi \mathrm{i}} \int_{-\infty}^{\infty} \frac{f(\xi)}{\xi - \bar{z}} \mathrm{d}\xi = 0$$

Subtracting these two equations gives Poisson's formula:

$$\begin{aligned} f(z) &= \frac{1}{2\pi \mathrm{i}} \int_{-\infty}^{\infty} \frac{f(\xi)}{\xi - z} \mathrm{d}\xi - \frac{1}{2\pi \mathrm{i}} \int_{-\infty}^{\infty} \frac{f(\xi)}{\xi - \bar{z}} \mathrm{d}\xi \\ &= \frac{y}{\pi} \int_{-\infty}^{\infty} \frac{f(\xi, 0)}{(\xi - x)^2 + y^2} \mathrm{d}\xi \end{aligned}$$

Proof complete. □

In Poisson's formula, if we let $f(z) = u(x, y) + \mathrm{i}v(x, y)$ and take the real part, we obtain

$$u(x, y) = \frac{y}{\pi} \int_{-\infty}^{\infty} \frac{u(\xi, 0)}{(\xi - x)^2 + y^2} \mathrm{d}\xi$$

Here, the function $u(x, y)$ satisfies the harmonic condition $\Delta u = 0$, which means that a harmonic function in the upper half-plane is fully determined by its values on the real axis.

2.3.2 *Integral representation of derivative*

Suppose $f(z)$ is analytic in a simply connected region B and continuous on $\bar{B}$, then for any point ζ in B, $f(z)$ has high-order derivatives given by

$$f^{(n)}(\zeta) = \frac{n!}{2\pi \mathrm{i}} \oint_L \frac{f(z)}{(z - \zeta)^{n+1}} \mathrm{d}z \tag{2.3.3}$$

Proof. Taking the nth derivative with respect to ζ on both sides of the equation gives the desired result. This equation can also be expressed as

$$f^{(n)}(z) = \frac{n!}{2\pi \mathrm{i}} \oint_L \frac{f(\zeta)}{(\zeta - z)^{n+1}} \mathrm{d}\zeta \tag{2.3.4}$$

It means that the value of the nth derivative of an analytic function at any point in a simply connected region is fully determined by the values of the function on the boundary. This formula also implies that if a single-valued function is analytic in a certain region, then it is infinitely differentiable in that region. □

Example 2.8. Compute the integral:

$$\oint_{|z-\mathrm{i}|=1} \frac{1}{(z^2 + 1)^2} \mathrm{d}z$$

Solution: As shown in Figure 2.13, the only singularity $z = \mathrm{i}$ lies inside the contour, so we can apply the high-order derivative formula to obtain

$$\oint_{|z-\mathrm{i}|=1} \frac{1}{(z^2+1)^2}\mathrm{d}z = \oint_{|z-\mathrm{i}|=1} \frac{\left[1/(z+\mathrm{i})^2\right]}{(z-\mathrm{i})^2}\mathrm{d}z$$

$$= 2\pi\mathrm{i}\frac{\mathrm{d}}{\mathrm{d}z}\left[\frac{1}{(z+\mathrm{i})^2}\right]\Big|_{z=\mathrm{i}} = \frac{\pi}{2}$$

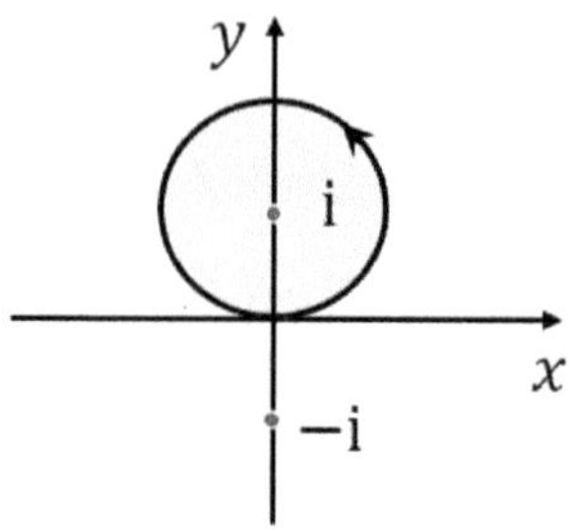

Figure 2.13.

Practice: Try calculating the integration using the parametric representation of a circle method.

2.3.3 *Multiple connected domains*

Let B be a region consisting of $C_0, C_1, C_2, \ldots, C_n$ that enclose a multiple connected domain. If $f(z)$ is analytic in B and continuous on $\bar{B}$, then for any point ζ in B, we have

$$f(\zeta) = \frac{1}{2\pi\mathrm{i}}\oint_{C_0} \frac{f(z)}{z-\zeta}\mathrm{d}z + \frac{1}{2\pi\mathrm{i}}\sum_{k=1}^{n}\oint_{C_k^-} \frac{f(z)}{z-\zeta}\mathrm{d}z \tag{2.3.5}$$

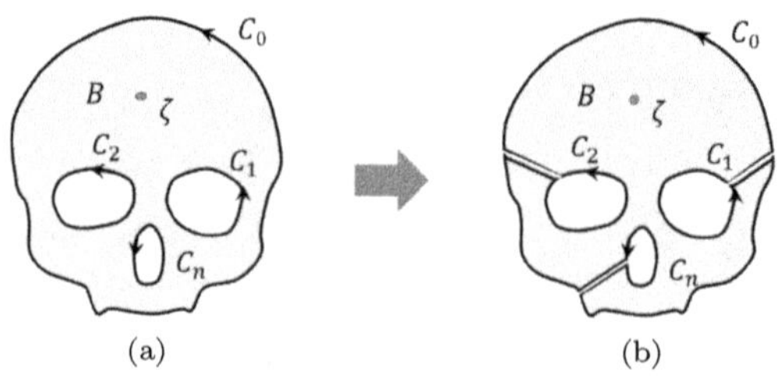

Figure 2.14.

Proof. Similar to the proof of Cauchy's theorem for multiply connected domains, we construct a bridge path as shown in Figure 2.14(b) to transform the multiple connected domain into a simply connected one. Then, we can use Cauchy's integral formula to prove the formula. □

This equation can also be expressed as

$$f(z) = \frac{1}{2\pi \mathrm{i}} \oint_{C_0} \frac{f(\zeta)}{\zeta - z} \mathrm{d}\zeta - \frac{1}{2\pi \mathrm{i}} \sum_{k=1}^{n} \oint_{C_k} \frac{f(\zeta)}{\zeta - z} \mathrm{d}\zeta \tag{2.3.6}$$

2.3.4 *The modulus theorems*

By applying the Cauchy integral formula, a series of theorems regarding analytic functions can be proven.

(1) **Maximum modulus theorem:** Let $f(z)$ be analytic in a region B and continuous on $\bar{B}$. Then, $|f(z)|$ can only attain its maximum value on the boundary curve L.

Proof. Consider the analytic function $[f(z)]^n$. If $\max|f(\zeta)| = M$ on the boundary curve and $|\zeta - z| \geq \delta$, where δ is a positive value, and l represents the total length of the boundary curve. By applying the Cauchy integral formula, it has

$$|[f(z)]^n| = \left|\frac{1}{2\pi \mathrm{i}} \oint_\Gamma \frac{[f(\zeta)]^n}{\zeta - z} \mathrm{d}\zeta\right| \leq \frac{1}{2\pi} \oint_\Gamma \left|\frac{[f(\zeta)]^n}{\zeta - z}\right| |\mathrm{d}\zeta| \leq \frac{1}{2\pi} \frac{M^n}{\delta} l$$

By taking the limit as $n \to \infty$, we obtain

$$|f(z)| \leq M \left(\frac{l}{2\pi\delta}\right)^{1/n} \leq M$$

Equality holds only when $f(z)$ is a constant function. □

A similar statement regarding the minimum modulus is not generally true. For example, in the closed disk $|z| \leq 1$, the function $f(z) = z^2$ has its minimum modulus not occurring on the circle. However, under certain constraints, the following theorem holds.

(2) **Minimum modulus theorem:** Let $f(z)$ be analytic in a region B and have no zeros in the region. Then, the minimum value of $|f(z)|$ can only occur on the boundary curve L, unless $f(z)$ is a constant function.

Proof. Let

$$g(z) = \frac{1}{f(z)}$$

Since $f(z)$ has no zeros, $g(z)$ is an analytic function in the region B. By the maximum modulus theorem, the maximum value of $|g(z)|$ occurs on the boundary curve. Therefore, the minimum value of $|f(z)|$ lies on the boundary. □

(3) **Liouville's theorem:** If $f(z)$ is an analytic function defined on the entire complex plane and is bounded, $|f(z)| \leq M$, then $f(z)$ must be a constant function.

Proof. Applying the Cauchy integral formula, we have

$$f'(z) = \frac{1}{2\pi \mathrm{i}} \oint_L \frac{f(\zeta)}{(\zeta - z)^2} d\zeta$$

Taking L as a circle centered at z with radius R, according to the maximum modulus theorem,

$$|f'(z)| \leq \frac{1}{2\pi} \frac{M}{R^2} 2\pi R = \frac{M}{R}$$

Since the radius R can be arbitrarily large, letting $R \to \infty$ implies $f'(z) = 0$ everywhere, indicating that $f(z)$ must be a constant. □

From Liouville's theorem, the following corollaries can be deduced:

Corollary 1. *If $f(z)$ is analytic on the entire complex plane and $|f(z)| \geq M$, then $f(z)$ must be a constant function.*

Corollary 2. *If $f(z)$ is analytic on the entire complex plane and $\lim_{z\to\infty} \frac{f(z)}{z} = 0$, then $f(z)$ must be a constant function.*

Corollary 3. *If an analytic function defined on the entire complex plane is not a constant function, it must have a singularity at $z = \infty$.*

The reader can practice proving these corollaries. In Chapter 1, the fundamental theorem of algebra was stated, and now a proof can be given.

(4) **Fundamental theorem of algebra:** Any polynomial equation

$$P(z) = a_n z^n + a_{n-1} z^{n-1} + \cdots + a_1 z + a_0 = 0$$

must have a complex root.

Proof. First, $P(z)$ is an analytic function on the complex plane, and as $z \to \infty$, $|P(z)| \to \infty$. We use a proof by contradiction: Suppose $P(z)$ has no complex roots, i.e., $P(z) \neq 0$. Then, $g(z) = \frac{1}{P(z)}$ is analytic on the entire complex plane. Taking a circle with the origin as its center and a radius of R, according to the maximum modulus theorem, the maximum value of $|g(z)|$ must occur on the circle. Since

$$|g(z)| = \frac{1}{|P(z)|} \xrightarrow{R \to \infty} 0$$

by Liouville's theorem, we conclude that $g(z) \equiv 0$, which contradicts our assumption. Hence, the proof is complete. □

Exercises

[1] Calculate the contour integration:

(a) $\oint_{|z-\mathrm{i}|=1} \frac{1}{(z^2+1)^2} \mathrm{d}z$;

(b) $\oint_{|z-1|+|z-\mathrm{i}|=4} \frac{1}{z(z^2+4)} \mathrm{d}z$

Answer: (a) $\frac{\pi}{2}$; (b) $\frac{\pi \mathrm{i}}{4}$

[2] Compute the integral:

$$\oint_{|z|=1} \frac{\sin z}{z^3} \mathrm{d}z$$

[3] Assume that $f(z)$ is analytic within a circle of radius a, and prove Poisson's formula for the circular domain:

$$f(z) = \frac{a^2 - r^2}{2\pi} \int_0^{2\pi} \frac{f(a\mathrm{e}^{\mathrm{i}\theta})}{a^2 + r^2 - 2ra\cos(\varphi - \theta)} \mathrm{d}\theta$$

Hint: Consider any point z within the circle and its conjugate $z_1 = a^2/\bar{z}$.

[4] Assume that $f(z)$ is analytic within a simply connected region B. Consider a circular boundary with z as the center and any radius ε and prove the mean value theorem:

$$f(z) = \frac{1}{2\pi}\int_0^{2\pi} f(z+\varepsilon e^{i\theta})d\theta$$

[5] For an analytic function $f(z) = u(x,y) + iv(x,y)$ in the upper half-plane, given $|f(z)| \xrightarrow{z\to\infty} 0$, prove the relationship between its real and imaginary parts:

$$\begin{cases} u(x,y) = \frac{x}{\pi}\int_{-\infty}^{\infty}\frac{v(\xi,0)}{(\xi-x)^2+y^2}d\xi \\ v(x,y) = -\frac{x}{\pi}\int_{-\infty}^{\infty}\frac{u(\xi,0)}{(\xi-x)^2+y^2}d\xi \end{cases}$$

Hint:

$$\begin{aligned} f(z) &= \frac{1}{2\pi i}\int_{-\infty}^{\infty}\frac{f(\xi)}{\xi - z}d\xi + \frac{1}{2\pi i}\int_{-\infty}^{\infty}\frac{f(\xi)}{\xi - \bar{z}}d\xi \\ &= \frac{x}{\pi i}\int_{-\infty}^{\infty}\frac{f(\xi,0)}{(\xi - x)^2 + y^2}d\xi \end{aligned}$$

[6] Given the function $\psi(t,x) = e^{2tx-t^2}$, where t is a complex variable, prove the formula:

$$\left.\frac{\partial^n}{\partial t^n}\psi(t,x)\right|_{t=0} = (-1)^n e^{x^2}\frac{d^n}{dx^n}e^{-x^2}$$

Hint:

$$\frac{n!}{2\pi i}\oint_L \frac{e^{2zx-z^2}}{z^{n+1}}dz = e^{x^2}\frac{n!}{2\pi i}\oint_L \frac{e^{-(z-x)^2}}{z^{n+1}}dz = (-1)^n e^{x^2}\frac{d^n}{dx^n}e^{-x^2}$$

[7] Prove Corollaries 1 and 2 of Liouville's theorem.
Hint: In Corollary 1, let $g(z) = \frac{1}{f(z)}$. Corollary 2 can be proven using a similar approach as the proof of Liouville's theorem.

2.4 Integration of Multivalued Functions

For multivalued functions, the definite integral is only well defined on a given single-valued branch. This means that when integrating along a path, we can only select a specific branch of the function on the integration path. The following example illustrates this.

Example 2.9. Consider the paths L_1 and L_2, as shown in Figure 2.15, Setting $\sqrt{z} = 1$ as $z = 1$, we want to calculate the integral of the multivalued functions:

(a) $I_1 = \int_{L_1} \frac{\mathrm{d}z}{\sqrt{z}}$; (b) $I_2 = \int_{L_2} \frac{\mathrm{d}z}{\sqrt{z}}$

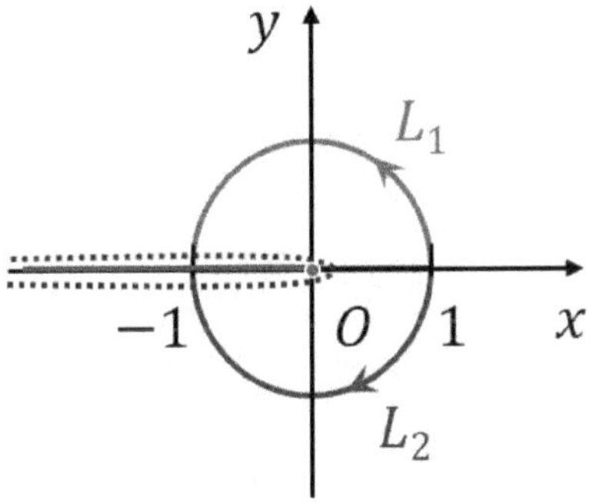

Figure 2.15.

Solution: Since $z = 1$, $\sqrt{z} = 1$, we know that $\arg z|_{z=1} = 0$.

(a) Along the path L_1, when reaching $z = -1$, $\arg z|_{z=-1} = \pi$. Hence,

$$\int_{L_1} \frac{\mathrm{d}z}{\sqrt{z}} = 2\,\sqrt{z}\Big|_{z=+1}^{z=-1} = 2(\sqrt{\mathrm{e}^{\pi\mathrm{i}}} - 1) = 2(\mathrm{i} - 1)$$

(b) Along the path L_2, when reaching $z = -1$, $\arg z|_{z=-1} = -\pi$. Hence,

$$\int_{L_2} \frac{\mathrm{d}z}{\sqrt{z}} = 2\sqrt{z}\Big|_{z=+1}^{z=-1} = 2(\sqrt{\mathrm{e}^{-\pi\mathrm{i}}} - 1) = -2(\mathrm{i} + 1)$$

The integral along the single-valued contour $L = L_1 + L_2^-$ is

$$\int_{L} \frac{\mathrm{d}z}{\sqrt{z}} = 4\mathrm{i}$$

The integral result does not satisfy Cauchy's theorem because the contour of integration is not closed.

Example 2.10. To calculate the integral along the contour C shown in Figure 2.16(a), taking the $\arg z = 0$ at the upper edge of the branch cut,

(a) $\oint_C \sqrt{z(z-1)}\mathrm{d}z$; (b) $\oint_C \frac{1}{\sqrt{z(z-1)}}\mathrm{d}z$

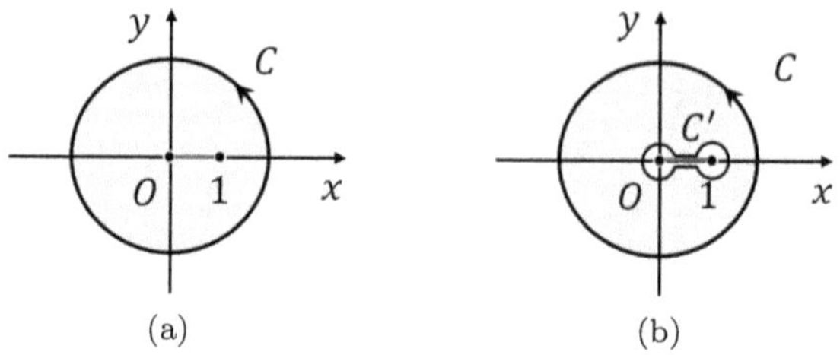

Figure 2.16.

Solution: (a) The function has two branch points at $z_1 = 0$ and $z_2 = 1$. Draw a branch cut between them. According to Cauchy's theorem for multiply connected domains, the integral along the contour C can be transformed into the integral along the contour C', as shown in Figure 2.16(b). It can be easily shown that the integrals along the two small circular arcs are zero. Therefore,

$$\begin{aligned}\oint_C \sqrt{z(z-1)}\mathrm{d}z &= \int_1^0 \sqrt{x|x-1|\mathrm{e}^{\pi\mathrm{i}}}\mathrm{d}x + \int_0^1 \sqrt{x\mathrm{e}^{2\pi\mathrm{i}}|x-1|\mathrm{e}^{\pi\mathrm{i}}}\mathrm{d}x\\ &= -2\mathrm{i}\int_0^1 \sqrt{x(1-x)}\mathrm{d}x = -2\mathrm{i}\mathrm{B}\left(\frac{3}{2},\frac{3}{2}\right)\\ &= -2\mathrm{i}\frac{\Gamma(3/2)\Gamma(3/2)}{\Gamma(3)} = -\frac{\pi\mathrm{i}}{4}\end{aligned}$$

(b) Using the same reasoning, we calculate the integral as follows:

$$\begin{aligned}\oint_C \frac{1}{\sqrt{z(z-1)}}\mathrm{d}z &= \int_1^0 \frac{1}{\sqrt{x(1-x)\mathrm{e}^{\pi\mathrm{i}}}}\mathrm{d}x + \int_0^1 \frac{1}{\sqrt{x\mathrm{e}^{2\pi\mathrm{i}}(1-x)\mathrm{e}^{\pi\mathrm{i}}}}\mathrm{d}x\\ &= 2\mathrm{i}\int_0^1 \frac{1}{\sqrt{x(1-x)}}\mathrm{d}x = 2\mathrm{i}\mathrm{B}\left(\frac{1}{2},\frac{1}{2}\right)\\ &= 2\mathrm{i}\frac{\Gamma(1/2)\Gamma(1/2)}{\Gamma(1)} = 2\pi\mathrm{i}\end{aligned}$$

Here, we have used the Betta function and the Gamma function which are addressed later. As we can see, after determining the branch cut, we can define the contour integral on a single-valued branch. The branch cut corresponds to a removed region where the function values on the upper and lower branches of the cut are distinct. The original integration region C becomes a multiply connected domain of a single-valued function. We can also interpret it as follows: Taking the radius of the circular contour C very large, it is as if we are looking from a far distance, where the two single branch points merge into a single pole, equivalent to integrating a single-valued function over a closed contour in a doubly connected domain.

Exercises

[1] For the paths L_1 and L_2 shown in Figure 2.15, with $\sqrt{z} = -1$ at $z = 1$, we calculate the following integrals:

$$I = \int_L \frac{\mathrm{d}z}{\sqrt{z}}$$

Answer: $2(1 - \mathrm{i})$; $2(1 + \mathrm{i})$

[2] Taking $\arg z = 2\pi$ along the upper edge of the branch cut $[0, 1]$, we calculate the integral along the contour shown in Figure 2.16(a):

$$\oint_C \sqrt{z(z-1)}\mathrm{d}z$$

2.5 Elliptic Integrals

2.5.1 *Integration of radical functions*

An integral of the form $\int R(u, \sqrt{p(u)})\mathrm{d}u$ is called an elliptic integral, where R is a rational function and $p(u)$ is a cubic or quartic polynomial. It is named elliptic integral because it first appeared in the calculation of the arc length of an ellipse. Consider the integral:

$$w(z) = \int_0^z \frac{\mathrm{d}u}{\sqrt{u(u-\alpha)(u-\beta)}} \tag{2.5.1}$$

Its integrand is a multivalued function with four branch points: 0, α, β, ∞. We make two branch cuts, as shown in Figure 2.17. The integral is well defined in each single-valued branch. When the path crosses the branch cuts, the integral becomes ambiguous: the function values along the integration path transition from one branch to another. However, if the path follows C_2 in Figure 2.17(b), where the dashed lines indicate a transition to another branch, the contour integral remains well defined and the integral value may be non-zero.

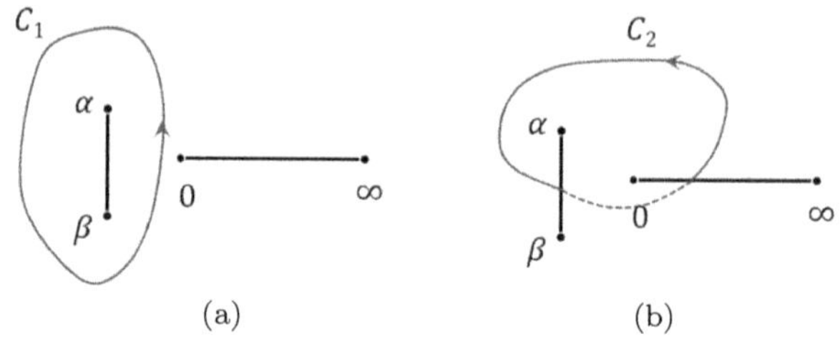

Figure 2.17.

Looking at this integral path from the perspective of Riemann surfaces provides further insights. According to Section 1.5, the multivalued function in the integral can be represented as a double-layer covering of the unit sphere, which is topologically equivalent to a torus with genus 1. The integration path from 0 to z can be regarded as a smooth curve on the torus. It is notable that there exist two classes of inequivalent closed curves on the torus, which do not form the boundaries of a region on the torus, such as C_1 and C_2 in Figure 2.18. Therefore, Green's theorem is no longer applicable. In fact, the contour integrals along these closed curves are non-zero:

$$\omega_1 = \oint_{C_1} \frac{du}{\sqrt{u(u-\alpha)(u-\beta)}}, \quad \omega_2 = \oint_{C_2} \frac{du}{\sqrt{u(u-\alpha)(u-\beta)}}$$

where ω_1 and ω_2 are complex numbers.

2.5.2 *Elliptic functions*

The function obtained by taking the inverse of an elliptic integral is called an elliptic function. This inverse function cannot be expressed in terms of elementary functions. For the integral value $\Phi^{-1} = w$ along a path L from 0 to z, we can add the windings by m times to

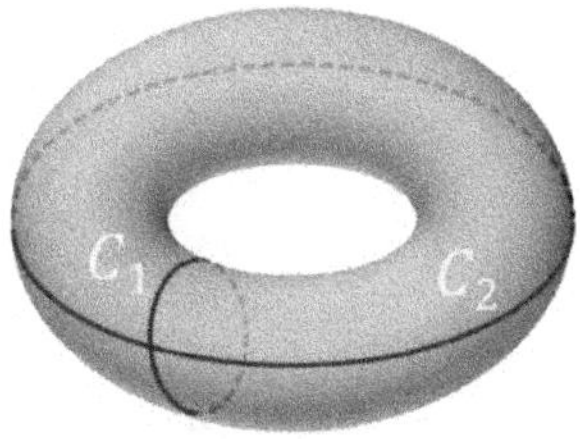

Figure 2.18.

C_1 and n times to C_2, respectively, i.e.,

$$\Phi^{-1} = w + m\omega_1 + n\omega_2 \ (m, n \in \mathbb{Z})$$

This implies that the inverse function of w exhibits translational periodicity in two different directions:

$$\Phi(w) = \Phi(w + m\omega_1 + n\omega_2)$$

Thus, elliptic functions are doubly periodic functions, where the complex numbers ω_1 and ω_2 in the w-plane are non-collinear.

General elliptic integrals can be expressed in one of the following three forms:

$$[1]\ F(z,k) = \int_0^z \frac{\mathrm{d}u}{\sqrt{(1-u^2)(1-k^2u^2)}}$$

$$[2]\ E(z,k) = \int_0^z \sqrt{\frac{1-k^2u^2}{1-u^2}}\mathrm{d}u$$

$$[3]\ \Pi(z,k,l) = \int_0^z \frac{\mathrm{d}u}{(1+lu^2)\sqrt{(1-u^2)(1-k^2u^2)}}$$

They are referred to as the first, second, and third kinds of Legendre elliptic integrals, where k and l are constants, and k is called the modulus of the elliptic integral. By substituting $u = \sin\theta$ as the independent variable, the three types of integrals become

$$[1]\ F(\phi,k) = \int_0^\phi \frac{\mathrm{d}\theta}{\sqrt{1-k^2\sin^2\theta}}$$

$$[2]\ E(\phi,k) = \int_0^\phi \sqrt{1-k^2\sin^2\theta}\mathrm{d}\theta$$

$$[3]\ \Pi(\phi,k,l) = \int_0^\phi \frac{\mathrm{d}\theta}{(1+l\sin^2\theta)\sqrt{1-k^2\sin^2\theta}}$$

The definite integral with upper limit $\phi = \pi/2$ is called a complete elliptic integral, denoted as

$$F\left(\frac{\pi}{2}, k\right) \equiv K(k), \quad E\left(\frac{\pi}{2}, k\right) \equiv E(k)$$

Geometrically, $4E(k)$ represents the circumference of an ellipse divided by its major axis and is somewhat similar to the role of π.

2.5.3 *Jacobi elliptic functions*

The first kind of elliptic integral is given by

$$w = \int_0^z \frac{du}{\sqrt{(1-u^2)(1-k^2u^2)}} = \int_0^\phi \frac{d\theta}{\sqrt{1-k^2\sin^2\theta}} \tag{2.5.2}$$

Its inverse function is called the Jacobi elliptic sine function, denoted by $\mathrm{sn}(w,k)$, which is

$$z = \mathrm{sn}(w,k) = \sin\phi$$

By letting $u = \cos\theta$, we introduce the Jacobi elliptic cosine function, denoted by $\mathrm{cn}(w,k)$, which is

$$w = \int_0^\phi \frac{d\theta}{\sqrt{1-k^2\cos^2\theta}} \longrightarrow \mathrm{cn}(w,k) = \cos\phi$$

The Jacobi elliptic sine and cosine functions satisfy the following relation:

$$\mathrm{sn}^2 w + \mathrm{cn}^2 w = 1$$

In addition, the following elliptic function is defined:

$$\mathrm{dn}(w,k) \stackrel{\mathrm{def}}{=} \sqrt{1-k^2\,\mathrm{sn}^2\,w}$$

It is customary to change the independent variable to z and use the symbols $\operatorname{sn} z$, $\operatorname{cn} z$, etc. to represent the Jacobi elliptic functions. They have the following fundamental properties:

(1) Double periodicity

$$\operatorname{sn}(z+4K)=\operatorname{sn} z, \quad \operatorname{sn}(z+2\mathrm{i}K')=\operatorname{sn} z \tag{2.5.3}$$

The two periods are given by the complete elliptic integrals

$$\omega_1=4K(k), \quad \omega_2=2\mathrm{i}K'(k') \tag{2.5.4}$$

where the complementary modulus is defined as

$$k' \stackrel{\text{def}}{=} \sqrt{1-k^2}, \quad K' \stackrel{\text{def}}{=} F\left(\frac{\pi}{2},k'\right)$$

We revisit the topic of periodicity of Jacobi elliptic functions later.

(2) Limit behavior

For $k\to 0$ since

$$\int_0^z \frac{\mathrm{d}u}{\sqrt{(1-u^2)(1-k^2u^2)}} \xrightarrow{k\to 0} \int_0^z \frac{\mathrm{d}u}{\sqrt{(1-u^2)}}$$

the Jacobi elliptic functions degenerate into trigonometric functions:

$$\operatorname{sn}(z,k)\to\sin z, \quad \operatorname{cn}(z,k)\to\cos z, \quad \operatorname{dn}(z,k)\to 1$$

For $k\to 1$ since

$$\int_0^z \frac{\mathrm{d}u}{\sqrt{(1-u^2)(1-k^2u^2)}} \xrightarrow{k\to 1} \int_0^z \frac{\mathrm{d}u}{1-u^2}$$

the Jacobi elliptic functions degenerate into hyperbolic functions:

$$\operatorname{sn}(z,k)\to\tanh z, \quad \operatorname{cn}(z,k)\sim\operatorname{dn}(z,k)\to\operatorname{sech} z$$

(3) Derivative relations

Taking the derivative of the first kind of elliptic integral,

$$\frac{\mathrm{d}z}{\mathrm{d}w} = \sqrt{(1-z^2)(1-k^2z^2)}$$

we have

$$\frac{\mathrm{d}}{\mathrm{d}w}\mathrm{sn}\, w = \sqrt{(1-\mathrm{sn}^2 w)(1-k^2\mathrm{sn}^2 w)} = \mathrm{cn}\, w\, \mathrm{dn}\, w$$

Similar relations can be derived for $\mathrm{cn}\, w$ and $\mathrm{dn}\, w$:

$$\frac{\mathrm{d}}{\mathrm{d}w}\mathrm{cn}\, w = -\mathrm{sn}\, w\, \mathrm{dn}\, w, \quad \frac{\mathrm{d}}{\mathrm{d}w}\mathrm{dn}\, w = -k^2\mathrm{sn}\, w\, \mathrm{cn}\, w$$

Figure 2.19 illustrates the characteristics of various curves when z takes a real variable, with $k = 0.7$.

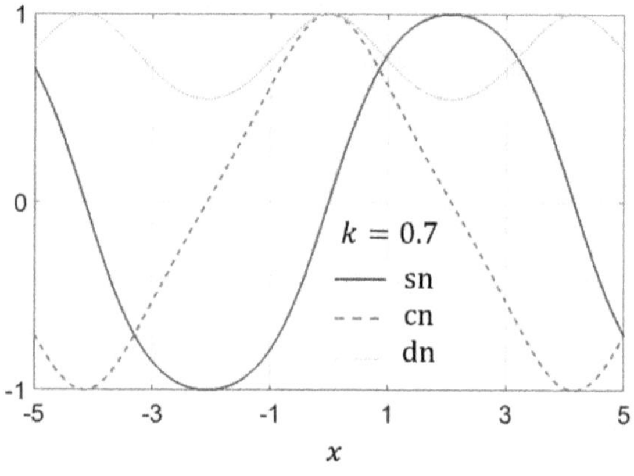

Figure 2.19.

(4) Addition formula

Just like trigonometric functions have addition formulas, such as

$$\sin(u \pm v) = \sin u \cos v \pm \cos u \sin v$$

$$\tan(u \pm v) = \frac{\tan u \pm \tan v}{1 \mp \tan u \tan v}$$

Jacobi elliptic functions also have similar addition formulas:

$$\operatorname{sn}(u+v) = \frac{\operatorname{sn} u \operatorname{cn} v \operatorname{dn} v + \operatorname{sn} v \operatorname{cn} u \operatorname{dn} u}{1 - k^2 \operatorname{sn}^2 u \operatorname{sn}^2 v}$$

$$\operatorname{cn}(u+v) = \frac{\operatorname{cn} u \operatorname{cn} v - \operatorname{sn} u \operatorname{sn} v \operatorname{dn} u \operatorname{dn} v}{1 - k^2 \operatorname{sn}^2 u \operatorname{sn}^2 v}$$

$$\operatorname{dn}(u+v) = \frac{\operatorname{dn} u \operatorname{dn} v - k^2 \operatorname{sn} u \operatorname{sn} v \operatorname{cn} u \operatorname{cn} v}{1 - k^2 \operatorname{sn}^2 u \operatorname{sn}^2 v}$$

The proof of the addition formulas is referred to literatures.[1]

ENDNOTE

By integrating the following geometric series term by term

$$\frac{1}{1+x} = 1 - x + x^2 - x^3 - \cdots$$

we obtain the logarithmic function:

$$\ln(1+x) = x - \frac{x^2}{2} + \frac{x^3}{3} - \frac{x^4}{4} + \cdots$$

In fact, most of the transcendental functions, including the logarithmic function, exponential function, trigonometric functions, and hyperbolic functions, can be obtained by integrating rational functions and inverting them. For example, for the equation of a circle $x^2 + y^2 = 1$, a segment of its arc length $\theta(x)$ is given by the integral

$$\theta(x) = \int_0^x \frac{\mathrm{d}t}{\sqrt{1-t^2}}$$

This is called the inverse sine function, and its inverse function $x = \theta^{-1}(x) \equiv \sin\theta$ is the sine function. Therefore, trigonometric functions are also known as circular functions. Through appropriate parameter transformations, the integrand can be expressed as a rational function. Another example is the parametric equations of an ellipse $x = a\sin\theta$, $y = b\cos\theta\ (a > b)$,

[1] Zhuxi Wang and Dunren Guo, *Introduction to Special Functions.*

the element of arc length is

$$\mathrm{d}s = \sqrt{(\mathrm{d}x)^2 + (\mathrm{d}y)^2} = \sqrt{a^2 \cos^2\theta + b^2 \sin^2\theta}\,\mathrm{d}\theta$$

Hence, a segment of the arc length of the ellipse is given by

$$\int \mathrm{d}s = a \int_0^{\phi} \sqrt{1 - k^2 \sin^2\theta}\,\mathrm{d}\theta, \quad k^2 = \frac{a^2 - b^2}{a^2} = \epsilon^2$$

This is the second kind of elliptic integral, where ϵ is the eccentricity of the ellipse, and the perimeter of the ellipse is $4aE(k)$.

For a function equation of the form $y'^2 = p(x)$, the inverse function relation for the integral (arc length of a cubic elliptic curve) is defined as

$$f^{-1}(x) = \int_0^x \frac{\mathrm{d}u}{\sqrt{p(u)}}$$

The difficulty of this integral lies in the multivalued nature of the integrand. Although attempts were made since the 17th century to integrate $p(x)$ for cubic or quartic polynomials, it was not until Jacobi's work that someone thought of inverting them, resulting in single-valued periodic functions.

Elliptic integrals arise from many important geometric and mechanical problems. Since their integrands cannot be expressed as rational functions, their inverse functions belong to a class of new transcendental functions. In mathematics, curves that require parameterization by elliptic functions are called elliptic curves. Interestingly, ellipses can be parameterized by rational functions, so they themselves do not belong to elliptic curves. When $p(x)$ is a polynomial of degree higher than four, it is referred to as a hyperelliptic integral.

Jacob Bernoulli was one of the earliest mathematicians to study the lemniscate equation, also known as the "lemniscate of Bernoulli". The equation is given by

$$(x^2 + y^2)^2 = 2a^2(x^2 - y^2)$$

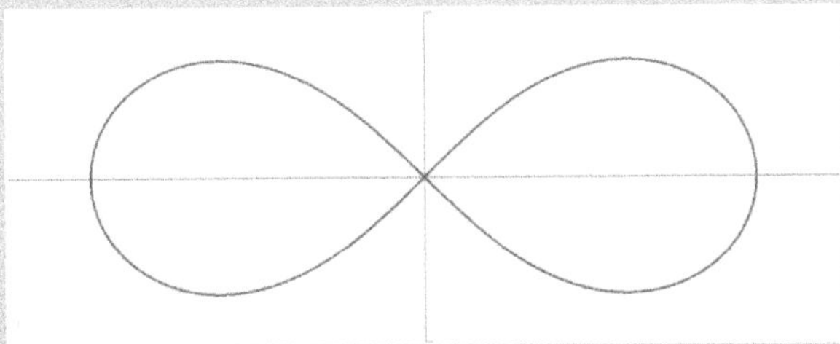

Figure 2.20.

As shown in Figure 2.20, its arc length can be expressed by elliptic integrals as

$$\int_0^x \frac{dt}{\sqrt{1-t^4}}$$

Gauss studied the inverse function of this integral, known as the lemniscatic sine function, $x = \mathrm{sl}(u)$, where

$$u = \int_0^x \frac{dt}{\sqrt{1-t^4}}$$

Gauss discovered that this function has periodicity similar to the sine function, with a period of

$$2\tilde{\omega} = 4\int_0^1 \frac{dt}{\sqrt{1-t^4}}$$

When extended to the complex variable case

$$\frac{d(it)}{\sqrt{1-(it)^4}} = \frac{idt}{\sqrt{1-t^4}}$$

it was found that the lemniscatic sine function also has a second period of $2i\tilde{\omega}$. This led to the discovery of one of the most important properties of elliptic functions, namely, their double periodicity.

Euler discovered the series expansion for trigonometric functions

$$\cot x = \sum_{n=-\infty}^{\infty} \frac{1}{x + n\pi}$$

showing that they have a period of π. Eisenstein further proved that all doubly periodic functions can be expressed using a similar series:

$$f(z) = \sum_{m,n=-\infty}^{\infty} \frac{1}{(z + m\omega_1 + n\omega_2)^2} \ (m, n \in \mathbb{Z}; \omega_1, \omega_2 \in \mathbb{C})$$

In fact, this series is equivalent to the $\mathcal{P}$-function of Weierstrass, which is the inverse function of the following elliptic integral:

$$\int_0^z \frac{\mathrm{d}u}{\sqrt{4u^3 - g_2 u - g_3}}$$

Abel proved that a meromorphic function (an analytic function with only poles) can have at most two independent periods. The geometric interpretation of Jacobi elliptic functions is discussed in Chapter 6.

According to the fundamental theorem of algebra, a polynomial of degree five has five roots. After Cardano obtained the solutions for cubic and quartic equations, mathematicians struggled for three hundred years until the arrival of the two tragic young geniuses, Abel and Galois, who ultimately proved that the roots of a general quintic polynomial cannot be expressed using elementary algebraic operations, namely, addition, subtraction, multiplication, division, and radicals. The conquest of this historical problem led to the birth of group theory and modern algebra.

In 1858, Hermite proved that the solutions of any quintic polynomial equation can be expressed using elliptic functions.

Exercises

[1] Derive the differential equation satisfied by the Jacobi elliptic functions:

(a) $w = \mathrm{sn}(z, k)$
(b) $w = \mathrm{cn}(z, k)$
(c) $w = \mathrm{dn}(z, k)$

Answer:

(a) $w'' + (1 + k^2)w - 2k^2 w^3 = 0$
(b) $w'' + (1 - 2k^2)w + 2k^2 w^3 = 0$
(c) $w'' - (2 - k^2)w + 2w^3 = 0$

[2] Prove that the following integral can be expressed as a rational function integral:

$$\int \frac{dt}{\sqrt{1 - t^2}}$$

Hint: Make the variable substitution $t = \frac{2v}{1+v^2}$.

[3] Make the variable substitution $t = 1/u$ to transform the function to

$$\frac{dt}{\sqrt{(t-a)(t-b)(t-c)}} \longmapsto \frac{du}{\sqrt{u(1-au)(1-bu)(1-cu)}}$$

[4] Find the circumference of the ellipse given by the equation

$$4x^2 + 9y^2 = 36$$

Answer: $12\mathrm{E}\left(\frac{\sqrt{5}}{3}\right) \approx 15.86$

[5] Find the arc length of the function $y = \sin x$ in the interval $[0, \pi]$.

Answer: $2\mathrm{E}\left(\frac{1}{\sqrt{2}}\right) \approx 3.82$

Chapter 3

Series Expansions

3.1 Convergence of Series

3.1.1 *Sequence and series*

The partial sums of a complex sequence $\{a_k\}_{k=1}^{\infty}$ constitute a new sequence $\{s_n = \sum_{k=1}^{n} a_k\}$. The infinite sum

$$\sum_{k=1}^{\infty} a_k = a_1 + a_2 + a_3 + \cdots + a_k + \cdots$$

is called a series. If the partial sequence has a finite limit s,

$$s_n = \sum_{k=1}^{n} a_k \overset{n\to\infty}{\longrightarrow} s$$

the series $\sum_{k=1}^{\infty} a_k$ is said to converge, denoted as

$$\lim_{n\to\infty} \sum_{k=1}^{n} a_k = s$$

The sequence $\{a_k\}_{k=1}^{\infty}$ has zero limit is a necessary condition for the convergence of the series $\sum_{k=1}^{\infty} a_k$. But it is not a sufficient condition. For example, the harmonic series

$$\sum_{k=1}^{\infty} \frac{1}{k} = 1 + \frac{1}{2} + \frac{1}{3} + \frac{1}{4} + \cdots + \frac{1}{k} + \cdots$$

diverges. Series that oscillate between two values for their partial sums also do not converge, for instance,

$$\sum_{k=0}^{\infty} a_k = 1 - 1 + 1 - 1 + \cdots + (-1)^k + \cdots$$

Cauchy convergence criterion: The sufficient and necessary condition for the convergence of the series $\sum_{k=1}^{\infty} a_k$ is $\lim_{m,n\to\infty} |s_m - s_n| \to$, which means that for any positive $\varepsilon > 0$, there exists a positive integer N such that for $k > N$, $|\sum_{k=N+1}^{N+p} a_k| < \varepsilon$, where p is any positive integer.

Since complex numbers consist of real and imaginary parts, the sufficient and necessary condition for the convergence of a complex series is that both the series formed by the real parts and the series formed by the imaginary parts converge.

3.1.2 *Function series*

If each term of a sequence is a complex function $\{w_k(z)\}$, then the infinite sum expression

$$\sum_{k=1}^{\infty} w_k(z) = w_1(z) + w_2(z) + \cdots + w_k(z) + \cdots$$

is called a function series. The convergence of a function series is also determined by the Cauchy criterion, which states that for any small positive number $\varepsilon > 0$, there exists a positive integer N such that for $k > N$, $|\sum_{k=N+1}^{N+p} w_k(z)| < \varepsilon$, where p is any positive integer.

Since each term is a function, the convergence of a function series may vary at different points z, leading to the issue of convergence domain. Additionally, the speed of convergence at different points may differ, which can affect the analytic properties of the series and requires further clarification. Below are several relevant concepts.

(1) *Absolute convergence*

In a function series, if the series formed by the absolute values of each term,

$$\sum_{k=1}^{\infty} |w_k(z)| = |w_1(z)| + |w_2(z)| + \cdots + |w_k(z)| + \cdots$$

converges at point z, the function series is said to be absolutely convergent at that point.

(2) *Uniform convergence*

The terms of a function series are functions of z. If the series converges at every point in the closed region $\bar{B}$, the series is said to converge in the region B. If the positive integer N in the Cauchy criterion is independent of z, then the complex function series uniformly converges in the domain B.

(3) *Absolute uniform convergence*

If the series formed by the absolute values of each function term uniformly converges for every point in a region B, the function series is said to be absolutely uniformly convergent.

Uniform convergence implies that a fixed positive integer N which is independent of z can be found such that the tail of the infinite series can be ignored, i.e., $|\sum_{k=N+1}^{\infty} w_k(z)| < \varepsilon$. Regarding the uniform convergence of function series, we have the following criterion.

Weierstrass M-test criterion: Let $\{M_k\}$ be a sequence of positive real numbers, and suppose the series $\sum_{k=1}^{\infty} M_k$ converges. If for all z in the closed region $\bar{B}$, $|w_k(z)| \leq M_k$ holds, then the function series $\sum_{k=1}^{\infty} w_k(z)$ uniformly converges in the region B.

Example 3.1. Prove the following series converges absolutely but non-uniformly:

$$f(x) = \sum_{k=1}^{\infty} \frac{x^2}{(1+x^2)^k} \quad (x \in \mathbb{R})$$

Proof. When $x = 0$, $f(0) = 0$. When $x \neq 0$, this is a geometric series, and by direct calculation, we obtain

$$f(x) = \sum_{k=1}^{\infty} \frac{x^2}{(1+x^2)^k} = 1$$

Hence, the series converges and is absolutely convergent for $-\infty < x < \infty$. Now let us consider the partial sum of the series starting from the $(N+1)$th term:

$$\left|\sum_{k=N+1}^{\infty} \frac{x^2}{(1+x^2)^k}\right| = \left|\frac{1}{(1+x^2)^N}\right| < \varepsilon$$

We observe that for small positive ε, the integer N depends on the value of x. Therefore, the series is not uniformly convergent.

It is worth noting that in this example, the function terms are continuous functions, but the sum of the series itself is not a continuous function. In other words, adding an infinite number of continuous functions results in a discontinuous function. This peculiar characteristic is deeply related to non-uniform convergence of the series. □

3.1.3 *Analyticity*

Let $\{w_k(z)\}$ be a function sequence, and each term of the sequence is analytic in the region B. The following theorems hold:

(1) **Derivative theorem:** *If the function series $f(z) = \sum_{k=1}^{\infty} w_k(z)$ uniformly converges in the region B, then $f(z)$ is analytic in B, and the derivative of $f(z)$ is equal to the sum of the derivatives of the series terms,*

$$f^{(n)}(z) = \sum_{k=1}^{\infty} w_k^{(n)}(z)$$

(2) **Integration theorem:** *Let L be a piecewise smooth path in the region B. If the function series $f(z) = \sum_{k=1}^{\infty} w_k(z)$ uniformly converges in B, then the path integral of $f(z)$ is equal to the sum of the path integrals of the series terms,*

$$\int_L f(z)\mathrm{d}z = \sum_{k=1}^{\infty} \int_L w_k(z)\mathrm{d}z$$

These two theorems indicate that if a series uniformly converges in a certain region, the derivative or integral of the series can be interchanged with the summation.

3.1.4 *Power series*

A series of the form $\sum_{k=0}^{\infty} a_k(z-z_0)^k$ is called a power series centered at z_0, where a_k are complex constants. If there exists a positive number R such that the power series converges when $|z-z_0| < R$ and diverges when $|z-z_0| > R$, then $|z-z_0| < R$ is called the convergence circle, and R is called the radius of convergence. The convergence circle and radius of convergence of a power series can be determined by several methods.

(1) *Ratio test*

If the ratio of consecutive terms of the power series satisfies

$$\lim_{k\to\infty} \frac{|a_{k+1}||z-z_0|^{k+1}}{|a_k||z-z_0|^k} = \lim_{k\to\infty} \frac{|a_{k+1}|}{|a_k|}|z-z_0| < 1$$

then the series converges, and the radius of convergence is given by

$$R = \lim_{k\to\infty} \left|\frac{a_k}{a_{k+1}}\right|$$

The ratio test is also known as the D'Alembert's test.

(2) *Root test*

If the power series terms satisfy

$$\lim_{k\to\infty} \sqrt[k]{|a_k||z-z_0|^k} < 1$$

then the series converges, and the radius of convergence is given by

$$R = \lim_{k\to\infty} \frac{1}{\sqrt[k]{|a_k|}}$$

The root test is also known as the Cauchy's test.

Example 3.2. Find the radius of convergence:

$$\sum_{k=0}^{\infty}(-1)^k z^{2k}$$

Solution: Using the ratio test, we have

$$\frac{|(-1)^{k+1}z^{2(k+1)}|}{|(-1)^k z^{2k}|} < 1$$

Therefore, the radius of convergence is $R = 1$.

In this problem, the series is a geometric series, and its sum can be directly calculated as

$$\sum_{k=0}^{\infty}(-1)^k z^{2k} = 1 - z^2 + z^4 - \cdots = \frac{1}{1+z^2}$$

This analytical expression shows that the function is analytic in the entire plane except at $z = \pm \mathrm{i}$. The radius of convergence of the series coincides with the expanding circle starting from the origin and reaching the singularities.

Question: When $|z - z_0| < R$, the series converges within the circle; but what about on the circle $|z - z_0| = R$? This is determined by the Gauss' test, as explained below.

(3) *The Gauss' test*

For power series $\sum_{k=1}^{\infty} w_k(z)$, if the limit of the ratio of consecutive terms is

$$\lim_{k\to\infty}\left|\frac{w_{k+1}}{w_k}\right| = 1$$

the ratio of consecutive terms can be expressed as

$$\left|\frac{w_k}{w_{k+1}}\right| = 1 + \frac{\lambda}{k} + \frac{\omega_k}{k^p}$$

Then the series converges if $\lambda > 1$ and diverges if $\lambda \leq 1$. The detailed proof of this test can be found in references.[1]

Example 3.3. Study the convergence of Legendre series:

$$y(x) = \sum_{k=0}^{\infty} a_{2k}x^{2k}$$

[1]Kunmiao Liang *et al.*, *Mathematical Methods for Physics.*

where the coefficients satisfy the recursive relation:

$$a_{2k+2} = \frac{(2k-l)(2k+l+1)}{(2k+2)(2k+1)} a_{2k}, \quad a_0 = 1$$

Solution: Using the Ratio test, the radius of convergence is given by

$$R = \lim_{k\to\infty} \left| \frac{a_{2k}}{a_{2k+2}} \right|^{1/2} = \lim_{k\to\infty} \left| \frac{(2k+2)(2k+1)}{(2k-l)(2k+l+1)} \right|^{1/2} = 1$$

Hence, the series converges for $|x| < 1$. Now let's examine the convergence of the series at $|x| = 1$. Since

$$\lim_{k\to\infty} \left| \frac{a_{2k}}{a_{2k+2}} \right| = \lim_{k\to\infty} \left| \frac{(2k+2)(2k+1)}{(2k-l)(2k+l+1)} \right| = \frac{4k^2+6k+2}{4k^2+2k-l(l+1)}$$

$$= 1 + \frac{1}{k} + \frac{1}{k^2} \frac{l(l+1)(l+1/k)}{4+2/k-l(l+1)/k^2}$$

Thus $\lambda = 1$. According to Gauss' test, the Legendre series diverges at $x = \pm 1$. Therefore, the convergence interval for the Legendre series is $x \in (-1, +1)$.

The analyticity of a power series is also related to the uniform convergence of the series. This can be determined using the Abel's theorem.

Abel's theorem: If a power series $\sum_{k=0}^{\infty} a_k (z-z_0)^k$ converges at $z = z_1$, then it must converge within the circle $|z-z_0| < |z_1 - z_0|$, and it uniformly converges within the closed circle $|z - z_0| \le \delta$, where $\delta < |z_1 - z_0|$.

Proof. Since the power series converges at $z = z_1$, we have

$$\lim_{k\to\infty} \frac{|a_{k+1}||z_1 - z_0|^{k+1}}{|a_k||z_1 - z_0|^k} = \lim_{k\to\infty} \frac{|a_{k+1}|}{|a_k|} |z_1 - z_0| < 1$$

Thus

$$\lim_{k\to\infty} \frac{|a_{k+1}|}{|a_k|} < \frac{1}{|z_1 - z_0|}$$

According to the ratio test,

$$\lim_{k\to\infty} \frac{|a_{k+1}||z - z_0|^{k+1}}{|a_k||z - z_0|^k} = \lim_{k\to\infty} \frac{|a_{k+1}|}{|a_k|} |z - z_0| < \frac{|z - z_0|}{|z_1 - z_0|} < 1$$

Therefore, the power series converges within the region $|z - z_0| < |z_1 - z_0|$. Furthermore, for any point z within the closed region $|z - z_0| \leq \delta$, where $\delta < |z_1 - z_0|$, we have $|a_k(z - z_0)^k| \leq |a_k \delta^k| \equiv M_k$. The series $\sum_{k=0}^{\infty} M_k$ is convergent due to

$$\lim_{k\to\infty} \frac{|M_{k+1}|}{|M_k|} = \lim_{k\to\infty} \frac{|a_{k+1}\delta^{k+1}|}{|a_k \delta^k|} < \frac{\delta}{|z_1 - z_0|} < 1$$

According to the Weierstrass M-test, the power series uniformly converges within the closed region $|z - z_0| \leq \delta$.
Corollary: If a power series $\sum_{k=0}^{\infty} a_k(z - z_0)^k$ diverges at $z = z_1$, then it diverges everywhere outside the circle $|z - z_0| > |z_1 - z_0|$.

It can be concluded from Abel's theorem that the power series is an analytic function within the convergence circle $|z - z_0| < R$. Thus, the convergence circle is a simply connected domain without singularities. Outside the convergence circle, the power series diverges at every point. □

ENDNOTE

The calculus established by Newton and Leibniz was based on intuition. In order to provide a solid foundation for calculus, it was necessary to give precise definitions for fundamental concepts such as real numbers, functions, and limits. This process led to the evolution of mathematical analysis over a span of nearly two hundred years.

Since the time of Newton, people have been discussing the concept of limits. By the late 18th century, S. F. Lacroix abandoned the traditional geometric tangent method and was the first to use limits to define derivatives. Cauchy was the first person to transform the vague notion of approximating a function to a particular value into rigorous mathematical terms. He used the language of $\delta - \varepsilon$ inequalities to express the concept of limits, defined the continuity of functions and the convergence of infinite series, and provided the definition of derivatives that is still used in today's mathematics textbooks. Prior to Cauchy, Lagrange's definition of derivatives was based on a questionable assumption that all functions can be represented by power series.

Regarding integration, the early popular approach was to consider it as the inverse operation of differentiation. Cauchy was the first to define integration as the limit of an infinite sum, thus opening up a vast new field. In 1840, Cauchy extended his integration definition to complex functions, introduced path integrals, and obtained a series of important results including Cauchy's theorem and the concept of residues. However, one of Cauchy's conclusions, "the limit of an infinite sequence of continuous functions must be a continuous function", is incorrect. Section 14.2 will give a counterexample. Weierstrass studied how to ensure the analyticity of function series in a given region and emphasized the uniform convergence of function series. This idea was first noticed by Stokes and Seidel, but Cauchy misinterpreted it. For example, Cauchy believed that a convergent series composed of continuous functions must be a continuous function. The counterexample given in this section disproves that assumption. Abel also provided a counterexample:

$$\sum_{k=1}^{\infty} \frac{(-1)^{k+1}}{k} \sin kx = \sin x - \frac{1}{2} \sin 2x + \frac{1}{3} \sin 3x - \cdots$$

Each term of this series is a continuously differentiable function, but the series function is discontinuous at $x = (2n+1)\pi (n \in \mathbb{Z})$, and therefore, it is not differentiable either. The theorem of analyticity states that for uniformly convergent series, the interchange of differentiation or integration with the summation of the series is valid. Non-uniformly convergent series may exhibit singularities in the corresponding region, and the interchange of differentiation or integration with the summation of the series is not permissible. For example, the Taylor series expansion

$$\mathrm{e}^{-x} = \sum_{k=1}^{\infty} \frac{(-1)^k}{k!} x^k = 1 - \frac{1}{1!}x + \frac{1}{2!}x^2 - \frac{1}{3!}x^3 + \cdots$$

The series on the right-hand side converges absolutely with a convergence radius of $R = \infty$. However, this does not imply that the following equation always holds:

$$\int_0^a \mathrm{e}^{-x} dx = \sum_{k=1}^{\infty} \int_0^a \frac{(-1)^k}{k!} x^k \mathrm{d}x$$

The equation holds when the integral limit a is a finite value. However, since $z = \infty$ is a singularity of the function $f(z) = \mathrm{e}^{-z}$, each term in the summation diverges as $a \to \infty$. Actually, the positive integer N in Cauchy's criterion depends on x:

$$\left| \sum_{k=N+1}^{N+p} \frac{(-1)^k}{k!} x^k \right| < \varepsilon$$

does not uniformly converge on $[0, \infty)$, so the interchange of integration and summation is illegal.

Exercises

[1] Prove the divergence of the harmonic series:

$$\sum_{k=1}^{\infty} \frac{1}{k} = 1 + \frac{1}{2} + \frac{1}{3} + \frac{1}{4} + \cdots\cdots$$

Hint: Consider $\frac{1}{k} \geq \int_k^{k+1} \frac{1}{x} \mathrm{d}x$.

[2] Study the convergence of the following series with the Gauss' test:

$$\text{(a)} \sum_{k=1}^{\infty} \frac{1}{k}; \quad \text{(b)} \sum_{k=1}^{\infty} \frac{1}{k\sqrt{k}}; \quad \text{(c)} \sum_{k=1}^{\infty} \frac{1}{k^2}$$

[3] Determine the uniform convergence region of the following power series:

$$\text{(a)} \sum_{k=0}^{\infty} z^k; \quad \text{(b)} \sum_{k=0}^{\infty} \frac{z^k}{k}; \quad \text{(c)} \sum_{k=0}^{\infty} \frac{z^{4k}}{k^2}$$

Hint: If the series converges uniformly for $|z| < 1$, then it converges at $|z| = 1$.

Answer: (a) $|z| \leq \delta < 1$; (b) $|z| \leq \delta < 1$; (c) $|z| < 1$

[4] If the series $\sum_{k=1}^{\infty} a_k$ and $\sum_{k=1}^{\infty} b_k$ both converge absolutely, prove that the series

$$\sum_{k=1}^{\infty}(a_k \cos kx + b_k \sin kx)$$

converges uniformly for $-\infty < x < \infty$.

[5] Find the convergence domain of the power series:

$$\text{(a)} \sum_{k=1}^{\infty} \frac{k!}{k^k} z^k; \quad \text{(b)} \sum_{k=1}^{\infty} k^{\ln k} z^k; \quad \text{(c)} \sum_{k=1}^{\infty}(-1)^k (z^2 + 2z + 2)^k$$

Answer: (a) $R = \mathrm{e}$; (b) $R = 1$; (c) $|z^2 + 2z + 2| < 1$

[6] Prove that

$$\ln(1 - z) = -z - \frac{z^2}{2} - \frac{z^3}{3} - \frac{z^4}{4} - \cdots (|z| < 1)$$

[7] Find the sum of the series:

$$\text{(a)} \sum_{k=1}^{\infty} kz^k (|z| < 1); \quad \text{(b)} \sum_{k=1}^{\infty} k^2 z^k (|z| < 1)$$

Answer:

$$\text{(a)} \frac{z}{(1 - z)^2}; \quad \text{(b)} \frac{z(1 + z)}{(1 - z)^3}$$

3.2 Taylor Series

3.2.1 *Taylor's theorem*

Let $f(z)$ be a function analytic in a region with center z_0 and radius R. For any point z within the circle, the function $f(z)$ can be expanded as a power series:

$$f(z) = \sum_{k=0}^{\infty} a_k (z - z_0)^k \tag{3.2.1}$$

where the coefficients a_k are given by

$$a_k = \frac{1}{2\pi \mathrm{i}} \oint_C \frac{f(\zeta)}{(\zeta - z_0)^{k+1}} d\zeta = \frac{1}{k!} f^{(k)}(z_0) \tag{3.2.2}$$

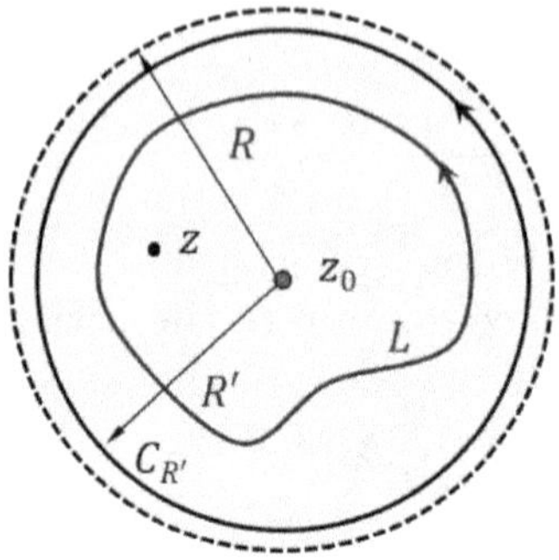

Figure 3.1.

Proof. As shown in Figure 3.1, consider a circle $C_{R'}$ with radius R' such that $R' < R$. According to Cauchy's integral formula, the value of the function at any point z within $C_{R'}$ reads

$$f(z) = \frac{1}{2\pi \mathrm{i}} \oint_{C_{R'}} \frac{f(\zeta)}{\zeta - z} d\zeta$$

Since $|z - z_0| < |\zeta - z_0|$ for any ζ on $C_{R'}$, we can expand it using the geometric series as

$$\frac{1}{\zeta - z} = \frac{1}{(\zeta - z_0) - (z - z_0)} = \frac{1}{\zeta - z_0} \cdot \frac{1}{1 - \frac{z - z_0}{\zeta - z_0}}$$

$$= \frac{1}{\zeta - z_0} \sum_{k=0}^{\infty} \left(\frac{z - z_0}{\zeta - z_0} \right)^k = \sum_{k=0}^{\infty} \frac{(z - z_0)^k}{(\zeta - z_0)^{k+1}}$$

Abel's theorem tells us that this geometric series uniformly converges within the closed region $C_{R'}$. Therefore,

$$f(z) = \sum_{k=0}^{\infty} (z - z_0)^k \cdot \frac{1}{2\pi \mathrm{i}} \oint_{C_{R'}} \frac{f(\zeta)}{(\zeta - z_0)^{k+1}} \mathrm{d}\zeta = \sum_{k=0}^{\infty} a_k (z - z_0)^k$$

Since the integrand is analytic within the closed region except z_0, the integral path $C_{R'}$ can be continuously deformed into any arbitrary path L within the region. Hence, the expansion coefficients are given by

$$a_k = \frac{1}{2\pi \mathrm{i}} \oint_L \frac{f(\zeta)}{(\zeta - z_0)^{k+1}} \mathrm{d}\zeta = \frac{1}{k!} f^{(k)}(z_0)$$

Representing an analytic function as a positive power series is known as the Taylor series expansion of the function. □

3.2.2 *Uniqueness theorem*

If function $f(z)$ is analytic in region B, then it can be uniquely expanded as a Taylor series in the neighborhood of any point within B.

Proof. Suppose the analytic function $f(z)$ can also be expanded as a different Taylor series at z_0,

$$f(z) = \sum_{k=0}^{\infty} b_k (z - z_0)^k$$

Then we have

$$\begin{aligned} & b_0 + b_1(z - z_0)^1 + b_2(z - z_0)^2 + \cdots + b_k(z - z_0)^k + \cdots \\ & = f(z_0) + \frac{f'(z_0)}{1!}(z - z_0)^1 + \frac{f''(z_0)}{2!}(z - z_0)^2 \\ & \quad + \cdots + \frac{f^{(k)}(z_0)}{k!}(z - z_0)^k + \cdots \end{aligned}$$

Since each term of the series is analytic within B and uniformly convergent within $|z - z_0| < R$, we can differentiate both sides term by term and then set $z = z_0$. This yields

$$b_0 = f(z_0), \quad b_1 = \frac{f'(z_0)}{1!}, \quad b_2 = \frac{f''(z_0)}{2!}, \ldots, b_k = \frac{f^{(k)}(z_0)}{k!}, \ldots$$

Therefore, the coefficients of the two series are the same. The uniqueness of the coefficients implies that the Taylor series expansion of $f(z)$ is unique in the neighborhood of any point in B. □

Example 3.4. Take the Taylor series expansion of function $f(z) = \sin z$ at $z_0 = 0$.

Solution: According to the Taylor expansion formula, we have

$$f(z) = \sum_{k=0}^{\infty} \frac{f^{(k)}(z_0)}{k!}(z - z_0)^k = z - \frac{1}{3!}z^3 + \frac{1}{5!}z^5$$
$$- \cdots + \frac{(-1)^k}{(2k+1)!}z^{2k+1} + \cdots$$

It can be observed that the expansion has the same form as that of a real variable function.

Example 3.5. Proof of Euler's formula:

$$\mathrm{e}^{\mathrm{i}\theta} = \cos\theta + \mathrm{i}\sin\theta$$

Solution: Consider the Taylor series expansion of

$$\mathrm{e}^{z} = 1 + \frac{1}{1!}z + \frac{1}{2!}z^2 + \frac{1}{3!}z^3 + \cdots$$

Substituting $z = \mathrm{i}\theta$, we have

$$\mathrm{e}^{\mathrm{i}\theta} = \left[1 - \frac{1}{2!}\theta^2 + \frac{1}{4!}\theta^4 - \cdots\right] + \mathrm{i}\left[\frac{1}{1!}\theta - \frac{1}{3!}\theta^3 + \frac{1}{5!}\theta^5 - \cdots\right]$$
$$= \cos\theta + \mathrm{i}\sin\theta$$

Example 3.6. Find the Taylor series expansion of function $f(z) = \ln z$ at $z_0 = 1$.

Solution: The function $f(z) = \ln z$ is multivalued, so we need to choose one branch for the Taylor series expansion. Since

$$f(1) = 2k\pi\mathrm{i}, \quad f'(1) = 1, \quad f''(1) = -1!,$$
$$f^{(3)}(1) = 2!, \quad f^{(4)}(1) = -3!, \cdots$$

We can express it as

$$\ln z = 2k\pi\mathrm{i} + (z-1) - \frac{1}{2}(z-1)^2 + \frac{1}{3}(z-1)^3 - \frac{1}{4}(z-1)^4 + \cdots$$

where $|z - 1| < 1$ and $k \in Z$.

3.2.3 *Zeros of analytic functions*

If a function $f(z)$ is analytic in the neighborhood of z_0 and $f(z_0) = 0$, then $z = z_0$ is called a zero of $f(z)$. Based on the Taylor expansion:

$$f(z) = \sum_{n=0}^{\infty} a_n (z - z_0)^n$$

If $z = z_0$ is a zero and

$$a_0 = a_1 = \cdots = a_{n-1} = 0$$

then z_0 is called an nth order zero of $f(z)$.

Isolation of zeros theorem: If $f(z)$ is analytic in a region containing $z = z_0$ and $f(z_0) = 0$, then it is possible to find a finite circle $|z - z_0| < \rho$ within which $f(z)$ has no other zeros.

Proof. Suppose $z = z_0$ is an nth order zero of $f(z)$, then we can express it as:

$$f(z) = (z - z_0)^n g(z)$$

where $g(z)$ is analytic in the neighborhood of $z = z_0$ and $g(z_0) \neq 0$. Since $g(z)$ is continuous at $z = z_0$, there exists a circle $|z - z_0| < \rho$ where $g(z)$ is nonzero for all z within that circle. Hence, $f(z)$ has no other zeros in the neighborhood of $z = z_0$. □

From the isolation of zeros theorem, we can derive the following corollaries:

Corollary 1. *If the zeros of an analytic function are not isolated, then the function is identically zero.*

Corollary 2. *If $f_1(z)$ and $f_2(z)$ are analytic in a region B, and there exists a continuous arc or a subregion $D \subset B$ where $f_1(z) = f_2(z)$, then $f_1(z) = f_2(z)$ in B.*

Zeros of an analytic function correspond to points where both the real and imaginary parts of the function are zero. If we plot the curves of the real and imaginary parts simultaneously, the intersection points represent the zeros of the complex function. Figure 3.2(a) depicts the zero lines of the function:

$$f(z) = (z^2 + 1)(z - 3 - \mathrm{i})^3$$

where $z = \pm\mathrm{i}$ and $z = 3 + \mathrm{i}$ are first-order and third-order zeros, respectively. The zeros are also singular points of the phase or argument of the function, which are undefined. When z completes a full revolution around a zero, the phase of the function increases by $2n\pi(n \in \mathbb{Z})$, as shown in Figure 3.2(b), where n is the order of the zero, known as the winding number. In superconductors or superfluids, the zeros of the macroscopic wave function correspond to quantized vortices.

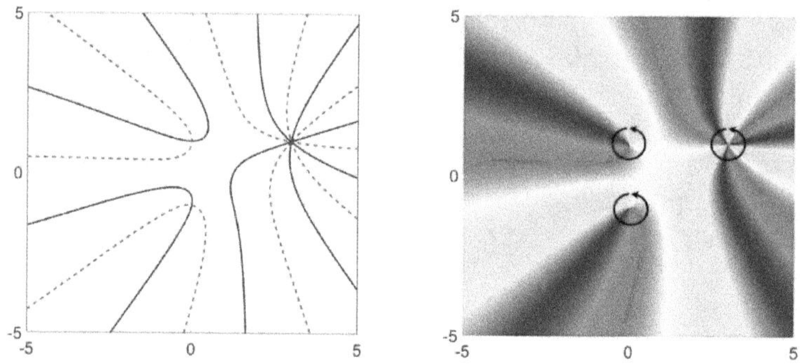

(a) (b)

Figure 3.2.

ENDNOTE

The isolation theorem for zeros can be addressed from the definition of the derivative of a complex function. Let z_0 be a zero of the analytic function $f(z)$, $f(z_0) = 0$. If there exists another zero z_1 in an infinitesimal neighborhood of z_0, i.e.,

$$\lim_{z_1 \to z_0} f(z_1) = f(z_0) = 0$$

Then,

$$\lim_{z_1 \to z_0} \frac{f(z_1) - f(z_0)}{z_1 - z_0} = 0$$

For analytic functions, the derivative at the point z_0 is independent of how you approach z_0. Therefore, $f'(z_0) = 0$. Similarly, $f'(z_1) = 0$. Hence, for any point z within the infinitesimal neighborhood of z_0, the second derivative is

$$f''(z_0) = \lim_{z \to z_0} \frac{f'(z) - f'(z_0)}{z - z_0} = \lim_{z_1 \to z_0} \frac{f'(z_1) - f'(z_0)}{z_1 - z_0} = 0$$

By induction, all higher-order derivatives of $f(z)$ at z_0 are zero. This implies that the analytic function $f(z)$ is identically zero in the neighborhood of z_0, which contradicts the assumption.

When a function $f(z)$ is analytic in the neighborhood of z_0 with $f'(z_0) = 0$ and $f''(z_0) \neq 0$, we can approximate the function using a Taylor series expansion and keep the terms up to the second order:

$$f(z) \approx f(z_0) + \frac{1}{2} f''(z_0)(z - z_0)^2$$

As $z \to z_0$, we can write $z - z_0 = r\mathrm{e}^{\mathrm{i}\theta}$ and $f''(z_0) = r_0 \mathrm{e}^{2\mathrm{i}\alpha}$, where for simplicity we omit the constant term $f(z_0)$. Thus, we have

$$f(z) \sim \frac{1}{2} r_0 r^2 \mathrm{e}^{\mathrm{i}(2\theta + 2\alpha)} = \frac{1}{2} r_0 r^2 [\cos 2(\theta + \alpha) + \mathrm{i} \sin 2(\theta + \alpha)]$$

The real and imaginary parts of the approximation are

$$u(r, \theta) \sim \frac{1}{2} r_0 r^2 \cos 2(\theta + \alpha), \quad v(r, \theta) \sim \frac{1}{2} r_0 r^2 \sin 2(\theta + \alpha)$$

When $\theta = -\alpha$, $u(r, \theta)$ represents an upward-opening parabola. When $\theta = \frac{\pi}{2} - \alpha$, it represents a downward-opening parabola. Therefore, $z = z_0$ is a saddle point for the real part function $u(r, \theta)$. Similarly, it is also a saddle point for the imaginary part function $v(r, \theta)$. The difference is that the saddle-shaped surface of $u(r, \theta)$ is rotated $\pi/4$ with respect to the surface of $v(r, \theta)$.

Hence, the zeros of the first derivative of a real function correspond to the extremal points of the function. In the case of a complex function, the zeros of the first derivative correspond to the common saddle points of the real and imaginary parts, which are the extrema in different directions.

Exercises

[1] Expand the following functions as a Taylor series around $z_0 = 0$:

$$\text{(a)}\ \tan z; \quad \text{(b)}\ \frac{1}{1 - 3z + 2z^2}$$

[2] Expand the following functions as a Taylor series around $z_0 = \mathrm{i}$:

$$\text{(a)}\ \sqrt[3]{z}; \quad \text{(b)}\ \mathrm{Ln}\, z$$

[3] Find the Taylor series expansion of the function around $z_0 = 0$:

$$f(z) = (1 + z)^m (m \notin \mathbb{Z})$$

[4] If the analytic functions $f(z)$ and $g(z)$ both have a first-order zero at $z = z_0$, i.e., the ratio

$$\left.\frac{f(z)}{g(z)}\right|_{z=z_0} = \frac{0}{0}$$

Prove L' Hopital's rule:

$$\lim_{z \to z_0} \frac{f(z)}{g(z)} = \lim_{z \to z_0} \frac{f'(z)}{g'(z)}$$

[5] Prove the integral representation of derivative using Taylor's theorem:

$$f^{(n)}(z) = \frac{n!}{2\pi \mathrm{i}} \oint_L \frac{f(\zeta)}{(\zeta - z)^{n+1}} \mathrm{d}\zeta$$

[6] Prove Liouville's theorem using Taylor's theorem.

[7] Prove the maximum modulus theorem using Taylor's theorem. *Hint*: Use proof by contradiction, assuming $|f(z_0)|$ is maximal, and consider the integral along the circular path centered at z_0.

3.3 Laurent Series

When $f(z)$ is analytic within the circle $|z - z_0| < R$, the Taylor theorem states that $f(z)$ can be expanded into a positive power series. Is it possible to have a convergent series with negative powers? And how does its convergence behave? This leads us to the discussion of the two-sided power series.

3.3.1 *Two-sided power series*

A series of the form:

$$\sum_{k=-\infty}^{\infty} a_k(z - z_0)^k = \cdots + a_{-k}(z - z_0)^{-k} + \cdots + a_{-1}(z - z_0)^{-1}$$
$$+ a_0 + a_1(z - z_0) + a_2(z - z_0)^2$$
$$+ \cdots a_k(z - z_0)^k + \cdots$$

is called a two-sided power series. Here,

$$S_+(z) = \sum_{k=0}^{\infty} a_k(z - z_0)^k$$

is referred to as the positive power or analytic part of the two-sided power series, and

$$S_-(z) = \sum_{k=-1}^{-\infty} a_k(z - z_0)^k$$

is referred to as the negative power or principal part of the two-sided power series.

Now, for the series to converge, both the positive power part and the negative power part must converge simultaneously. Let the convergence radius of the positive power part be R_1, meaning it converges within the disk $|z - z_0| < R_1$. Obviously, in order to ensure convergence of the negative power series, we must restrict z from approaching z_0 infinitely, as otherwise the negative power part will diverge. We can introduce a new variable $\zeta = \frac{1}{(z-z_0)}$, which, in terms of ζ, becomes a positive power series. Therefore, it has a convergence radius $|\zeta| < h \equiv \frac{1}{R_2}$. This implies that the negative power part converges outside the disk $|z - z_0| > R_2$. If $R_2 < R_1$, then the two-sided power series can only converge within the annular region:

$$R_2 < |z - z_0| < R_1$$

and this is referred to as the convergence annulus of the two-sided power series. If $R_2 > R_1$, the convergence domain of the two-sided power series is an empty set.

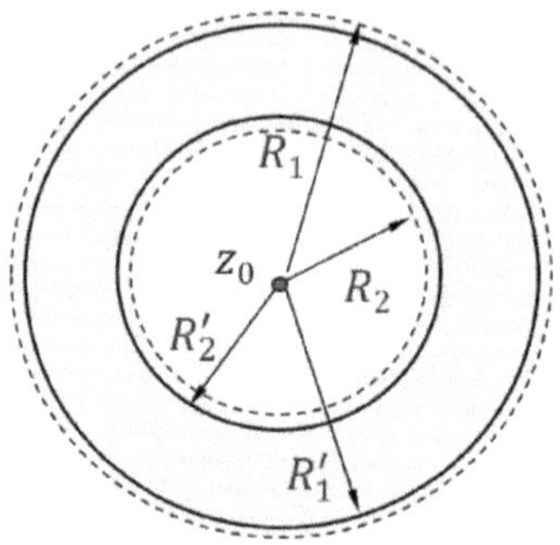

Figure 3.3.

If the two-sided power series converges within the annular region B, where $R_2 < |z - z_0| < R_1$, the Abel's theorem confirms that it uniformly converges within the closed region $R_2' < |z - z_0| < R_1'$, as shown in Figure 3.3. Analyticity theorems ensure the following properties:

(1) The series is continuous.
(2) The series is analytic within B and can be differentiated term by term.
(3) The series can be integrated term by term within B.

3.3.2 *Laurent's theorem*

When a complex function $f(z)$ has singularities within the region $|z - z_0| < R_1$, can it be expanded in a form similar to a Taylor series at the point z_0? The answer is yes, and we have the following Laurent's theorem:

Laurent's Theorem. *If the function $f(z)$ is single-valued and analytic within the annular region $R_2 < |z - z_0| < R_1$, then for any point z within the annulus, the function $f(z)$ can be expanded into a two-sided power series:*

$$f(z) = \sum_{k=-\infty}^{\infty} a_k (z - z_0)^k \tag{3.3.1}$$

where

$$a_k = \frac{1}{2\pi \mathrm{i}} \oint_L \frac{f(\zeta)}{(\zeta - z_0)^{k+1}} \mathrm{d}\zeta \tag{3.3.2}$$

The integration is taken along a closed path L within the annular region.

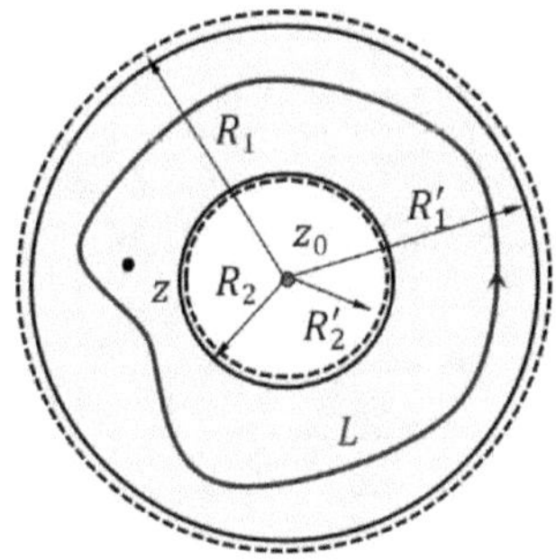

Figure 3.4.

Proof. Consider the circles $C_{R_1'}$ and $C_{R_2'}$ represented by solid lines shown in Figure 3.4. By applying the Cauchy's integral formula for multiply connected regions, for a point z within the annulus $f(z)$ reads

$$f(z) = \frac{1}{2\pi \mathrm{i}} \oint_{C_{R_1'}} \frac{f(\zeta)}{\zeta - z} \mathrm{d}\zeta - \frac{1}{2\pi \mathrm{i}} \oint_{C_{R_2'}} \frac{f(\zeta)}{\zeta - z} \mathrm{d}\zeta$$

For the integral along $C_{R_1'}$, due to $|z - z_0| < |\zeta - z_0|$, we have

$$\frac{1}{\zeta - z} = \frac{1}{(\zeta - z_0) - (z - z_0)} = \frac{1}{\zeta - z_0} \cdot \frac{1}{1 - \frac{z - z_0}{\zeta - z_0}}$$

$$= \frac{1}{\zeta - z_0} \sum_{k=0}^{\infty} \left(\frac{z - z_0}{\zeta - z_0} \right)^k = \sum_{k=0}^{\infty} \frac{(z - z_0)^k}{(\zeta - z_0)^{k+1}}$$

For the integral along $C_{R_2'}$, due to $|z-z_0| > |\zeta - z_0|$, we have

$$\frac{1}{\zeta - z} = \frac{1}{(\zeta - z_0) - (z - z_0)} = -\frac{1}{z - z_0} \cdot \frac{1}{1 - \frac{\zeta - z_0}{z - z_0}}$$
$$= -\frac{1}{z - z_0} \cdot \sum_{k=0}^{\infty} \left(\frac{\zeta - z_0}{z - z_0}\right)^k = -\sum_{k=0}^{\infty} \frac{(\zeta - z_0)^k}{(z - z_0)^{k+1}}$$

Substituting these two expressions into the integral representation and according to Abel's theorem, the geometric series uniformly converges within the closed region enclosed by $C_{R_1'}$ and $C_{R_2'}$. By interchanging the order of integration and summation, we have

$$f(z) = \frac{1}{2\pi \mathrm{i}} \sum_{k=0}^{\infty} (z - z_0)^k \oint_{C_{R_1'}} \frac{f(\zeta)}{(\zeta - z_0)^{k+1}} \mathrm{d}\zeta$$
$$+ \frac{1}{2\pi \mathrm{i}} \sum_{k=0}^{\infty} (z - z_0)^{-k-1} \oint_{C_{R_2'}} (\zeta - z_0)^k f(\zeta) \mathrm{d}\zeta$$
$$= \frac{1}{2\pi \mathrm{i}} \sum_{k=0}^{\infty} \left[\oint_L \frac{f(\zeta)}{(\zeta - z_0)^{k+1}} \mathrm{d}\zeta\right] (z - z_0)^k + \frac{1}{2\pi \mathrm{i}} \sum_{k=-1}^{-\infty}$$
$$\times \left[\oint_L \frac{f(\zeta)}{(\zeta - z_0)^{k+1}} \mathrm{d}\zeta\right] (z - z_0)^k = \sum_{k=-\infty}^{\infty} a_k (z - z_0)^k$$

where the analyticity of the integrand in the annular region allows us to deform the integration paths continually from $C_{R_1'} \to L$ and from $C_{R_2'} \to L$. As a result, the expansion coefficient is given by

$$a_k = \frac{1}{2\pi \mathrm{i}} \oint_L \frac{f(\zeta)}{(\zeta - z_0)^{k+1}} \mathrm{d}\zeta$$

Laurent series expansions is the representation of an analytic function within an annular region. It can be proved that the Laurent series expansion is also unique.

It should be mentioned that in the expansion series, although there are negative powers of $z - z_0$, the point z_0 may or may not be a singularity of the function $f(z)$. Therefore, it cannot be expressed as the kth derivative of $f(z)$ at z_0, i.e., $a_k \neq \frac{1}{k!} f^{(k)}(z_0)$. □

Example 3.7. Expand the function $f(z) = \mathrm{e}^{\frac{1}{z}}$ as Laurent series in the neighborhood of $z_0 = 0$.

Solution: First, expand the function e^{ζ} in a Taylor series around $\zeta = 0$:

$$\mathrm{e}^{\zeta} = 1 + \frac{1}{1!}\zeta + \frac{1}{2!}\zeta^2 + \frac{1}{3!}\zeta^3 + \cdots$$

Then, substitute $\zeta = \frac{1}{z}$ to yield the Laurent series expansion:

$$\mathrm{e}^{\frac{1}{z}} = 1 + \frac{1}{1!}\frac{1}{z} + \frac{1}{2!}\frac{1}{z^2} + \frac{1}{3!}\frac{1}{z^3} + \cdots$$

Example 3.8. Expand the function $f(z) = \frac{1}{z^2-1}$ in Laurent series in the following regions:

$$\text{(a) } 1 < |z| < \infty; \quad \text{(b) } 0 < |z-1| < 2$$

Solution: The expansion regions are shown in Figure 3.5.

(a) In the region $1 < |z| < \infty$, using the geometric series expansion, we have

$$f(z) = \frac{1}{z^2}\frac{1}{1-\frac{1}{z^2}} = \frac{1}{z^2}\sum_{k=0}^{\infty}\frac{1}{z^{2k}} = \sum_{k=1}^{\infty}\frac{1}{z^{2k}}$$

(b) In the region $0 < |z-1| < 2$, we have

$$\begin{aligned} f(z) = \frac{1}{z^2-1} &= \frac{1}{2}\frac{1}{z-1} - \frac{1}{2}\frac{1}{z+1} = \frac{1}{2}\frac{1}{z-1} - \frac{1}{4}\frac{1}{1+\frac{z-1}{2}} \\ &= \frac{1}{2}\frac{1}{z-1} - \frac{1}{4}\sum_{k=0}^{\infty}\frac{(-1)^k}{2^k}(z-1)^k \end{aligned}$$

From this example, we see that although the Laurent series has infinitely many negative power terms, $z_0 = 0$ is not a singularity of the function. The Taylor series expansion only needs to specify the neighborhood of a certain point. In contrast, the Laurent series expansion needs to indicate the region in which it is expanded, and its specific form depends on the specified expansion region. If only the center of the annulus, z_0, is a singularity of $f(z)$, then the inner convergence radius can be arbitrarily small. In this case, it is called

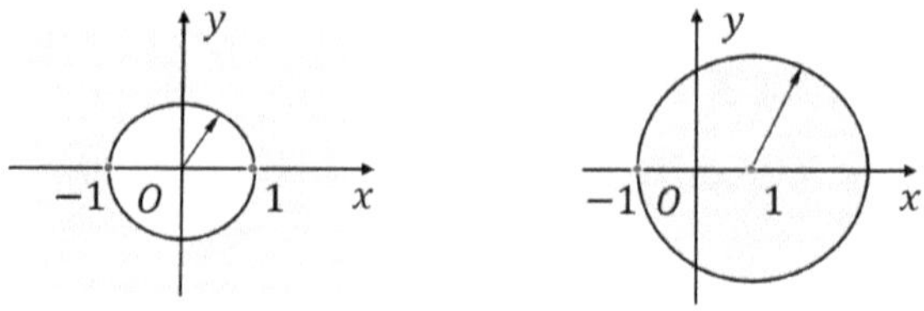

Figure 3.5.

a Laurent series expansion in the punctured neighborhood of the isolated singularity $z_0 = 0$.

ENDNOTE

Analyzing the general characteristics of power series helps us understand functions. However, not all functions can be expanded into power series. For example, the function $f(x) = \sqrt{x}$ has multivaluedness at $x = 0$ and is not strictly a function. Power series cannot reflect this behavior because they can only be single-valued. For bivariate algebraic functions that satisfy a polynomial equation $p(x, y) = 0$, Newton discovered that y can be expressed as a fractional power series in x:

$$y = a_0 + a_1 x^{r_1} + a_2 x^{r_2} + a_3 x^{r_3} + \cdots$$

where $r_1, r_2, r_3, \ldots$ are rational numbers. This equation can also be represented as a power series multiplied by a fractional power of x.

Fractional powers in the real number field are difficult to understand, and the fractional power series expansion of complex functions is called Puiseux expansions. It means that when expanding a multivalued function of order n in a series around a branch point $z = z_0$, a branch must be chosen first, and a new complex variable is introduced. For example, let us set

$$\zeta = \sqrt[n]{z - z_0}$$

An analytic function can be expanded as

$$f(z) = \sum_{k=-\infty}^{\infty} c_k \zeta^k = \sum_{k=-\infty}^{\infty} c_k (z - z_0)^{\frac{k}{n}}$$

Puiseux took Cauchy's work in function theory to the end of the first stage. Difficulties encountered in the integration of multivalued functions were still to be overcome, and the next stage of development in algebraic functions and their integration theory awaited the breakthrough by Riemann.

Sometimes, a point in the complex plane is a singularity in the principal branch of a multivalued function but not in other branches. Therefore, different results will be obtained when expanding the function in series on different branches. For example, the function $f(z) = \frac{1}{\ln z}$, in the principal branch, $z = 1$ is a pole of the function, but in other branches, it is not.

Exercises

[1] Let $z_0 = 1 + \frac{i}{2}$. Take the Laurent series expansion of the function $f(z) = \frac{1}{z^2-1}$ in the following regions:

$$\text{(a) } |z - z_0| < \frac{1}{2}; \quad \text{(b) } \frac{1}{2} < |z - z_0| < \frac{\sqrt{17}}{2}; \quad \text{(c) } |z - z_0| > \frac{\sqrt{17}}{2}$$

[2] Perform the Laurent series expansion of the function in the region of $|z| > 1$:

$$f(z) = \frac{1}{(z-1)^2(z+1)^2}$$

Answer:

$$\frac{1}{(z-1)^2(z+1)^2} = \sum_{k=0}^{\infty} \frac{k+1}{z^{2k+4}}$$

[3] Determine the convergence radius of the Laurent series expansion in the neighborhood of $z_0 = 0$ for the following functions:

$$\text{(a) } f(z) = \frac{z^2 - \pi^2}{\sin z}; \quad \text{(b) } f(z) = \frac{z^2 - \pi^2}{\sin^2 z}$$

[4] Prove the uniqueness of Laurent series expansion.
Hint: Multiply both sides of the series $f(z) = \sum_{k=-\infty}^{\infty} a_k(z - z_0)^k$ by $\frac{1}{2\pi i(z-z_0)^{k+1}}$ and perform a contour integration.

[5] Expand the complex function in the punctured neighborhood of $z_0 = 0$:

$$\text{(a) } f(z) = \sin\frac{1}{z}; \quad \text{(b) } f(z) = \cot z$$

Answer:

$$\text{(a) } \sum_{k=0}^{\infty}(-1)^k \frac{1}{(2k+1)!z^{2k+1}}; \quad \text{(b) } \frac{1}{z} - \frac{1}{3}z - \frac{1}{45}z^3 - \cdots$$

[6] Prove that $f(z) = \sqrt{z}$ cannot be expanded in power series at $z_0 = 0$.
Hint: Use proof by contradiction.

[7] Investigate the singularity of the multivalued function:

$$f(z) = \frac{1}{\sqrt{z} + 1}$$

3.4 Types of Singularities

3.4.1 *Singular point*

In mathematics, a point where a function is undefined is called a singular point. In real analysis, a singular point is a point where a function or its derivative is discontinuous. In complex analysis, singular points are defined as points where a function is not differentiable, and can be further classified into isolated singular points and non-isolated singular points.

(1) *Isolated singular point*

If a single-valued function $f(z)$ is not differentiable at a point z_0, but is differentiable at every point in the punctured neighborhood of z_0, then z_0 is called an isolated singular point of $f(z)$. For example, $z = 0$ or $z = \pm \mathrm{i}$ are isolated singular points of the following functions:

$$f(z) = \frac{1}{z}, \quad f(z) = \mathrm{e}^{\frac{1}{z}}, \quad f(z) = \frac{1}{1+z^2}$$

(2) *Non-isolated singular point*

If in any small neighborhood of z_0, there exists a non-differentiable point outside of z_0, then z_0 is called a non-isolated singular point of $f(z)$. The point $z = 0$ in the following function is a non-isolated singular point:

$$f(z) = \frac{1}{\sin(1/z)}$$

Because in any neighborhood of $z = 0$, there are always other singular points, as shown in Figure 3.6.

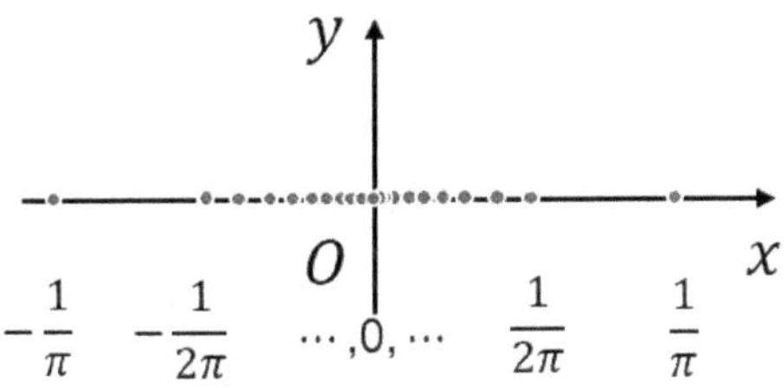

Figure 3.6.

Let us appreciate an example of a non-isolated singular point. For the function defined by an infinite series:

$$f(z) = \sum_{k=0}^{\infty} z^{2^k} = z + z^2 + z^4 + z^8 + z^{16} + \cdots$$

This series converges within the circular domain $|z| < 1$, so $f(z)$ is analytic inside the circle. However, as $z \to 1$, the series diverges, making $z = 1$ a singular point of the function. On the other hand,

$$f(z^2) = z^2 + z^4 + z^8 + z^{16} + \cdots = f(z) - z$$

As $z \to -1$, $f(z)$ also diverges, making $z = -1$ another singular point of the function. Similarly,

$$f(z^4) = z^4 + z^8 + z^{16} + \cdots = f(z) - z - z^2$$

More generally, for any natural number n, we have

$$f(z) = (z + z^2 + z^4 + z^8 + z^{16} + \cdots + z^{2^{n-1}}) + f(z^{2^n})$$

As $z^{2^n} \to 1$, all the series diverge, so $z = e^{2k\pi i/2^n}(k \in \mathbb{Z})$ are all singular points of $f(z)$. As $n \to \infty$, the gaps between singular points become infinitely close, forming a dense circular fence of non-isolated singular points, as shown in Figure 3.7. This dense set of singular points completely isolates the region inside and outside the unit circle.

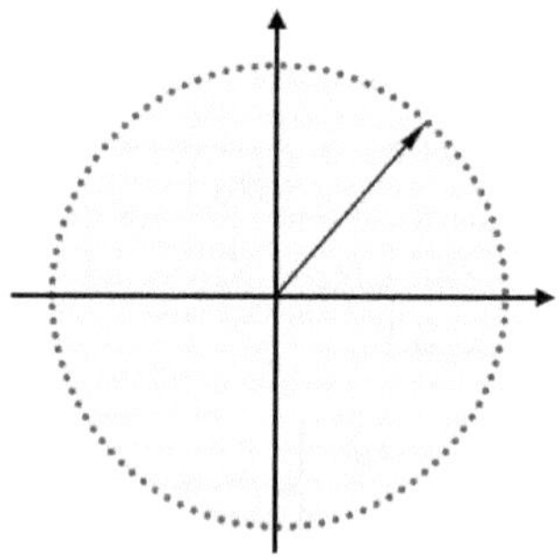

Figure 3.7.

Discussion: Non-isolated singular points are not rare. Functions like $f(z) = |z^2|$ are examples where the function is non-differentiable everywhere on the complex plane except at $z = 0$, making the entire complex plane a non-isolated singular point. If a complex function is arbitrarily defined and does not satisfy the Cauchy–Riemann conditions or is not a harmonic function, it means it is singular everywhere within all regions, such as the function $f(x, y) = (x^3y + 2x) + \mathrm{i}(y^4 + x^2)$. Since it lacks analyticity, functions with non-isolated singular points like this will not be considered in the future.

3.4.2 *Classification of isolated singular points*

The classification of isolated singular points of single-valued functions can be done using Laurent series expansion. In the punctured neighborhood of an isolated singular point, an analytic function can be expanded as follows:

$$f(z) = \sum_{k=-\infty}^{\infty} a_k (z - z_0)^k \tag{3.4.1}$$

where

$$a_k = \frac{1}{2\pi i} \oint_C \frac{f(\zeta)}{(\zeta - z_0)^{k+1}} d\zeta \tag{3.4.2}$$

Based on the highest negative power term, isolated singular points can be classified into three types.

(1) *Removable singularity*

If the Laurent series expansion of a function at point z_0 has no negative power terms, then z_0 is called a removable singularity. It can be deduced from the expansion that the function has a finite limit at the removable singularity:

$$\lim_{z \to z_0} f(z) = a_0$$

For example, $z = 0$ is a removable singularity of the function $f(z) = \frac{\sin z}{z}$.

(2) *Pole*

If the Laurent series expansion of a function at point z_0 has only a finite number of negative power terms, then z_0 is called a pole. The expansion can be written as

$$f(z) = a_{-m}(z - z_0)^{-m} + \cdots + a_{-2}(z - z_0)^{-2} + a_{-1}(z - z_0)^{-1} + a_0 + a_1(z - z_0) + a_2(z - z_0)^2 + \cdots + a_k(z - z_0)^k + \cdots$$

If the highest negative power term is of degree m, it is called an mth-order pole. Specifically, the pole for $m = 1$ is referred to as a simple pole. It is evident that the function approaches infinity at a pole:

$$\lim_{z \to z_0} f(z) = \infty$$

If the function $\varphi(z)$ is analytic in a neighborhood of z_0 and $\varphi(z_0) = 0$, then z_0 is a pole of the function $f(z) = 1/\varphi(z)$, and the order of the pole of $f(z)$ at z_0 is equal to the order of the zero of $\varphi(z)$ at z_0.

(3) *Essential singularity*

If the Laurent series expansion of a function at point z_0 has an infinite number of negative power terms, then z_0 is called an essential singularity. For example, $z_0 = 0$ is an essential singularity of $f(z) = e^{\frac{1}{z}}$, given by

$$e^{\frac{1}{z}} = \sum_{n=0}^{\infty} \frac{1}{n!} \frac{1}{z^n}$$

It can be proven that a function with an essential singularity neither has a finite limit nor approaches infinity at the singularity. The limit of the function varies depending on the approach of $z \to z_0$, indicating that there is no limit at an essential singularity. Any isolated singular point in a single-valued function that does not fall under removable singularities or poles is classified as an essential singularity.

3.4.3 *Classification of branch points*

In the case of multivalued functions, points where the function is not well-defined are called branch points, and they are also considered singular points. Due to the multivalued nature, it is not possible to define the function continuously in the punctured neighborhood of a branch point. Therefore, branch points are not isolated singular points and Laurent series expansion cannot be applied. Analogous to the classification of isolated singular points, branch points can be categorized into two types.

(1) *Algebraic branch points*

When z approaches a branch point, all branches of the function either tend to a finite limit or tend to infinity. For example, the square root function $f(z) = \sqrt[n]{z}$ has two algebraic branch points at $z = 0$ and $z = \infty$, respectively.

(2) *Transcendental branch points*

When z approaches a branch point, the limits of different branches do not exist. For example, the function $f(z) = e^{\frac{1}{\sqrt{z}}}$ has a transcendental branch point at $z = 0$, where the limits of its branches do not exist. However, $z = \infty$ is an algebraic branch point of this function. Branch points of logarithmic functions also fall under the category of transcendental branch points.

3.4.4 *Classification of analytic functions*

Based on the characteristics of isolated singular points, single-valued functions in a certain domain can be classified into two categories.

(1) *Holomorphic functions*

A holomorphic function is defined as a function that is analytic at every point within a region B. It is also known as an analytic function. If a function is analytic throughout the entire complex plane, it is called an entire function. The only isolated singular point of an entire function is $z = \infty$; otherwise, the function must be constant (Liouville's theorem). According to the Taylor series expansion, any entire function can be represented by a power series that converges throughout the entire plane.

(2) *Meromorphic function*

A meromorphic function is a function that is analytic in a region B except for a finite number of poles. It can be shown that a meromorphic function can be expressed as the ratio of two holomorphic functions, where the poles correspond to the zeros of the denominator function. Based on the definition, it can be inferred that a meromorphic function in a bounded domain can only have a finite number of poles, otherwise, the infinite number of poles must accumulate to a non-isolated singularity. However, in the closed complex plane, a meromorphic function can have an infinite number of poles.

ENDNOTE

Let us exhibit how peculiar essential singularities can be. Let $\alpha > 0$ be any real number, and consider the sequence of points:

$$z = \frac{1}{\ln \alpha + 2n\pi \mathrm{i}} \quad (n \in \mathbb{N})$$

As $n \to \infty$, $z \to 0$. Furthermore, we have

$$\mathrm{e}^{\frac{1}{z}} = \mathrm{e}^{\ln \alpha + 2n\pi \mathrm{i}} = \alpha \mathrm{e}^{2n\pi \mathrm{i}} = \alpha$$

It reveals that within any small neighborhood of the singularity $z = 0$, the function $f(z) = \mathrm{e}^{\frac{1}{z}}$ can take any real value. More generally, as $z \to 0$, the function can take any complex value. In fact, all essential singularities possess this property, as described by the Casorati-Weierstrass theorem: If z_0 is an essential singularity of the function $f(z)$, then for any complex number $A \in \bar{\mathbb{C}}$, there exists an infinite sequence of points $z_n \to z_0$ such that

$$\lim_{n \to \infty} f(z_n) = A$$

In order to gain a more intuitive understanding of essential singularities, Figure 3.8 depicts the function $f(z) = \mathrm{e}^{\frac{1}{z}}$ with the essential singularity at $z = 0$. Figure 3.8(a) represents the contour distribution of real and imaginary parts, while Figure 3.8(b) illustrates the distribution of phase angles.

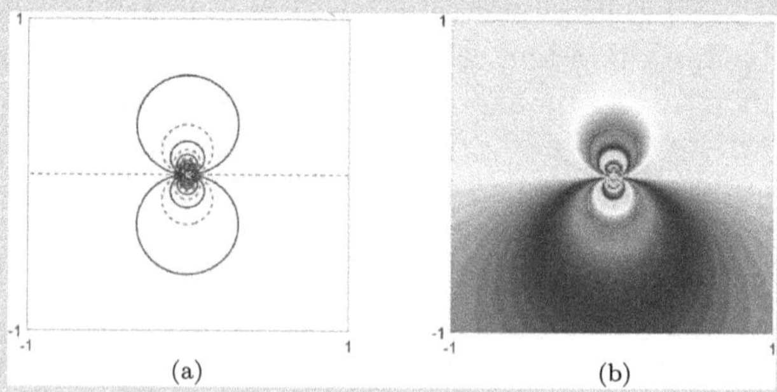

Figure 3.8.

Exercises

[1] Prove that $z = 0$ is a non-isolated singularity of the function:

$$f(z) = \frac{1}{e^{1/z^2} + 1}$$

[2] If z_0 is a zero of analytical function $\varphi(z)$, we need to prove that it is a pole of the function $f(z) = \frac{1}{\varphi(z)}$, and the pole order of $f(z)$ is equal to the zero order of $\varphi(z)$.

[3] Determine the isolated singularities and their types for the following functions:

$$\text{(a) } \frac{\cos z}{z^2 - \pi^2/4}; \quad \text{(b) } \frac{\ln(z+1)}{(z^2-1)^2}; \quad \text{(c) } \sqrt{z}\sin\frac{1}{z}; \quad \text{(d) } \frac{\ln z}{(z-1)^3}$$

Answer: (a)$z = \pm\frac{\pi}{2}$ is removable singularity and $z = \infty$ is an essential singularity; (b) $z = +1$ is a second-order pole; (c) no singularity; (d) $z = 1$ is a second-order pole.

[4] Determine whether $z = \infty$ is an isolated singularity and its type:

$$\text{(a) } \frac{z}{(z^2-2)^2}; \quad \text{(b) } e^z; \quad \text{(c) } z^2\sin\frac{1}{z}; \quad \text{(d) } \frac{\ln z}{(z-1)^3}$$

Answer: (a) removable singularity; (b) essential singularity; (c) simple pole; (d) non-isolated singularity.

[5] If the entire function $f(z)$ is not constant, prove that $z = \infty$ must be a essential singularity of the function $e^{f(z)}$.
Hint: The functions $e^{f(z)}$ and $e^{-f(z)}$ are both entire functions.

[6] Suppose that function $f(z)$ is analytic in a region B except at $z = z_0$, where z_0 is an essential singularity of $f(z)$. Prove the Casorati-Weierstrass theorem.
Hint: By using proof by contradiction, assume that for any complex number A and positive number δ, we always have $\lim_{z\to 0}|f(z) - A| > \delta$. Then $z = z_0$ must be a removable singularity of $g(z) = \frac{1}{f(z)-A}$, which means z_0 is a removable singularity or a pole of $f(z) = A + \frac{1}{g(z)}$, contradicting the given statement.

3.5 Singular Planar Fields

In Section 1.6 of Chapter 1, we discussed that under the conditions of source-free and circulation-free, a planar vector field $\boldsymbol{A} = (A_x, A_y)$

can be described by a complex potential, $f(z) = u(xy) + \mathrm{i}v(xy)$. The real and imaginary parts of the complex potential $f(z)$ are conjugate harmonic functions. The planar field is written as

$$\boldsymbol{A} \equiv A_x + \mathrm{i}A_y = \frac{\partial u}{\partial x} + \mathrm{i}\frac{\partial u}{\partial y} = \frac{\partial u}{\partial x} - \mathrm{i}\frac{\partial v}{\partial x} = \overline{f'(z)}$$

Now the question arises: Can we still describe a planar field using a complex potential when the field exhibits singularities, such as having sources or vortices?

3.5.1 *Source points and vortex points*

In a region B, if the divergence of a two-dimensional vector field $\boldsymbol{A} = (A_x, A_y)$ is non-zero, $\nabla \cdot \boldsymbol{A} \neq 0$, we call the field a source for $\nabla \cdot \boldsymbol{A} > 0$ or a sink for $\nabla \cdot \boldsymbol{A} < 0$. The points where $\nabla \cdot \boldsymbol{A} \neq 0$ are called source points, as shown in Figure 3.9(a). According to Stokes' theorem, the total flux through a closed loop L is given by

$$N = \oint_L \boldsymbol{A} \cdot \hat{n} \mathrm{d}l = \oint_L A_x \mathrm{d}y - A_y \mathrm{d}x \tag{3.5.1}$$

Here, $\hat{n}$ is the unit vector normal to the curve L in the outward direction, and we assume that outward flow is positive.

On the other hand, if the curl of a two-dimensional vector field is non-zero, we call the field a vortex, and the points where $\nabla \times \boldsymbol{A} \neq 0$ are called vortex points, as shown in Figure 3.9(b). The circulation around the loop L due to the vortex is given by

$$\Gamma = \oint_L \boldsymbol{A} \cdot \mathrm{d}\mathbf{l} = \oint_L A_x \mathrm{d}x + A_y \mathrm{d}y \tag{3.5.2}$$

In the case of a source-free and circulation-free field (divergence-free and curl-free), $A(z) = \overline{f'(z)} = A_x + \mathrm{i}A_y$, Therefore,

$$f'(z)\mathrm{d}z = (A_x \mathrm{d}x + A_y \mathrm{d}y) + \mathrm{i}(A_x \mathrm{d}y - A_y \mathrm{d}x)$$

Thus, the flux and circulation integrated along the contour L are given by

$$N = \mathrm{Im} \oint_L f'(z)\mathrm{d}z, \quad \Gamma = \mathrm{Re} \oint_L f'(z)\mathrm{d}z \tag{3.5.3}$$

Hence, if the planar vector field $\boldsymbol{A}$ has both sources and vortices, as shown in Figure 3.9(c), it implies that the complex potential is an analytic function with certain singularities. These singularities can be collectively represented as:

$$\Gamma + \mathrm{i}N = \oint_L f'(z)\mathrm{d}z \tag{3.5.4}$$

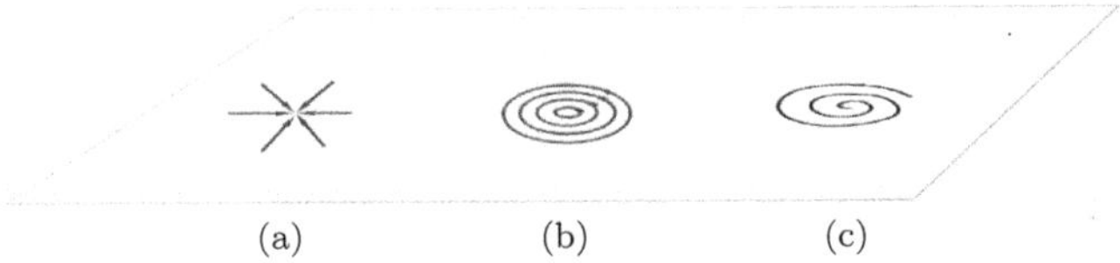

Figure 3.9.

3.5.2 *Complex potential functions*

(1) *Source field*

For simplicity, let's assume that the vector field $\boldsymbol{A}$ has only one source at the origin and no vortices. Due to the axisymmetric nature, the vector field takes a radial form: $\boldsymbol{A} = \varphi(r)\hat{\boldsymbol{r}}$. The total flux through a circular boundary with its center at the origin is given by

$$N = \oint_L A_x \mathrm{d}y - A_y \mathrm{d}x = \varphi(r) \cdot 2\pi r$$

Since the flux should be independent of the radius for flux conservation, we have $\varphi(r) = \frac{N}{2\pi r}$. By substituting

$$\boldsymbol{A} \equiv A_x + \mathrm{i}A_y = \varphi(r)\hat{\boldsymbol{r}} = \frac{N}{2\pi}\frac{1}{\bar{z}}$$

It can be recognized that

$$f'(z) = \frac{N}{2\pi}\frac{1}{z}$$

Thus, the complex potential is given by

$$f(z) = \frac{N}{2\pi}\ln z + C \tag{3.5.5}$$

This is a logarithmic function, and the source point of the field corresponds to the singularity of the logarithmic function. Since $z = \infty$ is also a singularity, it implies the existence of a sink at infinity. Physically, it means that sources and sinks must appear in pairs to ensure the stability of the planar field.

(2) *Vortex field*

If the vector field $\boldsymbol{A}$ has only one vortex point with circulation Γ, following similar reasoning, we can deduce $\boldsymbol{A} = \frac{\Gamma}{2\pi}\frac{\mathrm{i}}{\bar{z}}$, and thus the complex potential is

$$f(z) = \frac{\Gamma}{2\pi\mathrm{i}} \ln z + C \tag{3.5.6}$$

For a field with both sources and vortices as shown in Figure 3.9(c), we have $\boldsymbol{A} = \frac{N+\mathrm{i}\Gamma}{2\pi}\frac{1}{\bar{z}}$, and the complex potential can be expressed as

$$f(z) = \frac{N - \mathrm{i}\Gamma}{2\pi} \ln z \tag{3.5.7}$$

Surprisingly, for singular planar vector fields with source or vortex points, their complex potentials can be expressed by multivalued logarithmic functions. The source or vortex points correspond to the branch points of the logarithmic function, and the vortex at the origin manifests as an anti-vortex at infinity. The flux or circulation is related to the real or imaginary part of the complex potential.

Considering multiple sets of current lines perpendicular to the plane, according to Ampere's circuital law, the magnetic field generated by the current lines can be written as

$$H = \sum_{n=1}^{N} \frac{2\mathrm{i}I_n}{\bar{z} - \bar{\alpha}_n}$$

The corresponding complex potential is

$$f(z) = \sum_{n=1}^{N} 2I_n \ln(z - \alpha_n)$$

For the magnetic field generated by two identical current lines, the complex potential is

$$f(z) = 2I \ln(z^2 - \alpha^2)$$

Therefore, the equation describing the magnetic field lines reads

$$|z - \alpha| \cdot |z + \alpha| = \text{const}$$

Such equations define curves known as compressed egg-shaped lines, as shown in Figure 3.10.

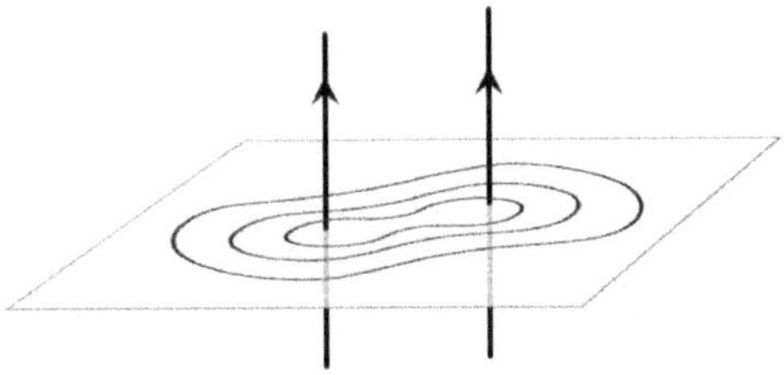

Figure 3.10.

Reflection: If $I_1 = \gamma I_2$, what would be the corresponding curve equation?

ENDNOTE

For continuous distributions of field sources, such as the static charge distribution on a metal surface, the surface of the metal forms a contour L with charge density $\rho(z)$. The potential distribution generated outside the conductor is given by

$$u(z) = \frac{1}{2\pi} \int_L \rho(\zeta) \ln |\zeta - z| \mathrm{d}\zeta$$

Considering it as the real part of a complex potential, we can see that the complex potential contains a singular boundary line formed by the metal surface, and the points on the boundary line are non-isolated singularities.

If a planar vector field contains only one source and one sink with strengths $\pm N$, separated by a distance h, this pair of positive and negative sources is called a dipole, and $p = Nh$ is called the dipole moment. The complex potential function of the planar field is given by

$$f(z) = \frac{N}{2\pi} \ln(z + h) - \frac{N}{2\pi} \ln z$$

Taking the limit as $h \to$, we have

$$f(z) = \lim_{h \to 0} \frac{Nh}{2\pi} \frac{\ln(z + h) - \ln z}{h} = \frac{p}{2\pi} \frac{\mathrm{d}}{\mathrm{d}z} \ln z = \frac{p}{2\pi z}$$

It can be seen that the dipole source forms a simple pole of the complex potential.

If there are pairs of dipoles at points $z = \alpha$ and $z = \alpha - h$, they together form a quadrupole. The complex potential of the planar vector field generated by the quadrupole is given by

$$f(z) = \lim_{h \to 0} \frac{ph}{2\pi} \frac{1}{h} \left[\frac{1}{z - \alpha + h} - \frac{1}{z - \alpha} \right] = -\frac{ph}{2\pi} \frac{1}{(z - \alpha)^2}$$

Therefore, the quadrupole source forms a second-order pole of the complex potential. Conversely, a meromorphic function with a second-order pole can be interpreted as a planar vector field generated by a quadrupole. Similarly, a meromorphic function with an nth order pole is considered to be generated by a planar vector field with $2n$ poles.

Exercises

[1] There are two parallel wires with a distance of $2d$ between them. The wires have uniform charge distributions with unit charge per unit length, $\pm q$. Find the complex potential function of the electric field perpendicular to the plane of the wires.
Answer: $f(z) = 2q \ln \frac{z+d}{z-d}$

[2] If there are closely spaced positive and negative vortices in the plane, write down their complex potential function.

[3] Consider a fluid velocity field $\mathbf{v} = \nabla \phi$, where the velocity potential is given by

$$\phi(z) = \ln \left| \frac{z - 1}{z + 1} \right|$$

Find the equation of the streamlines and calculate the flow rate from the singular points $z = \pm 1$.
Answer: $\arg(z - 1) - \arg(z - 1) = C; q = \pm \frac{1}{2}$

[4] Analyze the multipole structure of the complex potential:

$$f(z) = \frac{1}{(z + 1)^2 (z - 1)}$$

Chapter 4

Calculus of Residues

4.1 Residue Theorem

4.1.1 *Definition of residuc*

Let z_0 be an isolated singularity of the single-valued analytic function $f(z)$. In the punctured neighborhood of $z = z_0$, $f(z)$ can be expanded into a Laurent series:

$$f(z) = \sum_{n=-\infty}^{\infty} a_n (z - z_0)^n$$

Employing the integration formula:

$$I = \oint_L (z - z_0)^n \mathrm{d}z = \begin{cases} 0 & (n \neq -1) \\ 2\pi\mathrm{i} & (n = -1) \end{cases}$$

Integrate both sides along an closed contour L that encloses z_0. Due to the uniform convergence of the power series, the order of integration and summation can be interchanged, giving rise to

$$\oint_L f(z)\mathrm{d}z = \sum_{n=-\infty}^{\infty} \oint_L a_n (z - z_0)^n \mathrm{d}z = 2\pi\mathrm{i}a_{-1} \tag{4.1.1}$$

It can be observed that the result of the contour integral depends only on the coefficient a_{-1} of the $(z - z_0)^{-1}$ term in the Laurent series expansion. This coefficient is called the residue of $f(z)$ at z_0, denoted as

$$a_{-1} \equiv \mathrm{Res}[f(z)]|_{z=z_0} \equiv \underset{z=z_0}{\mathrm{Res}}[f(z)] \equiv \mathrm{Res}[f(z_0)]$$

Residue Theorem: Suppose that the function $f(z)$ is analytic within the region B enclosed by the closed contour L, except for a finite number of isolated singularities $z_1, z_2, \ldots, z_N$. Then, the contour is

$$\oint_L f(z)\mathrm{d}z = 2\pi \mathrm{i} \sum_{k=1}^{N} \mathrm{Res}[f(z_k)] \tag{4.1.2}$$

Proof. Take the integration contour as shown in Figure 4.1, and utilize the Cauchy's theorem for multiply connected domains along with Equation (4.1.1) to prove this theorem. □

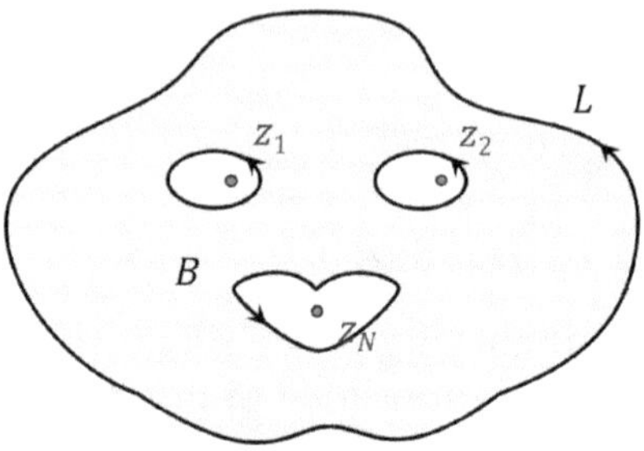

Figure 4.1.

4.1.2 *Calculation of residue*

Since the result of the contour integral depends only on the residues of the isolated singularities, we can calculate the contour integral by determining the residues. There are several common methods for calculating residues.

(1) *Laurent series expansion method*

Based on the definition of residues, expand the function $f(z)$ into a Laurent series and pick the coefficient of the first negative power term.

(2) *The first-order pole*

If z_0 is a simple pole of the single-valued analytic function $f(z)$, then

$$\underset{z=z_0}{\mathrm{Res}}[f(z)] = \lim_{z\to z_0}(z-z_0)f(z) \tag{4.1.3}$$

If $f(z) = \frac{P(z)}{Q(z)}$, where both $P(z)$ and $Q(z)$ are analytic functions, then

$$\operatorname*{Res}_{z=z_0}[f(z)] = \lim_{z\to z_0} \frac{P(z)}{Q'(z)} \tag{4.1.4}$$

(3) *The mth-order pole*

If z_0 is an mth-order pole of the single-valued analytic function $f(z)$, and

$$f(z) = \frac{a_{-m}}{(z-z_0)^m} + \frac{a_{-m+1}}{(z-z_0)^{m-1}} + \cdots + \frac{a_{-1}}{(z-z_0)} + a_0 + a_1(z-z_0) + \cdots$$

then the residue is given by

$$\operatorname*{Res}_{z=z_0} f(z) = a_{-1} = \frac{1}{(m-1)!} \lim_{z\to z_0} \frac{\mathrm{d}^{m-1}}{\mathrm{d}z^{m-1}}[(z-z_0)^m f(z)] \tag{4.1.5}$$

Example 4.1. Calculate the residue at $z = 1$:

$$f(z) = \frac{1}{z^n - 1} \quad (n \in \mathbb{N})$$

Solution: The singularity at $z = 1$ is a simple pole, so

$$\text{Res}[f(z)]|_{z=1} = \left.\frac{1}{nz^{n-1}}\right|_{z=1} = \frac{1}{n}$$

Example 4.2. Determine the poles of the function $f(z)$ and calculating the residues at these poles:

$$f(z) = \frac{z+2\mathrm{i}}{z^5 + 4z^3}$$

Solution: By simplifying the function, we have

$$f(z) = \frac{z+2\mathrm{i}}{z^5+4z^3} = \frac{1}{z^3(z-2\mathrm{i})}$$

It can be seen that the isolated singularities are $z = 0$ (the third-order pole) and $z = 2\mathrm{i}$ (the simple pole). The residues at these points are

$$\text{Res}[f(z)]|_{z=2\mathrm{i}} = \left.\frac{z+2\mathrm{i}}{5z^4+12z^2}\right|_{z=2\mathrm{i}} = \frac{\mathrm{i}}{8}$$

$$\text{Res}[f(z)]|_{z=0} = \frac{1}{2!}\frac{\mathrm{d}^2}{\mathrm{d}z^2}\left(\frac{1}{z-2\mathrm{i}}\right)\Bigg|_{z=0} = -\frac{\mathrm{i}}{8}$$

Example 4.3. Calculate the integral:

$$\oint_{|z-\mathrm{i}|=1} \frac{1}{(z^2+1)^2} \mathrm{d}z$$

Solution: Within the integration contour, $z = \mathrm{i}$ is a second-order pole of the function. The residue at this point is

$$\mathrm{Res}\left[\frac{1}{(z^2+1)^2}\right]\Bigg|_{z=\mathrm{i}} = \frac{1}{1!}\frac{\mathrm{d}}{\mathrm{d}z}\left[(z-\mathrm{i})^2\frac{1}{(z^2+1)^2}\right]\Bigg|_{z=\mathrm{i}} = -\frac{\mathrm{i}}{4}$$

Therefore, the integral is

$$\oint_{|z-\mathrm{i}|=1} \frac{1}{(z^2+1)^2} \mathrm{d}z = 2\pi\mathrm{i} \times \left(-\frac{\mathrm{i}}{4}\right) = \frac{\pi}{2}$$

4.1.3 *Residues at infinity*

The residues at infinity is defined as

$$\mathrm{Res}[f(\infty)] \stackrel{\mathrm{def}}{=} = \frac{1}{2\pi\mathrm{i}} \oint_{L}^{-} f(z)\mathrm{d}z \tag{4.1.6}$$

Here, L^- represents a counterclockwise contour around the point at infinity, with no other singularities inside the contour. The contour is equivalent to a counterclockwise contour around all singularities z_k on the complex plane. Hence,

$$\sum_{z_k} \mathrm{Res}[f(z_k)] + \mathrm{Res}[f(\infty)] = 0 \tag{4.1.7}$$

This conclusion is easily understood from the perspective of the Riemann sphere, where the contour around all finite singularities is equivalent to the contour around the north pole.

By performing a variable substitution $\zeta = \frac{1}{z}$, we have

$$\mathrm{Res}[f(\infty)] = \frac{1}{2\pi\mathrm{i}} \oint_{L^-} f(z)\mathrm{d}z = -\frac{1}{2\pi\mathrm{i}} \oint_{L} \frac{f(\zeta)}{\zeta^2}\mathrm{d}\zeta$$

It is worth noting that although the point at infinity may not be a singularity of the function, its residue is not necessarily zero. For example, consider the function $f(z) = \frac{1}{z}$, where $z = \infty$ is not a singularity, but its residue is $\text{Res}[f(\infty)] = -1$. Generally, if $f(z)$ is analytic for $|z| > R$ and its Laurent series expansion is given by

$$f(z) = \sum_{k=-\infty}^{\infty} a_k z^k \quad (|z| > R)$$

The residue at infinity is $\text{Res}[f(\infty)] = -a_{-1}$.

Example 4.4. Find the residue at infinity:

$$f(z) = \frac{\mathrm{e}^z}{z^3}$$

Solution: Expand the exponential function e^z as the Taylor series:

$$f(z) = \frac{1}{z^3} + \frac{1}{1!}\frac{1}{z^2} + \frac{1}{2!}\frac{1}{z} + \frac{1}{3!}z + \cdots$$

Hence,

$$\text{Res}[f(\infty)] = -\frac{1}{2}$$

Exercises

[1] Compute the residues at finite singularities:

(a) $f(z) = \dfrac{1}{z}\left[1 + \dfrac{1}{z+1} + \dfrac{1}{(z+1)^2} + \cdots + \dfrac{1}{(z+1)^n}\right]$;

(b) $\dfrac{\sqrt{z}}{\sinh\sqrt{z}}$; (c) $f(z) = \dfrac{\mathrm{e}^z - 1}{\sin^3 z}$; (d) $f(z) = z^3 \cos\dfrac{1}{z-2}$

Hint: (a) $z = -1$ is a nth-order pole; (b) The poles are $z = -n^2\pi^2 (n \in \mathbb{Z})$; (c) The poles are $z = n\pi (n \in \mathbb{Z})$, among which $n = 0$ is a second-order pole and others are third-order poles; (d) $z = 2$ is an essential singularity.

Answer:

(a) $\text{Res}[f(0)] = n + 1, \quad \text{Res}[f(-1)] = -n$

(b) $\text{Res}[f(-n^2\pi^2)] = 2(-1)^{n+1} n^2\pi^2 \ (n \in Z)$

(c) $\text{Res}[f(n\pi)] = \dfrac{1}{2}(-1)^{3n}\mathrm{e}^{n\pi} \ (n \in \mathbb{Z})$; (d) $\text{Res}[f(2)] = -\dfrac{143}{24}$

[2] Calculate the contour integrals:

$$(a) \oint_{|z|=2} \frac{z}{z^4-1} dz; \quad (b) \oint_{|z|=2} \frac{e^z}{z(z-1)^2} dz;$$

$$(c) \oint_{|z|=1} \cot z dz; \qquad (d) \oint_{|z-1|=1/2} \frac{\sin z}{\ln z} dz$$

[3] Find all singularities of the function $f(z) = \cot z^2$ and calculate their residues.

[4] Calculate the residues of the following functions at infinity:

$$(a)\ f(z) = e^{\frac{1}{1-z}}; \quad (b)\ f(z) = \frac{e^z}{(z-1)^n}\ (n \in \mathbb{N});$$

$$(c)\ f(z) = \sqrt{(z-1)(2-z)}; \quad (d)\ e^{\frac{1}{z}} \sin z$$

Hint: (a) Calculate the residue at $z = 1$: $\mathrm{Res}\, f(1) = -1$; (b) Calculate the residue at $z = 1$; (c) $z = 1$ and $z = 2$ are two branch points of the multivalued function. Perform the Taylor series expansion at $z = 0$ on one of the branch sheet; (d) $z = \infty$ is an essential singularity of the function $f(z) = \sin z$.

Answer:

$$(a)\ 1; \quad (b)\ -\frac{e}{(n-1)!}; \quad (c)\ \pm\frac{i}{8}; \quad (d)\ \sum_{n=1}^{\infty} \frac{(-1)^n}{(2n-1)!(2n)!}$$

[5] Calculate the contour integral:

$$\oint_{|z|=2} \frac{1}{(z^8+1)^2} dz$$

Hint: Calculate the residue at infinity.

Answer: $\mathrm{Res}\, f(\infty) = 0$

4.2 Integration of Real Functions

4.2.1 *Three typical integrals*

The residue theorem can be conveniently utilized to calculate the definite integrals of many real-valued functions. Below are three basic types of integrals.

(1) *Type I*

$$\int_0^{2\pi} R(\cos\theta, \sin\theta)\mathrm{d}\theta$$

Here the integrand $R(\cos\theta, \sin\theta)$ is a rational function of trigonometric functions. By making a change of variable $z = \mathrm{e}^{\mathrm{i}\theta}$, with the real variable θ ranging from 0 to 2π, the parameter integral becomes a contour integral of the complex variable z along the unit circular path $|z| = 1$:

$$I = \int_0^{2\pi} R(\cos\theta, \sin\theta)\mathrm{d}\theta = \oint_{|z|=1} R\left(\frac{z+z^{-1}}{2}, \frac{z-z^{-1}}{2\mathrm{i}}\right)\frac{\mathrm{d}z}{\mathrm{i}z} \tag{4.2.1}$$

Example 4.5. Calculate the definite integral:

$$\int_0^{2\pi} \frac{1}{1+\varepsilon\cos\theta}\mathrm{d}\theta \quad (0 < \varepsilon < 1)$$

Solution: Let $z = \mathrm{e}^{\mathrm{i}\theta}$, then

$$\begin{aligned}\int_0^{2\pi} \frac{1}{1+\varepsilon\cos\theta}\mathrm{d}\theta &= \oint_{|z|=1} \frac{1}{1+\varepsilon(z+z^{-1})/2}\frac{\mathrm{d}z}{\mathrm{i}z} \\ &= \oint_{|z|=1} \frac{-2\mathrm{i}}{\varepsilon z^2 + 2z + \varepsilon}\mathrm{d}z\end{aligned}$$

The simple poles of the integrand are given by

$$z_1 = \frac{-1+\sqrt{1-\varepsilon^2}}{\varepsilon}, \quad z_2 = \frac{-1-\sqrt{1-\varepsilon^2}}{\varepsilon}$$

They are a pair of conjugate points of the unit circle. It is easy to determine that z_1 lies inside the integration contour $|z| = 1$, while z_2 lies outside. Therefore, the residue is

$$\mathrm{Res}[f(z_1)] = \frac{-2\mathrm{i}}{2\varepsilon z_1 + 2} = \frac{-\mathrm{i}}{\sqrt{1-\varepsilon^2}}$$

Applying the residue theorem, the result of the integral is

$$\int_0^{2\pi} \frac{1}{1+\varepsilon\cos\theta}\mathrm{d}\theta = 2\pi\mathrm{i}\,\mathrm{Res}[f(z_1)] = \frac{2\pi}{\sqrt{1-\varepsilon^2}}$$

(2) *Type II*

$$\int_{-\infty}^{\infty} f(x)\mathrm{d}x$$

Here, it is assumed that the integrand $f(x)$ has no singularities on the real axis. The basic idea for calculating integrals of this type is to relate the definite integral of a real-valued function to the contour integral of a complex-valued function. The original integral along the real axis $(-\infty, \infty)$ is regarded as path L_1. By analytically extending the real-valued function to the complex plane, $f(x) \longrightarrow f(z)$, we introduce path L_2 in the upper half-plane to form a closed contour $L = L_1 + L_2$, as shown in Figure 4.2.

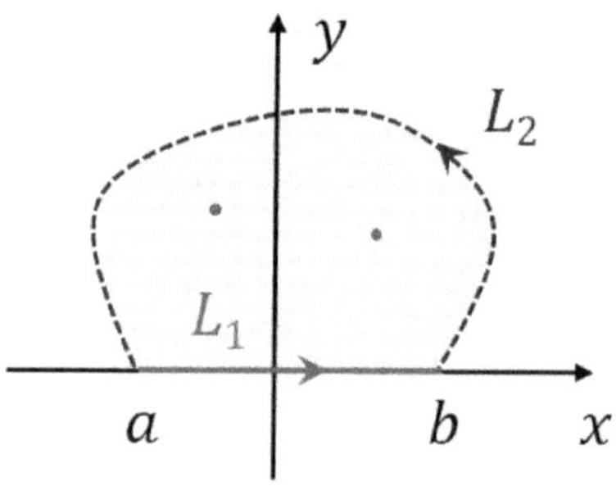

Figure 4.2.

$$\int_a^b f(x)\mathrm{d}x \longrightarrow \int_{L_1} f(z)\mathrm{d}z + \int_{L_2} f(z)\mathrm{d}z \longrightarrow \oint_{L_1+L_2} f(z)\mathrm{d}z$$

If we can compute the integral along path L_2, we obtain the integral of the real-valued function by applying the residue theorem to calculate the integral over the closed loop L.

Specifically, for the infinite integral $\int_{-\infty}^{\infty} f(x)\mathrm{d}x$, by analytically extending the integrand $f(x)$ to a complex-valued function $f(z)$, if $f(z)$ has a finite number of isolated singularities in the complex plane, we can introduce a semicircular path C_R to form the contour L as shown in Figure 4.3. Then we have

$$\oint_L f(z)\mathrm{d}z = \int_{-R}^{R} f(x)\mathrm{d}x + \int_{C_R} f(z)\mathrm{d}z$$

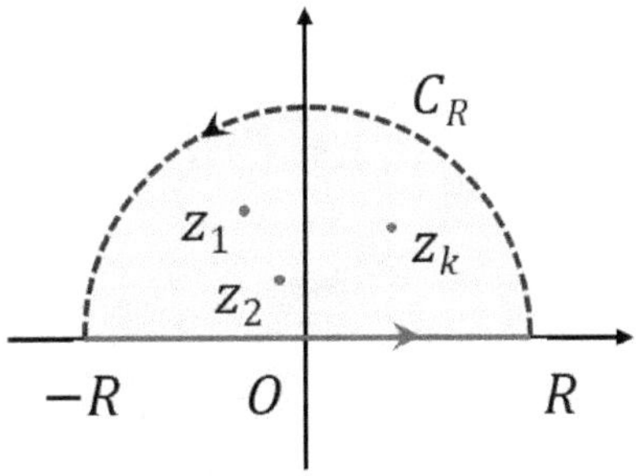

Figure 4.3.

We hope that the integral along the semicircle C_R vanishes,

$$\int_{C_R} f(z)\mathrm{d}z \xrightarrow{R\to\infty} 0$$

From this, we obtain

$$\int_{-\infty}^{\infty} f(x)\mathrm{d}x = 2\pi\mathrm{i} \sum_{z_k\in\mathbb{H}} \mathrm{Res}[f(z_k)] \tag{4.2.2}$$

where the symbol $\mathbb{H}$ denotes the upper half-plane: $\mathbb{H} = \{z \in \mathbb{C}|\mathrm{Im}\, z > 0\}$. It can be proven that as $R \to \infty$, as long as $|zf(z)|$ approaches zero uniformly, this condition will be satisfied.

Proof. Based on the properties of complex variable integration, for the path C_R in Figure 4.3, we have

$$\left|\int_{C_R} f(z)\mathrm{d}z\right| = \left|\int_{C_R} zf(z)\frac{\mathrm{d}z}{z}\right| \leq \int_{C_R} |zf(z)|\frac{|\mathrm{d}z|}{|z|}$$
$$\leq \max|zf(z)|\frac{\pi R}{R} = \pi \max|zf(z)| \xrightarrow{|z|\to\infty} 0$$

□

Example 4.6. Calculate the Integral:

$$\int_{-\infty}^{\infty} \frac{1}{(1+x^2)^2}\mathrm{d}x$$

Solution: We analytically extend the integrand to the complex plane as

$$f(z) = \frac{1}{(1+z^2)^2}$$

Since

$$|zf(z)| = \left|\frac{z}{(1+z^2)^2}\right| \stackrel{|z|\to\infty}{\longrightarrow} 0$$

We have

$$\int_{C_R} \frac{1}{(1+z^2)^2}\mathrm{d}z = 0$$

The function $f(z)$ has two second-order poles at $z = \pm i$. Among them, only $z = i$ lies in the upper half-plane, as shown in Figure 4.4. Therefore,

$$\int_{-\infty}^{\infty} \frac{1}{(1+x^2)^2}\mathrm{d}x = 2\pi\mathrm{i}\mathrm{Res}\left[\frac{1}{(1+z^2)^2}\right]\bigg|_{z=\mathrm{i}} = 2\pi\mathrm{i}\frac{\mathrm{d}}{\mathrm{d}z}\left.\frac{1}{(z+\mathrm{i})^2}\right|_{z=\mathrm{i}} = \frac{\pi}{2}$$

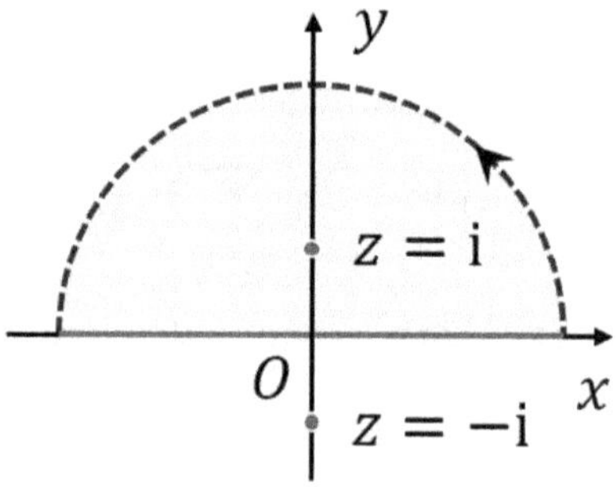

Figure 4.4.

(3) *Type III*

$$\int_{-\infty}^{\infty} f(x)\cos mx\mathrm{d}x, \quad \int_{-\infty}^{\infty} f(x)\sin mx\mathrm{d}x$$

These integrals involve algebraic and trigonometric functions. Presently, we assume that the integrand $f(z)$ has no singularities on the real axis. Consider the case of $m > 0$ and perform an analytic extension:

$$f(x)\mathrm{e}^{\mathrm{i}mx} \longrightarrow f(z)\mathrm{e}^{\mathrm{i}mz}$$

We construct a closed contour as shown in Figure 4.3 and still hope the integral along the semicircle C_R to be zero. This requires a certain condition for the integrand, known as Jordan's Lemma.

Jordan's Lemma. *If $m > 0$ and C_R is a semicircular path centered at the origin and located in the upper half-plane. If $f(z)$ approaches zero uniformly as z tends to infinity, then*

$$\lim_{R\to\infty}\int_{C_R} f(z)\mathrm{e}^{\mathrm{i}mz}\mathrm{d}z = 0$$

Proof. Parametrizing the upper half-plane semicircle C_R with $z = R\mathrm{e}^{\mathrm{i}\varphi}$, where $0 \leq \varphi \leq \pi$, one has

$$\begin{aligned}\left|\int_{C_R} f(z)\mathrm{e}^{\mathrm{i}mz}\mathrm{d}z\right| &= \left|\int_0^{\pi} f(R\mathrm{e}^{\mathrm{i}\varphi})\mathrm{e}^{-mR\sin\varphi}\mathrm{e}^{\mathrm{i}mR\cos\varphi}R\mathrm{e}^{\mathrm{i}\varphi}\mathrm{i}\mathrm{d}\varphi\right| \\ &\leq \max|f(z)|\int_0^{\pi} \mathrm{e}^{-mR\sin\varphi}R\mathrm{d}\varphi\end{aligned}$$

Since $\max|f(z)| \to 0$ as $R \to \infty$, it suffices to prove that the integral $\int_0^{\pi} \mathrm{e}^{-mR\sin\varphi}R\mathrm{d}\varphi$ is finite. Note that in the interval $0 \leq \varphi \leq \pi/2$, we have $0 \leq 2\varphi/\pi \leq \sin\varphi$, as shown in Figure 4.5. Therefore,

$$\begin{aligned}\int_0^{\pi} \mathrm{e}^{-mR\sin\varphi}R\mathrm{d}\varphi &= 2\int_0^{\frac{\pi}{2}} \mathrm{e}^{-mR\sin\varphi}R\mathrm{d}\varphi \\ < 2\int_0^{\frac{\pi}{2}} \mathrm{e}^{-\frac{2mR\varphi}{\pi}}R\mathrm{d}\varphi &= \frac{\pi}{m}(1-\mathrm{e}^{-mR}) \overset{R\to\infty}{\longrightarrow} \frac{\pi}{m}\end{aligned}$$

This completes the proof. □

According to Jordan's Lemma, Type III integrals can be expressed in terms of residues of the upper half-plane:

$$\int_{-\infty}^{\infty} f(x)\mathrm{e}^{\mathrm{i}mx}\mathrm{d}x = 2\pi\mathrm{i}\sum_{z_k\in\mathbb{H}} \mathrm{Res}[f(z_k)\mathrm{e}^{\mathrm{i}mz_k}] \tag{4.2.3}$$

When the integrand contains trigonometric functions, we use the following relationships:

$$\int_{-\infty}^{\infty} f(x)\mathrm{e}^{-\mathrm{i}mx}\mathrm{d}x \overset{x\to -x}{\longrightarrow} \int_{-\infty}^{\infty} f(-x)\mathrm{e}^{\mathrm{i}mx}\mathrm{d}x \quad (m > 0)$$

For even functions $f(x)$,

$$\int_0^{\infty} f(x)\cos mx\,\mathrm{d}x = \pi\mathrm{i}\sum_{z_k\in\mathbb{H}} \mathrm{Res}[f(z_k)\mathrm{e}^{\mathrm{i}mz_k}]$$

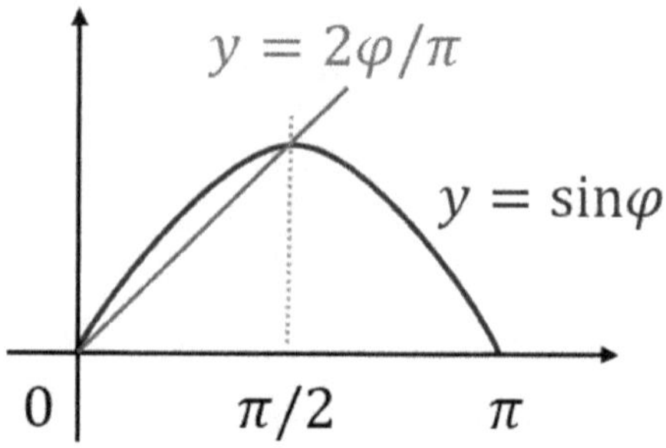

Figure 4.5.

For odd functions $f(x)$,

$$\int_0^{\infty} f(x) \sin mx \, \mathrm{d}x = \pi \sum_{z_k \in \mathbb{H}} \mathrm{Res}[f(z_k)\mathrm{e}^{\mathrm{i}mz_k}]$$

Example 4.7. Calculate the integral:

$$\int_0^{\infty} \frac{x \sin mx}{1+x^2} \mathrm{d}x \quad (m > 0)$$

Solution: Since the integrand is an even function, it can be written it in terms of exponential functions. After performing an analytic extension, we get

$$\int_0^{\infty} \frac{x \sin mx}{1+x^2} \mathrm{d}x = \frac{1}{2} \int_{-\infty}^{\infty} \frac{x \sin mx}{1+x^2} \mathrm{d}x = \frac{1}{2\mathrm{i}} \int_{-\infty}^{\infty} \frac{x\mathrm{e}^{\mathrm{i}mx}}{1+x^2} \mathrm{d}x$$
$$= 2\pi\mathrm{i} \times \frac{1}{2\mathrm{i}} \operatorname*{Res}_{z=i}[f(z)\mathrm{e}^{\mathrm{i}mz}] = \frac{\pi}{2}\mathrm{e}^{-m}$$

4.2.2 *Simple poles on the real axis*

Consider integrals of Type II or Type III for integrand $f(x)$ having a first-order pole $z = z_0$ on the real axis. We define the principal value integral as

$$\mathcal{P} \int_{-\infty}^{\infty} f(x) \mathrm{d}x = \lim_{\varepsilon \to 0} \left[\int_{-\infty}^{z_0 - \varepsilon} f(x) \mathrm{d}x + \int_{z_0 + \varepsilon}^{\infty} f(x) \mathrm{d}x \right]$$

A closed contour L is constructed as shown in Figure 4.6, and its integral reads

$$\oint_L f(z)\mathrm{d}z = \int_{-R}^{z_0-\varepsilon} f(x)\mathrm{d}x + \int_{z_0+\varepsilon}^{R} f(x)\mathrm{d}x + \int_{C_R} f(z)\mathrm{d}z + \int_{C_\varepsilon} f(z)\mathrm{d}z = 2\pi\mathrm{i}\sum_{z_k\in\mathbb{H}} \mathrm{Res}[f(z_k)]$$

Since $z = z_0$ is a simple pole, we write it as

$$f(z) = \frac{a_{-1}}{z - z_0} + g(z)$$

where $g(z)$ is analytic at $z = z_0$. Taking the limits $R \to \infty$ and $\varepsilon \to 0$, it has

$$\int_{C_R} f(z)\mathrm{d}z \to 0, \quad \int_{C_\varepsilon} f(z)\mathrm{d}z = -\pi\mathrm{i}a_{-1} = -\pi\mathrm{i}\mathrm{Res}[f(z_0)]$$

Therefore,

$$\mathbb{P}\int_{-\infty}^{\infty} f(x)\mathrm{d}x = 2\pi\mathrm{i}\sum_{z_k\in\mathbb{H}} \mathrm{Res}[f(z_k)] + \pi\mathrm{i}\sum_{z_0\in\mathbb{R}} \mathrm{Res}[f(z_0)] \qquad (4.2.4)$$

It means that the simple pole on the real axis contributes only half of its residue.

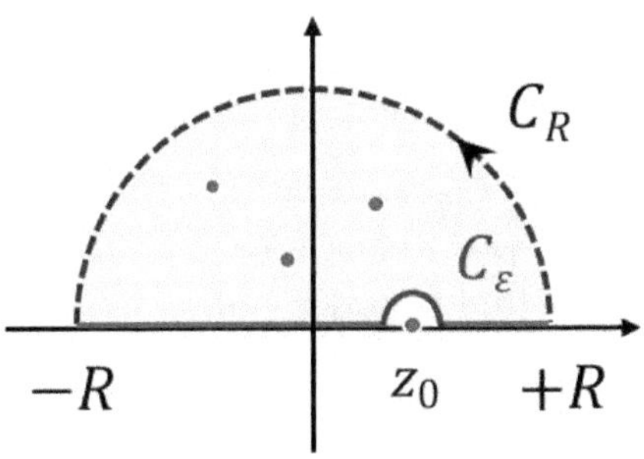

Figure 4.6.

Example 4.8. Calculate the integral:

$$\int_0^\infty \frac{\sin x}{x} \mathrm{d}x$$

Solution: Since the integrand is an even function, we rewrite it as the standard form:

$$\int_0^\infty \frac{\sin x}{x} \mathrm{d}x = \frac{1}{2} \int_{-\infty}^\infty \frac{\sin x}{x} \mathrm{d}x = \frac{1}{2} \int_{-\infty}^\infty \frac{\mathrm{e}^{\mathrm{i}x} - \mathrm{e}^{-\mathrm{i}x}}{2\mathrm{i}x} \mathrm{d}x = \frac{1}{2\mathrm{i}} \int_{-\infty}^\infty \frac{\mathrm{e}^{\mathrm{i}x}}{x} \mathrm{d}x$$

The function $z = \frac{\mathrm{e}^{\mathrm{i}z}}{z}$ has a simple pole at $z = 0$ on the real axis and no other singularities in the complex plane. Therefore,

$$\int_0^\infty \frac{\sin x}{x} \mathrm{d}x = \pi\mathrm{i} \times \frac{1}{2\mathrm{i}} \operatorname*{Res}_{z=0} \left[\frac{\mathrm{e}^{\mathrm{i}z}}{z} \right] = \frac{\pi}{2}$$

Example 4.9. Calculate the integral:

$$\int_0^\infty \frac{\sin x}{x(x^2 - 1)} \mathrm{d}x$$

Solution: Extend the integrand analytically to the complex plane. There are three simple poles on the real axis: $z_j = 0, \pm 1$. We construct the integration contour as shown in Figure 4.7. By calculating the residue at each pole, we obtain

$$\mathcal{P} \int_0^\infty \frac{\sin x}{x(x^2 - 1)} \mathrm{d}x = \pi\mathrm{i} \times \frac{1}{2\mathrm{i}} \sum_{z_j \in \mathbb{R}} \operatorname{Res} \left[\frac{\mathrm{e}^{\mathrm{i}z_j}}{z_j(z_j^2 - 1)} \right] = \frac{\pi}{2}(\cos 1 - 1)$$

Discussion:

(1) Singularities on the real axis can only be simple poles and cannot be second-order or higher-order poles.
(2) Can we choose the lower half of the circular arc for C_ε to perform the integration?
(3) If a simple pole is encountered on any straight line path during integration, it also contributes half of its residue.

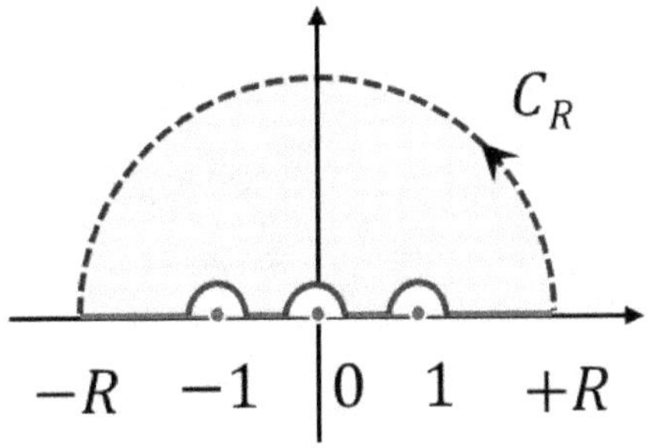

Figure 4.7.

4.2.3 *Poles on the lower half-plane*

For the Type III integrals, if $m < 0$, we sometimes perform a variable substitution by replacing $x \longrightarrow -x$ in the real-valued function and then proceed using the methods described earlier. However, sometimes it is more convenient to employ the integration along the half-circle path in the lower half-plane, which is denoted by $\bar{\mathbb{H}}$. We let $m' = |m|$ and establish the following conclusion:

Suppose $f(z)$ is a single-valued analytic function in the complex plane except for a finite number of isolated singularities at z_k, and $|f(z)|$ uniformly approaches zero as $|z| \to \infty$. Then, we have

$$\int_{-\infty}^{\infty} f(x)\mathrm{e}^{-\mathrm{i}m'x}\mathrm{d}x = -2\pi\mathrm{i} \sum_{z_k \in \bar{\mathbb{H}}} \mathrm{Res}[f(z_k)\mathrm{e}^{-\mathrm{i}m'z_k}] \tag{4.2.5}$$

Proof. Let C_R and C_R' be the semicircular paths in the upper and lower half-planes, respectively. Due to the uniform convergence of $|f(z)| \to 0$ as $|z| \to \infty$, the Jordan's Lemma holds for the infinite radius lower half-circle C_R', i.e.,

$$\lim_{R\to\infty} \int_{C_{R'}} f(z)\mathrm{e}^{-\mathrm{i}m'z}\mathrm{d}z = 0 \quad (m' > 0)$$

On the other hand, the integral along the upper half-circle C_R is non-zero. Considering the closed contour composed of the real axis and the lower half-circle C_R', the residue theorem gives rise to

$$\int_{-\infty}^{\infty} f(x)\mathrm{e}^{-\mathrm{i}m'x}\mathrm{d}x - \int_{C_{R'}} f(z)\mathrm{e}^{-\mathrm{i}m'z}\mathrm{d}z = -2\pi\mathrm{i} \sum_{z_k \in \bar{\mathbb{H}}} \mathrm{Res}[f(z_k)\mathrm{e}^{-\mathrm{i}m'z_k}]$$

The negative sign arises from integrating clockwise along the lower half-circle. Thus, we obtain

$$\int_{-\infty}^{\infty} f(x)\mathrm{e}^{-\mathrm{i}m'x}\mathrm{d}x = -2\pi\mathrm{i}\sum_{z_k\in\bar{\mathbb{H}}} \mathrm{Res}[f(z_k)\mathrm{e}^{-\mathrm{i}m'z_k}]$$

This completes the proof. □

We can further determine that the integral along the upper half-circle C_R is given by

$$\begin{aligned}\lim_{R\to\infty}\int_{C_R} f(z)\mathrm{e}^{-\mathrm{i}m'z}\mathrm{d}z &= \lim_{R\to\infty}\int_{C_R+C_{R'}} f(z)\mathrm{e}^{-\mathrm{i}m'z}\mathrm{d}z\\ &= 2\pi\mathrm{i}\sum_{z_k\in\mathbb{C}} \mathrm{Res}[f(z_k)\mathrm{e}^{-\mathrm{i}m'z_k}]\end{aligned}$$

Namely, the result of integrating along the upper half-circle C_R is related to the residues of isolated singularities in the entire complex plane.

Example 4.10. Calculate the integral:

$$\int_0^{\infty} \frac{\sin^3 x}{x^3}\mathrm{d}x$$

Solution: Consider the contour integral:

$$\oint_L \frac{\sin^3 z}{z^3}\mathrm{d}z$$

where the contour L is still the upper half-circle, avoiding the pole at $z = 0$ as shown in Figure 4.8. Since there are no singularities in the upper half-plane, we have

$$\oint_\Gamma \frac{\sin^3 z}{z^3}\mathrm{d}z = \int_{-R}^{R} \frac{\sin^3 x}{x^3}\mathrm{d}x + \int_{C_R} \frac{\sin^3 z}{z^3}\mathrm{d}z + \int_{C_\varepsilon} \frac{\sin^3 z}{z^3}\mathrm{d}z = 0$$

As $z = 0$ is a removable singularity, it can be shown that

$$\int_{C_\varepsilon} \frac{\sin^3 z}{z^3}\mathrm{d}z \xrightarrow{\varepsilon\to 0} 0$$

Taking the limit as $R \to \infty$, we have

$$\begin{aligned}\int_{-\infty}^{\infty} \frac{\sin^3 x}{x^3}\mathrm{d}x &= -\lim_{R\to\infty}\int_{C_R} \frac{\sin^3 z}{z^3}\mathrm{d}z \\ &= \frac{1}{8\mathrm{i}}\lim_{R\to\infty}\int_{C_R} \frac{\mathrm{e}^{3\mathrm{i}z} - 3\mathrm{e}^{\mathrm{i}z} + 3\mathrm{e}^{-\mathrm{i}z} - \mathrm{e}^{-3\mathrm{i}z}}{z^3}\mathrm{d}z\end{aligned}$$

The Jordan's Lemma indicates that

$$\lim_{R\to\infty}\int_{C_R} \frac{\mathrm{e}^{3\mathrm{i}z}}{z^3}dz = \lim_{R\to\infty}\int_{C_R} \frac{\mathrm{e}^{\mathrm{i}z}}{z^3}\mathrm{d}z = 0$$

Furthermore,

$$\begin{aligned}\lim_{R\to\infty}\int_{C_R} \frac{\mathrm{e}^{-\mathrm{i}z}}{z^3}\mathrm{d}z &= 2\pi\mathrm{i}\operatorname*{Res}_{z=0}\left[\frac{\mathrm{e}^{-\mathrm{i}z}}{z^3}\right] = -\pi\mathrm{i} \\ \lim_{R\to\infty}\int_{C_R} \frac{\mathrm{e}^{-3\mathrm{i}z}}{z^3}\mathrm{d}z &= 2\pi\mathrm{i}\operatorname*{Res}_{z=0}\left[\frac{\mathrm{e}^{-3\mathrm{i}z}}{z^3}\right] = -9\pi\mathrm{i}\end{aligned}$$

Hence,

$$\int_{0}^{\infty} \frac{\sin^3 x}{x^3}\mathrm{d}x = \frac{1}{2}\int_{-\infty}^{\infty} \frac{\sin^3 x}{x^3}\mathrm{d}x = \frac{1}{2} \times \frac{1}{8\mathrm{i}} \times 6\pi\mathrm{i} = \frac{3\pi}{8}$$

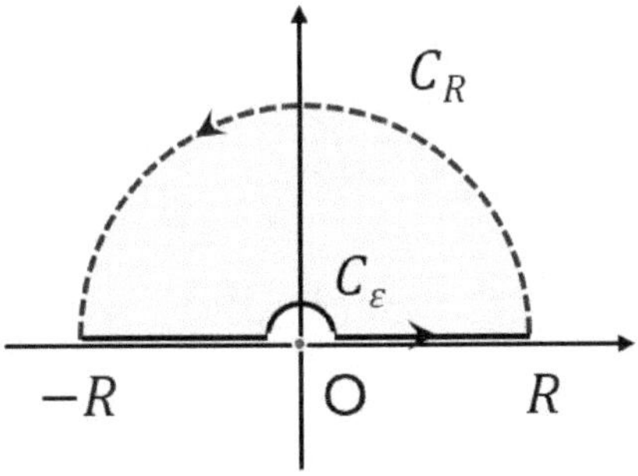

Figure 4.8.

ENDNOTE
It is not surprising that the integration of rational functions is related to the constant π, as lines are equivalent to circles and integrating along a circular path should be related to π. However, the infinite integration of rational functions mixed with trigonometric functions is related to another mathematic constant e, as shown in Example 4.7, may initially seem surprising. In fact, within the realm of real analysis, this is completely incomprehensible. However, from the perspective of complex analysis, one can appreciate the underlying reasoning. Due to the close relationship between trigonometric functions and exponential functions, the result of such integrals is simultaneously related to both π and e.

Exercises

[1] Calculate the integrals:

$$\text{(a)} \int_0^{2\pi} \frac{\sin^2\theta}{a+b\cos\theta}\mathrm{d}\theta \quad (a>b>0);$$

$$\text{(b)} \int_0^{2\pi} \frac{1}{(1+\varepsilon\cos\theta)^2}\mathrm{d}\theta \quad (1>\varepsilon>0);$$

$$\text{(c)} \int_0^{2\pi} \frac{1}{1+\sin^2\theta}\mathrm{d}\theta; \quad \text{(d)} \int_0^{\frac{\pi}{2}} \frac{1}{(a+\sin^2\theta)^2}\mathrm{d}\theta \quad (a>0);$$

$$\text{(e)} \int_0^{2\pi} \sin^{2n}\theta\mathrm{d}\theta; \quad \text{(f)} \int_0^{2\pi} \cos^{2n}\theta\mathrm{d}\theta \quad (n\geq 0)$$

Answer:

$$\text{(a)} \frac{2\pi(a-\sqrt{a^2-b^2})}{b^2}; \quad \text{(b)} \frac{2\pi}{(1-\varepsilon^2)^{3/2}};$$

$$\text{(c)} \frac{\pi}{\sqrt{2}}; \quad \text{(d)} \frac{\pi(2a+1)}{4(a^2+a)^{3/2}};$$

$$\text{(e)} \frac{(2n)!\pi}{2^{2n-1}(n!)^2}; \quad \text{(f)} \frac{(2n)!\pi}{2^{2n-1}(n!)^2}$$

[2] Prove the integrals:

$$\text{(a)} \int_{-\infty}^{\infty} \frac{x^2}{(1+x^2)^2} \mathrm{d}x = \frac{\pi}{2};$$

$$\text{(b)} \int_{-\infty}^{\infty} \frac{x}{(x^2+2x+2)(x^2+4)} \mathrm{d}x = -\frac{\pi}{10};$$

$$\text{(c)} \int_{-\infty}^{\infty} \frac{x-1}{x^5-1} \mathrm{d}x = \frac{4\pi}{5} \sin \frac{2\pi}{5}$$

[3] Calculate the integral ($t \in \mathbb{R}$):

$$\text{(a)} \int_{-\infty}^{\infty} \frac{1}{x^2+2tx+1} \mathrm{d}x; \quad \text{(b)} \int_{-\infty}^{\infty} \frac{1}{(x^2+2tx+1)^2} \mathrm{d}x$$

Answer:

$$\text{(a)} \begin{cases} \dfrac{\pi}{\sqrt{1-t^2}} & (t^2 < 1) \\ 0 & (t^2 > 1) \\ \text{NAN} & (t^2 = 1) \end{cases}; \quad \text{(b)} \begin{cases} \dfrac{\pi}{(1-t^2)^{3/2}} & (t^2 < 1) \\ \text{NAN} & (t^2 \geq 1) \end{cases}$$

[4] Let $z = z_0$ be a simple pole of the function $f(z)$. If the path C_ε around the pole z_0 is not a semicircle but a circular arc of angle α, prove the following result:

$$\lim_{\varepsilon \to 0} \int_{C_\varepsilon} f(z) \mathrm{d}z = \alpha \mathrm{i} \mathrm{Res}[f(z_0)]$$

[5] Prove the integrals:

$$\text{(a)} \int_{-\infty}^{\infty} \frac{\sin^2 x}{x^2+1} \mathrm{d}x = \frac{\pi}{2}\left[1 - \frac{1}{\mathrm{e}^2}\right]; \quad \text{(b)} \int_{-\infty}^{\infty} \frac{\cos x}{(1+x^2)^2} \mathrm{d}x = \frac{\pi}{\mathrm{e}}$$

[6] Utilizing the integral formula

$$\int_0^{2\pi} \frac{1}{1+\varepsilon \cos \theta} \mathrm{d}\theta = \frac{2\pi}{\sqrt{1-\varepsilon^2}}$$

to prove that

$$\int_0^{2\pi} \cos^{2n} \theta \mathrm{d}\theta = \frac{(2n)!\pi}{2^{2n-1}(n!)^2} \quad (n \geq 0)$$

Hint: Expand both sides as power series in ε.

[7] For a function $f(z) = u(x,y) + \mathrm{i}v(x,y)$ that is analytic in the upper half-plane and satisfies the conditions of Type II, with x_0 being a point on the real axis, prove the following dispersion relationship (named Hilbert transform) on the real axis:

$$\begin{cases} u(x_0) = \dfrac{1}{\pi}\mathbb{P}\displaystyle\int_{-\infty}^{\infty} \dfrac{v(x)}{x - x_0}\mathrm{d}x \\ v(x_0) = -\dfrac{1}{\pi}\mathbb{P}\displaystyle\int_{-\infty}^{\infty} \dfrac{u(x)}{x - x_0}\mathrm{d}x \end{cases}$$

4.3 Special Integrals

4.3.1 *Multiple-valued function integration*

In the context of integration, the multi-valued nature of functions is often taken as a source of trouble, as encountered in discussions of elliptic integrals. However, for many complicated real variable integrals, multiple-valued functions sometimes bring unexpected convenience.

Example 4.11. Calculate the integral:

$$\int_0^{\infty} \frac{x^{\alpha-1}}{1+x}\mathrm{d}x \quad (0 < \alpha < 1)$$

Solution: Extending the integrand analytically to the complex plane, one has

$$f(z) = \frac{z^{\alpha-1}}{z+1}$$

This is a multi-valued function with branch points at $z = 0$ and $z = \infty$. Along the positive real axis, a branch cut is made with the argument of z setting to zero on the upper edge and 2π on the lower edge. We choose a "keyhole" contour L as shown in Figure 4.9. Inside the contour, there is only one isolated singularity at $z = -1$. Therefore,

$$\oint_L \frac{z^{\alpha-1}}{1+z}\mathrm{d}z = \int_{\varepsilon}^{R} \frac{x^{\alpha-1}}{1+x}\mathrm{d}x + \int_{C_R} \frac{z^{\alpha-1}}{1+z}\mathrm{d}z + \int_{R}^{\varepsilon} \frac{x^{\alpha-1}\mathrm{e}^{2\pi\mathrm{i}(\alpha-1)}}{1+x}\mathrm{d}x$$
$$+ \int_{C_\varepsilon} \frac{z^{\alpha-1}}{1+z}\mathrm{d}z = 2\pi\mathrm{i}\operatorname*{Res}_{z=-1}[f(z)]$$

As $|z| \to \infty$, $|zf(z)| \to 0$, so the integral along the circular path C_R is zero. Furthermore, using the parameter method, it can be shown that as $\varepsilon \to 0$,

$$\int_{C_\varepsilon} \frac{z^{\alpha-1}}{1+z} \mathrm{d}z \to 0$$

Therefore,

$$(1 - \mathrm{e}^{2\pi \mathrm{i}\alpha}) \int_0^\infty \frac{x^{\alpha-1}}{1+x} \mathrm{d}x = 2\pi \mathrm{i} \mathrm{e}^{\pi \mathrm{i}(\alpha-1)}$$

Finally, we obtain the result of the integral as:

$$\int_0^\infty \frac{x^{\alpha-1}}{1+x} \mathrm{d}x = -\frac{2\pi \mathrm{i} \mathrm{e}^{\pi \mathrm{i}\alpha}}{1 - \mathrm{e}^{2\pi \mathrm{i}\alpha}} = \frac{\pi}{\sin \pi\alpha}$$

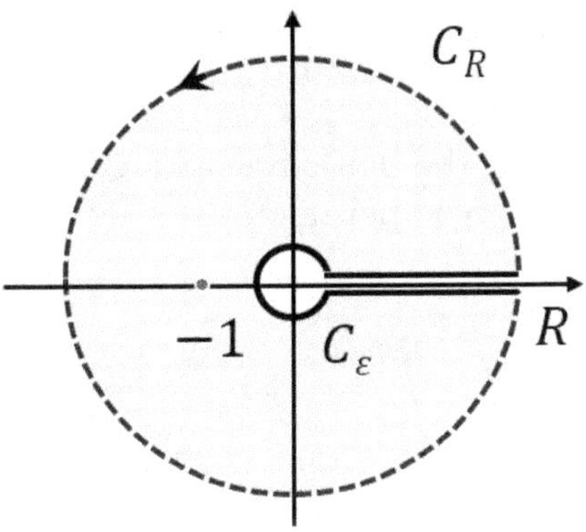

Figure 4.9.

This problem can also be solved by variable substitution $x + 1 = \frac{1}{y}$ and using the Beta function for calculation.

Example 4.12. Calculate the integral:

$$\int_0^\infty \frac{x^{\alpha-1}}{1-x} \mathrm{d}x \quad (0 < \alpha < 1)$$

Solution: Define the complex function:

$$f(z) = \frac{z^{\alpha-1}}{1-z}$$

This is a multi-valued function with no singularities in the complex plane except for the point $z = 1$ on the real axis. We consider a closed contour L as shown in Figure 4.10, which encloses no singularities inside the contour. Therefore,

$$\int_0^R \frac{x^{\alpha-1}}{1-x}\mathrm{d}x + \int_{C_1} \frac{z^{\alpha-1}}{1-z}\mathrm{d}z + \int_{C_R} \frac{z^{\alpha-1}}{1-z}\mathrm{d}z$$
$$+ \int_R^0 \frac{(x\mathrm{e}^{2\pi\mathrm{i}})^{\alpha-1}}{1-x}\mathrm{d}x + \int_{C_2} \frac{z^{\alpha-1}}{1-z}\mathrm{d}z + \int_{C_\varepsilon} \frac{z^{\alpha-1}}{1-z}\mathrm{d}z = 0$$

Since

$$|zf(z)| = \left|\frac{z^\alpha}{1-z}\right| \overset{z\to\infty}{\longrightarrow} 0$$

the integral along the circular path C_R is zero. Similarly, it can be shown that the integral along C_ε is also zero. The key now is to evaluate the integrals along the half-circle arcs C_1 and C_2. The infinitesimal upper half-circle C_1 is parameterized by equation $z - 1 = \varepsilon\mathrm{e}^{\mathrm{i}\theta}$. We have

$$\int_{C_1} \frac{z^{\alpha-1}}{1-z}\mathrm{d}z = \int_\pi^0 \frac{(1+\varepsilon\mathrm{e}^{\mathrm{i}\theta})^{\alpha-1}}{-\varepsilon\mathrm{e}^{\mathrm{i}\theta}}\mathrm{i}\varepsilon\mathrm{e}^{\mathrm{i}\theta}\mathrm{d}\theta \overset{\varepsilon\to 0}{\longrightarrow} \pi\mathrm{i}$$

It is notable that for the infinitesimal lower half-circle C_2, the parameter equation should be written as $z - \mathrm{e}^{2\pi\mathrm{i}} = \varepsilon\mathrm{e}^{\mathrm{i}\theta}$, which yields

$$\int_{C_2} \frac{z^{\alpha-1}}{1-z}\mathrm{d}z = \int_{2\pi}^\pi \frac{(\mathrm{e}^{2\pi\mathrm{i}}+\varepsilon\mathrm{e}^{\mathrm{i}\theta})^{\alpha-1}}{-\varepsilon\mathrm{e}^{\mathrm{i}\theta}}\mathrm{i}\varepsilon\mathrm{e}^{\mathrm{i}\theta}\mathrm{d}\theta \overset{\varepsilon\to 0}{\longrightarrow} \pi\mathrm{i}\mathrm{e}^{2\pi\alpha\mathrm{i}}$$

Finally, we get the integration

$$\mathcal{P}\int_0^\infty \frac{x^{\alpha-1}}{1-x}\mathrm{d}x = \frac{-\pi\mathrm{i}(1+\mathrm{e}^{2\pi\alpha\mathrm{i}})}{1-\mathrm{e}^{2\pi\alpha\mathrm{i}}} = \pi\cot\pi\alpha$$

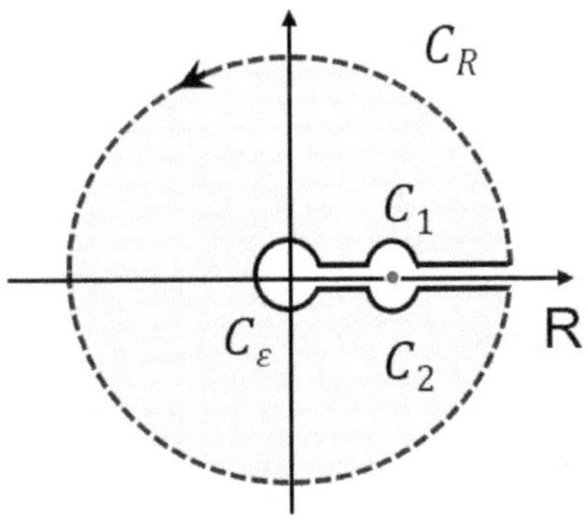

Figure 4.10.

4.3.2 *Special contour integration*

Example 4.13. Calculate the integral:

$$\int_{-\infty}^{\infty} \frac{e^{\alpha x}}{e^x + 1} dx \quad (0 < \alpha < 1)$$

Solution: The singularities of the integrand

$$f(z) = \frac{e^{\alpha z}}{1 + e^z}$$

are located at $z = (2k + 1)\pi i (k \in \mathbb{Z})$. The singularities are evenly distributed along the imaginary axis, making it unsuitable to choose a semicircular contour. Instead, we select a rectangular closed contour L as shown in Figure 4.11. The contour encloses only one singularity at $z = \pi i$. We have

$$\begin{aligned}\oint_L f(z)dz &= \int_{-R}^{R} \frac{e^{\alpha x}}{1 + e^x} dx + \int_0^{2\pi} \frac{e^{\alpha(R+iy)}}{1 + e^{R+iy}} i dy \\ &\quad + \int_R^{-R} \frac{e^{\alpha(x+2\pi i)}}{1 + e^{x+2\pi i}} dx + \int_{2\pi}^{0} \frac{e^{\alpha(-R+iy)}}{1 + e^{-R+iy}} i dy \\ &= 2\pi i \operatorname{Res}[f(z)]|_{z=\pi i}\end{aligned}$$

It can be easily shown that when $R \to \infty$, the second and fourth terms in the equation above tend to zero. Hence,

$$\int_{-\infty}^{\infty} \frac{e^{\alpha x}}{1+e^x} dx + e^{2\pi\alpha i} \int_{\infty}^{-\infty} \frac{e^{\alpha x}}{1+e^x} dx = 2\pi i \operatorname*{Res}_{z=\pi i} \left[\frac{e^{\alpha z}}{1+e^z} \right]$$

$$\int_{-\infty}^{\infty} \frac{e^{\alpha x}}{1+e^x} dx = 2\pi i \frac{e^{\pi(\alpha-1)i}}{1-e^{2\pi\alpha i}} = \frac{\pi}{\sin \alpha\pi}$$

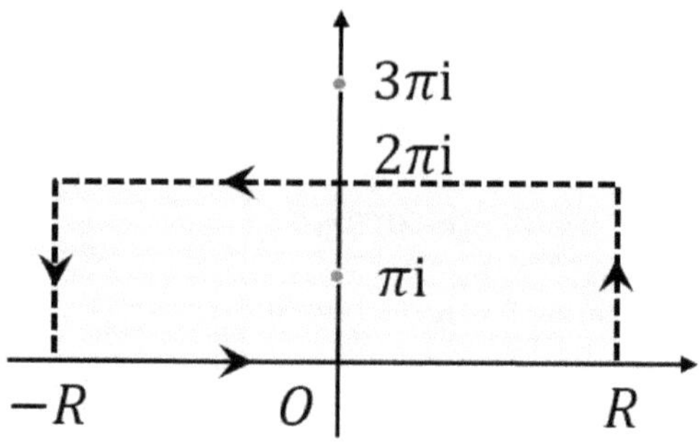

Figure 4.11.

Discussion: Example 4.11 can be transformed into this problem by the variable substitution $x = e^y$, which does not involve multi-valued function integration.
Practice: Perform the variable substitution $x = e^y$ on Example 4.12 and calculate the integral.

Example 4.14. Calculate Fresnel integrals:

$$\text{(a) } I_1 = \int_0^{\infty} \sin x^2 dx; \quad \text{(b) } I_2 = \int_0^{\infty} \cos x^2 dx$$

Solution: Consider the integral:

$$I_2 + iI_1 = \int_0^{\infty} e^{ix^2} dx$$

The function $f(z) = e^{iz^2}$ has no singularities in the entire plane. Taking the integration path L as shown in Figure 4.12, we get

$$\oint_L e^{iz^2} dz = \int_0^R e^{ix^2} dx + \int_{C_R} e^{iz^2} dz + \int_R^0 e^{i(\rho e^{\pi i/4})^2} d(\rho e^{\pi i/4}) = 0$$

Next, let us examine the integral along the circular arc. By substituting $\zeta = z^2$ and $\mathrm{d}z = \frac{\mathrm{d}\zeta}{2\sqrt{\zeta}}$, we have

$$\int_{C_R} \mathrm{e}^{\mathrm{i}z^2} dz = \int_{C_R}' \frac{\mathrm{e}^{\mathrm{i}\zeta}}{2\sqrt{\zeta}} \mathrm{d}\zeta$$

Applying the Jordan's lemma, this integral approaches zero as $R \to \infty$. Hence,

$$I_2 + \mathrm{i}I_1 = \int_0^\infty \mathrm{e}^{\mathrm{i}x^2} \mathrm{d}x = \mathrm{e}^{\pi\mathrm{i}/4} \int_0^\infty \mathrm{e}^{-\rho^2} \mathrm{d}\rho = \frac{\sqrt{\pi}}{2} \mathrm{e}^{\pi\mathrm{i}/4} = \sqrt{\frac{\pi}{8}}(1+\mathrm{i})$$

We obtain by comparing the real and imaginary parts,

$$\int_0^\infty \sin x^2 \mathrm{d}x = \int_0^\infty \cos x^2 \mathrm{d}x = \sqrt{\frac{\pi}{8}}$$

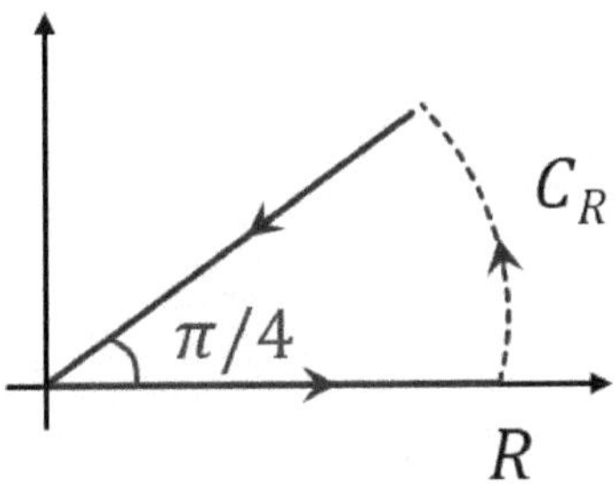

Figure 4.12.

Practice: By setting $y = x^2$, recalculate Example 4.14 using the integration path shown in Figure 4.13.

4.3.3 *Semi-infinite integrals*

For the semi-infinite integral $\int_0^\infty f(x)\mathrm{d}x$, if the integrand $f(x)$ is neither an odd function nor an even function, how to carry out the integration? we can multiply a logarithmic function to transform the integrand into a multi-valued function and then take the "keyhole" contour integration as shown in Figure 4.9. Assuming z_k are poles of complex function $f(z)$. If $f(x)$ has no positive real roots, we have the following conclusions:

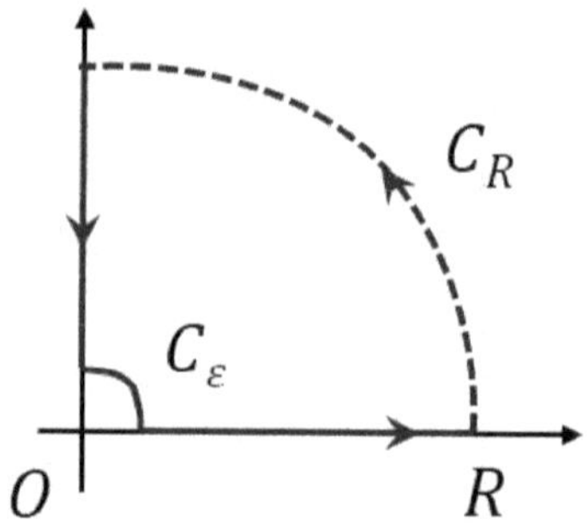

Figure 4.13.

(1) If when $z \to \infty$, $|zf(z)\ln z| \to 0$, then

$$\int_0^\infty f(x)\mathrm{d}x = -\sum_{z_k \in \mathbb{C}} \mathrm{Res}[f(z_k)\ln z_k] \tag{4.3.1}$$

(2) If when $z \to \infty$, $|zf(z)(\ln z)^2| \to 0$, then

$$\int_0^\infty f(x)\ln x \,\mathrm{d}x = -\frac{1}{2}\mathrm{Re}\sum_{z_k \in \mathbb{C}} \mathrm{Res}[f(z_k)(\ln z_k)^2] \tag{4.3.2}$$

Readers can refer to the following examples for proofs.

Example 4.15. Calculate the integral:

$$\int_0^\infty \frac{1}{1+x+x^2}\mathrm{d}x$$

Solution: Consider the logarithmic multi-valued function,

$$f(z) = \frac{\ln z}{1+z+z^2}$$

By making a cut along the real axis between the branch points $z = 0$ and $z = \infty$, we construct the closed "keyhole" contour L as shown in Figure 4.9. The singularities of $f(z)$ locate at $z_1 = \mathrm{e}^{2\pi\mathrm{i}/3}$ and

$z_2 = e^{4\pi i/3}$. Thus, we have

$$\oint_L f(z)dz = \int_0^\infty \frac{\ln x}{1+x+x^2}dx + \int_\infty^0 \frac{\ln x + 2\pi i}{1+x+x^2}dx$$
$$+ \int_{C_\varepsilon} f(z)dz + \int_{C_R} f(z)dz$$
$$= 2\pi i \sum_{z=z_1,z_2} \mathrm{Res}[f(z)]$$

The integrals along C_ε and C_R are both zero, so that

$$\int_0^\infty \frac{1}{1+x+x^2}dx = -\sum_{z=z_1,z_2} \mathrm{Res}\left[\frac{\ln z}{1+z+z^2}\right] = \frac{2\pi}{3\sqrt{3}}$$

Example 4.16. Calculate the integral:

$$\int_0^\infty \frac{\ln x}{1+x+x^2}dx$$

Solution: Consider the logarithmic multi-valued function,

$$f(z) = \frac{(\ln z)^2}{1+z+z^2}$$

We still take the "keyhole" contour L as shown in Figure 4.9. The singularities of $f(z)$ locate at $z_1 = e^{2\pi i/3}$ and $z_2 = e^{4\pi i/3}$. Since the integrals along C_ε and C_R are zero, we have

$$\oint_L f(z)dz = \int_0^\infty \frac{(\ln x)^2}{1+x+x^2}dx + \int_\infty^0 \frac{(\ln x + 2\pi i)^2}{1+x+x^2}dx$$
$$= -4\pi i \int_0^\infty \frac{\ln x}{1+x+x^2}dx + 4\pi^2 \int_0^\infty \frac{1}{1+x+x^2}dx$$
$$= 2\pi i \sum_{z=z_1,z_2} \mathrm{Res}\left[\frac{(\ln z)^2}{1+z+z^2}\right] = \frac{8\pi^3}{3\sqrt{3}}$$

Comparing the real and imaginary parts, we obtain

$$\int_0^\infty \frac{\ln x}{1+x+x^2}dx = 0$$

As a bonus, we obtain the integral result once again:

$$\int_0^\infty \frac{1}{1+x+x^2}dx = \frac{2\pi}{3\sqrt{3}}$$

4.3.4 "Dog bone" integrals

There are some integrals that do not fall into any of the three basic types mentioned above, where the integral along the large circular path is not equal to zero. In such cases, the residue at infinity needs to be used to evaluate the integral.

Example 4.17. Calculate the integral:

$$\int_{-1}^{1} \frac{1}{\sqrt[3]{(1-x)(1+x)^2}} \mathrm{d}x$$

Solution: The integrand is a multi-valued function given by

$$f(z) = \frac{1}{\sqrt[3]{(1-z)(1+z)^2}}$$

It has two branch points at $z_{1,2} = \pm 1$. We cut the plane between these two branch points and choose the branch where the function values on the cut are positive real numbers, i.e., $\arg f(z) = 0$. We construct the "dog bone" contour L as shown in Figure 4.14. As the variable z moves to the lower edge of the cut, the argument of the function $f(z)$ increases by $-\frac{4\pi}{3}$. Since there are no singularities enclosed by the contour, one has

$$\begin{aligned}\oint_L f(z)\mathrm{d}z = &\int_{-1}^{1} \frac{1}{\sqrt[3]{(1-x)(1+x)^2}} \mathrm{d}x + \mathrm{e}^{-\frac{4\pi \mathrm{i}}{3}} \\ &\times \int_{+1}^{-1} \frac{1}{\sqrt[3]{(1-x)(1+x)^2}} \mathrm{d}x \\ &- \int_{C_1} \frac{1}{\sqrt[3]{(1-z)(1+z)^2}} \mathrm{d}z - \int_{C_2} \frac{1}{\sqrt[3]{(1-z)(1+z)^2}} \mathrm{d}z \\ &+ \int_{C_R} \frac{1}{\sqrt[3]{(1-z)(1+z)^2}} \mathrm{d}z = 0\end{aligned}$$

It can be easily shown that the integrals around the two small circular paths C_1 and C_2 are both zero. Therefore,

$$\left(1-\mathrm{e}^{\frac{2\pi\mathrm{i}}{3}}\right)\int_{-1}^{1}\frac{1}{\sqrt[3]{(1-x)(1+x)^2}}\mathrm{d}x=-\int_{C_R}\frac{1}{\sqrt[3]{(1-z)(1+z)^2}}\mathrm{d}z$$

The integral along the contour C_R is equal to $2\pi\mathrm{i}$ times the residue at $z=\infty$. Note that as $z\to\infty$ along the real axis, the argument of the function $f(z)$ increases by $\mathrm{e}^{\frac{\pi\mathrm{i}}{3}}$. We have

$$f(z)=\frac{1}{\sqrt[3]{(1-z)(1+z)^2}}=\frac{1}{z\mathrm{e}^{-\pi\mathrm{i}/3}}\left(1-\frac{1}{z}\right)^{-\frac{1}{3}}\left(1+\frac{1}{z}\right)^{-\frac{2}{3}}$$

Expanding the right-hand side of the above equation in Laurent series with respect to $\frac{1}{z}$, we find the residue at $z=\infty$ to be

$$\mathrm{Res}[f(\infty)]=-\mathrm{e}^{\frac{\pi\mathrm{i}}{3}}$$

Therefore, the integral is

$$\int_{-1}^{1}\frac{1}{\sqrt[3]{(1-x)(1+x)^2}}\mathrm{d}x=2\pi\mathrm{i}\times\frac{-\mathrm{e}^{\pi\mathrm{i}/3}}{1-\mathrm{e}^{2\pi\mathrm{i}/3}}=\frac{2\pi}{\sqrt{3}}$$

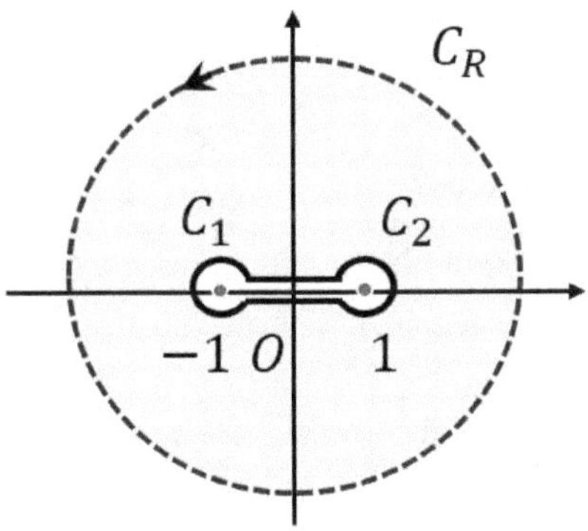

Figure 4.14.

In this example, the integration can also be evaluated using the Beta function method. The readers may try it as well.

ENDNOTE

By multiplying the integrand by a logarithmic function, we can transform the improper integral $\int_0^\infty f(x)\mathrm{d}x$ into a multi-valued function integral. We can also adopt different strategies. For example, consider the integral:

$$\int_0^\infty \frac{x^\alpha}{x^3+1}\mathrm{d}x \quad (\alpha < 2)$$

We consider the complex function $f(z) = \frac{z^\alpha}{z^3+1}$ and choose a closed contour shown in Figure 4.15, which forms a $\frac{2\pi}{3}$ sector. Within the contour, $f(z)$ has only one simple pole at $z_1 = \mathrm{e}^{\pi \mathrm{i}/3}$. Therefore,

$$\begin{aligned}\oint_L f(z)\mathrm{d}z &= \int_0^\infty \frac{x^\alpha}{x^3+1}\mathrm{d}x + \int_\infty^0 \frac{\rho^\alpha}{\rho^3+1}\mathrm{e}^{2\pi(\alpha+1)\mathrm{i}/3}\mathrm{d}\rho \\ &\quad + \int_{C_\varepsilon} f(z)\mathrm{d}z + \int_{C_R} f(z)\mathrm{d}z \\ &= 2\pi\mathrm{i}\sum_{z=z_1} \mathrm{Res}[f(z)]\end{aligned}$$

The integrals along C_ε and C_R are still zero, so we have

$$\left(1-\mathrm{e}^{\frac{2\pi(\alpha+1)\mathrm{i}}{3}}\right)\int_0^\infty \frac{x^\alpha}{x^3+1}\mathrm{d}x = 2\pi\mathrm{i} \times \frac{\mathrm{e}^{\pi(\alpha-2)\mathrm{i}/3}}{3}$$

This can be simplified to

$$\int_0^\infty \frac{x^\alpha}{x^3+1}\mathrm{d}x = \frac{\pi}{3\sin\frac{\pi(\alpha+1)}{3}}$$

Taking $\alpha = 0, 1$ and $\frac{1}{2}$, respectively, we get

$$\int_0^\infty \frac{1}{x^3+1}\mathrm{d}x = \frac{\pi}{3\sin\frac{\pi}{3}} = \frac{2\pi}{3\sqrt{3}}$$

$$\int_0^\infty \frac{x}{x^3+1}\mathrm{d}x = \frac{\pi}{3\sin\frac{2\pi}{3}} = \frac{2\pi}{3\sqrt{3}}$$

$$\int_0^\infty \frac{\sqrt{x}}{x^3+1}\mathrm{d}x = \frac{\pi}{3\sin\frac{\pi}{2}} = \frac{\pi}{3}$$

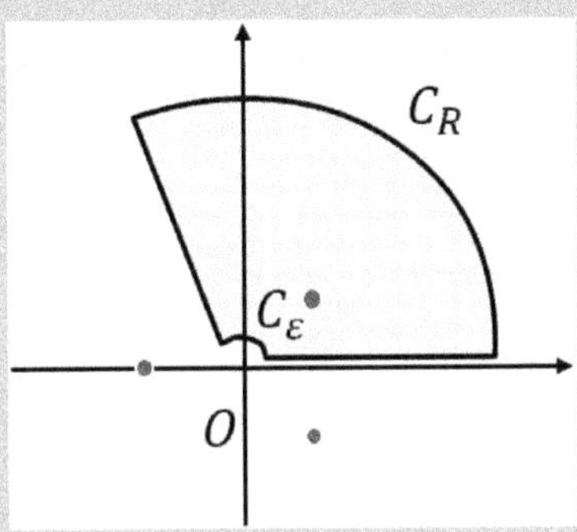

Figure 4.15.

In this way, it appears that all the definite integrals we have calculated involve infinite or semi-infinite integrals. For integrals over finite intervals (indefinite integrals), we can simply make a change of variable, transforming the finite interval into an infinite interval. For example, by setting $y = \frac{x-a}{x-b}$, we can transform the integral into the semi-infinite integral:

$$\int_a^b f(x)\mathrm{d}x \longrightarrow \int_0^\infty \tilde{f}(y)\mathrm{d}y$$

Exercises

[1] Prove the following integrals:

(a) $$\int_0^\infty \frac{1}{x^n+1}\mathrm{d}x = \frac{\pi/n}{\sin(\pi/n)} \quad (n \geq 2)$$

(b) $$\int_0^\infty \frac{\cosh bx}{\cosh x}\mathrm{d}x = \frac{\pi}{2\cos(b\pi/2)} \quad (|b| < 1)$$

Hint: (a) Consider a sector-shaped closed contour with $\theta = \frac{2\pi}{n}$. (b) Consider a rectangular contour containing only one singularity of $\cosh z$.

[2] Calculate the integrals ($0 < \alpha < 1$):

$$\text{(a)} \int_0^\infty x^{\alpha-1} \sin x \, \mathrm{d}x; \quad \text{(b)} \int_0^\infty x^{\alpha-1} \cos x \, \mathrm{d}x$$

Hint: Take a sector-shaped contour in the first quadrant and combine the two integrals.

Answer:

$$\text{(a) } \Gamma(\alpha) \sin \frac{\pi\alpha}{2}; \quad \text{(b) } \Gamma(\alpha) \cos \frac{\pi\alpha}{2}$$

[3] Calculate the integrals ($0 < a < b$):

$$\text{(a)} \int_0^\infty \frac{x^{a-1}}{x^b + 1} \mathrm{d}x; \quad \text{(b)} \int_0^\infty \frac{x^{a-1}}{x^b - 1} \mathrm{d}x$$

Hint: Take a sector contour with $0 \le \theta \le \frac{2\pi}{b}$ as shown in Figure 4.15, or directly substitute $y = x^b$.

Answer:

$$\text{(a)} \frac{\pi}{b \sin \frac{\pi a}{b}}; \quad \text{(b)} \ -\frac{\pi}{b} \cot \frac{\pi a}{b}$$

[4] Calculate the integrals:

$$\text{(a)} \int_0^\infty \frac{\ln x}{(x+a)(x+b)} \mathrm{d}x \quad (b > a > 0)$$

$$\text{(b)} \int_0^\infty \frac{\ln x}{x^a (x+1)} \mathrm{d}x \quad (1 > a > 0)$$

$$\text{(c)} \int_0^\infty \frac{x^a \ln x}{x^2 + 1} \mathrm{d}x \quad (1 > a > 0); \quad \text{(d)} \int_0^\infty \frac{(\ln x)^2}{x^3 + 1} \mathrm{d}x$$

Answer:

$$\text{(a)} \frac{(\ln a)^2 - (\ln b)^2}{2(a-b)}; \quad \text{(b)} \frac{\pi^2 \cos \pi a}{\sin^2 \pi a}; \quad \text{(c)} \frac{\pi^2 \sin \frac{a\pi}{2}}{4 \cos^2 \frac{a\pi}{2}}; \quad \text{(d)} \frac{10\pi^3}{81\sqrt{3}}$$

[5] Calculate the integrals:

$$\text{(a)} \int_0^1 \frac{1}{\sqrt{x(1-x)}}\mathrm{d}x; \quad \text{(b)} \int_0^1 \frac{x^4}{\sqrt{x(1-x)}}\mathrm{d}x$$

$$\text{(c)} \int_0^1 \frac{\sqrt[4]{x(1-x)^3}}{(1+x)^3}\mathrm{d}x$$

Answer: (a) π; (b) $\frac{35\pi}{128}$; (c) $\frac{3\sqrt[4]{2}}{64}\pi$

[6] Prove the integrals:

$$\int_0^1 \frac{1}{x^3+1}\mathrm{d}x = \frac{\pi}{3\sqrt{3}} + \frac{1}{3}\ln 2$$

Hint: Substitute $x = \frac{1}{y+1}$ to convert the integral to a semi-infinite integral.

[7] Prove the integral:

$$\int_0^1 \frac{x^\alpha(1-x)^{-\alpha}}{x^3+1}\mathrm{d}x = -\frac{\pi}{3}\frac{[2^{-\alpha} + 2\cos[(\alpha+2)\pi/3]]}{\sin \pi\alpha}$$

and prove that as $\alpha \to 0$,

$$\int_0^1 \frac{1}{x^3+1}\mathrm{d}x = \frac{\pi}{3\sqrt{3}} + \frac{1}{3}\ln 2$$

Hint: The two branch points are $z_1 = 0$ and $z_2 = 1$. Use the "dog bone" integral contour.

[8] Calculate the integral:

$$\int_0^\infty \frac{x\cos mx}{1+x^2}\mathrm{d}x \quad (m > 0)$$

Hint: Take a quarter sector-shaped contour in the first quadrant, where the simple pole on the path contributes half of the residue.
Answer: $-\frac{1}{2}[\mathrm{e}^{-m}\mathrm{Ei}(m) + \mathrm{e}^{m}\mathrm{Ei}(-m)]$, where $\mathrm{Ei}(x)$ is the exponential integral function which is defined in Appendix A.2.

[9] Prove the integrals:

$$\text{(a)} \int_0^\infty \frac{x}{e^x - 1} dx = \frac{\pi^2}{6};$$

$$\text{(b)} \int_0^\infty \frac{\sinh ax}{e^x - 1} dx = \frac{1}{2a} - \frac{\pi}{2} \cot \pi a \quad (a \notin \mathbb{Z})$$

Hint: Consider respectively the contours shown in the following figure:

$$\text{(a)} \oint_L \frac{z^2}{e^z - 1} dz, \quad \text{(b)} \oint_L \frac{e^{\pm az}}{e^z - 1} dz$$

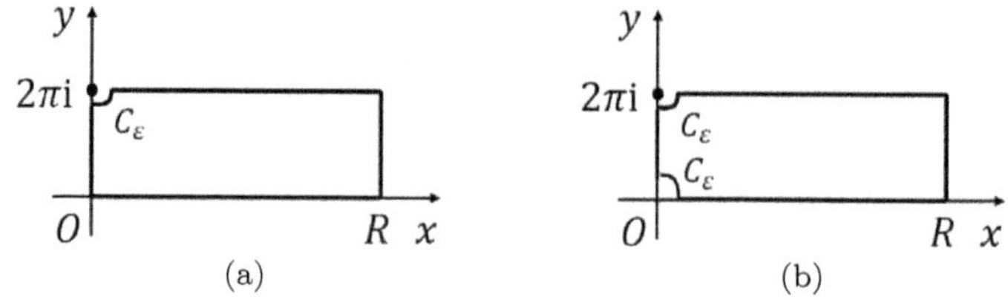

4.4 Series Summation

Consider the Bose distribution function in quantum statistics:

$$f_B(z) = \frac{1}{e^{\beta z} - 1}$$

where $\beta = 2\pi$. This function has simple poles at $z_n = n\text{i}$, all of which are located on the imaginary axis as shown in Figure 4.16(a), and the residues of all poles are the same:

$$\text{Res}[f_B(z_n)] = \frac{1}{2\pi}$$

For a complex variable function $F(z)$, we temporarily assume it has no singularities on the imaginary axis. According to the residue theorem for the contour C and Γ shown in Figure 4.16(b), we have

$$\oint_C F(z) f_B(z) dz + \oint_\Gamma F(z) f_B(z) dz = 2\pi\text{i} \sum_{\text{poles of } F(z)} \text{Res}[F(z) f_B(z)]$$

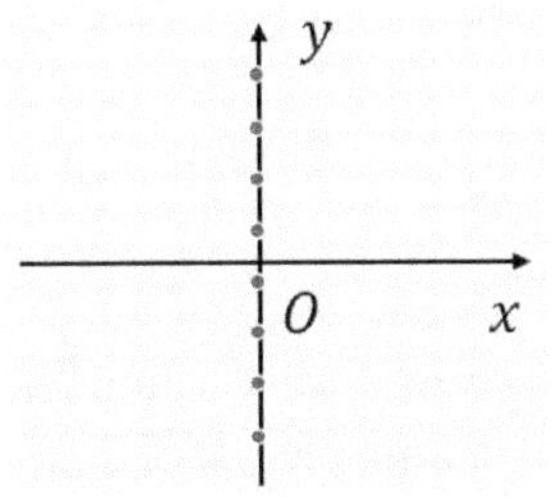

(a)

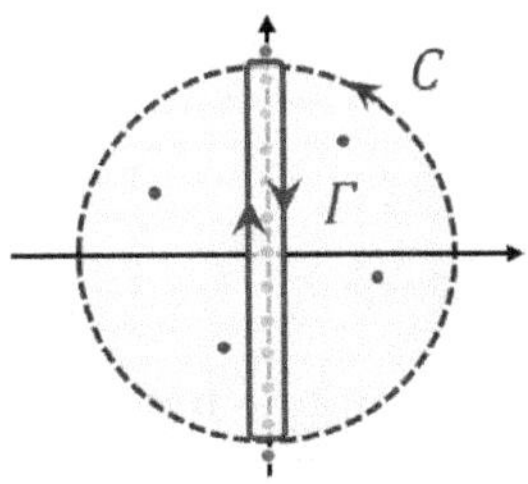

(b)

Figure 4.16.

and

$$\oint_{\Gamma} F(z)f_{\mathrm{B}}(z)\mathrm{d}z = -2\pi\mathrm{i} \sum_{\text{poles of } f_{\mathrm{B}}(z)} \mathrm{Res}[F(z)f_{\mathrm{B}}(z)]$$

Since the pole positions of $f_{\mathrm{B}}(z)$ being at $z_n = n\mathrm{i}$, we choose the radius of contour C to be $R = \left(n + \frac{1}{2}\right)$, meaning that the integral path passes between two adjacent poles. This ensures that the values of $|f_{\mathrm{B}}(z)|$ along the integration contour remain finite. If we choose a function $F(z)$ such that on the contour C the value $|zF(z)| \to 0$ as $R \to \infty$, it must then satisfy

$$|zF(z)f_{\mathrm{B}}(z)| \xrightarrow{z\in C} 0$$

Then the integrand belong to the Type II. Consequently, the integral

$$\oint_{C} F(z)f_{\mathrm{B}}(z)\mathrm{d}z \xrightarrow{R\to\infty} 0$$

Since

$$2\pi\mathrm{i}\sum_{z_n} \mathrm{Res}[F(z_n)f_{\mathrm{B}}(z_n)] = \mathrm{i}\sum_{z_n} F(z_n)$$

the series sum is given by

$$\sum_{z_n=n\mathrm{i}} F(z_n) = -2\pi \sum_{\text{poles of } F(z)} \mathrm{Res}[F(z)f_{\mathrm{B}}(z)] \tag{4.4.1}$$

Similar results can be obtained for the Fermi distribution function:

$$\sum_{z_n=\left(n+\frac{1}{2}\right)\mathrm{i}} F(z_n) = 2\pi \sum_{\text{poles of } F(z)} \mathrm{Res}[F(z)f_{\mathrm{F}}(z)]$$

where

$$f_{\mathrm{F}}(z)=\frac{1}{\mathrm{e}^{2\pi z}+1}$$

Discussion:

(1) If $F(z)$ has singularities on the imaginary axis, then calculate their residues separately, as shown in Example 4.18.
(2) The Bose distribution function $f_{\mathrm{B}}(z)$ has infinitely many singularities on the imaginary axis, uniformly spaced. Therefore, there will always be singularities outside the circular contour C. What radius R should be taken for the circular contour? This is the particularity of this problem. Due to the contour C passing between two adjacent poles and as long as the radius R of the circle is sufficiently large and the series is convergent, according to the Cauchy convergence criterion, it is known that the residues of singularities outside the circle C contribute negligibly to the infinite series. Therefore, this method is feasible.

Example 4.18. Find the series summation:

$$\sum_{n=1}^{\infty}\frac{1}{n^p}\quad(p=2,4,6,\ldots)$$

Solution: Consider the function

$$F(z)=\frac{1}{z^p}$$

Then $z=0$ is a $(p+1)$th order pole of the function $F(z)f_{\mathrm{B}}(z)$. According to the sum formula (4.4.1), we have

$$\sum_{n\neq 0}\frac{1}{n^p}=-2\pi\mathrm{i}^p\operatorname*{Res}_{z=0}\left[F(z)f_{\mathrm{B}}(z)\right]=-\frac{2\pi\mathrm{i}^p}{p!}\frac{\mathrm{d}^p}{\mathrm{d}z^p}\left(\frac{z}{\mathrm{e}^{2\pi z}-1}\right)\bigg|_{z=0}$$

By calculating the derivatives, it reveals that

$$p=2\longrightarrow\sum_{n=1}^{\infty}\frac{1}{n^2}=\frac{\pi^2}{6}$$

$$p=4\longrightarrow\sum_{n=1}^{\infty}\frac{1}{n^4}=\frac{\pi^4}{90}$$

$$p = 6 \longrightarrow \sum_{n=1}^{\infty} \frac{1}{n^6} = \frac{\pi^6}{945}$$

$$p = 8 \longrightarrow \sum_{n=1}^{\infty} \frac{1}{n^8} = \frac{\pi^8}{9450}$$

$$p = 10 \longrightarrow \sum_{n=1}^{\infty} \frac{1}{n^{10}} = \frac{\pi^{10}}{93555}$$

ENDNOTE

The Bose and Fermi distribution functions in quantum statistics seem to have special significance in mathematics. For simplicity, let us take $\beta = 1$. We have

$$f_{\rm B}(z) = \frac{1}{{\rm e}^z - 1}, \quad f_{\rm F}(z) = \frac{1}{{\rm e}^z + 1}$$

(1) Bose distribution function

Consider the function $G(x, z)$ expanded as a Taylor series with respect to the variable z:

$$G(x, z) = \frac{z{\rm e}^{xz}}{{\rm e}^z - 1} = \sum_{n=0}^{\infty} \frac{{\rm B}_n(x)}{n!} z^n$$

is called the generating function or generating series of Bernoulli polynomials ${\rm B}_k(x)$. When $x = 0$, we have

$$\frac{z}{{\rm e}^z - 1} = \sum_{n=0}^{\infty} \frac{{\rm B}_n}{n!} z^n$$

where ${\rm B}_n$ is called the Bernoulli numbers,

$${\rm B}_n = \frac{{\rm d}^n}{{\rm d}z^n} \left(\frac{z}{{\rm e}^z - 1} \right) \bigg|_{z=0}$$

By expanding, we can find $B_0 = 0$, $B_1 = -1/2$, $B_{2n+1} = 0$, and B_{2n} satisfies the recursive relation

$$\sum_{k=0}^{[n/2]} \frac{n!}{(n-2k+1)!} \frac{B_{2k}}{(2k)!} = \frac{1}{2}$$

Setting $n = 2, 4, 6, \ldots$ successively allows us to calculate the values of B_{2n}. The values for the first few terms are shown in Table 4.1.

Table 4.1.

B_2	B_4	B_6	B_8	B_{10}	B_{12}	B_{14}	B_{16}
$\frac{1}{6}$	$-\frac{1}{30}$	$\frac{1}{42}$	$-\frac{1}{30}$	$\frac{5}{66}$	$-\frac{691}{2730}$	$\frac{7}{6}$	$-\frac{3617}{510}$

According to the Cauchy integral formula, the Bernoulli numbers can be expressed as contour integrals:

$$B_n = \frac{n!}{2\pi i} \oint_L \frac{z}{e^z - 1} \frac{dz}{z^{n+1}}$$

where the contour L encloses the origin and lies inside a circular boundary with radius less than 2π. In principle, all orders of Bernoulli numbers can be calculated using this integral, but when n is large, $z = 0$ becomes a $(n+1)$th order pole, making the calculation of residues very cumbersome. To solve this, we use a contour as shown in Figure 4.17. Since the integrand is a single-valued function, the integrals along L_1 and L_2 cancel each other out, and the integral along the large circle C_R is zero. Thus, the integral around $z = 0$ is equal to the sum of residues at all single poles $z_k = 2k\pi i$ on the imaginary axis, i.e.,

$$\begin{aligned} B_n &= \frac{n!}{2\pi i} \oint_L \frac{z}{e^z - 1} \frac{dz}{z^{n+1}} = -\frac{n!}{2\pi i} \times 2\pi i \operatorname*{Res}_{z \neq 0} \left[\frac{z^{-n}}{e^z - 1} \right] \\ &= -\frac{n!}{(2\pi i)^n} \sum_{k=1}^{\infty} \left[\frac{1}{k^n} + \frac{1}{(-k)^n} \right] \end{aligned}$$

Therefore,

$$B_{2n+1} = 0$$

$$B_{2n} = (-1)^{n+1} \frac{(2n)!}{(2\pi)^{2n}} \sum_{k=1}^{\infty} \frac{2}{k^{2n}} = (-1)^{n+1} \frac{2(2n)!}{(2\pi)^{2n}} \zeta(2n)$$

where $\zeta(z) = \sum_{k=1}^{\infty} \frac{1}{k^z}$ is known as the Riemann zeta function.

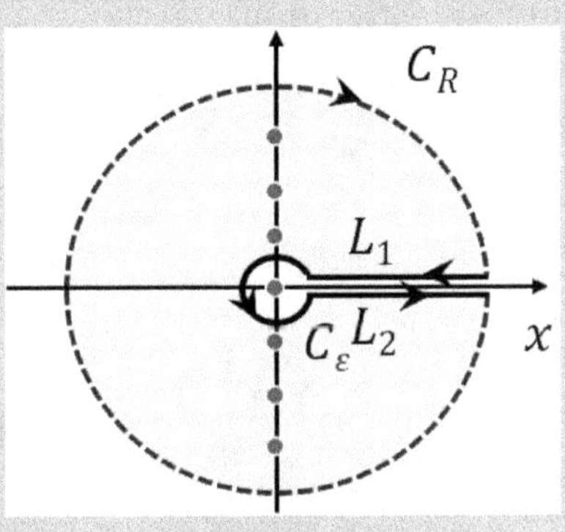

Figure 4.17.

When $x \neq 0$, we have

$$G(x, z) = \frac{ze^{xz}}{e^z - 1} = \sum_{k=0}^{\infty} \frac{B_k}{k!} z^k \cdot \sum_{l=0}^{\infty} \frac{x^l}{l!} z^l = \sum_{n=0}^{\infty} \frac{z^n}{n!} \cdot \sum_{k=0}^{n} \binom{n}{k} B_k x^{n-k}$$

So the Bernoulli polynomials can be expressed as

$$B_n(x) = \sum_{k=0}^{n} \binom{n}{k} B_k x^{n-k}$$

For example,

$$B_0(x) = 1$$

$$B_1(x) = x - \frac{1}{2}$$

$$B_2(x) = x^2 - x + \frac{1}{6}$$

$$B_3(x) = x^3 - \frac{3}{2}x^2 + \frac{1}{2}x$$

$$\cdots\cdots$$

(2) Fermi distribution function
Consider the generating function

$$G(x, z) = \frac{2e^{xz}}{e^z+1} = \sum_{n=0}^{\infty} \frac{E_n(x)}{n!} z^n$$

where $E_n(x)$ is called the Euler polynomials. Let $x = \frac{1}{2}$, then we have

$$\frac{2e^{z/2}}{e^z+1} = \sec h\frac{z}{2} = \sum_{n=0}^{\infty} \frac{E_n}{n!}\left(\frac{z}{2}\right)^n \quad (|z| < \pi)$$

Here, E_n is called the Euler numbers. Since the generating function $G(x, z)$ is an even function with respect to z, we have $E_{2n+1} = 0$, and E_{2n} satisfies the recursive relation

$$\sum_{n=0}^{k} \frac{(2k)!}{(2n)!(2k-2n)!} E_{2n} = 0 \ (k \geq 1)$$

as well as

$$E_{2n} = (-1)^n 2^{2n} E_{2n}\left(\frac{1}{2}\right)$$

The first few Euler numbers are shown in Table 4.2.

Table 4.2.

E_2	E_4	E_6	E_8	E_{10}	E_{12}	E_{14}	E_{16}
−1	5	−61	1385	−50521	2702765	−199360981	19391512145

For a general x, the Euler polynomials can be derived as

$$\mathrm{E}_n(x) = \sum_{k=0}^{[n/2]} \frac{(-1)^k \mathrm{E}_{2k}}{2^{2k}} \binom{n}{2k} \left(x - \frac{1}{2}\right)^{n-2k}$$

For example,

$$\begin{aligned} &\mathrm{E}_0(x) = 1 \\ &\mathrm{E}_1(x) = x - \frac{1}{2} \\ &\mathrm{E}_2(x) = x(x-1) \\ &\mathrm{E}_3(x) = \left(x - \frac{1}{2}\right)\left(x^2 - x - \tfrac{1}{2}\right) \\ &\cdots \end{aligned}$$

Let us briefly examine why some series are related to Bose or Fermi distribution functions. Because when $\operatorname{Re} z > 0$, $|\mathrm{e}^{-\beta z}| < 1$, the Bose distribution can be expanded in series as

$$f_{\mathrm{B}}(z) = \frac{1}{\mathrm{e}^{\beta z} - 1} = \mathrm{e}^{-\beta z} \frac{1}{1 - \mathrm{e}^{-\beta z}} = \mathrm{e}^{-\beta z} \sum_{n=0}^{\infty} \mathrm{e}^{-n\beta z}$$

Let $t = \mathrm{e}^{-\beta z}$, then we have

$$f_{\mathrm{B}}(t) = \sum_{n=1}^{\infty} t^n \quad (|t| < 1)$$

Likewise, when $\operatorname{Re} z < 0$, $|\mathrm{e}^{\beta z}| < 1$, we have

$$f_{\mathrm{B}}(z) = \frac{1}{\mathrm{e}^{\beta z} - 1} = -\sum_{n=0}^{\infty} \mathrm{e}^{n\beta z} = -\sum_{n=0}^{\infty} t^{-n} \quad (|t| > 1)$$

This is the simplest geometric series. Hence, it is not surprising that certain series are related to the Bose distribution function.

The Fermi distribution function also has a similar geometric series representation.

When p is an even number, the infinite series $\sum_{n=1}^{\infty} \frac{1}{n^p}$ has a concise expression. However, when p is an odd number, there doesn't seem to be a similar concise expression. R. Apéry proved that the sum of the series for $p = 3$ is irrational, but whether it is related to π remains unknown. For other odd values of p, it is even unknown whether the sum of the series is irrational. It is truly a perplexing and fascinating aspect of mathematics, isn't it?

Exercises

[1] Proof:

$$\text{(a)} \sum_{n=1}^{\infty} \frac{1}{n^2 + a^2} = \frac{\pi}{2a} \coth \pi a - \frac{1}{2a^2};$$

$$\text{(b)} \frac{1}{e^z - 1} = -\frac{1}{2} + \frac{1}{z} + \sum_{n=1}^{\infty} \frac{2z}{z^2 + 4n^2\pi^2}$$

Hint: (a) Take $F(z) = \frac{1}{z^2 - a^2}$; (b) Make use the result of (a) and set $a = \frac{z}{2\pi}$.

[2] Calculate the series summation:

$$\text{(a)} \sum_{n=1}^{\infty} \frac{1}{n(n+2)}; \quad \text{(b)} \sum_{n=-\infty}^{\infty} \frac{(-1)^n}{(2n+1)^3}$$

Answer: (a) $\frac{3}{4}$; (b) $\frac{\pi^3}{32}$

[3] Proof:

$$\frac{1}{\cosh \frac{\pi}{2}} - \frac{1}{3 \cosh \frac{3\pi}{2}} + \frac{1}{5 \cosh \frac{5\pi}{2}} - \frac{1}{7 \cosh \frac{7\pi}{2}} + \cdots = \frac{\pi}{8}$$

Hint: Take $F(z) = \frac{e^{\pi z}}{z \cos \pi z}$ and employ the form of Fermion distribution function.

[4] Proof:

$$\sum_{n=1}^{\infty} \frac{1}{n^p} = -\frac{(2\pi i)^p}{2p!} B_p$$

where p is a positive even number.

Chapter 5

Analytic Theory

5.1 Analytic Continuation

5.1.1 *Definition*

Analytic continuation is a concept in complex analysis that deals with the extension of the definition domain of an analytic function. Let us examine the function $f(z) = \frac{1}{1+z^2}$ and its infinite series representation:

$$\frac{1}{1+z^2} = 1 - z^2 + z^4 - z^6 + \cdots$$

The function $f(z)$ on the left-hand side is analytic on the entire complex plane except for the singularities at $z = \pm \mathrm{i}$. The infinite series $g(z) = 1 - z^2 + z^4 - z^6 + \cdots$ on the right-hand side converges and is analytic within the disk B with radius $|z| < 1$. Beyond this convergence region, the series diverges and loses its meaning. In the overlapping region, the two functions are completely equal. Comparatively, the function $f(z)$ is analytical in a larger region compared to $g(z)$. We say that $f(z)$ is the analytic continuation of the function $g(z)$.

In general, if there exists a single-valued analytic function $g(z)$ in a region B, and another analytic function $f(z)$ is defined in a larger region G that preserves its analytic properties, such that $f(z) = g(z)$ in the common region B, then $f(z)$ is called the analytic continuation of $g(z)$.

Conversely, considering the function $f(z) = \frac{1}{1+z^2}$ expanded in a Taylor series around $z = \frac{3}{4}$, its convergence domain is $\left|z - \frac{3}{4}\right| < \frac{5}{4}$. Two convergence domains B and B' are shown in Figure 5.1(a). The two series functions are equal within the intersection of the convergence regions, defining an analytic function on the region $G = B \bigcup B'$. By continuously applying this method, except for some isolated singularities, the function can be analytically continued from the domain B to the entire complex plane, as shown in Figure 5.1(b).

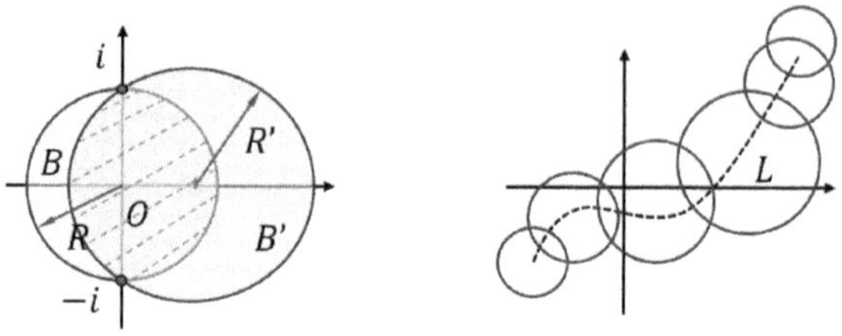

(a) (b)

Figure 5.1.

For example, the function $f_1(z) = \int_0^\infty \mathrm{e}^{-zt}\mathrm{d}t$ has an analytic domain in the right half-plane $S_1 : \mathrm{Re}\, z > 0$, while the function $f_2(z) = \mathrm{i}\sum_{n=0}^{\infty}\left[\frac{z+\mathrm{i}}{\mathrm{i}}\right]^n$ has an analytic domain in the circular region $S_2 : |z + \mathrm{i}| < 1$. Within the common domain, they are both equal to $\frac{1}{z}$. We can define an analytic continuation function as

$$f(z) = \begin{cases} f_1(z) & (z \in S_1) \\ f_2(z) & (z \in S_2) \end{cases}$$

5.1.2 *Fundamental theorems*

In regards to analytic continuation, the following fundamental properties are observed:

(1) **Continuous extension lemma:** Consider two non-overlapping regions B_1 and B_2 with a common boundary γ. Let $f_1(z)$ and $f_2(z)$ be analytic functions in B_1 and B_2, respectively, and continuous on the regions $B_1 \bigcup \gamma$ and $B_2 \bigcup \gamma$. If for all $z \in \gamma$ we have $f_1(z) = f_2(z)$, then the function

$$f(z) \stackrel{\text{def}}{=} \begin{cases} f_1(z) & (z \in B_1 \bigcup \gamma) \\ f_2(z) & (z \in B_2 \bigcup \gamma) \end{cases}$$

is analytic in the region $B \equiv B_1 \bigcup \gamma \bigcup B_2$. We say that $f_1(z)$ and $f_2(z)$ are analytic continuations of each other.

(2) **Uniqueness theorem:** The analytic continuation of a function is unique.

Proof. Suppose a function $f(z)$ is analytic in a domain B, and there exist two analytic continuation functions $g_1(z)$ and $g_2(z)$ in a larger domain $G \supset B$, such that $g_1(z) = g_2(z) \equiv f(z)$ within B. Let

$$h(z) = g_1(z) - g_2(z)$$

Since $h(z) \equiv 0$ in the region B by assumption, the corollary of the isolated zeros theorem implies that $h(z) \equiv 0$ within the whole domain G. □

(3) **Schwarz reflection principle:** If $f(z)$ is an analytic function in a region B, where a segment of its boundary lies on the real axis, and $f(z)$ is a real function on this segment, then its analytic continuation can be reflected symmetrically with respect to the real axis, giving rise to a new region B' reflected with respect to the real axis. The analytic continuation function $g(z)$ satisfies:

$$g(z) = \overline{f(\bar{z})} = \bar{f}(\bar{z}) \tag{5.1.1}$$

(4) **Frobenius theorem:** The complex number field is the only possible algebraic extension of the real number field that preserves its algebraic properties.

According to the Frobenius theorem, the most common type of analytic continuation is the direct extension from real variables to complex variables, extending the function from the real axis to the complex plane. This property has been utilized in Chapter 4, where the residue theorem was used to evaluate integrals of real functions.

ENDNOTE

For a real geometric series, the sum can be calculated as

$$1 + x + x^2 + x^3 + x^4 + \cdots = \frac{1}{1-x} \quad (x \in \mathbb{R})$$

The series converges for $|x| < 1$. However, the function $f(x) \equiv \frac{1}{1-x}$ on the right-hand side is defined beyond the range $|x| < 1$. In fact, it is meaningful for the entire real axis except for $x = 1$, and the two are equal in the interval $-1 < x < 1$. The convergence radius of the series can be determined from the singularity position of $f(x)$ as $R = 1$, as shown in Figure 5.2(a).

For the series

$$1 - x^2 + x^4 - x^6 + \cdots = \frac{1}{1+x^2}$$

it is not immediately apparent why its convergence range is also $|x| < 1$, as the function on the right-hand side has no singularities on the entire real axis (Figure 5.2(b)). However, from the perspective of the complex domain,

$$1 - z^2 + z^4 - z^6 + \cdots = \frac{1}{1+z^2}$$

the singularity of the function $f(z) = \frac{1}{1+z^2}$ is located at $z = \pm\mathrm{i}$, and the distance from it to $z = 0$ is exactly the convergence radius $R = 1$.

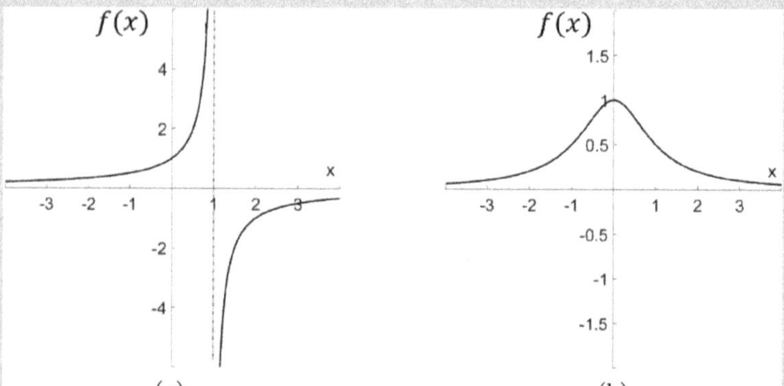

Figure 5.2.

Cauchy established the theory of analytic functions based on derivatives and integrals, while Weierstrass opened up another path from an analytical perspective. Weierstrass was a systematic and hardworking person who did not rely on intuition but focused on establishing mathematics on a solid foundation of reasoning. He developed the theory of analytic functions starting from power series and developed methods for analytic continuation.

During the process of extending functions by increasing convergence circles, it is possible for the function values in the overlapping region of successive circles to be distinct, as shown in Figure 5.3. This is a situation encountered by multivalued functions, and the singularities that occur in this process must be located on the boundary of the convergence circles.

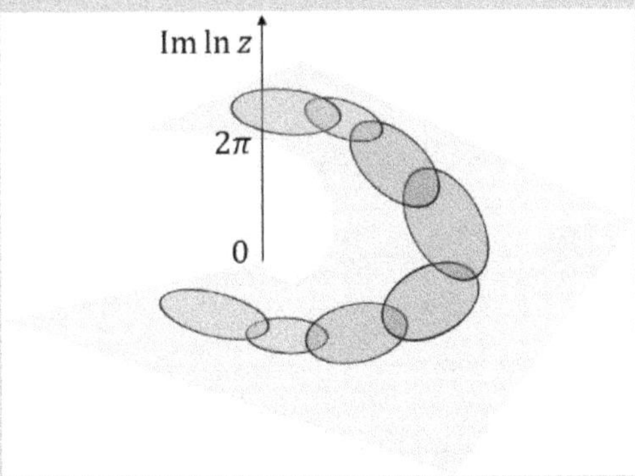

Figure 5.3.

An alternative way of defining complex functions is to analytically extend the real functions. For example, based on the Taylor series expansions of real functions:

$$\sin x = x - \frac{1}{3!}x^3 + \frac{1}{5!}x^5 - \cdots$$

$$\cos x = 1 - \frac{1}{2!}x^2 + \frac{1}{4!}x^4 - \cdots$$

By replacing the real variable x with a complex variable z, we can define complex trigonometric functions through analytic continuation as

$$\sin z \stackrel{\text{def}}{=} = z - \frac{1}{3!}z^3 + \frac{1}{5!}z^5 - \cdots$$

$$\cos z \stackrel{\text{def}}{=} = 1 - \frac{1}{2!}z^2 + \frac{1}{4!}z^4 - \cdots$$

According to the Taylor expansion of the exponential function:

$$\mathrm{e}^x = 1 + \frac{1}{1!}x + \frac{1}{2!}x^2 + \frac{1}{3!}x^3 + \frac{1}{4!}x^4 + \frac{1}{5!}x^5 - \cdots$$

We define the complex exponential function as

$$\mathrm{e}^z \stackrel{\text{def}}{=} 1 + \frac{1}{1!}z + \frac{1}{2!}z^2 + \frac{1}{3!}z^3 + \frac{1}{4!}z^4 + \frac{1}{5!}z^5 - \cdots$$

From this, we directly obtain Euler's formula:

$$\begin{aligned}\mathrm{e}^{\mathrm{i}x} &= \left[1 - \frac{1}{2!}x^2 + \frac{1}{4!}x^4 - \cdots\right] + \mathrm{i}\left[x - \frac{1}{3!}x^3 + \frac{1}{5!}x^5 - \cdots\right]\\ &= \cos x + \mathrm{i}\sin x\end{aligned}$$

and find the relationships between trigonometric functions and exponential functions:

$$\cos z = \frac{\mathrm{e}^{\mathrm{i}z} + \mathrm{e}^{-\mathrm{i}z}}{2}, \quad \sin z = \frac{\mathrm{e}^{\mathrm{i}z} - \mathrm{e}^{-\mathrm{i}z}}{2\mathrm{i}}$$

As for real functions, given a function $f(x)$ defined on an interval $[x_1 x_2]$, as shown in Figure 5.4(a), it seems that there are numerous ways to smoothly extend the function from the endpoints x_1 and x_2 to the whole real axis, as depicted in Figures 5.4(b) and 5.4(c). However, complex analysis theory tells us that there is only one unique way to extend the function while preserving its analyticity. This is because if a complex function is analytic at some point, it is infinitely differentiable there. Applying this property to real functions, it becomes a sufficient and necessary condition for a function curve to be smooth. In other words, if you have a smooth real function defined on a certain interval, then the function extended to the entire real axis while maintaining smoothness is unique.

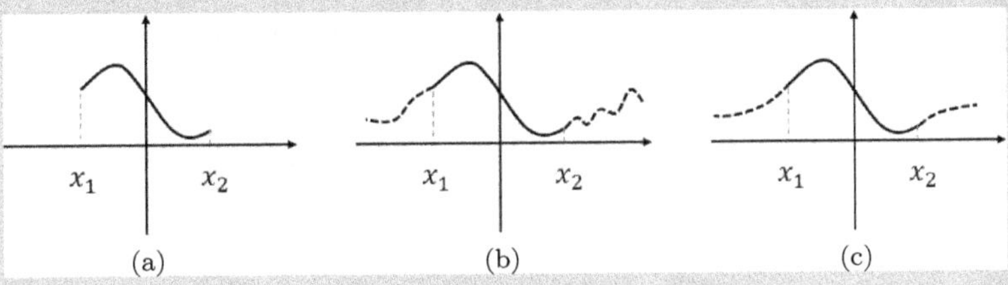

Figure 5.4.

The geometric series can be analytically continued from the interior of the convergence circle to the exterior. In Section 3.4, we mentioned an infinite series:

$$\sum_{k=0}^{\infty} z^{2^k} = z + z^2 + z^4 + z^8 + z^{16} + \cdots$$

It has a unit circle consisting of dense non-isolated singular points. The series converges uniformly inside the circle, making it analytic. Can this function be analytically continued outside the unit circle? The answer is negative. This raises the question: What kind of infinite series can or cannot be analytically continued from the interior of the convergence circle to the exterior?

Examine the decimal part of any real number, which is denoted as $a_1a_2a_3a_4\ldots$. For example, $\sqrt{2} = 1.414213562\ldots$, where $a_1 = 4$, $a_2 = 1$, $a_3 = 4$, $a_4 = 2, \ldots$. We construct a series with these numbers:

$$\sum_{k=0}^{\infty} a_k z^k$$

For finite decimal rational numbers, this series degenerates into a polynomial, making it analytic in the entire plane. For rational numbers with a repeating cycle of length p, the series can be expressed as

$$\sum_{k=1}^{\infty} a_k z^k = \sum_{k=0}^{\infty}(a_0 + a_1 z + a_2 z^2 + \cdots a_p z^{p-1}) z^{pk}$$

Its convergence domain is $|z| < 1$. The sum of the series within the convergence domain reads

$$\sum_{k=1}^{\infty} a_k z^k = \frac{a_0 + a_1 z + a_2 z^2 + \cdots a_{p-1} z^{p-1}}{1 - z^p} \tag{5.1.2}$$

This is a fractional rational function with p simple poles evenly distributed on the unit circle. The series can be analytically continued from within the unit circle to the entire plane. In the complex plane, it has $p-1$ zeros and is analytic at $z = \infty$, making it a meromorphic function.

For irrational numbers, by contrast, we still find that the convergence domain of the infinite series $\sum_{k=1}^{\infty} a_k z^k$ is $|z| < 1$. Taking account of the fact that the repeating cycle of $p \to \infty$ for irrational numbers, we surmise there are numerous non-isolated singularities distributed on the unit circle. If the surmise is true, then the irrational number series cannot be analytically continued from the interior to the exterior of the unit circle.

Reflection:

(1) For an irrational number series, can we prove that all points on the unit circle are dense non-isolated singularities?
(2) If the irrational number series cannot be continued to the exterior of the unit circle, is it possible to construct an analytical function meaningful outside the circle, analogous to the rational number series?

Exercises

[1] Prove that the functions $f_1(z)$ and $f_2(z)$ defined by series are analytic continuations of each other:

$$f_1(z) = 1 + az + a^2z^2 + \cdots$$

$$f_2(z) = \frac{1}{1-z} - \frac{(1-a)z}{(1-z)^2} + \frac{(1-a)^2z^2}{(1-z)^3} + \cdots$$

[2] The two power series,

$$f_1(z) = \sum_{n=1}^{\infty} \frac{z^n}{n}, \quad f_2(z) = \mathrm{i}\pi + \sum_{n=1}^{\infty}(-1)^n \frac{(z-2)^n}{n}$$

have no common convergence region. We need to prove that they are still analytic continuations of each other.

[3] Prove the Schwarz reflection principle: $g(z) = \bar{f}(\bar{z})$ is an analytic function.
Hint: Prove that the real and imaginary parts of $g(z)$ satisfy the Cauchy–Riemann conditions.

[4] Suppose the analytic function $f(z) = u + \mathrm{i}v$ satisfies the Schwarz reflection condition. Prove that u and v are even and odd functions of y, respectively.

5.2 Analytic Continuation Functions

5.2.1 *Gamma function*

(1) *Euler integral*

Consider the factorial function of integers: $f(n) = n!$, with the results shown as the data points in Figure 5.5. The German mathematician Goldbach posed the question: Can we find a continuous function that smoothly connects these points?

Euler indeed found such a function, but it is not an ordinary function. It is defined as the integral of a real variable,

$$\Gamma(x) \stackrel{\text{def}}{=} \int_0^{\infty} \mathrm{e}^{-t} t^{x-1} \mathrm{d}t \quad (x > 0) \tag{5.2.1}$$

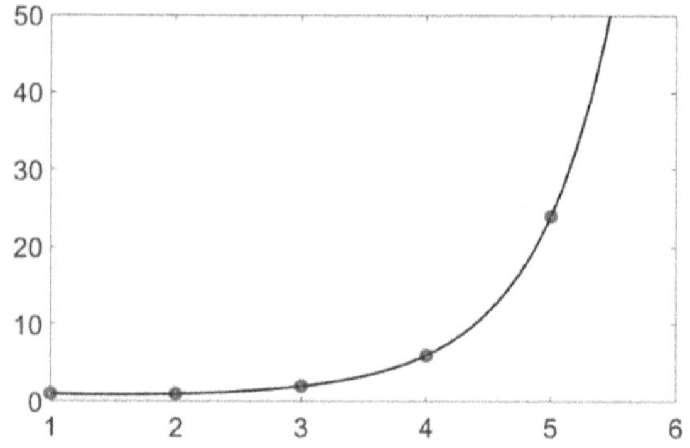

Figure 5.5.

This is called the Gamma function, also known as the second Euler integral. This integral converges for $x > 0$, meaning that the Gamma function is only defined on the positive real axis. Its fundamental properties are listed as follows:

$$\Gamma(1) = 1; \quad \Gamma(n+1) = n\Gamma(n) = n!$$

$$\Gamma\left(\frac{1}{2}\right) = \sqrt{\pi}; \quad \Gamma\left(n+\frac{1}{2}\right) = \frac{(2n-1)!!}{2^n}\sqrt{\pi}$$

$$\Gamma(x+1) = x\Gamma(x);$$

$$\Gamma(x)\Gamma(1-x) = \frac{\pi}{\sin \pi x}; \quad \Gamma(x)\Gamma(-x) = -\frac{\pi}{x \sin \pi x}$$

These properties can be directly proven from the definition of the Gamma function.

(2) *Analytical continuation*

If we extend the function analytically to the entire complex plane, then

$$\Gamma(z) \stackrel{\text{def}}{=} \int_0^\infty e^{-t} t^{z-1} dt \tag{5.2.2}$$

The relations still hold:

$$\Gamma(z+1) = z\Gamma(z); \quad \Gamma(z)\Gamma(1-z) = \frac{\pi}{\sin \pi z} \tag{5.2.3}$$

Figure 5.6(a) depicts the behavior of the Gamma function in the entire real domain, showing that the integral remains meaningful for $x < 0$. Figure 5.6(b) illustrates the modulus distribution of the Gamma function in the complex plane.

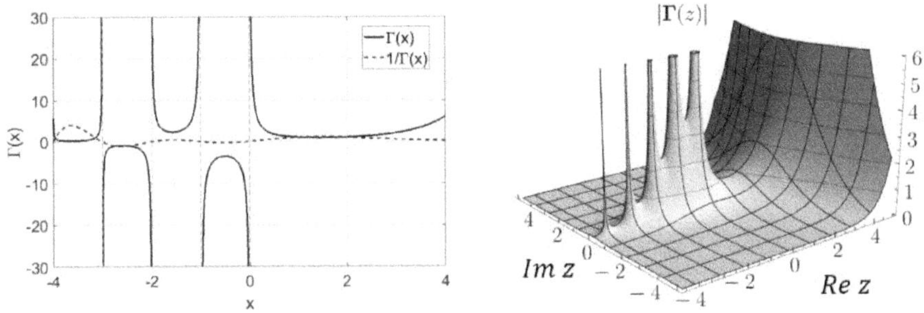

(a) (b)

Figure 5.6.

(3) *Poles and zeros*

Since $\Gamma(z) = \frac{\Gamma(z+1)}{z}$, it can be inferred that all singularities of the Gamma function occur at $z_k = 0, -1, -2, -3, \ldots$, which are all negative integer values. All singularities are first-order poles and have the following limit behavior:

$$\Gamma(z)|_{z\to 0} = \left.\frac{\Gamma(z+1)}{z}\right|_{z\to 0} \sim (-1)^0 \frac{1}{0!} \cdot \frac{1}{z}$$

$$\Gamma(z)|_{z\to -1} = \left.\frac{\Gamma(z+2)}{z(z+1)}\right|_{z\to -1} \sim (-1)^1 \frac{1}{1!} \cdot \frac{1}{z+1}$$

$$\Gamma(z)|_{z\to -2} = \left.\frac{\Gamma(z+3)}{z(z+1)(z+2)}\right|_{z\to -2} \sim (-1)^2 \frac{1}{2!} \cdot \frac{1}{z+2}$$

$$\Gamma(z)|_{z\to -n} = \left.\frac{\Gamma(z+n+1)}{z(z+1)\cdots(z+n)}\right|_{z\to -n} \sim (-1)^n \frac{1}{n!} \cdot \frac{1}{z+n}$$

It can be proven that the Gamma function has no zeros in the complex plane. Therefore, the function $\frac{1}{\Gamma(z)}$ has no singularities in the entire plane, as shown by the dashed line in Figure 5.6(a). It can be represented as an infinite product of zeros in Weierstrass form:

$$\frac{1}{\Gamma(z)} = z e^{\gamma z} \prod_{n=1}^{\infty} \left(1 + \frac{z}{n}\right) e^{-\frac{z}{n}} \tag{5.2.4}$$

where

$$\gamma = \lim_{n\to\infty} \left(\sum_{k=1}^{n} \frac{1}{k} - \ln n\right) = 0.577216\cdots$$

is called Euler's constant. From this, the formulas follow:

$$\sin \pi z = \frac{\pi}{\Gamma(z)\Gamma(1-z)} = \pi z \prod_{n=1}^{\infty} \left(1 - \frac{z^2}{n^2}\right) \tag{5.2.5}$$

$$\cos \pi z = \frac{\sin 2\pi z}{2 \sin \pi z} = \prod_{n=1}^{\infty} \left[1 - \frac{4z^2}{(2n-1)^2}\right] \tag{5.2.6}$$

These formulas imply that trigonometric functions can be decomposed into product factors, similar to polynomials. This is not a special case; later on, it will be proven that this is a general property of entire functions.

(4) *Stirling's formula*

As $|z| \to \infty$, the Stirling asymptotic formula holds:

$$\Gamma(z) \sim \sqrt{2\pi} z^{z-\frac{1}{2}} e^{-z} \tag{5.2.7}$$

In particular, as $n \to \infty$, we have the Stirling approximation for factorial,

$$n! = n\Gamma(n) \approx \sqrt{2\pi n} n^n e^{-n} \tag{5.2.8}$$

This Stirling approximation for factorial is frequently met in the statistical physics.

(5) *Integral representation*

The Gamma function can also be represented by a path integral, known as the Schlaefli integral,

$$\Gamma(z) = \frac{1}{e^{2\pi i z} - 1} \int_L e^{-t} t^{z-1} dt \tag{5.2.9}$$

The integration path L is shown in Figure 5.7.

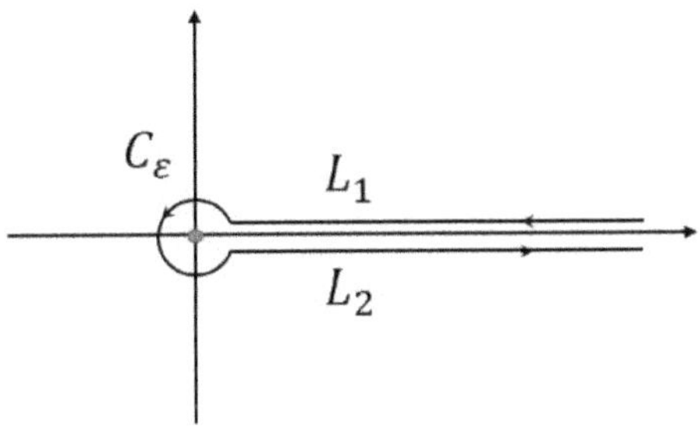

Figure 5.7.

Discussion: Little is known about Euler's constant γ. It is unsure whether γ is related to the mathematical constant π or e. Some speculate that it is a transcendental number, but in fact, it is even unknown whether it is an irrational number.

5.2.2 *Beta function*

The Beta function, also known as the first Euler integral, is defined as

$$B(p, q) \stackrel{\text{def}}{=} \int_0^1 t^{p-1}(1-t)^{q-1} dt \quad (p, q > 0) \tag{5.2.10}$$

It can be easily proven that $B(p, q) = B(q, p)$. Substituting the variable $t = \sin^2\theta$, we obtain an alternative representation,

$$B(p, q) = 2 \int_0^{\frac{\pi}{2}} \sin^{2p-1}\theta \cos^{2q-1}\theta \, d\theta \tag{5.2.11}$$

The real variables p and q can also be analytically extended to the complex plane, resulting in a bivariate complex function.

The Beta function and gamma function are intimately related by the following important formula:

$$\mathrm{B}(p,q)=\frac{\Gamma(p)\Gamma(q)}{\Gamma(p+q)} \tag{5.2.12}$$

Proof. Let $t=x^2$, the gamma function is rewritten as

$$\Gamma(p)=\int_0^\infty \mathrm{e}^{-t}t^{p-1}\mathrm{d}t=2\int_0^\infty \mathrm{e}^{-x^2}x^{2p-1}\mathrm{d}x$$

We have

$$\Gamma(p)\Gamma(q)=4\int_0^\infty \mathrm{e}^{-x^2}x^{2p-1}\mathrm{d}x\int_0^\infty \mathrm{e}^{-y^2}y^{2q-1}\mathrm{d}y$$

The integration domain is the first quadrant. By employing the polar coordinates, with $x=r\cos\theta$, $y=r\sin\theta$ and $\mathrm{d}x\mathrm{d}y=r\mathrm{d}r\mathrm{d}\theta$, it yields

$$\Gamma(p)\Gamma(q)=4\int_0^\infty \mathrm{e}^{-r^2}r^{2p+2q-1}\mathrm{d}r\int_0^{\frac{\pi}{2}}\sin^{2p-1}\theta\cos^{2q-1}\theta\,\mathrm{d}\theta$$

Taking $s=r^2$, and making use of formula (5.2.11), we get

$$\Gamma(p)\Gamma(q)=\int_0^\infty \mathrm{e}^{-s}s^{p+q-1}\mathrm{d}s\mathrm{B}(p,q)=\Gamma(p+q)\mathrm{B}(p,q)$$

Proof completed. □

Example 5.1. Compute the Integral:

$$\int_{-1}^1(1-x^2)^{z-1}\mathrm{d}x \quad (\mathrm{Re}\,z>0)$$

Solution: Let $t=x^2$, then we have

$$\int_{-1}^1(1-x^2)^{z-1}\mathrm{d}x=\int_0^1(1-t)^{z-1}t^{-\frac{1}{2}}\mathrm{d}t=\mathrm{B}\left(z,\frac{1}{2}\right)$$
$$=\frac{\Gamma(z)\Gamma\left(\frac{1}{2}\right)}{\Gamma\left(z+\frac{1}{2}\right)}$$

Example 5.2. Proof:

$$\Gamma(z)\Gamma(1-z)=\frac{\pi}{\sin\pi z}$$

Proof. Using formula (5.2.12) and letting $u = \tan\theta$, we have

$$\Gamma(z)\Gamma(1-z) = \mathrm{B}(z, 1-z) = 2\int_0^\infty \frac{u^{2z-1}}{u^2+1}\mathrm{d}u \quad (0 < \mathrm{Re}\, z < 1)$$

By utilizing the results of the multivalued function integral from Example 4.10 in Chapter 4, the proof can be obtained. □

5.2.3 *Polygamma functions*

(1) *Digamma function*

The digamma function, also known as the ψ function, is defined as the logarithmic derivative of the gamma function:

$$\psi(z) \stackrel{\text{def}}{=} \frac{\Gamma'(z)}{\Gamma(z)} = \frac{\mathrm{d}}{\mathrm{d}z}\ln\Gamma(z) \tag{5.2.13}$$

The digamma function also has first-order poles at $z_k = 0, -1, -2, -3, \ldots$, with residues $\mathrm{Res}[\psi(z_k)] = -1$. The digamma function has the following fundamental properties:

$$\psi(z+n) = \psi(z) + \frac{1}{z} + \frac{1}{z+1} + \frac{1}{z+2} + \cdots + \frac{1}{z+n-1}$$

$$\psi(z) = -\gamma + \sum_{n=0}^{\infty}\left(\frac{1}{n+1} - \frac{1}{n+z}\right)$$

$$\psi(z) - \psi(-z) = -\frac{1}{z} - \pi\cot\pi z$$

Below are some special values of the ψ function:

$$\psi(1) = -\gamma, \quad \psi'(1) = \frac{\pi^2}{6}$$

$$\psi\left(\frac{1}{2}\right) = -\gamma - 2\ln 2, \quad \psi'\left(\frac{1}{2}\right) = \frac{\pi^2}{2},$$

$$\psi\left(\frac{1}{4}\right) = -\gamma - \frac{\pi}{2} - 3\ln 2, \quad \psi'\left(\frac{1}{3}\right) = -\gamma - \frac{\pi}{2\sqrt{3}} - \frac{3}{2}\ln 3$$

The digamma function can be utilized to calculate the summation of some kind of rational terms series. Assume that

$$\sum_{n=0}^{\infty} u_n = \sum_{n=0}^{\infty} \frac{p(n)}{q(n)}$$

where $p(n)$ and $q(n)$ are polynomials in n. If α_k are all the first-order zeros of $q(n)$, then

$$q(n) = (n+\alpha_1)(n+\alpha_2)\cdots(n+\alpha_m)$$

Therefore,

$$u_n = \frac{p(n)}{q(n)} \equiv \sum_{k=1}^{m} \frac{b_k}{n+\alpha_k}$$

To ensure the convergence of the series, it is required that $\lim_{n\to\infty} n \cdot u_n \to 0$, implying

$$\sum_{k=1}^{m} b_k = 0$$

Under this condition, we have the following summation formula:

$$\sum_{n=0}^{\infty} u_n = -\sum_{k=1}^{m} b_k \psi(\alpha_k) \tag{5.2.14}$$

Proof. Using the formula

$$\psi(z) = -\gamma + \sum_{n=0}^{\infty} \left(\frac{1}{n+1} - \frac{1}{n+z} \right)$$

We have

$$b_k \psi(\alpha_k) = -\gamma b_k + b_k \sum_{n=0}^{\infty} \left(\frac{1}{n+1} - \frac{1}{n+\alpha_k} \right)$$

Summing both sides over k and make use of the equality $\sum_{k=1}^{m} b_k =$, we get

$$\sum_{n=0}^{\infty} \sum_{k=1}^{m} \frac{b_k}{n+\alpha_k} = -\sum_{k=1}^{m} b_k \psi(\alpha_k)$$

Example 5.3. Find the summation of series:

$$\sum_{n=0}^{\infty} \frac{1}{n^2 + a^2}$$

Solution: Using formula (5.2.14) and properties of the ψ function, we have

$$\begin{aligned}\sum_{n=0}^{\infty} \frac{1}{n^2 + a^2} &= \frac{\mathrm{i}}{2a} \sum_{n=0}^{\infty} \left(\frac{1}{n + \mathrm{i}a} - \frac{1}{n - \mathrm{i}a} \right) = -\frac{\mathrm{i}}{2a} [\psi(\mathrm{i}a) - \psi(-\mathrm{i}a)] \\ &= -\frac{\mathrm{i}}{2a} \left[-\frac{1}{\mathrm{i}a} - \pi \cot(\mathrm{i}\pi a) \right] = \frac{1}{2a^2} [1 + \pi a \coth(\pi a)]\end{aligned}$$

The result is also obtained by using the method of residues in Section 4.4, Exercise 1. □

(2) *Polygamma functions*

Polygamma functions are the higher-order Gamma functions which are defined as the mth derivative of the ψ function:

$$\psi^{(m)}(z) \overset{\text{def}}{=} \frac{\mathrm{d}^m}{\mathrm{d}z^m} \psi(z) = \frac{\mathrm{d}^{m+1}}{\mathrm{d}z^{m+1}} \ln \Gamma(z) \tag{5.2.15}$$

They have $(m + 1)$th order poles at $z = 0, -1, -2, -3, \ldots$. The polygamma functions have properties such as integral representations and recurrence relations:

$$\psi^{(m)}(z) = (-1)^{m+1} \int_0^{\infty} \frac{t^m \mathrm{e}^{-zt}}{1 - \mathrm{e}^{-t}} \mathrm{d}t = -\int_0^1 \frac{t^{z-1}}{1 - t} (\ln t)^m \mathrm{d}t$$

$$\psi^{(m)}(z + 1) = \psi^{(m)}(z) + \frac{(-1)^m m!}{z^{m+1}}$$

$$\psi^{(m)}(z) = (-1)^{m+1} m! \sum_{k=0}^{\infty} \frac{1}{(z + k)^{m+1}}$$

The graphs in Figure 5.8 illustrate the curves of polygamma functions for real variables, while Figure 5.9 shows the distribution of their arguments when analytically extended to the complex plane.

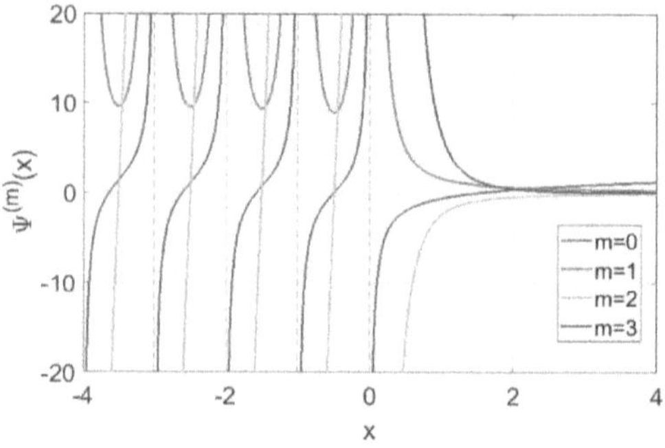

Figure 5.8. (images from Wikipedia).

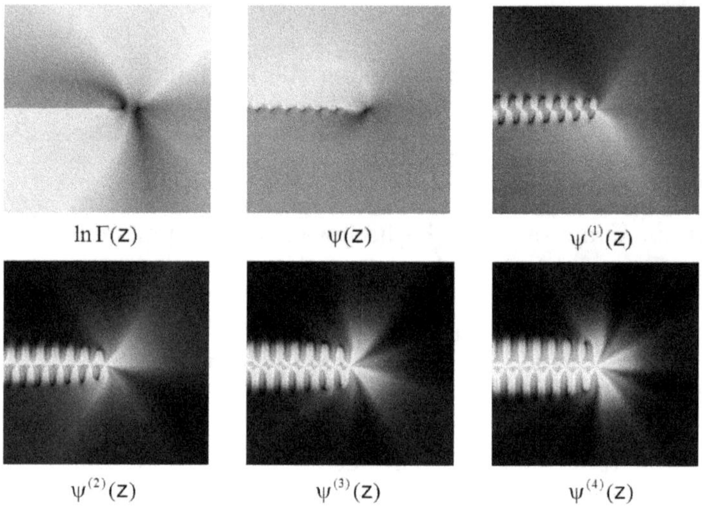

Figure 5.9. (images from Wikipedia).

5.2.4 *Riemann zeta function*

Consider the infinite series:

$$\zeta(s) = 1 + \frac{1}{2^s} + \frac{1}{3^s} + \frac{1}{4^s} + \cdots \quad (s > 1)$$

When $s = 2n$ is even, we can find the sum of $\zeta(2n)$, for example,

$$\zeta(2) = 1 + \frac{1}{2^2} + \frac{1}{3^2} + \frac{1}{4^2} + \cdots = \frac{\pi^2}{6}$$

As the real number s is extended to the complex variable z, it is called the Riemann series. When its analytic domain is extended to the entire complex plane, it is called the Riemann zeta function. Figure 5.10 depicts the two-dimensional distribution of $\zeta(z)$, with colors indicating the phase. The figure shows that $z = 1$ is the only simple pole of the function, and all zeros are distributed along the lines $z = -2k(k \in \mathbb{N})$ and $\operatorname{Re} z = \frac{1}{2}$.

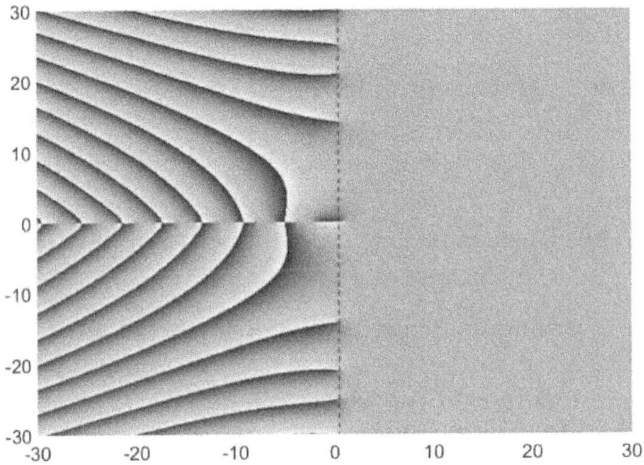

Figure 5.10.

The integral

$$F(z) = \int_0^\infty \varphi(t) t^{z-1} \mathrm{d}t \tag{5.2.16}$$

is called the Mellin transformation of the function $\varphi(t)$. It has the property of multiplication transformation invariance, which means

$$\int_0^\infty f(at)\frac{\mathrm{d}t}{t} = \int_0^\infty f(t)\frac{\mathrm{d}t}{t}$$

Let $\varphi(t) = \mathrm{e}^{-nt}$, then

$$\int_0^\infty \mathrm{e}^{-nt} t^s \frac{\mathrm{d}t}{t} = \frac{1}{n^s}\int_0^\infty \mathrm{e}^{-t} t^s \frac{\mathrm{d}t}{t} \equiv \frac{1}{n^s}\Gamma(s) \quad (\operatorname{Re} s > 1)$$

Summing both sides of the above equation over n, the right-hand side becomes the Riemann zeta function. Therefore,

$$\Gamma(s)\zeta(s) = \int_0^\infty \sum_{n=1}^\infty \mathrm{e}^{-nt} t^s \frac{\mathrm{d}t}{t} = \int_0^\infty \frac{t^{s-1}}{\mathrm{e}^t - 1} \mathrm{d}t$$

This gives the integral representation of the zeta function:

$$\zeta(z) = \frac{1}{\Gamma(z)} \int_0^\infty \frac{t^{z-1}}{\mathrm{e}^t - 1} \mathrm{d}t$$

From this, the following formula can be derived,

$$\zeta(1-z) = \frac{2}{(2\pi)^z} \cos\left(\frac{\pi z}{2}\right) \Gamma(z)\zeta(z) \tag{5.2.17}$$

Note that the Bose distribution function appears again in this formula.

Riemann Hypothesis: Riemann Hypothesis states that apart from certain trivial zeros at $z = -2k (k \in \mathbb{N})$, all non-trivial zeros of the ζ function are located on the vertical line with the real part $x = \frac{1}{2}$. Figure 5.11(a) depicts the distribution of several zeros near the line $z = \frac{1}{2} + \mathrm{i}y$. Figure 5.11(b) shows the two-dimensional curves of the real part u and the imaginary part v of $\zeta(z) = u + \mathrm{i}v$ as they vary along the line $z = \frac{1}{2} + \mathrm{i}y$. From the figure, it can be observed that the curve repeatedly passes through the origin, indicating that $\zeta(z)$ has zeros along the line $z = \frac{1}{2} + \mathrm{i}y$.

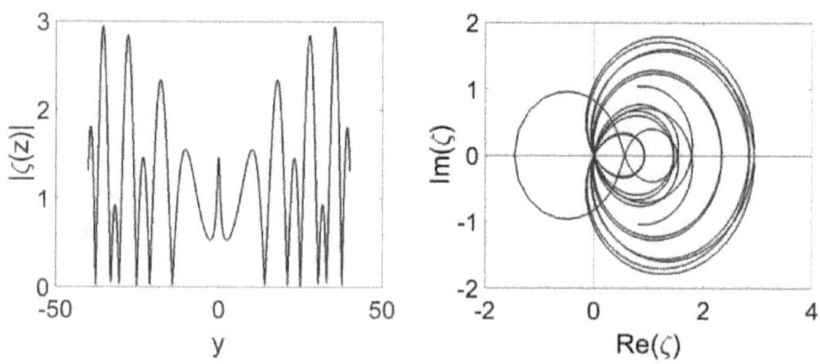

(a) (b)

Figure 5.11.

ENDNOTE

The Gamma function can also be introduced in another insightful way. Consider the complex variable function:

$$\Gamma_n(z) \overset{\text{def}}{=} \int_0^n \left(1 - \frac{t}{n}\right)^n t^{z-1}\mathrm{d}t \quad (\operatorname{Re} z > 0, n \geq 1)$$

Then we have $\Gamma(z) = \lim_{n\to\infty} \Gamma_n(z)$. By making the variable substitution $s = \frac{t}{n}$, we obtain

$$\Gamma_n(z) = n^z \int_0^1 (1-s)^n s^{z-1}\mathrm{d}s$$

Hence

$$\Gamma_1(z) = \int_0^1 (1-s)s^{z-1}\mathrm{d}s = \frac{1}{z(z+1)}$$

$$\Gamma_n(z) = \frac{n^z}{z}\int_0^1 (1-s)^n \mathrm{d}s^z = \left(\frac{n}{n-1}\right)^z \frac{n}{z}\Gamma_{n-1}(z+1)$$

Repeating the recursive process, we get

$$\begin{aligned}\Gamma_n(z) &= \frac{n^z n!}{z(z+1)\cdots(z+n-2)}\Gamma_1(z+n-1) \\ &= \frac{n^z n!}{z(z+1)\cdots(z+n)}\end{aligned}$$

Therefore, the Gamma function can be represented as

$$\Gamma(z) = \lim_{n\to\infty} \frac{n^z n!}{z(z+1)\cdots(z+n)}$$

This expression is known as the Gauss product representation. A direct result of this representation is

$$\Gamma(z+1) = z\Gamma(z)$$

Furthermore,

$$\frac{1}{\Gamma_n(z)} = z\mathrm{e}^{\gamma_n z}\prod_{k=1}^{n}\left(1+\frac{z}{k}\right)\mathrm{e}^{-\frac{z}{k}}$$

where

$$\gamma_n = 1 + \frac{1}{2} + \cdots + \frac{1}{n} - \ln n$$

As $n \to \infty$, γ_n converges to γ, which is the Euler's constant. Therefore,

$$\lim_{n\to\infty}\frac{1}{\Gamma_n(z)} = \frac{1}{\Gamma(z)} = z\mathrm{e}^{\gamma z}\prod_{k=1}^{\infty}\left(1+\frac{z}{k}\right)\mathrm{e}^{-\frac{z}{k}} \quad (z \in \mathbb{C})$$

From this, we can deduce that

$$\frac{1}{\Gamma(z)\Gamma(-z)} = -z^2\prod_{k=1}^{\infty}\left(1-\frac{z^2}{k^2}\right) = -\frac{z\sin\pi z}{\pi}$$

Using the relation $\Gamma(1-z) = -z\Gamma(-z)$, we obtain the formula:

$$\Gamma(z)\Gamma(1-z) = \frac{\pi}{\sin\pi z}$$

Since Euler established the relationship between trigonometric and exponential functions (Euler's formula) in the 18th century, the theory of complex analysis has undergone rapid development. However, it was not until the 19th century that the foundations of complex analysis were solidified, with the notable contributions of three individuals: Cauchy, Weierstrass, and Riemann. Cauchy developed the theory of complex integration, Weierstrass formalized algebraic theory based on the convergence of power series, and Riemann made pioneering contributions in the field of geometry. His ideas are crucial to the entire mathematical edifice.

The study of number theory first appeared around 300 BC when Euclid proved the existence of infinitely many prime numbers in his book "Elements". However, there was a gap of two thousand years until Euler reignited people's interest in prime number research in 1737 by proving the divergence of the sum of reciprocals of all prime numbers $p \in \mathbb{P}$, denoted as $\sum_{p\in\mathbb{P}} \frac{1}{p}$. Euler used the formula:

$$\sum_{n=1}^{\infty} \frac{1}{n^s} = \prod_{p\in\mathbb{P}} \left(1 - \frac{1}{p^s}\right)^{-1} \qquad (s > 1)$$

Euler was the first person to apply analytical methods to solve arithmetic problems. Many mathematicians felt uneasy about this, and it wasn't until a hundred years later that Dirichlet attempted to prove the prime number theorem using real analysis.

In 1859, Riemann published his only work in number theory, a paper of fewer than 10 pages. This paper caused a huge impact in various areas of pure mathematics, and its influence continues to this day, possibly lasting for several hundred years in the future. The starting point of the paper was the aforementioned formula proven by Euler. Riemann realized that only by considering s as a complex variable could he explore the properties of prime number distribution more deeply. Since then, this function has been referred to as the Riemann zeta function. Riemann stated several properties of the ζ function, one of which is the Riemann hypothesis. On the eve of the 20th century, Hadamard and de la Vallee Poussin finally proved Gauss's conjecture, the prime number theorem, using the Riemann zeta function: As the value of n increases, the density of prime numbers decreases proportionally to $\ln n$.

Through computer simulations, researchers have already discovered as many as 10^{13} non-trivial zeros of the Riemann zeta function, and they all invariably lie on the critical line $z = \frac{1}{2} + iy$, as depicted in Figure 5.10. It is reasonable to acknowledge that the Riemann hypothesis is undoubtedly correct! However, mathematicians as a group, cannot be considered reasonable in such matters. The rigorous proof of the Riemann hypothesis remains the Mount Everest challenging human intelligence, awaiting the emergence of another extraordinary genius.

Exercises

[1] Prove that the Gamma function has no zeros in the entire complex plane.

[2] Prove the integrals for $a > 1$:

$$\text{(a)} \int_0^\infty \cos x^a \mathrm{d}x = \frac{1}{a}\Gamma\left(\frac{1}{a}\right)\cos\frac{\pi}{2a};$$

$$\text{(b)} \int_0^\infty \sin x^a \mathrm{d}x = \frac{1}{a}\Gamma\left(\frac{1}{a}\right)\sin\frac{\pi}{2a}$$

Hint: Consider a quarter sector in the first quadrant.

[3] Using the formula

$$\ln n! = \ln 1 + \ln 2 + \ln 3 + \cdots + \ln n = \sum_{k=1}^{n} \ln k$$

and the formula for the integral of a continuous function,

$$\int_1^n \ln x \, \mathrm{d}x = (x\ln x - x)|_1^n = n\ln n - n + 1$$

prove when n is large, the Stirling's approximation formula holds:

$$\ln n! \approx n\ln n - n$$

[4] Us the formula

$$n! = \Gamma(n+1) = \int_0^\infty x^n \mathrm{e}^{-x} \mathrm{d}x = \int_0^\infty \mathrm{e}^{n\ln x - x} \mathrm{d}x$$

to prove when n is large, the Stirling's approximation formula holds:

$$n! \approx \sqrt{2\pi n} n^n \mathrm{e}^{-n}$$

Hint: Substitute $x = n + y \longrightarrow \ln x = \ln n + \ln\left(1 + \frac{y}{n}\right) \approx \ln n + \frac{y}{n} - \frac{y^2}{2n^2}$

[5] Prove the integral representation of the Gamma function known as the Schlafli integral:

$$\Gamma(z) = \frac{1}{\mathrm{e}^{2\pi \mathrm{i} z} - 1} \int_L \mathrm{e}^{-t} t^{z-1} \mathrm{d}t$$

[6] Calculate the integrals:

$$\int_{-1}^{1} (1-x)^p (1+x)^q \mathrm{d}x \quad (\mathrm{Re}\, p > -1, \mathrm{Re}\, q > -1)$$

[7] Proof:

$$\mathrm{B}(p,q) = \int_0^\infty \frac{x^{p-1}}{(1+x)^{p+q}} \mathrm{d}x$$

[8] Calculate the integrals:

$$\text{(a)} \quad \int_0^{\frac{\pi}{2}} \tan^\alpha \theta \mathrm{d}\theta \quad (-1 < \alpha < 1);$$

$$\text{(b)} \quad \int_0^\infty \frac{\sinh^a x}{\cosh^b x} \mathrm{d}x \quad (-1 < a < b)$$

Hint: Make substitution: (a) $x = \tan^2 \theta$; (b) $t = \sinh^2 x$

Answer:

$$\text{(a)} \quad \frac{\pi}{2\cos(\pi\alpha/2)}; \quad \text{(b)} \quad \frac{1}{2}\mathrm{B}\left(\frac{a+1}{2}, \frac{b-a}{2}\right)$$

[9] Find the sum of infinite series:

$$\text{(a)} \quad \sum_{n=0}^{\infty} \frac{1}{(3n+1)(3n+2)(3n+3)};$$

$$\text{(b)} \quad \sum_{n=0}^{\infty} \frac{1}{(n+1)^2(2n+1)^2}$$

Hint:

$$\text{(a)} \sum_{n=0}^{\infty} \frac{1}{(3n+1)(3n+2)(3n+3)}$$

$$= \frac{1}{6}\frac{1}{n+1/3} - \frac{1}{3}\frac{1}{n+2/3} + \frac{1}{6}\frac{1}{n+1}$$

$$= -\frac{1}{6}\left[\psi\left(\frac{1}{3}\right) - 2\psi\left(\frac{2}{3}\right) + \psi(1)\right]$$

$$\text{(b)} \sum_{n=0}^{\infty} \frac{1}{(n+1)^2(2n+1)^2}$$

$$= \left[\frac{4}{n+1} + \frac{1}{(n+1)^2}\right] - \left[\frac{4}{n+1/2} - \frac{1}{(n+1/2)^2}\right]$$

$$= -[4\psi(1) - \psi'(1)] + \left[4\psi\left(\frac{1}{2}\right) + \psi'\left(\frac{1}{2}\right)\right]$$

Answer:

$$\text{(a)}\ \frac{1}{4}\left(\frac{\pi}{\sqrt{3}} - \ln 3\right); \quad \text{(b)}\ \frac{2}{3}\pi^2 - 8\ln 2$$

[10] Prove Euler's prime number formula:

$$\zeta(s) = \frac{1}{1-\frac{1}{2^s}} \cdot \frac{1}{1-\frac{1}{3^s}} \cdot \frac{1}{1-\frac{1}{5^s}} \cdot \cdots \cdot \frac{1}{1-\frac{1}{p^s}} \cdot \cdots \qquad (s \in \mathbb{N}, p \in \mathbb{P})$$

Hint: Use the geometric series formula for each term:

$$\frac{1}{1-\frac{1}{p^s}} = 1 + \frac{1}{p^s} + \frac{1}{p^{2s}} + \frac{1}{p^{3s}} + \cdots$$

5.3 Logarithmic Integrals

5.3.1 *Zero and pole theorems*

Theorem 1. *Let α and β be an n-order zero and an m-order pole of the function $f(z)$ respectively. Then α and β are both first-order poles* *of the logarithmic derivative function* $[\ln f(z)]'$, *and their residues are given by*:

$$\operatorname*{Res}_{z=\alpha}\left[\frac{\mathrm{d}}{\mathrm{d}z}\ln f(z)\right] = \operatorname*{Res}_{z=\alpha}\left[\frac{f'(z)}{f(z)}\right] = n$$

$$\operatorname*{Res}_{z=\beta}\left[\frac{\mathrm{d}}{\mathrm{d}z}\ln f(z)\right] = \operatorname*{Res}_{z=\beta}\left[\frac{f'(z)}{f(z)}\right] = -m$$

Proof. Since $f(z) = (z-\alpha)^n h(z)$, where $h(z)$ is analytic in the neighborhood of α and $h(\alpha) \neq 0$, we have

$$\frac{f'(z)}{f(z)} = \frac{n}{z-\alpha} + \frac{h'(z)}{h(z)}$$

Thus, α is a first-order pole of function $\frac{f'(z)}{f(z)}$, and its residue is

$$\operatorname*{Res}_{z=\alpha}\left[\frac{f'(z)}{f(z)}\right] = n$$

On the other hand, β is an m-order pole of the function $f(z)$, so that

$$f(z) = \frac{g(z)}{(z-\beta)^m}$$

where $g(z)$ is analytic in the neighborhood of β and $g(\beta) \neq 0$. Hence,

$$\frac{f'(z)}{f(z)} = -\frac{m}{z-\beta} + \frac{g'(z)}{g(z)}$$

Therefore, β is also a first-order pole of function $\frac{f'(z)}{f(z)}$, and its residue is

$$\operatorname*{Res}_{z=\beta}\left[\frac{f'(z)}{f(z)}\right] = -m$$

□

Theorem 2. *Let* $f(z)$ *be a meromorphic function that is analytic within the closed contour* C *and non-zero on* C. *Then the integral of its logarithmic derivative is given by*

$$\frac{1}{2\pi \mathrm{i}}\oint_C \frac{f'(z)}{f(z)}\mathrm{d}z = \frac{1}{2\pi \mathrm{i}}\oint_C \mathrm{d}[\ln f(z)] = N(f,C) - P(f,C) \qquad (5.3.1)$$

where $N(f,C)$ *and* $P(fC)$ *represent the number of zeros and poles of* $f(z)$ *inside* C, *respectively, where* n*-order zeros are counted as* n

individual zeros and m-order poles as m individual poles. Generally, if $\varphi(z)$ *is analytic within the closed region bounded by* C*, then*

$$\frac{1}{2\pi \mathrm{i}} \oint_C \varphi(z) \frac{f'(z)}{f(z)} \mathrm{d}z = \sum_k n_k \varphi(\alpha_k) - \sum_j m_j \varphi(\beta_j) \tag{5.3.2}$$

The proof is straightforward and can be completed by the reader.

5.3.2 *Argument principle*

When a closed contour encircles a simple zero or pole, the argument increases or decreases by 2π, as shown in Figure 5.12 for the function.

$$f(z) = \frac{(z^2+1)(z-3-2\mathrm{i})^3}{(z+2+2\mathrm{i})^2}$$

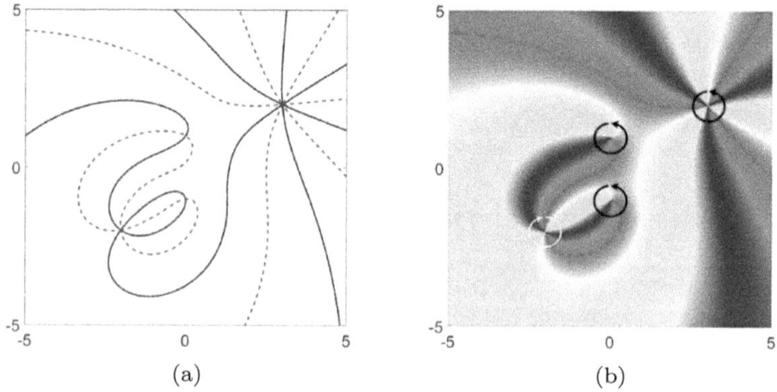

Figure 5.12.

If a meromorphic function $f(z)$ has $N(f, C)$ zeros and $P(f, C)$ poles inside the contour C and is analytic and non-zero on C, then the total change in the argument around the contour is given by

$$\Delta_C \arg[f(z)] = 2\pi[N(f, C) - P(f, C)] \tag{5.3.3}$$

The argument principle is, in fact, a geometric interpretation of the zero and pole counting theorem. From Equation (5.3.1), we have

$$\Delta_C \arg[f(z)] = -\mathrm{i} \oint_C \frac{f'(z)}{f(z)} \mathrm{d}z$$

For example, in Figure 5.12(b), the total change in argument is

$$\Delta_C \arg[f(z)] = 2\pi[(2+3) - 2] = 6\pi$$

Rouché's theorem. *If functions $f(z)$ and $g(z)$ are analytic inside and on a contour C and $|f(z)| > |g(z)|$ on C, then $f(z) + g(z)$ and $f(z)$ have the same number of zeros inside C,*

$$N(f+g, C) = N(f, C)$$

Proof. As shown in Figure 5.13(a), since on the contour C we have

$$|f(z) + g(z)| \geq |f(z)| - |g(z)| > 0$$

Both $f(z) + g(z)$ and $f(z)$ are analytic inside C and have no zeros on C. Rewriting

$$f(z) + g(z) = f(z)\left[1 + \frac{g(z)}{f(z)}\right]$$

the change in the argument is given by

$$\Delta_C \arg[f(z) + g(z)] = \Delta_C \arg[f(z)] + \Delta_C \arg\left[1 + \frac{g(z)}{f(z)}\right]$$

Considering the function $\zeta(z) = 1 + \frac{g(z)}{f(z)}$, since on the contour C we have

$$\left|\frac{g(z)}{f(z)}\right| = |\zeta(z) - 1| < 1$$

The mapping of the contour C to the ζ plane, denoted as $C \longmapsto \Gamma$, always lies inside the unit circle $|\zeta(z) - 1| = 1$, as shown in Figure 5.13(b). Therefore, the change in argument around Γ, $\Delta_\Gamma \arg[\zeta(z)] = 0$, i.e.,

$$\Delta_C \left[1 + \frac{g(z)}{f(z)}\right] = 0$$

Hence,

$$\Delta_C[f(z) + g(z)] = \Delta_C[f(z)]$$

According to the argument principle, we get

$$N(f+g, C) = N(f, C) \qquad \square$$

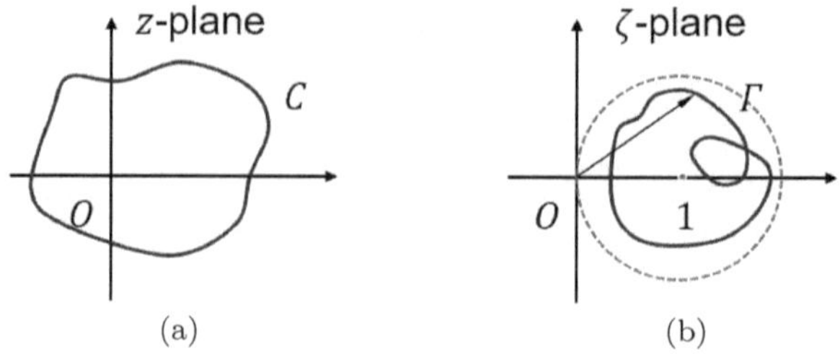

Figure 5.13.

Reflection: If $|f(z)| > |g(z)|$ on contour C, do $f(z)+g(z)$ and $f(z)$ have the same number of poles?

Example 5.4. Proof of the fundamental theorem of algebra: Every nth-degree polynomial equation has exactly n roots.

Proof. Consider an nth-degree polynomial:

$$P_n(z) = a_0 z^n + a_1 z^{n-1} + \cdots + a_{n-1} z + a_n \quad (a_0 \neq 0)$$

Let us define $f(z) = a_0 z^n$ and consider the function:

$$\varphi(z) \equiv P_n(z) - f(z) = a_1 z^{n-1} + \cdots + a_{n-1} z + a_n$$

Since

$$\lim_{z\to\infty} \frac{|\varphi(z)|}{|f(z)|} = 0$$

we have $|f(z)| > |\varphi(z)|$ on the contour of the circle with radius R. According to Rouché's theorem, $P_n(z)$ and $f(z)$ should have the same number of zeros inside the contour C. Besides, the equation $f(z) = a_0 z^n = 0$ has n-fold root $z = 0$ inside the circle $|z| < R$. Therefore, the polynomial equation $P_n(z) = 0$ has exactly n roots inside the circle $|z| < R$. □

Example 5.5. Consider an nth-degree polynomial:

$$P_n(z) = a_0 z^n + a_1 z^{n-1} + \cdots + a_{n-1} z + a_n \quad (a_0 \neq 0)$$

that satisfies the condition $|a_k| > |a_0| + \cdots + |a_{k-1}| + |a_{k+1}| + \cdots + |a_n|$. We want to prove that $P_n(z) = 0$ has $n - k$ roots inside the unit circle $|z| < 1$.

Proof. Let

$$f(z) = a_k z^{n-k}, \quad g(z) = P_n(z) - a_k z^{n-k}$$

It can be easily shown that $|f(z)| > |g(z)|$ on the unit circle $|z| = 1$. According to Rouché's theorem, we know that $P_n(z)$ and $f(z)$ have the same number of zeros inside the unit circle. Therefore, the polynomial equation $P_n(z) = 0$ has $n - k$ roots inside the circle $|z| < 1$. □

Exercises

[1] Prove Theorem 2 in this section.

[2] Prove that all the roots of the equation $z^7 - z^3 + 12 = 0$ lie within the annular region $1 < |z| < 2$.

[3] Determine the number of roots of the following equations inside the unit circle $|z| < 1$:

$$\text{(a) } 2z^5 - z^3 + z^2 - 2z + 8 = 0; \quad \text{(b) } z^4 - 5z + 1 = 0$$

Answer: (a) 0; (b) 1

[4] How many roots does the equation $z^9 + z^5 - 8z^3 + 2z + 1 = 0$ have within the annular region $1 < |z| < 2$?
Answer: 6

[5] How many zeros does the polynomial $P(z) = z^6 + 4z^4 + z^3 + 2z^2 + z + 5$ have in the first quadrant of the complex plane?
Hint: Study the change in argument of the mapped contour in the ζ-plane under the function $\zeta = P(z)$ for a sector contour in the first quadrant of the z-plane.
Answer: 2

5.4 Decomposition of Meromorphic Functions

5.4.1 *Pole decomposition*

Let a_n $(n = 1, 2, 3, \ldots)$ be all the poles of the meromorphic function $f(z)$ on the complex plane, with p_n the order of the poles. First, we expand $f(z)$ in a Laurent series in the punctured neighborhood of

a_1,

$$f(z) = f_1(z) + \sum_{\nu=1}^{p_1} \frac{c_{-\nu}^{(1)}}{(z-a_1)^\nu}$$

Here, $c_{-\nu}^{(1)}$ denotes the expansion coefficient of each negative power term of the pole a_1, and p_1 is the order of the pole. The positive part of the Laurent series, $f_1(z)$, is analytic at the point a_1, but $a_2, a_3, a_4, \ldots$, remain its poles. Therefore, we continue by expanding $f_1(z)$ in a Laurent series around a_2 in a punctured neighborhood,

$$f_1(z) = f_2(z) + \sum_{\nu=1}^{p_2} \frac{c_{-\nu}^{(2)}}{(z-a_2)^\nu}$$

At this point, $a_3a_4, \ldots$ still remain as poles of $f_2(z)$. We continue this process, and eventually separate out all the poles of $f(z)$:

$$f(z) = f_N(z) + \sum_{n=1}^{N} g_n(z)$$

Here, $f_N(z)$ no longer contains any poles in the entire plane, and the principal part (negative power part) of the Laurent series at all the poles is given by

$$g_n(z) = \sum_{\nu=1}^{p_n} \frac{c_{-\nu}^{(n)}}{(z-a_n)^\nu}$$

If $f(z)$ has only a finite number of poles on the complex plane, $h(z) \equiv f_N(z)$ will be an entire function. Thus, the meromorphic function $f(z)$ can be decomposed into the sum of an entire function and the principal parts obtained by Laurent series expansion at all its poles:

$$f(z) = h(z) + g(z)$$

$$g(z) = \sum_{n=1}^{N} g_n(z)$$

According to Liouville's theorem, $h(z) = c$, where c is a constant. Therefore, the meromorphic function can be decomposed into the ratio of two polynomial functions.

If $f(z)$ has an infinite number of poles on the complex plane, the point at infinity must be a non-isolated singularity, and the sum of the principal parts,

$$g(z) = \sum_{n=1}^{\infty} g_n(z)$$

forms an infinite series, giving rise to convergence issues in this case.

5.4.2 *Mittag–Leffler expansion*

In general, the infinite series formed by the principal parts is divergent for all z. To obtain an effective partial fraction representation, some modifications need to be introduced to eliminate the divergence of the series. Assuming that $z = z_0$ is not a pole of $f(z)$, it can be proven that these modifications can be chosen as a truncation of the Taylor series expansion of the principal part at z_0.

Mittag–Leffler theorem: For any sequence of points $a_n \in \mathbb{C}$, with $\lim_{z\to\infty} a_n = \infty$, and a sequence of functions $g_n(z)$, there exists a meromorphic function $f(z)$ whose poles are at $z = a_n$, and each pole has a principal part given by $g_n(z)$.

The proof of this theorem can be found in the literatures.[1]

Corollary: Any meromorphic function $f(z)$ can be expanded as a series:

$$f(z) = h(z) + \sum_{n=1}^{\infty} [g_n(z) - P_n(z)] \tag{5.4.1}$$

This series converges uniformly on any compact set, where $h(z)$ is an entire function. $g_n(z)$ is the principal part of $f(z)$, and $P_n(z)$ is a truncated polynomial obtained by expanding $g_n(z)$ in a Taylor series at $z = 0$:

$$P_n(z) = \sum_{k=0}^{m_n} \frac{g_n^{(k)}(0)}{k!} z^k \tag{5.4.2}$$

The choice of m_n ensures that

$$|g_n(z) - P_n(z)| < \frac{1}{2^n}$$

[1]Zhuxi Wang and Dunren Guo, *Introduction to Special Functions.*

This meromorphic function decomposition in terms of poles is known as the Mittag–Leffler expansion.

Example 5.6. Take the Mittag–Leffler expansion of the meromorphic function:

$$f(z) = \frac{1}{\sin^2 z}$$

Solution: The function has second-order poles at $z_n = n\pi$ $(n \in \mathbb{Z})$, and their principal parts are given by

$$g_n(z) = \frac{1}{(z - n\pi)^2}$$

The series formed by the principal parts is

$$\sum_{n=-\infty}^{\infty} \frac{1}{(z - n\pi)^2} \quad (n \in \mathbb{Z})$$

This infinite series uniformly converges on any closed set where $z \neq n\pi$, there is no need for a modification polynomial $P_n(z)$. The entire function part is

$$h(z) = \frac{1}{\sin^2 z} - \frac{1}{z^2} - \sum_{n \neq 0} \frac{1}{(z - n\pi)^2}$$

It is a periodic function with a period of π along the real axis. Considering the strip region $\{0 \leq Rez \leq \pi\}$, the infinite series on the right-hand side is convergent. The first two terms tend to zero as $\operatorname{Im} z \to \infty$ and remain finite as $z \to 0$. Therefore, the entire function $h(z)$ is bounded in the strip region. Moreover, since $h(z)$ is a periodic function along the real axis, it is bounded on the entire complex plane. By Liouville's theorem, $h(z)$ must be constant. Taking $z = 0$ or $\operatorname{Im} z \to \infty$, we can determine that $h(z) \equiv 0$. Hence, the Mittag–Leffler expansion is

$$\frac{1}{\sin^2 z} = \sum_{-\infty}^{\infty} \frac{1}{(z - n\pi)^2} \tag{5.4.3}$$

Example 5.7. Expand the meromorphic function using Mittag–Leffler expansion:

$$f(z) = \cot z$$

Solution: The function has simple poles at $z_n = n\pi$ ($n \in \mathbb{Z}$), and the series formed by the principal parts is

$$g(z) = \sum_{n=-\infty}^{\infty} \frac{1}{z - n\pi}$$

This series is divergent. We can take the zeroth-order term of the Taylor series expansion of $g(z)$ at $z = 0$ as $P_n(z)$, which is given by

$$\sum_{n\neq 0} \left(\frac{1}{z - n\pi} + \frac{1}{n\pi} \right) = \sum_{n\neq 0} \frac{z}{(z - n\pi)n\pi}$$

It is evident that this infinite summation uniformly converges on any closed set. Using the same reasoning as in Example 5.6, we can prove that the entire function $h(z) \equiv 0$. Therefore,

$$\cot z = \frac{1}{z} + \sum_{n\neq 0} \left(\frac{1}{z - n\pi} + \frac{1}{n\pi} \right) = \frac{1}{z} + \sum_{n=1}^{\infty} \frac{2z}{z^2 - n^2\pi^2} \tag{5.4.4}$$

Discussion: In this case, the divergent cancellation term is actually zero because $\sum_{n\neq 0} \frac{1}{n\pi} = 0$. If taking the zeroth-order term of Taylor series expansion is not sufficient to eliminate divergence, higher-order expansion terms need to be considered as modifications.

Practice: Prove the following relation:

$$\coth z = \frac{1}{z} + \sum_{n=1}^{\infty} \frac{2z}{z^2 + n^2\pi^2}$$

ENDNOTE

The basic idea of eliminating the divergence of a series can be understood from the divergence of the series at $z = 0$. Let us write

$$g(0) = \sum_{n=1}^{\infty} g_n(0) = \sum_{n=1}^{\infty} \sum_{\nu=1}^{p_n} \frac{c_{-\nu}^{(n)}}{(-a_n)^\nu} \quad (\nu = 1, 2, \ldots, p_n)$$

The reason for the divergence of the series $g(0)$ is that as $n \to \infty$, the terms of the series do not decay fast enough. To eliminate the divergence of the zeroth power in each term of the series, we subtract the constant $\sum_{\nu=1}^{p_n} \frac{c_{-\nu}^{(n)}}{(-a_n)^\nu}$ from $g(z)$. If the resulting series after subtraction still diverges at $g'(0)$, then we need to subtract the divergence of the first power of z as well,

$$\sum_{n=1}^{\infty}\left\{\left[\sum_{\nu=1}^{p_n}\frac{c_{-\nu}^{(n)}}{(z-a_n)^\nu}-\sum_{\nu=1}^{p_n}\frac{c_{-\nu}^{(n)}}{(-a_n)^\nu}\right]-\sum_{\nu=1}^{p_n}\frac{-\nu c_{-\nu}^{(n)}}{(-a_n)^{\nu+1}}z\right\}$$

This process continues by progressively subtracting the terms that cause divergence, which is the Taylor polynomial,

$$P_n(z)=\sum_{k=0}^{m_n}\frac{g_n^{(k)}(0)}{k!}z^k$$

until all the divergences of the derivative $g^{(n)}(0)$ are eliminated, thus ensuring the finiteness of $g(z)$ at $z=0$.

If $z=0$ is also a pole of the meromorphic function, we set aside the negative powers of the Laurent series expansion at that point and perform a Taylor series expansion correction for the divergences caused by other poles at $z=0$.

According to Section 3.5, a plane scalar field with singularity can be described using a complex potential with singularity. The field generated by a dipole corresponds to a simple pole, while the field generated by a quadrupole corresponds to a pole of order two, and so on. Thus, a meromorphic function describes the complex potential generated by the distribution of multiple poles in the plane. From a physical perspective, the Coulomb potential generated by a dipole is inversely proportional to distance and belongs to a long-range potential. The total potential is the superposition of the potentials produced by different point charges. If there are infinitely many dipoles in the plane, the potential at any point will be infinitely large. In physics, dealing with such potentials often involves subtracting this infinite background. In the Mittag–Leffler expansion, the infinite contribution caused by the superposition of the negative first power terms from the Laurent series expansion should also be subtracted, namely the residues of all the poles. Since divergence occurs at every point in the plane, resulting in a uniform infinite background, we can choose any point for the subtraction. As for poles of order two or higher, if there is no negative first power term, they belong to short-range potentials, and the superposition does not lead to infinity. Therefore, no subtraction correction is necessary, which is the difference between Example 5.5 and Example 5.6. However, if the Laurent series expansion of higher-order poles contains a negative first power term, subtraction correction is also required.

Consider two non-collinear complex numbers ω_1 and ω_2 on the complex plane. We can construct the principal part of a meromorphic function with second-order poles at the periodic lattice points $z_{m,n} = m\omega_1 + n\omega_2$ $(m, n \in Z)$ by using the series:

$$\sum_{m,n=-\infty}^{\infty} \frac{1}{(z - m\omega_1 - n\omega_2)^2}$$

This infinite series does not converge, so we need to subtract the divergent part at $z = 0$ to obtain a meromorphic function. The result is

$$\mathcal{P}(z) = \frac{1}{z^2} + \sum_{m,n\neq 0} \left[\frac{1}{(z - m\omega_1 - n\omega_2)^2} - \frac{1}{(m\omega_1 + n\omega_2)^2} \right]$$

This function is known as the Weierstrass P-function, which exhibits double periodicity: $\mathcal{P}(z + m\omega_1 + n\omega_2) = \mathcal{P}(z)$. Expanding the $\mathcal{P}$-function as a Laurent series at $z = 0$, we can observe that all odd power terms are zero. By comparing coefficients, it can be verified that $\mathcal{P}(z)$ satisfies the differential equation:

$$[\mathcal{P}'(z)]^2 = 4[\mathcal{P}(z)]^3 - g_2\mathcal{P}(z) - g_3$$

Here, g_2 and g_3 are constants known as the modular invariants:

$$g_2 = 60 \sum_{m,n\neq 0} \frac{1}{(m\omega_1 + n\omega_2)^4}, \quad g_3 = 140 \sum_{m,n\neq 0} \frac{1}{(m\omega_1 + n\omega_2)^6}$$

Therefore, the $\mathcal{P}$-function exhibits characteristics of elliptic integrals:

$$\int_0^z \frac{1}{\sqrt{4u^3 - g_2 u - g_3}} \mathrm{d}u$$

Its domain forms a torus-shaped surface.

Exercises

[1] Prove the following:

$$\text{(a)}\ \tan z = \sum_{n=1}^{\infty} \frac{8z}{(2n-1)^2\pi^2 - 4z^2}$$

$$\text{(b)}\ \frac{1}{\sin z} = \frac{1}{z} + \sum_{n=1}^{\infty} (-1)^n \frac{2z}{z^2 - n^2\pi^2}$$

Hint: Use the formulas

$$\text{(a)}\ \tan z = \cot z - 2\cot 2z; \quad \text{(b)}\ \frac{1}{\sin z} = \frac{1}{2}\left(\cot\frac{z}{2} + \tan\frac{z}{2}\right)$$

[2] Proof:

$$\text{(a)}\ \frac{1}{\mathrm{e}^z - 1} = -\frac{1}{2} + \frac{1}{z} + \sum_{n=1}^{\infty} \frac{2z}{z^2 + 4n^2\pi^2}$$

$$\text{(b)}\ \frac{1}{\mathrm{e}^z + 1} = \frac{1}{2} - \sum_{n=0}^{\infty} \frac{2z}{z^2 + (2n+1)^2\pi^2}$$

[3] Expand the meromorphic functions as Mittag–Leffler series:

$$f(z) = \frac{1}{\sin z^2}$$

Hint: Consider the first-order correction term in the Taylor expansion.

[4] Find the poles and principal parts of the following series and express them in terms of trigonometric functions:

$$\sum_{n=-\infty}^{\infty} \frac{1}{z^3 - n^3}$$

Answer: $\frac{\pi}{3z^2}[\cot(\pi z) + \omega \cot(\lambda \pi z) + \omega^2 \cot(\lambda^2 \pi z)], \omega = \mathrm{e}^{\frac{2\pi \mathrm{i}}{3}}$

[5] Study the logarithmic derivative of the Γ function:

$$\psi(z) \equiv \frac{\mathrm{d}}{\mathrm{d}z}[\ln \Gamma(z)]$$

(a) Prove that $z = -n(n = 0, 1, 2, \ldots)$ are first-order poles of $\psi(z)$, and the residue at each pole is

$$\mathrm{Res}[\psi(-n)] = -1$$

(b) Expand $\psi(z)$ using the poles at $z = -n$ $(n = 0, 1, 2, \ldots)$ as Mittag–Leffler series and eliminate the divergence of the series:

$$\psi(z) = h(z) - \frac{1}{z} + \sum_{n=1}^{\infty} \left(\frac{1}{n} - \frac{1}{z+n} \right)$$

Prove that the entire function $h(z)$ is a constant: $h(z) \equiv -\gamma = \psi(1)$.

Hint: Consider a closed contour C_n with the origin as the center and $n < R_n < n+1$ as the radius. Evaluate the contour integral $\oint_{C_n} h(z)\mathrm{d}z = 0$. Let $R_n \to \infty$ to show that $h(\infty)$ must be finite.

(c) Prove the Euler's constant

$$\gamma = \lim_{n\to\infty} \left(1 + \frac{1}{2} + \frac{1}{3} + \cdots + \frac{1}{n} - \ln n \right)$$

Hint: Integrate both sides of the expansion and use the identity: $\Gamma(z+1) = z\Gamma(z)$.

(d) Prove the inverse Γ function can be represented as

$$\frac{1}{\Gamma(z)} = z\mathrm{e}^{\gamma z} \prod_{n=1}^{\infty} \left(1 + \frac{z}{n} \right) \mathrm{e}^{-\frac{z}{n}}$$

(e) Proof:

$$\psi(z) = -\gamma - \frac{1}{z} - \sum_{k=1}^{\infty} (-1)^k \zeta(k+1) z^k$$

$$\zeta(k) = \sum_{n \in N} \frac{1}{n^k} = \frac{(-1)^k}{k!} \frac{\mathrm{d}^k}{\mathrm{d}z^k} [z\psi(z)]|_{z=0}$$

5.5 Decomposition of Entire Functions

5.5.1 *Factorization of entire functions*

Any polynomial of degree n, denoted as $P_n(z)$, has n zeros α_k ($k = 1, 2, \ldots, n$). The polynomial can be expressed in factored form as

$$P_n(z) = A' \prod_{k=1}^{n} (z - \alpha_k) = A \prod_{k=1}^{n} \left(1 - \frac{z}{\alpha_k}\right)$$

Similar properties apply to entire functions, which can have either no zeros or infinitely many zeros. It can be easily shown that an entire function with no zeros can be represented as $f(z) = \mathrm{e}^{g(z)}$, where $g(z)$ is also an entire function. If an entire function has n zeros α_k ($k = 1, 2, \ldots n$), it can be formally expressed as

$$f(z) = A\mathrm{e}^{g(z)} \prod_{k=1}^{n} \left(1 - \frac{z}{\alpha_k}\right)$$

However, when an entire function has infinitely many countable zeros, it involves infinite products, and the issue of convergence arises.

5.5.2 *Convergence of infinite products*

For an infinite product of complex terms:

$$\prod_{n=1}^{\infty} q_n = \prod_{n=1}^{\infty} (1 + c_n) \tag{5.5.1}$$

If the limit Π of its partial products is a non-zero finite value, i.e.,

$$\Pi_N = \prod_{n=1}^{N} (1 + c_n) \overset{N \to \infty}{\longrightarrow} \Pi$$

it is said to be convergent and denoted as $\Pi = \lim_{n\to\infty} \Pi_n$. Since $1 + c_n = \frac{\Pi_n}{\Pi_{n-1}}$, it follows that $c_n \to 0$ as $n \to \infty$ is a necessary condition for the convergence of an infinite product. The following convergence criteria exist:

(1) A necessary and sufficient condition for the convergence of an infinite product is that the series $\sum_{n=1}^{\infty} \ln q_n$ converges.
(2) A necessary and sufficient condition for the convergence of an infinite product is that the series $\sum_{n=1}^{\infty} c_n$ converges.

The infinite product can be rewritten as

$$\prod_{n=1}^{\infty} q_n = \prod_{n=1}^{N} q_n \exp\left[\sum_{n=N+1}^{\infty} \mathrm{Ln}(1 + c_n)\right]$$

where Ln denotes the principal value of the logarithmic function. If the corresponding series of logarithms is absolutely convergent, the infinite product is said to be absolutely convergent, and the criterion for this is stated as:

An infinite product $\prod_{n=1}^{\infty} q_n$ is absolutely convergent if and only if the infinite series $\sum_{n=1}^{\infty} |c_n|$ is absolutely convergent.

Proof. For a given positive integer N, it can be shown that $|c_n| < \frac{1}{2}$ for $n > N$. Therefore, there exists a positive real number $\alpha > 0$ such that

$$|Ln(1 + c_n)| \leq \alpha |c_n|$$

Since the infinite series $\sum_{n=1}^{\infty} |c_n|$ is absolutely convergent, the series of logarithms $\sum_{n=1}^{\infty} \mathrm{Ln}(1+c_n)$ is absolutely convergent, implying the absolute convergence of the infinite product. □

Example 5.8. Studying the convergence of infinite products:

$$\text{(a)} \prod_{n=1}^{\infty}\left[1 + (-1)^{n+1}\frac{1}{n}\right]; \quad \text{(b)} \prod_{n=1}^{\infty}\left[1 + \frac{\mathrm{i}}{n}\right]$$

Solution: (a) Since the harmonic series $\sum_{n=1}^{\infty} \frac{1}{n} = \infty$ diverges, according to the convergence criteria, $\prod_{n=1}^{\infty}\left[1 + \frac{(-1)^{n+1}}{n}\right]$ does not converge absolutely. On the other hand, the series $\sum_{k=1}^{\infty}(-1)^{n+1}\frac{1}{n}$

converges, so the infinite product converges but not absolutely. In fact,

$$\prod_{n=1}^{\infty}\left[1+(-1)^{n+1}\frac{1}{n}\right] = (1+1)\left(1-\frac{1}{2}\right)\left(1+\frac{1}{3}\right)\left(1-\frac{1}{4}\right)\cdots$$
$$= 2\cdot\frac{1}{2}\cdot\frac{4}{3}\cdot\frac{3}{4}\cdot\frac{6}{5}\cdot\frac{5}{6}\cdot\cdots\to 1$$

(b) Since $\sum_{k=1}^{\infty}\operatorname{Ln}\left[1+\frac{\mathrm{i}}{n}\right] = \sum_{k=1}^{\infty}\left[\frac{\mathrm{i}}{n}+\mathcal{O}\left(\frac{1}{n^2}\right)\right]$ does not converge, $\prod_{n=1}^{\infty}\left[1+\frac{\mathrm{i}}{n}\right]$ does not converge. Moreover, because

$$0 < \operatorname{Ln}\left|1+\frac{\mathrm{i}}{n}\right| < \frac{1}{2}\ln\left(1+\frac{1}{n^2}\right) < \frac{1}{n^2}$$

it follows that $\prod_{n=1}^{\infty}\left|1+\frac{\mathrm{i}}{n}\right|$ converges. Therefore, the infinite product does not converge, but the product of its moduli converges.

Conclusion: The absolute convergence of an infinite product is not equivalent to the convergence of the product of its moduli.

5.5.3 *Weierstrass factorization theorem*

For any sequence of points $\alpha_n \in \mathbb{C}$ satisfying $\lim_{n\to\infty}\alpha_n = \infty$, there exists an entire function $f(z)$ whose zeros are located at $z = \alpha_n$, and the order of each zero is equal to the number of times α_n is repeated in the sequence.

Corollary. *Any entire function can be decomposed into an infinite product of its zeros:*

$$f(z) = z^m \mathrm{e}^{g(z)}\prod_{n=1}^{\infty}\left(1-\frac{z}{\alpha_n}\right)\mathrm{e}^{\frac{z}{\alpha_n}+\frac{1}{2}\left(\frac{z}{\alpha_n}\right)^2+\cdots+\frac{1}{m_n}\left(\frac{z}{\alpha_n}\right)^{p_n}} \tag{5.5.2}$$

Here, $m \geq 0$ is the order of the zero of $f(z)$ at $z = 0, g(z)$ is an entire function, and p_n is chosen such that the series $\sum_{n=1}^{\infty}\left(\frac{z}{\alpha_n}\right)^{p_n+1}$ converges absolutely and uniformly on any compact set.

The factorization of an entire function $f(z)$ is closely related to the Mittag–Leffler expansion of meromorphic functions. This can be understood as follows: take the logarithmic derivative $h(z) =$

$[\ln f(z)]'$ of the entire function, where the zeros of $f(z)$ are transformed into simple poles of $h(z)$, and then apply the Mittag–Leffler expansion to it.

Example 5.9. Proof:

$$\sin z = z\prod_{n=1}^{\infty}\left(1-\frac{z^2}{n^2\pi^2}\right)$$

Proof. Using the Mittag–Leffler expansion (5.4.4), we have

$$\frac{\mathrm{d}}{\mathrm{d}z}\ln(\sin z) = \cot z = \frac{1}{z} + \sum_{n=1}^{\infty}\frac{2z}{z^2-n^2\pi^2}$$

Integrating both sides gives

$$\sin z = z\prod_{n=1}^{\infty}\left(1-\frac{z^2}{n^2\pi^2}\right)$$

Example 5.10. Constructing an entire function with all its zeros at negative integer values $z_n = -n (n \in N)$.
Solution: In this case, the infinite product is given by

$$\prod_{n=1}^{\infty}\left(1+\frac{z}{n}\right)$$

This infinite product does not converge and needs to be regularized. Since

$$\left|\mathrm{Ln}\left(1+\frac{z}{n}\right)-\frac{z}{n}\right| \leq C\frac{|z|^2}{n^2} \quad (|z| \leq R, n \geq 2R)$$

where the right side represents the convergent series $\sum_{n=1}^{\infty}\frac{|z|^2}{n^2}$, we can define the desired entire function as

$$f(z) = \prod_{n=1}^{\infty}\left(1+\frac{z}{n}\right)\mathrm{e}^{-\frac{z}{n}}$$

which converges absolutely in the entire plane and has the specified zeros.

ENDNOTE

According to the fundamental theorem of algebra, an n-degree polynomial (which is also an entire function) has exactly n roots. Multiple roots are treated as distinct single roots, and the polynomial can be uniquely expressed as a factorization product:

$$P_n(z) = (z - \alpha_1)(z - \alpha_2) \cdots (z - \alpha_n)$$

This is also known as the factorization expansion of the polynomial with respect to its zeros.

An entire function has no singularities in the complex plane but may have zeros, $z = \alpha_n$ $(n = 1, 2, \ldots)$, roots of the function equation. The Weierstrass theorem states that an entire function can also be expressed in the factorization form of its zeros:

$$f(z) = \mathrm{e}^{\tilde{g}(z)} \prod_{n=1}^{N} (z - \alpha_n) = \mathrm{e}^{g(z)} z^m \prod_{n=1}^{N} \left(1 - \frac{z}{\alpha_n}\right)$$

where $g(z)$ is also an entire function. It can be observed that the zeros reflect the main characteristics of the entire function. Consider the case of an infinite number of zeros in the complex plane, we can take the logarithm of both sides of the equation:

$$\ln f(z) = g(z) + \sum_{n=1}^{\infty} \ln \left(1 - \frac{z}{\alpha_n}\right)$$

Here, $\ln f(z)$ is still an entire function. As discussed in Section 3.5, the term $\ln(z - \alpha_n)$ represents a complex potential generated by a point source. Based on the knowledge of electrostatics, $\ln(z - \alpha_n)$ corresponds to a two-dimensional long-range Coulomb potential, and the superposition of an infinite number of electric charges will result in divergence. Therefore, this divergence must be subtracted from the above infinite series.

The subtraction is carried out using the Mittag–Leffler scheme, which involves expanding each term of the series

$$\sum_{n=1}^{\infty} \ln\left(1 - \frac{z}{\alpha_n}\right)$$

as a Taylor series around $z = 0$ and subtracting a truncated Taylor polynomial,

$$P_n(z) = -\sum_{k=1}^{m_n} \frac{1}{k}\left(\frac{z}{\alpha_n}\right)^k$$

Hence,

$$\ln f(z) = g(z) + \sum_{n=1}^{\infty}\left[\ln\left(1 - \frac{z}{\alpha_n}\right) + \sum_{k=1}^{m_n} \frac{1}{k}\left(\frac{z}{\alpha_n}\right)^k\right]$$

If $z = 0$ is an mth order zero of $f(z)$, it can be expressed in the Weierstrass factorization form as

$$f(z) = z^m e^{g(z)} \prod_{n=1}^{\infty}\left(1 - \frac{z}{\alpha_n}\right) e^{\frac{z}{\alpha_n} + \frac{1}{2}\left(\frac{z}{\alpha_n}\right)^2 + \cdots + \frac{1}{m_n}\left(\frac{z}{\alpha_n}\right)^{m_n}}$$

Exercises

[1] Determine the convergence of the following infinite products and find their limits:

$$\text{(a)} \prod_{n=2}^{\infty}\left[1 - \frac{1}{n}\right]; \quad \text{(b)} \prod_{n=2}^{\infty}\left[1 - \frac{1}{n^2}\right];$$

$$\text{(c)} \prod_{n=2}^{\infty}\left[1 - \frac{2}{n(n+1)}\right]; \quad \text{(d)} \prod_{n=2}^{\infty}\left[1 - \frac{2}{n^3+1}\right]$$

Answer: (a) Not absolutely convergent; (b) $\frac{1}{2}$; (c) $\frac{1}{3}$; (d) $\frac{2}{3}$

[2] Express the hyperbolic functions as infinite products:

$$\text{(a)}\ \sinh z = z\prod_{n=1}^{\infty}\left(1+\frac{z^2}{n^2\pi^2}\right)$$

$$\text{(b)}\ \cosh z = z\prod_{n=1}^{\infty}\left[1+\frac{4z^2}{(2n-1)^2\pi^2}\right]$$

[3] Proof:

$$\cos z = \prod_{n=1}^{\infty}\left[1-\frac{4z^2}{(2n-1)^2\pi^2}\right]$$

Hint: Use the identity $\cos z = \frac{\sin 2z}{2\sin z}$ and factor the numerator into two infinite products based on parity.

[4] Proof:

$$\text{(a)}\ \prod_{n=2}^{\infty}\left(1-\frac{1}{n^2}\right)=\frac{1}{2};\quad \text{(b)}\ \prod_{n=2}^{\infty}\left(1+\frac{1}{n^2}\right)=\frac{\mathrm{e}^{\pi}-\mathrm{e}^{-\pi}}{2\pi}$$

[5] Proof of Wallis's formula:

$$\frac{2\cdot 2}{1\cdot 3}\cdot\frac{4\cdot 4}{3\cdot 5}\cdot\cdots\cdot\frac{2n\cdot 2n}{(2n-1)\cdot(2n+1)}\cdots=\frac{\pi}{2}$$

[6] Construct an entire function with all its zeros as single zeros on the real axis, given by $z_n = \pm\sqrt[4]{n}\ (n\in\mathbb{N})$.

Answer:

$$f(z)=\prod_{n=1}^{\infty}\left(1-\frac{z^2}{\sqrt{n}}\right)\mathrm{e}^{\frac{z^2}{\sqrt{n}}}$$

Chapter 6

Conformal Mappings

6.1 Conformal Transformation

6.1.1 *Invariance of harmonic equation*

The analytic function $\zeta(z) = \xi(x, y) + i\eta(x, y)$ defines a mapping that corresponds to a variable substitution $(x, y) \longmapsto (\xi, \eta)$, which transforms the two-dimensional Laplace's equation in the z-plane,

$$\frac{\partial^2 u}{\partial x^2} + \frac{\partial^2 u}{\partial y^2} = 0 \tag{6.1.1}$$

into the corresponding equation in the ζ-plane:

$$(\xi_x^2 + \xi_y^2)u_{\xi\xi} + 2(\xi_x\eta_x + \xi_y\eta_y)u_{\xi\eta} + (\eta_x^2 + \eta_y^2)u_{\eta\eta}$$
$$+ (\xi_{xx} + \xi_{yy})u_\xi + (\eta_{xx} + \eta_{yy})u_\eta = 0$$

Since the real and imaginary parts of $\zeta(z)$ are both harmonic functions and satisfy the Cauchy–Riemann conditions,

$$\frac{\partial \xi}{\partial x} = \frac{\partial \eta}{\partial y}, \quad \frac{\partial \xi}{\partial y} = -\frac{\partial \eta}{\partial x}$$

we have

$$|\zeta'(z)|^2(u_{\xi\xi} + u_{\eta\eta}) = 0$$

This equation implies that if $\zeta(z)$ is an analytic function, then the Laplace's equation remains invariant under its mapping. In other words, a harmonic function u in a certain region of the z-plane is mapped to the corresponding harmonic function in the ζ-plane:

$$u_{\xi\xi} + u_{\eta\eta} = 0 \tag{6.1.2}$$

The transformation also applies to the two-dimensional Poisson's equation:

$$(u_{xx} + u_{yy}) = \rho(x, y)$$

which transforms into

$$u_{\xi\xi} + u_{\eta\eta} = \rho'(\xi, \eta)$$

where

$$\rho'(\xi, \eta) \equiv \frac{1}{|\zeta'(z)|^2}\rho(x(\xi, \eta), y(\xi, \eta)) \tag{6.1.3}$$

Here, $\rho'(\xi, \eta)$ represents the modified distribution of the external source $\rho(x, y)$ after the transformation. Despite this change, the total charge of the external source remains invariant, as

$$|\zeta'(z)|^2 = \left(\frac{\partial \xi}{\partial x}\right)^2 + \left(\frac{\partial \xi}{\partial y}\right)^2 = \begin{vmatrix} \frac{\partial \xi}{\partial x} & \frac{\partial \xi}{\partial y} \\ \frac{\partial \eta}{\partial x} & \frac{\partial \eta}{\partial y} \end{vmatrix} \equiv \det J$$

where $\det J$ is the Jacobian determinant of the coordinate transformation. Hence,

$$Q = \int_{D'} \rho'(\xi, \eta) \mathrm{d}\xi \mathrm{d}\eta = \int_D \rho'(x, y) \det J \mathrm{d}x \mathrm{d}y \equiv \int_D \rho(x, y) \mathrm{d}x \mathrm{d}y$$

The density distribution of a point source can be represented using the Dirac δ function:

$$\rho(x, y) = q\delta(x - x_0)\delta(y - y_0)$$

Therefore, the Poisson's equation

$$\frac{\partial^2 u}{\partial x^2} + \frac{\partial^2 u}{\partial y^2} = q\delta(x - x_0)\delta(y - y_0)$$

remains in the same form after the transformation:

$$\frac{\partial^2 u}{\partial \xi^2} + \frac{\partial^2 u}{\partial \eta^2} = q\delta(\xi - \xi_0)\delta(\eta - \eta_0)$$

The Dirac delta function will be specifically addressed later, but for now, just bear in mind the basic properties:

$$\int_{-\infty}^{\infty} \delta(x)\mathrm{d}x = 1, \quad \int_{-\infty}^{\infty} \delta(x)f(x)\mathrm{d}x = f(0)$$

6.1.2 *Geometric interpretation of derivatives*

Let there be a parameterized curve $C(t)$ on the z-plane: $z = z(t)$. It is mapped to a curve $\zeta(C)$ on the ζ-plane by the analytic function $\zeta(z)$: $C(t) \longmapsto \zeta(C)$. According to $\zeta(t) = \zeta(z(t))$, we have

$$\left.\frac{\mathrm{d}\zeta(t)}{\mathrm{d}t}\right|_{t=t_0} = \left.\frac{\mathrm{d}\zeta(z)}{\mathrm{d}z}\right|_{z=z_0} \cdot \left.\frac{\mathrm{d}z(t)}{\mathrm{d}t}\right|_{t=t_0} \tag{6.1.4}$$

where $z_0 = z(t_0)$. Thus, the argument of the derivative satisfies the relation:

$$\arg\left[\zeta'(t_0)\right] = \arg\left[\zeta'(z_0)\right] + \arg\left[z'(t_0)\right]$$

Since $\frac{\mathrm{d}z(t)}{\mathrm{d}t}$ and $\frac{\mathrm{d}\zeta(t)}{\mathrm{d}t}$ respectively reflect the slopes of the curve C at the point z_0 and the image curve $\zeta(C)$ at the point $\zeta_0 = \zeta(z_0)$, the argument of the derivative $\arg\left[\zeta'(z_0)\right]$ represents the increment of the inclination angle of the curve $\zeta(C)$ relative to the curve C at the point z_0, as shown in Figure 6.1. Using polar coordinates: $\zeta'(z_0) = r\mathrm{e}^{\mathrm{i}\alpha}$, the argument $\alpha = \arg\left[\zeta'(z_0)\right]$ represents the angle by which the tangent line turns. The magnitude of the derivative $r = |\zeta'(z_0)|$ reflects the

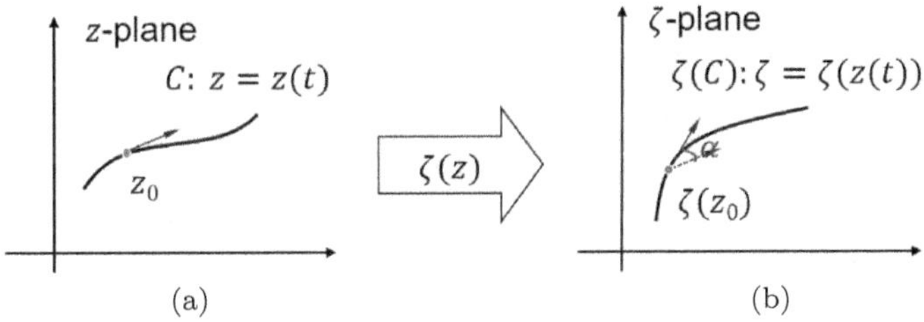

Figure 6.1.

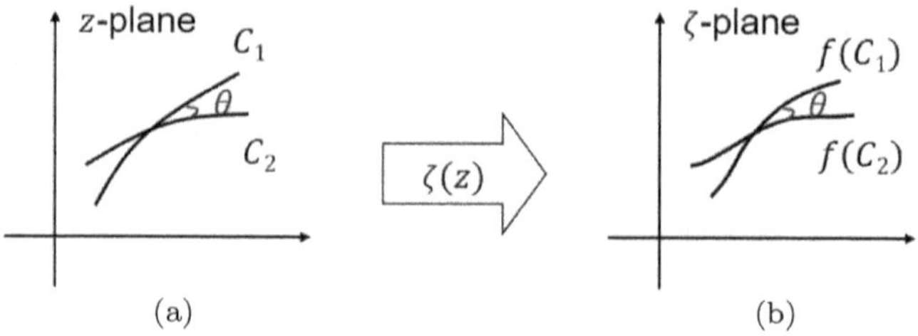

Figure 6.2.

scaling factor of the differential element passing through the point z_0 after the mapping.

If there are two curves C_1 and C_2 intersecting at z_0 on the z-plane and they are mapped by the function $\zeta(z)$ to the corresponding intersecting curves $\zeta(C_1)$ and $\zeta(C_2)$ on the ζ-plane. Since the derivative is independent of the approach of z_0, each curve's tangent line at z_0 turns by the same angle $\alpha = \arg[\zeta'(z_0)]$. Therefore, the angle between $\zeta(C_1)$ and $\zeta(C_2)$ remains unchanged compared to the angle between C_1 and C_2, as shown in Figure 6.2. This is the property of conformal invariance of analytic functions.

6.1.3 *Conformal mapping*

For a polygon on the z-plane, when it is mapped by an analytic function, it becomes a polygon on the ζ-plane, and the interior angles of the polygon remain unchanged. However, since the magnitude of the derivative is different at each point, the scaling factor of the differential element at each point is variable. Therefore, the polygon at a finite scale will undergo noticeable shape changes, as shown in Figure 6.3.

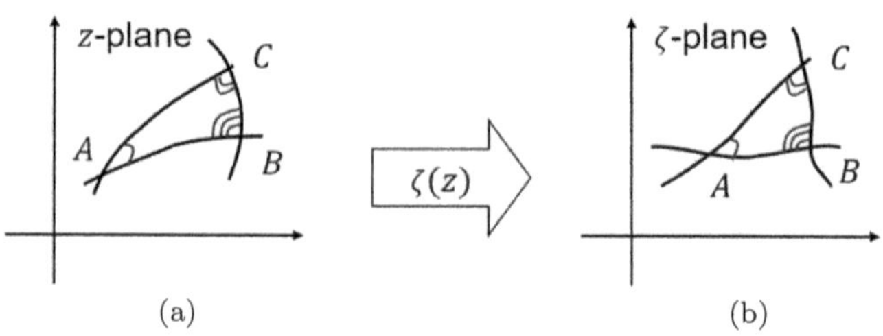

Figure 6.3.

In general, if a mapping $\zeta(z)$ preserves the angles between curves passing through z_0 and their orientations, it is called angle-preserving or locally conformal at z_0. Conformal mappings preserve the angles and local shapes of objects, but they do not necessarily preserve the similarity of their finite sizes (Figure 6.4).

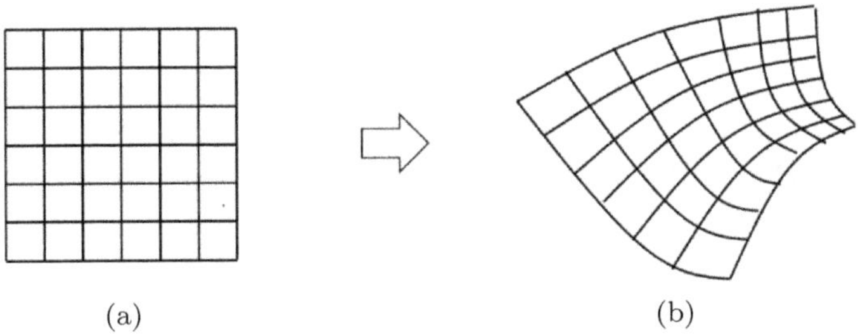

Figure 6.4.

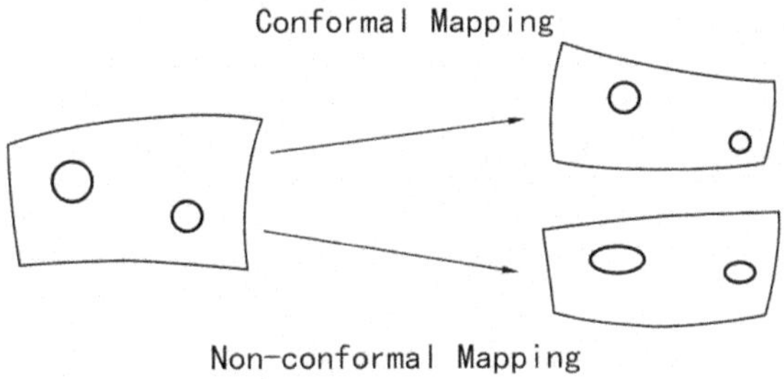

Figure 6.5.

Under a conformal mapping, circles can either enlarge or shrink, but they cannot be transformed into ellipses or vice versa, as shown in Figure 6.5. The fundamental problem in the theory of conformal mapping is to construct a function that maps one region to another given region, for a given pair of regions B and B'. Thus, it is necessary to determine the existence and uniqueness of conformal mappings. In the complex analysis, any simply connected region can be mapped to another simply connected region by an appropriate analytic function.

(1) **Riemann mapping theorem:** Through an appropriate analytic function transformation, any simply connected region is conformally equivalent to a unit circle.

(2) **Principle of boundary correspondence:** Suppose there are two regions B and B' with boundary curves C and C', respectively, and B' is bounded. The function $\zeta(z)$ is analytic within region B and continuous on $\bar{B}$, and implements a bijective mapping that preserves the orientation of traversal from C to C'. Then it establishes a single-valued conformal mapping from region B to B'.

For a region B, its boundary curve C can be represented by a real parameter t: $C = C(t)$. Assuming there is a conformal mapping $\zeta(z)$ from B to B', and B' has no infinite branches, then $\zeta(C(t))$ represents a continuous and bijective mapping between the boundary curves of region B and B'. In the following sections, several basic transformations of analytic functions will be discussed.

ENDNOTE

Prior to the 18th century, people were already familiar with the stereographic projection on a sphere. Hipparchus, Ptolemy, and even earlier ancient Egyptians may have used the stereographic projection to create star charts or navigation charts. In 1590, Thomas Harriot noticed that the stereographic projection possesses conformal invariance, as shown in Figure 6.6(a), meaning that it preserves angles or maintains local similarity. However, objects undergo distortion on a finite scale, as depicted in Figure 6.6(b) with a polar projection map. Although such maps suffer from scale distortions, they were widely used in early Western navigation due to their convenience in determining the direction of navigation based on the positions of stars. In 1695, Edmond Halley mathematically proved the conformal invariance of the stereographic projection. Euler, Lagrange, and others contributed to the development of conformal theory, using complex numbers. Gauss extended Lagrange's theory to conformal mappings from arbitrary surfaces to the plane. Riemann seems to be the first to consider conformal mappings as the foundation of the theory of analytic functions, and his Riemann mapping theorem provides a fundamental description of conformal mapping theory.

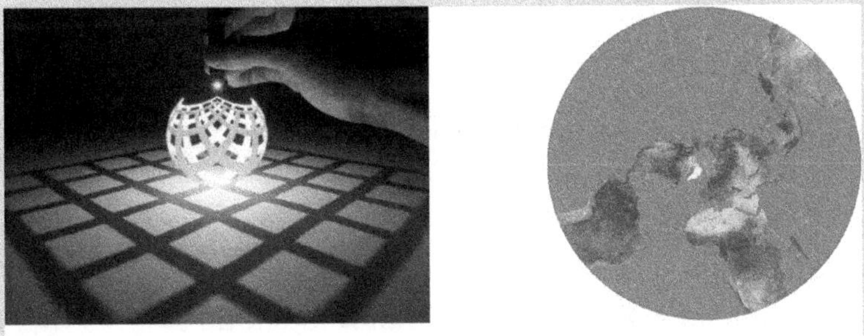

(a) (b)

Figure 6.6. (image from Wikipedia).

An important property of the stereographic projection is that it maps circles on the sphere to circles or straight lines on the plane. From the perspective of Riemann spheres, the real axis is just a circle, indistinguishable from the unit circle. Spherical motions (rigid rotations) are conformal, representing conformal mappings from the Riemann sphere to itself, which can be achieved through a fractional linear transformation.

Exercises

[1] Determine the region to which the function $\zeta = \sqrt{z}$ maps the interior of the circle $r = \cos\phi\left(-\frac{\pi}{2} < \phi \le \frac{\pi}{2}\right)$, and draw the curve.
Answer: The curve is a lemniscate given by $\rho = \sqrt{\cos 2\theta}$.

[2] Determine the region to which the function $\zeta = z^2$ maps the interior of the circle $r = \cos\phi\left(-\frac{\pi}{2} < \phi \le \frac{\pi}{2}\right)$, and draw the curve.
Answer: The curve is a cardioid given by $\rho = \cos^2\left[\frac{\theta}{2}\right]$.

[3] Proof: The stereographic projection maps circles on the sphere to circles in the plane.
Hint: Use a plane to intersect the unit sphere, given by the equation $aX + bY + cZ = d$. The intersection of the plane and the sphere is a circle on the sphere.

[4] Let P be a point on the Riemann sphere, and z be its stereographic projection point on the complex plane. Prove the following:

(a) The diameter of the sphere passing through P, the point $-P$, corresponds to the stereographic projection point $-\frac{1}{\bar{z}}$ in the complex plane.
(b) Under stereographic projection, a 180° rotation of the sphere around the x-axis corresponds to inversion: $z \longmapsto \frac{1}{z}$.
Hint: The stereographic projection of a point $P = (X, Y, Z)$ on the sphere is given by $z = \frac{X+\mathrm{i}Y}{1-Z}$.

6.2 Elementary Function Transformations

6.2.1 *Power function transformation*

The linear function $\zeta(z) = az + b$ (where a and b are complex constants) has a derivative $\zeta'(z) = a$, which is a constant. This means that the magnification factor $|a|$ of the linear function remains constant, and different parts of the graph are scaled proportionally while maintaining the same shape. The argument of the constant a represents the rotation angle of the graph in the complex plane.

The power function $\zeta(z) = z^n$ has a derivative $\zeta'(z) = nz^{n-1}$. Since the derivative at the origin $z = 0$ is zero, the angle between the two curves at the origin is no longer preserved; it is magnified by a factor of n. The transformation $\zeta(z) = z^{\frac{1}{n}}$ reduces the angle at the origin by a factor of n. Outside the origin, power functions preserve conformal invariance.

Example 6.1. Consider a large metal conductor with a missing dihedral angle $\alpha = \frac{\pi}{3}$, and the electric potential on the conductor is V_0. We want to determine the electric potential distribution in the space of the dihedral angle.

Solution: Let us take the transformation $\zeta(z) = z^3$, which maps the region of the dihedral angle $\alpha = \frac{\pi}{3}$ to the upper half-plane (Figure 6.7). In the upper half-plane of ζ, the electric potential distribution is independent of the real axis ξ and proportional to η, given by

$$u = A\eta + V_0 = A\,\mathrm{Im}\,\zeta + V_0$$

where A is a real constant depending on the charge density on the surface of the conductor. Considering the potential as the imaginary part, the complex potential becomes

$$w = A\zeta + \mathrm{i}V_0 = Az^3 + \mathrm{i}V_0$$

Returning to the original z-plane, the electric potential distribution in the region of the dihedral angle is given by

$$u = \operatorname{Im} w = A \operatorname{Im} z^3 + V_0 = A(3x^2y - y^3) + V_0$$

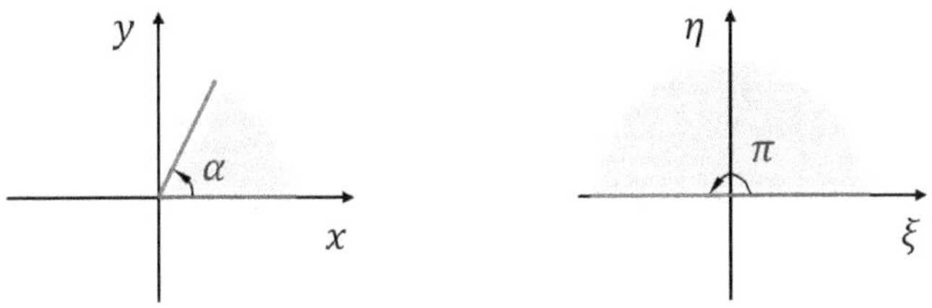

(a) (b)

Figure 6.7.

Reflection: If α is an arbitrary angle, can we determine the electric potential distribution?

Example 6.2. Study of water flow in a flat-bottomed tank with a vertical thin plate blocking the flow, as shown in Figure 6.8(a).

Solution: First, we apply the transformation $\zeta_1 = z^2$, which maps the boundary of the tank from Figure 6.8(a) to the boundary shown in Figure 6.8(b). The negative real axis and positive real axis are folded together. Then, we take $\zeta_2 = \zeta_1 + h^2$ to shift the folded horizontal boundary to the right by h^2, aligning the endpoints to the origin as shown in Figure 6.8(c). Finally, we apply the transformation $\zeta = \sqrt{\zeta_2}$ to unfold the folded positive real axis into the complete real axis as depicted in Figure 6.8(d). The combined transformation function is given by

$$\zeta = \sqrt{z^2 + h^2}$$

On the flat horizontal plane, the water flow is uniform, and the velocity potential is a linear function in the horizontal direction:

$$u = A\xi = A \operatorname{Re} \zeta$$

The corresponding complex potential is

$$w = A\zeta = A\sqrt{z^2 + h^2}$$

Thus, the velocity field distribution in the z-plane is

$$v_x + \mathrm{i}v_y = \frac{\partial u}{\partial x} + i\frac{\partial u}{\partial y} = \overline{w'(z)} = \frac{A\bar{z}}{\sqrt{\bar{z}^2 + h^2}}$$

To determine the constant A, let us assume that the flow velocity at infinity is uniformly distributed in the horizontal direction, i.e.,

$$\lim_{z\to\infty} \frac{A\bar{z}}{\sqrt{\bar{z}^2 + h^2}} = \lim_{z\to\infty} \frac{A}{\sqrt{1 + (h/\bar{z})^2}} \xrightarrow{z\to\infty} v_x = A \equiv v_0$$

Therefore, the velocity field distribution is given by

$$v_x = \mathrm{Re}\left[\frac{v_0\bar{z}}{\sqrt{\bar{z}^2 + h^2}}\right], \quad v_y = \mathrm{Im}\left[\frac{v_0\bar{z}}{\sqrt{\bar{z}^2 + h^2}}\right]$$

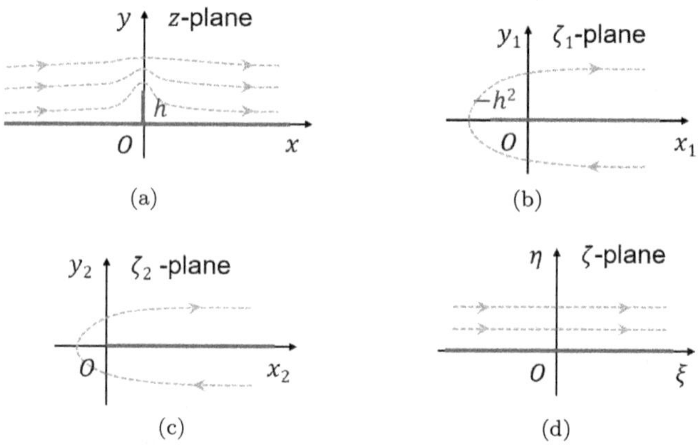

Figure 6.8.

6.2.2 *Exponential function and logarithmic function transformations*

(1) *Exponential function*

$$\zeta(z) = \mathrm{e}^z = \mathrm{e}^x\mathrm{e}^{\mathrm{i}y}$$

Due to

$$|\zeta| = \mathrm{e}^x, \quad \arg\zeta = y$$

The exponential function maps the z-plane's vertical lines (with constant x) to circles in the ζ-plane with a radius of $r = \mathrm{e}^x$, as shown by the solid line in Figure 6.9. It also maps the z-plane's horizontal lines (with constant y) to rays in the ζ-plane emanating from the origin with an angle $\phi = \arg\zeta = y$, as shown by the dashed line in Figure 6.9. The orthogonality between the solid and dashed lines remains unchanged. If we consider the principal value of the argument, $0 \leq \arg\zeta \leq 2\pi$, then the exponential function maps a region in the z-plane that is parallel to the real axis and has a width of $0 \leq y \leq 2\pi$ to the entire ζ-plane.

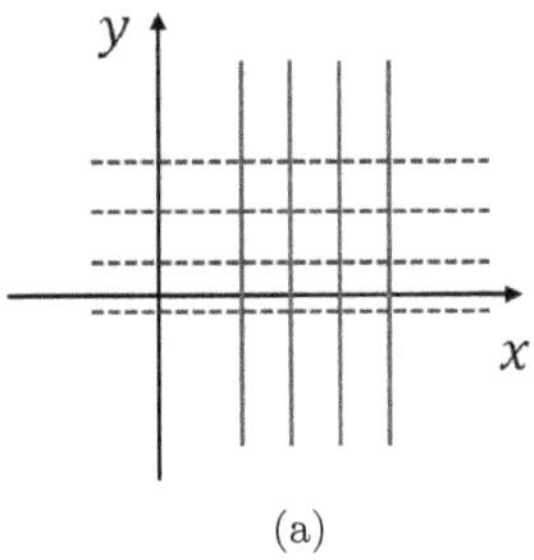

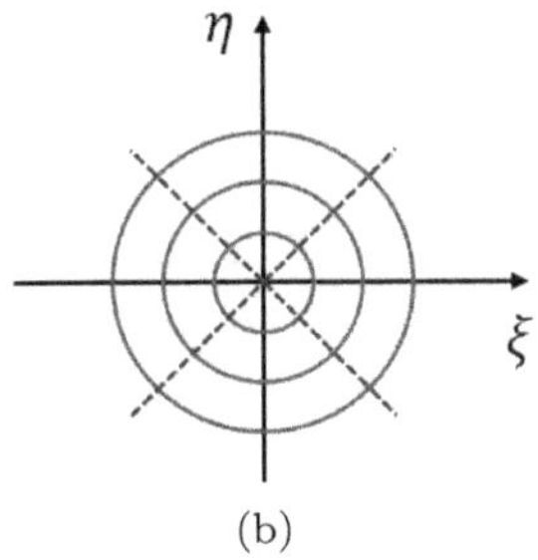

Figure 6.9.

(2) *Logarithmic function*

$$\zeta(z) = \ln z = \ln|z| + \mathrm{i}\arg z + 2k\pi\mathrm{i} \quad (k \in \mathbb{Z})$$

It is the inverse function of the exponential function. The logarithmic function maps circles centered at the origin in the z-plane to lines parallel to the imaginary axis in the ζ-plane. It also maps rays in the z-plane with constant arguments to lines parallel to the real axis in the ζ-plane. This can be observed by comparing Figures 6.9(b) to 6.9(a).

Example 6.3. Consider a capacitor consisting of two concentric cylinders, with radii R_1 and R_2. We want to calculate the capacitance per unit length of the cylindrical capacitor.

Solution: We apply the logarithmic function transformation

$$\zeta(z) = \ln z = \ln|z| + \mathrm{i}\arg z \quad (0 \leq \arg z < 2\pi)$$

This transformation maps the z-plane's concentric circles with radii $|z| = R_1$ and R_2 to parallel lines in the ζ-plane, where the real part is a constant value $\xi = \ln R_1$ and $\xi = \ln R_2$, respectively, and the imaginary part spans a width of $0 \leq \eta < 2\pi$, as shown in Figure 6.10(b). This forms a parallel-plate capacitor. The capacitance per unit length of this parallel-plate capacitor can be calculated as

$$C = \frac{\varepsilon_0 A}{d} = \frac{2\pi\varepsilon_0}{\ln\left(\frac{R_2}{R_1}\right)}$$

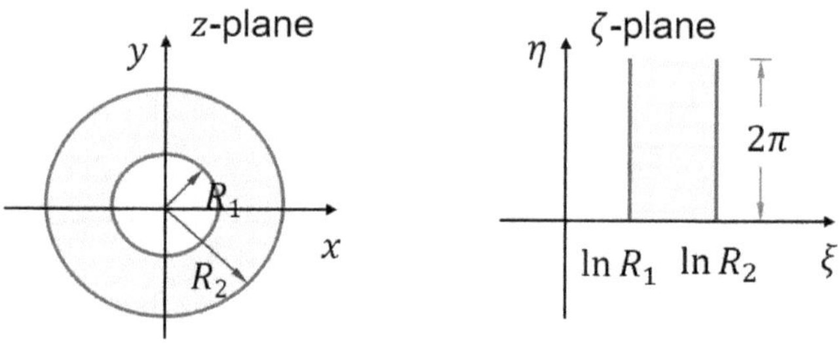

(a) (b)

Figure 6.10.

6.2.3 *Fractional linear transformations*

(1) Inversion transformation

$$\zeta(z) = \frac{R^2}{z}$$

Taking the polar coordinate representation, $z = \rho e^{\mathrm{i}\phi}$, we have

$$\zeta(z) = \frac{R^2}{\rho} e^{-\mathrm{i}\phi}$$

This transformation can be achieved by applying two consecutive transformations:

$$z_1 = \frac{R^2}{\bar{z}}, \quad \zeta = \bar{z}_1$$

As a result, circles in the z-plane are mapped to circles in the ζ-plane, the regions inside the circle are mapped to the regions outside the circle, pairs of conjugate points on the circle remain conjugate, and the center of the circle is mapped to infinity, as shown in Figure 6.11. The pair of points z and z_1 in the figure is called conjugate points of the circle which satisfy $|z| \cdot |z_1| = R^2$. For any point inside the circle, there exists a unique conjugate point outside the circle, and vice versa.

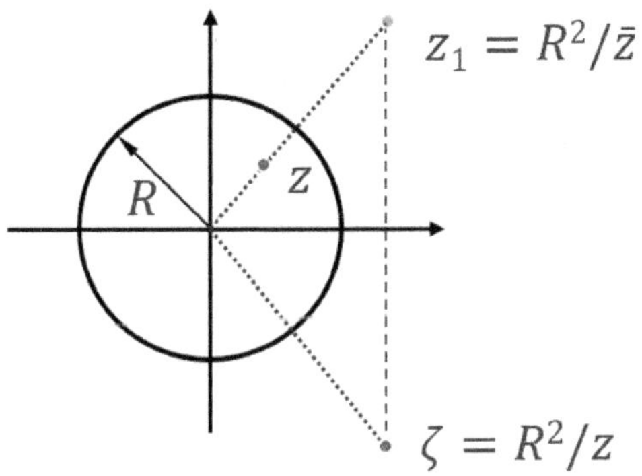

Figure 6.11.

(2) Fractional linear transformation

$$\zeta(z) = \frac{az+b}{cz+d} \quad (ad - bc \neq 0) \tag{6.2.1}$$

The numerator and denominator in this expression are both linear functions, hence it is called a fractional linear transformation or sometimes a Möbius transformation. The fractional linear transformation can be decomposed into three successive transformations: translation, inversion, and another translation.

$$\zeta = \frac{a}{c} + \frac{b - ad/c}{cz+d} = \frac{a}{c} + \frac{(bc-ad)/c^2}{z + d/c}$$

Therefore,

$$\left|\zeta - \frac{a}{c}\right| = \frac{|(bc-ad)/c^2|}{|z+d/c|} \xrightarrow{p=-d/c} \left|\zeta - \frac{a}{c}\right| = \frac{|(b+ap)/c|}{|z-p|}$$

Fractional linear transformations have the following fundamental properties:

[1] They map circles to circles, with conjugate points on a circle remaining conjugate. They can transform lines in the z-plane to circles or lines in the ζ-plane, and vice versa.
[2] Fractional linear transformations implement mappings from the Riemann sphere to itself.
[3] Fractional linear transformations are non-commutative, and the set of all fractional linear transformations forms a group.

A fractional linear transformation is determined by four complex constants: a, b, c and d, with the condition that $ad - bc \neq 0$. However, removing a common constant from the numerator and denominator leaves three independent constants that determine the transformation properties. Therefore, there exists a unique fractional linear transformation that maps three distinct points z_1, z_2, z_3 to three distinct points $\zeta_1, \zeta_2, \zeta_3$ in the ζ-plane, i.e.,

$$(z_1, z_2, z_3) \longmapsto (\zeta_1, \zeta_2, \zeta_3)$$

Example 6.4. There is a large grounded conducting plane, and there is a long wire parallel to the conducting plane, with a distance of a between them. If the wire is uniformly charged with a charge per unit length of Q, the electric potential distribution is to be determined.

Solution: We can choose a suitable coordinate system such that the conducting plane is located on the real axis and the long wire is at $z = \mathrm{i}a$. In the two-dimensional case, the electric potential satisfies the Poisson's equation:

$$\begin{cases} \dfrac{\partial^2 u}{\partial x^2} + \dfrac{\partial^2 u}{\partial y^2} = \dfrac{Q}{2\pi\varepsilon_0}\delta(y-a)\delta(x) \\ u|_{y=0} = 0 \end{cases}$$

Now, we attempt to find a transformation that maps the real axis to a circle, the wire to the center of the circle, and the mirror image of the wire with respect to the real axis to the infinity point, which is the conjugate point of the circle, as shown in Figure 6.12.

$$z = \mathrm{i}a \longmapsto \zeta = 0, \quad z = -\mathrm{i}a \longmapsto \zeta = \infty, \quad y = 0 \longmapsto |\zeta| = R$$

Using a fractional linear transformation, we can achieve this goal by taking

$$\zeta = R\frac{x - \mathrm{i}a}{x + \mathrm{i}a}$$

Clearly, the real axis $z = x$ of the conducting plane becomes a circle with center at the origin and radius R:

$$|\zeta| = \left| R\frac{x - \mathrm{i}a}{x + \mathrm{i}a} \right| = R$$

The electric potential distribution for this concentric cylindrical system is given by

$$u = \frac{Q}{2\pi\varepsilon_0} \ln \frac{\rho}{R}$$

Returning to the z-plane, we obtain the electric potential distribution as

$$\begin{aligned} u &= \frac{Q}{2\pi\varepsilon_0} \ln \frac{\rho}{R} = \frac{Q}{2\pi\varepsilon_0} \ln \frac{|\zeta|}{R} = \frac{Q}{2\pi\varepsilon_0} \ln \frac{|z - \mathrm{i}a|}{|z + \mathrm{i}a|} \\ &= \frac{Q}{2\pi\varepsilon_0} \ln \frac{x^2 + (y - a)^2}{x^2 + (y + a)^2} \end{aligned}$$

This result can also be obtained by superposition of the potentials produced by the wire and its mirror image with respect to its horizontal plane.

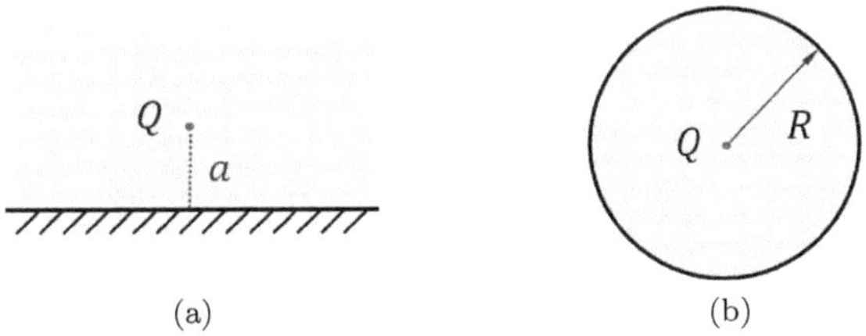

Figure 6.12.

Example 6.5. Two parallel cylindrical conductors with radii R_1 and R_2 are considered, and their axes are separated by a distance $L(L > R_1 + R_2)$. The goal is to find the capacitance per unit length between the cylinders.

Solution: To transform the parallel cylinders into concentric cylinders, we need to find a transformation that utilizes the unique pair of common conjugate points on the line connecting their centers. Let A and B be these conjugate points, as shown in Figure 6.13(a). We have the relationships:

$$|O_1A| \cdot |O_1B| = R_1^2, \quad |O_2A| \cdot |O_2B| = R_2^2$$

Taking O_1 as the origin, $|O_1A| = x_1$, and $|O_1B| = x_2$, it is rewritten as

$$x_1 \cdot x_2 = R_1^2, \quad (L - x_1) \cdot (L - x_2) = R_2^2$$

Solving the equations, we obtain

$$x_{1,2} = \frac{1}{2L}\left[(L^2 + R_1^2 - R_2^2) \pm \sqrt{(L^2 + R_1^2 - R_2^2)^2 - 4R_1^2L^2}\right]$$

We then consider the fractional linear transformation

$$\zeta = \frac{z - x_1}{z - x_2}$$

This transformation maps point A to the origin and point B to infinity. It can be verified that the two cylinders are transformed into concentric cylinders with radii:

$$R_1' = \left|\frac{-R_1 - x_1}{-R_1 - x_2}\right|, \quad R_2' = \left|\frac{L + R_2 - x_1}{L + R_2 - x_2}\right|$$

Therefore, the capacitance per unit length, which is the capacitance of the coaxial capacitors, is given by

$$C = \frac{2\pi\varepsilon_0}{\ln\left(\frac{R_2'}{R_1'}\right)}$$

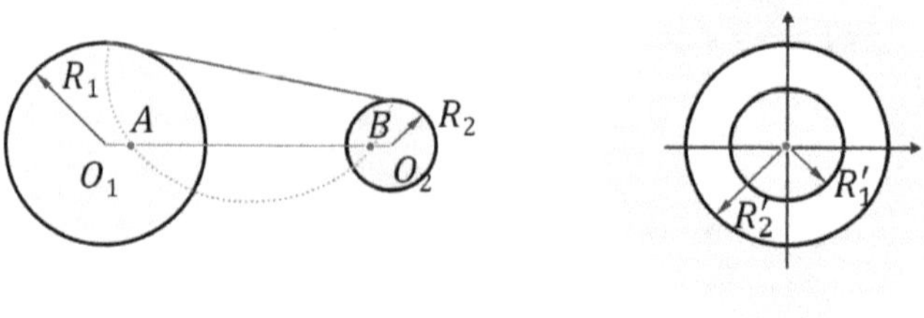

(a) (b)

Figure 6.13.

Example 6.6. In Figure 6.13(a), there is a semicircular protrusion on the real axis with a radius of $R = 1$. The goal is to find a conformal transformation that flattens the semicircular protrusion.

Solution: To achieve this, we can proceed in several steps. First, we perform a fractional linear transformation:

$$z_1 = \frac{z-1}{z+1}$$

This transformation maps point C in Figure 6.14(a) to the origin C_1 in Figure 6.14(b), point A to infinity, and point B to the point B_1 with $z_1 = \mathrm{i}$ on the imaginary axis. Thus, the semicircular arc ABC is mapped to the upper half of the z_1-plane. On the other hand, it maps the infinite points D and E on the real axis to the same point D_1 with $z_1 = +1$. Therefore, the upper plane with the semicircular protrusion becomes the first quadrant of the z_1-plane. Next, we apply the transformation:

$$z_2 = z_1^2$$

This maps the first quadrant to the upper half-plane. Note that the arrangement of points on the real axis becomes

$$A_2 \to B_2 \to C_2 \to D_2(E_2) \to A_2$$

which is different from the original order of points on the protruded real axis:

$$E \to A \to B \to C \to D$$

To align the points properly, we apply another fractional linear function:

$$\zeta = \frac{z_2+1}{-z_2+1}$$

This transformation maps the circular arc ABC back to the line segment $[-1, 0, +1]$ (Figure 6.14(c)). The final complete transformation is given by

$$\zeta(z) = \frac{z_2+1}{-z_2+1} = \frac{z_1^2+1}{-z_1^2+1} = \frac{1}{2}\left(z + \frac{1}{z}\right)$$

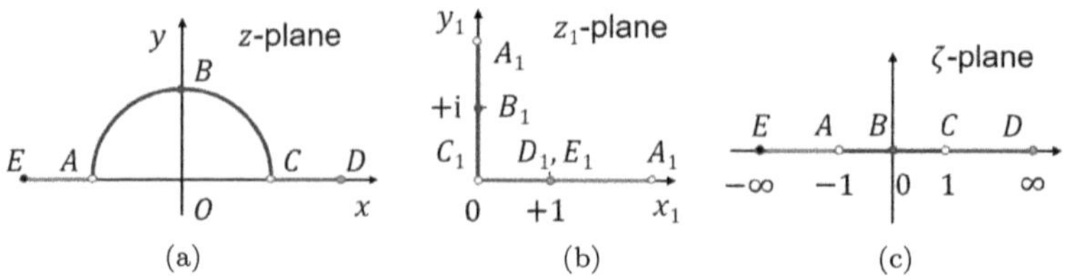

Figure 6.14.

ENDNOTE

Consider two fractional linear transformations:

$$\boldsymbol{L}_1\colon z \longmapsto \frac{a_1 z + b_1}{c_1 z + d_1} \quad (a_1 d_1 - b_1 c_1 \neq 0)$$

$$\boldsymbol{L}_2\colon z \longmapsto \frac{a_2 z + b_2}{c_2 z + d_2} \quad (a_2 d_2 - b_2 c_2 \neq 0)$$

The composition of these two mappings, $\boldsymbol{L} = \boldsymbol{L}_2 \circ \boldsymbol{L}_1$, is also a fractional linear transformation:

$$z \longmapsto \frac{a_1\left(\frac{a_2 z + b_2}{c_2 z + d_2}\right) + b_1}{c_1\left(\frac{a_2 z + b_2}{c_2 z + d_2}\right) + d_1} = \frac{(a_1 a_2 + b_1 c_2) z + (a_1 b_2 + b_1 d_2)}{(c_1 a_2 + d_1 c_2) z + (c_1 b_2 + d_1 d_2)}$$

We note that a fractional linear transformation

$$\boldsymbol{L}\colon z \longmapsto \frac{az + b}{cz + d}$$

can be described by a 2×2 matrix:

$$\boldsymbol{L} = \begin{pmatrix} a & b \\ c & d \end{pmatrix}$$

The condition for a fractional linear transformation to be invertible is $ad - bc \neq 0$, which means the corresponding matrix $\boldsymbol{L}$ is invertible. The composition of two successive mappings is equivalent to the product of the matrices:

$$\begin{pmatrix} a_1 & b_1 \\ c_1 & d_1 \end{pmatrix} \begin{pmatrix} a_2 & b_2 \\ c_2 & d_2 \end{pmatrix} = \begin{pmatrix} a_1 a_2 + b_1 c_2 & a_1 b_2 + b_1 d_2 \\ c_1 a_2 + d_1 c_2 & c_1 b_2 + d_1 d_2 \end{pmatrix}$$

It can be verified that fractional linear transformations form a group, which is a closed algebraic set under multiplication, with the following properties:

[1] Associativity: $\boldsymbol{L}_1 \circ (\boldsymbol{L}_2 \circ \boldsymbol{L}_3) = (\boldsymbol{L}_1 \circ \boldsymbol{L}_2) \circ \boldsymbol{L}_3$;
[2] Identity element: The identity mapping $z \longmapsto z$;
[3] Existence of inverse element $\boldsymbol{L}^{-1}$: $z \longmapsto \frac{dz-b}{a-cz}$.

Fractional linear transformations are generally non-commutative, i.e., $\boldsymbol{L}_1 \circ \boldsymbol{L}_2 \neq \boldsymbol{L}_2 \circ \boldsymbol{L}_1$, so it is a non-Abelian group.

Due to the conformal invariance of the stereographic projection between the extended complex plane $\bar{\mathbb{C}}$ and the Riemann sphere, Gauss discovered that any continuous rotation of the sphere can be represented by the following fractional linear transformation:

$$\boldsymbol{L} : z \longmapsto \frac{az + \bar{b}}{-bz + \bar{a}} \quad (|a|^2 + |b|^2 = 1)$$

Its coefficient matrix is

$$\boldsymbol{U} = \begin{pmatrix} a & \bar{b} \\ -b & \bar{a} \end{pmatrix} \longrightarrow \begin{cases} \boldsymbol{U}^{+}\boldsymbol{U} = 1 \\ \det \boldsymbol{U} = 1 \end{cases}$$

This is a unitary transform, forming the $SU(2)$ group, which is isomorphic to the three-dimensional space rotation group $SO(3)$. Thus, the motion of a rigid body (sphere rotation) can be described using fractional linear transformations, involving a pair of complex numbers a and b. By taking $a = \alpha + \mathrm{i}\beta$ and $b = \gamma + \mathrm{i}\delta$, with $\alpha, \beta, \gamma, \delta \in \mathbb{R}$, the matrix representation is given by

$$\begin{aligned}\begin{pmatrix} a & b \\ -\bar{b} & \bar{a} \end{pmatrix} &= \alpha \begin{pmatrix} 1 & 0 \\ 0 & 1 \end{pmatrix} + \beta \begin{pmatrix} \mathrm{i} & 0 \\ 0 & -\mathrm{i} \end{pmatrix} + \gamma \begin{pmatrix} 0 & 1 \\ -1 & 0 \end{pmatrix} + \delta \begin{pmatrix} 0 & \mathrm{i} \\ \mathrm{i} & 0 \end{pmatrix} \\ &= \alpha \boldsymbol{I} + \beta \boldsymbol{i} + \gamma \boldsymbol{j} + \delta \boldsymbol{k}\end{aligned}$$

where the matrices $\boldsymbol{i}, \boldsymbol{j}, \boldsymbol{k}$ satisfy

$$\boldsymbol{i}^2 = \boldsymbol{j}^2 = \boldsymbol{k}^2 = \boldsymbol{ijk} = \mathbf{-1}$$

These matrices are exactly the basis of quaternions, and the rotation matrix U corresponds to a quaternion. Therefore, specific quaternion multiplication describes the three-dimensional point rotation of a rigid body.

In quantum mechanics, Pauli matrices are usually defined as Hermitian matrices:

$$\mathrm{i}\sigma_x \longleftrightarrow -\boldsymbol{k}, \quad \mathrm{i}\sigma_y \longleftrightarrow -\boldsymbol{j}, \quad \mathrm{i}\sigma_z \longleftrightarrow -\boldsymbol{i}$$

They describe the intrinsic quantum states of spin-$\frac{1}{2}$ particles in non-relativistic quantum mechanics.

A similar story happened in the remarkable history of developing relativistic quantum mechanics. In order to incorporating the Schrödinger equation into the framework of relativistic covariance, Dirac attempted to linearize the relativistic energy-momentum equation with quadratic terms

$$E^2 = m^2c^4 + p^2c^2$$

with the following assumption:

$$\sqrt{x^2 + y^2} = \alpha x + \beta y$$

where α and β are undetermined "coefficients", satisfying

$$x^2 + y^2 = (\alpha x + \beta y) \cdot (\alpha x + \beta y) = \alpha^2 x^2 + \beta^2 y^2 + (\alpha\beta + \beta\alpha)xy$$

Hence,

$$\alpha^2 = \beta^2 = 1, \quad \alpha\beta + \beta\alpha = 0$$

It means α and β must be non-commutative numbers! Doesn't this sound familiar? Yes, you see the shadow of quaternions! If all three components of momentum are taken account in, we will precisely obtain the Hamiltonian quaternions. Quaternions thus make their incredible way into quantum physics, where the electron spin, as a relativistic effect, directly manifests itself in the Dirac equation.

Dirac's extraordinary insight not only established relativistic quantum mechanics but also led to predictions about positrons. It becomes an immortal discovery alongside Newton's theory of universal gravitation, Maxwell's electromagnetic wave theory, and Einstein's theory of gravity.

Exercises

[1] Find fractional linear transformations to implement the following point-to-point mappings:

$$(a)\ (1, \mathrm{i}, -1) \longmapsto (\mathrm{i}, -1, 1); \quad (b)(0, 1, \infty) \longmapsto (-1, -\mathrm{i}, 1)$$

[2] On the real axis, the electric potential is V_0 for $x \in (0, a)$, and the potential is for $x > a$ and along the positive imaginary axis. Find the potential distribution in the first quadrant.
Hint: Perform successively the transformations $\zeta_1 = z^2, \zeta_2 = \frac{\zeta_1 - a^2}{\zeta_1}$, and $\zeta = \ln \zeta_2$, to map the first quadrant to a parallel-plate capacitor.
Answer:

$$u = \frac{V_0}{\pi}\eta = \frac{V_0}{\pi} \arg \frac{z^2 - a^2}{z^2}$$

[3] There is a long conductor with a radius of a near an infinite metal plane, with the axis of the conductor at a distance of b from the plane. Find a conformal transformation that transforms this system into concentric cylinders.

[4] A hollow conducting cylindrical shell with a radius of R_1 surrounding another conducting cylindrical core with a radius of R_2, where both cylinders are parallel and separated by a distance of $L(L < R_1 - R_2)$. Find the capacitance per unit length.
Answer:

$$C = \frac{2\pi\varepsilon_0}{\operatorname{arccos} h \left[\frac{(R_1^2 + R_2^2 - L^2)}{2R_1 R_2}\right]}$$

[5] Map the curved region to the unit circle: $|z| \leq 2, \mathrm{Im}\, z \geq 1$.
Answer:

$$\zeta_1 = \frac{z+\sqrt{3}-\mathrm{i}}{-z+\sqrt{3}+\mathrm{i}}, \quad \zeta_2 = -\zeta_1^3, \quad \zeta = \frac{\zeta_2 - \mathrm{i}}{\zeta_2 + \mathrm{i}}$$

[6] Find a transformation that maps the $\frac{1}{4}$ sector of radius $R = 1$ to the upper half-plane.
Answer:

$$\zeta = \left(\frac{1+z^2}{1-z^2}\right)^2$$

[7] There is a strip-shaped region D in the z-plane: $-\frac{\pi}{4} < \mathrm{Re}\, z < \frac{\pi}{4}$. Find a function $\zeta(z)$ that conformally maps it to the unit circle $|\zeta| < 1$, such that the three mapping points are

$$z = \pm\frac{\pi}{4} \longmapsto \zeta = \pm 1, \quad z = \mathrm{i}\infty \longmapsto \zeta = \mathrm{i}$$

Hint: $\zeta_1 = \mathrm{e}^{\mathrm{i}z}$ maps the two edges of the strip region $\mathrm{Re}\, z = \pm\frac{\pi}{4}$ to the rectangular region $\theta = \pm\frac{\pi}{4}$, then rotate counterclockwise by $\frac{\pi}{4}$ to the first quadrant.
Answer: $\zeta(z) = \tan z$

[8] Prove that the transformations $z \longmapsto z+1$ and $z \longmapsto \frac{1}{z}$ are not commutative.

[9] For a half-disc of radius R, with the circular boundary at temperature T_1 and the base at temperature T_2, find the stable temperature distribution on the half-disc.
Hint: Perform the following transformations to map the half-disc to a horizontal plate:

$$\zeta_1 = \frac{z+R}{z-R}, \quad \zeta_2 = \mathrm{e}^{-\frac{3\pi\mathrm{i}}{2}}\zeta_1, \quad \zeta_3 = \zeta_2^2, \quad \zeta = \ln \zeta_3$$

Answer:

$$\frac{2}{\pi}(T_1 - T_2)[\mathrm{Arg}(R+z) - \mathrm{Arg}(R-z)] + T_2$$

[10] Map a quarter-circle arc of radius R to the interior of the unit circle.

[11] Programming exercise: Use the hyperbolic function $\zeta(z) = \sinh z$ as the mapping function, we are need to determine

(a) What curves do lines parallel to the y-axis and x-axis in the z-plane respectively map to?

(b) What curves do concentric circles centered at the origin in the z-plane map to?

Answer: (a) Ellipses and hyperbolas (b) The images of circles with radii $R = 2, 4, 8$, respectively, which rotate around the origin with winding numbers $n = 1, 3, 5$.

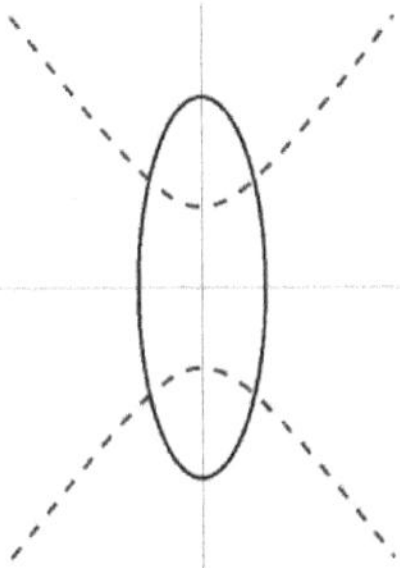

(a)

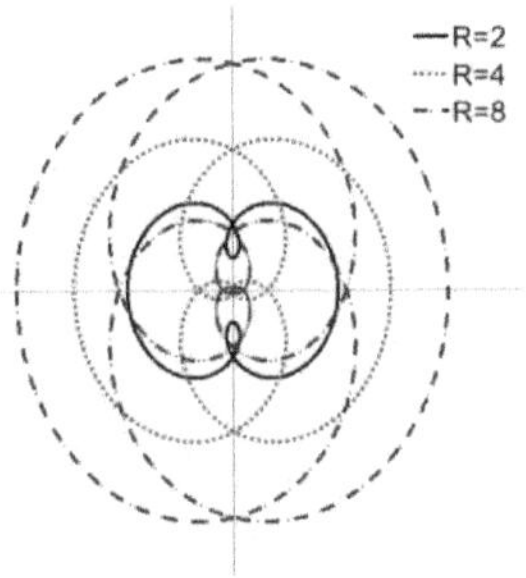

(b)

6.3 Joukowsky Transformation

In Section 6.2, we obtained a transformation function given by

$$\zeta(z) = \frac{1}{2}\left(z + \frac{1}{z}\right) \tag{6.3.1}$$

This transformation is known as the Joukowsky transformation and possesses several special properties. In this section, we will examine it in more detail.

Considering the real and imaginary parts of $\zeta(z) = \xi + \mathrm{i}\eta$, let $z = \rho e^{\mathrm{i}\phi}$, we have

$$\xi = \frac{1}{2}\left(\rho + \frac{1}{\rho}\right)\cos\phi, \quad \eta = \frac{1}{2}\left(\rho - \frac{1}{\rho}\right)\sin\phi$$

After eliminating the parameter ϕ, we obtain

$$\frac{\xi^2}{\left(\frac{1}{2}\rho + \frac{1}{2\rho}\right)^2} + \frac{\eta^2}{\left(\frac{1}{2}\rho - \frac{1}{2\rho}\right)^2} = 1 \tag{6.3.2}$$

Therefore, the Joukowsky transformation maps the circle $|z| = \rho$ on the z-plane to an ellipse on the ζ-plane. The lengths of the major and minor axes of the ellipse are given by

$$a = \frac{1}{2}\left|\rho + \frac{1}{\rho}\right|, \quad b = \frac{1}{2}\left|\rho - \frac{1}{\rho}\right|$$

The focal length $c = \sqrt{a^2 - b^2} = 1$ remains a constant. Hence, circles of different radii ρ are transformed into different ellipses, all sharing the common foci $\zeta = \pm 1$. Specifically, the unit circle on the z-plane is mapped to a line segment on the ζ-plane, with $\xi \in [-1, +1]$, forming an ellipse with a minor axis of zero.

As the radius ρ of the circle increases infinitely starting from 1, the major and minor semiaxes a and b also increase infinitely. This maps the exterior of the unit circle on the z-plane to the entire ζ-plane. Conversely, as ρ shrinks from 1 to zero, the major and minor semiaxes a and b still increase infinitely. Therefore, the interior of the unit circle on the z-plane is also mapped to the entire ζ-plane. Consequently, this is a multivalued mapping, expressed as a square root function

$$z = \zeta + \sqrt{\zeta^2 - 1}$$

Along the real axis from $\zeta = -1$ to $\zeta = +1$, a branch cut corresponds to the unit circle on the z-plane, as shown in Figure 6.15.

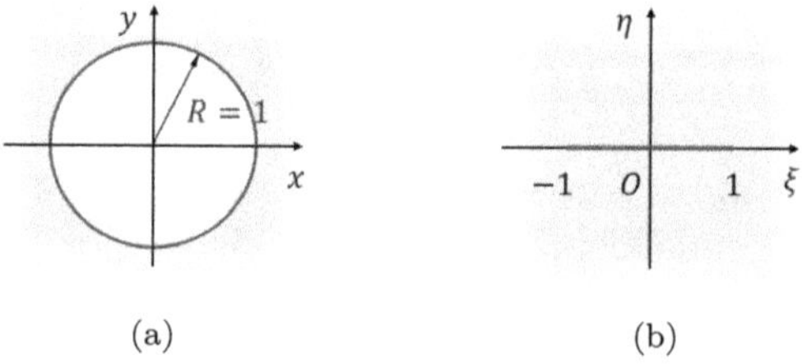

Figure 6.15.

On the other hand, by eliminating the parameter ρ from Equation (6.3.2), the rays on the z-plane emanating from the origin with $\arg z = \phi$ are mapped to hyperbolas on the ζ-plane given by

$$\frac{\xi^2}{\cos^2\phi} - \frac{\eta^2}{\sin^2\phi} = 1 \tag{6.3.3}$$

The real and imaginary semiaxes of these hyperbolas are given by $a = |\cos\phi|$ and $b = |\sin\phi|$. This family of hyperbolas is confocal, with focal length $c = \sqrt{a^2 + b^2} = 1$, and the foci are located at $\xi = \pm 1$.

The Joukowsky transformation $\zeta(z) = \frac{1}{2}\left(z + \frac{1}{z}\right)$ maps circles to ellipses, the unit circle to the line segment $[-1, +1]$, and rays originating from the origin to hyperbolas. The families of ellipses and hyperbolas are orthogonal to each other.

It should be noted that the Joukowsky transformation has singularities at $z = 0$. Additionally, due to the derivative of the transformation function:

$$\frac{\mathrm{d}}{\mathrm{d}z}\zeta = \frac{1}{2}\left(1 - \frac{1}{z^2}\right)$$

the transformation function is not conformal at $z = 0$, $z = \infty$ and the critical points $z = \pm 1$. Rewriting the Joukowsky transformation as

$$\frac{\zeta - 1}{\zeta + 1} = \left(\frac{z - 1}{z + 1}\right)^2$$

it can also be observed that the argument at $z = \pm 1$ is magnified by a factor of 2.

Example 6.7. Find the static electric field distribution generated by a long elliptical cylindrical conductor, where the major and minor semi-axes of the elliptical cross-section are a and b, respectively.

Solution: The focal length of the ellipse is $c = \sqrt{a^2 - b^2}$. Using the Joukowsky transformation:

$$z = \frac{c}{2}\left(\zeta + \frac{1}{\zeta}\right)$$

the ellipse on the z-plane is transformed into a circle on the ζ-plane with a radius $R = \frac{a+b}{c} = \frac{c}{a-b}$. Therefore, the potential distribution of a long elliptical cylinder is given by

$$u = C_1 \ln|\zeta| + C_2$$

where C_1 and C_2 are real constants. The corresponding complex potential is

$$w = C_1 \ln\zeta + C_2 = C_1 \ln\left(z + \sqrt{z^2 - c^2}\right) + C_2$$

Hence, the electric potential distribution produced by the elliptical cylinder is

$$u = C_1 \mathrm{Re}\left[\ln\left(z + \sqrt{z^2 - c^2}\right)\right] + C_2$$

ENDNOTE

The Joukowsky transformation can also map circles to airfoil profiles. Consider a circle on the z-plane centered at ih and passing through the points $z = \pm 1$, as shown in Figure 6.16(a).

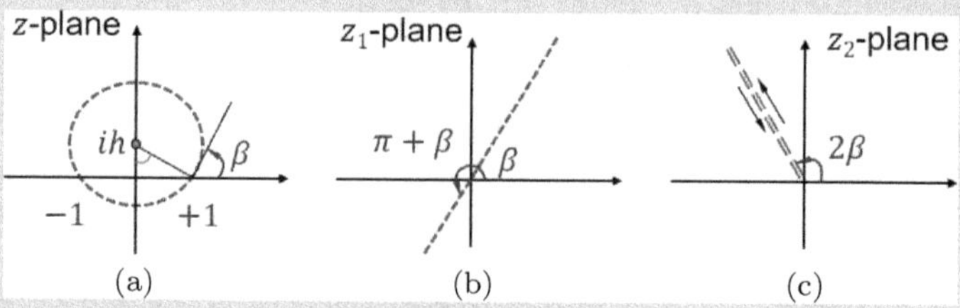

Figure 6.16.

The fractional linear transformation

$$z_1 = \frac{z-1}{z+1}$$

implement the mapping, $z = 1 \longmapsto 0$ and $z = -1 \longmapsto \infty$. The real axis is still mapped to the real axis, while the circle on the z-plane is transformed into a line passing through the origin on the z_1-plane, as shown in Figure 6.16(b). The angle it forms with the real axis is given by

$$\beta = \frac{\pi}{2} - \arctan h$$

Then, applying the transformation:

$$z_2 = z_1^2$$

It maps the line in Figure 6.16(b) to the rays on the z_2-plane that fold back and forth at an angle of $\alpha = 2\beta$, as shown in Figure 6.16(c).

On the other hand, considering a circular arc on the ζ-plane passing through $+1$, going through ih, and ending at -1, as shown by the dotted line in Figure 6.17(b), the fractional linear transformation

$$\zeta_1 = \frac{\zeta - 1}{\zeta + 1}$$

implements the mapping $+1 \longmapsto 0$ and $-1 \longmapsto \infty$. It also maps the segment of the circular arc to the folding rays passing through the origin on the ζ_1-plane, with the same angle:

$$\alpha = \pi - 2 \arctan h \equiv 2\beta$$

Therefore, the transformation $\zeta_1 = z_2$ relates the circle on the z-plane (Figure 6.16(a)) to the segment of the circular arc on the ζ-plane (dotted line in Figure 6.17(b)), expressed as

$$\frac{\zeta - 1}{\zeta + 1} = \left(\frac{z - 1}{z + 1}\right)^2 \longrightarrow \zeta = \frac{1}{2}\left(z + \frac{1}{z}\right)$$

This is precisely the Joukowsky transformation. Since the dotted line circle on the z-plane transforms into the back-and-forth circular arc on the ζ-plane, any circle tangent to the dotted line at $z = +1$ (as shown in Figure 6.17(a)) will be mapped to a closed curve on the ζ-plane tangent to the dotted line at $z = +1$, forming the airfoil profile shown in Figure 6.17(b). In the era before computers, this mapping provided a theoretical model for systematically analyzing the aerodynamic properties of wings.

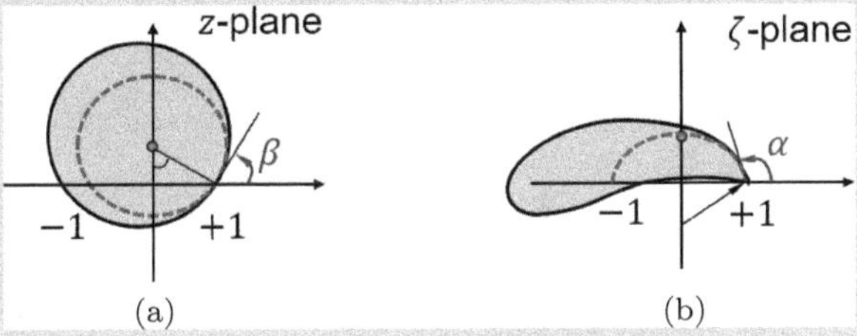

Figure 6.17.

Exercises

[1] Find the distribution of the electric field in space produced by an infinitely long charged conductor with a thin strip of width $2a$.
Answer:

$$z = \frac{a}{2}\left(\zeta + \frac{1}{\zeta}\right), \quad u = C_1 \ln|\zeta| + C_2$$

[2] Find a map to transform a unit disk with a cut of depth h to a perfect unit disk (Figure 6.18).
Answer:

$$z_1 = \mathrm{e}^{-\mathrm{i}\theta} z, \quad z_2 = -\mathrm{i}\frac{z_1 - 1}{z_1 + 1}, \quad \zeta = \sqrt{z_2^2 + \left(\frac{h}{2-h}\right)^2}$$

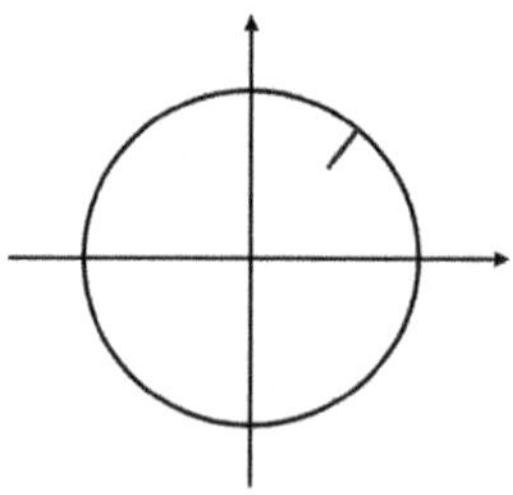

Figure 6.18.

[3] The external region of the hexagonal star with length 1 (Figure 6.19) is mapped to the external region of the unit circle.

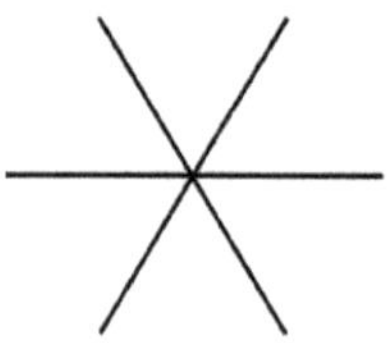

Figure 6.19.

Hint: Let $z_1 = z^3, \zeta_1 = \zeta^3$, and find the transformation between z_1 and ζ_1.
Answer: $z^3 = \frac{1}{2}\left(\zeta^3 + \frac{1}{\zeta^3}\right)$

[4] Find the transformation function $\zeta(z)$ that maps the region $|\mathrm{Re}\, z| \le \frac{\pi}{2}$, $\mathrm{Im}\, z \ge 0$ to the upper half-plane.

Hint: Take $\zeta_1 = \mathrm{e}^{\mathrm{i}z}, \zeta_2 = -\mathrm{i}\zeta_1, \zeta(z) = \frac{1}{2}\left(\zeta_2 + \frac{1}{\zeta_2}\right)$

Answer: $\zeta(z) = \sin z$

6.4 Polygon Mapping

Let us consider the mapping between a polygonal region and the upper half-plane. Suppose there is an n-sided polygon on the z-plane with vertices $a_1, a_2, a_3, \ldots, a_n$, and the external angles at these vertices are given by $\theta_1, \theta_2, \theta_3, \ldots, \theta_n$ (counterclockwise being positive). The total sum of these angles is

$$\theta_1 + \theta_2 + \theta_3 + \cdots + \theta_n = 2\pi$$

We want to find a transformation that maps the vertices $a_1, a_2, a_3, \ldots$ to points $b_1, b_2, b_3, \ldots$ on the real axis of the ζ-plane, i.e.,

$$a_k \longmapsto b_k = \zeta(a_k)$$

This transformation should also map the interior region of the polygon on the z-plane to the upper half-plane of ζ (Figure 6.20).

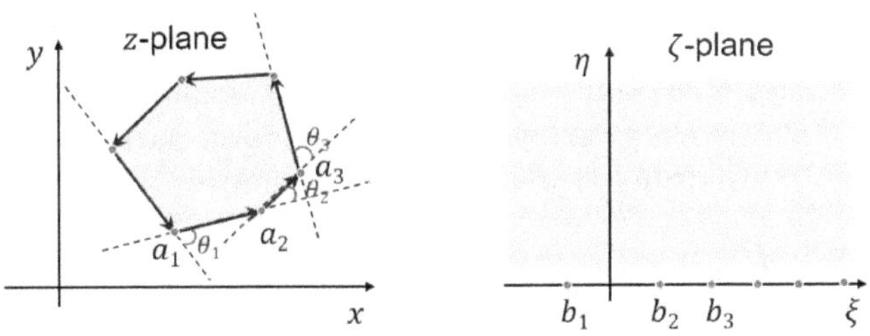

(a) (b)

Figure 6.20.

Since the different vertex angles are transformed into points on the real axis, and the angle between adjacent sides becomes π, the transformation does not preserve angles at these points. Therefore, it should be described by a power function, and $\zeta'(z)$ should be zero or infinite at these points. Consider the relationship between the external angles of the polygon:

$$\phi_k = \arg \left.\frac{\mathrm{d}z}{\mathrm{d}\zeta}\right|_{\zeta=b_k-0}, \quad \phi_{k+1} = \arg \left.\frac{\mathrm{d}z}{\mathrm{d}\zeta}\right|_{\zeta=b_k+0}$$

$$\longrightarrow \theta_k = \phi_{k+1} - \phi_k = \arg\left[\left.\frac{\mathrm{d}z}{\mathrm{d}\zeta}\right|_{\zeta=b_k+0} - \left.\frac{\mathrm{d}z}{\mathrm{d}\zeta}\right|_{\zeta=b_k-0}\right]$$

When ζ goes from the left side to the right side of b_k, the angle turned on the ζ-plane is $-\pi$. Corresponding to the angle θ_k on the z-plane, the mapping between the angles is $-\pi \longmapsto \theta_k$. Namely, the angle is magnified/reduced by $\frac{\theta_k}{\pi}$. To achieve this, we use the power function transformation,

$$z'(\zeta) = A(\zeta - b_k)^{-\frac{\theta_k}{\pi}}$$

This transformation can accomplish this objective. Implementing this mapping for all vertices, we have

$$z'(\zeta) = A(\zeta - b_1)^{-\frac{\theta_1}{\pi}}(\zeta - b_2)^{-\frac{\theta_2}{\pi}} \cdots (\zeta - b_n)^{-\frac{\theta_n}{\pi}}$$

After integrating the above equation, we obtain the Schwarz–Christoffel transformation:

$$z(\zeta) = A\int_{\zeta_0}^{\zeta} (\zeta - b_1)^{-\frac{\theta_1}{\pi}}(\zeta - b_2)^{-\frac{\theta_2}{\pi}} \cdots (\zeta - b_n)^{-\frac{\theta_n}{\pi}}\mathrm{d}\zeta \qquad (6.4.1)$$

If we set one of the points, b_1, as infinity, the mapping starts from $b_2 \longmapsto a_2$. In this case, we can ignore the first term of (6.4.1), and the Schwarz–Christoffel transformation becomes

$$z(\zeta) = A\int_{\zeta_0}^{\zeta} (\zeta - b_2)^{-\frac{\theta_2}{\pi}}(\zeta - b_3)^{-\frac{\theta_3}{\pi}} \cdots (\zeta - b_n)^{-\frac{\theta_n}{\pi}}\mathrm{d}\zeta \qquad (6.4.2)$$

It should be noted that this transformation is independent of the order of the vertices, so it does not uniquely describe the mapping from the polygon to the real axis. This is mainly because the transformation only involves derivatives, i.e., the slopes of the lines, and does not directly provide the relative positions of the vertices. When considering a specific problem, it is necessary to specify the correspondence between points.

Example 6.8. Solve Example 6.2 using the Schwarz–Christoffel transformation.

Solution: Considering the mapping between the vertices $a_k \longmapsto b_k$ from (a) to (b) in Figure 6.21, we set b_1 as an infinitely distant point, and b_2 and b_4 as ± 1. Observing the change in angles between adjacent a_k, we can choose the Schwarz–Christoffel transformation as follows:

$$z(\zeta) = z_0 + A \int (\zeta + h)^{-\frac{1}{2}} \zeta^{+1} (\zeta - h)^{-\frac{1}{2}} \mathrm{d}\zeta$$
$$= z_0 + A \int \frac{\zeta}{\sqrt{\zeta^2 - h^2}} \mathrm{d}\zeta = z_0 + A\sqrt{\zeta^2 - h^2}$$

From $a_3 \longmapsto b_3$, we have $A = 1$. Therefore, the transformation function is

$$\zeta = \sqrt{z^2 + h^2}$$

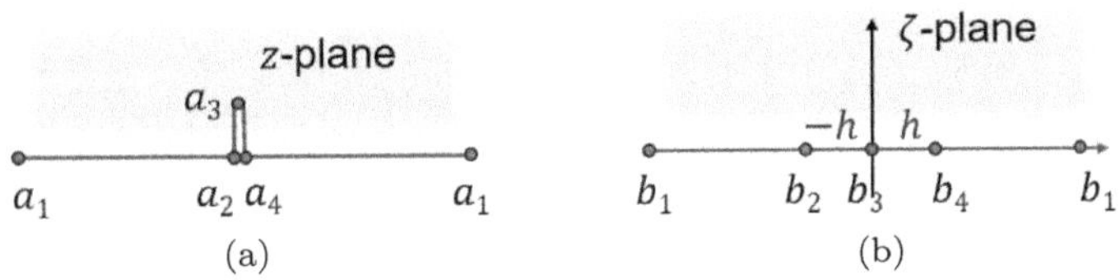

Figure 6.21.

Example 6.9. Mapping of a rectangle region $A_1A_2A_3A_4$ on the plane to the upper half-plane.

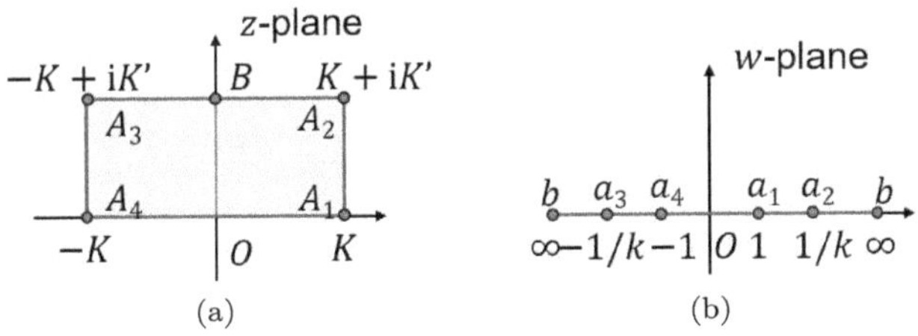

Figure 6.22.

Solution: We map each vertex A_1, A_2, A_3, A_4 of the rectangle to points a_1, a_2, a_3, a_4 on the real axis, and map point B to the point at infinity (Figure 6.22). The corresponding Schwarz–Christoffel transform is given by

$$\begin{aligned} z &= C_1 \int_0^w (w-1)^{-\frac{1}{2}} \left(w - \frac{1}{k}\right)^{-\frac{1}{2}} \left(w + \frac{1}{k}\right)^{-\frac{1}{2}} (w+1)^{-\frac{1}{2}} dw + C_2 \\ &= C \int_0^w \frac{1}{\sqrt{(1-w^2)(1-k^2w^2)}} dw \end{aligned}$$

Since $O \longmapsto O$, we have $C_2 = 0$. Taking the constant $C = 1$ and using the mapping relation $A_1 \longmapsto a_1 = 1$, we have

$$K = \int_0^1 \frac{1}{\sqrt{(1-t^2)(1-k^2t^2)}} dt$$

Here, $K(k)$ represents the complete elliptic integral of the first kind. Next, considering $A_2 \longmapsto a_2 = \frac{1}{k}$ $(0 < k < 1)$, we have

$$\begin{aligned} K + \mathrm{i}K' &= \int_0^{\frac{1}{k}} \frac{1}{\sqrt{(1-t^2)(1-k^2t^2)}} \mathrm{d}t \\ &= \int_0^1 \frac{1}{\sqrt{(1-t^2)(1-k^2t^2)}} \mathrm{d}t + \int_1^{\frac{1}{k}} \frac{1}{\sqrt{(1-t^2)(1-k^2t^2)}} \mathrm{d}t \end{aligned}$$

Let $k'^2 = 1 - k^2$, and perform the variable substitution,

$$t \longmapsto \frac{1}{\sqrt{1-k'^2\tau^2}}$$

We get

$$\mathrm{i}K' = \mathrm{i} \int_0^1 \frac{1}{\sqrt{(1-\tau^2)(1-k'^2\tau^2)}} \mathrm{d}\tau$$

Therefore,

$$K' = \int_0^1 \frac{1}{\sqrt{(1-t^2)(1-k'^2t^2)}} \mathrm{d}t \equiv K(k')$$

The inverse function of the Schwarz–Christoffel transformation, which maps the upper half-plane to the rectangle, is the Jacobi elliptic sine function:

$$w = sn(z, k)$$

It implements the mapping from the rectangle to the upper half-plane.

The Schwarz–Christoffel transformation also maps the interior of the unit circle to the interior of the polygon:

$$z(\zeta) = A \int_{\zeta_0}^{\zeta} (\zeta - b_1)^{-\frac{\theta_1}{\pi}} (\zeta - b_2)^{-\frac{\theta_2}{\pi}} \cdots (\zeta - b_n)^{-\frac{\theta_n}{\pi}} \mathrm{d}\zeta \qquad (6.4.3)$$

where b_k are the points on the unit circle corresponding to the vertices of the polygon. Additionally, assuming that the points at infinity in the z-plane and ζ-plane correspond to each other, this is omitted here.

ENDNOTE

Chapter 2 introduced elliptic integrals and elliptic functions. Now let us specifically examine the mapping properties of the first kind of elliptic integrals,

$$F(z, k) = \int_0^z \frac{\mathrm{d}u}{\sqrt{(1-u^2)(1-k^2u^2)}}$$

Here, $0 < k < 1$ is the modulus of the elliptic integral. We consider the branch of the square root in the integral as the positive value on the x-axis interval $[0, 1]$ to study the mapping of the upper half-plane.

When $z = x$ moves along the real axis from left to right over $[0, 1]$ (interval I), the integral $F(z, k)$ takes positive values and increases from to

$$K = F(1, k) = \int_0^1 \frac{\mathrm{d}x}{\sqrt{(1-x^2)(1-k^2x^2)}}$$

Thus, it maps $[0, 1]$ to $[0, K]$. When $z = x$ passes through $\left[1, \frac{1}{k}\right]$ (interval II), one of the four linear factors in the square root changes sign, resulting in a purely imaginary integral with an argument of $\frac{\pi}{2}$,

$$\mathrm{i}K' = \mathrm{i} \int_1^{\frac{1}{k}} \frac{\mathrm{d}x}{\sqrt{(x^2-1)(1-k^2x^2)}}$$

It maps to the line segment $[KK + \mathrm{i}K']$. As $z = x$ continues moving to the right after passing through the point $\frac{1}{k}$ (interval III), the sign of $1 - kx$ in the square root also changes, increasing the argument by $\frac{\pi}{2}$. Therefore, the integral becomes negative, and the total argument becomes π,

$$F(x,k) = \int_{\frac{1}{k}}^{x} \frac{\mathrm{d}x}{\sqrt{(1-x^2)(1-k^2x^2)}}$$
$$= -\int_{\frac{1}{k}}^{x} \frac{\mathrm{d}x}{\sqrt{(x^2-1)(k^2x^2-1)}}$$

By making the variable substitution $x \to \frac{1}{kx}$, we have

$$\int_{\frac{1}{k}}^{\infty} \frac{\mathrm{d}x}{\sqrt{(x^2-1)(k^2x^2-1)}} = \int_{0}^{1} \frac{\mathrm{d}x}{\sqrt{(1-x^2)(1-k^2x^2)}} = K$$

Thus, it maps to the line segment $[K + \mathrm{i}K', \mathrm{i}K']$. By following the same process, the mapping process corresponding to integration along the negative real axis can be obtained, as shown in Figure 6.23.

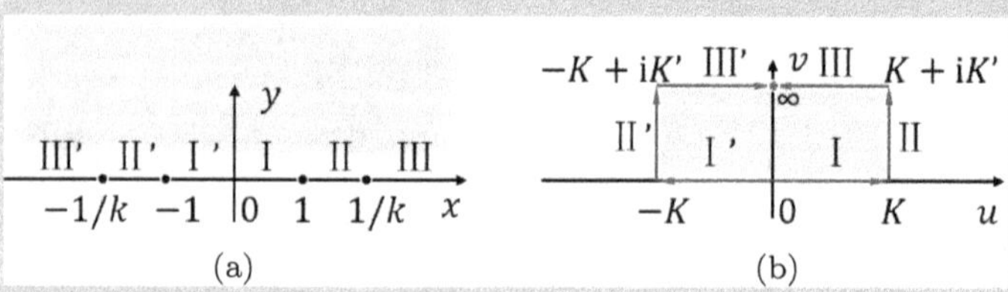

Figure 6.23.

So, the first kind of elliptic integral $F(zk)$ implements a conformal mapping from the upper half-plane H to the rectangular region B:

$$H \longmapsto B\colon [-K, K; K + \mathrm{i}K', -K + \mathrm{i}K']$$

Conversely, its inverse function is the Jacobi elliptic sine function $sn(z,k)$, which implements a mapping from the rectangular region B to the upper half-plane H. The function

$sn(z,k)$ defined in the rectangular region B can be analytically extended using the Schwarz reflection principle. For example, by symmetrically extending it across the edge I of the rectangle to the rectangular region B' shown in Figure 6.24, the extended analytic function $sn(z,k)$ maps B' to the lower half-plane of the complex plane. Similarly, by extending the function analytically across the edge II of the rectangle, we obtain the region B'', which is also mapped to the lower half-plane. Likewise, B'' can be analytically extended to the region B''', and the extended function $sn(z,k)$ maps B''' back to the upper half-plane. By extending $sn(z,k)$ analytically to the entire z-plane, the shaded rectangular regions are all mapped to the upper half-plane, while the white rectangular regions are all mapped to the lower half-plane. The double periodicity of the Jacobi elliptic sine function becomes clear: it is the translational periodicity of the rectangular region along the real and imaginary axes of the complex plane, with translation periods of $4K$ and $2\mathrm{i}K'$, respectively:

$$sn(z + 4Km + 2iK'n) = sn\, z \quad (m, n \in \mathbb{N})$$

Since $(0, \mathrm{i}K') \longmapsto \infty$, at positions $z = 2mK + i(2n+1)K'$, the function snz has simple poles, represented by the circular points on the edges of the rectangle, as shown in Figure 6.24. The basic properties of Jacobi elliptic functions are listed in Table 6.1.

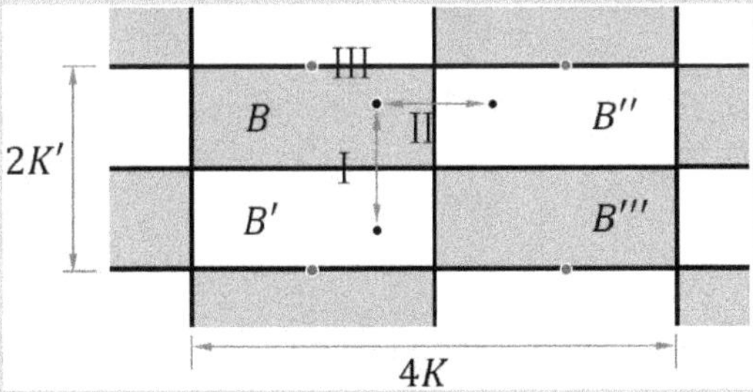

Figure 6.24.

Table 6.1.

	Period (ω_1, ω_2)	Zeros	Poles
snz	$4K; 2iK'$	$2mK + 2niK'$	$2mK + (2n+1)iK'$
cnz	$4K; 2K + 2iK'$	$(2m+1)K + 2niK'$	$2mK + (2n+1)iK'$
dnz	$2K; 4iK'$	$(2m+1)K + (2n+1)iK'$	$2mK + (2n+1)iK'$

For general elliptic functions, their double periods in the complex plane form a parallelogram region, equivalent to being defined on a torus, as shown in Figure 6.25. Their analytic properties are described by Liouville's theorem:

[1] **Liouville's first theorem.** *A double periodic function without singularities must be constant.*
Therefore, non-constant double periodic functions must have singularities. Only double periodic functions with poles (meromorphic functions) are called elliptic functions, and the number of poles within a parallelogram period (counting mth-order poles as m individual poles) is referred to as the order of the elliptic function.

[2] **Liouville's second theorem.** *The sum of the residues of an elliptic function at the poles within a parallelogram period is zero.*

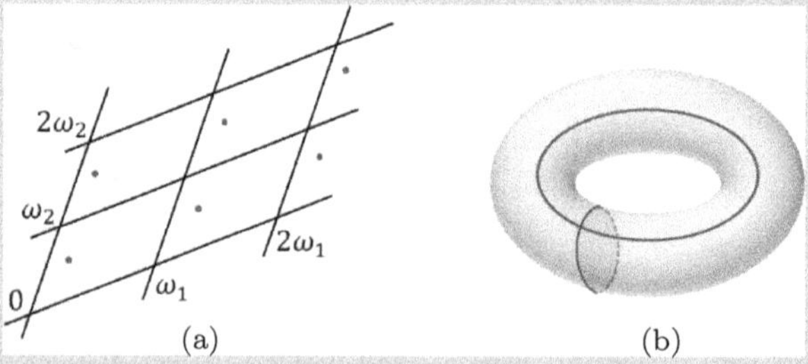

Figure 6.25.

This theorem indicates that elliptic functions are at least of second order: Jacobi elliptic functions have two simple poles within one period, while the Weierstrass elliptic function ($\mathbb{P}$-function) has a second-order pole with a residue of zero.

[3] **Liouville's third theorem.** *The number of zeros of an elliptic function within a parallelogram period is equal to the number of poles.*

Exercises

[1] Construct a conformal mapping that maps the upper half-plane $\operatorname{Im} z > 0$ to the half-strip region:

$$\{\operatorname{Im} w > 0, \quad -a < \operatorname{Re} w < a\}$$

Hint: Use the Schwarz–Christoffel transformation to implement the mapping: $-a \mapsto -1, a \mapsto +1, \infty \mapsto \infty$.
Answer: $z = \sin\left(\frac{\pi w}{2a}\right)$

[2] Construct a conformal mapping that maps the upper half-plane to the upper half-plane with a secant line $[ib, i\infty)$ on the imaginary axis.
Hint: Apply the Schwarz–Christoffel transformation with the mapping: $0 \mapsto -\infty, \infty \mapsto +\infty, -1 \mapsto ib$.
Answer: $\frac{b}{2}\left(\sqrt{z} - \frac{1}{\sqrt{z}}\right)$

[3] Construct a conformal mapping to flatten the protrusions in Figure 6.26:
(a) the rectangle;
(b) the inclined plate.

Figure 6.26.

Answer:

$$\text{(a) } z = C_1 \int_0^w \sqrt{\frac{1-k^2u^2}{1-u^2}}\mathrm{d}u + C_2$$

$$\text{(b) } w = (z-1)^\alpha \left(1 + \frac{\alpha}{1-\alpha} z\right)^{1-\alpha}$$

[4] Two long, thin conductor strips of width b are placed parallel on the same horizontal line with a distance of $2a$ between their nearby ends. Determine the capacitance per unit length.
Answer:

$$C = \frac{\varepsilon_0 K(k')}{K(k)} = \frac{\varepsilon_0 K\left(\frac{\sqrt{(a+b)^2-a^2}}{a+b}\right)}{K\left(\frac{a}{a+b}\right)}$$

[5] Proof:

(a) A double periodic function without singularities must be constant.

(b) The sum of the residues of an elliptic function at the poles within a parallelogram period is zero.
Hint: Use periodicity to show that the sum of integrals along two pairs of opposite edges of the parallelogram is zero.

(c) The number of zeros of an elliptic function within a parallelogram period is equal to the number of poles.
Hint: The logarithmic derivative of an elliptic function is still a double periodic function with simple poles.

(d) The sum of zeros minus the sum of poles of an elliptic function within a parallelogram period is equal to its period.
Hint: Let L be the boundary loop of the parallelogram period. The sum of the differences between the zeros and poles of an elliptic function is given by

$$\sum_{k=1}^{p} (\alpha_k - \beta_k) = \frac{1}{2\pi \mathrm{i}} \oint_L z \frac{f'(z)}{f(z)} \mathrm{d}z$$

The logarithmic derivative $\frac{f'(z)}{f(z)}$ has simple poles at the zeros and poles of $f(z)$, and its argument changes by $2\pi p$ over a period.

6.5 Conformal Self-Maps

6.5.1 *Region self-maps*

For an analytic function $f(z)$, if there exists a point $z = z_0$ such that $f(z_0) = z_0$, then z_0 is called a fixed point of $f(z)$. A non-identity fractional linear transformation can have either one or two fixed points. For example, $z = \infty$ is a fixed point of the mapping $f(z) = az + b$.

A one-to-one conformal mapping from an open set U to itself is called a conformal self-map of U: $U \longmapsto U$. In fact, fractional linear transformations are the self-maps of the Riemann sphere $\mathbb{S}$ onto itself. The set of all self-maps of U forms a group, which means that if f and g are self-maps of U, then $f \circ g$ is also a self-map of U; the inverse mapping of a conformal self-map is also a conformal self-map; and $g(z) = z$ is the identity mapping.

Let D be the open unit disk on the complex plane $D = \{|z| < 1\}$. Obviously, the rotation $z \longmapsto \mathrm{e}^{\mathrm{i}\theta} z (\theta \in \mathbb{R})$ constitutes a conformal self-map of the unit disk. Generally, all self-maps of the open unit disk D have the following fractional linear form:

$$\psi(z) = \mathrm{e}^{\mathrm{i}\theta} \frac{z - a}{1 - \bar{a}z} \quad (|a| < 1)$$

Proof. Considering the boundary of the unit disk $z = \mathrm{e}^{\mathrm{i}\phi}$, we have

$$|\psi(z)| = \left| \frac{\mathrm{e}^{\mathrm{i}\phi} - a}{1 - \bar{a}\mathrm{e}^{\mathrm{i}\phi}} \right| = \left| \frac{1 - \bar{a}\mathrm{e}^{\mathrm{i}\phi}}{1 - \bar{a}\mathrm{e}^{\mathrm{i}\phi}} \right| = 1$$

According to the boundary correspondence principle, we can conclude that $\psi(z)$ is a conformal self-map of the unit disk.

An important property of the function $\psi(z)$ is that it implements a reciprocal mapping: $\psi(0) = a, \psi(a) = 0$, and the inverse mapping of $f(z)$ is itself: $\psi(z) = \psi^{-1}(z)$.

All fractional linear mappings of the following form:

$$z \longmapsto \frac{az + b}{cz + d} \quad (a, b, c, d \in \mathbb{R},\ ad - bc = 1)$$

implement self-maps of the upper half-plane $H = \{\mathrm{Im}\, z > 0\}$. These mappings form a special matrix group called the modular group.
Practice: Prove $\psi(z) = \psi^{-1}(z)$. □

6.5.2 *Hyperbolic geometry*

Let the analytic function $w = \psi(z)$ be a conformal self-map of the open unit disk D. Then we have

$$\frac{|\mathrm{d}w|}{1-|w|^2} = \frac{|\mathrm{d}z|}{1-|z|^2} \tag{6.5.1}$$

If L is a smooth curve inside D and L' is the image of L under the self-map, then

$$\rho \equiv 2\int_{L'} \frac{|\mathrm{d}w|}{1-|w|^2} = 2\int_{L} \frac{|\mathrm{d}z|}{1-|z|^2}$$

Using the hyperbolic metric, the distance between two points z_0 and z_1 on the complex plane is defined as

$$\rho(z_0, z) \overset{\text{def}}{=} 2\int_{z_0}^{z} \frac{|\mathrm{d}z|}{1-|z|^2} \tag{6.5.2}$$

The hyperbolic metric distance between two points remains invariant under conformal self-maps, i.e.,

$$\rho(\psi(z_0), \psi(z)) = \rho(z_0, z)(z_0 z \in D)$$

For any two points z_0 and z_1 inside the open unit disk D, there exists a unique shortest curve with respect to the hyperbolic metric, which is an arc passing through z_0 and z_1 and orthogonal to the unit circle, as shown in Figure 6.27(a). This curve is called a geodesic, and the hyperbolic metric distance from the origin to z is given by

$$\rho(0, z) = 2\int_0^z \frac{|\mathrm{d}z|}{1-|z|^2} = \int_0^{|z|} \left[\frac{1}{1-t} + \frac{1}{1+t}\right] \mathrm{d}t = \ln\frac{1+|z|}{1-|z|}$$

It can be observed that as z approaches the boundary of the unit circle, the distance from the origin to z tends to infinity:

$$\rho(0, z) \overset{|z|\to 1}{\longrightarrow} \infty$$

Due to the invariance of conformal self-maps, any arc orthogonal to the unit circle has an infinite hyperbolic distance between its two intersection points, which is equivalent to a straight line in hyperbolic geometry. A geodesic triangle is formed by three hyperbolic

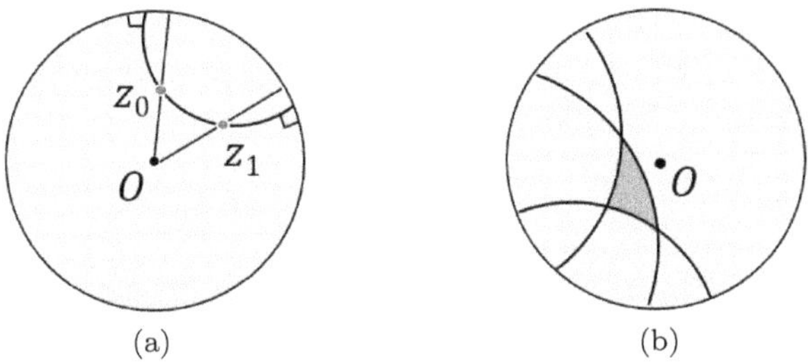

Figure 6.27.

geodesics, as shown in Figure 6.27(b), and its sum of interior angles is less than π. It may seem unfortunate since the sum of interior angles being equal to π reflects the elegant simplicity of Euclidean geometry. However, hyperbolic geometry also exhibits its unique aspect, where the area of a hyperbolic triangle is equal to the sum of its three interior angles and is independent of its side lengths.

6.5.3 *Julia set*

Let us assume that the analytic function $f(z)$ implements a conformal self-map of a region U in the complex plane, $U \longmapsto U$. Consider its nth iteration:

$$z \longmapsto f(z) \longmapsto f(f(z)) \longmapsto \cdots \longmapsto f(f(\cdots f(f(z))\cdots))$$

which is denoted as $f^n(z)$. The set of self-maps of the entire function $f(z)$ is called the Julia set $J[f]$, defined as

$$f(J[f]) = J[f] = f^{-1}(J[f]) \tag{6.5.3}$$

In other words, $z_0 \in J[f]$ if and only if $f(z_0) \in J[f]$. For example, for the function $f(z) = z^2$, the Julia set $J[f]$ is the unit disk D, which remains invariant under iterative applications of $f(z)$. Similarly, for the function $f(z) = z^2 - 2$, the Julia set $J[f]$ is the interval $[-2, 2]$, which also remains invariant under iterative applications of $f(z)$.

It is important not to confuse the nth iteration of $f(z)$ with the nth power of $f(z)$, denoted as $[f(z)]^n$. For example, for $f(z) = z + 1$, the nth iteration is $f^n(z) = z + n$, while for $f(z) = z^d$, the nth iteration is $f^n(z) = z^{d^n}$. If $f(z)$ is a d-degree polynomial, then $f^n(z)$ is a d^n-degree polynomial.

In general, it is difficult to anticipate the geometric structure of the Julia set $J[f]$ when repeatedly iterating an analytic function. For example, consider the quadratic function:

$$f_c(z) = z^2 + c \quad (c \in \mathbb{C})$$

Given a complex number c, repeatedly applying the iteration $z \longmapsto z^2 + c$ starting from an initial $z = 0$, yield the sequence:

$$\longmapsto c \longmapsto c^2 + c \longmapsto (c^2 + c)^2 + c \longmapsto ((c^2 + c)^2 + c)^2 + c \longmapsto \cdots$$

If the results of iterating any number of times remain within a finite range, then the set of all these points forms a region invariant under self-maps, forming the Julia set:

$$J[f_c] = \{z \in \mathbb{C} : \forall n \in N, |f_c^n(z)| \leq 2\}$$

where $f_c^n(z)$ represents the nth iteration of $f_c(z)$. On the other hand, points that escape to infinity under repeated iteration form the Fatou set. Clearly, the Julia set and the Fatou set are complementary. By choosing different parameters c and applying suitable coloring schemes, Figure 6.28 showcases the mesmerizing Julia sets.

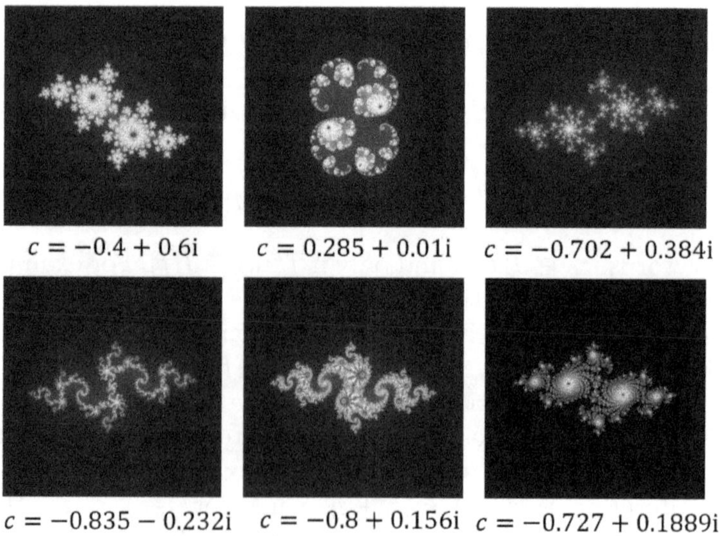

Figure 6.28. (image from Wikipedia).

This concept applies to higher-degree polynomial functions as well, as illustrated in Figure 6.29. Additionally, Julia sets can be defined for multivariable complex functions.

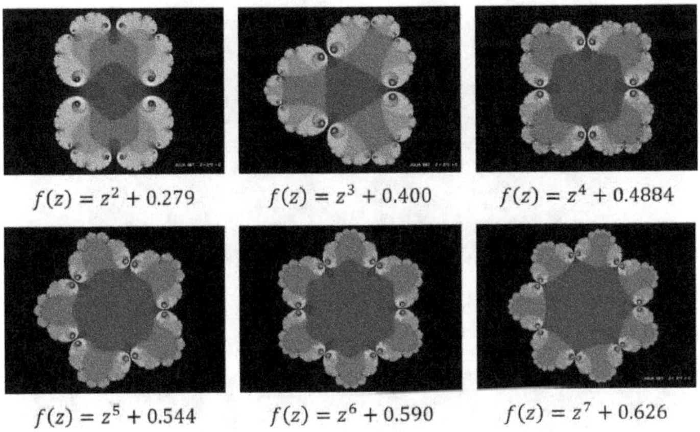

Figure 6.29. (image from Wikipedia).

6.5.4 *Mandelbrot set*

The Mandelbrot set M is defined as the set of parameters c in the complex plane for which the corresponding Julia set $J[f_c]$ is connected. The Julia sets $J[f_c]$ are formed by the quadratic polynomial functions:

$$f_c(z) = z^2 + c \quad (c \in \mathbb{C})$$

It can be proven that a parameter c belongs to the Mandelbrot set if and only if $|f_c^n(0)| \leq 2$ for all $n \geq 1$. The Mandelbrot set is a closed subset of the disk $|c| \leq 2$. Figure 6.30 displays the famous Mandelbrot set for the function $f_c(z)$.

The Mandelbrot set has a remarkable characteristic of not having a clear boundary. It can be infinitely zoomed in, revealing intricate and self-similar structures, as depicted in Figure 6.31. It empirically embodies the Chinese maxim: "unknown beginning, unknown end, unknown gate, unknown extremity, unknown source". A finite area can have an infinitely long boundary, and the length of the boundary depends on the scale of measurement, giving rise to a new branch of mathematics called fractal geometry.

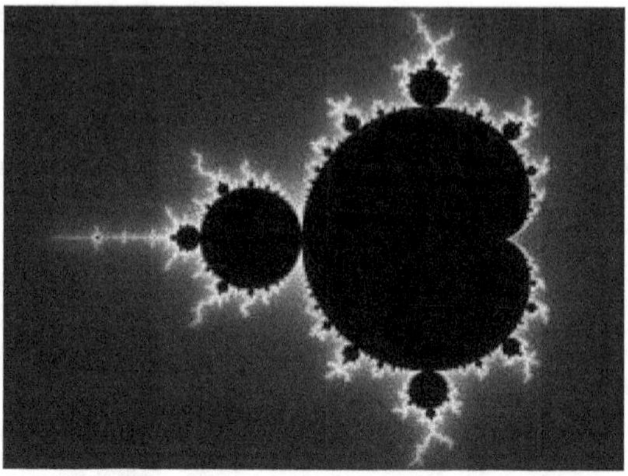

Figure 6.30. (image from Wikipedia).

Figure 6.31. (image from Wikipedia).

ENDNOTE

In 1882, J. H. Poincaré provided a natural interpretation of Lobachevskian geometry using fractional linear transformations. In different cases, these transformations can represent the rigid motions of Euclidean geometry, spherical geometry, and hyperbolic geometry. This insight led to a fundamental breakthrough in non-Euclidean geometry. The three different geometries are closely related through complex analysis:

(1) **Euclidean metric**

The infinitesimal length in the complex plane is given by

$$|\mathrm{d}z| = |z_1 - z_2|$$

The geodesics are straight lines, and the sum of the interior angles of a triangle is π.

(2) **Hyperbolic metric**

The infinitesimal length in the open unit disk is given by

$$\frac{2\mathrm{d}|z|}{1 - |z|^2}$$

The geodesics are arcs orthogonal to the unit circle, and the sum of the interior angles of a triangle is less than π.

(3) **Spherical metric**

The infinitesimal length on the Riemann sphere is given by

$$\frac{2\mathrm{d}|z|}{1 + |z|^2}$$

The geodesics are great circle arcs on the Riemann sphere, and the sum of the interior angles of a spherical triangle is greater than π.

The Poincaré disk, shown in Figure 6.27, is a hyperbolic geometry model proposed by E. Beltrami. It considers a unit open disk D, where the size of an object is proportional to $1-|z|^2$. As one moves farther from the center, the object becomes smaller. If a person's field of view is limited and they cannot see far away, they would not perceive themselves as becoming smaller or larger. Thus, as the distance from the center increases, people appear smaller, and relative to them, the perceived space also becomes larger. When an object approaches the boundary, its size approaches zero, and the space becomes infinitely large. Therefore, an object can never reach the boundary.

Figure 6.32 is an artwork called "Circle Limit" by M. C. Escher (1959), which visually embodies the concept of hyperbolic geometry. Intuitively, the entire world of the hyperbolic plane is squeezed inside the Euclidean circle, making the fish near the edge appear very crowded. However, this is an illusion. According to the conformal invariance property of hyperbolic geometry, fish of the same color in the image are identical in size and shape. From the perspective of Euclidean metric, objects near the edge appear to be getting smaller. But from the viewpoint of hyperbolic geometry, the boundary circle is always at an infinite distance, so there is no boundary circle.

Figure 6.32. (image from Wikipedia).

On the other hand, it has long been known that the sum of interior angles of a spherical triangle is greater than π, and no squares can be drawn on a sphere. It is impossible to flatten an entire spherical surface without creating cracks or wrinkles, and so on. However, people have not considered it as a different geometry from Euclidean geometry because they are accustomed to viewing the sphere as an object embedded in three-dimensional Euclidean space, thereby overlooking its non-Euclidean properties. Imagine a world where the perception of space is limited to only the two dimensions of the sphere, and the curvature of this world is positive. In such a geometry, the concept of parallel lines does not exist. In comparison, hyperbolic geometry corresponds to spherical geometry with imaginary radius, where circular functions like $\sin x$ and $\cos x$ are replaced by hyperbolic functions $\sinh x$ and $\cosh x$. Gauss may have been the first person to understand the concept of changing curvature in two-dimensional space, and Riemann extended this concept to higher dimensions. Their work laid the mathematical foundation for Albert Einstein's theory of general relativity in the 20th century.

In 1967, the French mathematician Benoit Mandelbrot raised a seemingly trivial question: How long is the coastline of Britain? When measured in kilometers and then converted to meters, the lengths obtained should be the same, right? In fact, when measured in meters, the length will be longer than when measured in kilometers because when measuring in kilometers, the intricate details of the coastline that are shorter than one kilometer are ignored. By measuring in meters, the winding and intricate nature of the coastline can be captured, resulting in a longer measured length.

Now, does the length have a maximum value? Mandelbrot discovered that as the measuring unit tends to infinitesimally small, more previously ignored curve details are revealed, and theoretically, the length obtained may tend to infinity! Therefore, he claimed that the length of the coastline is uncertain and depends on the measuring scale. This seems paradoxical because it assumes that the coastline is an irregular and non-smooth curve, different in nature from the regular geometric shapes or smooth curves studied in classical geometry. Mandelbrot called such curves fractals, meaning "fragments". In other words, these curves are continuous but not differentiable everywhere. Mandelbrot discovered that these structures have a fascinating characteristic called self-similarity, where the structure of any small region is essentially similar to that of a larger region. To describe this self-similarity, Mandelbrot introduced the concept of fractal dimension, which can take fractional values, in contrast to the usual integer values of geometric dimensions.

In fact, fractals are ubiquitous in nature. Figure 6.33 shows self-similar fractal structures in landforms (a), plant root systems (b), lightning (c), and the Yangtze River system (d). However, it was in the abstract realm of complex analysis theory that people truly began to notice fractal phenomena. It seems that mathematicians are more adept at exploring new territories through abstract thinking than finding inspiration from the real world. This surprising phenomenon is not unique in mathematics. For example, symmetries are abundant, but the impetus to study group theory came from the problem of solving quintic equations. Another famous example is the Seven Bridges of Königsberg problem, which Euler successfully solved but considered it not a mathematical problem "worthy of wasting the precious time of mathematicians".

When we suddenly encounter completely different natural processes that exhibit such striking similarities, does our dormant curiosity triggered with a shiver of lightning?

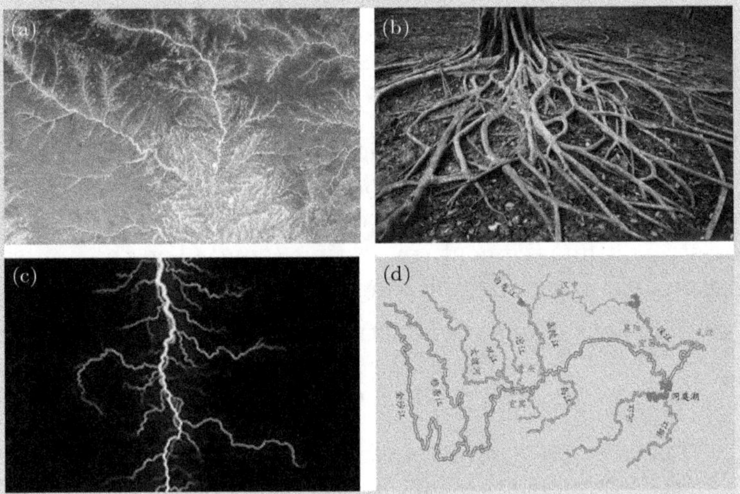

Figure 6.33. (image from networks).

Exercises

[1] Let T be a Möbius transformation with two fixed points z_1 and z_2. If S is also a Möbius transformation, prove that the fixed points of the transformation $S^{-1}TS$ are $S^{-1}z_1$ and $S^{-1}z_2$.

[2] If two non-constant Möbius transformations T and S have the same fixed points, prove that $TS = ST$.

[3] Prove that the fractional linear transformation

$$w = \mathrm{e}^{\mathrm{i}\theta}\frac{z-a}{1-\bar{a}z} \quad (|z|<1)$$

preserves hyperbolic metric length:

$$\frac{|\mathrm{d}w|}{1-|w|^2} = \frac{|\mathrm{d}z|}{1-|z|^2}$$

[4] Prove that the fractional linear transformation

$$w = \frac{az+\bar{b}}{-bz+\bar{a}}, \quad |a|^2+|b|^2=1$$

preserves spherical metric length:

$$\frac{|dw|}{1+|w|^2} = \frac{|dz|}{1+|z|^2}$$

[5] On the sphere, consider two points P and Q, and let their stereographic projections on the complex plane be z_1 and z_2, respectively. We define the chordal distance $d(z_1, z_2)$ as the length of the straight line segment connecting P and Q on the unit sphere. We need to prove the following properties:

(a) Chordal distance is a metric: $d(z_1, z_2) = d(z_2, z_1)$, and

$$d(z_1, z_2) \leq d(z_1, z_3) + d(z_3, z_2)$$

(b) The chordal distance between two points z_1 and z_2 on the complex plane is given by

$$d(z_1, z_2) = \frac{2|z_1 - z_2|}{\sqrt{1+|z_1|^2}\sqrt{1+|z_2|^2}} \quad (z_1 z_2 \in \mathbb{C})$$

As $z_1 \to z_2$, the chordal distance becomes equal to the distance on the unit sphere (the metric length).

(c) $d(z, \infty) = \frac{2}{\sqrt{1+|z|^2}}$

(d) $d(z_1, z_2) = d\left(\frac{1}{z_1}, \frac{1}{z_2}\right)$ $(z_1, z_2 \neq 0)$

(e) $d(z, 0) = d\left(\frac{1}{z}, \infty\right)$ $(z \neq 0)$
Hint: $[d(z_1, z_2)]^2 = (X_1 - X_2)^2 + (Y_1 - Y_2)^2 + (Z_1 - Z_2)^2$

[6] Find a transformation that maps the unit disk $|z| < 1$ to the unit disk $|w| < 1$ and transforms $z = a$ to $w = 0$.

Chapter 7

Fourier Analysis

In music theory, the interval between two audio frequencies that differ by a factor of 2 is divided into octaves. When the frequency ratio between two sounds is a rational number such as $\frac{1}{2}, \frac{2}{3}, \frac{3}{4}$ etc., the sounds are perceived as harmonious. This is known as the Pythagorean scale. In ancient times, people couldn't understand the mathematical principles behind it, let alone the physical principles. They simply revered rational numbers as the embodiment of nature's perfection, believing that "everything is number". The mystery behind this is related to Fourier series.

7.1 Fourier Series

7.1.1 *Orthogonal set of trigonometric functions*

Consider an infinite set composed of the following family of trigonometric functions:

$$\left\{1, \cos\frac{2\pi x}{T}, \cos\frac{4\pi x}{T}, \ldots, \cos\frac{2n\pi x}{T}, \ldots; \sin\frac{2\pi x}{T}, \ldots, \sin\frac{2n\pi x}{T}, \ldots\right\} \tag{7.1.1}$$

Except for the first constant term, each term is a sine or cosine function with a frequency that is an integer multiple of a basic frequency $\omega_0 = \frac{2\pi}{T}$. Importantly, these trigonometric functions with harmonic relationships are orthogonal to each other. This means that the integral of the product of any two different functions over one period (inner product) is zero. Let $T = 2l$, we have

$$\frac{1}{l}\int_{-l}^{l} \sin\frac{2m\pi x}{T}\cos\frac{2n\pi x}{T}\,\mathrm{d}x = 0$$

$$\frac{1}{l}\int_{-l}^{l} \cos\frac{2m\pi x}{T}\cos\frac{2n\pi x}{T}\,\mathrm{d}x = \delta_{mn}$$

$$\frac{1}{l}\int_{-l}^{l} \sin\frac{2m\pi x}{T}\sin\frac{2n\pi x}{T}\,\mathrm{d}x = \delta_{mn}$$

Here, δ_{mn} denotes the Kronecker delta symbol:

$$\delta_{mn} = \begin{cases} 1 & (m = n) \\ 0 & (m \neq n) \end{cases}$$

Considering these functions in set (7.1.1) as the basis functions, if $f(x)$ is a function with a period of T, $f(x+T) = f(x)$, it can be expanded into a Fourier series:

$$f(x) = \frac{a_0}{2} + \sum_{n=1}^{\infty}\left(a_n\cos\frac{2n\pi x}{T} + b_n\sin\frac{2n\pi x}{T}\right) \tag{7.1.2}$$

The expansion coefficients are given by

$$a_n = \frac{1}{l}\int_{-l}^{l} f(\xi)\cos\frac{2n\pi\xi}{T}\,\mathrm{d}\xi, \quad b_n = \frac{1}{l}\int_{-l}^{l} f(\xi)\sin\frac{2n\pi\xi}{T}\,\mathrm{d}\xi$$

The reason this result holds is that besides the orthogonality property of the basis functions, they also form a complete set.

7.1.2 *Dirichlet's theorem*

The Fourier expansion formula is an infinite series that requires convergence. For this, it is necessary to impose certain conditions on the periodic function $f(x)$. This leads to Dirichlet's theorem, which states that if the function $f(x)$ is continuous on the interval $[-l, l]$, or has only a finite number of first-class discontinuities and a finite number of extremal points, then the Fourier series converges, and we have

$$\frac{a_0}{2}+\sum_{n=1}^{\infty}\left(a_n\cos\frac{n\pi x}{l}+b_n\sin\frac{n\pi x}{l}\right)=\frac{1}{2}[f(x-0)+f(x+0)] \tag{7.1.3}$$

For functions defined on finite intervals, the function can be transformed into a periodic function by periodic translation, and then the Fourier series expansion can be performed. The choice of periodic extension depends on the given constraints or physical conditions.

(1) If the function $f(x)$ is odd, it can be expanded as the Fourier sine series:

$$\begin{cases} f(x)=\sum\limits_{n=1}^{\infty} b_n\sin\dfrac{n\pi x}{l} \\ b_n=\dfrac{2}{l}\displaystyle\int_0^l f(\xi)\sin\frac{n\pi\xi}{l}\,\mathrm{d}\xi \end{cases}$$

(2) If the function $f(x)$ is even, it can be expanded as the Fourier cosine series:

$$\begin{cases} f(x)=\dfrac{a_0}{2}+\sum\limits_{n=1}^{\infty} a_n\cos\dfrac{n\pi x}{l} \\ a_n=\dfrac{2}{l}\displaystyle\int_0^l f(\xi)\cos\frac{n\pi\xi}{l}\,\mathrm{d}\xi \end{cases}$$

Figure 7.1 illustrates the synthesis of a periodic function using harmonics with integer multiples of frequencies. The bold line represents the synthesized periodic function, while the thin lines represent the harmonic components. The second row represents the proportion (amplitude) a_n or b_n of each frequency component. Conversely, a periodic function contains different harmonic components, with harmonics having frequency multiples, and the lowest frequency is called the fundamental frequency. The Fourier series expansion separates these harmonic components from the periodic function.

Discussion: Both the upper and lower rows in Figure 7.1 describe the properties of the same function: the upper row represents the curve in the configuration space (sometimes referred to as the time

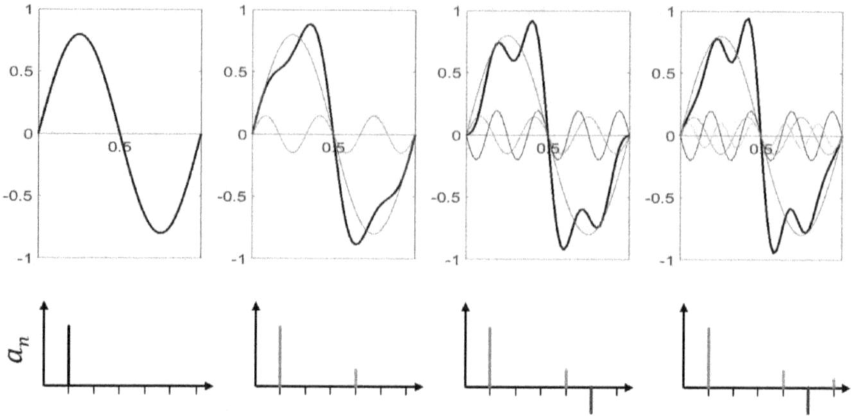

Figure 7.1.

domain), while the lower row represents the curve in the frequency spectrum space (sometimes referred to as the frequency domain), which consists of discrete points for periodic functions. These two representations are completely equivalent. By knowing the distribution of the spectrum and combining it with the corresponding harmonic functions, the function in the configuration space can be fully described, and vice versa. This situation is generally applicable, such as in the case of a function that is continuously differentiable at $x = 0$, which can be expanded into a Taylor series:

$$f(x) = \sum_{k=0}^{\infty} a_k x^k, \quad a_k = \frac{1}{k!} f^{(k)}(0)$$

This expansion is unique. Treating $\{x^k\}_{k=0}^{\infty}$ as a set of linearly independent basis functions, the function $f(x)$ in the configuration space can be represented by a set of discrete coefficients a_k:

$$f(x) \longleftrightarrow \{a_k\}_{k=0}^{\infty}$$

Example 7.1. Expand the following function in finite interval into a Fourier series:

$$f(x) = x + x^2 \quad (-\pi \leq x \leq \pi)$$

Solution: Periodically shift the function defined on $x \in [-\pi, \pi]$ to obtain a periodic function as shown in Figure 7.2. It is a smooth

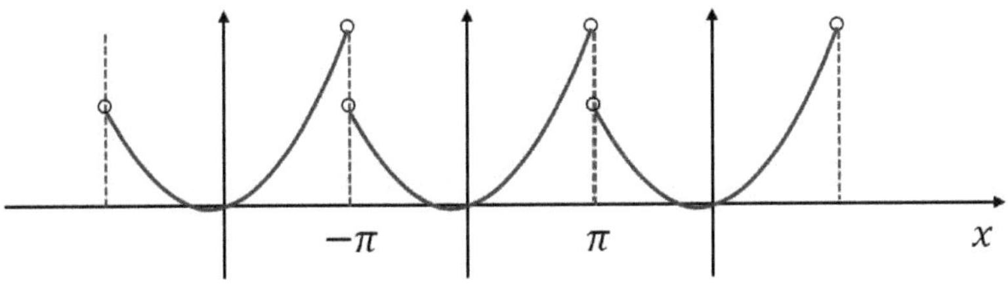

Figure 7.2.

function with breakpoints. Expand it into a Fourier series as

$$x + x^2 = \frac{1}{3}\pi^2 + 4\sum_{n=1}^{\infty} \frac{(-1)^n}{n^2}\cos nx + 2\sum_{n=1}^{\infty} \frac{(-1)^{n+1}}{n}\sin nx$$

Based on the Fourier series expansion result in this example, we can obtain the following useful arithmetic formulas:

[1] Setting $x = 0$, we have

$$1 - \frac{1}{2^2} + \frac{1}{3^2} - \frac{1}{4^2} + \cdots = \frac{\pi^2}{12}$$

[2] Setting $x = \pi$, we have

$$1 + \frac{1}{2^2} + \frac{1}{3^2} + \frac{1}{4^2} + \cdots = \frac{\pi^2}{6}$$

[3] Adding the two equations, we have

$$1 + \frac{1}{3^2} + \frac{1}{5^2} + \frac{1}{7^2} + \cdots = \frac{\pi^2}{8}$$

When a function defined on a finite interval is periodically shifted, the resulting periodic function often has breakpoints. According to Dirichlet's theorem, the Fourier series value at the breakpoint is equal to the average of the function values to the left and right of the breakpoint.

Example 7.2. Find the Fourier series expansion of the sawtooth wave function shown in Figure 7.3(a).

Solution: The Fourier series expansion of the sawtooth wave function is

$$f(x) = \frac{h}{2} + \sum_{n=1}^{\infty} a_n \sin n\omega x$$

where the expansion coefficients are given by

$$a_n = -\frac{h}{n\pi}, \qquad \omega = \frac{2\pi}{d}$$

Figure 7.3(b) illustrates the frequency spectrum components $|a_n|$. It can be observed that the high-frequency components decrease as n increases. This carries significant physical meaning. The high-frequency components correspond to the sharp or highly curved parts of the function in Figure 7.3(a). Figure 7.4 shows the process of the Fourier series approximation to the sawtooth wave function for different values of n ($n = 5, 10, 20, 100$). It demonstrates that the low-frequency components reflect the overall characteristics of the function curve, while the high-frequency components capture the local features of the function curve.

Since the function has discontinuous breakpoints, it implies that the high-frequency components are difficult to fully describe with a finite spectrum. This leads to the occurrence of the Gibbs phenomenon, where the Fourier series expansion, even with a large number of terms, results in an approximately 18% deviation at the breakpoints of the function. This is shown in Figure 7.4 for $n = 100$, and this deviation cannot be eliminated by increasing n.

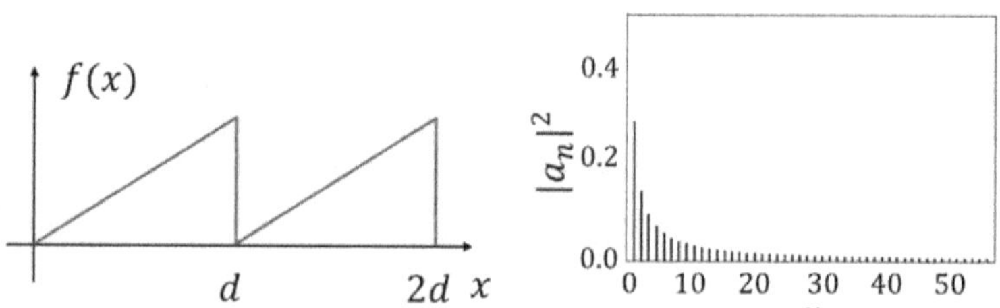

(a) (b)

Figure 7.3.

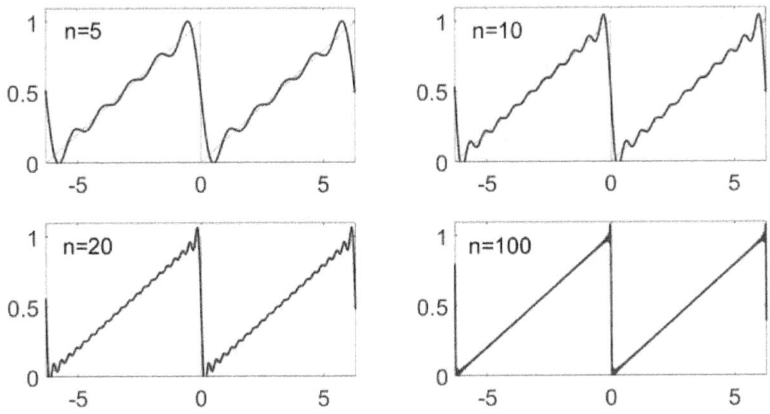

Figure 7.4.

7.1.3 *The complex exponential Fourier series*

The complex exponential Fourier series is a form of expansion using complex exponential basis functions, which are given by

$$\left\{1, \mathrm{e}^{\pm\frac{\pi \mathrm{i}x}{l}}, \mathrm{e}^{\pm\frac{2\pi \mathrm{i}x}{l}}, \ldots, \mathrm{e}^{\pm\frac{\mathrm{i}n\pi x}{l}}, \ldots\right\} \tag{7.1.4}$$

These basis functions satisfy the orthogonality and completeness conditions:

$$\frac{1}{2l}\int_{-l}^{l} \mathrm{e}^{\frac{\mathrm{i}m\pi x}{l}}[\mathrm{e}^{\frac{\mathrm{i}n\pi x}{l}}]^{*}\,\mathrm{d}x = \delta_{mn} \tag{7.1.5}$$

By performing a periodic translation of the function $f(x)$ defined on the interval $[-l, l]$, it can be expanded using the complex exponential Fourier series as follows:

$$f(x) = \sum_{n=-\infty}^{\infty} c_n \mathrm{e}^{\frac{\mathrm{i}n\pi x}{l}}, \quad c_n = \frac{1}{2l}\int_{-l}^{l} f(\xi)\mathrm{e}^{-\frac{\mathrm{i}n\pi\xi}{l}}\,\mathrm{d}\xi \tag{7.1.6}$$

The relationship between complex exponential basis functions and trigonometric basis functions is simply a linear combination of the basis functions, which can be seen as an orthogonal transformation from one set of basis functions to another:

$$\cos\frac{n\pi x}{l} = \frac{\mathrm{e}^{\mathrm{i}n\pi x/l} + \mathrm{e}^{-\mathrm{i}n\pi x/l}}{2}, \quad \sin\frac{n\pi x}{l} = \frac{\mathrm{e}^{\mathrm{i}n\pi x/l} - \mathrm{e}^{-\mathrm{i}n\pi x/l}}{2\mathrm{i}}$$

Take $z = \mathrm{e}^{\mathrm{i}x}$, and consider a real function $f(x)$ represented as $f(z)$ by extending it analytically to the complex plane. The Fourier series expansion of $f(z)$ on the unit circle $x \in [-\pi, \pi]$ becomes

$$f(z) = \sum_{n=-\infty}^{\infty} c_n z^n$$

$$c_n = \frac{1}{2\pi} \int_{-\pi}^{\pi} f(\xi) \mathrm{e}^{-\mathrm{i}n\xi} \, \mathrm{d}\xi = \frac{1}{2\pi \mathrm{i}} \oint_{|z|=1} \frac{f(z)}{z^{n+1}} \, \mathrm{d}z$$

It can be observed that the Fourier series expansion of the function $f(x)$ on the unit circle corresponds to the Laurent series expansion of the analytically extended function $f(z)$ centered at $z = 0$. However, it is important to note that the convergence domain of this Laurent series, i.e., the intersection of the convergence domains of the positive and negative power parts, is usually an empty set.

In order to obtain a non-empty convergence domain for the Laurent series, the function $f(x)$ and its higher-order derivatives must be continuous everywhere and satisfy a "closed" periodic boundary condition, such as $f(-\pi) = f(\pi)$.

Example 7.3. Find the Fourier series expansion of the function $f(x)$ on the interval $x \in [-\pi, \pi]$:

$$f(x) = \frac{a \sin x}{1 - 2a \cos x + a^2} \quad (|a| < 1)$$

Solution: By letting $z = \mathrm{e}^{\mathrm{i}x}$ and extending the function analytically to the complex plane, we have

$$f(z) = \frac{1 - z^2}{2\mathrm{i}\left[z^2 - \left(a + \frac{1}{a}\right) z + 1\right]}$$

Its Laurent series expansion is

$$f(z) = \frac{1}{2\mathrm{i}} \sum_{n=0}^{\infty} a^n \left(z^n - \frac{1}{z^n}\right)$$

This Laurent series converges in the annular region $|a| < |z| < 1/|a|$, and $f(z)$ is analytic within the convergence annulus. The unit circle of the trigonometric series lies within this annular region. Returning to the original real parameter, we obtain the Fourier sine series expansion of $f(x)$ as follows:

$$\frac{a \sin x}{1 - 2a \cos x + a^2} = \sum_{n=1}^{\infty} a^n \sin nx$$

7.1.4 *Fourier series sum*

Consider the series sum:

$$\sum_{n=1}^{\infty}\frac{\cos nx}{n}\quad (0<x<2\pi)$$

Since this series diverges at $x=0$ and $x=2\pi$, we take

$$\sum_{n=1}^{\infty}\frac{\cos nx}{n}=\lim_{r\to 1}\sum_{n=1}^{\infty}\frac{r^n\cos nx}{n}$$

The series on the right-hand side converges absolutely for $|r|<1$. Due to

$$\begin{aligned}\sum_{n=1}^{\infty}\frac{r^n\cos nx}{n}&=\frac{1}{2}\sum_{n=1}^{\infty}\frac{r^n\mathrm{e}^{\mathrm{i}nx}}{n}+\frac{1}{2}\sum_{n=1}^{\infty}\frac{r^n\mathrm{e}^{-\mathrm{i}nx}}{n}\\&=-\frac{1}{2}[\ln(1-r\mathrm{e}^{\mathrm{i}x})+\ln(1-r\mathrm{e}^{-\mathrm{i}x})]\\&=-\ln[(1+r^2)-2r\cos x]^{\frac{1}{2}}\end{aligned}$$

Taking $r=1$, we have

$$\sum_{n=1}^{\infty}\frac{\cos nx}{n}=-\ln(2-2\cos x)^{\frac{1}{2}}=-\ln\left(2\sin\frac{x}{2}\right)\quad (0<x<2\pi)$$

Here are several commonly used Fourier series sum formulas. Readers are encouraged to verify them:

$$\sum_{n=1}^{\infty}\frac{\sin nx}{n}=\begin{cases}-\frac{1}{2}(\pi+x) & (-\pi\le x<0)\\ \frac{1}{2}(\pi-x) & (0\le x<\pi)\end{cases}$$

$$\sum_{n=1}^{\infty}(-1)^{n+1}\frac{\sin nx}{n}=\frac{x}{2}\quad (-\pi\le x<\pi)$$

$$\sum_{n=0}^{\infty}\frac{\sin(2n+1)x}{2n+1}=\begin{cases}-\frac{\pi}{4} & (-\pi\le x<0)\\ \frac{\pi}{4} & (0\le x<\pi)\end{cases}$$

$$\sum_{n=1}^{\infty}\frac{\cos nx}{n}=-\ln\left(2\sin\frac{x}{2}\right)\quad (0<x<2\pi)$$

$$\sum_{n=1}^{\infty}(-1)^n\frac{\cos nx}{n} = -\ln\left(2\cos\frac{x}{2}\right) \quad (-\pi < x < \pi)$$

$$\sum_{n=0}^{\infty}\frac{\cos(2n+1)x}{2n+1} = \frac{1}{2}\ln\left(2\cot\frac{|x|}{2}\right) \quad (0 < x < 2\pi)$$

ENDNOTE

Even if the function diverges at $x = 1$, mathematicians consider $f(x) = \frac{1}{1-x}$ to be a legitimate function. However, until the early 19th century, classical mathematicians could not accept functions that were discontinuous or had derivative discontinuities, such as $f(x) = |x|$. Euler believed that such functions should not be considered as functions, perhaps because they could not be represented as power series at the points of discontinuity (Lagrange used power series to define derivatives). In people's minds, a function had to be representable by a formula, so when Fourier presented periodic functions with discontinuities, such as the sawtooth wave or square wave, that could be expressed as a series of trigonometric functions, it caused a stir among the traditional forces in mathematics: how could a discontinuous function be represented as a sum of continuous functions?

According to the result in Example 7.2, the periodic sawtooth wave can be represented by a trigonometric series as follows:

$$f(x) = \frac{h}{2} - \sum_{n=1}^{\infty} a_n \sin n\omega x, \quad a_n = -\frac{h}{n\pi}$$

It can be viewed as a real function defined on the unit circle. Let $z = \mathrm{e}^{\mathrm{i}\omega x}$, then

$$f(z) = \frac{h}{2} + \sum_{n=1}^{\infty}\frac{ih}{2n\pi}\left(z^n - \frac{1}{z^n}\right)$$

The convergence domain of the positive power terms $S_+(z)$ is $R_1 < 1$, and the convergence domain of the negative power terms $S_-(z)$ is $R_2 > 1$. Therefore, the convergence ring of its Laurent series is an empty set. In fact, the analytic expression for the sum of the series can be written as

$$\begin{cases} S_+(z) = \sum\limits_{n=1}^{\infty} \dfrac{ih}{2n\pi} z^n = -\dfrac{ih}{2\pi} \ln(1-z) & (|z| < 1) \\ S_-(z) = -\sum\limits_{n=1}^{\infty} \dfrac{ih}{2n\pi} \dfrac{1}{z^n} = \dfrac{ih}{2\pi} \ln\left(1 - \dfrac{1}{z}\right) & (|z| > 1) \end{cases}$$

Thus, $f(z)$ is analytically extended to the entire complex plane. The breakpoints $x = 0$ and $x = d$ of the real function $f(x)$ correspond to the branch points $z = 1$ of the complex logarithmic function $S_\pm(z)$. As shown in Figure 7.5, the function values on the unit circle are given by

$$f(z) = \frac{h}{2} + S_+(e^{i\theta}) + S_-(e^{i\theta}) = \frac{h}{2\pi}\theta$$

As z crosses the point $z = 1$ on the unit circle from the upper half-plane, the argument of $(1-z)$ jumps from $-\frac{\pi}{2}$ to $\frac{\pi}{2}$, while the argument of $\left(1 - \frac{1}{z}\right)$ jumps from $\frac{\pi}{2}$ to $-\frac{\pi}{2}$. Therefore, the value of the function $f(z)$ jumps by h at $z = 1$, indicating that the Fourier series indeed describes the sawtooth wave function.

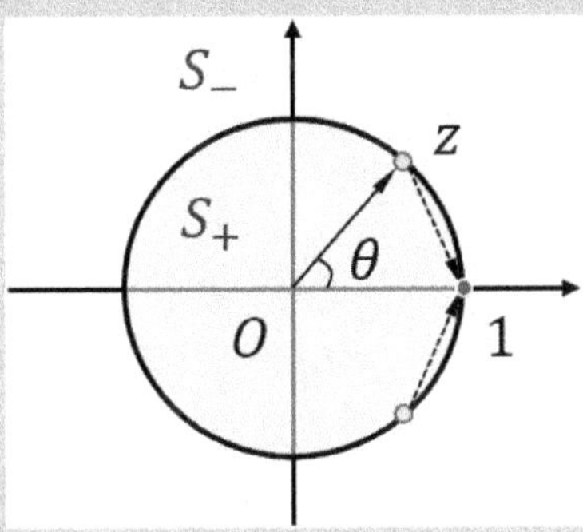

Figure 7.5.

It is evident that understanding the properties of the function $f(x) = |x|$ is quite challenging within the realm of real numbers. It is not differentiable at $x = 0$ and cannot be expanded into a Taylor series. However, when it is expanded into a Fourier series over the interval $x \in [-\pi, \pi]$ and extended analytically to the complex plane by letting $z = e^{ix}$, it can be expressed in the form of a Laurent series:

$$f(z) = \frac{\pi}{2} - \sum_{k=0}^{\infty} \frac{4}{(2k+1)^2\pi} \left[z^{2k+1} + \frac{1}{z^{2k+1}} \right]$$

The singularity of the function at $x = 0$ or $z = 1$ becomes apparent, and in this sense, it possesses the same legitimacy as the ordinary function $f(x) = x$. In the future, it will be discussed that functions with breakpoints are actually necessary elements of the completeness of inner product spaces.

Fourier's work freed people from the notion that functions had to be analytic or expandable into Taylor series. The Taylor series only represents an analytic function in a neighborhood of a point, although the convergence radius may be infinite. On the other hand, the Fourier series represents a function over a finite interval. The fact that a square wave can be represented by a Fourier series made series become functions in the traditional sense, and mathematicians gradually realized that functions represented by series do not guarantee differentiability. It is commonly believed that continuous functions are differentiable — at least, the set of non-differentiable points is not dense. However, Weierstrass constructed the following function:

$$f(x) = \sum_{k=0}^{\infty} a^k \cos(b^k \pi x) \quad (0 < a < 1)$$

where b is a positive odd integer and $ab > 1+3\pi/2$. Weierstrass proved that this function is continuous everywhere but non-differentiable everywhere. This "pathological" function is not smooth and exhibits irregularities. Physically, it corresponds to the typical case of the Brownian motion trajectory of particles. Another example of a pathological function is

$$f(x) = \lim_{\substack{m \to \infty \\ n \to \infty}} \cos^{2n}(m!\pi x) = \begin{cases} 1 & (x \in \mathbb{Q}) \\ 0 & (x \notin \mathbb{Q}) \end{cases}$$

Dirichlet successfully solved the convergence problem of Fourier series using Cauchy's analytical methods. However, does this series converge uniformly? The various subtle issues regarding the convergence of Fourier series prompted people to explore the properties of real numbers themselves, ultimately leading to the development of set theory by Cantor.

The establishment of Fourier series marks a milestone in the conceptualization of functions: they are input–output rules, where any input value has a specific output value. These input–output relationships do not necessarily need to be expressed as formulas and can even be non-numerical. In the past, mathematicians had no interest in studying heartbeat or stock price fluctuation curves, but now they are also considered as functions. The frequency distribution of these curves contains previously unimaginable properties. By manipulating the Fourier spectrum, people can generate new functions. For example, music synthesizers use Fourier series to simulate the sounds of violins or flutes, microphones extend the range of human vocalization, and even create sounds that have never existed in nature. Fourier optics has long been a specialized discipline for dealing with spectral characteristics of light.

Exercises

[1] Find the Fourier trigonometric series expansion for the following functions in the interval $x \in [-\pi, \pi]$, where α is not an integer:

$$\text{(a) } f(x) = \cos \alpha x; \quad \text{(b) } f(x) = \cosh \alpha x$$

Answer:

$$\text{(a)}\ \frac{2\sin\pi\alpha}{\pi}\left[\frac{1}{2\alpha}+\sum_{k=1}^{\infty}(-1)^{k-1}\frac{\alpha\cos kx}{k^2-\alpha^2}\right]$$

$$\text{(b)}\ \frac{2\sinh\pi\alpha}{\pi}\left[\frac{1}{2\alpha}+\sum_{k=1}^{\infty}(-1)^{k}\frac{\alpha\cos kx}{k^2+\alpha^2}\right]$$

[2] Proof:

$$|\sin x| = \frac{2}{\pi} - \frac{4}{\pi}\sum_{k=1}^{\infty}\frac{\cos 2kx}{4k^2-1}$$

[3] Develop the Fourier trigonometric series expansion for the function $f(x) = |x|$ in the periodic interval $x \in [-\pi, \pi]$, and prove that

$$\sum_{k=0}^{\infty}\frac{1}{(2k+1)^2} = \frac{\pi^2}{8}$$

Answer:

$$|x| = \frac{\pi}{2} - \sum_{k=0}^{\infty}\frac{4}{(2k+1)^2\pi}\cos{(2k+1)x}$$

[4] Expand the following functions in the Fourier exponential series in the interval $x \in [-\pi, \pi]$, where $\alpha \neq \beta$:

$$f(x) = \begin{cases} \alpha x & (0 < x \leq \pi) \\ \beta x & (-\pi \leq x < 0) \end{cases}$$

Answer:

$$f(x) = \frac{(\alpha-\beta)\pi}{4} + \sum_{n\neq 0}^{\infty}\left[\frac{(\alpha+\beta)\mathrm{i}}{2n}(-1)^n + \frac{\beta-\alpha}{2n^2\pi}(1-(-1)^n)\right]\mathrm{e}^{nx\mathrm{i}}$$

[5] Given the Taylor series expansion:

$$\ln(1+z) = \sum_{n=1}^{\infty}(-1)^{n+1}\frac{z^n}{n} \quad (|z|<1)$$

we need to prove the following Fourier series summation formulas:

(a) $\displaystyle\sum_{n=1}^{\infty}\frac{\cos nx}{n} = -\ln\left(2\sin\frac{x}{2}\right) \quad (0 < x < 2\pi)$

(b) $\displaystyle\sum_{n=1}^{\infty}(-1)^n\frac{\cos nx}{n} = -\ln\left(2\cos\frac{x}{2}\right) \quad (-\pi < x < \pi)$

(c) $\displaystyle\sum_{n=0}^{\infty}\frac{\cos(2n+1)x}{2n+1} = \frac{1}{2}\ln\left(2\cot\frac{|x|}{2}\right) \quad (0 < x < 2\pi)$

7.2 Fourier Transform

Section 7.1 discusses the Fourier series expansion of periodic functions. Functions defined on a finite interval can be extended to periodic functions through translation, allowing for the Fourier series expansion of that interval. However, can the Fourier series expansion be applied to non-periodic functions defined on the infinite interval $x \in (-\infty, \infty)$?

7.2.1 *Fourier integral*

First, consider a non-periodic function $f(x)$ defined on the infinite interval as the limiting case of a periodic function $g(x)$ with a period of $2l$. The Fourier series expansion of $g(x)$ is given by

$$g(x) = \frac{a_0}{2} + \sum_{n=1}^{\infty}\left(a_n\cos\frac{n\pi x}{l} + b_n\sin\frac{n\pi x}{l}\right)$$

As l approaches infinity, the above equation becomes the Fourier expansion of the non-periodic function $f(x)$. Introducing the discontinuous variable:

$$k_n = \frac{n\pi}{l} \quad (n = 0, 1, 2, \ldots)$$

$$\Delta k_n = k_n - k_{n-1} = \frac{\pi}{l}$$

Then, we have

$$f(x) = \frac{a_0}{2} + \sum_{n=1}^{\infty}(a_n\cos k_n x + b_n\sin k_n x)$$

If the integral of $f(x)$ over the interval $(-l, +l)$ is finite, utilizing the integrability condition of the function, we have

$$\lim_{l\to\infty} a_0 = \lim_{l\to\infty} \frac{1}{l}\int_{-l}^{l} f(\xi)\mathrm{d}\xi = 0$$

The first term on the right-hand side becomes

$$\begin{aligned}
&\lim_{l\to\infty}\sum_{n=1}^{\infty}\left[\frac{1}{l}\int_{-l}^{l} f(\xi)\cos k_n\xi\,\mathrm{d}\xi\right]\cos k_n x \\
&= \lim_{l\to\infty}\sum_{n=1}^{\infty}\left[\frac{1}{\pi}\int_{-l}^{l} f(\xi)\cos k_n\xi\,\mathrm{d}\xi\right]\cos k_n x\Delta k_n \\
&\xrightarrow{\Delta k_n\to 0} \int_0^{\infty}\left[\frac{1}{\pi}\int_{-\infty}^{\infty} f(\xi)\cos k\xi\,\mathrm{d}\xi\right]\cos kx\,\mathrm{d}k \equiv \int_0^{\infty} A(k)\cos kx\,\mathrm{d}k
\end{aligned}$$

Similarly, the expression for the second term on the right-hand side can be obtained. Thus, we obtain the Fourier integral representation:

$$f(x) = \int_0^{\infty} A(k)\cos kx\,\mathrm{d}k + \int_0^{\infty} B(k)\sin kx\,\mathrm{d}k \tag{7.2.1}$$

where

$$A(k) = \frac{1}{\pi}\int_{-\infty}^{\infty} f(\xi)\cos k\xi\,\mathrm{d}\xi, \quad B(k) = \frac{1}{\pi}\int_{-\infty}^{\infty} f(\xi)\sin k\xi\,\mathrm{d}\xi \tag{7.2.2}$$

In contrast to the discrete spectrum of periodic functions, the integral coefficients $A(k)$ and $B(k)$ $(k > 0)$ for non-periodic functions also reflect the frequency components of the function's spectrum, although this spectrum is generally continuous.

Fourier integral theorem: If a function $f(x)$ satisfies the Dirichlet conditions on a finite interval and is absolutely integrable over the interval $(-\infty, \infty)$, that is, the integral $\int_{-\infty}^{\infty} |f(x)|\mathrm{d}x$ is finite, then $f(x)$ can be represented in the form of a Fourier integral as follows:

$$\frac{1}{2}[f(x-0) + f(x+0)] = \int_0^{\infty} [A(k)\cos kx + B(k)\sin kx]\mathrm{d}k \tag{7.2.3}$$

Proof omitted.

Using the same approach, we can derive the complex exponential form of the Fourier integral formula for non-periodic functions, which has a more compact form:

$$f(x) = \sum_{n=-\infty}^{\infty} c_n \mathrm{e}^{\frac{\mathrm{i}n\pi x}{l}} = \frac{1}{2l} \sum_{n=-\infty}^{\infty} \mathrm{e}^{\frac{\mathrm{i}n\pi x}{l}} \int_{-l}^{l} f(\xi)\mathrm{e}^{-\frac{\mathrm{i}n\pi\xi}{l}}\,\mathrm{d}\xi$$

$$\xrightarrow{l\to\infty} \frac{1}{2\pi} \int_{-\infty}^{\infty} \left[\int_{-\infty}^{\infty} f(\xi)\mathrm{e}^{-\mathrm{i}k\xi}\,\mathrm{d}\xi\right] \mathrm{e}^{\mathrm{i}kx}\,\mathrm{d}k \equiv \int_{-\infty}^{\infty} F(k)\mathrm{e}^{\mathrm{i}kx}\,\mathrm{d}k$$

where

$$F(k) = \frac{1}{2\pi} \int_{-\infty}^{\infty} f(x)\mathrm{e}^{-\mathrm{i}kx}\,\mathrm{d}x \tag{7.2.4}$$

$F(k)$ is called the Fourier transform of the function $f(x)$, denoted as $F(k) = \mathfrak{F}[f(x)]$, and

$$f(x) = \int_{-\infty}^{\infty} F(k)\mathrm{e}^{\mathrm{i}kx}\,\mathrm{d}k \tag{7.2.5}$$

is called the inverse transform of $F(k)$, denoted as $f(x) = \mathfrak{F}^{-1}[F(k)]$. They form a pair of mappings, where $F(k)$ is often referred to as the image function of $f(x)$, and $f(x)$ as the original function of $F(k)$. Some physics books define the Fourier transform in a symmetric form:

$$\begin{cases} F(k) = \mathfrak{F}[f(x)] \stackrel{\text{def}}{=} \dfrac{1}{\sqrt{2\pi}} \displaystyle\int_{-\infty}^{\infty} f(x)\mathrm{e}^{-\mathrm{i}kx}\,\mathrm{d}x \\ f(x) = \mathfrak{F}^{-1}[F(k)] \stackrel{\text{def}}{=} \dfrac{1}{\sqrt{2\pi}} \displaystyle\int_{-\infty}^{\infty} F(k)\mathrm{e}^{\mathrm{i}kx}\,\mathrm{d}k \end{cases} \tag{7.2.6}$$

(1) Fourier cosine transform: If $f(x)$ is an even function, then

$$F_c(k) = \mathfrak{F}_c[f(x)] \stackrel{\text{def}}{=} \frac{1}{\pi} \int_0^{\infty} f(x)\cos kx\,\mathrm{d}x \tag{7.2.7}$$

(2) Fourier sine transform: If $f(x)$ is an odd function, then

$$F_s(k) = \mathfrak{F}_s[f(x)] \stackrel{\text{def}}{=} \frac{1}{\pi} \int_0^{\infty} f(x)\sin kx\,\mathrm{d}x \tag{7.2.8}$$

Since $F_c(k)$ and $F_s(k)$ are even and odd functions of k, respectively, for $0 < k < \infty$, their inverse transforms are given by

$$f(x) = \mathfrak{F}_c^{-1}[F_c(k)] = 2\int_0^{\infty} F_c(k)\cos kx\, \mathrm{d}k \tag{7.2.9}$$

$$f(x) = \mathfrak{F}_s^{-1}[F_s(k)] = 2\int_0^{\infty} F_s(k)\sin kx\, \mathrm{d}k \tag{7.2.10}$$

The Fourier cosine transform and Fourier sine transform are collectively referred to as the Trigonometric transforms.

Example 7.4. Try expressing the Haar function as a Fourier integral:

$$f(x) = \begin{cases} h & (|x| \le d) \\ 0 & (|x| > d) \end{cases}$$

Solution: To solve this, note that $f(x)$ is an even function. Thus, its Fourier transform $F(k)$ will also be an even function. According to the definition of the Fourier transform:

$$F(k) = \frac{1}{2\pi}\int_{-\infty}^{\infty} f(x)\mathrm{e}^{-ikx}\,\mathrm{d}x = \frac{1}{\pi}\int_0^{d} f(x)\cos kx\,\mathrm{d}x = \frac{h}{\pi}\frac{\sin kd}{k}$$

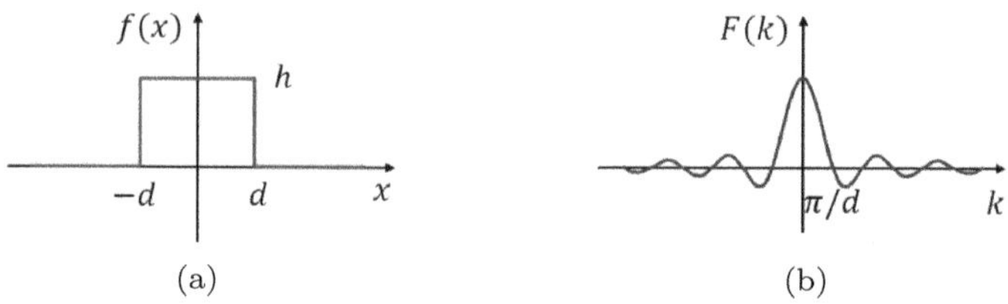

(a) (b)

Figure 7.6.

Discussion: The relationship between the function $f(x)$ and its Fourier transform $F(k)$ is analogous to the relationship between a diffraction slit and its diffraction pattern in optics. This experimental setup in optics is known as the Fraunhofer diffraction. Figure 7.6 illustrates the relationship between the width of the square wave and the width of its frequency spectrum: $\Delta k \sim \pi/d$, indicating that functions with wider spatial widths have narrower frequency spectra.

7.2.2 *Fundamental properties*

Regarding the properties of the Fourier transform, the following theorems hold:

[1] Derivative theorem

$$\mathfrak{F}[f'(x)] = \mathrm{i}kF(k)$$

[2] Integration theorem

$$\mathfrak{F}\left[\int_0^x f(\xi)\mathrm{d}\xi\right] = \frac{1}{\mathrm{i}k}F(k)$$

[3] Scaling theorem

$$\mathfrak{F}[f(ax)] = \frac{1}{|a|}F\left(\frac{k}{a}\right)$$

[4] Shift theorem

$$\mathfrak{F}[f(x - x_0)] = \mathrm{e}^{-\mathrm{i}kx_0}F(k)$$

[5] Phase shift theorem

$$\mathfrak{F}[\mathrm{e}^{\mathrm{i}k_0x}f(x)] = F(k - k_0)$$

[6] Multiplication theorem

$$\int_{-\infty}^{\infty} f_1(x)\bar{f_2}(x)\mathrm{d}x = 2\pi\int_{-\infty}^{\infty} F_1(k)\bar{F_2}(k)\mathrm{d}k$$

When $f_1(x) = f_2(x) \equiv f(x)$, we have

$$\int_{-\infty}^{\infty} |f(x)|^2\,\mathrm{d}x = 2\pi\int_{-\infty}^{\infty} |F(k)|^2\,\mathrm{d}k$$

These theorems can all be directly proven from the definition of the Fourier transform. Readers are encouraged to practice the proofs on their own.

Derivative properties for Fourier cosine and sine transforms:

$$\mathfrak{F}_c[f'(x)] = \frac{1}{\pi}\int_0^{\infty} f'(x)\cos kx\,dx = -\frac{1}{\pi}f(0) + kF_s(k)$$

$$\mathfrak{F}_c[f''(x)] = \frac{1}{\pi}\int_0^{\infty} f''(x)\cos kx\,\mathrm{d}x = -\frac{1}{\pi}f'(0) - k^2F_c(k)$$

$$\mathfrak{F}_s[f'(x)] = \frac{1}{\pi}\int_0^{\infty} f'(x)\sin kx\,\mathrm{d}x = -kF_c(k) \tag{7.2.11}$$

$$\mathfrak{F}_s[f''(x)] = \frac{1}{\pi}\int_0^{\infty} f''(x)\sin kx\,\mathrm{d}x = \frac{1}{\pi}kf(0) - k^2F_s(k)$$

It is worth noting that the trigonometric transforms of the derivatives of a function are related to the boundary values at $x = 0$. If $f'(0)$ is given on the boundary, the Fourier cosine transform should be used. If $f(0)$ is given, the Fourier sine transform should be used.

Discussion: The fundamental properties of Fourier transforms have corresponding manifestations in the physical world:

(1) The scaling theorem reflects the relationship between the original function $f(x)$ and its transform $F(k)$, showing that the width of the diffracting slit is inversely proportional to the width of the diffraction pattern.
(2) The delay theorem's translation corresponds to moving the diffraction slit up or down. In this case, the diffraction pattern remains unchanged. What is the role of the factor e^{-ikx_0}? How to reveal its physical meaning?
(3) In the shift theorem, the complex factor e^{ik_0x} corresponds to oblique incidence in the diffraction experiment. The frequency spectrum shift, $F(k - k_0)$, indicates a shift in the diffraction pattern. What does it imply, and is there any connection between the shift theorem and the Doppler effect?
(4) The multiplication theorem reflects energy conservation, indicating that the total energy distributed in the spatial domain is equal to the total energy distributed in the frequency domain.

Example 7.5. Finding the Fourier Transform of an N-Periodic Function:

$$g(x) = \sum_{n=0}^{N-1} f(x - nd)$$

where $f(x + d) = f(x)$.

Solution: Let $\mathfrak{F}[f(x)] = F(k)$. Applying the delay theorem to the periodic function $f(x-nd) = f(x)$, we have

$$\mathfrak{F}[f(x-nd)] = \mathrm{e}^{-inkd}F(k)$$

Thus,

$$G(k) \equiv \mathfrak{F}[g(x)] = \sum_{n=0}^{N-1} \mathrm{e}^{-inkd}F(k) = \left(\frac{1-\mathrm{e}^{-\mathrm{i}Nkd}}{1-\mathrm{e}^{-\mathrm{i}kd}}\right)F(k)$$

This gives the distribution formula for the spectral intensity:

$$|G(k)| = \left|\frac{\sin\left(\frac{Nkd}{2}\right)}{\sin\left(\frac{kd}{2}\right)}\right| \times |F(k)|$$

Taking $f(x)$ as the Haar function, the term $|G(k)|$ corresponds to the intensity distribution of optical grating diffraction in optics. The second term on the right, $|F(k)|$, represents the single-slit diffraction factor, while the first term represents the interference factor between the slits. For normal incidence, the change in wave vector of the diffracted light is given by $k = q\sin\theta$, where θ is the diffraction angle and $q = \frac{2\pi}{\lambda}$.

7.2.3 *Three-dimensional Fourier transform*

For a three-dimensional function $f(x,y,z)$, its Fourier transform involves taking the Fourier transform of each variable (x,y,z) separately and can be represented as

$$f(x,y,z) = \int_{-\infty}^{\infty}\int_{-\infty}^{\infty}\int_{-\infty}^{\infty} F(k_x,k_y,k_z)\mathrm{e}^{\mathrm{i}(k_x x+k_y y+k_z z)}\,\mathrm{d}k_x\,\mathrm{d}k_y\,\mathrm{d}k_z \tag{7.2.12}$$

$$F(k_x,k_y,k_z) = \frac{1}{(2\pi)^3}\int_{-\infty}^{\infty}\int_{-\infty}^{\infty}\int_{-\infty}^{\infty} f(x,y,z)\mathrm{e}^{-\mathrm{i}(k_x x+k_y y+k_z z)}\,\mathrm{d}x\mathrm{d}y\mathrm{d}z \tag{7.2.13}$$

In a more compact vector form, this can be written as

$$f(\boldsymbol{r}) = \iiint_{-\infty}^{\infty} F(\boldsymbol{k})\mathrm{e}^{\mathrm{i}\boldsymbol{k}\cdot\boldsymbol{r}}\,\mathrm{d}\boldsymbol{k}$$

$$F(\boldsymbol{k}) = \frac{1}{(2\pi)^3}\iiint_{-\infty}^{\infty} f(\boldsymbol{r})\mathrm{e}^{-\mathrm{i}\boldsymbol{k}\cdot\boldsymbol{r}}\,\mathrm{d}\boldsymbol{r}$$

When computing the three-dimensional Fourier transform, the spherical coordinate system is often used.

Example 7.6. Finding the Fourier Transform of the Yukawa Potential:

$$v_\alpha(\boldsymbol{r}) = \frac{q\mathrm{e}^{-\alpha r}}{r} \qquad (\alpha > 0)$$

Solution: Using the definition of the three-dimensional Fourier transform, we have

$$V_\alpha(\boldsymbol{k}) = \frac{1}{(2\pi)^3}\iiint_{-\infty}^{\infty} \frac{qe^{-\alpha r}}{r}\mathrm{e}^{-\mathrm{i}\boldsymbol{k}\cdot\boldsymbol{r}}\,\mathrm{d}\boldsymbol{r}$$

In spherical coordinates with the direction of $\boldsymbol{k}$ as the polar axis and $|\boldsymbol{k}| = k$, we have

$$\begin{aligned} V_\alpha(\boldsymbol{k}) &= \frac{q}{(2\pi)^3}\int_0^\infty r\mathrm{d}r \int_0^\pi \sin\theta\,\mathrm{d}\theta \int_0^{2\pi} \mathrm{e}^{-\alpha r}\mathrm{e}^{-ikr\cos\theta}\,\mathrm{d}\phi \\ &= \frac{q}{(2\pi)^2}\int_0^\infty \frac{\mathrm{e}^{-\alpha r}}{ik}(\mathrm{e}^{ikr} - \mathrm{e}^{-ikr})\mathrm{d}r = \frac{q}{2\pi^2}\frac{1}{k^2+\alpha^2} \end{aligned}$$

In the limit of $\alpha \to 0$, we obtain the Fourier transform of the Coulomb potential $v_{\mathrm{Coul}}(\boldsymbol{r}) = \frac{q}{r}$, which is

$$V_{\mathrm{Coul}}(\boldsymbol{k}) = \frac{q}{2\pi^2}\frac{1}{k^2}$$

ENDNOTE

Mapping and transformation: A function is a point-to-point mapping from one set to another, as shown in Figure 7.7(a). In contrast, each point in the Fourier transform's image space contains all the information of the original set, as shown in Figure 7.7(b). The reverse is also true.

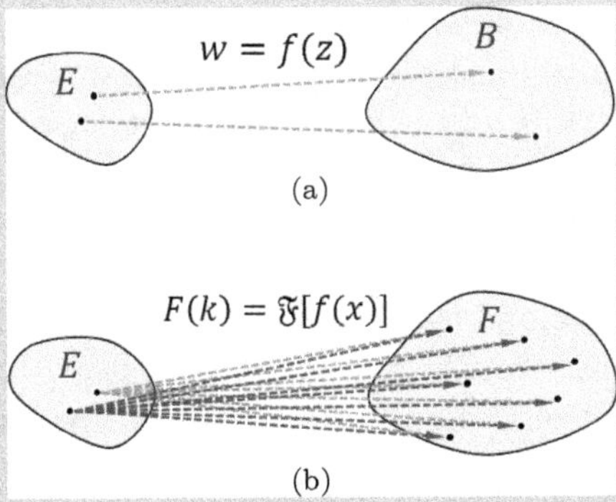

Figure 7.7.

This relationship between function mapping and Fourier transformation is similar to the relationship between lens imaging and holographic imaging in optics. Lens imaging has a one-to-one correspondence and forms a mapping relationship, as shown in Figure 7.8(a), where the image plane corresponds to the function space. Integral transformation, on the other hand, corresponds to the holographic imaging process, where each point on the original image is mapped to the entire holographic plane, and conversely, each point on the image plane also contains all the information of the original function, as shown in Figure 7.8(b). The holographic plane corresponds to the transformed image space or frequency domain.

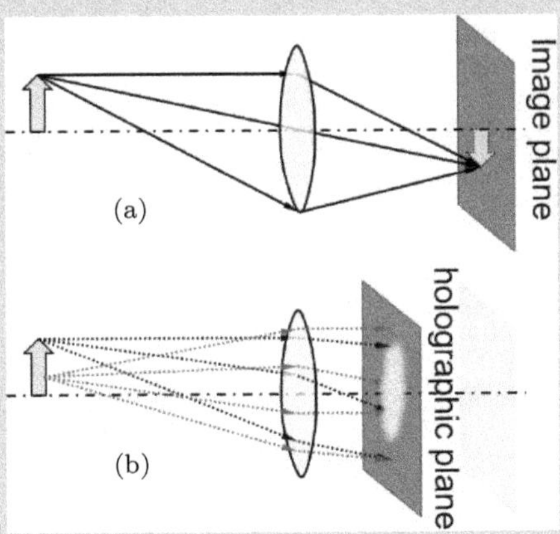

Figure 7.8.

Holographic imaging devices essentially belong to a special category of Fraunhofer diffraction, the latter being an extension of Fresnel diffraction with the addition of a convex lens. It can be demonstrated that a Fraunhofer diffraction setup constitutes a standard Fourier transformer, as shown in Figure 7.9(a). In this setup, the diffraction screen or object represents the function $f(x)$ in configuration space, while the diffraction pattern on the receiving screen represents the modulus of the Fourier image function $|F(k)|$ in transform space. Due to the nature of optical imaging, the phase information of the image function is lost. Based on this observation, we can explore certain physical scenarios involving the effects of Fourier transform properties.

(1) **Scaling Theorem**

$$\mathfrak{F}[f(ax)] = \frac{1}{|a|}F\left(\frac{k}{a}\right)$$

If the diffraction screen is a single slit, the original function is a Haar function:

$$f(x) = \begin{cases} 1 & (|x| \leq d/2) \\ 0 & (|x| > d/2) \end{cases}$$

where d is the slit width. According to the scaling theorem, when the width of the slit is magnified or reduced by a factor a, the width of the diffraction fringes is correspondingly reduced or magnified by the same factor. The slit width and the width of the zero-order diffraction fringe satisfy: $\Delta x \cdot \Delta k \sim 1$, as shown in Figures 7.9(a) and 7.9(b).

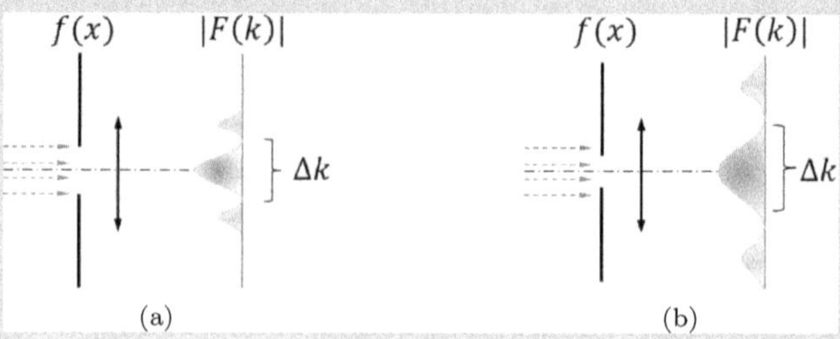

Figure 7.9.

The scaling relationship of Fourier transforms is also reflected in quantum mechanics, where the wave function in momentum space and the wave function in position space satisfy a Fourier transform relationship (see ENDNOTE in Section 14.4):

$$\varphi(p) = \frac{1}{\sqrt{2\pi\hbar}} \int_{-\infty}^{\infty} \psi(x) \mathrm{e}^{-\frac{ipx}{\hbar}} \, \mathrm{d}x \equiv \mathfrak{F}[\psi(x)]$$

and the typical widths of the original function and the image function satisfy:

$$\Delta x \cdot \Delta p \sim \Delta x \cdot \hbar \Delta k \sim \hbar$$

where Δx and Δp are inversely proportional, a relationship known as the Heisenberg uncertainty principle. This is not truly a general principle but merely a common property of Fourier transforms.

(2) **Translation Theorem**

$$\mathfrak{F}[f(x - x_0)] = \mathrm{e}^{-ikx_0} F(k)$$

As shown in Figures 7.10(a) and 7.10(b), when the diffraction screen is shifted vertically, $f(x) \to f(x - x_0)$, how does the diffraction pattern on the receiving screen change? No changes at all! According to the translation theorem of Fourier transforms, the image function gains a phase factor, $F(k) \to \mathrm{e}^{-ikx_0} F(k)$. However, since the observed diffraction pattern Quenches the phase information, $|\mathrm{e}^{-ikx_0} F(k)| \equiv |F(k)|$, no change in the pattern is seen. To reveal the effect of this phase factor, one can illuminate the diffraction screen with an additional coherent beam, as shown in Figure 7.10(c). In this case, interference fringes appear on the receiving screen. When the diffraction screen moves, the interference fringes also shift correspondingly, as shown in Figure 7.10(d).

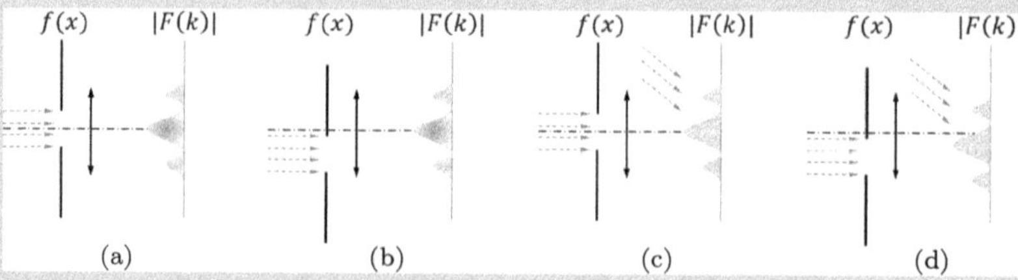

Figure 7.10.

(3) **Phase Shift Theorem**

$$\mathfrak{F}[\mathrm{e}^{ik_0x}f(x)] = F(k - k_0)$$

What happens if the incident light in Figure 7.9 is changed to oblique incidence? Experimentally, it is observed that the diffraction fringes shift, as shown in Figure 7.11. The shift distance is

$$k_0 = \frac{2\pi}{\lambda}\sin\theta_0$$

where θ_0 is the angle of oblique incidence and λ is the wavelength of light. The phase factor attached to the original function can be understood as follows: oblique incidence causes a phase difference in the plane wave arriving at point x on the diffraction screen,

$$\Delta\phi = \frac{2\pi x}{\lambda}\sin\theta_0 \equiv k_0x$$

thereby adding a phase factor e^{ik_0x} to the original function $f(x)$. From the phase shift theorem, the displacement of the spectrum function k_0 arises from this phase difference.

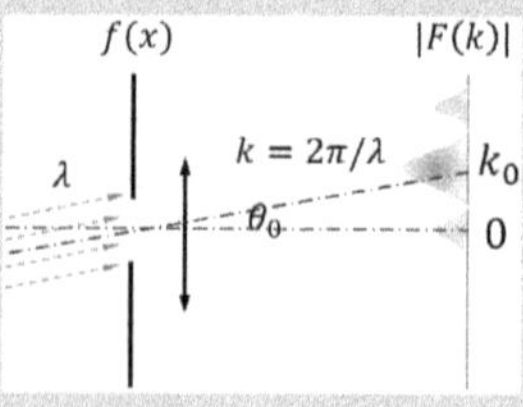

Figure 7.11.

The phase shift theorem also appears in another physical scenario. As shown in Figure 7.12, when a plane wave of frequency ω_0 propagates at a velocity c along the positive x-direction, if an observer moves at a velocity v along the x-direction, he will observe the frequency of the wave changes. This is the Doppler effect. The effect can be interpreted by the phase shift theorem as follows: A plane wave can be expressed as

$$f(x,t) = \mathrm{e}^{\mathrm{i}\left(\omega_0 t - \frac{x}{c}\right)}$$

In the moving reference of frame, $x = x' - vt$, so the wave function is expresses as

$$f(x',t) = \mathrm{e}^{\mathrm{i}\left(\omega_0 t - \frac{x'-vt}{c}\right)} = \mathrm{e}^{\mathrm{i}\frac{\omega_0 v}{c}t}\mathrm{e}^{\mathrm{i}\left(\omega_0 t - \frac{x'}{c}\right)}$$

Suppose the Fourier transform with respective to time t yields the frequency spectrum function $F(x,\omega) = \mathfrak{F}[f(x,t)]$. For a moving observer, the spectrum in the moving frame of reference is

$$F(x',\omega) = F\left(x, \omega - \frac{\omega_0 v}{c}\right)$$

According to the phase shift theorem, the frequency spectrum relation is

$$\omega' = \omega - \frac{\omega_0 v}{c}$$

As the plane wave has single frequency $\omega = \omega_0$, the moving observer detects a shifted frequency of

$$\omega' = \frac{c-v}{c}\omega_0 \quad \text{or} \quad \nu' = \frac{c-v}{c}\nu_0$$

This is exactly the Doppler shift formula.

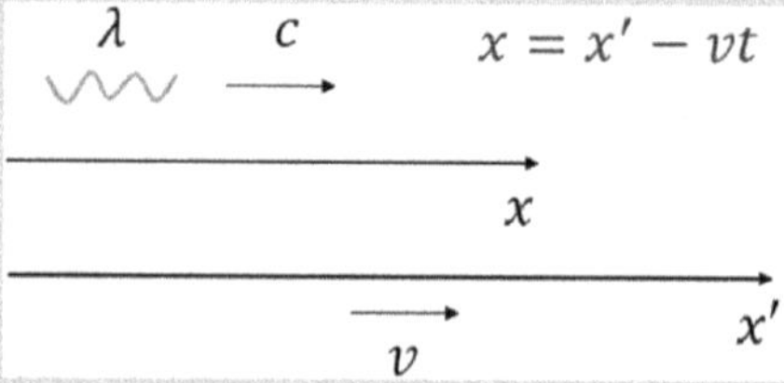

Figure 7.12.

The scattering amplitude of light in a material is given by

$$\langle \boldsymbol{q}'|\hat{U}|\boldsymbol{q}\rangle \stackrel{\text{def}}{=} \frac{1}{(2\pi)^3}\iiint_{-\infty}^{\infty} u(\boldsymbol{r})\mathrm{e}^{-\mathrm{i}(\boldsymbol{q}'-\boldsymbol{q})\cdot\boldsymbol{r}}\,\mathrm{d}\boldsymbol{r}$$

where $\boldsymbol{q}$ and $\boldsymbol{q}'$ are the wave vectors of the incident and scattered waves, and $u(\boldsymbol{r})$ represents the scattering potential of the material. Let $\boldsymbol{k} = \boldsymbol{q}' - \boldsymbol{q}$, then we have

$$\langle \boldsymbol{q}'|\hat{U}|\boldsymbol{q}\rangle = \mathfrak{F}[u(\boldsymbol{r})] \equiv U(\boldsymbol{k})$$

If the material has a periodic structure, meaning that the scattering potential satisfies three-dimensional translational periodicity,

$$u(\boldsymbol{r}+\boldsymbol{R}) = u(\boldsymbol{r})$$

where $\boldsymbol{R} = m_1\boldsymbol{a}_1 + m_2\boldsymbol{a}_2 + m_3\boldsymbol{a}_3$ ($m_j \in \mathbb{Z}$), and $\boldsymbol{a}_j (j = 1, 2, 3)$ are the Bravais lattice vectors of the crystal, then we can express the scattering potential as

$$u(\boldsymbol{r}+\boldsymbol{R}) = \iiint_{-\infty}^{\infty} U(\boldsymbol{k})\mathrm{e}^{\mathrm{i}\boldsymbol{k}\cdot(\boldsymbol{r}+\boldsymbol{R})}\,\mathrm{d}\boldsymbol{k} = \iiint_{-\infty}^{\infty} U(\boldsymbol{k})\mathrm{e}^{\mathrm{i}\boldsymbol{k}\cdot\boldsymbol{r}}\mathrm{e}^{\mathrm{i}\boldsymbol{k}\cdot\boldsymbol{R}}\,\mathrm{d}\boldsymbol{k}$$

This shows that the periodicity of the scattering potential requires the wave vectors to satisfy $\mathrm{e}^{\mathrm{i}\boldsymbol{k}\cdot\boldsymbol{R}} = 1$, which implies that $\boldsymbol{k}\cdot\boldsymbol{R}$ must be a multiple of 2π. Therefore, $\boldsymbol{k}$ can only take a series of discrete values.

Introducing the reciprocal lattice vectors $\boldsymbol{b}_j (j = 1, 2, 3)$ of the three-dimensional lattice:

$$\boldsymbol{b}_1 = \frac{2\pi(\boldsymbol{a}_2 \times \boldsymbol{a}_3)}{\boldsymbol{a}_1 \cdot (\boldsymbol{a}_2 \times \boldsymbol{a}_3)}, \quad \boldsymbol{b}_2 = \frac{2\pi(\boldsymbol{a}_3 \times \boldsymbol{a}_1)}{\boldsymbol{a}_1 \cdot (\boldsymbol{a}_2 \times \boldsymbol{a}_3)}, \quad \boldsymbol{b}_3 = \frac{2\pi(\boldsymbol{a}_1 \times \boldsymbol{a}_2)}{\boldsymbol{a}_1 \cdot (\boldsymbol{a}_2 \times \boldsymbol{a}_3)}$$

where they satisfy the orthogonality relation $\boldsymbol{b}_i \cdot \boldsymbol{a}_j = 2\pi\delta_{ij}$ We can express $\boldsymbol{k}$ as $\boldsymbol{k} = n_1\boldsymbol{b}_1 + n_2\boldsymbol{b}_2 + n_3\boldsymbol{b}_3$ $(n_j \in \mathbb{Z})$. It can be easily verified that $\cdot\boldsymbol{R} = 2\pi N$ $(N \in \mathbb{Z})$. In terms of wave vectors, this gives

$$(\boldsymbol{q}' - \boldsymbol{q}) \cdot \boldsymbol{a}_j = 2n\pi \quad (n \in \mathbb{Z})$$

This is the Bragg scattering formula in optics. For elastic scattering, $|\boldsymbol{q}| = |\boldsymbol{q}'| = \frac{2\pi}{\lambda}$, and the equation can be simplified to the familiar form:

$$2d \sin\theta = n\lambda$$

where d is the spacing between the incident crystal planes, as shown in Figure 7.13(a). Scattered light that does not satisfy the Bragg formula will undergo coherent cancellation. Figure 7.13(b) shows a crystal diffraction pattern, which appears as discrete and regular spots. The diffraction pattern has a one-to-one correspondence with the three-dimensional Fourier transform of the lattice, and by analyzing X-ray diffraction patterns, we can understand the structure of the crystal.

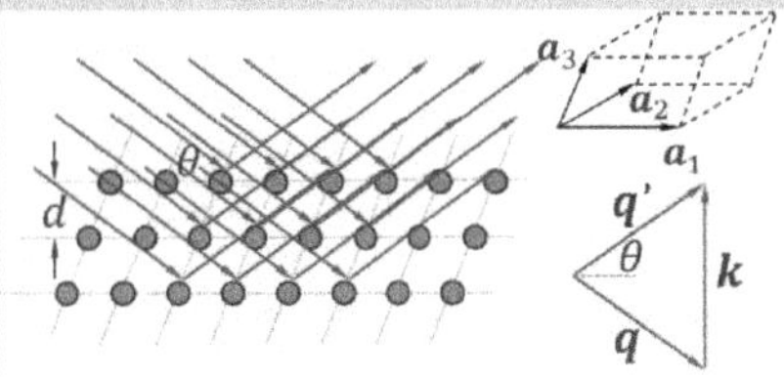

(a)

(b)

Figure 7.13. (image from network).

Exercises

[1] Find the Fourier transform of the functions:

$$\text{(a) } f(x) = \mathrm{e}^{-a|x|} \quad (a > 0); \quad \text{(b) } f(x) = \frac{1}{|x|^a} \quad (0 < a < 1)$$

Hint: (b) After converting it into a semi-infinite integral, using the residue theorem, taking a quarter of the first quadrant as the contour of integration, and transforming the integral into a Gamma function integral, we discuss the cases of $k > 0$ and $k < 0$, respectively.

Answer:

$$\text{(a) } \frac{1}{\pi a} \frac{a^2}{a^2 + k^2}; \quad \text{(b) } \frac{\Gamma(1-a)\sin(a\pi/2)}{\pi |k|^{1-a}}$$

[2] Find the Fourier transform of the functions:

$$\text{(a) } f(x) = \frac{\sin \pi x}{x}; \quad \text{(b) } f(x) = \operatorname{sec\,h}(ax) \quad (a > 0)$$

Answer:

$$\text{(a) } \begin{cases} \frac{1}{2} & |k| \le \pi \\ 0 & |k| > \pi \end{cases}; \quad \text{(b) } \frac{1}{2a} \operatorname{sec\,h}\left(\frac{\pi k}{2a}\right)$$

[3] Prove that the Fourier transform satisfying the Parseval's relation:

$$\int_{-\infty}^{\infty} f_1(x)\bar{f_2}(x)\mathrm{d}x = 2\pi \int_{-\infty}^{\infty} F_1(k)\bar{F_2}(k)\mathrm{d}k$$

Hint: Make use of the properties of Dirac function:

$$\delta(k - k') = \frac{1}{2\pi}\int_{-\infty}^{\infty} \mathrm{e}^{\mathrm{i}(k-k')x}\,\mathrm{d}x, \quad \int_{-\infty}^{\infty} f(k')\delta(k - k')\mathrm{d}k' = f(k)$$

[4] Solve the Poisson equation via three-dimensional Fourier transform:

$$\nabla^2 u(\boldsymbol{r}) = \varphi(\boldsymbol{r})$$

Answer:

$$u(\boldsymbol{r}) = \frac{1}{4\pi}\iiint \frac{\varphi(\boldsymbol{r}')}{|\boldsymbol{r} - \boldsymbol{r}'|}\mathrm{d}\boldsymbol{r}'$$

[5] Solve the Airy equation by Fourier transform:

$$y''(x) - xy(x) = 0 \quad (-\infty < x < \infty)$$

Answer:

$$y(x) = \frac{C}{2\pi}\int_{-\infty}^{\infty} e^{\frac{i}{3}k^3 + ikx}\, dk$$

[6] Calculate the definite integral:

$$I = \int_0^{\infty} e^{-(a^2x^2 + b^2x^{-2})}\, dx$$

Hint: Perform Fourier transform with respect to the parameter b.
Answer: $\frac{1}{\sqrt{2}a} e^{-2ab}$

[7] Perform Fourier sine and cosine transform for the function:

$$f(x) = \frac{1}{\sqrt{x}}$$

(a) Prove it poses self-reciprocal:

$$\mathfrak{F}_c\left[\frac{1}{\sqrt{x}}\right] = \frac{1}{\pi}\int_0^{\infty} \frac{1}{\sqrt{x}} \cos xt\, dx = \frac{1}{\sqrt{2\pi t}}$$

$$\mathfrak{F}_s\left[\frac{1}{\sqrt{x}}\right] = \frac{1}{\pi}\int_0^{\infty} \frac{1}{\sqrt{x}} \sin xt\, dx = \frac{1}{\sqrt{2\pi t}}$$

(b) Make use of the results to calculate the Fresnel integral:

$$\int_0^{\infty} \cos x^2\, dx; \quad \int_0^{\infty} \sin x^2\, dx$$

7.3 Convolution Theorem

7.3.1 *Convolution function*

Let us first define the convolution of two functions $f_1(x)$ and $f_2(x)$,

$$f_1(x) * f_2(x) \stackrel{\text{def}}{=} \int_{-\infty}^{\infty} f_1(\xi) f_2(x - \xi) d\xi \tag{7.3.1}$$

The convolution has the following properties:

$$\begin{aligned}
&f_1(x) * f_2(x) = f_2(x) * f_1(x)\\
&f_1(x) * [f_2(x) * f_3(x)] = [f_1(x) * f_2(x)] * f_3(x)\\
&f_1(x) * [f_2(x) + f_3(x)] = f_1(x) * f_2(x) + f_1(x) * f_3(x)\\
&[f_1(x) * f_2(x)]' = f_1'(x) * f_2(x) + f_1(x) * f_2'(x)
\end{aligned}$$

These properties can be directly derived from the definition. Thus, it can be concluded that convolution satisfies the commutative, associative, and distributive laws. Regarding the Fourier transform of the convolution function, there is the Convolution Theorem:

$$\mathfrak{F}[f_1(x) * f_2(x)] = 2\pi F_1(k)F_2(k) \tag{7.3.2}$$

Proof. According to the definition of the Fourier transform, we have

$$\begin{aligned}
\mathfrak{F}[f_1(x) * f_2(x)] &= \frac{1}{2\pi}\int_{-\infty}^{\infty} f_1(x) * f_2(x)\mathrm{e}^{-ikx}\,\mathrm{d}x\\
&= \frac{1}{2\pi}\int_{-\infty}^{\infty}\left[\int_{-\infty}^{\infty} f_1(\xi)f_2(x-\xi)\mathrm{d}\xi\right]\mathrm{e}^{-ikx}\,\mathrm{d}x\\
&= \frac{1}{2\pi}\int_{-\infty}^{\infty} f_1(\xi)\left[\int_{-\infty}^{\infty} f_2(x-\xi)\mathrm{e}^{-ikx}\,\mathrm{d}x\right]\mathrm{d}\xi\\
&= \frac{1}{2\pi}\int_{-\infty}^{\infty} f_1(\xi)\left[\int_{-\infty}^{\infty} f_2(y)\mathrm{e}^{-iky-ik\xi}\,\mathrm{d}y\right]\mathrm{d}\xi\\
&= \frac{1}{2\pi}\int_{-\infty}^{\infty} f_1(\xi)\mathrm{e}^{-ik\xi}\,\mathrm{d}\xi\int_{-\infty}^{\infty} f_2(y)\mathrm{e}^{-iky}\,\mathrm{d}y\\
&= 2\pi F_1(k)F_2(k)
\end{aligned}$$

In addition, the area under the convolution function is equal to the product of the areas under the two functions, that is

$$\int_{-\infty}^{\infty}[f_1(x) * f_2(x)]\mathrm{d}x = \int_{-\infty}^{\infty}\int_{-\infty}^{\infty} f_1(\xi)f_2(x-\xi)\mathrm{d}\xi\,\mathrm{d}x$$

$$= \int_{-\infty}^{\infty} f_1(\xi) \left[\int_{-\infty}^{\infty} f_2(x-\xi)\mathrm{d}x \right] \mathrm{d}\xi$$

$$= \int_{-\infty}^{\infty} f_1(\xi)\mathrm{d}\xi \int_{-\infty}^{\infty} f_2(x)\mathrm{d}x \tag{7.3.3}$$

Convolution is commonly encountered in linear response systems. When a physical system is subjected to an external impulse, it responds accordingly. For example, when an external force $f(t)$ is applied, it causes a displacement $x(t)$. Typically, the response exhibits a time delay. Assuming the response is linear, the entire process can be described as follows:

$$x(t) = \int_{-\infty}^{\infty} f(t')\kappa(t-t')\mathrm{d}t'$$

Here, $\kappa(t-t')$ is called the response function, which relates the cumulative effect of the external force to the system's response. Taking the Fourier transform, we have

$$X(\omega) = \frac{1}{2\pi} \int_{-\infty}^{\infty} x(t)\mathrm{e}^{-\mathrm{i}\omega t}\, dt$$

This yields

$$X(\omega) = K(\omega)F(\omega)$$

This equation indicates that the response function is represented simply as a product in the frequency domain. The response function $K(\omega)$ is typically a complex function, where the real part represents the in-phase displacement with the external force, and the imaginary part reflects the system's dissipative capacity. □

Example 7.7. Find the response function for a damped oscillator subjected to an external force:

$$mx''(t) + \alpha x'(t) + kx(t) = f(t)$$

Solution: By taking the Fourier transform of both sides of the equation, we can directly obtain the response function as follows:

$$K(\omega) = \frac{X(\omega)}{F(\omega)} = \frac{1}{m} \frac{1}{\omega_0^2 - \omega^2 - \mathrm{i}\omega\gamma}$$

Here $\omega_0^2 = \frac{k}{m}$ and $\gamma = \frac{\alpha}{m}$. The imaginary part of the response function is given by

$$\operatorname{Im} K(\omega) = \frac{1}{m} \frac{\omega\gamma}{(\omega_0^2 - \omega^2)^2 + (\omega\gamma)^2}$$

This expression exhibits a resonance peak near $\omega = \omega_0$, indicating that the system has a strong response or energy absorption capability in the frequency range around ω_0.

7.3.2 *Correlation function*

In physics, the cross-correlation function of two functions $f_1(x)$ and $f_2(x)$ is commonly defined as the correlation product:

$$\begin{aligned} R_{12}(x) = f_1(x) \circ f_2(x) &\stackrel{\text{def}}{=} \int_{-\infty}^{\infty} f_1(\xi + x)\bar{f}_2(\xi)\mathrm{d}\xi \\ &= \int_{-\infty}^{\infty} f_1(\xi)\bar{f}_2(\xi - x)\mathrm{d}\xi \end{aligned} \tag{7.3.4}$$

From the definition, the cross-correlation and convolution have a similar definition: both involve multiplying two sequences with a time shift and then summing. The difference lies in the fact that the cross-correlation function involves shifting or delaying one of the sequences before multiplication and summation, while convolution involves flipping one of the sequences, shifting it, and then multiplying and summing. It can be seen that the cross-correlation function of $f_1(x)$ and $f_2(x)$ is equivalent to the convolution of the functions $\bar{f}_1(-x)$ and $f_2(x)$, with no substantial difference between them.

Using the Fourier transform representation, it can be easily proven using the Wiener-Khinchin theorem:

$$R_{12}(x) = 2\pi \int_{-\infty}^{\infty} F_1(k)\bar{F}_2(k)\mathrm{e}^{ikx}\,\mathrm{d}k \tag{7.3.5}$$

If $f_1(x) = f_2(x)$, it is called the autocorrelation function:

$$R(x) = \int_{-\infty}^{\infty} f(\xi + x)\bar{f}(\xi)\mathrm{d}\xi \tag{7.3.6}$$

It is evident that $R(-x) = \bar{R}(x)$, and

$$R(x) = 2\pi \int_{-\infty}^{\infty} F(k)\bar{F}(k)\mathrm{e}^{ikx}\,\mathrm{d}k$$

The power spectral density function is defined as

$$S(k) = 2\pi F(k)\bar{F}(k)$$

It can be shown that it is the Fourier transform of the autocorrelation function $R(x)$:

$$\begin{cases} R(x) = \displaystyle\int_{-\infty}^{\infty} S(k)\mathrm{e}^{ikx}\,\mathrm{d}k \\ S(k) = \dfrac{1}{2\pi}\displaystyle\int_{-\infty}^{\infty} R(x)\mathrm{e}^{-ikx}\,\mathrm{d}x \end{cases}$$

Based on the energy integral, it can be derived that

$$\int_{-\infty}^{\infty} |R(x)|^2\,\mathrm{d}x = 2\pi\int_{-\infty}^{\infty} |S(k)|^2\,\mathrm{d}k \tag{7.3.7}$$

It can be proven that the cross-correlation function of the sum of two independent functions is equal to the sum of their individual autocorrelation functions. For continuous-time white noise signals, their autocorrelation function is $R(x) = \delta_{t0}$, meaning that they are completely uncorrelated in time except at $t = 0$.

ENDNOTE

Correlation measures the similarity between two functions. In the context of random processes, when two signal sequences or functions have similar frequency characteristics, the correlation function exhibits a peak. However, for general non-periodic signal sequences or functions, it is not easy to visually determine their similarity. This raises a question: in what way are they similar, and how can their degree of similarity be quantified?

In experimental measurements, variance is often used to describe the fluctuations of a random process. Similarly, for two real-valued signal sequences or real functions $f_1(x)$ and $f_2(x)$, to compare their similarity, we consider the squared difference:

$$\Delta \overset{\text{def}}{=} \int_{-\infty}^{\infty} |f_1(\xi) - f_2(\xi)|^2\,\mathrm{d}\xi$$

This involves squaring and summing the deviations at each point. Obviously, $\Delta \geq 0$, and a smaller value of the squared difference indicates a higher degree of similarity between the two functions. However, this formula has a drawback: different signal sequences may not be synchronized in time but can still be highly similar. To address this, we can compare the two signal sequences by introducing different shifts or delays, defining the function:

$$\Delta(x) \stackrel{\text{def}}{=} \int_{-\infty}^{\infty} |f_1(\xi + x) - f_2(\xi)|^2 \, d\xi$$

When $\Delta(x)$ reaches its minimum value, it indicates the highest level of matching between the two functions. Expanding the squared modulus inside the integral, we have

$$\begin{aligned}\Delta(x) &= \int_{-\infty}^{\infty} \{|f_1(\xi + x)|^2 + |f_2(\xi)|^2 - f_1(\xi + x)\bar{f}_2(\xi) \\ &\quad - \bar{f}_1(\xi + x)f_2(\xi)\} \, d\xi \\ &= C - \int_{-\infty}^{\infty} [f_1(\xi + x)\bar{f}_2(\xi) + \bar{f}_1(\xi + x)f_2(\xi)] d\xi \\ &= C - [R_{12}(x) + \overline{R_{12}}(x)]\end{aligned}$$

The first two terms in the integral are constants and can be ignored.

Hence, the similarity between two functions can be characterized by the correlation function $R_{12}(x)$. If the correlation function $R_{12}(x)$ is relatively flat, it indicates a lower degree of similarity between $f_1(x)$ and $f_2(x)$. Conversely, if the correlation function $R_{12}(x)$ exhibits peaks, it suggests the presence of some correlation between them. From a wave perspective, the correlation function is similar to the coherence coefficient of two waveforms. If the two waveforms are incoherent, the coherence coefficient is zero. Similarly, if the correlation function of two functions is zero, it means they are uncorrelated or do not have overlapping frequency spectra.

We can also examine the correlation of spectral functions by defining the spectral variance:

$$\tilde{\Delta} \overset{\text{def}}{=} \int_{-\infty}^{\infty} |F_1(k) - F_2(k)|^2 \, \mathrm{d}k$$

A smaller value of $\tilde{\Delta}$ indicates a higher similarity between the spectra of the two functions. However, since $F_1(k)$ and $F_2(k)$ are usually complex functions, this formula has a limitation. Complex numbers can be represented as vectors in a two-dimensional space, where the vector's length represents the magnitude and the direction represents the phase angle. Therefore, complex functions can be represented by a three-dimensional curve along the k-axis, as shown in Figure 7.14. To address this, we introduce a "spiral" spectral variance:

$$\tilde{\Delta}(\alpha) \overset{\text{def}}{=} \int_{-\infty}^{\infty} |F_1(k)\mathrm{e}^{\mathrm{i}k\alpha} - F_2(k)|^2 \, \mathrm{d}k$$

Here, α describes the degree of spiral rotation, and the additional phase factor $\mathrm{e}^{\mathrm{i}k\alpha}$ does not alter the distribution of the spectral functions' magnitudes. If $\tilde{\Delta}(\alpha)$ reaches its minimum value for certain α values, it indicates that the spectral functions $F_1(k)$ and $F_2(k)$ are closest. Correspondingly, we have

$$\tilde{\Delta}(\alpha) = C - \int_{-\infty}^{\infty} [F_1(k)\bar{F}_2(k)\mathrm{e}^{\mathrm{i}k\alpha} + F_2(k)\bar{F}_1(k)\mathrm{e}^{-\mathrm{i}k\alpha}]\mathrm{d}k$$

By letting $\alpha \to x$, the second term in the above equation essentially becomes the correlation function $R_{12}(x)$, so $\tilde{\Delta}(x) \approx \Delta(x)$.

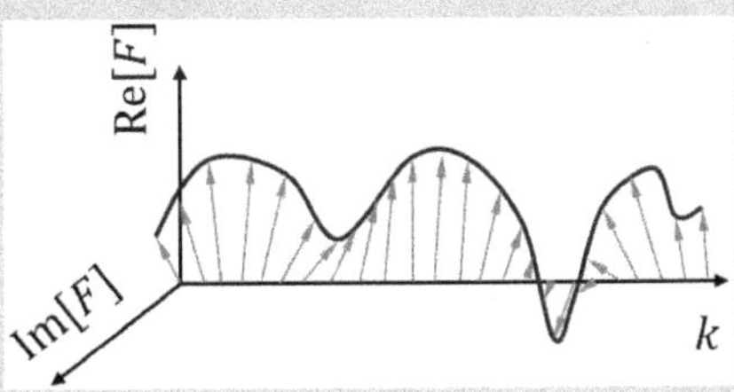

Figure 7.14.

The autocorrelation function represents the self-similarity of a sequence and can be referred to as temporal correlation in the time domain. It helps identify repeated patterns in a sequence that may be obscured by noise. In the spatial domain, it is referred to as spatial correlation, indicating the correlation between density fluctuations at different points in a system at equilibrium. When the density distribution exhibits some periodicity, the autocorrelation function also exhibits corresponding periodicity.

Since convolution and correlation functions share a similar concept, convolution can also be used to describe the similarity between two functions. The convolution function is an important reference for pattern recognition and multiple convolutions can characterize the common features of multiple signal sequences:

$$R_n(x) = f_1(x) * f_2(x) * f_3(x) * \cdots * f_n(x)$$

Now let's consider the following questions:
(1) If we define a "frequency-shift" correlation function:

$$\Delta_{12}(\lambda, x) = \int_{-\infty}^{\infty} |F_1(k - \lambda)\mathrm{e}^{ikx} - F_2(k)|^2\, \mathrm{d}k$$

What property does it reveal about the functions? What is the significance of the "frequency-shift" autocorrelation function?
(2) If a sequence exhibits self-similarity, what measure should be used to characterize it?

Exercises

[1] Prove the properties of convolution:

(a) $f_1(x) * f_2(x) = f_2(x) * f_1(x)$
(b) $f_1(x) * [f_2(x) * f_3(x)] = [f_1(x) * f_2(x)] * f_3(x)$
(c) $f_1(x) * [f_2(x) + f_3(x)] = f_1(x) * f_2(x) + f_1(x) * f_3(x)$

[2] Prove the properties of correlation function:

$$\text{(a) } f_1(x) \circ f_2(x) \neq f_2(x) \circ f_1(x)$$

$$\text{(b) } f_1(x) \circ [f_2(x) \circ f_3(x)] \neq [f_1(x) \circ f_2(x)] \circ f_3(x)$$

[3] Prove that the correlation function of the sum of two unrelated functions is equal to the sum of their individual correlation functions.

[4] Prove:

$$\text{(a) } \mathfrak{F}[f_1(x) * f_2(x) * \cdots * f_n(x)] = (2\pi)^{n-1} F_1(k) F_2(k) \cdots F_n(k)$$

$$\text{(b) } \mathfrak{F}[f_1(x) \circ f_2(x) \circ \cdots \circ f_n(x)] = (2\pi)^{n-1} F_1(k) F_2(k) \cdots F_n(k)$$

7.4 Poisson Summation Formula

In this section, we introduce another important relation related to the Fourier transform called the Poisson summation formula. If a function $f(x)$ satisfies the Dirichlet conditions and is absolutely integrable over the interval $(-\infty, \infty)$, and we define $F(k) = \mathfrak{F}[f(x)]$, then the formula is given by

$$\sum_{n\in\mathbb{Z}} f(2\pi n) = \sum_{n\in\mathbb{Z}} F(n) \tag{7.4.1}$$

Proof. In Section 4.4, we discussed the properties of the Bose distribution formula $f_B = \frac{1}{\mathrm{e}^{\mathrm{i}z}-1}$. It has simple poles at $z = 2\pi n$ with residues of $-\mathrm{i}$. Consider the contour Γ shown in Figure 7.15. According to the residue theorem, we have

$$2\pi \sum_{n\in\mathbb{Z}} f(2\pi n) = \oint_\Gamma \frac{f(z)}{\mathrm{e}^{\mathrm{i}z} - 1}\,\mathrm{d}z = \int_{L_1} \frac{f(z)}{\mathrm{e}^{\mathrm{i}z} - 1}\,\mathrm{d}z + \int_{L_2} \frac{f(z)}{\mathrm{e}^{\mathrm{i}z} - 1}\,\mathrm{d}z$$

For the path L_1, where the imaginary part of z, $y < 0$, we have $|\mathrm{e}^{\mathrm{i}z}| = |\mathrm{e}^{-y}| > 1$, and we can expand it as a geometric series:

$$\frac{1}{\mathrm{e}^{\mathrm{i}z} - 1} = \mathrm{e}^{-\mathrm{i}z} \sum_{n=0}^{\infty} \mathrm{e}^{-\mathrm{i}nz}$$

For the path L_2, where the imaginary part of z, $y > 0$, we have $|\mathrm{e}^{\mathrm{i}z}| = |\mathrm{e}^{-y}| < 1$, and we can still expand it as a geometric series:

$$\frac{1}{\mathrm{e}^{\mathrm{i}z} - 1} = -\sum_{n=0}^{\infty} \mathrm{e}^{\mathrm{i}nz}$$

Therefore,

$$\begin{aligned} 2\pi \sum_{n \in Z} f(2\pi n) &= \int_{L_1} f(z) \sum_{n=0}^{\infty} \mathrm{e}^{-\mathrm{i}(n+1)z} \mathrm{d}z - \int_{L_2} f(z) \sum_{n=0}^{\infty} \mathrm{e}^{\mathrm{i}nz} \, \mathrm{d}z \\ &= \sum_{n=0}^{\infty} \int_{-\infty}^{\infty} f(x) \mathrm{e}^{-\mathrm{i}(n+1)x} \, \mathrm{d}x - \sum_{n=0}^{\infty} \int_{\infty}^{-\infty} f(x) \mathrm{e}^{\mathrm{i}nx} \, \mathrm{d}x \\ &= \sum_{n=0}^{\infty} 2\pi F(n+1) + \sum_{n=0}^{\infty} 2\pi F(-n) \end{aligned}$$

Thus,

$$\sum_{n \in \mathbb{Z}} f(2\pi n) = \sum_{n \in \mathbb{Z}} F(n)$$

□

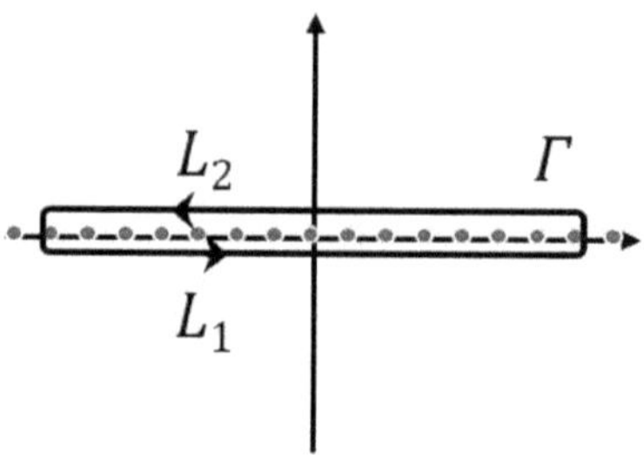

Figure 7.15.

Poisson summation formula has several far-reaching consequences. Here are some notable ones:

[1] The Fourier transform of Gaussian function is

$$\mathfrak{F}[\mathrm{e}^{-\alpha^2 x^2}] = \frac{1}{2\sqrt{\pi}\alpha} \mathrm{e}^{-\frac{k^2}{4\alpha^2}}$$

Take $\alpha^2 = \frac{t}{4\pi}(t > 0)$, we have

$$\mathfrak{F}[\mathrm{e}^{-\frac{tx^2}{4\pi}}] = \frac{1}{\sqrt{t}}\mathrm{e}^{-\frac{\pi k^2}{t}}$$

By applying the Poisson summation formula, it follows that

$$\sum_{n=-\infty}^{\infty} \mathrm{e}^{-\pi n^2 t} = \sum_{n=-\infty}^{\infty} \frac{\mathrm{e}^{-\pi n^2/t}}{\sqrt{t}} \tag{7.4.2}$$

Defining the ϑ-function by

$$\vartheta(t) \stackrel{\text{def}}{=} \sum_{n=-\infty}^{\infty} \mathrm{e}^{-\pi n^2 t} \quad (t > 0)$$

It means that

$$\vartheta(t) = \frac{1}{\sqrt{t}}\vartheta\left(\frac{1}{t}\right) \tag{7.4.3}$$

This ϑ-function has intimate relation with Riemann's ζ-function.

[2] From Exercise [2] of Section 7.2, the Fourier transformation of function $f(x) = \frac{1}{\cosh \pi x}$ is:

$$\mathfrak{F}\left[\frac{1}{\cosh(ax)}\right] = \frac{1}{2a\cosh\left(\frac{\pi k}{2a}\right)}$$

With the phase shift theorem, we have

$$\mathfrak{F}\left[\frac{\mathrm{e}^{-ibx}}{\cosh(ax)}\right] = \frac{1}{2a\cosh\left(\frac{\pi(k+b)}{2a}\right)}$$

By taking $a = \frac{1}{2t}$, and applying the Poisson summation formula, it follows that

$$\sum_{n=-\infty}^{\infty} \frac{\mathrm{e}^{-2\pi ibn}}{\cosh\left(\frac{\pi n}{t}\right)} = \sum_{n=-\infty}^{\infty} \frac{t}{\cosh[\pi(n+b)t]} \tag{7.4.4}$$

[3] Assuming $a > 0$, the Fourier transform of function $f(x) = \frac{a}{x^2+a^2}$ is

$$\mathfrak{F}\left[\frac{a}{x^2+a^2}\right] = \pi\mathrm{e}^{-a|k|}$$

Using the scaling theorem and the Poisson summation formula, we have

$$\sum_{n=-\infty}^{\infty} \frac{a}{n^2+a^2} = \pi \sum_{n=-\infty}^{\infty} \mathrm{e}^{-2\pi a|n|}$$

Since $\mathrm{e}^{-2\pi a|n|} < 1$, It can be simplified as

$$2a\sum_{n=0}^{\infty} \frac{1}{n^2+a^2} + \frac{1}{a} = \pi + 2\pi \sum_{n=1}^{\infty} \mathrm{e}^{-2\pi a n}$$

$$= -\pi + \frac{2\pi}{1-\mathrm{e}^{-2\pi a}} = \pi \frac{1+\mathrm{e}^{-2\pi a}}{1-\mathrm{e}^{-2\pi a}}$$

or

$$\sum_{n=1}^{\infty} \frac{1}{n^2+a^2} = -\frac{1}{2a^2} + \frac{\pi}{2a} \coth \pi a$$

These are previously derived results.

Exercises

[1] Assume $\operatorname{Im} a > 0$, prove that

$$\sum_{n=-\infty}^{\infty} \frac{1}{(n+a)^2} = \frac{\pi^2}{\sin^2(\pi a)}$$

ENDNOTE

We can understand the Fourier transform from the mapping relationship shown in Figure 7.16: the original function space is discretized into many small segments x_j ($j \in \mathbb{Z}$), and the corresponding function values are $f(x_j)$. Similarly, the image space is also discretized into segments k_n ($n \in \mathbb{Z}$), with corresponding function values $F(k_n)$ (represented as real functions in the figure). We can consider $f(x_j)\Delta x_j$ as the amplitudes of a series of subwave sources. The mapping from the position space to the frequency domain can be imagined as a propagation process with a wave vector k_n. The jth subwave source has a propagation distance of x_j relative to the subwave source at $x = 0$, and the propagation function is $\mathrm{e}^{\mathrm{i}k_n \cdot x_j}$. Therefore, the amplitude of propagation is $f(x_j)\mathrm{e}^{\mathrm{i}k_n \cdot x_j}\Delta x_j$. The total amplitude propagated with wave vector k_n is the superposition of amplitudes from all subwave sources reaching that point:

$$F(k_n) = \sum_j f(x_j)e^{ik_n \cdot x_j}\Delta x_j$$

Taking the limits as $j, n \to \infty$, we have

$$F(k) = \lim_{j\to\infty} \sum_j f(x_j)\mathrm{e}^{\mathrm{i}k \cdot x_j}\Delta x_j = \int_{-\infty}^{\infty} f(x)\mathrm{e}^{\mathrm{i}k \cdot x}\, dx$$

From this perspective, the Fourier transform shares similarities with the Huygens principle in function space.

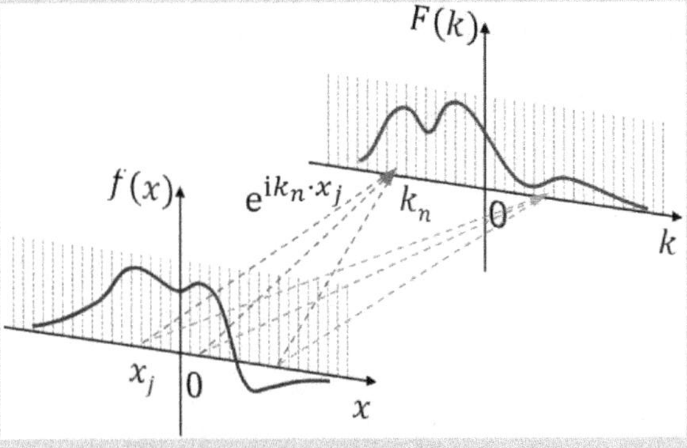

Figure 7.16.

Since the Fourier transform is a global transformation, meaning that any point in the frequency domain requires all the information from the position space to be represented, it usually involves a large computational cost. We consider an approximate approach:

$$F(k_n) = \sum_j f(x_j)e^{ik_n \cdot x_j}\Delta x_j = \sum_{|x_j|<\Delta} f(x_j)e^{ik_n \cdot x_j}\Delta x_j + \sum_{|x_j|>\Delta} f(x_j)e^{ik_n \cdot x_j}\Delta x_j$$

Here, we set the range of the window Δ as

$$|k_n \cdot x_j| \lesssim \frac{\pi}{2} \longrightarrow \Delta \sim \frac{\pi}{2|k_n|}$$

The propagation function $e^{ik_n \cdot x_j}$ can be represented by a two-dimensional unit vector, where the phase $k_n \cdot x_j$ indicates the direction of the vector. For the subwave sources within the window, the vector directions are relatively close. However, for the subwave sources outside the window, the vector directions vary significantly. Since $f(x)$ satisfies absolute integrability, when x is large, the variation of $f(x)$ tends to be flat. Therefore, the effects of these vectors cancel each other out:

$$\sum_{|x_j|>\Delta} f(x_j)e^{ik_n \cdot j dx}\,dx \sim A \sum_{|x_j|>\Delta} e^{ik_n \cdot j dx}\,dx \to 0$$

Hence, $F(k)$ can be approximated by the Fourier transform of a function within the range of the window $\Delta_k \sim \frac{\pi}{|k|}$:

$$F(k) \approx \sum_{|x_j|<\Delta_k} f(x_j)e^{ik \cdot x_j}\Delta x_j = \int_{-\Delta_k}^{\Delta_k} f(x)e^{ik \cdot x}\,dx$$

It can be seen that the low-frequency components with smaller k require a wider window, while the high-frequency components with larger k only need to be approximated within a narrower window. This idea of using functions with a finite range as an approximation for the Fourier transform ultimately developed into the wavelet transform method.

Chapter 8

Function Transformations

8.1 Laplace Transform

8.1.1 *Absolute integrability*

The condition for the existence of the Fourier transform is that the original function $f(t)$ satisfies the Dirichlet conditions on any finite interval and is absolutely integrable over the interval $(-\infty, \infty)$. This is a very strong restriction, and most functions do not satisfy this condition. In order to ensure the absolute integrability of the function, we introduce a convergence factor defined as

$$g(t) = f(t)\mathrm{e}^{-\sigma t}H(t) \quad (\sigma > 0)$$

where the Heaviside step function is defined as

$$H(t) \stackrel{\text{def}}{=} \begin{cases} 1 & (t \geq 0) \\ 0 & (t < 0) \end{cases}$$

The function $g(t)$ generally satisfies the condition of absolute integrability. The Fourier transform of $g(t)$ is

$$G(\omega) = \frac{1}{2\pi}\int_{-\infty}^{\infty} g(t)\mathrm{e}^{-\mathrm{i}\omega t}\mathrm{d}t = \frac{1}{2\pi}\int_{0}^{\infty} f(t)\mathrm{e}^{-(\sigma+\mathrm{i}\omega)t}\mathrm{d}t$$

Let $p = \sigma + \mathrm{i}\omega$ and $F(p) = 2\pi G(\omega)$, we define

$$F(p) \stackrel{\text{def}}{=} \int_{0}^{\infty} f(t)\mathrm{e}^{-pt}\mathrm{d}t \tag{8.1.1}$$

Function $F(p)$ is called the Laplace transform of the function $f(t)$. It should be noted that p is a complex number with $\mathrm{Re}\, p = \sigma > 0$.

The inverse Fourier transform of the function $G(\omega)$ is given by

$$g(t) = f(t)\mathrm{e}^{-\sigma t}H(t) = \int_{-\infty}^{\infty} G(\omega)\mathrm{e}^{\mathrm{i}\omega t}\mathrm{d}\omega$$

Therefore, the inverse Laplace transform is given by

$$f(t) = \frac{1}{2\pi\mathrm{i}}\int_{\sigma-\mathrm{i}\infty}^{\sigma+\mathrm{i}\infty} F(p)\mathrm{e}^{pt}\mathrm{d}p \quad (t \geq 0) \tag{8.1.2}$$

The Laplace transform and its inverse transform are denoted by the symbol $\mathcal{L}$ and $\mathcal{L}^{-1}$:

$$\begin{cases} F(p) = \mathcal{L}[f(t)] \\ f(t) = \mathcal{L}^{-1}[F(p)] \end{cases}$$

$F(p)$ is referred to as the image function of $f(t)$, and $f(t)$ is referred to as the original function of $F(p)$.

The conditions for the existence of the Laplace transform are as follows:

(1) For any finite interval in the range $\infty > t \geq 0$, except for a finite number of first-class discontinuities, the function $f(t)$ and its derivative are continuous.
(2) There exist constants $M > 0$ and $\sigma \geq 0$ such that for any $\infty > t \geq 0$, $|f(t)| < M\mathrm{e}^{\sigma t}$, where the lower bound of σ is called the convergence abscissa.

The definition and integrability conditions of the Laplace transform indicate that the image function $F(p)$ is analytic in the right half-plane $\mathrm{Re}\, p > \sigma$. Here are the Laplace transform functions of several elementary functions:

$$\mathcal{L}[1] = \frac{1}{p} \quad (\mathrm{Re}\, p > 0)$$

$$\mathcal{L}[t] = \frac{1}{p^2}, \quad \mathcal{L}[t^n] = \frac{n!}{p^{n+1}}$$

$$\mathcal{L}[\mathrm{e}^{st}] = \frac{1}{p-s} \quad (\mathrm{Re}\, p > \mathrm{Re}\, s)$$

$$\mathcal{L}[te^{st}] = \frac{1}{(p-s)^2}, \quad \mathcal{L}[t^n e^{st}] = \frac{n!}{(p-s)^{n+1}}$$

$$\mathcal{L}[\delta(t-a)] = e^{-ap} \quad (a > 0)$$

$$\mathcal{L}[\sin \omega t] = \frac{\omega}{p^2+\omega^2}, \quad \mathcal{L}[\cos \omega t] = \frac{p}{p^2+\omega^2}$$

These results can be directly proven using the definition of the Laplace transform.

8.1.2 *Basic properties*

[1] **Linearity theorem:**

$$\mathcal{L}[c_1 f_1(t) + c_2 f_2(t)] = c_1 L[f_1(t)] + c_2 \mathcal{L}[f_2(t)]$$

[2] **Derivative theorem:**

$$\mathcal{L}[f'(t)] = pF(p) - f(0)$$
$$\mathcal{L}[f^{(n)}(t)] = p^n F(p) - p^{n-1} f(0) - \cdots - p f^{(n-2)}(0) - f^{(n-1)}(0)$$

[3] **Integration theorem:**

$$\mathcal{L}\left[\int_0^t f(\tau)\mathrm{d}\tau\right] = \frac{F(p)}{p}$$

[4] **Scaling theorem:**

$$\mathcal{L}\left[f\left(\frac{t}{a}\right)\right] = aF(ap) \quad (a > 0)$$

[5] **Delay theorem:**

$$\mathcal{L}[f(t-t_0)] = e^{-pt_0} F(p)$$

[6] **"Phase shift" theorem:**

$$\mathcal{L}[e^{-\lambda t} f(t)] = F(p+\lambda)$$

[7] **Convolution theorem:**

$$\mathcal{L}[f_1(t) * f_2(t)] = F_1(p) F_2(p)$$

Here, the convolution is defined as

$$f_1(t) * f_2(t) \equiv \int_0^t f_1(\tau) f_2(t-\tau)\mathrm{d}\tau \qquad (8.1.3)$$

Note that the convolution definition here differs slightly from the convolution definition in the Fourier transform.

We will only prove the convolution theorem. By definition, it has

$$\mathcal{L}[f_1(t) * f_2(t)] = \int_0^\infty f_1(t) * f_2(t)\mathrm{e}^{-pt}\mathrm{d}t$$
$$= \int_0^\infty \left[\int_0^t f_1(\tau)f_2(t-\tau)\mathrm{d}\tau\right]\mathrm{e}^{-pt}\mathrm{d}t$$

This is a double integral, and the integration region is shown as the shaded area in Figure 8.1. By interchanging the order of integration and noting that the integration range for t becomes $[\tau, \infty]$, we have

$$\mathcal{L}[f_1(t) * f_2(t)] = \int_0^\infty \left[\int_\tau^\infty f_2(t-\tau)\mathrm{e}^{-pt}\mathrm{d}t\right] f_1(\tau)\mathrm{d}\tau$$
$$= \int_0^\infty \left[\int_0^\infty f_2(\xi)\mathrm{e}^{-p\xi}\mathrm{d}\xi\right] f_1(\tau)\mathrm{e}^{-p\tau}\mathrm{d}\tau$$
$$= \int_0^\infty f_1(\tau)\mathrm{e}^{-p\tau}\mathrm{d}\tau \int_0^\infty f_2(\xi)\mathrm{e}^{-p\xi}\mathrm{d}\xi = F_1(p)F_2(p)$$

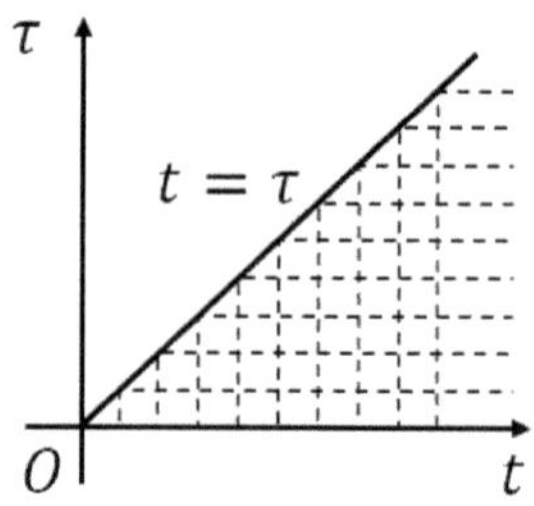

Figure 8.1.

Example 8.1. Proof:

$$\mathcal{L}[\mathrm{e}^{-\lambda t}\sin\omega t] = \frac{\omega}{(p+\lambda)^2+\omega^2}$$

Proof. We start with the identity,

$$\mathcal{L}[\sin\omega t] = \frac{\omega}{p^2+\omega^2}$$

Applying the shifting theorem, we substitute p with $p+\lambda$ to get

$$\mathcal{L}[\mathrm{e}^{-\lambda t}\sin\omega t] = \frac{\omega}{(p+\lambda)^2+\omega^2}$$ □

Example 8.2. Find the Laplace transform of the function:

$$f(t) = t^a \quad (a > -1)$$

Solution 1: According to the definition of the Laplace transform, we have

$$\mathcal{L}[t^a] = \int_0^\infty t^a \mathrm{e}^{-pt}\mathrm{d}t$$

Let $z = pt$, then

$$\mathcal{L}[t^a] = \frac{1}{p^{a+1}} \int_L z^a \mathrm{e}^{-z}\mathrm{d}z$$

Since $\mathrm{Re}\, p > 0$, when $\mathrm{Im}\, p > 0$ the integration path L lies in the first quadrant, as shown in Figure 8.2(a). By cutting along the positive real axis and setting $\arg z = 0$ along the upper edge of the cut, we construct a sector-shaped contour Γ. Since there are no singularities within the contour, the loop integration vanishes:

$$\oint_\Gamma z^a \mathrm{e}^{-z}\mathrm{d}z = \int_L z^a \mathrm{e}^{-z}\mathrm{d}z + \int_{C_R} z^a \mathrm{e}^{-z}\mathrm{d}z + \int_\infty^0 x^a \mathrm{e}^{-x}\mathrm{d}x + \int_{C_\varepsilon} z^a \mathrm{e}^{-z}\mathrm{d}z = 0$$

It can be proven that

$$\int_{C_R} z^a \mathrm{e}^{-z}\mathrm{d}z \xrightarrow{R\to\infty} 0, \quad \int_{C_\varepsilon} z^a \mathrm{e}^{-z}\mathrm{d}z \xrightarrow{\varepsilon\to 0} 0$$

Therefore,

$$\int_L z^a \mathrm{e}^{-z}\mathrm{d}z = \int_0^\infty x^a \mathrm{e}^{-x}\mathrm{d}x = \Gamma(a+1)$$

Thus, we have

$$\mathcal{L}[t^a] = \frac{\Gamma(a+1)}{p^{a+1}}$$

When $\mathrm{Im}\, p < 0$, the integration path L lies in the fourth quadrant. By constructing a contour as shown in Figure 8.2(b), we can obtain the same result as in the case of $\mathrm{Im}\, p > 0$.

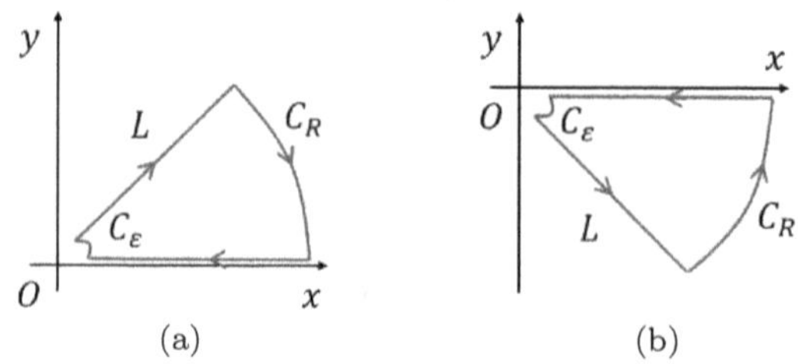

Figure 8.2.

Solution 2: According to the definition of the Gamma function in Section 5.2,

$$\Gamma(a) \overset{\text{def}}{=} \int_0^{\infty} \mathrm{e}^{-t} t^{a-1} \mathrm{d}t \quad (a > 0)$$

For a positive real variable $x > 0$, we make the variable substitution $t \longmapsto xt$, which gives

$$\int_0^{\infty} t^a \mathrm{e}^{-xt} \mathrm{d}t = \frac{\Gamma(a+1)}{x^{a+1}}$$

By analytically continuing the function from the real variable x to the complex plane, we have

$$\mathcal{L}[t^a] = \int_0^{\infty} t^a \mathrm{e}^{-pt} \mathrm{d}t = \frac{\Gamma(a+1)}{p^{a+1}}$$

ENDNOTE

Pierre–Simon Laplace, known as the French Newton, proved in his work *Celestial Mechanics* that the solar system is a stable dynamical system. Newton had sought divine intervention for this. It is said that Napoleon once asked why the book did not mention God creating the universe, and Laplace replied, "Your Majesty, I had no need of that hypothesis". *Celestial Mechanics* extensively discusses potential theory and has had a profound impact on physics to the extent that the fundamental differential equation in potential theory is called the Laplace equation, although Laplace never mentioned that the concept of potential was invented by Lagrange.

Laplace's other major work, *An Analytical Theory of Probability*, is his greatest contribution to probability theory. He skillfully used Laplace transforms, generating functions, and many other mathematical techniques. It is somewhat controversial that Laplace rarely mentioned the work of his predecessors in both books as if he had invented everything himself.

If we express a power series as

$$\sum_{n=0}^{\infty} a(n)x^n$$

it has a formal resemblance to the integral:

$$\int_0^{\infty} a(t)x^t \mathrm{d}t$$

In a restatement of notation, taking $x = \mathrm{e}^{-p}$, the integral becomes

$$\int_0^{\infty} a(t)\mathrm{e}^{-pt}\mathrm{d}t$$

This is the Laplace transform of the function $a(t)$. Therefore, the Laplace transform is an analog of a power series in the context of a continuous limit. Since power series are important in analysis, it can be anticipated that the Laplace transform is also important.

Exercises

[1] Find the Laplace transform of the following functions:

$$\text{(a) } f(t) = t^2 + t\mathrm{e}^t; \quad \text{(b) } f(t) = \mathrm{e}^{-2t}\sin 6t - 5\mathrm{e}^{-2t}$$

$$\text{(c) } f(t) = a^t; \quad \text{(d) } f(t) = \frac{\cos at}{\sqrt{t}}$$

[2] Find the Laplace transform of the following functions:

$$\text{(a) } \frac{1}{\sqrt{\pi t}}; \quad \text{(b) } \sin(\omega t + \alpha)$$

[3] Derive the formula of the Laplace inverse transform:

$$f(t) = \frac{1}{2\pi \mathrm{i}} \int_{\sigma - \mathrm{i}\infty}^{\sigma + \mathrm{i}\infty} F(p)\mathrm{e}^{pt}\mathrm{d}p \quad (t \geq 0)$$

[4] Assuming $f(t)$ is a periodic function with $f(t+\alpha) = f(t)$, prove that

$$F(p) = \frac{1}{1 - \mathrm{e}^{-\alpha p}} \int_0^{\alpha} \mathrm{e}^{-pt} f(t)\mathrm{d}t$$

[5] If $f(t)$ is bounded as $t \to \infty$, we need to prove the initial value theorem and the final value theorem:

$$\text{(a)} \ \lim_{t\to 0} f(t) = \lim_{p\to\infty} pF(p); \quad \text{(b)} \ \lim_{t\to\infty} f(t) = \lim_{p\to 0} pF(p)$$

[6] Find the Laplace transform of the ϑ function:

$$\vartheta(t) = \sum_{n=-\infty}^{\infty} \mathrm{e}^{-n^2\pi^2 t}$$

Answer: $\bar{\vartheta}(p) = \frac{\coth\sqrt{p}}{\sqrt{p}}$

8.2 Laplace Inverse Transform

After applying the Laplace transform to a function, it is necessary to invert it back to the original function. There are several basic methods for performing the inverse transform as follows.

8.2.1 *Partial fraction decomposition*

For the inversion of a composite rational function, it can be simplified by decomposing the fraction. If the function contains exponential factors, the shifting theorem can be used to find the original function.

Example 8.3. Given the Laplace transform function $F(p)$, find the original function:

$$F(p) = \frac{p^3 + 2p^2 - 9p + 36}{p^4 - 81}$$

Solution: Decompose the rational function into simple fractions:

$$F(p) = \frac{p^3 + 2p^2 - 9p + 36}{(p-3)(p+3)(p^2+9)}$$
$$= \frac{1}{2}\frac{1}{p-3} - \frac{1}{2}\frac{1}{p+3} + \frac{p}{p^2+9} - \frac{1}{3}\frac{3}{p^2+9}$$

Therefore,

$$f(t) = L^{-1}[F(p)] = \frac{1}{2}\mathrm{e}^{3t} - \frac{1}{2}\mathrm{e}^{-3t} + \cos 3t - \frac{1}{3}\sin 3t$$

Example 8.4. Perform inversion of Laplace transform of the function:

$$F(p) = \frac{\mathrm{e}^{-\alpha p}}{p(p+b)}$$

Solution: Simplify the rational function:

$$F(p) = \frac{\mathrm{e}^{-\alpha p}}{p(p+b)} = \frac{1}{b}\mathrm{e}^{-\alpha p}\left[\frac{1}{p} - \frac{1}{p+b}\right]$$

Using the shifting theorem, we have

$$f(t) = \mathcal{L}^{-1}[F(p)] = \frac{1}{b}[1 - \mathrm{e}^{-b(t-\alpha)}]H(t-\alpha)$$

Note: Due to the time delay involved, the original function needs to be multiplied by the unit step function $H(t)$.

8.2.2 *Convolution theorem*

For the inversion of the product of two Laplace-transformed functions, the convolution theorem, along with the Laplace transform function table provided in Appendix A.2, can be utilized for calculations.

Example 8.5. Given the Laplace transform function, find the original function:

$$F(p) = \frac{1}{p^2(p^2+\lambda)^3}$$

Solution: Let us break down the function into two parts: $F_1(p) = \frac{1}{p^2}$ which has the original function $f_1(t) = t$ and $F_2(p) = \frac{1}{(p^2+\lambda)^3}$ with original function $f_2(t)$. Considering that

$$\mathcal{L}^{-1}\left[\frac{1}{p^2+\lambda}\right]=\frac{1}{\sqrt{\lambda}}\sin\sqrt{\lambda}t$$

and

$$\frac{1}{(p^2+\lambda)^3}=\frac{1}{2}\frac{\partial^2}{\partial\lambda^2}\frac{1}{p^2+\lambda}$$

the original function $f_2(t)$ is computed as

$$\begin{aligned}f_2(t)&=\frac{1}{2}\frac{\partial^2}{\partial\lambda^2}\left[\frac{1}{\sqrt{\lambda}}\sin\sqrt{\lambda}t\right]\\&=\frac{1}{8}\left[\frac{3}{\lambda^{5/2}}\sin\sqrt{\lambda}t-\frac{3t}{\lambda^2}\cos\sqrt{\lambda}t-\frac{t^2}{\lambda^{3/2}}\sin\sqrt{\lambda}t\right]\end{aligned}$$

Using the convolution theorem, we have

$$\begin{aligned}\mathcal{L}^{-1}\left[\frac{1}{p^2(p^2+\lambda)^3}\right]&=f_2(t)*f_1(t)\\&=\frac{1}{8}\int_0^t\left[\frac{3}{\lambda^{5/2}}\sin\sqrt{\lambda\tau}-\frac{3\tau}{\lambda^2}\cos\sqrt{\lambda\tau}\right.\\&\quad\left.-\frac{t^2}{\lambda^{3/2}}\sin\sqrt{\lambda\tau}\right](t-\tau)\mathrm{d}\tau\end{aligned}$$

The result can be further simplified, but the details are dropped here.

8.2.3 *Riemann–Mellin inversion formula*

For more complex Laplace-transformed functions, the Riemann–Mellin inversion method can be used to directly invert them by employing contour integration. The approach is as follows:

According to the definition of the Laplace transform, the original function can be expressed as

$$f(t)=\mathcal{L}^{-1}[F(p)]=\frac{1}{2\pi\mathrm{i}}\int_{\sigma-\mathrm{i}\infty}^{\sigma+\mathrm{i}\infty}F(p)\mathrm{e}^{pt}\mathrm{d}p$$

where the integration path is a straight line parallel to the imaginary axis with a real part of σ. To form a closed contour, we consider the left half-circle arc, C_R, as shown in Figure 8.3. As $|p|\to\infty$, if $|F(p)|$

uniformly approaches zero within the range $\frac{\pi}{2} - \delta \leq \arg p \leq \frac{3\pi}{2} + \delta$, then it can be proven that

$$\int_{C_R} F(p)\mathrm{e}^{pt}\mathrm{d}p \overset{R\to\infty}{\longrightarrow} 0$$

Proof. The integral along C_R can be divided into several segments:

$$\int_{C_R} F(p)\mathrm{e}^{pt}\mathrm{d}p = \int_{AB} F(p)\mathrm{e}^{pt}\mathrm{d}p + \int_{BCD} F(p)\mathrm{e}^{pt}\mathrm{d}p + \int_{DE} F(p)\mathrm{e}^{pt}\mathrm{d}p$$

Here, segment AB denotes the integral path from point A to point B, segment BCD denotes the integral path along the semi-circular arc from point B to point D, and segment DE denotes the integral path from point D to point E. For the integral along segment BCD, we can perform a variable substitution by letting $p = \mathrm{i}z$. This substitution satisfies the Jordan lemma, which implies that the integral along this segment is zero. For the integral along segment AB, we substitute $p = R\mathrm{e}^{\mathrm{i}\theta}$, where R is the radius and θ is the angle parameter. Thus

$$\begin{aligned}\int_{AB} F(p)\mathrm{e}^{pt}\mathrm{d}p &\leq \int_{AB} |F(p)||\mathrm{e}^{pt}||\mathrm{d}p| = \int_{AB} |F(p)|\mathrm{e}^{\sigma t}R\mathrm{d}\theta \\ &\leq \max|F(p)|\mathrm{e}^{\sigma t}R\alpha \leq \max|F(p)|e^{\sigma t}\sigma \longrightarrow 0\end{aligned}$$

Similarly, the integral along segment DE is also zero. According to the residue theorem, the original function can be expressed as

$$f(t) = \sum_{\text{poles of } F(p)} \mathrm{Res}[F(p)\mathrm{e}^{pt}] \tag{8.2.1}$$

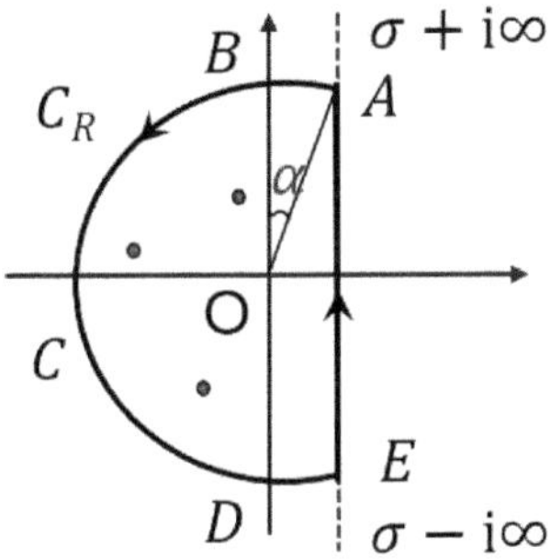

Figure 8.3.

If $F(p)$ is a multi-valued function, it is necessary to draw branch cuts and select an appropriate closed contour before applying the inversion method. □

Example 8.6. Find the original function of the Laplace transform:

$$F(p) = \frac{1}{\sqrt{p}}$$

Solution: We need to use the Riemann–Mellin inversion method. The image function $F(p)$ is a multi-valued function with branch points at $z = 0$ and $z = \infty$. To ensure that the integration path does not cross the branch cut, we choose the integration contour as shown in Figure 8.4. Since there are no singularities within the contour, we can apply the Riemann–Mellin inversion formula:

$$f(t) = \frac{1}{2\pi \mathrm{i}} \int_{\sigma-\mathrm{i}\infty}^{\sigma+\mathrm{i}\infty} \frac{\mathrm{e}^{pt}}{\sqrt{p}} \mathrm{d}p = -\frac{1}{2\pi \mathrm{i}} \left(\int_{l_1} + \int_{l_2} + \int_{C_\varepsilon} + \int_{C_R} \right) \frac{\mathrm{e}^{pt}}{\sqrt{p}} \mathrm{d}p$$

Since the integral along C_R is zero, it can be proven that

$$\int_{C_\varepsilon} \frac{\mathrm{e}^{pt}}{\sqrt{p}} \mathrm{d}p \xrightarrow{\varepsilon \to 0} 0$$

For the paths l_1 and l_2, we take $p = x\mathrm{e}^{\mp\pi \mathrm{i}}$. Thus,

$$\begin{aligned} f(t) &= -\frac{1}{2\pi \mathrm{i}} \left(\int_{l_1} \frac{\mathrm{e}^{pt}}{\sqrt{p}} \mathrm{d}p + \int_{l_2} \frac{\mathrm{e}^{pt}}{\sqrt{p}} \mathrm{d}p \right) \\ &= \frac{1}{2\pi \mathrm{i}} \int_0^\infty \frac{\mathrm{e}^{-xt}}{-\mathrm{i}\sqrt{x}} \mathrm{d}x + \frac{1}{2\pi \mathrm{i}} \int_\infty^0 \frac{\mathrm{e}^{-xt}}{\mathrm{i}\sqrt{x}} \mathrm{d}x = \frac{1}{\pi} \int_0^\infty \frac{\mathrm{e}^{-xt}}{\sqrt{x}} \mathrm{d}x \end{aligned}$$

By making the variable substitution $u = \sqrt{tx}$, the integral becomes a Gaussian integral and the final result is

$$f(t) = \frac{2}{\pi\sqrt{t}} \int_0^\infty \mathrm{e}^{-u^2} \mathrm{d}u = \frac{1}{\sqrt{\pi t}}$$

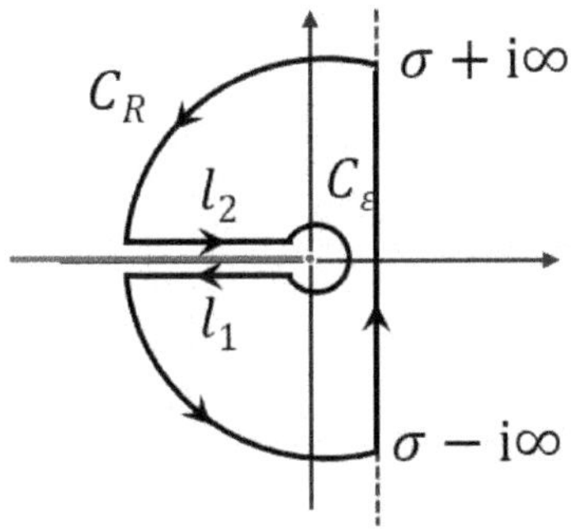

Figure 8.4.

Exercises

[1] Find the original function of the Laplace transform for the following functions:

$$(\text{a})\ F(p) = \frac{6}{(p+1)^4}; \quad (\text{b})\ F(p) = \frac{1}{(p^2+2p+2)^2}$$

[2] Use the Riemann–Mellin inversion method to find the original function for the following functions:

$$(\text{a})\ F(p) = \frac{1}{\sqrt{p}}\mathrm{e}^{-a\sqrt{p}}; \quad (\text{b})\ F(p) = \frac{1}{p^2(p^2+\lambda)^3}$$

[3] Find the Laplace inverse transform for the following functions:

$$(\text{a})\ F(p) = \frac{\cosh(ap)}{p^4\cosh p}; \quad (\text{b})\ F(p) = \frac{\sinh(ap)}{p^2\sinh p} \quad (0 < a < 1)$$

[4] Find the original function of the Laplace transform:

$$F(p) = \frac{\coth\sqrt{p}}{\sqrt{p}}$$

Hint: Making use of the expansion series,

$$\frac{1}{1-\mathrm{e}^{-2\sqrt{p}}} = 1 + \mathrm{e}^{-2\sqrt{p}} + \mathrm{e}^{-4\sqrt{p}} + \cdots$$

8.3 Some Applications

8.3.1 *Solving differential equations*

By using Laplace transforms, differential equations can be transformed into algebraic equations. The original function can then be obtained by applying inverse transforms. When performing Laplace transforms, the initial values are directly substituted using the derivative theorem, making it convenient for solving initial value problems in ordinary differential equations.

(1) *Constant coefficient differential equation*

Example 8.7. Solve the following ordinary differential equation:

$$\begin{cases} y'' + 2y' - 3y = \mathrm{e}^{-t} \\ y(0) = 0, \quad y'(0) = 1 \end{cases}$$

Solution: Let us assume the Laplace transform $L[y(t)] = Y(p)$. Applying Laplace transforms to both sides of the equation, we have

$$[p^2Y(p) - py(0) - y'(0)] + 2[pY(p) - y(0)] - 3Y(p) = \frac{1}{p+1}$$

Using the initial conditions, then

$$Y(p) = \frac{p+2}{(p+1)(p-1)(p+3)} = \frac{-\frac{1}{4}}{p+1} + \frac{\frac{3}{8}}{p-1} + \frac{-\frac{1}{8}}{p-3}$$

Hence,

$$y(t) = \mathcal{L}^{-1}[Y(p)] = -\frac{1}{4}\mathrm{e}^{-t} + \frac{3}{8}\mathrm{e}^{t} - \frac{1}{8}\mathrm{e}^{-3t}$$

Example 8.8. Solve systems of ordinary differential equations:

$$\begin{cases} y'' - x'' + x' - y = \mathrm{e}^{-t} - 2 \\ 2y'' - x'' - 2y' + x = -t \\ y(0) = y'(0) = 0, \quad x(0) = x'(0) = 0 \end{cases}$$

Solution: Let us define

$$\mathcal{L}[y(t)] = Y(p), \quad \mathcal{L}[x(t)] = X(p)$$

By applying Laplace transforms to both sides of the system and using the initial conditions, we obtain

$$\begin{cases} p^2Y(p) - p^2X(p) + pX(p) - Y(p) = \dfrac{1}{p-1} - \dfrac{2}{p} \\ 2p^2Y(p) - p^2X(p) + X(p) - 2pY(p) = -\dfrac{1}{p^2} \end{cases}$$

Solving for $X(p)$ and $Y(p)$, and then applying inverse Laplace transforms, we get

$$\begin{cases} x(t) = \mathcal{L}^{-1}[X(p)] = -t + t\mathrm{e}^t \\ y(t) = \mathcal{L}^{-1}[Y(p)] = 1 - \mathrm{e}^{-t} + t\mathrm{e}^t \end{cases}$$

(2) *Linear coefficient differential equations*

If the coefficients of a differential equation are linear functions, Laplace transforms can also be employed to solve them. Let

$$\mathcal{L}[y(t)] = \int_0^{\infty} y(t)\mathrm{e}^{-pt}\mathrm{d}t = Y(p)$$

since

$$\mathcal{L}[ty(t)] = \int_0^{\infty} ty(t)\mathrm{e}^{-pt}\mathrm{d}t = -\frac{\partial}{\partial p}\int_0^{\infty} y(t)\mathrm{e}^{-pt}\mathrm{d}t = -Y'(p)$$

$$\begin{aligned} \mathcal{L}[ty'(t)] &= -\frac{\partial}{\partial p}\int_0^{\infty} y'(t)\mathrm{e}^{-pt}\mathrm{d}t = -\frac{\partial}{\partial p}[pY(p) - y(0)] \\ &= -pY'(p) - Y(p) \\ \mathcal{L}[ty''(t)] &= -\frac{\partial}{\partial p}\int_0^{\infty} y''(t)\mathrm{e}^{-pt}\mathrm{d}t = -\frac{\partial}{\partial p}[p^2Y(p) - py(0) - y'(0)] \\ &= -p^2Y'(p) - 2pY(p) + y(0) \end{aligned}$$

Using these transformations, we can convert a high-order linear coefficient differential equation into a first-order differential equation in terms of the transformed function $Y(p)$, which can then be directly integrated to solve.

Example 8.9. To solve the following ordinary differential equation,

$$\begin{cases} (2t-1)y'' + 3ty' + (t+1)y = 0 \\ y(0) = 0, \quad y'(0) = 0 \end{cases}$$

Solution: Applying Laplace transforms to both sides of the equation and simplifying, we have

$$(-2p^2 - 3p - 1)Y'(p) - (p^2 + 4p + 2)Y(p) = 0$$

or

$$\frac{d \ln Y(p)}{dp} = -\frac{p^2 + 4p + 2}{2p^2 + 3p + 1} = -\frac{1}{2} - \frac{1}{4}\frac{1}{(p+1/2)} - \frac{1}{(p+1)}$$

Solving for $Y(p)$, we obtain

$$Y(p) = \frac{\mathrm{e}^{-p/2}}{(p+1/2)^{1/4}(p+1)}$$

Using the results of inverse Laplace transforms,

$$\frac{1}{(p+1/2)^{1/4}} \longmapsto \frac{1}{\Gamma(1/2)} t^{-\frac{3}{4}} \mathrm{e}^{-\frac{t}{2}}, \qquad \frac{1}{p+1} \longmapsto \mathrm{e}^{-t}$$

and applying the convolution theorem and time delay theorem, we get

$$y(t) = \frac{1}{\Gamma\left(\frac{1}{2}\right)} \mathrm{e}^{-t+\frac{1}{2}} \int_0^{t-\frac{1}{2}} \tau^{-\frac{3}{4}} \mathrm{e}^{\frac{\tau}{2}} \mathrm{d}\tau$$

8.3.2 *Solving integral equations*

Integral equations involving convolutions can be solved using Laplace transforms, assuming convolutions are applicable in Laplace transforms. For example, consider the integral equation

$$y(t) + \lambda \int_0^t g(t-\tau) y(\tau) \mathrm{d}\tau = f(t)$$

where λ is a constant. Applying Laplace transforms to both sides of the equation, we have

$$[1+\lambda G(p)]Y(p) = F(p)$$

yielding

$$Y(p) = \frac{F(p)}{1+\lambda G(p)}$$

By applying inverse transforms, we can solve the equation.

Example 8.10. Solve the following integral equation:

$$y(t) + \lambda \int_0^t \mathrm{e}^{-(t-\tau)} y(\tau)\mathrm{d}\tau = f(t)$$

Solution: Let $g(t) = \mathrm{e}^{-t}$, its Laplace transform function is

$$G(p) = \frac{1}{p+1}$$

Therefore,

$$Y(p) = \frac{(p+1)F(p)}{p+\lambda+1} = F(p) - \lambda\frac{F(p)}{p+\lambda+1}$$

Applying inverse Laplace transforms, we get

$$y(t) = f(t) - \lambda \int_0^t \mathrm{e}^{-(\lambda+1)(t-\tau)} f(\tau)\mathrm{d}\tau$$

8.3.3 *Evaluating real function integral*

Laplace transform can also be used to compute integrals of certain real-valued functions, sometimes providing a more convenient approach compared to the direct application of the residue theorem.

Example 8.11. Compute the integral:

$$I = \int_0^\infty \frac{\cos 7x}{x^2+a^2}\mathrm{d}x \quad (a>0)$$

Solution: Consider the integral for the parameterized function,

$$I(t) = \int_0^\infty \frac{\cos tx}{x^2 + a^2} \mathrm{d}x = \frac{1}{2} \int_{-\infty}^\infty \frac{\cos tx}{x^2 + a^2} \mathrm{d}x$$

Applying the Laplace transform with respect to the variable t, we have

$$\begin{aligned} \bar{I}(p) &= \int_0^\infty I(t) \mathrm{e}^{-pt} \mathrm{d}t = \frac{1}{2} \int_{-\infty}^\infty \frac{1}{x^2 + a^2} \cdot \frac{p}{x^2 + p^2} \mathrm{d}x \\ &= \pi \mathrm{i} \operatorname*{Res}_{z_k \in \mathbb{H}} \left(\frac{1}{z^2 + a^2} \cdot \frac{p}{z^2 + p^2} \right) = \frac{\pi}{2a} \cdot \frac{1}{p + a} \end{aligned}$$

Taking the inverse transform, we obtain

$$I(t) = \mathcal{L}^{-1}[\bar{I}(p)] = \frac{\pi}{2a} \mathrm{e}^{-at}$$

When setting $t = 7$, it yields

$$I = \frac{\pi}{2a} \mathrm{e}^{-7a}$$

By differentiating $I(t)$ with respect to t and evaluating at $t = m$, we obtain the integral result once again:

$$\int_0^\infty \frac{x \sin mx}{x^2 + a^2} \mathrm{d}x = \frac{\pi}{2} \mathrm{e}^{-m}$$

Example 8.12. Calculate the integral:

$$I = \int_0^1 \frac{1}{x^\alpha (1 - x)^{1-\alpha}} \mathrm{d}x \quad (\alpha > 0)$$

Solution: Consider the integral in the form of a convolution:

$$I(t) = \int_0^t \frac{1}{x^\alpha (t - x)^{1-\alpha}} \mathrm{d}x = \frac{1}{t^\alpha} * \frac{1}{t^{1-\alpha}}$$

Applying the Laplace transform and using the identities

$$\mathcal{L}\left[\frac{1}{t^\alpha}\right] = \frac{\Gamma(1-\alpha)}{p^{1-\alpha}}, \quad \mathcal{L}\left[\frac{1}{t^{1-\alpha}}\right] = \frac{\Gamma(\alpha)}{p^\alpha}$$

we have

$$I(p) = \frac{\Gamma(\alpha)\Gamma(1-\alpha)}{p}$$

Performing the inverse transform, we get

$$\int_0^1 \frac{1}{x^\alpha(1-x)^{1-\alpha}}\mathrm{d}x = \Gamma(\alpha)\Gamma(1-\alpha)$$

The result shows that the integral is independent of t. This fact can also be directly observed by substituting $x \to tx$ in the integral.

8.3.4 *Summation of series*

Similar to the Fourier transform, the Laplace transform can also be used to compute certain expressions for the summation of series. The approach is to apply the Laplace transform for function $f(t)$:

$$F(p) = \int_0^\infty f(t)\mathrm{e}^{-pt}\mathrm{d}t$$

Setting $p = n$ and summing over n, considering that $\mathrm{e}^{-t} < 1$ for $t > 0$, we have

$$\sum_{n=1}^\infty F(n) = \sum_{n=1}^\infty \int_0^\infty f(t)\mathrm{e}^{-nt}\mathrm{d}t = \int_0^\infty f(t)\sum_{n=1}^\infty \mathrm{e}^{-nt}\mathrm{d}t = \int_0^\infty \frac{f(t)}{\mathrm{e}^t - 1}\mathrm{d}t$$

If we choose $f(t) = t^{s-1}$, where $s > -2$, then

$$F(p) = \int_0^\infty t^{s-1}\mathrm{e}^{-pt}\mathrm{d}t = \frac{\Gamma(s)}{p^s}$$

Therefore,

$$\sum_{n=1}^\infty \frac{\Gamma(s)}{n^s} = \int_0^\infty \frac{t^{s-1}}{\mathrm{e}^t - 1}\mathrm{d}t$$

By performing analytic continuation, we once again obtain the integral expression for the Riemann zeta function:

$$\zeta(z) = \frac{1}{\Gamma(z)}\int_0^\infty \frac{t^{z-1}}{\mathrm{e}^t - 1}\mathrm{d}t$$

Example 8.13. Calculate the summation of series:

$$\sum_{n=1}^{\infty} \frac{1}{n^2 - a^2} \quad (a \notin \mathbb{Z},\ \mathrm{Re}\, a > 0)$$

Solution: We can make use of the Laplace transform formula,

$$\int_0^\infty \sinh at\, \mathrm{e}^{-pt} \mathrm{d}t = \frac{a}{p^2 - a^2}$$

By setting $p = n$ and summing over n, we get

$$\sum_{n=1}^{\infty} \frac{1}{n^2 - a^2} = \frac{1}{a} \int_0^\infty \frac{\sinh at}{\mathrm{e}^t - 1} \mathrm{d}t$$

The term on the right-handed side can be integrated as

$$\sum_{n=1}^{\infty} \frac{1}{n^2 - a^2} = \frac{1}{a} \int_0^\infty \frac{\sinh at}{\mathrm{e}^t - 1} \mathrm{d}t = \frac{1}{2a^2} - \frac{\pi}{2a} \cot \pi a$$

Setting $a \to \mathrm{i}a$, we obtain the summation formula once again:

$$\sum_{n=1}^{\infty} \frac{1}{n^2 + a^2} = -\frac{1}{2a^2} + \frac{\pi}{2a} \coth \pi a$$

For $a = 0$, we have

$$\sum_{n=1}^{\infty} \frac{1}{n^2} = \sum_{n=1}^{\infty} \int_0^\infty t\mathrm{e}^{-nt} \mathrm{d}t = \int_0^\infty \frac{t}{\mathrm{e}^t - 1} \mathrm{d}t = \frac{\pi^2}{6}$$

When calculating the summation of series, the following formulas are frequently used:

$$\int_0^\infty \mathrm{e}^{-\alpha t} \mathrm{e}^{-pt} \mathrm{d}t = \frac{1}{p - \alpha}, \quad \int_0^\infty t^{\alpha - 1} \mathrm{e}^{-pt} \mathrm{d}t = \frac{\Gamma(\alpha)}{p^\alpha}$$

$$\int_0^\infty \sin \omega t \mathrm{e}^{-pt} \mathrm{d}t = \frac{\omega}{p^2 + \omega^2}, \quad \int_0^\infty \cos \omega t \mathrm{e}^{-pt} \mathrm{d}t = \frac{p}{p^2 + \omega^2}$$

$$\int_0^\infty \sinh \omega t \mathrm{e}^{-pt} \mathrm{d}t = \frac{\omega}{p^2 - \omega^2}, \quad \int_0^\infty \cosh \omega t \mathrm{e}^{-pt} \mathrm{d}t = \frac{p}{p^2 - \omega^2}$$

Exercises

[1] Solve the initial value problem for ordinary differential equations:

$$\text{(a)}\ \frac{d^3y}{dt^3}+3\frac{d^2y}{dt^2}+3\frac{dy}{dt}+y=6e^{-t},\quad y(0)=y'(0)=y''(0)=0;$$

$$\text{(b)}\ \begin{cases}\dfrac{dy}{dt}+2y+2x=10e^{2t}, \\[2ex] \dfrac{dx}{dt}-2y+x=7e^{2t},\end{cases}\qquad \begin{cases}y(0)=1\\ x(0)=3\end{cases}$$

[2] Use Laplace transform to evaluate the integrals:

$$\text{(a)}\ \int_0^\infty \frac{\sin^2 x}{x^2}dx;\qquad \text{(b)}\ \int_0^\infty \frac{\sin x}{x(x^2+1)}dx$$

$$\text{(c)}\ \int_0^\infty \frac{e^{-x}}{x^2+a^2}dx;\quad \text{(d)}\ \int_0^\infty \frac{e^{-x^2}}{x^2+a^2}dx$$

[3] Calculate the integrals using the convolution theorem:

$$\text{(a)}\ \int_0^1 \frac{1}{\sqrt{x(1-x)}}dx;\quad \text{(b)}\ \int_0^1 \frac{x^4}{\sqrt{x(1-x)}}dx$$

Answer: (a) π; (b) $\frac{35\pi}{128}$

[4] The zero-order Bessel equation is

$$xy''+y'+xy=0$$

Find the solution that satisfies $y(0)=1$.
Hint: Apply the Laplace transform, expand the transform of the function as a Laurent series, and then find the series form of the original function.
Answer:

$$Y(p)=\frac{c}{p\sqrt{1+\frac{1}{p^2}}}=c\sum_{n=0}^{\infty}\frac{(2n)!}{2^{2n}(n!)^2}\frac{(-1)^n}{p^{2n+1}}$$

$$y(x)=1-\frac{x^2}{2^2}+\frac{x^4}{2^2\cdot 4^2}-\frac{x^6}{2^2\cdot 4^2\cdot 6^2}+\cdots$$

[5] Use Laplace transform to evaluate the series sums:

$$\text{(a)} \sum_{n=0}^{\infty} \frac{1}{(3n+1)(3n+2)(3n+3)}; \quad \text{(b)} \sum_{n=-\infty}^{\infty} \frac{1}{(n^2+1)^2}$$

Hint:

$$\text{(a)}\ \mathcal{L}^{-1}\left[\frac{1}{(3p+1)(3p+2)(3p+3)}\right]$$

$$= \frac{1}{6}(\mathrm{e}^{-\frac{t}{3}} - 2\mathrm{e}^{-\frac{2t}{3}} + \mathrm{e}^{-t}) = \frac{1}{6}(\mathrm{e}^{\frac{t}{3}} - 1)^2 \mathrm{e}^{-t}$$

$$\longrightarrow \sum_{n=0}^{\infty} \frac{1}{(3n+1)(3n+2)(3n+3)} = \frac{1}{6}\int_0^{\infty} \frac{(\mathrm{e}^{\frac{t}{3}} - 1)^2}{\mathrm{e}^t - 1}\mathrm{d}t$$

$$= \frac{1}{2}\int_0^{\infty} \frac{x}{(x+1)(x^2+3x+3)}\mathrm{d}x$$

$$\text{(b)} \sum_{n=1}^{\infty} \frac{1}{n^2+a^2} = -\frac{1}{2a^2} + \frac{\pi}{2a}\coth \pi a$$

Differentiate the above expression with respect to the parameter a and then let $a \to 1$.

Answer:

$$\text{(a)}\ \frac{\pi}{4\sqrt{3}} - \frac{1}{4}\ln 3; \quad \text{(b)}\ \frac{\pi}{4}\coth \pi + \frac{\pi^2}{4}\operatorname{csch}^2 \pi - \frac{1}{2}$$

[6] Solving Abel equations:

$$\text{(a)} \int_0^t \frac{u(\tau)}{\sqrt{t-\tau}}\mathrm{d}\tau = f(t); \quad \text{(b)} \int_0^t \frac{u(\tau)}{\sqrt{t^2-\tau^2}}\mathrm{d}\tau = f(t)$$

Answer:

$$\text{(a)}\ u(t) = \frac{1}{\pi}\frac{\mathrm{d}}{\mathrm{d}t}\int_0^t \frac{f(\tau)}{\sqrt{t-\tau}}\mathrm{d}\tau; \quad \text{(b)}\ u(t) = \frac{2}{\pi}\frac{\mathrm{d}}{\mathrm{d}t}\int_0^t \frac{f(\tau)\tau}{\sqrt{t^2-\tau^2}}\mathrm{d}\tau$$

[7] Let $F(p) = \mathcal{L}[f(x)]$, prove the following formulas:

$$\text{(a)}\ \mathcal{L}\left[\frac{f(x)}{x}\right] = \int_p^{\infty} F(p)\mathrm{d}p; \quad \text{(b)} \int_0^{\infty} \frac{f(x)}{x}\mathrm{d}x = \int_0^{\infty} F(p)\mathrm{d}p$$

Hint: The condition $\mathcal{L}\left[\frac{f(x)}{x}\right] \to 0$ must be satisfied as $p \to \infty$.

[8] Calculate integrals ($a, b > 0$):

$$\text{(a)} \int_0^{\infty} \frac{\sin x}{x} \mathrm{d}x; \qquad \text{(b)} \int_0^{\infty} \frac{\cos ax - \cos bx}{x} \mathrm{d}x$$

$$\text{(c)} \int_0^{\infty} \frac{\mathrm{e}^{-ax} \sin bx}{x} \mathrm{d}x; \quad \text{(d)} \int_0^{\infty} \frac{\mathrm{e}^{-ax} - \mathrm{e}^{-bx}}{x} \mathrm{d}x$$

Hint: Use the previous formula.
Answer:

$$\text{(a)}\ \frac{\pi}{2}; \quad \text{(b)}\ \ln \frac{b}{a}; \quad \text{(c)}\ \text{atan}\left(\frac{b}{a}\right); \quad \text{(d)}\ \ln \frac{b}{a}$$

8.4 Z-Transform

Previously, we introduced the Fourier transform and the Laplace transform, both of which are integral transforms. In mathematics, there are many similar integral transforms, such as the Euler transform, the Mellin transform, and the Hankel transform, all of which have a general form:

$$F(k) = \int K(k, x) f(x) \mathrm{d}x \tag{8.4.1}$$

Here, $K(k, x)$ is referred to as the kernel of the integral transform, as listed in Table 8.1.

Table 8.1.

Transform Type	Fourier Transform	Laplace Transform	Euler Transform	Mellin Transform	Hankel Transform
$K(k, x)$	e^{ikx}	e^{-kx}	$(k - x)^{\nu}$	x^{k-1}	$xJ_n(kx)$

Each integral transform has its special applications, which will not be discussed individually. In this section, we briefly introduce a transform method for handling discrete sequences, known as the Z-transform.

8.4.1 Definition of the Z-transform

Let $\{f_k\}(k \in \mathbb{Z})$ be a discrete signal data. We introduce a complex variable z and define

$$F(z) = \sum_{k=-\infty}^{\infty} f_k z^{-k} \tag{8.4.2}$$

$F(z)$ is called the two-sided Z-transform of the sequence $\{f_k\}$ denoted as

$$F(z) \equiv Z[f_k]$$

$F(z)$ is referred to as the image function of the sequence $\{f_k\}$. If the summation is only taken over non-negative values of k, it is called the one-sided Z-transform of the sequence $\{f_k\}_{k=0}^{\infty}$ as follows:

$$F(z) = \sum_{k=0}^{\infty} f_k z^{-k} \tag{8.4.3}$$

The existence condition for the above-defined Z-transform is that the power series must converge, i.e.,

$$\sum_{k=-\infty}^{\infty} |f_k z^{-k}| < \infty$$

Existence theorem: If the sequence $\{f_k\}$ is bounded within a finite interval $M < k < N$ and satisfies the following conditions for positive real numbers α and β:

$$\lim_{k\to-\infty} |f_k|\beta^k = 0, \quad \lim_{k\to\infty} |f_k|\alpha^{-k} = 0$$

then the two-sided Z-transform $F(z)$ of $\{f_k\}$ converges absolutely and uniformly within the annular region $\alpha < |z| < \beta$.

Example 8.14. Find the one-sided Z-transform of the sequence $\{1, 1, 1, 1, \ldots\}$.
Solution:

$$F(z) = \sum_{k=0}^{\infty} z^{-k} = 1 + \frac{1}{z} + \frac{1}{z^2} + \cdots = \frac{z}{z-1} \quad (|z| > 1)$$

Discussion: The discrete Fourier transform can be seen as a special case of the Z-transform: Let $z = \mathrm{e}^{2\pi \mathrm{i} n/N}$, then we have

$$F_n = \frac{1}{2\pi}\sum_{k=0}^{N-1} f_k \mathrm{e}^{-\frac{2\pi \mathrm{i} kn}{N}}, \quad f_k = \sum_{k=0}^{N-1} F_k \mathrm{e}^{\frac{2\pi \mathrm{i} kn}{N}}$$

Furthermore, by setting $z = \mathrm{e}^{\frac{2\pi \mathrm{i} n\alpha}{N}}$, where $\alpha \neq \pm 1$, we obtain the discrete fractional Fourier transform:

$$F_n = \frac{1}{2\pi}\sum_{k=0}^{N-1} f_k \mathrm{e}^{-\frac{2\pi \mathrm{i}\alpha kn}{N}}, \quad f_k = \sum_{k=0}^{N-1} F_k \mathrm{e}^{\frac{2\pi \mathrm{i}\alpha kn}{N}}$$

8.4.2 *Basic properties*

[1] **Linearity theorem**

$$\mathcal{Z}[af_k + bf_k] = a\mathcal{Z}[f_k] + b\mathcal{Z}[f_k]$$

[2] **Scaling theorem**

$$\mathcal{Z}[a^k f_k] = F\left(\frac{z}{a}\right)$$

[3] **Shifting theorem**

A. Two-sided Z-transform:

$$\mathcal{Z}[f_{k\pm m}] = z^{\pm m} F(z)$$

B. One-sided Z-transform:

$$\begin{cases} \mathcal{Z}[f_{k-m}] = z^{-m}F(z) + \sum\limits_{k=0}^{m-1} f_{k-m} z^{-k} \\ \mathcal{Z}[f_{k+m}] = z^{m}F(z) - \sum\limits_{k=0}^{m-1} f_{k} z^{m-k} \end{cases}$$

[4] **Differentiation theorem**

$$\mathcal{Z}[kf_k] = -z\frac{\mathrm{d}}{\mathrm{d}z}F(z)$$

$$\mathcal{Z}[k^m f_k] = (-1)^m \left(z\frac{\mathrm{d}}{\mathrm{d}z}\right)^m F(z)$$

[5] **Convolution theorem**

$$\mathcal{Z}[f_k * g_k] = \mathcal{Z}[f_k] \cdot \mathcal{Z}[g_k]$$

The convolution of two sequences is defined as

$$f_k * g_k \stackrel{\text{def}}{=} \sum_{l=-\infty}^{\infty} f_l g_{k-l}$$

The above theorems can all be proven directly from the definitions, and we omit the proofs here.

8.4.3 *Inverse transform*

For the Z-transform,

$$F(z) = \sum_{k=-\infty}^{\infty} f_k z^{-k}$$

If $F(z)$ is an analytic function, according to the residue calculus method, its inverse transform can be represented using a contour integral,

$$f_k = \mathcal{Z}^{-1}[F(z)] = \frac{1}{2\pi \mathrm{i}} \oint_L F(z) z^{k-1} \mathrm{d}z \tag{8.4.4}$$

where the contour L is a closed curve that encloses all the singularities of $F(z)$.

In practical applications, the one-sided Z-transform is often more important. This is because the generating function $G(z)$ of the sequence $\{f_k\}_{k=0}^{\infty}$ precisely corresponds to its Z-transform. In other words, the inverse Z-transform of the image function $F(z)$ is given by the expansion coefficients of $F\left(\frac{1}{z}\right)$.

Example 8.15. To find the inverse Z-transform of a function,

$$F(z) = \frac{z(z+1)}{(z-1)^3}$$

Solution 1: Let $z = y^{-1}$, then we have

$$\frac{z(z+1)}{(z-1)^3} = \frac{y^{-1}(y^{-1}+1)}{(y^{-1}-1)^3} = \sum_{k=0}^{\infty} k^2 y^k = \sum_{k=0}^{\infty} k^2 z^{-k}$$

Therefore, the corresponding discrete sequence is

$$\{f_k\} = \{k^2\}_{k=0}^{\infty}$$

Solution 2: Using the inverse transform formula, we have

$$f_k = \frac{1}{2\pi \mathrm{i}} \oint_L F(z) z^{k-1} \mathrm{d}z = \frac{1}{2\pi \mathrm{i}} \oint_L \frac{z^{k+1} + z^k}{(z-1)^3} \mathrm{d}z$$

Since $z = 1$ is the only third-order pole of the integrated function, we obtain

$$f_k = \frac{1}{2!} \frac{\mathrm{d}^2}{\mathrm{d}z^2} (z^{k+1} + z^k) \bigg|_{z=1} = k^2$$

This problem can also be solved using the results from Example 8.14 and the differentiation theorem.

8.4.4 *Application examples*

Example 8.16. Finding the closed-form expression of the Fibonacci sequence,

$$f_{k+2} = f_{k+1} + f_k, \quad f_0 = 0, \ f_1 = 1$$

Solution: Let $F(z) \equiv Z[f_k]$. Applying the Z-transform to both sides of the equation and making use of the shifting theorem, we have

$$z^2[F(z) - f_0 - f_1 z^{-1}] = z[F(z) - f_0] + F(z)$$

Solving for $F(z)$, we obtain

$$\begin{aligned} F(z) &= \frac{z}{z^2 - z - 1} = \frac{1}{\sqrt{5}} \frac{z}{z - \frac{1+\sqrt{5}}{2}} - \frac{1}{\sqrt{5}} \frac{z}{z - \frac{1-\sqrt{5}}{2}} \\ &= \frac{1}{\sqrt{5}} \sum_{k=0}^{\infty} \left(\frac{1+\sqrt{5}}{2} \right)^k z^{-k} - \frac{1}{\sqrt{5}} \sum_{k=0}^{\infty} \left(\frac{1-\sqrt{5}}{2} \right)^k z^{-k} \end{aligned}$$

Therefore, the closed-form expression for the Fibonacci sequence is given by

$$f_k = \frac{1}{\sqrt{5}} \left(\frac{1+\sqrt{5}}{2} \right)^k - \frac{1}{\sqrt{5}} \left(\frac{1-\sqrt{5}}{2} \right)^k$$

8.4.5 *Relationship with Laplace transform*

By treating a discrete sequence as a discontinuous function, we can observe the inherent connection between the Z-transform and the Laplace transform. For example, if we divide time into intervals of Δt and define a function $f(t)$ such that its value is a constant f_k for $t \in (t_k, t_{k+1})$, then $f(t)$ can be expressed as

$$f(t) = f_0[H(t) - H(t-1)] + f_1[H(t-1) - H(t-2)] + \cdots$$

Here, $H(t)$ is the Heaviside step function since

$$\mathcal{L}[H(t-k) - H(t-(k+1))] = \int_k^{k+1} \mathrm{e}^{-pt}\mathrm{d}t = \frac{\mathrm{e}^{-kp}}{p}(1-\mathrm{e}^{-p})$$

Applying the Laplace transform to $f(t)$, we have

$$\mathcal{L}[f(t)] = \frac{1}{p}(1-\mathrm{e}^{-p})\sum_{k=0}^{\infty} f_k \mathrm{e}^{-kp}$$

Let $z = \mathrm{e}^p$, then we have

$$\mathcal{L}[f(t)] = \frac{1-\frac{1}{z}}{\ln z} Z[f_k] \tag{8.4.5}$$

Therefore, in many cases, the Z-transform can also be obtained using the Laplace transform.

ENDNOTE

A field is a set in which addition, subtraction, multiplication, and division can be performed. In addition to closure, a field specifies several fundamental rules. First, it requires an element "0" that remains unchanged when other numbers are added to it. It also requires an element "1" that remains unchanged when other numbers are multiplied by it.

Furthermore, addition and multiplication must be commutative. Therefore, quaternion algebra is just a division algebra and does not constitute a field. The sets $\mathbb{N}$ and $\mathbb{Z}$ are not fields, but $\mathbb{Q}$ and $\mathbb{Z}$, $\mathbb{C}$ form fields. People can also construct fields with finite elements, such as $\{1, 0, -1\}$, which also forms a field. Another type of field is an extension field, where we can add an element that was not previously present in a given field. For example, by adding the element $\sqrt{2}$ to the rational number field $\mathbb{Q}$ and performing addition, subtraction, multiplication, and division operations, we obtain an extension field of the form $a + b\sqrt{2}$ $(a, b \in \mathbb{Q})$:

$$\frac{a + b\sqrt{2}}{c + d\sqrt{2}} = \frac{ac - 2bd}{c^2 - 2d^2} + \frac{bc - ad}{c^2 - 2d^2}\sqrt{2}$$

The arithmetic operations on these numbers are similar to complex numbers. Note that this field is not the same as the real number field $\mathbb{R}$; it does not include irrational numbers like $\sqrt{3}$ or π.

Although the set of integers $\mathbb{Z}$ is not a field, it forms a ring, which means it supports addition, subtraction, and multiplication, but not necessarily division. Every integer can be uniquely decomposed into a product of prime numbers, a property known as the unique factorization theorem or the fundamental theorem of arithmetic.

Suppose p is an odd prime number, and $p = 1 \pmod 4$. Then the equation $x^2 + y^2 = p$ has a positive integer solution $(x, y) \in \mathbb{N}$, e.g., $5 = 1^2 + 2^2$. However, for some prime numbers like $p = 3$, there are no integer solutions. The reason is that $5 = 1^2 + 2^2 = (2 + \mathrm{i})(2 - \mathrm{i})$, meaning that $p = 5$ can be factored into "composite" numbers in a generalized "prime factorization" sense, making it a "composite" in this sense. On the other hand, $p = 3$ cannot be factorized in this generalized prime factorization. This process involves adding the imaginary unit i to the ring of integers $\mathbb{Z}$, creating an extended ring called $\mathbb{Z}(\mathrm{i})$, which consists of all complex numbers of the form $m + n\mathrm{i}$. Gauss first studied these complex numbers with both real and imaginary parts as integers in his work "Disquisitiones Arithmeticae", and they are now known as Gaussian integers.

Is the factorization of Gaussian integers unique like the fundamental theorem of arithmetic? The answer is no! For example, $65 = (7+4\mathrm{i})(7-4\mathrm{i}) = (8+\mathrm{i})(8-\mathrm{i})$. In order to restore the uniqueness of factorization, Kummer introduced the concept of "ideal numbers" in 1846, which allows factorization into "ideal" prime numbers. This idea led Dedekind to define ideals in algebraic rings. Dedekind found that Kummer's ideal numbers were not actual numbers but rather subsets of the ring. If non-zero elements of the integral domain $\mathbb{Z}$ are allowed to be divided, it gives rise to a new set called the field of rational numbers $\mathbb{Q}$, and the extension of $\mathbb{Q}$ is denoted as $\mathbb{Q}(\mathrm{i})$.

Let K be a finite extension field of $\mathbb{Q}$, with its integral domain denoted as R. An ideal is a closed subring of R, where any element multiplied by an element from the parent ring R belongs to the subring similar to the concept of invariant subgroups in group theory. Elements of the ideal subring can be uniquely factorized just like the prime factorization in the set of integers $\mathbb{Z}$. An element with only one factor is called a prime ideal. This discovery led to the shift in number theory research from numbers to sets.

Algebraic numbers are solutions to polynomial equations of the form:

$$a_n x^n + a_{n-1} x^{n-1} + \cdots + a_1 x + a_0 = 0$$

where $a_0, a_1, \ldots, a_n \in \mathbb{Z}$ (integers). In particular, when $a_n = 1$, the solutions are called algebraic integers. For example, $\sqrt{-2}$ is an algebraic integer that satisfies the equation:

$$x^2 + 2 = 0$$

The concept of algebraic numbers is a natural extension of rational numbers, which is the case when $n = 1$. Cantor proved that the set of all algebraic numbers is countable, whereas the set of real numbers is uncountable. This implies the existence of real numbers that are not algebraic, known as transcendental numbers, as they "transcend" algebraic methods. π is an example of a transcendental number, and a famous story in the mathematics is the proof that $2^{\sqrt{2}}$ is transcendental.

The Fibonacci sequence, where each term is an integer, can only be expressed using irrational numbers in its closed-form history of expression. This is an example of using algebraic numbers to explain the nature of integers. The underlying idea is that the Fibonacci sequence defines the irrational number $\sqrt{5}$ because as $k \to \infty$,

$$\frac{f_{k+1}}{f_k} \to \varphi \equiv \frac{1+\sqrt{5}}{2}$$

This is known as the golden ratio. The general term of the sequence is derived based on the overall behavior of the Fibonacci sequence as a whole reflected in each individual term. This fact is reminiscent of Cardano's formula, which states that real roots of a cubic algebraic equation require the use of imaginary numbers.

Based on the closed-form expression of the Fibonacci sequence,

$$\begin{aligned} f_k &= \frac{1}{\sqrt{5}}\left(\frac{1+\sqrt{5}}{2}\right)^k - \frac{1}{\sqrt{5}}\left(\frac{1-\sqrt{5}}{2}\right)^k \\ &\equiv \frac{1}{\sqrt{5}}[\varphi^k - \psi^k] \quad (k \in \mathbb{N}) \end{aligned}$$

where φ is the golden ratio. If we replace k with continuous real numbers, we obtain a continuous smooth function that connects the Fibonacci sequence. The real part of this function is shown in Figure 8.5(a), which oscillates and extends to negative values, forming a bidirectional Fibonacci sequence $f_{-k} = (-1)^{k+1} f_k$. The function is analytically extended to the complex plane, resulting in

$$f(z) = \frac{1}{\sqrt{5}}(\varphi^z - \psi^z)$$

This is the spiral function described in Section 1.2, with $\alpha = 1$, $\beta = \sqrt{5}$. It is an entire function, and Figure 8.5(b) illustrates the distribution of its argument (phase). All the zeros of the function are evenly spaced on a straight line, parameterized by

$$z(t) = \frac{2\pi t}{\pi + 2\mathrm{i}\ln\varphi} \quad (-\infty < t < \infty)$$

with a slope of $k = \tan\theta = -\frac{2}{\pi}\ln\varphi$. The zeros lie at $t = n$ $(n \in \mathbb{N})$. The line of zeros is mapped by $f(z)$ to the curve shown in Figure 8.5(c), which repeatedly passes through the zeros. Figure 8.5(d) depicts the images of two lines parallel to the line of zeros, forming an infinitely extending spiral without endpoints. Figure 8.6 shows the spiral curve at different scales parameterized by

$$F(t) = \frac{1}{\sqrt{5}}\varphi^{\frac{2\pi^2 t}{\omega}}\mathrm{e}^{-\frac{4\pi\mathrm{i}t}{\omega}\ln^2\varphi}[\varphi^h - \mathrm{e}^{2\pi\mathrm{i}t}\psi^h] \quad (-\infty < t < \infty)$$

where $\omega = \pi^2 + 4\ln^2\varphi$ and h is an arbitrary complex constant. This curve is a smooth golden spiral.

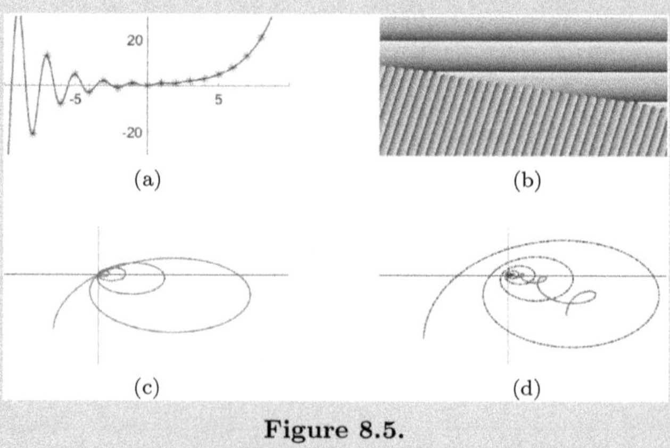

Figure 8.5.

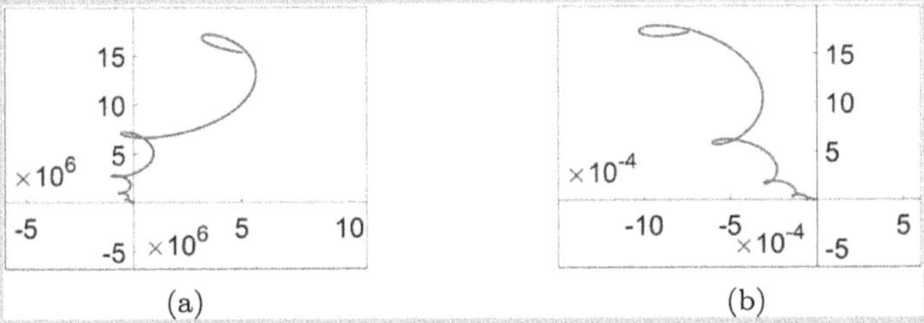

Figure 8.6.

It can be proven that the Fibonacci function exhibits the following properties:

(a) $f(k) = f_k$;
(b) $f(z+2) = f(z+1) + f(z)$;
(c) $f(z+k+1) = f_{k+1}f(z+1) + f_k f(z)$;
(d) $f(\alpha+\beta+1) = f(\alpha+1)f(\beta+1) + f(\alpha)f(\beta)(\alpha, \beta \in \mathbb{C})$

Exercises

[1] Find the one-sided Z-transform of the sequence:

$$\{1, 2, 3, 4, \ldots\}$$

[2] Find the Z-transform of the following sequences ($k \in \mathbb{N}$):

$$\text{(a) } \{\sin \beta k\}; \quad \text{(b) } \{ka^k\};$$
$$\text{(c) } \left\{\frac{1}{k!}a^k\right\}; \quad \text{(d) } \left\{\frac{1}{k+1}\right\}$$

[3] Prove the Z-transform convolution theorem:

$$\mathscr{Z}[f_k * g_k] = \mathscr{Z}[f_k] \cdot \mathscr{Z}[g_k]$$

[4] Find the inverse Z-transform:

$$F(z) = \frac{z}{(z+1)(z-1)^2}$$

[5] Find the general term representation of the difference equation:

$$u_{n+2} - (b+c)u_{n+1} + bcu_n = 0$$

Answer:

$$u_n = \frac{u_1 - cu_0}{b-c}b^n - \frac{u_1 - bu_0}{b-c}c^n$$

[6] Solve the difference equation:

$$\begin{aligned}&\text{(a) } u_{n+1} - au_n = nb^n, \quad u_0 = 1;\\ &\text{(b) } u_{n+2} - 2bu_{n+1} + b^2u_n = b^n, \quad u_0 = u_1 = 0\end{aligned}$$

[7] Prove:

$$\frac{1+\sqrt{5}}{2} = 1 + \frac{1}{1+\frac{1}{1+\frac{1}{1+\cdots}}}$$

Hint: Use $\varphi^2 - \varphi - 1 = 0$ or the relationship of the Fibonacci sequence:

$$\frac{f_{k+1}}{f_k} = 1 + \frac{1}{\frac{f_k}{f_{k-1}}}$$

[8] Prove the relationships for the Fibonacci function:

$$\begin{aligned}&\text{(a) } f(z+2) = f(z+1) + f(z);\\ &\text{(b) } f(z+n) = f_n f(z+1) + f_{n-1} f(z);\\ &\text{(c) } f(\alpha+\beta+1) = f(\alpha+1)f(\beta+1) + f(\alpha)f(\beta) \quad (\alpha, \beta \in \mathbb{C})\end{aligned}$$

[9] Prove that $\sqrt{2} + \sqrt{3}$ is an algebraic number satisfying the equation:

$$x^4 - 10x^2 + 1 = 0$$

Chapter 9

General Solutions of Differential Equations

This chapter introduces several methods for finding general solutions to linear ordinary differential equations (ODEs) and partial differential equations (PDEs). The general solution refers to the solution of an ODE without imposing any constraints. Since the equations are linear, the general solution is a linear combination of all linearly independent solutions. The last two sections of this chapter provide a brief introduction to some methods for finding solitary wave solutions to nonlinear equations.

9.1 Constant Coefficient Ordinary Differential Equations

We first introduce the basic methods for finding general solutions of constant coefficient linear ODEs. A general nth-order linear ODE can be expressed as

$$\hat{L}y \equiv y^{(n)}(x) + a_{n-1}y^{(n-1)}(x) + \cdots + a_1 y^{(1)}(x) + a_0 y(x) = r(x) \tag{9.1.1}$$

where a_k are constants, $r(x)$ is the non-homogeneous term, and $\hat{L}$ represents the linear differential operator acting on the function $y(x)$, defined as

$$\hat{L} = \frac{\mathrm{d}^n}{\mathrm{d}x^n} + a_{n-1}\frac{\mathrm{d}^{n-1}}{\mathrm{d}x^{n-1}} + \cdots + a_1\frac{\mathrm{d}}{\mathrm{d}x} + a_0$$

9.1.1 *Homogeneous equations*

When $r(x) = 0$, the equation is called a homogeneous linear differential equation:

$$y^{(n)} + a_{n-1}y^{(n-1)} + \cdots + a_1 y^{(1)} + a_0 y = 0 \tag{9.1.2}$$

Assuming the solution has the form of an exponential function, $y = \mathrm{e}^{\lambda x}$, substitute it into the equation and obtain

$$[\lambda^n + a_{n-1}\lambda^{n-1} + \cdots + a_1\lambda + a_0]\mathrm{e}^{\lambda x} = 0$$

Thus, we have

$$p(\lambda) \equiv \lambda^n + a_{n-1}\lambda^{n-1} + \cdots + a_1\lambda + a_0$$

This expression is called the characteristic polynomial of the ODE, and its roots are referred to as characteristic roots. According to the fundamental theorem of algebra, an nth-degree polynomial equation has n complex roots, including k_j $(j = 1, 2, \ldots m)$ roots with multiplicity k_j (repeated roots). Hence,

$$p(\lambda) = (\lambda - \lambda_1)^{k_1}(\lambda - \lambda_2)^{k_2} \cdots (\lambda - \lambda_m)^{k_m}$$

where $k_1 + k_2 + \cdots + k_m = n$. The set of functions:

$$\{\mathrm{e}^{\lambda_j x}, x\mathrm{e}^{\lambda_j x}, x^2\mathrm{e}^{\lambda_j x}, \ldots, x^{k_j-1}\mathrm{e}^{\lambda_j x}\}_{j=1}^{m} \tag{9.1.3}$$

are all linearly independent solutions of the homogeneous linear ODE.

Example 9.1. To solve the ODE:

$$\frac{\mathrm{d}^2y}{\mathrm{d}x^2} + a\frac{\mathrm{d}y}{\mathrm{d}x} + by = 0 \quad (a, b \in \mathbb{R})$$

Solution: The characteristic polynomial of the equation is

$$p(\lambda) \equiv \lambda^2 + a\lambda + b$$

It has two roots:

$$\lambda_{1,2} = \frac{1}{2}(-a \pm \sqrt{a^2 - 4b})$$

We need to discuss the properties of the solutions in three cases:

(1) $a^2 > 4b$, it has two distinct real roots. Letting $\gamma = \frac{1}{2}\sqrt{a^2 - 4b}$, the general solution is

$$y(x) = c_1 e^{\lambda_1 x} + c_2 e^{-\lambda_2 x} = (c_1 e^{\gamma x} + c_2 e^{-\gamma x}) e^{-\frac{ax}{2}}$$

(2) $a^2 = 4b$, it has one repeated root. the general solution is

$$y(x) = c_0 e^{-\frac{ax}{2}} + c_1 x e^{-\frac{ax}{2}}$$

(3) $a^2 < 4b$, it has two distinct complex roots. Letting $\omega = \frac{1}{2}\sqrt{4b - a^2}$, the general solution is

$$y(x) = c_1 e^{-\frac{1}{2}ax + i\omega x} + c_2 e^{-\frac{1}{2}ax - i\omega x} = (d_1 \cos \omega x + d_2 \sin \omega x) e^{-\frac{ax}{2}}$$

9.1.2 *Non-homogeneous equations*

For non-homogeneous equations where $r(x) \neq 0$, there is no general analytical method to find solutions for higher-order ODEs. However, for specific forms of the non-homogeneous term, such as the form:

$$r(x) = e^{\alpha x} S(x)$$

We can look for particular solutions based on the form of $S(x)$ in two cases:

(1) **$S(x)$ is a polynomial:**

$$S(x) = b_m x^m + b_{m-1} x^{m-1} + \cdots + b_0$$

The particular solution of the equation is

$$\tilde{y}(x) = e^{\alpha x} x^k q_m(x) \tag{9.1.4}$$

Here, $q_m(x)$ is an mth-degree polynomial, and k is equal to the multiplicity of a specific characteristic root λ_k, where $\lambda_k = \alpha$.

Example 9.2. Solve the ODE:

$$y''' + 3y'' + 3y' + y = e^{-x}(x - 5)$$

Solution: The characteristic equation is

$$\lambda^3 + 3\lambda^2 + 3\lambda + 1 = 0$$

It has a triple root $\lambda_1 = -1$. Therefore, the general solution of the corresponding homogeneous equation is

$$y(x) = (c_0 + c_1 x + c_2 x^2)\mathrm{e}^{-x}$$

Since $\lambda_1 = \alpha = -1$ and $S(x)$ is a linear polynomial, the particular solution of the non-homogeneous ODE can be written as

$$\tilde{y}(x) = x^3(b_1 x + b_0)\mathrm{e}^{-x}$$

By substituting the particular solution into the original equation, we find

$$b_0 = -\frac{5}{6}, \quad b_1 = \frac{1}{24}$$

Hence, the general solution of the equation is

$$y(x) = (c_0 + c_1 x + c_2 x^2)\mathrm{e}^{-x} + \frac{1}{24}x^3(x - 20)\mathrm{e}^{-x}$$

(2) **$S(x)$ is a trigonometric function:**

$$S(x) = P_m(x)\cos\beta x + Q_n(x)\sin\beta x$$

Here, $P_m(x)$ and $Q_n(x)$ are real coefficient polynomials of degrees m and n, respectively. The particular solution of the equation is

$$\tilde{y}(x) = \mathrm{e}^{\alpha x}x^k[p_s(x)\cos\beta x + q_s(x)\sin\beta x] \tag{9.1.5}$$

Here k represents the multiplicity of the characteristic root λ_k, $p_s(x)$ and $q_s(x)$ are s-degree polynomials, where $s = \max[m, n]$.

Example 9.3. Find the general solution for the equation:

$$y'' + 4y' + 4y = \cos 2x$$

Solution: Solving the characteristic equation

$$\lambda^2 + 4\lambda + 4 = 0$$

it has only one double real root $\lambda = -2$. Therefore, the general solution of the homogeneous equation is

$$y(x) = (c_1 + c_2 x)\mathrm{e}^{-2x}$$

Here, c_1 and c_2 are arbitrary constants. Since $\alpha = \pm 2\mathrm{i}$ are not characteristic roots of the equation, the particular solution reads

$$\tilde{y}(x) = a\cos 2x + b\sin 2x$$

Substituting it into the equation:

$$8b\cos 2x - 8a\sin 2x = \cos 2x$$

we get $a = 0, b = \frac{1}{8}$. Hence, the general solution of the equation is

$$y(x) = (c_1 + c_2 x)\mathrm{e}^{-2x} + \frac{1}{8}\sin 2x$$

(3) **General non-homogeneous term:**
For a second-order linear ODE with a general non-homogeneous term $r(x)$ of the form:

$$\frac{\mathrm{d}^2 y}{\mathrm{d}x^2} + a\frac{\mathrm{d}y}{\mathrm{d}x} + by = r(x) \tag{9.1.6}$$

where a and b are constants, we can find a particular solution by using the transformation

$$y(x) = \mathrm{e}^{-\frac{1}{2}ax}u(x) \tag{9.1.7}$$

After eliminating the first derivative term, the equation is transformed into the following form:

$$\frac{\mathrm{d}^2 u}{\mathrm{d}x^2} - pu = f(x) \tag{9.1.8}$$

where

$$f(x) = \mathrm{e}^{\frac{1}{2}ax}r(x), \quad p = \frac{1}{4}a^2 - b$$

The general solution of the corresponding homogeneous equation is

$$u(x) = A\mathrm{e}^{\sqrt{p}x} + B\mathrm{e}^{-\sqrt{p}x}$$

To obtain a particular solution of the non-homogeneous equation, we use the method of undetermined coefficients. Let A and B be functions of x. We assume the particular solution as

$$\tilde{u}(x) = A(x)\mathrm{e}^{\sqrt{p}x} + B(x)\mathrm{e}^{-\sqrt{p}x}$$

Differentiating $u(x)$, we have

$$\frac{\mathrm{d}u}{\mathrm{d}x} = A'(x)\mathrm{e}^{\sqrt{p}x} + \sqrt{p}A(x)\mathrm{e}^{\sqrt{p}x} + B'(x)\mathrm{e}^{-\sqrt{p}x} - \sqrt{p}B(x)\mathrm{e}^{-\sqrt{p}x}$$

$$\frac{\mathrm{d}^2u}{\mathrm{d}x^2} = \frac{d}{dx}[A'(x)\mathrm{e}^{\sqrt{p}x} + B'(x)\mathrm{e}^{-\sqrt{p}x}] + \sqrt{p}[A'(x)\mathrm{e}^{\sqrt{p}x} - B'(x)\mathrm{e}^{-\sqrt{p}x}]$$
$$+ p[A(x)\mathrm{e}^{\sqrt{p}x} + B(x)\mathrm{e}^{-\sqrt{p}x}]$$

Substituting these expressions into the equation, and since we only need to find one particular solution, we take

$$\begin{cases} A'(x)\mathrm{e}^{\sqrt{p}x} + B'(x)\mathrm{e}^{-\sqrt{p}x} = 0 \\ \sqrt{p}[A'(x)\mathrm{e}^{\sqrt{p}x} - B'(x)\mathrm{e}^{-\sqrt{p}x}] = f(x) \end{cases}$$

Solving these equations to obtain

$$A(x) = \frac{1}{2\sqrt{p}}\int^{(x)} f(\xi)\mathrm{e}^{-\sqrt{p}\xi}\mathrm{d}\xi, \quad B(x) = -\frac{1}{2\sqrt{p}}\int^{(x)} f(\xi)\mathrm{e}^{\sqrt{p}\xi}\mathrm{d}\xi$$

Therefore, the general solution of Equation (9.1.8) is

$$u(x) = A_0\mathrm{e}^{\sqrt{p}x} + B_0\mathrm{e}^{-\sqrt{p}x} + \frac{\mathrm{e}^{\sqrt{p}x}}{2\sqrt{p}}\int^{(x)} f(\xi)\mathrm{e}^{-\sqrt{p}\xi}\mathrm{d}\xi$$
$$- \frac{\mathrm{e}^{-\sqrt{p}x}}{2\sqrt{p}}\int^{(x)} f(\xi)\mathrm{e}^{\sqrt{p}\xi}\mathrm{d}\xi$$

Therefore, the solution of Equation (9.1.6) is

$$y(x) = A_0\mathrm{e}^{(\sqrt{p}-\frac{1}{2}a)x} + B_0\mathrm{e}^{-(\sqrt{p}+\frac{1}{2}a)x}$$
$$+ \frac{1}{2\sqrt{p}}\mathrm{e}^{(\sqrt{p}-\frac{a}{2})x}\int^{(x)} r(\xi)\mathrm{e}^{-(\sqrt{p}-\frac{a}{2})\xi}\mathrm{d}\xi$$
$$- \frac{1}{2\sqrt{p}}\mathrm{e}^{-(\sqrt{p}+\frac{a}{2})x}\int^{(x)} r(\xi)\mathrm{e}^{(\sqrt{p}+\frac{a}{2})\xi}\mathrm{d}\xi$$

9.1.3 *Euler-type equation*

A Euler-type equation is a variable coefficient differential equation of the form:

$$x^n \frac{\mathrm{d}^n y}{\mathrm{d}x^n} + a_{n-1} x^{n-1} \frac{\mathrm{d}^{n-1} y}{\mathrm{d}x^{n-1}} + \cdots + a_1 x \frac{\mathrm{d}y}{\mathrm{d}x} + a_0 y = 0 \tag{9.1.9}$$

where a_j $(j = 1, 2, \ldots, n-1)$ are constants. By making the variable substitution $x = \mathrm{e}^t$, we have

$$\frac{\mathrm{d}y}{\mathrm{d}x} = \mathrm{e}^{-t} \frac{\mathrm{d}y}{\mathrm{d}t}, \quad \frac{\mathrm{d}^2 y}{\mathrm{d}x^2} = \mathrm{e}^{-2t} \left(\frac{\mathrm{d}^2 y}{\mathrm{d}t^2} - \frac{\mathrm{d}y}{\mathrm{d}t} \right)$$

$$\frac{\mathrm{d}^k y}{\mathrm{d}x^k} = \mathrm{e}^{-kt} \left(\frac{\mathrm{d}^k y}{\mathrm{d}t^k} + \beta_1 \frac{\mathrm{d}^{k-1} y}{\mathrm{d}t^{k-1}} + \cdots + \beta_{k-1} \frac{\mathrm{d}y}{\mathrm{d}t} \right)$$

where β_j are constants. Thus, the equation can be transformed into a constant coefficient homogeneous linear differential equation:

$$\frac{\mathrm{d}^n y}{\mathrm{d}t^n} + b_{n-1} \frac{\mathrm{d}^{n-1} y}{\mathrm{d}t^{n-1}} + \cdots + b_1 \frac{\mathrm{d}y}{\mathrm{d}t} + b_0 y = 0$$

Another more direct method is to assume that the Euler's equation has a solution of the form: $y(x) \sim x^\alpha$. Substituting it into the equation, we obtain an algebraic equation of degree n for α. By solving this equation, we can find n linearly independent power function solutions. If there are m repeated roots, then the corresponding m linearly independent solutions are given by

$$\{x^{\alpha_j}, x^{\alpha_j} \ln x, x^{\alpha_j} \ln^2 x, \ldots, x^{\alpha_j} \ln^{m-1} x\} \tag{9.1.10}$$

Exercises

[1] Find the general solution to the homogeneous equation:

$$y^{(4)} - 5y'' + 4y = 0$$

[2] Find the general solution to the homogeneous ODEs:

(a) $y'' - 2y' - 3y = 3x + 1$

(b) $y'' - y = \mathrm{e}^x \sin 2x$

(c) $y'' - 4y' + 4y = \mathrm{e}^x + x\mathrm{e}^{2x}$

(d) $y^{(6)} - y^{(4)} = x^2$

(e) $y''' + 3y'' - 4y = x\mathrm{e}^{-2x}$

[3] Find the general solution to the Euler-type equation:

$$x^3y''' - x^2y'' + 2xy' - 2y = x^3$$

Hint: First, assume the general solution to the homogeneous differential equation is $y(x) \sim x^\alpha$. Solve for the roots α_j of the characteristic equation. If there are repeated roots, the linearly independent solutions are $\{x^{\alpha_j}, x^{\alpha_j}\ln x, x^{\alpha_j}\ln^2 x, \ldots\}$. Then, assume the particular solution to the non-homogeneous equation is $\tilde{y}(x) = Ax^3$. Substitute it into the equation and solve for A.

9.2 Variable Coefficient Ordinary Differential Equations

Euler-type equations belong to a special class of variable coefficient ODEs. For general variable coefficient ODEs, there is no universal analytical method for finding solutions. The typical approach is to use power series solutions, where the solution of the equation in a neighborhood of a chosen point is represented as a power series with undetermined coefficients. Substituting this series into the equation leads to recurrence relations among the coefficients. It is essential to ensure the convergence of the infinite series when using the power series method.

In this section, we mainly discuss second-order linear variable coefficient ODEs. Without loss of generality, we first consider the complex variable second-order homogeneous ODEs:

$$y''(z) + p(z)y' + q(z)y(z) = 0 \tag{9.2.1}$$

In a neighborhood of a point z_0, if both $p(z)$ and $q(z)$ are analytic, then z_0 is called a regular point of the equation. If either $p(z)$ or $q(z)$ is singular, then z_0 is called a singular point of the equation. At the regular point, solutions to the ODE are analytic and their series representations have no negative power terms. At the singular point, however, solutions to the ODE may also be singular and hence their series representations have negative power terms. We will discuss these two cases separately.

9.2.1 *Regular point*

If the coefficient functions $p(z)$ and $q(z)$ of a linear second-order ODE are analytic functions in the neighborhood $|z - z_0| < R$, then the solution of the equation is also analytic at z_0 and can be represented by a Taylor series:

$$y(z) = \sum_{k=0}^{\infty} a_k (z - z_0)^k$$

Example 9.4. Find the series solution of the Legendre equation:

$$(1 - x^2)\frac{\mathrm{d}^2 y}{\mathrm{d}x^2} - 2x\frac{\mathrm{d}y}{\mathrm{d}x} + l(l+1)y = 0$$

Solution: The coefficient functions of the equation are

$$p(x) = -\frac{2x}{1 - x^2}, \quad q(x) = \frac{l(l+1)}{1 - x^2}$$

It can be observed that $x_0 = 0$ is a regular point of the equation, as the functions $p(z)$ and $q(z)$ are analytic at that point. Assume the series solution of the equation as

$$y(x) = \sum_{k=0}^{\infty} a_k x^k$$

By substituting this series into the equation and combining terms with the same powers of x, we obtain a recurrence relation among the coefficients a_k:

$$a_{k+2} = \frac{(k-l)(k+l+1)}{(k+2)(k+1)} a_k$$

It can be seen that for a second-order differential equation, only two coefficients are independent. We can choose them as a_0 and a_1. The general solution of the equation can be represented as

$$y(x) = a_0 y_0(x) + a_1 y_1(x)$$

It is a superposition of two linearly independent solutions, $y_0(x)$ and $y_1(x)$, where

$$\begin{aligned} y_0(x) = 1 &+ \frac{(-l)(l+1)}{2!}x^2 + \frac{(2-l)(-l)(l+1)(l+3)}{4!}x^4 + \cdots \\ &+ \frac{(2k-2-l)(2k-4-l)\cdots(-l)(l+1)\cdots(l+2k-1)}{(2k)!} \\ &\times x^{2k} + \cdots \end{aligned}$$

and

$$y_1(x) = x + \frac{(1-l)(l+2)}{3!}x^3 + \frac{(3-l)(1-l)(l+2)(l+4)}{5!}x^5 + \cdots$$
$$+ \frac{(2k-1-l)(2k-3-l)\cdots(1-l)(l+2)\cdots(l+2k)}{(2k+1)!}$$
$$\times\, x^{2k+1} + \cdots$$

By applying the ratio test, it can be easily proved that the convergence radius of both series solutions is

$$R = \lim_{k\to\infty}\sqrt{\left|\frac{a_k}{a_{k+2}}\right|} = \lim_{k\to\infty}\sqrt{\frac{(k+2)(k+1)}{(k-l)(k+l+1)}} = 1$$

This convergence radius can also be determined from the positions of the singular points of the coefficient functions $p(z)$ and $q(z)$ at $z = \pm 1$.

9.2.2 *Regular singularity*

In most cases, the solutions to the ODEs are singular at the singular points. For a physical problem, it is meaningless if both the two solutions are singular. We should try to avoid this serious situation to occur. A somehow less serious situation for singular ODEs is that while one solution is singular, the other solution may be finite. Hence it is physically acceptable. This type of ODEs imposes special constraints on the coefficient function $p(z)$ and $q(z)$, and it is known as the regular singularity.

For a second order ODE, if z_0 is a pole of the coefficient function $p(z)$ up to the first order and a pole of the coefficient function $q(z)$ up to the second order, i.e.,

$$p(z) = \sum_{k=-1}^{\infty} p_k(z-z_0)^k, \quad q(z) = \sum_{k=-2}^{\infty} q_k(z-z_0)^k$$

then it is called a regular singular point of the ODE. Mathematically, it can be proven that the series solution in the neighborhood of a regular singularity has a finite number of negative powers. Below we will only consider the situation of regular singularity.

Fuchs' theorem states that for a regular singular point z_0, the equation has a series solution of the following form:

$$y_1(z) = \sum_{k=0}^{\infty} a_k (z - z_0)^{s_1 + k} \tag{9.2.2}$$

This form of series is called a Frobenius series.

The second linearly independent solution of the equation has two forms:

(1) In case of $s_1 - s_2 \notin \mathbb{Z}$,

$$y_2(z) = \sum_{k=0}^{\infty} b_k (z - z_0)^{s_2 + k} \quad (s_1 - s_2 \notin \mathbb{Z}) \tag{9.2.3}$$

(2) in case of $s_1 - s_2 \in \mathbb{Z}$,

$$y_2(z) = A y_1(z) \ln(z - z_0) + \sum_{k=0}^{\infty} b_k (z - z_0)^{s_2 + k} \quad (s_1 - s_2 \in \mathbb{Z}) \tag{9.2.4}$$

Here, s_1 and s_2 are called characteristic indices, with $\operatorname{Re} s_1 > \operatorname{Re} s_2$. They are the roots of the following indicial equation:

$$s(s-1) + s p_{-1} + q_{-2} = 0 \tag{9.2.5}$$

The indicial equation arises when we substitute the Frobenius series solution into the differential equation and let the coefficients of the zeroth power of $(z - z_0)$ vanish.

Example 9.5. Find two linearly independent series solutions in the neighborhood of $x = 0$ for the ODE:

$$xy'' - xy' + y = 0$$

Solution: We first identify the regular singular point at $x = 0$ and solve the indicial equation which is given by

$$s(s-1) + s p_{-1} + q_{-2} = 0$$

For the given equation, $p_{-1} = -1$ and $q_{-2} = 1$, we get the two roots: $s_1 = 1$ and $s_2 = 0$. Now, let us assume the series solution for the equation as

$$y(x) = x^{s_1} \sum_{k=0}^{\infty} a_k x^k$$

Substituting into the ODE, we obtain the first solution:

$$y_1(x) = x$$

As to the second solution for $s_2 = 0$, we suppose

$$y_2(x) = A y_1(x) \ln x + \sum_{k=0}^{\infty} b_k x^k$$

Substituting $y_2(x)$ back into the equation, we get

$$A = -b_0, b_k = -\frac{1}{(k-1)k!} b_0 \quad (k \geq 2)$$

Hence,

$$y_2(x) = x \ln x - 1 + \sum_{k=2}^{\infty} \frac{1}{(k-1)k!} x^k + b_1 x$$

As the last term in $y_2(x)$ is identical to $y_1(x)$, we can choose the second solution as

$$y_2(x) = x \ln x - 1 + \sum_{k=2}^{\infty} \frac{1}{(k-1)k!} x^k$$

Example 9.6. Finding series solutions of Bessel's equation:

$$x^2 y'' + x y' + (x^2 - \nu^2) y = 0$$

Solution: Since $x_0 = 0$ is a first-order pole of $p(x) = \frac{1}{x}$ and a second-order pole of $q(x) = 1 - \frac{\nu^2}{x^2}$, it is a regular singular point of the Bessel's equation. Solving the indicial equation:

$$s^2 - \nu^2 = 0$$

we obtain two characteristic indices, $s_1 = \nu$ and $s_2 = -\nu$. We need to consider several cases:

(1) $\boldsymbol{s_1 - s_2} = 2\nu \notin \mathbb{Z}$

If the difference between the two roots is not an integer or a half-integer, then the two linearly independent solutions of the equation can be directly taken as

$$y(x) = a_0 x^s + a_1 x^{s+1} + a_2 x^{s+2} + \cdots + a_k x^{s+k} + \cdots$$

Substituting this series into the Bessel's equation, combining like terms, and equating the coefficients of x^k to zero, we obtain the following relations:

$$\begin{aligned}
&[s^2 - \nu^2]a_0 = 0 \\
&[(s+1)^2 - \nu^2]a_1 = 0 \\
&\vdots \\
&[(s+k)^2 - \nu^2]a_k + a_{k-2} = 0
\end{aligned}$$

Assuming $a_0 \neq 0$, we can deduce that $a_1 = 0$, and obtain the recurrence relation:

$$a_k = \frac{-1}{(s+k+\nu)(s+k-\nu)} a_{k-2}$$

Taking $s_1 = \nu$, we obtain a particular solution of the equation:

$$\begin{aligned}
y_1(x) = a_0 x^\nu \Bigg[1 &- \frac{1}{1!(\nu+1)} (\frac{x}{2})^2 + \frac{1}{2!(\nu+1)(\nu+2)} \left(\frac{x}{2}\right)^4 \\
&- \cdots + (-1)^k \frac{1}{k!(\nu+1)(\nu+2)\cdots(\nu+k)} \left(\frac{x}{2}\right)^{2k} + \cdots \Bigg]
\end{aligned}$$

The convergence radius of this series is given by

$$R = \lim_{k\to\infty} \left| \frac{a_{k-2}}{a_k} \right| = \lim_{k\to\infty} 2^k k(2\nu + k) \to \infty$$

Usually, $a_0 = \frac{1}{2^\nu \Gamma(\nu+1)}$ is chosen, and this solution is called the νth order Bessel function, denoted as $\mathrm{J}_\nu(x)$, which reads

$$\mathrm{J}_\nu(x) = \sum_{k=0}^{\infty} (-1)^k \frac{1}{k!\Gamma(\nu+k+1)} \left(\frac{x}{2}\right)^{\nu+2k}$$

Similarly, the other solution of the equation corresponding to $s_2 = -\nu$ is denoted as $\mathrm{J}_{-\nu}(x)$, given by

$$\mathrm{J}_{-\nu}(x) = \sum_{k=0}^{\infty} (-1)^k \frac{1}{k!\Gamma(-\nu + k + 1)} \left(\frac{x}{2}\right)^{-\nu+2k}$$

The general solution of the equation can be expressed as

$$y(x) = c_1 \mathrm{J}_\nu(x) + c_2 \mathrm{J}_{-\nu}(x)$$

Thus, we have obtained the solutions for non-integer or non-half-integer order Bessel's equation, which consist of two linearly independent solutions that converge globally.

(2) **Integer-order Bessel's equation ($\boldsymbol{\nu = m}$)**
From the indicial equation, we have $s_1 = m$, $s_2 = -m$, and $s_1 - s_2 = 2m$, which is zero or a positive integer. For the larger root $s_1 = m$, the first solution is the mth order Bessel function:

$$\mathrm{J}_m(x) = \sum_{k=0}^{\infty} (-1)^k \frac{1}{k!\Gamma(m + k + 1)} \left(\frac{x}{2}\right)^{m+2k}$$

For the smaller root $s_2 = -m$, if we choose the second solution as $\mathrm{J}_{-\nu}(x) \longrightarrow \mathrm{J}_{-m}(x)$, we will find that $\mathrm{J}_{-m}(x)$ and $\mathrm{J}_m(x)$ are linearly dependent. The proof is as follows:

Since the Γ function diverges for $k < m$, the series $\mathrm{J}_{-m}(x)$ can only start from $k = m$. Let $n = k - m$, we have

$$\begin{aligned}\mathrm{J}_{-m}(x) &= \sum_{k=m}^{\infty} (-1)^k \frac{1}{k!\Gamma(-m + k + 1)} \left(\frac{x}{2}\right)^{-m+2k} \\ &= \sum_{n=0}^{\infty} (-1)^{n+m} \frac{1}{(n + m)!\Gamma(n + 1)} \left(\frac{x}{2}\right)^{m+2n} = (-1)^m \mathrm{J}_m(x)\end{aligned}$$

Hence $\mathrm{J}_{-m}(x)$ and $\mathrm{J}_m(x)$ are linearly dependent. Therefore, we need to find another solution for the equation, which should have the following form:

$$y_2(x) = A\mathrm{J}_m(x) \ln x + \sum_{k=0}^{\infty} b_k x^{-m+k}$$

For the Bessel's equation, we no longer follow this approach to solve for the coefficients b_k. Alternatively, the second solution is usually taken as the Neumann function, which is a combination of the two linearly independent solutions corresponding to non-integer ν:

$$\mathrm{N}_\nu(x) = \alpha \mathrm{J}_\nu(x) + \beta \mathrm{J}_{-\nu}(x)$$

Taking $\alpha = \cot \nu\pi$ and $\beta = -\csc \nu\pi$, we obtain a new function known as the νth order Neumann function:

$$\mathrm{N}_\nu(x) = \frac{\mathrm{J}_\nu(x)\cos \nu\pi - \mathrm{J}_{-\nu}(x)}{\sin \nu\pi} \tag{9.2.6}$$

When $\nu = m$ is an integer, using L'Hôpital's rule, we have

$$\begin{aligned} \mathrm{N}_m(x) &= \lim_{\nu \to m} \mathrm{N}_\nu(x) = \lim_{\nu \to m} \frac{\mathrm{J}_\nu(x)\cos \nu\pi - \mathrm{J}_{-\nu}(x)}{\sin \nu\pi} \\ &= \frac{2}{\pi}\mathrm{J}_m(x) \ln\left(\frac{x}{2}\right) - \frac{1}{\pi}\sum_{k=0}^{m-1} \frac{(m-k-1)!}{k!}\left(\frac{x}{2}\right)^{2k-m} \\ &\quad - \frac{1}{\pi}\sum_{k=0}^{\infty} \frac{(-1)^k}{k!(m+k)!}[\psi(k+1) + \psi(m+k+1)]\left(\frac{x}{2}\right)^{2k+m} \end{aligned}$$

where ψ is the digamma function.

Therefore, the two linearly independent solutions for the integer-order Bessel's equation are $\mathrm{J}_m(x)$ and $\mathrm{N}_m(x)$. The general solution of the equation can be expressed as

$$y(x) = c_1 \mathrm{J}_m(x) + c_2 \mathrm{N}_m(x)$$

Example 9.7. Find the series solution of the hypergeometric equation:

$$x(x-1)y'' + [(\alpha + \beta + 1)x - \gamma]y' + \alpha\beta y = 0$$

Solution: The equation has three regular singularities: $x = 0$, $x = 1$ and $x = \infty$. Its indicial equation has two roots at $x_0 = 0$:

$$s_1 = 0, \quad s_2 = 1 - \gamma$$

Consider the series solution:

$$y(x) = a_0 x^s + a_1 x^{s+1} + a_2 x^{s+2} + \cdots + a_k x^{s+k} + \cdots$$

Substituting this into the hypergeometric equation, we obtain the recursive relation for the coefficient corresponding to $s_1 = 0$:

$$a_{k+1} = \frac{(k+\alpha)(k+\beta)}{(k+\gamma)(k+1)} a_k \quad (k = 0, 1, 2, \ldots)$$

Therefore, one solution of the equation is given by

$$y_1(x) = F(\alpha, \beta, \gamma; x)$$

where $\mathrm{F}(\alpha, \beta, \gamma; x)$ is known as the hypergeometric function or Gaussian hypergeometric function:

$$F(\alpha, \beta, \gamma; x) = \frac{\Gamma(\gamma)}{\Gamma(\alpha)\Gamma(\beta)} \sum_{k=0}^{\infty} \frac{\Gamma(\alpha+k)\Gamma(\beta+k)}{k!\Gamma(\gamma+k)} x^k \quad (|x| < 1) \tag{9.2.7}$$

The radius of convergence of the series is $R = 1$. When $\gamma \notin \mathbb{Z}$, the other linearly independent solution of the equation is given by

$$y_2(x) = x^{1-\gamma}\mathrm{F}(\alpha - \gamma + 1, \beta - \gamma + 1, 2 - \gamma; x) \tag{9.2.8}$$

If $\gamma \in \mathbb{Z}$, the procedure for finding the second linearly independent solution $\tilde{y}_2(x)$ follows the case where the difference in characteristic indices is an integer. The details are omitted here.

The hypergeometric equation, also known as the Gaussian equation, can be transformed into the confluent hypergeometric equation, also known as the Kummer equation, by making the variable substitution: $x \longrightarrow \frac{x}{b}$, and then letting $b = \beta \longrightarrow \infty$. The confluent hypergeometric equation is given by

$$xy'' + (\gamma - x)y' - \alpha y = 0 \tag{9.2.9}$$

It has a regular singular point at $x = 0$, while the other two regular singular points at $x = 1$ and $x = \infty$ converge to a non-regular singular point at $x = \infty$, hence the name "confluent". One series solution of the Kummer equation is given by

$$F(\alpha, \gamma, x) = \sum_{k=0}^{\infty} \frac{\Gamma(k+\alpha)\Gamma(\gamma)}{k!\Gamma(\alpha)\Gamma(k+\gamma)} x^k \quad (|x| < \infty) \tag{9.2.10}$$

It is known as the confluent hypergeometric function or Kummer function.

Discussion: If a differential equation does not satisfy the conditions of the Fuchs' theorem, in general, its solutions cannot be expressed in terms of a Frobenius series. For example, the following equation has a non-regular singular point at $z = 0$:

$$z^4 y'' + y = 0$$

This equation has two linearly independent solutions:

$$y_1 = z \sin \frac{1}{z}, \quad y_2 = z \cos \frac{1}{z}$$

It is observed that $z = 0$ is an essential singularity, and they cannot be expanded in the form of a Frobenius series.

9.2.3 *Second solution of the ODE*

As mentioned earlier, the second solution of the Bessel equation can be taken as the Neumann function. In general, for a homogeneous second-order ODE of the form:

$$y'' + p(x)y' + q(x)y = 0$$

If one solution $y_1(x)$ is known, the other linearly independent solution can be directly obtained as

$$y_2(x) = y_1(x) \int_\alpha^x \frac{1}{[y_1(s)]^2} \exp\left[-\int_c^s p(t)\mathrm{d}t\right] \mathrm{d}s \tag{9.2.11}$$

Proof. Let $y_1(x)$ and $y_2(x)$ be two linearly independent solutions of the ODE. Define the Wronskian determinant as

$$W(y_1, y_2; x) = \begin{vmatrix} y_1(x) & y_1'(x) \\ y_2(x) & y_2'(x) \end{vmatrix} = y_1(x)y_2'(x) - y_1'(x)y_2(x)$$

Taking the derivative of the above equation, we have

$$W'(y_1, y_2; x) = y_1 y_2'' - y_1'' y_2 = p(x)(y_1' y_2 - y_1 y_2') = -p(x)W(y_1, y_2; x)$$

Thus, we obtain

$$W(y_1, y_2; x) = W(y_1, y_2; c) \exp\left[-\int_c^x p(\xi)\mathrm{d}\xi\right] \tag{9.2.12}$$

Dividing both sides of Equation (9.2.12) by $y_1^2(x)$, the left side becomes $\left[\frac{y_2(x)}{y_1(x)}\right]'$. Integrating both sides complete the proof. $\square$

From Equation (9.2.12), it is observed that the Wronskian determinant of any two solutions of a second-order ODE does not change sign within the defined interval $[a, b]$. If it is zero at a certain point, it remains zero throughout the entire interval $[a, b]$. We have the following proposition:

Two solutions y_1 and y_2 of a second-order ODE are linearly independent if and only if their Wronskian determinant is non-zero,

$$W(y_1, y_2; x) \neq 0$$

Example 9.8. For the equation $y'' - k^2 y = 0$, given one solution $y_1(x) = \mathrm{e}^{kx}$, we need to find the second solution.
Solution: Since $p(x) = 0$, the second solution is given by

$$y_2(x) = \mathrm{e}^{kx} \int_\alpha^x \frac{1}{\mathrm{e}^{2ks}} \mathrm{d}s = -\frac{1}{2k}\mathrm{e}^{-kx} + \frac{\mathrm{e}^{-2k\alpha}}{2k}\mathrm{e}^{kx}$$

The second term is the same as the first solution, so we can omit it. Taking the linearly independent solution, we have

$$y_2(x) = \mathrm{e}^{-kx}$$

Example 9.9. For the Legendre equation, given one solution $\mathrm{P}_l(x)$, we need to find the second solution:

$$(1 - x^2)\frac{\mathrm{d}^2 y}{\mathrm{d}x^2} - 2x\frac{\mathrm{d}y}{\mathrm{d}x} + l(l+1)y = 0$$

Solution: Since

$$p(x) = -\frac{2x}{1 - x^2}, \quad q(x) = \frac{l(l+1)}{1 - x^2}$$

the second solution is

$$\begin{aligned}
\mathrm{Q}_l(x) &= \mathrm{P}_l(x) \int_\alpha^x \frac{1}{[\mathrm{P}_l(s)]^2} \exp\left[-\int_0^s \frac{2t}{1 - t^2}\mathrm{d}t\right] \mathrm{d}s \\
&= \mathrm{P}_l(x) \int_\alpha^x \frac{1}{[P_l(s)]^2} \frac{1}{1 - s^2}\mathrm{d}s
\end{aligned}$$

(1) For $l = 0$, $\mathrm{P}_0(x) = 1$. We have

$$\mathrm{Q}_0(x) = \int_\alpha^x \frac{1}{1-s^2}\mathrm{d}s = \frac{1}{2}\left(\ln\frac{1+x}{1-x} - \ln\frac{1+\alpha}{1-\alpha}\right)$$

Taking $\alpha = 0$, we get

$$\mathrm{Q}_0(x) = \frac{1}{2}\ln\frac{1+x}{1-x} \quad (|x| < 1)$$

(2) For $l = 1$, $\mathrm{P}_1(x) = x$. We have

$$\mathrm{Q}_1(x) = x\int_\alpha^x \frac{1}{s^2(1-s^2)}\mathrm{d}s = Ax + Bx\ln\frac{1+x}{1-x} + C$$

Taking $A = 0$, $B = \frac{1}{2}$, and $C = -1$, we get

$$\mathrm{Q}_1(x) = \frac{1}{2}x\ln\frac{1+x}{1-x} - 1$$

9.2.4 *Particular solution for the non-homogeneous ODEs*

For a general variable coefficient non-homogeneous second-order ODE,

$$\hat{L}[y] = y''(x) + p(x)y'(x) + q(x)y(x) = r(x)$$

As mentioned earlier, if we know one solution of the corresponding homogeneous equation, we can find another linearly independent solution. Additionally, we can find a particular solution for the non-homogeneous equation:

$$\tilde{y}(x) = y_2(x)\int^{(x)} \frac{r(\xi)y_1(\xi)}{W(y_1, y_2; \xi)}\mathrm{d}\xi - y_1(x)\int^{(x)} \frac{r(\xi)y_2(\xi)}{W(y_1, y_2; \xi)}\mathrm{d}\xi$$

where $W(y_1, y_2; x)$ is the Wronskian determinant.

Proof. Let us assume that the two solutions of the homogeneous equation are $y_1(x)$ and $y_2(x)$:

$$\begin{cases} y_1'' + py_1' + qy_1 = 0 \\ y_2'' + py_2' + qy_2 = 0 \end{cases}$$

Using the method of undetermined coefficients, we assume that the particular solution of the non-homogeneous equation is

$$\tilde{y}(x) = A(x)y_1(x) + B(x)y_2(x)$$

Substitute it into the non-homogeneous equation and rearrange to yield,

$$\begin{aligned}(A'y_1 + B'y_2)' &+ p(A'y_1 + B'y_2)\\ &+ [(Ay_1' + By_2')' + p(Ay_1' + By_2') + q(Ay_1 + By_2)] = r(x)\end{aligned}$$

By setting $A'y_1 + B'y_2 = 0$, we have

$$A'y_1' + B'y_2' = r(x)$$

Solving the two equations, we find

$$A(x) = -C_1 \int^{(x)} \frac{r(\xi)y_2(\xi)}{W(y_1, y_2; \xi)} \mathrm{d}\xi, \quad B(x) = C_2 \int^{(x)} \frac{r(\xi)y_1(\xi)}{W(y_1, y_2; \xi)} \mathrm{d}\xi$$

Choosing $C_1 = C_2 = 1$, we obtain a particular solution,

$$\tilde{y}(x) = y_2(x) \int^{(x)} \frac{r(\xi)y_1(\xi)}{W(y_1, y_2; \xi)} \mathrm{d}\xi - y_1(x) \int^{(x)} \frac{r(\xi)y_2(\xi)}{W(y_1, y_2; \xi)} \mathrm{d}\xi \qquad \square$$

Example 9.10. To find the solution to the non-homogeneous differential equation:

$$y'' + y = \csc x$$

Solution: The corresponding homogeneous equation has two linearly independent solutions:

$$y_1(x) = \sin x, \quad y_2(x) = \cos x$$

The Wronskian determinant is given by

$$W(y_1, y_2; x) = y_1 y_2' - y_1' y_2 = -\sin^2 x - \cos^2 x = -1$$

Therefore, a particular solution of the equation is

$$\begin{aligned}\tilde{y}(x) &= y_2(x)\int^{(x)}\frac{r(\xi)y_1(\xi)}{W(y_1,y_2;\xi)}\mathrm{d}\xi + y_1(x)\int^{(x)}\frac{r(\xi)y_2(\xi)}{W(y_1,y_2;\xi)}\mathrm{d}\xi\\ &= -\cos x\int^{(x)}\sin\xi\csc\xi\mathrm{d}\xi + \sin x\int^{(x)}\cos\xi\csc\xi\,\mathrm{d}\xi\\ &= -x\cos x + \sin x\ln(\sin x)\end{aligned}$$

The general solution is

$$y(x) = c_1\sin x + c_2\cos x - x\cos x + \sin x\ln(\sin x)$$

ENDNOTE

Knowing the form of one solution to a second-order ODE is sufficient to determine the other solution. This indicates that the two solutions of the equation cannot be arbitrary; they have a certain relationship between them, which is familiar in polynomial equations. For example, in a quadratic equation:

$$x^2 + a_1x + a_2 = 0$$

there exists the following relationship between the roots and coefficients:

$$x_1 + x_2 = -a_1, \quad x_1\cdot x_2 = a_2$$

This relationship exhibits symmetry when exchanging the two roots. Vieta discovered that in a cubic equation

$$x^3 + a_1x^2 + a_2x + a_3 = 0$$

there also exists a relationship between the roots and coefficients:

$$\begin{cases} x_1 + x_2 + x_3 = -a_1\\ x_1\cdot x_2 + x_2\cdot x_3 + x_3\cdot x_1 = a_2\\ x_1\cdot x_2\cdot x_3 = -a_3\end{cases}$$

Similar relationships exist for quartic and quintic equations. G. Desargues extended this rule to arbitrary-degree polynomial equations

$$x^n + a_1x^{n-1} + a_2x^{n-2} + \cdots + a_{n-1}x + a_n = 0$$

where its n roots and coefficients satisfy

$$\begin{cases} x_1 + x_2 + \cdots + x_n = -a_1 \\ x_1x_2 + x_1x_3 + \cdots + x_1x_n + x_2x_3 + \cdots + x_{n-1}x_n = a_2 \\ x_1x_2x_3 + x_1x_2x_4 + \cdots + x_1x_{n-1}x_n + x_2x_3x_4 + \cdots \\ \quad + x_{n-2}x_{n-1}x_n = -a_3 \\ \cdots \\ x_1x_2\cdots x_{n-1}x_n = (-1)^n a_n \end{cases}$$

These relationships exhibit permutation symmetry for all the roots.

Although mathematical giants like Newton and Euler had noticed this intriguing symmetry, no one understood its profound significance. A seemingly trivial deduction by A. Van der Monde sparked a revolution. He proposed expressing each root using symmetric functions of all the roots of the equation. For example, in the case of a quadratic equation, the roots can be written as

$$x_{1,2} = \frac{(x_1 + x_2) \pm \sqrt{x_1^2 - 2x_1x_2 + x_2^2}}{2}$$

The significance lies in the symmetric functions $R_1 = x_1 + x_2$ and $R_2 = x_1^2 - 2x_1x_2 + x_2^2$, which, upon introducing radicals, generate two asymmetric roots:

$$x_1 = \frac{R_1 + \sqrt{R_2}}{2}, \quad x_2 = \frac{R_1 - \sqrt{R_2}}{2}$$

This expression can be generalized to cubic and quartic equations and holds promise for extending to quintic equations.

However, things didn't go as expected. Lagrange attempted to introduce a series of symmetric functions, denoted as R_n, formed by combining all the roots of the equation. He hoped to represent all the roots of an nth-degree polynomial equation using these functions. This method worked for cubic and quartic equations but failed for quintic equations. The construction of the function R_n seemed to lack a discernible pattern.

Abel proved that polynomial equations of degree four or higher cannot have algebraic solutions. However, he did not indicate which equations can be solved algebraically. For instance, Gauss had proven that binomial equations of the form $x^p = 1$ (where p is a prime number) have algebraic solutions. It became necessary to establish a general proof for determining which high-degree equations can have algebraic solutions. This task was brilliantly accomplished by Galois. The key idea of Galois' theory is that there exists a polynomial equation with coefficients from a certain field that has no solutions within that field. To obtain the remaining solutions, the coefficient field must be extended to a larger field called the field of roots. The form of the equation's solutions depends on the relationship between the coefficient field and the field of roots. Galois discovered that this relationship can be expressed using symmetry groups. Permutation operations on the field of roots, forming a permutation group, are capable of preserving the coefficient field. These operations constitute a subgroup known as the Galois group. A polynomial equation belonging to a particular coefficient field may have solutions in an extension field, and the relationship between the coefficient field and the field of roots is characterized by the structure of the Galois group.

For a second-order ODE:

$$y'' + p(x)y' + q(x) = 0$$

its solutions are also subject to certain symmetry constraints, manifested in the Wronskian determinant:

$$W(y_1, y_2; x) = W(y_1, y_2; c) \exp \int_c^x p(\xi) \mathrm{d}\xi$$

where c is a constant. The relationship between the two solutions and the coefficient functions is given by

$$p(x) = \frac{y_1 y_2'' - y_2 y_1''}{W(y_1, y_2)}, \quad q(x) = \frac{y_1' y_2'' - y_2' y_1''}{W(y_1, y_2)}$$

For higher-order differential equations, this kind of symmetry is more subtle and not easily discovered by humans.

Exercises

[1] Find the series solution of the equation $y'' - x^2 y = 0$ in the neighborhood of $x = 0$.
Answer:

$$y_1(x) = \sum_{k=0}^{\infty} \frac{\Gamma(3/4)}{k!\Gamma(k + 3/4)} \left(\frac{x}{2}\right)^{4k}$$

$$y_2(x) = \sum_{k=0}^{\infty} \frac{\Gamma(5/4)}{k!\Gamma(k + 5/4)} \left(\frac{x}{2}\right)^{4k+1}$$

[2] Find two linearly independent solutions in the neighborhood of $x = 0$:

(a) $x(x+1)y'' - (x-1)y' + y = 0$; (b) $x(x-1)^2 y'' - 2y = 0$

Answer:

(a) $\begin{cases} y_1(x) = 1 - x \\ y_2(x) = (1-x)\ln x + 4 \end{cases}$; (b) $\begin{cases} y_1(x) = \dfrac{x}{1-x} \\ y_2(x) = \dfrac{x}{1-x}\ln x + \dfrac{1+x}{2} \end{cases}$

[3] Find the series solution of the confluent hypergeometric equation in the neighborhood of $x = 0$:

$$xy'' + (\gamma - x)y' - \alpha y = 0$$

[4] Find the series solution of the Legendre equation in the neighborhood of $x = +1$:

$$(1 - x^2)\frac{\mathrm{d}^2 y}{\mathrm{d}x^2} - 2x\frac{\mathrm{d}y}{\mathrm{d}x} + l(l+1)y = 0$$

Answer:

$$y(x) = a_0 \left[1 - \frac{\Gamma(l+2)}{\Gamma(l)} \cdot \frac{1}{(1!)^2}\left(\frac{1-x}{2}\right) + \frac{\Gamma(l+3)}{\Gamma(l-1)} \cdot \frac{1}{(2!)^2}\right.$$

$$\left. \times \left(\frac{1-x}{2}\right)^2 - \frac{\Gamma(l+4)}{\Gamma(l-2)} \cdot \frac{1}{(3!)^2}\left(\frac{1-x}{2}\right)^3 + \cdots\right]$$

[5] Given that $y_1(x) = x$ is a solution of the differential equation:

$$x^2 y'' + xy' - y = 0$$

find its general solution.
Answer: $y(x) = c_1 x + c_2 x^{-1}$

[6] Given the Bessel equation with two series solutions $\mathrm{J}_\nu(x)$ and $\mathrm{J}_{-\nu}(x)$,

(a) Prove that the Wronskian determinant is

$$W[\mathrm{J}_\nu(x), \mathrm{J}_{-\nu}(x); x] = \frac{A}{x}$$

(b) Compute $A = -\frac{2}{\pi}\sin\nu\pi$ using the series representation of Bessel functions and prove that $\mathrm{J}_\nu(x)$ and $\mathrm{J}_{-\nu}(x)$ are linearly independent when ν is not an integer, while they are linearly dependent when ν is an integer.

(c) Introduce the linear superposition function $\mathrm{N}_\nu(x) = c_1\mathrm{J}_\nu(x) + c_2\mathrm{J}_{-\nu}(x)$ and prove that

$$W[\mathrm{J}_\nu(x), \mathrm{N}_\nu(x); x] = -\frac{2c_2}{\pi x}\sin\nu\pi$$

When taking $c_2 = -\frac{1}{\sin\nu\pi}$, one has $W[\mathrm{J}_\nu(x), \mathrm{N}_\nu(x); x] = \frac{2}{\pi x} \neq 0$. Hence, $\mathrm{J}_\nu(x)$ and $\mathrm{N}_\nu(x)$ are linearly independent.
Hint: We need only consider the first term of the Bessel series and make use of the formula: $\Gamma(\nu)\Gamma(1-\nu) = \frac{\pi}{\sin\nu\pi}$.

[7] For the Bessel equation, given $J_\nu(x)$ as the first solution, prove that

(a) The second solution can be expressed as

$$\mathrm{N}_\nu(x) = \mathrm{J}_\nu(x) \int \frac{1}{\xi[\mathrm{J}_\nu(\xi)]^2} \mathrm{d}\xi$$

(b) For the zeroth-order Bessel function, $\mathrm{J}_0(x) = 1 - \frac{x^2}{4} + \frac{x^4}{64} - \mathcal{O}(x^6)$, find $\mathrm{N}_0(x)$ in the series representation:

$$\mathrm{N}_0(x) = \mathrm{J}_0(x)\left[\ln x + \frac{x^2}{4} + \frac{5x^4}{128} + \cdots\right]$$

[8] Solve the non-homogeneous differential equation:

$$(1-x)y'' + xy' - y = (1-x)^2$$

Hint: Find one solution to the homogeneous equation, $y = x$, and then find the second solution and a particular solution.
Answer: $y = Ax + B\mathrm{e}^x + x^2 + 1$

[9] For a non-homogeneous second-order constant coefficient ODE,

$$\frac{\mathrm{d}^2 y}{\mathrm{d}x^2} + a\frac{\mathrm{d}y}{\mathrm{d}x} + by = r(x)$$

where a and b are constants, prove that a particular solution is given by

$$\begin{aligned}\tilde{y}(x) = &\frac{1}{2\sqrt{p}}\mathrm{e}^{\left(\sqrt{p}-\frac{a}{2}\right)x} \int^{(x)} r(\xi)\mathrm{e}^{-\left(\sqrt{p}-\frac{a}{2}\right)\xi}\mathrm{d}\xi \\ &- \frac{1}{2\sqrt{p}}\mathrm{e}^{-\left(\sqrt{p}+\frac{a}{2}\right)x} \int^{(x)} r(\xi)\mathrm{e}^{\left(\sqrt{p}+\frac{a}{2}\right)\xi}\mathrm{d}\xi\end{aligned}$$

where $p = \frac{a^2}{4} - b$.
Hint: Use the transformation $y(x) = \mathrm{e}^{-\frac{1}{2}ax}u(x)$ to eliminate the first-order derivative term and rewrite the equation as

$$\frac{\mathrm{d}^2 u}{\mathrm{d}x^2} - pu = \mathrm{e}^{\frac{1}{2}ax}r(x)$$

[10] Prove that the two non-zero solutions y_1 and y_2 of a second-order homogeneous ODE are linearly dependent if and only if the Wronskian determinant is zero, i.e.,

$$W(y_1, y_2; x) = 0$$

[11] Consider the second-order ODE:

$$x^2y'' + (x+1)y' - y = 0$$

(a) Prove that $x = 0$ is an irregular singular point.
(b) Given that $y_1(x) = x + 1$ is a series solution, find the second solution and verify that it cannot be expanded as a Frobenius series.
Answer: $y_2(x) = x\mathrm{e}^{\frac{1}{x}}$

[12] Find a particular solution for the following differential equations:

$$\text{(a) } y'' + 2y' + y = \mathrm{e}^{-x}\ln x; \quad \text{(b) } y'' + 2y' + 5y = \mathrm{e}^{-x}\sec 2x$$

Answer:

$$\text{(a) } \tilde{y} = \frac{1}{2}x^2\mathrm{e}^{-x}\ln x - \frac{3}{4}x^2\mathrm{e}^{-x};$$

$$\text{(b) } \tilde{y} = \frac{1}{2}x\mathrm{e}^{-x}\sin 2x + \frac{1}{4}\mathrm{e}^{-x}\cos 2x\ln(\cos 2x)$$

9.3 Constant Coefficient Partial Differential Equations

In this section, we consider PDEs with two independent variables (x, y) as follows:

$$a_0\frac{\partial^n u}{\partial x^n} + a_1\frac{\partial^n u}{\partial x^{n-1}\partial y} + \cdots + a_n\frac{\partial^n u}{\partial y^n} + b_0\frac{\partial^{n-1} u}{\partial x^{n-1}} + b_1\frac{\partial^{n-1} u}{\partial x^{n-2}\partial y}$$
$$+ \cdots + b_{n-1}\frac{\partial^{n-1} u}{\partial y^{n-1}} + \cdots + c\frac{\partial u}{\partial x} + d\frac{\partial u}{\partial y} + fu = r(x, y)$$

Generally, the coefficients $a_0, a_1, \ldots, b_0, b_1, \ldots, c, d, f$ are functions of the independent variables (x, y). Introducing the operator symbols:

$$\hat{\mathrm{D}}_x = \frac{\partial}{\partial x}, \quad \hat{\mathrm{D}}_y = \frac{\partial}{\partial y}$$

The PDEs can be expressed as

$$\hat{L}(\hat{\mathrm{D}}_x, \hat{\mathrm{D}}_y)u = r(x, y)$$

with differential operator

$$\hat{L}(\hat{\mathrm{D}}_x, \hat{\mathrm{D}}_y) \equiv a_0\hat{\mathrm{D}}_x^n + a_1\hat{\mathrm{D}}_x^{n-1}\hat{\mathrm{D}}_y + \cdots + a_n\hat{\mathrm{D}}_y^n + b_0\hat{\mathrm{D}}_x^{n-1} + b_1\hat{\mathrm{D}}_x^{n-2}\hat{\mathrm{D}}_y + \cdots + b_{n-1}\hat{\mathrm{D}}_y^{n-1} + \cdots + c\hat{\mathrm{D}}_x + d\hat{\mathrm{D}}_y + f$$

In general, PDEs do not have universally applicable solution methods. Below, we provide some general solutions specifically for PDEs with constant coefficients.

9.3.1 *Homogeneous partial differential equations*

Consider the homogeneous PDE with constant coefficients: $r(x, y) = 0$.

(1) $\hat{L}(\hat{\mathrm{D}}_x, \hat{\mathrm{D}}_y)$ is a homogeneous operator in terms of $\hat{\mathrm{D}}_x$ and $\hat{\mathrm{D}}_y$. Suppose

$$\hat{L}(\hat{\mathrm{D}}_x, \hat{\mathrm{D}}_y)u = [a_0\hat{\mathrm{D}}_x^n + a_1\hat{\mathrm{D}}_x^{n-1}\hat{\mathrm{D}}_y + \cdots + a_n\hat{\mathrm{D}}_y^n]u = 0$$

Assuming the solution of the equation has the form $u = \phi(y + \alpha x)$, we have

$$\hat{\mathrm{D}}_x^k u = \alpha^k \phi^{(k)}(y + \alpha x), \quad \hat{\mathrm{D}}_y^k u = \phi^{(k)}(y + \alpha x)$$
$$\hat{\mathrm{D}}_x^r \hat{\mathrm{D}}_y^s u = \alpha^r \phi^{(r+s)}(y + \alpha x)$$

Substituting these into the equation, we get

$$(a_0\alpha^n + a_1\alpha^{n-1} + \cdots + a_n)\phi^{(n)}(y + \alpha x) = 0$$

If the characteristic equation

$$a_0\alpha^n + a_1\alpha^{n-1} + \cdots + a_n = 0$$

has n distinct roots $(\alpha_1, \alpha_2, \ldots, \alpha_n)$, then the solution of the equation is

$$u = \phi_1(y + \alpha_1 x) + \phi_2(y + \alpha_2 x) + \cdots + \phi_n(y + \alpha_n x)$$

where ϕ_j $(j = 1, 2, \ldots, n)$ are any arbitrary functions. If α_j is a root of multiplicity k, then the solution of the PDE is

$$u = \phi_{j,1}(y + \alpha_j x) + x\phi_{j,2}(y + \alpha_j x) + \cdots + x^{k-1}\phi_{j,k}(y + \alpha_j x)$$

Example 9.11. Find the general solution of the equation:

$$[\hat{\mathrm{D}}_x^2 - 2\hat{\mathrm{D}}_x\hat{\mathrm{D}}_y + \hat{\mathrm{D}}_y^2]u = 0$$

Solution: The equation is a homogeneous PDE. From the characteristic equation,

$$\alpha^2 - 2\alpha + 1 = 0$$

we obtain a double root $\alpha = 1$. Thus, the solution is

$$u = \psi(x + y) + x\phi(x + y)$$

The specific forms of the unknown functions $\psi(x, y)$ and $\phi(x, y)$ need to be determined by the boundary or initial conditions of the equation, which will be discussed separately.

(2) $\hat{L}(\hat{\mathrm{D}}_x, \hat{\mathrm{D}}_y)$ is not a homogeneous operator in terms of $\hat{\mathrm{D}}_x$ and $\hat{\mathrm{D}}_y$.
Let us consider a first-order PDE:

$$[\hat{\mathrm{D}}_x - \alpha\hat{\mathrm{D}}_y - \beta]u = 0$$

When $\beta = 0$, the solution of the equation is

$$u = \phi(y + \alpha x)$$

When $\beta \neq 0$, we assume the solution has the form $u = g(x)\phi(y+\alpha x)$. Substituting this into the equation, we can solve for $g(x)$ to obtain $g(x) = \mathrm{e}^{\beta x}$. Therefore, the solution of the equation is

$$u = \mathrm{e}^{\beta x}\phi(y + \alpha x)$$

In general, higher-order PDEs can be factored into first-order equations in terms of $\hat{\mathrm{D}}_x$ and $\hat{\mathrm{D}}_y$, allowing us to find their general solutions. If the equation has multiple factors, such as

$$[\hat{\mathrm{D}}_x - \alpha\hat{\mathrm{D}}_y - \beta]^2 u = 0$$

then the general solution is

$$u = \mathrm{e}^{\beta x}\phi(y + \alpha x) + x\mathrm{e}^{\beta x}\psi(y + \alpha x)$$

Example 9.12. Find the general solution of the PDE:

$$\frac{\partial^2 u}{\partial x^2} - \frac{\partial^2 u}{\partial x \partial y} - 2\frac{\partial^2 u}{\partial y^2} + 2\frac{\partial u}{\partial x} + 2\frac{\partial u}{\partial y} = 0$$

Solution: We factorize the equation as

$$[\hat{\mathrm{D}}_x^2 - \hat{\mathrm{D}}_x\hat{\mathrm{D}}_y - 2\hat{\mathrm{D}}_y^2 + 2\hat{\mathrm{D}}_x + 2\hat{\mathrm{D}}_y]u = (\hat{\mathrm{D}}_x + \hat{\mathrm{D}}_y)(\hat{\mathrm{D}}_x - 2\hat{\mathrm{D}}_y + 2)u = 0$$

The general solution of the equation is

$$u = \phi(x - y) + \mathrm{e}^{-2x}\psi(y + 2x)$$

9.3.2 *Non-homogeneous partial differential equations*

For a non-homogeneous PDE with $r(x, y) \neq 0$,

$$\hat{L}(\hat{\mathrm{D}}_x, \hat{\mathrm{D}}_y)u = r(x, y)$$

We can first find the general solution of the corresponding homogeneous equation and then look for a particular solution. The particular solution can be expressed in the form:

$$\tilde{u}(x, y) = \frac{1}{\hat{L}(\hat{\mathrm{D}}_x, \hat{\mathrm{D}}_y)} r(x, y)$$

As an example, let us discuss a few special non-homogeneous terms.
(1) Exponential form:

$$r(x, y) = \mathrm{e}^{ax+by}$$

Since

$$\hat{\mathrm{D}}_x \mathrm{e}^{ax+by} = a\mathrm{e}^{ax+by}, \quad \hat{\mathrm{D}}_y \mathrm{e}^{ax+by} = b\mathrm{e}^{ax+by}$$

Therefore

$$\frac{1}{\hat{L}(\hat{\mathrm{D}}_x, \hat{\mathrm{D}}_y)} r(x, y) = \frac{1}{\hat{L}(a, b)} r(x, y)$$

Example 9.13. Find the general solution of the equation:

$$(\hat{\mathrm{D}}_x + \hat{\mathrm{D}}_y)(2\hat{\mathrm{D}}_x - 3\hat{\mathrm{D}}_y)u = 5\mathrm{e}^{x-y}$$

Solution: The general solution of the homogeneous PDE is

$$u = \phi(y - x) + \psi(2y + 3x)$$

For the particular solution of the non-homogeneous PDE, we have

$$\tilde{u}(x,y) = \frac{5}{(\hat{\mathrm{D}}_x + \hat{\mathrm{D}}_y)(2\hat{\mathrm{D}}_x - 3\hat{\mathrm{D}}_y)}\mathrm{e}^{x-y}$$
$$= \frac{5}{(\hat{\mathrm{D}}_x + \hat{\mathrm{D}}_y)[2 - 3(-1)]}\mathrm{e}^{x-y}$$

Therefore

$$(\hat{\mathrm{D}}_x + \hat{\mathrm{D}}_y)\tilde{u}(x,y) = \mathrm{e}^{x-y} \longrightarrow \tilde{u}(x,y) = x\mathrm{e}^{x-y}$$

The general solution of the PDE is

$$u(x,y) = \phi(y-x) + \psi(2y+3x) + x\mathrm{e}^{x-y}$$

(2) Polynomial form

$$r(x,y) = x^m y^n$$

Example 9.14. Find the general solution of the equation:

$$[\hat{\mathrm{D}}_x^2 - 2\hat{\mathrm{D}}_x\hat{\mathrm{D}}_y + \hat{\mathrm{D}}_y^2]u = 12xy$$

Solution: The general solution of the homogeneous PDE is

$$u = \psi(x+y) + x\phi(x+y)$$

For the particular solution of the non-homogeneous equation, we can choose

$$\tilde{u}(x,y) = \frac{12}{\hat{\mathrm{D}}_x^2 - 2\hat{\mathrm{D}}_x\hat{\mathrm{D}}_y + \hat{\mathrm{D}}_y^2}xy = \frac{12}{(\hat{\mathrm{D}}_x - \hat{\mathrm{D}}_y)^2}xy$$
$$= \frac{12}{\hat{\mathrm{D}}_x^2}\left(1 - \frac{\hat{\mathrm{D}}_y}{\hat{\mathrm{D}}_x}\right)^{-2} xy = \frac{12}{\hat{\mathrm{D}}_x^2}\left(1 + 2\frac{\hat{\mathrm{D}}_y}{\hat{\mathrm{D}}_x} + \cdots\right)xy$$
$$= \frac{12}{\hat{\mathrm{D}}_x^2}\left(xy + \frac{2}{\hat{\mathrm{D}}_x}x\right) = 12\left(y\frac{1}{\hat{\mathrm{D}}_x^2}x + \frac{2}{\hat{\mathrm{D}}_x^3}x\right)$$
$$= 12\left(\frac{1}{6}x^3y + \frac{1}{12}x^4\right) = x^4 + 2x^3y$$

Therefore, the general solution of the PDE is

$$u(x, y) = \psi(x + y) + x\phi(x + y) + x^4 + 2x^3y$$

In this example, the non-homogeneous term is a polynomial, so we can guess that the particular solution is also a fourth-order polynomial. By substituting it into the equation, we can determine the coefficients of the polynomial.

ENDNOTE

In mathematics, there are two fundamental types of equations: algebraic equations, which have been extensively studied by Abel and Galois, and differential equations, which originate from physics and describe how physical quantities change with respect to time, providing information about their rates of change. Differential equations arise in physics to describe planetary motion, wave propagation, fluid flow, heat transfer, electromagnetic interactions, and more. Newton realized that by focusing on the rates of change of physical quantities rather than the quantities themselves, the laws of nature can be described concisely.

Is there a theory of differential equations similar to Galois theory for algebraic equations? Is there a way to determine if a differential equation can be solved using specific methods? Norwegian mathematician Sophus Lie recognized that once a solution of a differential equation is obtained, it is possible to apply certain transformations to generate other forms of solutions. Multiple solutions obtained from a single solution are linked by a transformation group, determined by the symmetry of the differential equation.

This clue suggests the existence of a hidden theory waiting to be discovered. Recall the impact of Galois on symmetry in the context of algebraic equations. What would happen if a similar approach were applied to differential equations? Galois's research focused on discontinuous finite groups, such as the group of all permutations of the five roots of a fifth-degree equation, which has 120 elements. The simplest infinite group is the symmetry group of a circle, which contains all continuous rotational transformations and is denoted as $U(1)$. When two successive rotations occur, the angles are simply added together. Since only one independent variable is required to determine an angle, the $U(1)$ group is a one-dimensional continuous group known as a Lie group.

A Lie group is both a group and a manifold or multidimensional space. The structure of the group can be linearized, meaning that the curved manifold of a Lie group can be replaced with a flat Euclidean space. The linearized group structure gives the space an algebraic structure that describes the behavior of infinitesimal transformations. This structure is called the Lie algebra of the group, and it has the same dimension as the group manifold but is much simpler.

At the beginning of the 20th century, a theory of differential equations analogous to Galois theory was finally established. The theory of Lie groups and Lie algebras became more than just a tool for determining solvability of differential equations; it permeated almost every branch of mathematics. The root of this development lies in symmetry. Today, symmetry is deeply ingrained in all areas of mathematics and serves as the foundation for many physical concepts. Symmetry describes the laws hidden in the world, and continuous symmetries such as rotations are closely related to the properties of space, time, and matter, hinting at the existence of various conservation laws. The connection between symmetry and conservation laws was discovered by Emmy Noether.

When solving inhomogeneous PDEs, the differential operators are treated as algebraic variables for multiplication and division operations, and even for Taylor series expansions. This method was invented by Oliver Heaviside, but its essence remains somewhat elusive. Heaviside could not provide a theoretical explanation for it; he simply relied on physical intuition. One can assume that by applying Fourier transforms or Laplace transforms to differential equations, transforming them into algebraic equations in an image space, solving algebraically, and then transforming back to the original space, the rationale behind this method can be roughly explained.

Exercises

[1] Find the general solution to the homogeneous PDEs:

$$\text{(a) } [\hat{\mathrm{D}}_x^2 - 2\hat{\mathrm{D}}_x\hat{\mathrm{D}}_y - 3\hat{\mathrm{D}}_y^2]u = 0; \quad \text{(b) } [\hat{\mathrm{D}}_x^2 - 2\hat{\mathrm{D}}_x\hat{\mathrm{D}}_y + 2\hat{\mathrm{D}}_y^2]u = 0$$

Answer:

$$\begin{aligned}&\text{(a) } u = \phi(x-y) + \psi(3x+y);\\&\text{(b) } u = \phi((1+\mathrm{i})x - y) + \psi((1-\mathrm{i})x + y)\end{aligned}$$

[2] Find the general solution to the non-homogeneous PDEs:

$$\text{(a) } [\hat{\mathrm{D}}_x^2 - 6\hat{\mathrm{D}}_x\hat{\mathrm{D}}_y + 9\hat{\mathrm{D}}_y^2]u = 6x+2y; \quad \text{(b) } [\hat{\mathrm{D}}_x^2 + \hat{\mathrm{D}}_y^2]u = 12(x+y)$$

Answer:

$$\begin{aligned}&\text{(a) } u = x^2(y+3x) + x\phi(y+3x) + \psi(y+3x)\\&\text{(b) } u = 2x^3 + 2y^3 + \phi(x+\mathrm{i}y) + \psi(x-\mathrm{i}y)\end{aligned}$$

[3] Solve the PDE:

$$[x^2\hat{\mathrm{D}}_x^2 - 2xy\hat{\mathrm{D}}_x\hat{\mathrm{D}}_y + y^2\hat{\mathrm{D}}_y^2 + x\hat{\mathrm{D}}_x + y\hat{\mathrm{D}}_y]u = 0$$

Hint: Let $x = \mathrm{e}^t$ and $y = \mathrm{e}^s$, and assume a trial solution $u = f(\alpha t + s)$.
Answer: $u = f_1(\ln xy) + \ln x f_2(\ln xy)$

[4] For a constant coefficient PDE with non-homogeneous term $r(x,y) = \mathrm{e}^{ax+by}g(x,y)$, prove that

$$\frac{1}{\hat{L}(\hat{\mathrm{D}}_x, \hat{\mathrm{D}}_y)} r(x,y) = \mathrm{e}^{ax+by} \frac{1}{\hat{L}(\hat{\mathrm{D}}_x + a, \hat{\mathrm{D}}_y + b)} g(x,y)$$

9.4 Nonlinear Equations

A nonlinear differential equation refers to an equation that contains nonlinear terms of the unknown function or its differentials. For nonlinear equations, properties such as uniqueness, singularity, and finiteness of solutions do not generally exist. The principle of superposition fails to apply, and therefore, there is no general solution method. Specific methods need to be employed based on the particular problem. Nonlinear equations that can be rigorously solved are very limited, and in some cases, stable solitary wave solutions exist. This section only briefly introduces solitary wave solutions for several types of nonlinear equations with constant coefficients.

9.4.1 *Wave dispersion*

Waves propagate at a certain velocity while maintaining their shape during propagation. Consider the simplest linear wave equation:

$$u_t + au_x = 0$$

Its general solution is $f(x - at)$. During the propagation of the wave, the waveform remains unchanged and propagates at a constant velocity.

For slightly more complex linear wave equations:

$$u_t + u_x + u_{xxx} = 0$$

The plane wave solution is

$$u(xt) = \mathrm{e}^{\mathrm{i}(kx-\omega t)}$$

Substituting this into the equation yields $\omega = k - k^3$, which indicates that the wave propagation velocity depends on the wave vector k:

$$c = \frac{\omega}{k} = 1 - k^2$$

This is known as wave dispersion.

For nonlinear wave equations:

$$u_t + u_x + uu_x = 0$$

The solution can still be expressed in the following form:

$$u(xt) = f[x - (1 + u)t]$$

In this case, the wave propagation velocity will depend on the wave displacement, meaning that the waveform will change during propagation. However, for certain special nonlinear equations, there also exists a class of special traveling wave solutions where both the waveform and propagation velocity remain constant during propagation. The following section will introduce the typical example of a solitary wave.

9.4.2 *First-order nonlinear equations*

Here, we will only discuss the solution methods for two simple first-order nonlinear differential equations.

(1) *Bernoulli equation*

The Bernoulli equation is given by

$$y'(x) = p(x)y(x) + q(x)[y(x)]^n \quad (n \neq 0, 1)$$

It is a homogeneous nonlinear equation. By making the variable substitution $u(x) = [y(x)]^{1-n}$, we can transform it into a first-order linear differential equation:

$$u'(x) = (1-n)[p(x)u(x) + q(x)]$$

This equation can be solved by first obtaining the general solution of the homogeneous equation and then using the method of variation of parameters to find a particular solution. Details are omitted here.

(2) *Riccati equation*

The Riccati equation is a second-degree nonlinear equation for $y(x)$:

$$y'(x) = p(x)[y(x)]^2 + q(x)y(x) + r(x)$$

In fact, there is no general method for solving the Riccati equation. However, if a particular solution $\tilde{y}(x)$ of the equation can be guessed, a more general solution can be written as $y(x) = \tilde{y}(x) + u(x)$, where $u(x)$ satisfies the Bernoulli equation:

$$u'(x) = [2p(x)\tilde{y}(x) + q(x)]u(x) + p(x)[u(x)]^2$$

9.4.3 *Solitary wave solutions*

These types of solutions are typically localized within a finite spatial range and exhibit stable shapes and definite propagation velocities. When they encounter each other, they pass through each other unaffected, similar to classical particles. Therefore, they are also referred to as soliton solutions. Here, we provide a few examples of solitary wave solutions for common nonlinear equations.

(1) *KdV equation*

The Korteweg–de Vries (KdV) equation is a model for water surface solitary waves, in its simplified form given as

$$u_\tau + u_\xi + 12uu_\xi + u_{\xi\xi\xi} = 0 \tag{9.4.1}$$

We aim to find solitary wave solutions that resemble traveling waves. Let us make the transformation $\theta = a\xi - \omega\tau + \delta$, where a and δ are constants. We require that as $|\theta| \to \infty$, both u and its derivatives tend to zero. By setting $\omega = a + a^3$, the equation becomes

$$u_{\theta\theta\theta} + \frac{12}{a^2}uu_\theta - u_\theta = 0$$

This simplifies to

$$u_{\theta\theta} + \frac{6}{a^2}u^2 - u + c_1 = 0$$

Integrating this equation yields

$$\frac{1}{2}u_\theta^2 + \frac{2}{a^2}u^3 - \frac{1}{2}u^2 + c_1u + c_2 = 0$$

From the asymptotic stability, we find that $c_1 = c_2 = 0$. Hence, we have

$$u_\theta^2 = u^2\left(1 - \frac{4}{a^2}u\right)$$

This leads to

$$\frac{a\mathrm{d}u}{u\sqrt{a^2 - 4u}} = \mathrm{d}\theta$$

Integrating further, we obtain

$$\theta = \ln\frac{a - \sqrt{a^2 - 4u}}{a + \sqrt{a^2 - 4u}}$$

This results in the solitary wave solution:

$$u = \frac{a^2\mathrm{e}^\theta}{(\mathrm{e}^\theta + 1)^2} = \frac{1}{4}a^2\sec^2\frac{a}{2}\left[\xi - (1 + a^2)\tau + \frac{\delta}{a}\right]$$

This is a solitary wave packet that only protrudes within a small range. It moves forward at a velocity of $v = 1 + a^2$, while maintaining its shape and amplitude. This type of solitary wave is sometimes referred to as a bright soliton. The speed of the wave packet depends on its amplitude, with higher crests moving faster.

(2) *Sine-Gordon equation*

The sine-Gordon equation is given by

$$\phi_{xx} - \phi_{tt} = \sin\phi \tag{9.4.2}$$

We make a coordinate change by

$$\xi = \frac{x-t}{2}, \quad \tau = \frac{x+t}{2}$$

The equation becomes

$$\phi_{\xi\tau} = \sin\phi$$

Let's assume that the equation has two solutions represented by two functions:

$$\phi = u + v, \quad \rho = u - v$$

Considering the equation expressed in terms of the variables $u(\xi,\tau)$ and $v(\xi,\tau)$:

$$\begin{cases} u_\xi = f(v) \\ v_\tau = g(u) \end{cases} \longrightarrow \begin{cases} u_{\xi\tau} = g(u)f'(v) \\ v_{\xi\tau} = g'(u)f(v) \end{cases}$$

Thus, we have

$$\begin{aligned} &\phi_{\xi\tau} = (u+v)_{\xi\tau} = g(u)f'(v) + g'(u)f(v) \\ &\sin\phi = \sin(u+v) = \sin u\cos v + \cos u\sin v \end{aligned}$$

Choosing

$$g(u)f'(v) = \sin u\cos v$$

which leads to

$$\frac{g(u)}{\sin u} = \frac{\cos v}{f'(v)} = \alpha$$

We obtain

$$\begin{cases} v_\tau = g(u) = \alpha\sin u \\ u_\xi = f(v) = \dfrac{1}{\alpha}\sin v \end{cases}$$

Thus

$$\begin{cases} \frac{1}{2}(\phi+\rho)_\xi = \frac{1}{\alpha}\sin\frac{1}{2}(\phi-\rho) \\ \frac{1}{2}(\phi-\rho)_\tau = \alpha\sin\frac{1}{2}(\phi+\rho) \end{cases}$$

Considering a particular solution $\rho = 0$, the other solution ϕ satisfies the equations:

$$\phi_\xi = \frac{2}{\alpha}\sin\frac{\phi}{2}, \quad \phi_\tau = 2\alpha\sin\frac{\phi}{2}$$

Solving these equations, we obtain

$$\phi = 4\arctan\left\{\exp\left[\frac{1}{\alpha}\xi + c(\tau)\right]\right\} = 4\arctan[e^{a(x-bt)+\delta}]$$

where

$$a = \frac{1}{2}\left(\alpha + \frac{1}{\alpha}\right), \quad b = \frac{1-\alpha^2}{1+\alpha^2}, \quad c(\tau) = \alpha\tau + \delta$$

This solution is called the kink solution of the sine-Gordon equation, which is also a type of solitary wave solution, as shown in Figure 9.1. It is sometimes referred to as a dark soliton solution. It is worth noting that the wave speed b depends on the initial wave shape.

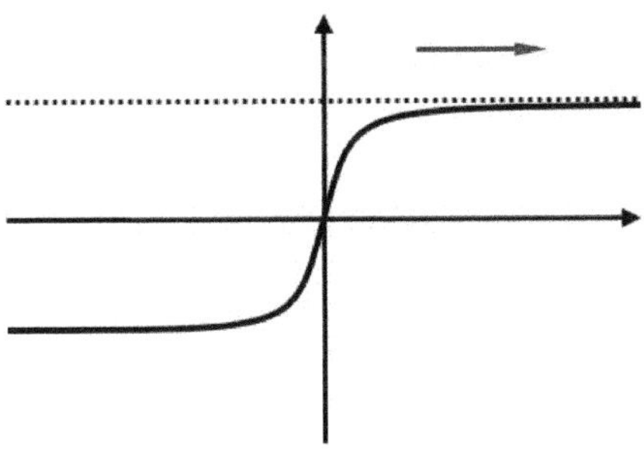

Figure 9.1.

Example 9.15. Find the traveling solution of the sine-Gordon equation:

Solution: Making substitution

$$\xi = x - at, \quad \phi(xt) = \phi(\xi)$$

The equation is changed into

$$(a^2 - 1)\frac{\mathrm{d}^2\phi}{\mathrm{d}\xi^2} + \sin\phi = 0$$

Multiplying $\frac{\mathrm{d}\phi}{\mathrm{d}\xi}$ for both sides and integrating, we obtain

$$\frac{1}{2}(a^2 - 1)\left(\frac{\mathrm{d}\phi}{\mathrm{d}\xi}\right)^2 + c - \cos\phi = 0$$

Hence

$$\frac{\mathrm{d}\phi}{\mathrm{d}\xi} = \sqrt{\frac{2[\cos\phi(\xi) - c]}{a^2 - 1}}$$

Take $c = 1$, one has

$$\xi - \xi_0 = \sqrt{1 - a^2}\ln\left[\tan\frac{\phi(\xi)}{4}\right]$$

We finally obtain the traveling solitary wave solution:

$$\phi(\xi) = 4\arctan\left[\exp\frac{\xi - \xi_0}{\sqrt{1 - a^2}}\right] = 4\arctan\left[\exp\frac{x - at}{\sqrt{1 - a^2}}\right]$$

According to the inverse tangent function form of the kink solution, we conjecture the following solution:

$$\phi(x, t) = \arctan\left[\frac{T(t)}{X(x)}\right]$$

By substituting it into the sine-Gordon equation, we obtain a new class of soliton solutions:

$$\phi(x, t) = 4\arctan\left[\sqrt{\frac{1 - \beta}{\beta}}\frac{\sin\sqrt{\beta}t}{\cosh\sqrt{1 - \beta}x}\right] \quad (0 < \beta < 1)$$

Where β is an integration constant. Since this solution does not propagate in time but oscillates in place, it is called a "breather soliton".

(3) *Nonlinear Schrödinger equation*

The nonlinear Schrödinger equation is given by

$$i\phi_t + \phi_{xx} + \beta\phi|\phi|^2 = 0 \tag{9.4.3}$$

We are looking for asymptotically stable traveling wave solutions with amplitude modulation:

$$\phi = \mathrm{e}^{\mathrm{i}(kx-vt)}u(\theta), \quad \theta = x - bt$$

The equation becomes

$$u_{\theta\theta} + \mathrm{i}(2k-b)u_\theta + (v-k^2)u + \beta u^3 = 0$$

Taking $k = \frac{1}{2}b$ and $v = \frac{1}{4}b^2 - a^2$, the equation simplifies to

$$u_{\theta\theta} - a^2u - \beta u^3 = 0 \longrightarrow u_\theta^2 = a^2u^2 - \frac{1}{2}\beta u^4$$

Integrating, we obtain

$$\mathrm{d}\theta = \frac{\mathrm{d}u}{u\sqrt{a^2 - \frac{\beta u^2}{2}}} \longrightarrow \theta = -\frac{1}{a}\ln\frac{a + \sqrt{a^2 - \beta u^2/2}}{\sqrt{\beta/2}u}$$

Finally, the solution is rewritten as

$$u(\theta) = a\sqrt{\frac{2}{\beta}}\sec h(a\theta)$$

Consequently,

$$\phi(x,t) = a\sqrt{\frac{2}{\beta}}\exp\left\{\mathrm{i}\left[\frac{1}{2}bx - \left(\frac{1}{4}b^2 - a^2\right)t\right]\right\}\sec h[a(x-bt)]$$

9.4.4 *Elliptic equation solutions*

In general, elliptic equations have a nonlinear form:

$$(y')^2 = a_0 + a_1y + a_2y^2 + a_3y^3 + a_4y^4 \tag{9.4.4}$$

or

$$y'' = a_0 + a_1y + a_2y^2 + a_3y^3$$

Let us discuss several special types:

(1) **Type I**

$$(y')^2 = a + by^2 + cy^4 \tag{9.4.5}$$

or

$$y'' = by + 2cy^3$$

The solutions of this equation can be expressed using the Weierstrass elliptic function $\wp(x)$:

$$y = \sqrt{\frac{1}{c}\left[\wp(x - x_0; g_2, g_3) - \frac{b}{3}\right]}$$

where $g_2 = \frac{4}{3}(b^2 - 3ac)$ and $g_3 = \frac{4b}{27}(9ac - 2b^2)$. Specifically,

$$y'' = -\omega^2(1 + k^2)y + \frac{2\omega^2 k^2}{A^2}y^3 \longrightarrow y = A\,\mathrm{sn}(\omega x, k)$$

$$y'' = \omega^2(2k^2 - 1)y - \frac{2\omega^2 k^2}{A^2}y^3 \longrightarrow y = A\,\mathrm{cn}(\omega x, k)$$

$$y'' = \omega^2(2 - k^2)y - \frac{2\omega^2}{A^2}y^3 \longrightarrow y = A\,\mathrm{dn}(\omega x, k)$$

$$y'' = -\omega^2(1 + k^2)y + \frac{2\omega^2}{A^2}y^3 \longrightarrow y = \frac{A}{\mathrm{sn}(\omega x, k)}$$

$$y'' = \omega^2(2k^2 - 1)y + \frac{2\omega^2 k'^2}{A^2}y^3 \longrightarrow y = \frac{A}{\mathrm{cn}(\omega x, k)}$$

$$y'' = \omega^2(2 - k^2)y - \frac{2\omega^2 k'^2}{A^2}y^3 \longrightarrow y = \frac{A}{\mathrm{dn}(\omega x, k)}$$

and so on.

(2) **Type II**

$$(y')^2 = ay + by^2 + cy^3 \tag{9.4.6}$$

or

$$y'' = \frac{a}{2} + by + \frac{3c}{2}y^2$$

This equation can also be represented using the Weierstrass elliptic function:

$$y = -\frac{b}{3c} + \wp\left(\frac{\sqrt{c}}{2}(x - x_0); g_2, g_3\right)$$

where $g_2 = \frac{4}{3c^2}(b^2 - 3ac)$ and $g_3 = \frac{4b}{27c^3}(9ac - 2b^2)$. Specifically,

$$y'' = 2\omega^2 A - 4\omega^2(1 + k^2)y + \frac{6\omega^2 k^2}{A}y^2 \longrightarrow y = A\,\mathrm{sn}^2(\omega x, k)$$

$$y'' = 2\omega^2 A k'^2 + 4\omega^2(2k^2 - 1)y - \frac{6\omega^2 k^2}{A}y^2 \longrightarrow y = A\,\mathrm{cn}^2(\omega x, k)$$

$$y'' = -2\omega^2 A k'^2 + 4\omega^2(2 - k^2)y - \frac{2\omega^2}{A}y^2 \longrightarrow y = A\,\mathrm{dn}^2(\omega x, k)$$

$$y'' = 2\omega^2 A k^2 - 4\omega^2(1 + k^2)y + \frac{6\omega^2}{A}y^2 \longrightarrow y = \frac{A}{\mathrm{sn}^2(\omega x, k)}$$

$$y'' = -2\omega^2 A k^2 + 4\omega^2(2k^2 - 1)y + \frac{6\omega^2 k'^2}{A}y^2 \longrightarrow y = \frac{A}{\mathrm{cn}^2(\omega x, k)}$$

$$y'' = -2\omega^2 A + 4\omega^2(2 - k^2)y - \frac{6\omega^2 k'^2}{A}y^2 \longrightarrow y = \frac{A}{\mathrm{dn}^2(\omega x, k)}$$

and so on.

(3) **Type III**

$$(y')^2 = a + by + cy^2 + dy^3 \tag{9.4.7}$$

or

$$y'' = \frac{b}{2} + cy + \frac{3d}{2}y^2$$

This equation has solutions in terms of the Weierstrass elliptic function:

$$y = -\frac{c}{3d} + \frac{4}{d}\wp(x - x_0; g_2, g_3)$$

where $g_2 = \frac{1}{12}(c^2 - 3bd)$ and $g_3 = \frac{1}{432}(9bcd - 27ad^2 - 2c^3)$. Specifically,

$$y'^2 = B(y - \alpha)(y - \beta)(y - \gamma) \quad (B > 0,\ \alpha > \beta > \gamma)$$

$$\longrightarrow y = \gamma + (\beta - \gamma)sn^2(\sqrt{B(\alpha - \gamma)/4}x, k), \quad k = \sqrt{\frac{\beta - \gamma}{\alpha - \gamma}}$$

$$y'^2 = -B(y - \alpha)(y - \beta)(y - \gamma) \quad (B > 0, \alpha > \beta > \gamma)$$

$$\longrightarrow y = \beta + (\alpha - \beta)cn^2(\sqrt{B(\alpha - \gamma)/4}x, k), \quad k = \sqrt{\frac{\alpha - \beta}{\alpha - \gamma}}$$

and so on.

Example 9.16. Solve the Duffing equation:

$$\frac{\mathrm{d}^2 x}{\mathrm{d}t^2} + \omega^2 x + \varepsilon\beta_0^2 x^3 = 0$$

Solution: The equation has periodic solutions in the form of Jacobi elliptic functions. When $\varepsilon > 0$, the solution is given by

$$x(t) = A\,\mathrm{cn}(\omega x, k)$$

where

$$\omega^2 = \omega_0^2 + \varepsilon\beta_0^2 A^2, \quad k^2 = \frac{\varepsilon\beta_0^2 A^2}{2\omega^2}$$

The period of the Jacobi elliptic cosine function is $4\mathrm{K}(k)$. Therefore, the oscillation period of the Duffing equation depends on the amplitude and is given by

$$T = \frac{4\mathrm{K}(k)}{\omega} = \frac{2\pi}{\omega}\mathrm{F}\left(\frac{1}{2}, \frac{1}{2}, 1, k^2\right)$$

$$= \frac{2\pi}{\omega}\left[1 + \left(\frac{1}{2}\right)^2 k^2 + \left(\frac{3}{8}\right)^2 k^4 + \cdots\right]$$

When $\varepsilon < 0$, the solution is given by

$$x(t) = A\,\mathrm{sn}(\omega x, k)$$

where

$$\omega^2 = \omega_0^2 + \frac{1}{2}\varepsilon\beta_0^2 A^2, \quad k^2 = -\frac{1}{2}\frac{\varepsilon\beta_0^2 A^2}{2\omega^2}$$

The period of the Jacobi elliptic sine function is also $4\mathrm{K}(k)$.

Example 9.17. Determine the motion of the circular pendulum shown in Figure 9.2.

Solution: For a gravity pendulum with relatively large oscillation amplitude, the trajectory of the small ball forms an arc of a circle. The equation of motion for the pendulum is given by

$$m\frac{\mathrm{d}^2 s}{\mathrm{d}t^2} = -\frac{\partial V(s)}{\partial s}$$

where the gravitational potential $V(s) = mgy(s)$. As $y(s) = l(1 - \cos\theta)$, we introduce $\omega^2 = \frac{g}{l}$, yielding

$$\frac{\mathrm{d}^2\theta}{\mathrm{d}t^2} + \omega^2 \sin\theta = 0$$

This is a nonlinear equation. Expanding in a Taylor series around $\theta = 0$, we have

$$\frac{\mathrm{d}^2\theta}{\mathrm{d}t^2} + \omega^2\left(\theta - \frac{1}{3!}\theta^3 + \frac{1}{5!}\theta^5 - \cdots\right) = 0$$

When the amplitude of the oscillation is small, under the first-order linear approximation, we get the equation of motion for the simple pendulum as

$$\frac{\mathrm{d}^2\theta}{\mathrm{d}t^2} + \omega^2\theta = 0$$

The period of the simple pendulum is independent of the amplitude:

$$T = \frac{2\pi}{\omega} = \sqrt{\frac{l}{g}}$$

When the amplitude is large, we cannot solely rely on linear approximation, but the nonlinear equation is still solvable. By multiplying both sides of the equation by $\frac{\mathrm{d}\theta}{\mathrm{d}t}$ and then integrating, we obtain

$$\left(\frac{\mathrm{d}\theta}{\mathrm{d}t}\right)^2 = 2\omega^2\cos\theta + c$$

Assuming the pendulum starts at rest with $\theta = \theta_0$, we find $c = -2\omega^2 \cos\theta_0$. Integrating the equation, we get

$$t = \frac{1}{\sqrt{2}\omega} \int_0^{\theta_0 - \theta} \frac{\mathrm{d}\theta}{\sqrt{\cos\theta - \cos\theta_0}}$$

The integral can be solved with Jacobian elliptic function as

$$\sin\frac{\theta - \theta_0}{2} = \sin\frac{\theta_0}{2} \operatorname{sn}\omega t$$

This leads to the period of the circular pendulum as

$$T = \frac{2\sqrt{2}}{\omega} \int_0^{\theta_0} \frac{\mathrm{d}\theta}{\sqrt{\cos\theta - \cos\theta_0}} = \frac{4\mathrm{K}(k)}{\omega}$$

where $k = \sin\frac{\theta_0}{2}$, indicating that the period of the circular pendulum is dependent on the amplitude θ_0. As $\theta_0 \to \pi$, the oscillating period $T \to \infty$.

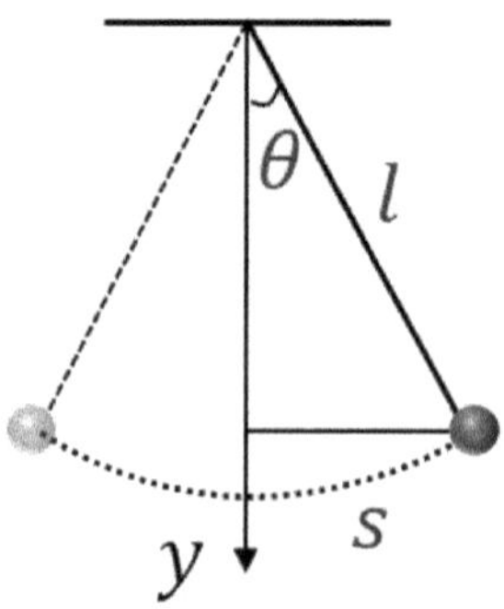

Figure 9.2.

Reflection: Besides the simple pendulum, what other types of pendulum have periods that are independent of the amplitude?

ENDNOTE

In the ocean, ships sometimes encounter a strange phenomenon: calm and windless waters suddenly rise with high waves and then disappear abruptly. These waves are called rogue waves, which appear and vanish without a trace. Their peak height is significantly higher than the surrounding background, posing a significant threat to ships. Here, we use the nonlinear Schrödinger equation as an example to demonstrate the peculiar solution of rogue waves:

$$i\frac{\partial \phi}{\partial t} + \frac{\partial^2 \phi}{\partial x^2} + 2g|\phi|^2\phi = 0$$

Assuming the equation has a rational fraction form solution:

$$\phi = \left(1 + \frac{A + 2B}{C}\right) \mathrm{e}^{it}$$

We choose

$$A = a_0 + a_1 x + a_2 t, \quad B = b_0 + b_1 x + b_2 t$$

and

$$C = c_0 + c_1(x - \alpha)^2 + c_2(t - \beta)^2$$

where c_1 and c_2 are non-zero constants. Substituting these into the equation and setting the coefficients of the cross terms $x^m t^n (m, n \geq 0)$ to zero, we obtain

$$\phi(xt) = \left\{1 - \frac{4[1 + 2i(t - \beta)]}{1 + 4(x - \alpha)^2 + 4(t - \beta)^2}\right\} \mathrm{e}^{it}$$

The center position is determined by the values of α and β. Figure 9.3 shows the spatiotemporal distribution $|\phi(x, t)|$ of this rogue wave solution, which can be seen as a type of spatiotemporal soliton.

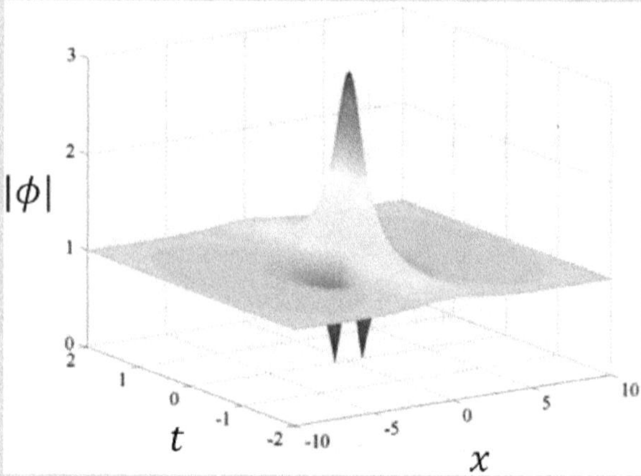

Figure 9.3.

The origin of rogue waves in the ocean is complex, and there is no definitive conclusion to date. It may be related to wind, currents, water wave diffraction, or nonlinear effects. Similar phenomena resembling rogue waves have also been observed in liquid helium and nonlinear optics.

There is a fundamental difference between singularities in linear and nonlinear differential equations. For example, in the nonlinear equation:

$$(x^2 - y)y' = x^2 + y^2$$

The curve $y = x^2$ is a set of singularities, making it impossible to construct solutions for the equation in the domain $[ab]$ because there will always be a y value that is undefined. Linear differential equations do not have this problem since the coefficients are only functions of x.

If a physical system follows a linear differential equation and initial conditions, its future behavior can be predicted. When the initial conditions undergo small changes, the solution of the equation will also undergo small changes, meaning that linear differential equations are continuous functions of their initial conditions. However, for nonlinear differential equations, even small changes in initial conditions will lead to completely different solutions. Since it is mathematically impossible to precisely determine the initial conditions in practical operations, nonlinear differential equations will result in unpredictable outcomes or the system entering a chaotic state, which refers to deterministic nonlinear dynamical systems that appear to be in a random state. Although there are densely packed nearly periodic orbits in certain parts of the phase space, forming attractors (Figure 9.4). The state of the system's motion is highly sensitive to initial conditions, exhibiting what is known as the "butterfly effect". M. Tabor and F. Calogero suggested interpreting chaos as motion on Riemann surfaces.

In the 1980s and 1990s, research on chaos phenomena was in full swing, with some enthusiasts even calling it the third pillar of modern physics, alongside relativity and quantum mechanics. However, as the excitement subsided, it was realized that chaos theory did not provide revolutionary insights into the physical world, like relativity or quantum mechanics did.

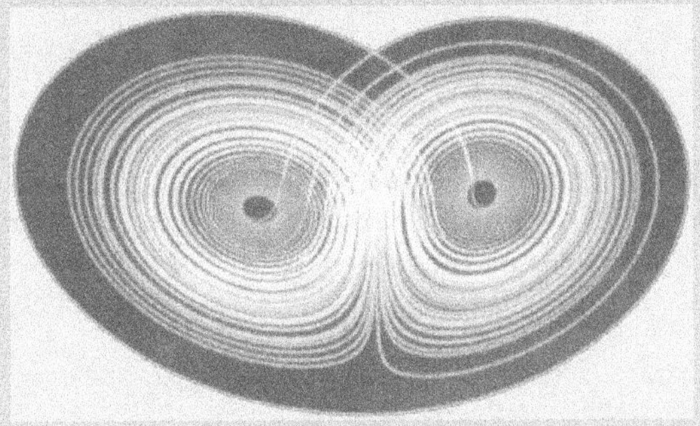

Figure 9.4.

Exercises

[1] Solve the Bernoulli equation:

$$y' + xy = xy^3$$

[2] Given that a particular solution of the Riccati equation is $\tilde{y} = 2$, find the general solution of the equation:

$$y' = y^2 - y - 2$$

[3] Find a traveling wave solution of the ϕ^4 equation:

$$\phi_{xx} - \phi_{tt} = \lambda\phi^3 - m^2\phi(\lambda > 0)$$

Answer:

$$\phi(x,t) = \pm\tanh\left[\frac{1}{\sqrt{2}}\frac{1}{\sqrt{1-v^2}}(x - vt) + \delta\right]$$

[4] Proof of the period of a circular pendulum:

$$T = \frac{2\sqrt{2}}{\omega}\int_0^{\theta_0}\frac{\mathrm{d}\theta}{\sqrt{\cos\theta - \cos\theta_0}} = \frac{4\mathrm{K}(k)}{\omega}$$

Hint: Make the variable substitution

$$\sin u = \frac{\sin\left(\frac{\theta}{2}\right)}{\sin\left(\frac{\theta_0}{2}\right)}$$

[5] Hamiltonian of a non-rigid pendulum:

$$H = \frac{\lambda}{2}z^2 - \sqrt{1 - z^2}\cos\phi$$

Solve for the motion of the pendulum.
Hint: The Hamiltonian $H \equiv H_0$ is a conserved quantity. The canonical equations of motion are

$$\dot{z} = -\frac{\partial H}{\partial \phi}, \quad \dot{\phi} = \frac{\partial H}{\partial z}$$

Answer:

$$z(t) = \begin{cases} A\,\mathrm{cn}[A\lambda k(t - t_0), k] & (0 < k < 1) \\ A\,\mathrm{dn}\left[A\lambda(t - t_0), \dfrac{1}{k}\right] & (k > 1) \end{cases}$$

$$k^2 = \frac{1}{2}\left[1 + \frac{\lambda H_0 - 1}{\sqrt{\lambda^2 + 1 - 2\lambda H_0}}\right]$$

Chapter 10

Differential Equations and Boundary Conditions

10.1 Mathematical Physics Equations

Physics is the science that studies the motion and changes of matter. Since the establishment of Newtonian mechanics and calculus, almost all mathematical descriptions of physical problems are presented in the form of differential equations (including integral equations).

Single-variable motion equations: These describe the relationship between physical quantities and a single independent variable, usually in the form of ordinary differential equations. Examples include Newton's equations of motion for particles, circuit equations, and so on. To obtain a specific trajectory or curve of change, it is necessary to know the initial values of the physical quantities at a given time, which are referred to as initial conditions of the equation.

Equations of motion for continuous media systems: In physics and engineering applications, there is another class of problems where physical quantities vary continuously in both time and space. These equations involve multiple independent variables, and thus are described by partial differential equations. Examples include the transmission equation of electromagnetic waves in space, the heat diffusion equation, Schrödinger's equation in quantum mechanics, and so on. To obtain an accurate solution for a specific system, it is not only necessary to know the initial values of the physical quantities at the initial time (initial conditions in terms of time), but also the distribution of the system on its boundaries, known as boundary conditions. In mathematics, the combination of initial and boundary conditions is referred to as the boundary value problem.

Taking the example of the motion of two heavy objects connected by a rope passing over a pulley: If the pulley is lightweight and the rope is weightless and inextensible, as shown in Figure 10.1, the entire system can be treated as a single body, and the equations of motion for the system can be derived using Newton's second law. However, if the rope has mass and is extensible, the problem becomes more complex. The motion state of each point on the rope, such as velocity, acceleration, and direction, varies with position and time. In this case, a differential analysis of each point on the rope is required to derive the corresponding equations of motion. This method is known as the method of elements.

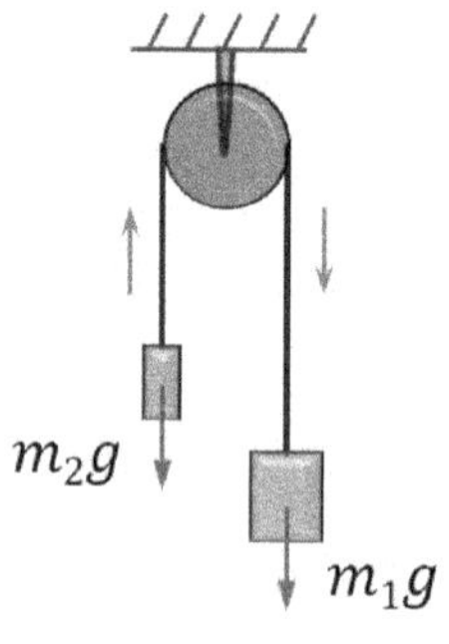

Figure 10.1.

The essence of the method of elements is to divide a continuous system in space into many interconnected small parts (elements). Each element is treated as a particle, and its equations of motion are analyzed based on the corresponding physical laws when it interacts with adjacent elements. To obtain the equations of motion for the elements, it is generally necessary to know the form of interaction between adjacent elements. This requires the use of fundamental physical laws, such as Hooke's law, the heat conduction law, and so on. In most cases, linear approximations can be used to describe these physical laws. After obtaining the equations of motion for the elements, taking a continuous limit allows us to derive the equations of motion for all points in the system, which are generally partial differential equations.

This section will derive several common mathematical physics equations, including the wave equation, transport equation, and steady-state field equations.

10.1.1 *Transverse vibration equation of a string*

When a uniform and soft string is stretched and subjected to small transverse vibrations near its equilibrium position, we seek to find the motion of each point on the string (see Figure 10.2).

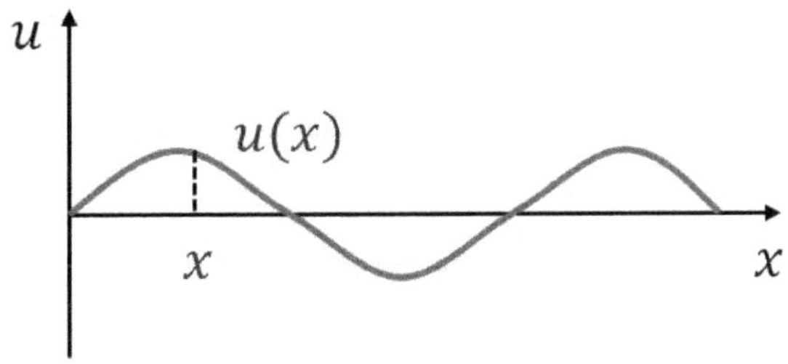

Figure 10.2.

To understand the relationship between the displacement $u(x, t)$ of each point on the string and time, we will approach the problem in several steps.

(1) *Simplification of model*

In order to simplify the problem, we make the following assumptions:

- **Soft string:** The tension in the string is always along the tangent direction, meaning there is no shear stress.
- **Light string:** The weight of the string itself can be neglected compared to the tension in the string.
- **Small amplitude vibrations:** Each point on the string only undergoes transverse vibrations near its equilibrium position, and the relative displacement $\mathrm{d}u = u(x + \mathrm{d}x) - u(x)$ between adjacent points x and $x + \mathrm{d}x$ is much smaller than the distance $\mathrm{d}x$.

(2) *Analysis of elements*

Let $u(x)$ and $u(x + \mathrm{d}x)$ be the displacements from the equilibrium position at x and $x + \mathrm{d}x$, respectively. The length of this segment of the string is given by

$$\mathrm{d}s = \sqrt{(\mathrm{d}x)^2 + (\mathrm{d}u)^2} \approx \mathrm{d}x$$

Considering $\mathrm{d}s$ as a particle, we can analyze its motion by applying Newton's second law. As shown in Figure 10.3, we can write the equation of motion for the element as follows:

$$\begin{aligned} T_2 \cos\alpha_2 - T_1 \cos\alpha_1 &= 0 \\ T_2 \sin\alpha_2 - T_1 \sin\alpha_1 &= (\rho \mathrm{d}s) u_{tt} \end{aligned}$$

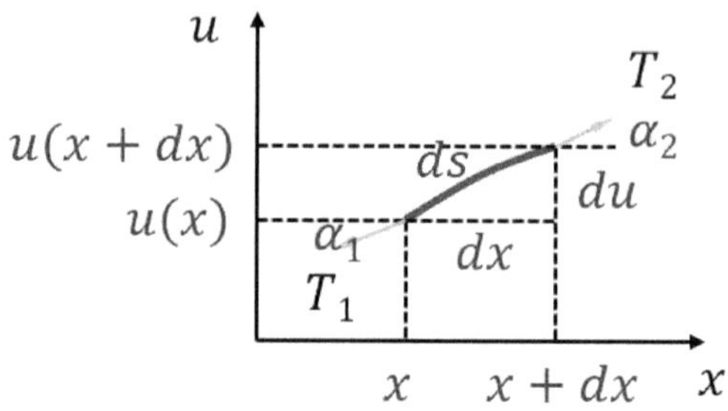

Figure 10.3.

(3) *Linearization of equation*

Assume that the amplitude of the string's vibration is small, i.e., $\alpha_1, \alpha_2 \ll 1$. We make a linear approximation by using $\cos\alpha_1 \approx \cos\alpha_2 \approx 1$. It implies that the tensions T_1 and T_2 in the string are equal, denoted as T. Additionally, we have $\sin\alpha \approx \alpha \approx \tan\alpha = \frac{\mathrm{d}u}{\mathrm{d}x}$. Therefore, we rewrite the equation as

$$T_2 \sin\alpha_2 - T_1 \sin\alpha_1 = T\left(\frac{\mathrm{d}u}{\mathrm{d}x}\right)_{x+\mathrm{d}x} - T\left(\frac{\mathrm{d}u}{\mathrm{d}x}\right)_x = T\frac{\mathrm{d}^2 u}{\mathrm{d}x^2}\mathrm{d}x$$

This leads to the vibration equation of the string:

$$u_{tt} - a^2 u_{xx} = 0 \tag{10.1.1}$$

Here, the parameter $a = \sqrt{\frac{T}{\rho}}$. From the dimensional analysis of Equation (10.1.1), we can deduce that the parameter a has the dimension of velocity. Later, we will see that a is exactly the speed of waves propagate along the string.

If the string is applied by an external force during vibration, according to the above deduction, we can prove that the string satisfies the nonhomogeneous equation:

$$u_{tt} - a^2 u_{xx} = f(x,t)$$

where $f(x,t)$ is the external force per unit length.

10.1.2 *Longitudinal vibration equation of a rod*

(1) *Analysis of elements*

By dividing the rod into small elements as shown in Figure 10.4, let $u(x)$ and $u(x+\mathrm{d}x)$ be the longitudinal displacements of points x and $x+\mathrm{d}x$, respectively. At a given time t, the relative deformation of the element is given by

$$\left.\frac{u(x+\mathrm{d}x) - u(x)}{\mathrm{d}x}\right|_t = u_x$$

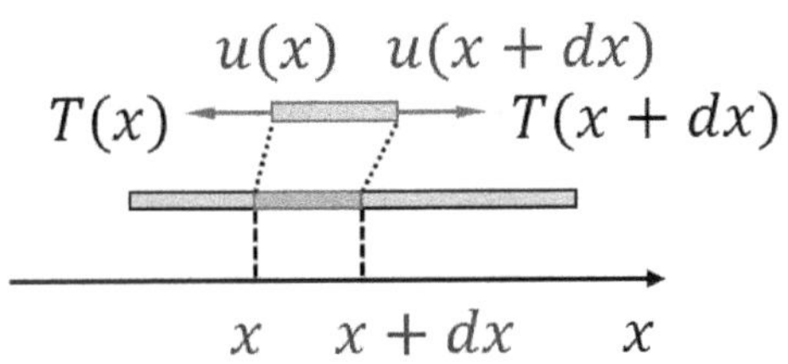

Figure 10.4.

(2) *Equation of motion*

According to Hooke's law, the force acting on the element is proportional to its relative deformation:

$$T(x) = ESu_x$$

where E is the Young's modulus and S is the cross-sectional area of the rod. Applying Newton's second law, we have

$$\begin{aligned} T(x+\mathrm{d}x) - T(x) &= ESu_x(x+\mathrm{d}x) - ESu_x(x) \\ &= ES\left(\frac{\partial u_x}{\partial x}\right)\mathrm{d}x = \rho S \mathrm{d}x u_{tt} \end{aligned}$$

The vibration equation of the rod is given by

$$u_{tt} - a^2 u_{xx} = 0$$

where $a = \sqrt{\frac{E}{\rho}}$. It can be observed that the longitudinal vibration equation of a rod has exactly the same form as the transverse vibration equation of a string.

10.1.3 *Diffusion Equation*

(1) *Background knowledge*

In a system where the concentration of a substance is not uniform in space, the substance will diffuse from areas of high concentration to areas of low concentration. We want to understand the rate of diffusion and how the concentration distribution changes with time during the diffusion process. Experimental evidence shows that when the spatial variation of concentration $u(\boldsymbol{r}, t)$ is small, diffusion obeys Fick's law:

$$\boldsymbol{q}(\boldsymbol{r}, t) = -D\nabla u(\boldsymbol{r}, t)$$

where $\boldsymbol{q}(\boldsymbol{r}, t)$ represents the diffusion flux, i.e., the number or mass of particles passing through a unit cross-sectional area per unit time, and D is the diffusion coefficient. Fick's law is independent of Newton's laws of motion and states that the diffusion flux is proportional to the gradient of concentration $u(\boldsymbol{r}, t)$.

(2) *Analysis of elements*

Consider a small cubic element in the system, as shown in Figure 10.5. Based on the conservation of particles, the change in the number of particles inside the cube $\mathrm{d}Q(\boldsymbol{r}, t)$ is equal to the number of particles flowing into the boundary. For simplicity, let's consider the variation of particle flux along the one-dimensional x-axis:

$$\mathrm{d}Q(x,t) = [\boldsymbol{q}(x,t)\mathrm{d}s - \boldsymbol{q}(x+\mathrm{d}x,t)\mathrm{d}s]\mathrm{d}t = -\frac{\partial \boldsymbol{q}}{\partial x}\mathrm{d}x\mathrm{d}s\mathrm{d}t$$

where ds is the cross-sectional area and the change in concentration inside the cube is given by

$$\mathrm{d}u = \frac{\mathrm{d}Q}{\mathrm{d}V} = \frac{\mathrm{d}Q}{\mathrm{d}s\mathrm{d}x}$$

Using Fick's law, we can obtain the one-dimensional diffusion equation:

$$u_t - a^2 u_{xx} = 0 \tag{10.1.2}$$

Similarly, the three-dimensional diffusion equation can be derived as

$$u_t - a^2 \Delta u = 0 \tag{10.1.3}$$

Here, $a^2 = D$, and

$$\Delta \overset{\text{def}}{=} \partial_x^2 + \partial_y^2 + \partial_z^2 \equiv \nabla^2$$

is called the Laplacian operator.

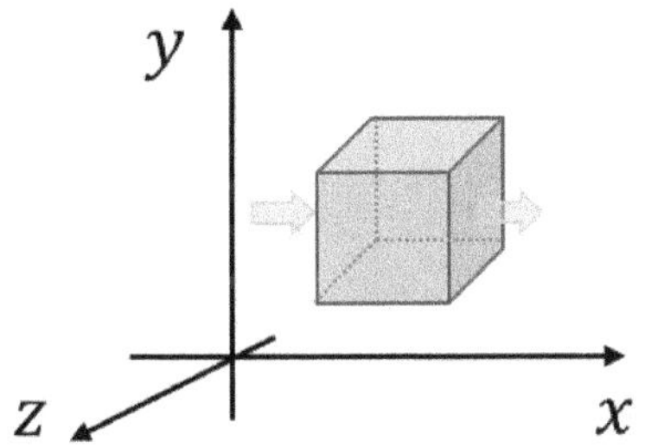

Figure 10.5.

10.1.4 *Heat conduction equation*

(1) *Background knowledge*

When the temperature of an object is not uniform, heat will transfer from areas of higher temperature to areas of lower temperature. Let's assume the temperature distribution in the object is $u(\boldsymbol{x}, t)$. When the spatial variation of temperature $u(\boldsymbol{r}, t)$ is small, heat conduction follows Fourier's law empirically:

$$\boldsymbol{q}(\boldsymbol{r}, t) = -\kappa \nabla u(\boldsymbol{r}, t)$$

where $\boldsymbol{q}(\boldsymbol{r}, t)$ represents the heat flux, i.e., the amount of heat passing through a unit cross-sectional area per unit time, and κ is the thermal conductivity. Fourier's law states that the heat flux is proportional to the gradient of temperature $u(\boldsymbol{r}, t)$. The process of heat conduction is also independent of Newton's laws of motion.

(2) *Derivation of equation*

Similar to the derivation of the diffusion equation, consider a small cubic element in the system. Based on the conservation of heat, the change in heat within a portion of the object $\mathrm{d}Q(\boldsymbol{r}, t)$ is equal to the net heat flowing into the boundary. For one-dimension case it is given by

$$\mathrm{d}Q(x, t) = [\boldsymbol{q}(x, t)\mathrm{d}s - \boldsymbol{q}(x + \mathrm{d}x, t)\mathrm{d}s]\mathrm{d}t = -\frac{\partial \boldsymbol{q}}{\partial x}\mathrm{d}x\mathrm{d}s\mathrm{d}t$$

The change in temperature is related to the specific heat capacity c and density ρ of the material:

$$\mathrm{d}u = \frac{\mathrm{d}Q}{c\rho \mathrm{d}s\mathrm{d}x}$$

Using Fourier's law of heat conduction, we obtain the one-dimensional heat conduction equation:

$$u_t - a^2 u_{xx} = 0 \tag{10.1.4}$$

Similarly, the three-dimensional heat conduction equation can be derived as

$$u_t - a^2 \Delta u = 0 \tag{10.1.5}$$

Here, $a^2 = \frac{\kappa}{c\rho}$. If there is a heat source $f(\boldsymbol{r}, t)$ present in the object, such as self-heating or heat dissipation in a specific manner, then the heat conduction process satisfy the nonhomogeneous equation:

$$u_t - a^2 \Delta u = f(\boldsymbol{r}, t) \tag{10.1.6}$$

10.1.5 *Other physics equation*

Below are listed some equations in physics:

- Transmission line equation

$$\begin{cases} j_{tt} - a^2 j_{xx} = 0, \quad v_{tt} - a^2 v_{xx} = 0 \\ a^2 = \dfrac{1}{LC} \end{cases}$$

- Small vibrations of uniform thin films

$$\begin{cases} u_{tt} - a^2 \Delta_2 u = 0 \\ a^2 = \dfrac{T}{\rho} \end{cases}$$

- Electrostatic field equations

$$\Delta\varphi = -\frac{\rho}{\varepsilon_0}$$

- Electromagnetic wave equations

$$\boldsymbol{E}_{tt} - a^2 \Delta \boldsymbol{E} = 0, \quad \boldsymbol{H}_{tt} - a^2 \Delta \boldsymbol{H} = 0$$

$$a^2 = \frac{1}{\mu_0 \varepsilon_0}$$

- Time-dependent Schrödinger equation

$$\mathrm{i}\hbar \frac{\partial}{\partial t}\psi = -\frac{\hbar^2}{2m}\Delta\psi + V\psi$$

- Stationary Schrödinger equation

$$-\frac{\hbar^2}{2m}\Delta\psi + V\psi = E\psi$$

These equations have the following fundamental characteristics:

(1) They are all linear differential equations and can be represented using linear differential operators $\hat{L}$ as $\hat{L}u = f(\boldsymbol{r}, t)$, where $f(\boldsymbol{r}, t)$ is independent of u and is referred to as the nonhomogeneous term of the equation. The equation $\hat{L}u = 0$ is called the homogeneous equation.

(2) The highest order of spatial and temporal derivatives in these equations does not exceed the second order.
(3) They all contain terms in the form of the Laplace operator or its higher-dimensional counterparts.

ENDNOTE

From Newtonian mechanics to Maxwell's equations, physical laws exhibit time-inversion symmetry. Observing the above differential equations, it is evident that equations containing second-order time derivatives possess time-inversion symmetry, while equations containing first-order time derivatives, such as the heat conduction equation or diffusion equation, do not possess time-inversion symmetry. This is not surprising, as heat conduction or diffusion processes are irreversible, with thermodynamic entropy playing a decisive role. However, the Schrödinger equation in quantum mechanics also contains a first-order time derivative. Does it violate time-inversion symmetry as well?

Here, the imaginary unit i plays a miraculous role and ensures time-inversion symmetry in an incredible way. Applying a time-reversal transformation $t \to -t$ to the Schrödinger equation, we have

$$-\mathrm{i}\hbar\frac{\partial}{\partial t}\psi(\boldsymbol{r},-t) = -\frac{\hbar^2}{2m}\Delta\psi(\boldsymbol{r},-t) + V(\boldsymbol{r},-t)\psi(\boldsymbol{r},-t)$$

Taking the complex conjugate of this equation, we obtain

$$\mathrm{i}\hbar\frac{\partial}{\partial t}\psi^*(\boldsymbol{r},-t) = -\frac{\hbar^2}{2m}\Delta\psi^*(\boldsymbol{r},-t) + V^*(\boldsymbol{r},-t)\psi^*(\boldsymbol{r},-t)$$

Assuming that the external potential satisfies $V^*(\boldsymbol{r},-t) = V(\boldsymbol{r},t)$, the equation remains invariant under time inversion if

$$\psi^*(\boldsymbol{r},-t) = \psi(\boldsymbol{r},t)$$

The probability density of the electron exhibits time-inversion symmetry:

$$\rho(\boldsymbol{r}, t) = |\psi(\boldsymbol{r}, t)|^2 \equiv |\psi(\boldsymbol{r}, -t)|^2$$

In Chapter 12, we will show that a one-dimensional Gaussian wave packet evolves in time t as

$$\psi(x, t) = \frac{1}{\sqrt{2\pi(\sigma^2 + \mathrm{i}\hbar t/m)}} \exp\left[-\frac{x^2}{2(\sigma^2 + \mathrm{i}\hbar t/m)}\right]$$

Unlike the motion of mechanical waves, the motion of an electron wave packet closely resembles the diffusion behavior of classical particles, and time seems to acquire a specific direction. But this is an illusion! Despite the irreversible diffusion of the electron wave packet, it still satisfies

$$\rho(\boldsymbol{r}, t) = \rho(\boldsymbol{r}, -t)$$

In other words, even if time is reversed and flows backward, the electron wave packet continues to diffuse rather than converge. To give an imprecise analogy, imagine filming the diffusion processes of classical particles and quantum wave packets and then playing the film in reverse. Classical diffusion would appear as convergence, while quantum diffusion would still be diffusion. This scenario has a significant impact on our worldview!

The Copenhagen interpretation considers the wave function as a kind of probability, indicating the existence of uncertainty. According to traditional thermodynamic theory, this uncertainty brings about entropy, and the increase in entropy during the diffusion of the wave packet gives time a direction. However, the Schrödinger equation preserves time-inversion symmetry, implying that there is no issue of entropy increase during the diffusion of the wave packet. The wave function does not contain any inherent uncertainty. Mysterious!

Exercises

[1] The string is vibrating in water, assuming that the resistance experienced per unit length is proportional to the velocity of vibration: $f = -\alpha u_t$. Derive the wave equation for the string.
Answer: $u_{tt} - a^2 u_{xx} + \frac{\alpha}{\rho} u_t = 0$

[2] A uniform wire has a resistivity of r and carries a constant current with a uniform distribution of current density $\boldsymbol{j}$. Derive the heat conduction equation for the wire.
Answer: $c\rho u_t - \kappa \Delta u = j^2 r$

[3] It an external force of $f(x,t)$ per unit length transversely applied on the soft string, prove the equation of motion is

$$u_{tt} - a^2 u_{xx} = f(x,t)$$

[4] Derive the wave equation under isothermal conditions for sound waves.

10.2 Boundary Value Problems

10.2.1 *Physical conditions of equations*

Boundary value problems arise from the general differential equations, where the specific distribution of physical quantities depends on the initial values of the given system and the values on the boundaries. These initial values and boundary conditions, collectively known as boundary value conditions, complement the general motion equations. Equations without specified initial or boundary conditions are referred to as generalized equations. Chapter 9 introduces some general solution methods for such equations.

For instance, the propagation of waves in a medium can be either traveling waves or standing waves. Although the electrostatic field generated by charges in space satisfies the Poisson equation, the specific distribution of the electric field depends not only on the distribution of free charges but also on the presence of other substances in the vicinity, which are the boundary conditions.

(1) *Initial conditions*

The distribution of physical quantities is given at the initial time: $u(x,t)|_{t=0} = \varphi(x)$, indicating that the values of physical quantities at various spatial points are specified at $t = 0$. Since some motion equations involve second-order derivatives with respect to time, it is also necessary to know the distribution of the "velocity" at the initial time, i.e., the values of the first-order derivatives of the physical quantity, $u_t(x,t)|_{t=0} = \psi(x)$.

When the physical quantities of the system do not change with time and reach a steady state, no initial conditions are required.

(2) *Boundary conditions*

Newton's equations of motion for particles only involve second-order derivatives with respect to time, so only the initial position and initial velocity of the particle are needed. For the motion equations of continuous media, which involve derivatives with respect to both time and space, the values and derivative values on the spatial boundaries, known as boundary conditions, are required. Mathematically, boundary conditions can be classified into three types, considering the system's boundary denoted as Σ.

- **Type 1:** Dirichlet boundary condition: The values on the boundary are given, also known as the Dirichlet boundary condition:

$$u(\boldsymbol{r},t)|_\Sigma = f(t)$$

- **Type 2:** Neumann boundary condition: The values of the normal derivatives on the boundary are given, also known as the Neumann boundary condition:

$$\left.\frac{\partial u(\boldsymbol{r},t)}{\partial n}\right|_\Sigma = g(t)$$

- **Type 3:** Cauchy boundary condition: A combination of the above two, known as the Cauchy boundary condition:

$$\left.\left[u + \alpha\frac{\partial u(\boldsymbol{r},t)}{\partial n}\right]\right|_\Sigma = h(t)$$

Here, f, g, and h are known functions, and α is a constant. The functions f, g, and h in the above boundary conditions are independent of the physical quantity u itself. When , g, and h are all zero, the boundary conditions are referred to as homogeneous boundary conditions. The following examples illustrate the boundary value conditions.

[1] Vibration of a string
There are three types of constraints on the displacement function $u(x,t)$ at the boundaries or endpoints $x = 0$ and $x = l$, respectively:

- **Type 1:** Dirichlet boundary condition:

$$u(0,t) = f_1(t), \quad u(l,t) = f_2(t)$$

It specifies the displacement changes at the boundaries. When $f(t) = 0$, it indicates the endpoints are fixed.

- **Type 2:** Neumann boundary condition:

$$u_x(0,t) = g_1(t), \quad u_x(l,t) = g_2(t)$$

It specifies the force changes at the boundaries. When $g(t) = 0$, it indicates the endpoints are free.

- **Type 3:** Cauchy boundary condition:

$$u(0,t) + \alpha_1 u_x(0,t) = h_1(t), \quad u(l,t) + \alpha_2 u_x(l,t) = h_2(t)$$

It specifies a combination of displacement and force changes.
The initial conditions are

$$u(x,0) = \varphi(x), \quad u_t(x,0) = \psi(x)$$

[2] Heat conduction equation
There are three types of constraints on the temperature function $u(\boldsymbol{x},t)$ at the surface Σ of an object.

- **Type 1:** Dirichlet boundary condition:

$$u(\boldsymbol{r},t)|_\Sigma = f(t)$$

It specifies the temperature changes on the surface. When $f(t) = 0$, it indicates a constant temperature on the boundary.

- **Type 2:** Neumann boundary condition:

$$\left.\frac{\partial u(\boldsymbol{r},t)}{\partial n}\right|_\Sigma = g(t)$$

It specifies the heat flux changes on the surface. When $g(t)= 0$, it indicates adiabatic conditions on the boundary.

- **Type 3:** Cauchy boundary condition:

$$\left[u + \alpha\frac{\partial u(\boldsymbol{r}, t)}{\partial n}\right]\bigg|_{\Sigma} = h(t)$$

It specifies a combination of both temperature and heat flux changes.

10.2.2 *Connection conditions*

Sometimes, systems are non-uniform and composed of different media with closely connected interfaces. For example, when two rods of different materials are joined, there will be a sudden change in phase or velocity as a wave propagates to the interface. Therefore, it is necessary to understand the connection relationship of physical quantities at the interface.

For the connection of two rods shown in Figure 10.6(a), with Young's moduli E_1 and E_2 respectively, and subjected to equal forces at the connection point $x = x_0$, the following conditions hold:

$$u_1(x_0, t) = u_2(x_0, t), \quad E_1 u_{1x}(x_0, t) = E_2 u_{2x}(x_0, t)$$

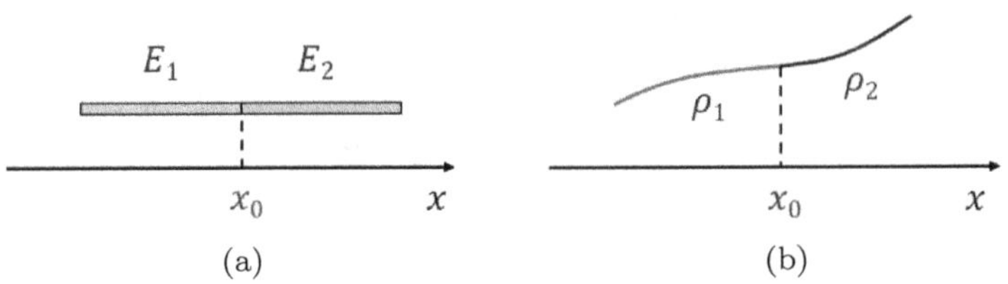

Figure 10.6.

Reflection: the connection of two segments of lightweight and flexible strings, as shown in Figure 10.6(b), with densities ρ_1 and ρ_2, respectively. What are the connection conditions in this case?

ENDNOTE

If one end of the rod at $x = l$ undergoes free cooling with an ambient temperature of θ, the heat flux leaving the rod is related to the temperature difference according to Newton's law of cooling:

$$-\kappa \frac{\partial u}{\partial x}\bigg|_{x=l} = h(u|_{x=l} - \theta)$$

This corresponds to a Type 3 boundary condition.

In some boundary conditions, the functions f, g and h themselves involve u or its derivative terms. As long as these terms are of the first degree, they belong to a linear system. There are also cases where nonlinear boundary conditions occur. For example, in the case of thermal radiation, the surface of an object radiates heat outward according to Stefan's law, where the radiative heat flux density is proportional to the fourth power of the temperature:

$$\boldsymbol{q} \propto u^4|_\Sigma$$

Exercises

[1] For a uniform rod of length l, with a constant heat flux entering at both ends with intensity q_0, write down the boundary conditions.

[2] Apply pressure to both ends of a uniform rod of length l, causing it to contract to a length $l(1 - 2\varepsilon)$. After removing the external forces, let the rod undergo free vibrations. Write down the boundary conditions and initial conditions for the vibration equation.

[3] Find the connection conditions for two different materials in close contact, assuming a slender rod. The thermal conductivity, specific heat, and density of the two materials are κ_1, c_1, ρ_1, and κ_2, c_2, ρ_2, respectively.

[4] In the scenario shown in Figure 10.7, where a mass m is attached to the middle of a horizontal string, determine the connection conditions for the string.

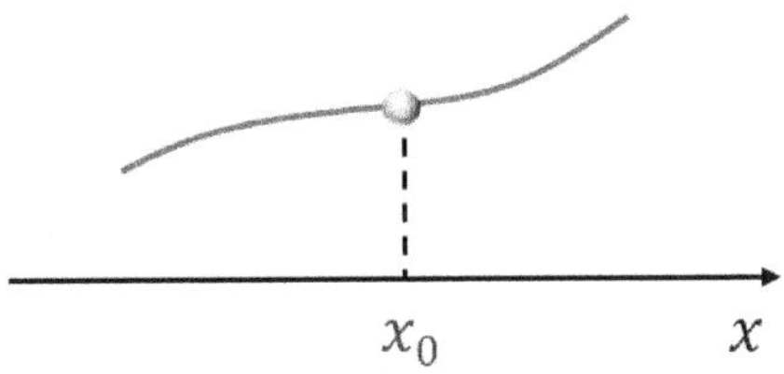

Figure 10.7.

[5] Determine the electrostatic connection conditions between two different dielectrics, as shown in Figure 10.8.

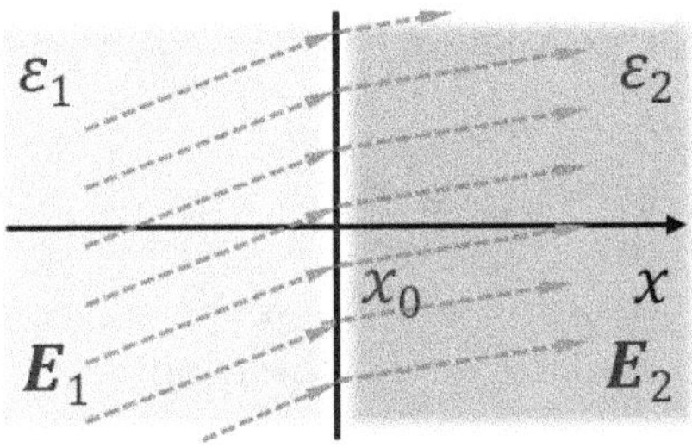

Figure 10.8.

[6] Three rods made of different materials are connected as shown in Figure 10.9. The thermal conductivity, specific heat, and density of the two materials are κ_i, c_i, and ρ_i $(i = 1, 2, 3)$, respectively. Write out the joining condition for the heat conduction equation.

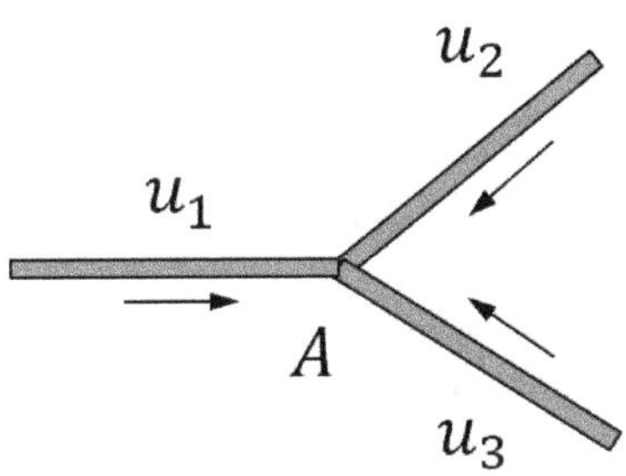

Figure 10.9.

Answer:

$$u_1|_A = u_2|_A = u_3|_A, \quad \kappa_1 u_{1x}|_A + \kappa_2 u_{2x}|_A + \kappa_3 u_{3x}|_A = 0$$

10.3 D'Alembert's Formula

This section discusses a special type of initial value problem that studies how initial conditions determine the behavior of equations.

10.3.1 *Wave equation for an infinite string*

$$\begin{cases} u_{tt} - a^2 u_{xx} = 0 & (-\infty < x < \infty) \\ u|_{t=0} = \phi(x), & u_t|_{t=0} = \psi(x) \end{cases}$$

Based on the algorithm for the general solution of constant coefficient PDEs in Section 9.3, we assume that the solution of the equation has the form:

$$u = f(x + \lambda t)$$

The characteristic equation is $\lambda^2 - a^2 = 0$, which yields $\lambda = \pm a$. Hence, the general solution of the equation is

$$u = f_1(x + at) + f_2(x - at)$$

Here, f_1 and f_2 are undetermined functions related to the initial conditions.

We can also make a variable substitution:

$$\xi = x + at, \quad \eta = x - at$$

This transforms the equation to

$$\frac{\partial^2 u}{\partial \xi \partial \eta} = 0$$

Integrating with respect to η first, we obtain $\frac{\partial u}{\partial \xi} = f(\xi)$. Then integrating with respect to ξ, we arrive at the solution of the equation:

$$u = \int f(\xi)\mathrm{d}\xi + f_2(\eta) \equiv f_1(\xi) + f_2(\eta) = f_1(x + at) + f_2(x - at)$$

It is observed that $f_2(x - at)$ describes a traveling wave propagating in the positive x direction with a velocity of a. The shape of the wave remains unchanged during the propagation. Similarly, $f_1(x+at)$ describes a traveling wave propagating in the negative x direction with a velocity of a. Therefore, $u = f_1(x+at) + f_2(x-at)$ represents the superposition of waves propagating in the positive and negative directions with a velocity of a. The shapes of both waves remain constant, and the overall waveform only changes when they overlap.

The forms of the functions $f_1(x)$ and $f_2(x)$ are determined by the initial conditions:

$$\begin{cases} f_1(x) + f_2(x) = \phi(x) \\ af_1'(x) - af_2'(x) = \psi(x) \end{cases}$$

Integrating the second equation, we have

$$f_1(x) - f_2(x) = \frac{1}{a}\int_{x_0}^{x} \psi(\xi)\mathrm{d}\xi + f_1(x_0) - f_2(x_0)$$

From this, we can solve for

$$f_1(x) = \frac{1}{2}\phi(x) + \frac{1}{2a}\int_{x_0}^{x} \psi(\xi)\mathrm{d}\xi + \frac{1}{2}[f_1(x_0) - f_2(x_0)]$$

$$f_2(x) = \frac{1}{2}\phi(x) - \frac{1}{2a}\int_{x_0}^{x} \psi(\xi)\mathrm{d}\xi - \frac{1}{2}[f_1(x_0) - f_2(x_0)]$$

Finally, we obtain the particular solution of the wave equation for an infinite string, known as D'Alembert's formula:

$$u(x,t) = \frac{1}{2}[\phi(x+at) + \phi(x-at)] + \frac{1}{2a}\int_{x-at}^{x+at} \psi(\xi)\mathrm{d}\xi \qquad (10.3.1)$$

Example 10.1. Studying the propagation of a triangle wave:

$$\begin{cases} u_{tt} - a^2u_{xx} = 0 \quad (-\infty < x < \infty) \\ u|_{t=0} = \phi(x) = \begin{cases} 1+x & (x \in [-1,0]) \\ 1-x & (x \in [0,1]) \\ 0 & (|x| > 1) \end{cases} \\ u_t|_{t=0} = \psi(x) = 0 \end{cases}$$

Solution: Since $\psi(x) = 0$, according to D'Alembert's formula, we have

$$u(x,t) = \frac{1}{2}[\phi(x+at) + \phi(x-at)]$$

This formula indicates that the initial waveform $\phi(x)$ is decomposed into two equal subwaves:

$$f_1(x) = f_2(x) = \frac{1}{2}\phi(x)$$

They propagate in opposite directions, and the waveform remains unchanged during the propagation. Figure 10.10 visually illustrates the shape and position of the wave at different moments in time.

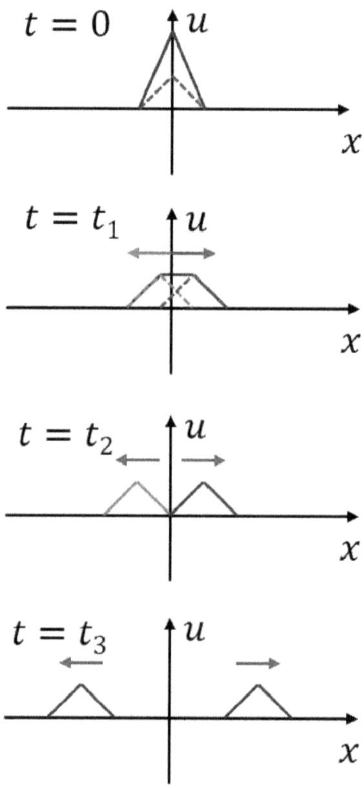

Figure 10.10.

10.3.2 *Endpoint reflection*

Consider the free vibration of a semi-infinite string with the assumption that $x = 0$ is a fixed endpoint:

$$\begin{cases} u_{tt} - a^2 u_{xx} = 0 & (0 \le x < \infty) \\ u(x,t)|_{x=0} = 0 \\ u(x,t)|_{t=0} = \phi(x), & u_t(x,t)|_{t=0} = \psi(x) \end{cases}$$

Since the point $x = 0$ is always fixed, let's assume that the semi-infinite string is the positive half of an infinite string. The displacement of the infinite string must be an odd function, so we need to extend the initial displacement and initial velocity as odd functions. This is referred to as odd extension:

$$\Phi(x) = \begin{cases} \phi(x) & (x \ge 0) \\ -\phi(-x) & (x < 0) \end{cases}$$

$$\Psi(x) = \begin{cases} \psi(x) & (x \ge 0) \\ -\psi(-x) & (x < 0) \end{cases}$$

Thus, the semi-infinite problem is transformed into an infinite problem,

$$\begin{cases} \tilde{u}_{tt} - a^2 \tilde{u}_{xx} = 0 & (-\infty < x < \infty) \\ \tilde{u}_{tt} - a^2 \tilde{u}_{xx} = 0 & (-\infty < x < \infty) \end{cases}$$

Applying D'Alembert's formula, we have

$$\begin{aligned} \tilde{u}(x,t) &= \frac{1}{2}[\Phi(x-at) + \Phi(x+at)] + \frac{1}{2a}\int_{x-at}^{x+at} \Psi(\xi)\mathrm{d}\xi \\ &= \frac{1}{2}\left[\Phi(x-at) - \frac{1}{a}\int_{-\infty}^{x-at} \Psi(\xi)\mathrm{d}\xi\right] \\ &\quad + \frac{1}{2}\left[\Phi(x+at) + \frac{1}{a}\int_{-\infty}^{x+at} \Psi(\xi)\mathrm{d}\xi\right] \end{aligned}$$

Finally, the determined shape of the semi-infinite string is

$$u(x,t) = \tilde{u}(x,t)|_{x \ge 0}$$

Example 10.2. Investigating the motion of a triangle wave on a semi-infinite string with one end fixed:

$$\begin{cases} u_{tt} - a^2 u_{xx} = \quad (0 \leq x < \infty) \\ u|_{x=0} = 0 \\ u|_{t=0} = \begin{cases} x - d + 1 & (x \in [d-1, d]) \\ d + 1 - x & (x \in [d, d+1]) \\ 0 & (|x - d| > 1) \end{cases} \\ u_t|_{t=0} = 0 \end{cases}$$

Solution: Since the point $x = 0$ is fixed, we extend the semi-infinite string as an odd extension of an infinite string. Since the initial velocity is $u_t|_{t=0} = \psi(x) = 0$, the D'Alembert's solution is given by

$$u(x,t) = \frac{1}{2}[\Phi(x + at) + \Phi(x - at)]_{x \geq 0}$$

Figure 10.11 visually illustrates the variation of the waveform over time. The initial triangle wave is divided into two subwaves, labeled as ① and ②: $f_1(x) = f_2(x) = \frac{1}{2}\phi(x)$, propagating in opposite directions. In the $-x$ direction, there are two non-physical subwaves, ③ and ④, which are odd symmetrical to ① and ②, respectively, also propagating in opposite directions. Specifically,

$$\begin{cases} \frac{1}{2}\Phi(x - at) = \textcircled{1} + \textcircled{3} \\ \frac{1}{2}\Phi(x + at) = \textcircled{2} + \textcircled{4} \end{cases}$$

In the physical region ($x > 0$), when the backward propagating subwave ② reaches the endpoint $x = 0$, reflection occurs. According to the odd extension property shown in the figure, subwave ② is replaced by subwave ③, and the reflected wave has the opposite phase (negative displacement) compared to the incident wave. The physical effect is that the two subwaves, ① and ③, with opposite phases propagate together, and this phenomenon of phase reversal in the reflected wave is known as half-wave loss.

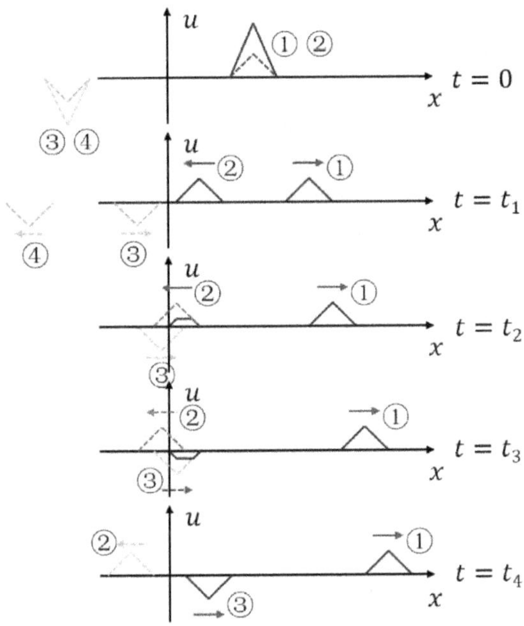

Figure 10.11.

If the initial velocity $\psi(x) \neq 0$, the wave can still be decomposed into two independent subwaves propagating in opposite directions, but the shapes of the two subwaves will be different. They are respectively given by

$$u_1(x) = \frac{1}{2}\varphi(x) + \frac{1}{2}\int_{-\infty}^{x} \psi(\xi)\mathrm{d}\xi$$

$$u_2(x) = \frac{1}{2}\varphi(x) - \frac{1}{2}\int_{-\infty}^{x} \psi(\xi)\mathrm{d}\xi$$

ENDNOTE

D'Alembert's formula can be used to solve the wave equation on an infinite string, providing a concise visualization of wave propagation and reflection. However, this method is not universally applicable, and for higher-dimensional unbounded problems, other methods such as Fourier transform are employed, which will be covered in subsequent chapters. D'Alembert's formula demonstrates that the propagation of one-dimensional waves follows the Huygens' principle, but for two-dimensional or three-dimensional spaces, whether wave propagation also adheres to Huygens' principle remains a subject for future investigation.

Exercises

[1] Solving free vibrations of an infinite string, with the initial displacement be represented by $\varphi(x)$, and the initial velocity by $-a\varphi'(x)$.
Answer: $u(x,t) = \varphi(x - at)$

[2] Solving longitudinal vibration of a semi-infinite rod, assuming one end is subjected to a longitudinal force $F = A\sin\omega t$, and the cross-sectional area S and Young's modulus. The initial displacement and velocity of the rod are both zero.
Hint: Use the Laplace transform.
Answer:

$$u(x,t) = \frac{Aa}{ES\omega}\left[\cos\omega\left(t - \frac{x}{a}\right) - 1\right]\mathrm{H}\left(t - \frac{x}{a}\right)$$

[3] Given a semi-infinite string with both initial displacement and velocity being zero, and one end of the string undergoing small vibrations, $u|_{x=0} = A\sin\omega t$, find the solution for the string's vibration.
Hint: One approach to solve this problem is to directly guess a particular solution for the equation using the boundary condition or to apply the Laplace transform to the equation.
Answer:

$$u(x,t) = A\sin\omega\left(t - \frac{x}{a}\right)\mathrm{H}\left(t - \frac{x}{a}\right)$$

[4] Study the motion of an infinite string:

$$\begin{cases} u_{tt} - a^2 u_{xx} = 0 \quad (-\infty < x < \infty) \\ u|_{t=0} = \phi(x) = 0 \\ u_t|_{t=0} = \psi(x) = \begin{cases} \psi_0 & (x \in [x_1, x_2]) \\ 0 & (x \notin [x_1, x_2]) \end{cases} \end{cases}$$

[5] Investigating the motion of a triangle wave on a semi-infinite string with one end free:

$$\begin{cases} u_{tt} - a^2 u_{xx} = 0 \quad (0 \le x < \infty) \\ u_x|_{x=0} = 0 \\ u|_{t=0} = \begin{cases} x - d + 1 & (x \in [d-1, d]) \\ d + 1 - x & (x \in [d, d+1]) \\ 0 & (|x - d| > 1) \end{cases} \\ u_t|_{t=0} = 0 \end{cases}$$

10.4 Orthogonal Curvilinear Coordinates

10.4.1 *Coordinate transformation method*

Let there be an orthogonal curvilinear coordinate system (q_1, q_2, q_3), and at each point in space, there are three mutually orthogonal basis vectors $(\boldsymbol{e}_1, \boldsymbol{e}_2, \boldsymbol{e}_3)$ corresponding to the three tangent directions. The direction of the basis vectors generally depends on the position of the point in space. The curvilinear coordinate system (q_1, q_2, q_3) is related to the Cartesian coordinate system (x, y, z) through an orthogonal transformation:

$$\begin{cases} x = x(q_1, q_2, q_3) \\ y = y(q_1, q_2, q_3) \\ z = z(q_1, q_2, q_3) \end{cases}$$

Using the orthogonality of the coordinate basis vectors, $\boldsymbol{e}_i \cdot \boldsymbol{e}_j = \delta_{ij}$, the line element length in three-dimensional space is given by

$$
\begin{aligned}
(\mathrm{d}s)^2 &= (\mathrm{d}x)^2 + (\mathrm{d}y)^2 + (\mathrm{d}z)^2 \\
&= \sum_j \left(\frac{\partial x}{\partial q_j}\mathrm{d}q_j\right)^2 + \sum_j \left(\frac{\partial y}{\partial q_j}\mathrm{d}q_j\right)^2 + \sum_j \left(\frac{\partial z}{\partial q_j}\mathrm{d}q_j\right)^2 \\
&= \sum_j \left[\left(\frac{\partial x}{\partial q_j}\right)^2 + \left(\frac{\partial y}{\partial q_j}\right)^2 + \left(\frac{\partial z}{\partial q_j}\right)^2\right](\mathrm{d}q_j)^2 \\
&\equiv (\mathrm{d}s_1)^2 + (\mathrm{d}s_2)^2 + (\mathrm{d}s_3)^2
\end{aligned}
$$

Thus, the line element length along the direction of the curvilinear coordinate q_j is given by

$$
\mathrm{d}s_j = \sqrt{\left(\frac{\partial x}{\partial q_j}\right)^2 + \left(\frac{\partial y}{\partial q_j}\right)^2 + \left(\frac{\partial z}{\partial q_j}\right)^2}\,\mathrm{d}q_j \equiv h_j \mathrm{d}q_j
$$

where h_j denote the Lamé coefficients or scale factors of the orthogonal curvilinear coordinate system:

$$
h_j(q_1, q_2, q_3) \stackrel{\text{def}}{=} \sqrt{\left(\frac{\partial x}{\partial q_j}\right)^2 + \left(\frac{\partial y}{\partial q_j}\right)^2 + \left(\frac{\partial z}{\partial q_j}\right)^2} \tag{10.4.1}
$$

(1) *Gradient*

The rate of change of a scalar function $u(q_1, q_2, q_3)$ along a curve coordinate q_j is given by

$$
\frac{\partial u}{\partial s_j} = \frac{1}{h_j}\frac{\partial u}{\partial q_j}
$$

Therefore, the gradient of the scalar function $u(q_1, q_2, q_3)$ is given by

$$
\nabla u = \frac{1}{h_1}\frac{\partial u}{\partial q_1}\boldsymbol{e}_1 + \frac{1}{h_2}\frac{\partial u}{\partial q_2}\boldsymbol{e}_2 + \frac{1}{h_3}\frac{\partial u}{\partial q_3}\boldsymbol{e}_3 \tag{10.4.2}
$$

(2) *Divergence*

The change in flux of a vector function $\boldsymbol{A} = A_1\boldsymbol{e}_1 + A_2\boldsymbol{e}_2 + A_3\boldsymbol{e}_3$ along the $\boldsymbol{e}_1$ direction is given by

$$\begin{aligned}
&(A_1 \mathrm{d}s_2 \mathrm{d}s_3)|_{q_1+\mathrm{d}q_1} - (A_1 \mathrm{d}s_2 \mathrm{d}s_3)|_{q_1} \\
&\quad = [(A_1 h_2 h_3)|_{q_1+\mathrm{d}q_1} - (A_1 h_2 h_3)|_{q_1}]\mathrm{d}q_2 \mathrm{d}q_3 \\
&\quad = \frac{\partial}{\partial q_1}(A_1 h_2 h_3)\mathrm{d}q_1 \mathrm{d}q_2 \mathrm{d}q_3
\end{aligned}$$

Therefore, the divergence of the vector function $\boldsymbol{A}(q_1, q_2, q_3)$ is given by

$$\nabla \cdot \boldsymbol{A} = \frac{1}{h_1 h_2 h_3}\left[\frac{\partial}{\partial q_1}(A_1 h_2 h_3) + \frac{\partial}{\partial q_2}(A_2 h_3 h_1) + \frac{\partial}{\partial q_3}(A_3 h_1 h_2)\right] \tag{10.4.3}$$

(3) *Curl*

Similarly, the curl of a vector function can be obtained as

$$\nabla \times \boldsymbol{A} = \frac{1}{h_1 h_2 h_3}\begin{vmatrix} h_1\boldsymbol{e}_1 & h_2\boldsymbol{e}_2 & h_3\boldsymbol{e}_3 \\ \dfrac{\partial}{\partial q_1} & \dfrac{\partial}{\partial q_2} & \dfrac{\partial}{\partial q_3} \\ h_1 A_1 & h_2 A_2 & h_3 A_3 \end{vmatrix} \tag{10.4.4}$$

10.4.2 *3D Laplace operator*

Based on $\Delta u = \nabla \cdot \nabla u$, the expression for the Laplace equation in an orthogonal curvilinear coordinate system is given by

$$\begin{aligned}
\Delta u = \frac{1}{h_1 h_2 h_3}&\left[\frac{\partial}{\partial q_1}\left(\frac{h_2 h_3}{h_1}\frac{\partial u}{\partial q_1}\right) + \frac{\partial}{\partial q_2}\left(\frac{h_3 h_1}{h_2}\frac{\partial u}{\partial q_2}\right)\right. \\
&\left. + \frac{\partial}{\partial q_3}\left(\frac{h_1 h_2}{h_3}\frac{\partial u}{\partial q_3}\right)\right]
\end{aligned} \tag{10.4.5}$$

(1) *Spherical coordinate system*

The coordinate transformation relationship is

$$x = r\sin\theta\cos\varphi, \quad y = r\sin\theta\sin\varphi, \quad z = r\cos\theta$$

The Lamé coefficients are

$$h_r = 1, \quad h_\theta = r, \quad h_\varphi = r\sin\theta$$

In spherical coordinate system, the Laplace operator is given by

$$\Delta u = \frac{1}{r^2}\frac{\partial}{\partial r}\left(r^2\frac{\partial u}{\partial r}\right) + \frac{1}{r^2\sin\theta}\frac{\partial}{\partial\theta}\left(\sin\theta\frac{\partial u}{\partial\theta}\right) + \frac{1}{r^2\sin^2\theta}\frac{\partial^2 u}{\partial\varphi^2} \tag{10.4.6}$$

(2) *Cylindrical coordinate system*

The coordinate transformation relationship is

$$x = \rho\sin\varphi, \quad y = \rho\sin\varphi, \quad z = z$$

It yields the Lamé coefficients:

$$h_\rho = 1, \quad h_\varphi = \rho, \quad h_z = 1$$

In cylindrical coordinate system, the Laplace operator is given by

$$\Delta u = \frac{1}{\rho}\frac{\partial}{\partial\rho}\left(\rho\frac{\partial u}{\partial\rho}\right) + \frac{1}{\rho^2}\frac{\partial^2 u}{\partial\varphi^2} + \frac{\partial^2 u}{\partial z^2} \tag{10.4.7}$$

10.4.3 *n-Dimensional Laplace operator*

In the generalized n-dimensional hyperspherical coordinate system $(q_1, q_2, \ldots, q_n)$, the Lamé coefficients are given by

$$h_j = \sqrt{\sum_{k=1}^{n}\left(\frac{\partial x_k}{\partial q_k}\right)^2}$$

The Laplace operator in this coordinate system is expressed as

$$\Delta u = \frac{1}{h_1h_2\cdots h_n}\sum_{k=1}^{n}\frac{\partial}{\partial q_k}\left(\frac{h_1h_2\cdots h_n}{h_k^2}\frac{\partial}{\partial q_k}\right) \tag{10.4.8}$$

Using the hyperspherical coordinate system, we have

$$x_k = r\left(\prod_{j=1}^{k-1}\sin\varphi_j\right)\cos\varphi_k \quad (k=1,2,\ldots,n)$$

where $\prod_{j=1}^{0}\sin\varphi_j = 1$, with $0 \le \varphi_k \le \pi$ $(k = 1,2,\ldots,n-2)$ and $0 \le \varphi_{n-1} \le 2\pi$. Specifically,

$$\begin{aligned}
x_1 &= r\cos\varphi_1\\
x_2 &= r\sin\varphi_1\cos\varphi_2\\
x_3 &= r\sin\varphi_1\sin\varphi_2\cos\varphi_3\\
&\vdots\\
x_{n-1} &= r\sin\varphi_1\sin\varphi_2\cdots\sin\varphi_{n-2}\cos\varphi_{n-1}\\
x_n &= r\sin\varphi_1\sin\varphi_2\cdots\sin\varphi_{n-2}\sin\varphi_{n-1}
\end{aligned}$$

The corresponding Lamé coefficients are

$$\begin{aligned}
&h_1 = 1, \quad h_2 = r, \quad h_3 = r\sin\varphi_1, \ldots\\
&\vdots\\
&h_{n-1} = r\sin\varphi_1\sin\varphi_2\cdots\sin\varphi_{n-3}\\
&h_n = r\sin\varphi_1\sin\varphi_2\cdots\sin\varphi_{n-2}
\end{aligned}$$

Therefore, the Laplace operator in the n-dimensional spherical coordinate system is given by

$$\begin{aligned}
\Delta u = {}& \frac{1}{r^{n-1}}\frac{\partial}{\partial r}\left(r^{n-1}\frac{\partial u}{\partial r}\right) + \frac{1}{r^2\sin^{n-2}\varphi_1}\frac{\partial}{\partial\varphi_1}\left(\sin^{n-2}\varphi_1\frac{\partial u}{\partial\varphi_1}\right) + \cdots\\
&+ \frac{1}{r^2\sin^2\varphi_1\cdots\sin^2\varphi_{k-1}\sin^{n-(k+1)}\varphi_k}\frac{\partial}{\partial\varphi_k}\\
&\times\left(\sin^{n-(k+1)}\varphi_k\frac{\partial u}{\partial\varphi_k}\right) + \cdots\\
&+ \frac{1}{r^2\sin^2\varphi_1\cdots\sin^2\varphi_{k-1}\sin^2\varphi_{n-2}}\frac{\partial^2 u}{\partial\varphi_{n-1}^2}
\end{aligned} \tag{10.4.9}$$

It can be easily shown that the Jacobian determinant of the coordinate transformation is

$$\det J = r^{n-1}(\sin\varphi_1)^{n-2}(\sin\varphi_2)^{n-3}\cdots\sin\varphi_{n-2}$$

The volume element transformation is given by

$$\mathrm{d}^n x = \det J \mathrm{d}r\mathrm{d}\varphi_1\cdots\mathrm{d}\varphi_{n-1} = r^{n-1}\mathrm{d}r\mathrm{d}\Omega_n$$

$$\mathrm{d}\Omega_n = (\sin\varphi_1)^{n-2}(\sin\varphi_2)^{n-3}\cdots\sin\varphi_{n-2}\mathrm{d}\varphi_1\cdots\mathrm{d}\varphi_{n-1}$$

ENDNOTE

We can also intuitively derive the Laplace operator in curvilinear coordinate systems from a geometric perspective.

(1) Cylindrical coordinates

$$\begin{cases} x = \rho\cos\varphi \\ y = \rho\sin\varphi, \\ z = z \end{cases} \qquad \begin{cases} \boldsymbol{e}_\rho = \boldsymbol{i}\cos\varphi + \boldsymbol{j}\sin\varphi \\ \boldsymbol{e}_\varphi = -\boldsymbol{i}\sin\varphi + \boldsymbol{j}\cos\varphi \\ \boldsymbol{e}_z = \boldsymbol{k} \end{cases}$$

Taking the partial derivatives of the basis vectors, we have

$$\frac{\partial \boldsymbol{e}_\rho}{\partial \varphi} = \boldsymbol{e}_\varphi, \quad \frac{\partial \boldsymbol{e}_\varphi}{\partial \varphi} = -\boldsymbol{e}_\rho, \quad \frac{\partial \boldsymbol{e}_z}{\partial \varphi} = 0$$

$$\frac{\partial \boldsymbol{e}_\rho}{\partial \rho} = \frac{\partial \boldsymbol{e}_\varphi}{\partial \rho} = \frac{\partial \boldsymbol{e}_z}{\partial \rho} = 0$$

$$\frac{\partial \boldsymbol{e}_\rho}{\partial z} = \frac{\partial \boldsymbol{e}_\varphi}{\partial z} = \frac{\partial \boldsymbol{e}_z}{\partial z} = 0$$

$$\nabla = \boldsymbol{e}_\rho\frac{\partial}{\partial\rho} + \boldsymbol{e}_\varphi\frac{1}{\rho}\frac{\partial}{\partial\varphi} + \boldsymbol{e}_z\frac{\partial}{\partial z} = \boldsymbol{i}\frac{\partial}{\partial x} + \boldsymbol{j}\frac{\partial}{\partial y} + \boldsymbol{k}\frac{\partial}{\partial z}$$

$$\nabla^2 = \nabla\cdot\nabla = \left(\boldsymbol{e}_\rho\frac{\partial}{\partial\rho} + \boldsymbol{e}_\varphi\frac{1}{\rho}\frac{\partial}{\partial\varphi} + \boldsymbol{e}_z\frac{\partial}{\partial z}\right)$$

$$\cdot\left(\boldsymbol{e}_\rho\frac{\partial}{\partial\rho} + \boldsymbol{e}_\varphi\frac{1}{\rho}\frac{\partial}{\partial\varphi} + \boldsymbol{e}_z\frac{\partial}{\partial z}\right)$$

The partial derivatives of the basis vectors can also be obtained directly from the geometric representation, which can be seen in Figure 10.12. Therefore, the Laplace operator is given by

$$\Delta u = \frac{1}{\rho}\frac{\partial}{\partial \rho}\left(\rho\frac{\partial u}{\partial \rho}\right) + \frac{1}{\rho^2}\frac{\partial^2 u}{\partial \varphi^2} + \frac{\partial^2 u}{\partial z^2}$$

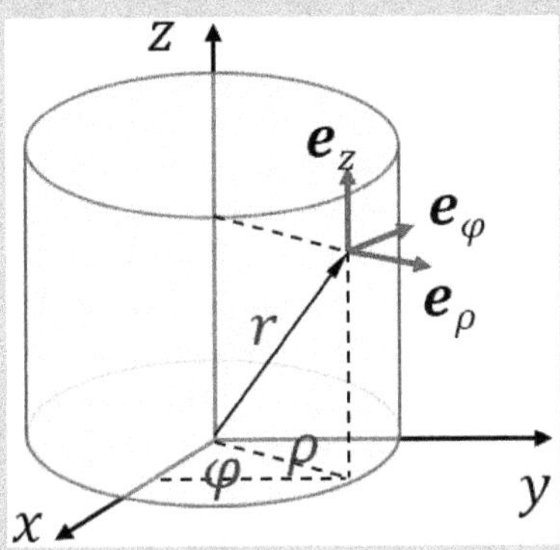

Figure 10.12.

(2) Spherical coordinates

$$\begin{cases} x = r\sin\theta\cos\varphi \\ y = r\sin\theta\sin\varphi, \\ z = r\cos\theta \end{cases} \quad \begin{cases} \boldsymbol{e}_r = \boldsymbol{i}\sin\theta\sin\varphi + \boldsymbol{j}\sin\theta\cos\varphi + \boldsymbol{k}\cos\theta \\ \boldsymbol{e}_\theta = \boldsymbol{i}\cos\theta\cos\varphi + \boldsymbol{j}\cos\theta\sin\varphi - \boldsymbol{k}\sin\theta \\ \boldsymbol{e}_\varphi = -\boldsymbol{i}\sin\varphi + \boldsymbol{j}\cos\varphi \end{cases}$$

Taking the partial derivatives of the basis vectors:

$$\frac{\partial \boldsymbol{e}_r}{\partial \theta} = \boldsymbol{e}_\theta, \quad \frac{\partial \boldsymbol{e}_\theta}{\partial \theta} = -\boldsymbol{e}_r, \quad \frac{\partial \boldsymbol{e}_\varphi}{\partial \theta} = 0$$

$$\frac{\partial \boldsymbol{e}_r}{\partial \varphi} = \boldsymbol{e}_\varphi \sin\theta, \quad \frac{\partial \boldsymbol{e}_\theta}{\partial \varphi} = -\boldsymbol{e}_\varphi\cos\theta$$

$$\frac{\partial \boldsymbol{e}_\varphi}{\partial \varphi} = -\boldsymbol{e}_r\sin\theta - \boldsymbol{e}_\theta\cos\theta$$

$$\frac{\partial \boldsymbol{e}_r}{\partial r} = \frac{\partial \boldsymbol{e}_\theta}{\partial r} = \frac{\partial \boldsymbol{e}_\varphi}{\partial r} = 0$$

$$\nabla = \boldsymbol{e}_r\frac{\partial}{\partial r} + \boldsymbol{e}_\theta\frac{1}{r}\frac{\partial}{\partial \theta} + \boldsymbol{e}_\varphi\frac{1}{r\sin\theta}\frac{\partial}{\partial \varphi}$$

$$\nabla^2 = \nabla \cdot \nabla = \left(\boldsymbol{e}_r \frac{\partial}{\partial r} + \boldsymbol{e}_\theta \frac{1}{r} \frac{\partial}{\partial \theta} + \boldsymbol{e}_\varphi \frac{1}{r \sin\theta} \frac{\partial}{\partial \varphi} \right)$$
$$\cdot \left(\boldsymbol{e}_r \frac{\partial}{\partial r} + \boldsymbol{e}_\theta \frac{1}{r} \frac{\partial}{\partial \theta} + \boldsymbol{e}_\varphi \frac{1}{r \sin\theta} \frac{\partial}{\partial \varphi} \right)$$

Similarly, the partial derivatives of the basis vectors can be obtained directly from the geometric representation, which can be seen in Figure 10.13. Therefore, the Laplace operator is given by

$$\Delta u = \frac{1}{r^2} \frac{\partial}{\partial r} \left(r^2 \frac{\partial u}{\partial r} \right) + \frac{1}{r^2 \sin\theta} \frac{\partial}{\partial \theta} \left(\sin\theta \frac{\partial u}{\partial \theta} \right) + \frac{1}{r^2 \sin^2\theta} \frac{\partial^2 u}{\partial \varphi^2}$$

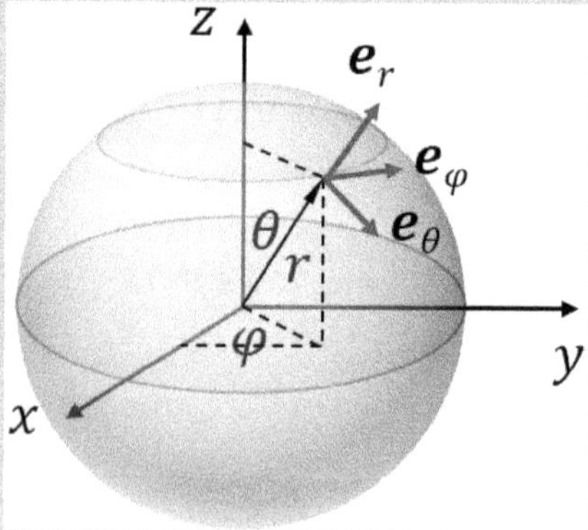

Figure 10.13.

Absolutely, hands-on derivation is quite necessary for anyone studying mathematics or physics. And it is enough to practice once.

In addition to cylindrical and spherical coordinate systems, there are other coordinate systems like conical coordinates and ellipsoidal coordinates. These coordinate systems are used to describe various physical phenomena and simplify problem-solving in different contexts. For those interested in further exploring these coordinate systems and their mathematical properties, the book *Introduction to Special Functions* by Wang Zhuxi and Guo Dunren is a great resource. It covers a variety of special functions and coordinate systems, offering insights into advanced mathematical concepts often encountered in physics and engineering.

Exercises

[1] Prove the vector curl formula for an orthogonal curvilinear coordinate system:

$$\nabla \times \boldsymbol{A} = \frac{1}{h_1 h_2 h_3} \begin{vmatrix} h_1 \boldsymbol{e}_1 & h_2 \boldsymbol{e}_2 & h_3 \boldsymbol{e}_3 \\ \dfrac{\partial}{\partial q_1} & \dfrac{\partial}{\partial q_2} & \dfrac{\partial}{\partial q_3} \\ h_1 A_1 & h_2 A_2 & h_3 A_3 \end{vmatrix}$$

10.5 Classification of Partial Differential Equations

10.5.1 *Characteristic equation*

A general second-order linear partial differential equation can be represented as

$$\sum_{i,j} a_{ij}(x) u_{x_i x_j} + \sum_i b_i(x) u_{x_i} + c(x) u + f(x) = 0$$

Here, $a_{ij}(x), b_i(x), c(x_k)$, and $f(x_k)$ are functions of the independent variables $x_1, x_2, \ldots, x_n$, and are independent of the function u. This equation is called a linear partial differential equation (PDE). If $f(x) = 0$, the equation is homogeneous; otherwise, it is non-homogeneous.

Now, let's focus on the classification of partial differential equations with two independent variables (x, y),

$$a_{11} u_{xx} + 2a_{12} u_{xy} + a_{22} u_{yy} + b_1 u_x + b_2 u_y + cu + f = 0 \quad (10.5.1)$$

Here, $a_{11}, a_{12}, a_{22}, b_1, c$, and f are all real functions of x and y. We can perform a change of variables as follows,

$$\begin{cases} x = x(\xi, \eta) \\ y = y(\xi, \eta) \end{cases} \longmapsto \begin{cases} \xi = \xi(x, y) \\ \eta = \eta(x, y) \end{cases}$$

The equation becomes

$$A_{11}u_{\xi\xi} + 2A_{12}u_{\xi\eta} + A_{22}u_{\eta\eta} + B_1u_\xi + B_2u_\eta + Cu + F = 0$$

Here,

$$A_{11}(\xi, \eta) = a_{11}\xi_x^2 + 2a_{12}\xi_x\xi_y + a_{22}\xi_y^2 = 0$$
$$A_{22}(\xi, \eta) = a_{11}\eta_x^2 + 2a_{12}\eta_x\eta_y + a_{22}\eta_y^2 = 0$$
$$A_{12}(\xi, \eta) = a_{11}\xi_x\eta_x + a_{12}(\xi_x\eta_y + \xi_y\eta_x) + a_{22}\xi_y\eta_y = 0$$

To simplify the equation, we choose new independent variables (ξ, η) such that $A_{11} = 0$ or $A_{22} = 0$, satisfying the condition:

$$a_{11}z_x^2 + 2a_{12}z_xz_y + a_{22}z_y^2 = 0$$

Alternatively,

$$a_{11}\left(-\frac{z_x}{z_y}\right)^2 - 2a_{12}\left(-\frac{z_x}{z_y}\right) + a_{22} = 0$$

For the curve equation $z(x, y) = C$, we have $\frac{\mathrm{d}y}{\mathrm{d}x} = -\frac{z_x}{z_y}$. The above condition becomes a first-order differential equation:

$$a_{11}\left(\frac{\mathrm{d}y}{\mathrm{d}x}\right)^2 - 2a_{12}\left(\frac{\mathrm{d}y}{\mathrm{d}x}\right) + a_{22} = 0 \tag{10.5.2}$$

This equation is called the characteristic equation of the second-order partial differential equation, and its integral curves $z(x, y) = C$ are called the characteristic lines of the equation. The two solutions of the characteristic equation are

$$\begin{aligned} \frac{\mathrm{d}y}{\mathrm{d}x} &= \frac{a_{12} + \sqrt{a_{12}^2 - a_{11}a_{22}}}{a_{11}} \\ \frac{\mathrm{d}y}{\mathrm{d}x} &= \frac{a_{12} - \sqrt{a_{12}^2 - a_{11}a_{22}}}{a_{11}} \end{aligned} \tag{10.5.3}$$

10.5.2 *Classification of partial differential equations*

Based on the form of the solutions to the characteristic equation, PDEs can be classified into three types. Letting $\Delta = a_{12}^2 - a_{11}a_{22}$, the PDEs can be classified into the following three types:

(1) *Hyperbolic equations:* $\Delta > 0$

For this type of equation, the characteristic equation has two real characteristic lines: $\xi(x,y) = C_1$ and $\eta(x,y) = C_2$. By taking (ξ,η) as new independent variables, we have $A_{11} = 0$ and $A_{22} = 0$. The second-order PDE becomes

$$u_{\xi\eta} = -\frac{1}{2A_{12}}[B_1 u_\xi + B_2 u_\eta + Cu + F] = 0$$

Further substitution:

$$\begin{cases} \xi = \alpha + \beta \\ \eta = \alpha - \beta \end{cases}$$

transforms the equation into the standard hyperbolic form:

$$u_{\alpha\alpha} - u_{\beta\beta} = -\frac{1}{A_{12}}[(B_1+B_2)u_\alpha + (B_1-B_2)u_\beta + 2Cu + 2F] \quad (10.5.4)$$

Examples of hyperbolic PDEs include the wave equation.

(2) *Parabolic equations:* $\Delta = 0$

In this case, the characteristic equation has one combined real characteristic line: $\xi(x,y) = C$. Taking (ξ,η) as new independent variables, we have

$$a_{12} = \pm\sqrt{a_{11}a_{22}}, \quad \frac{\xi_x}{\xi_y} = -\frac{dy}{dx} = -\frac{a_{12}}{a_{11}}$$

Consequently, we have

$$A_{11} = A_{22} = 0, \quad A_{22} \neq 0$$

The equation can be transformed into the standard parabolic form:

$$u_{\eta\eta} = -\frac{1}{A_{22}}[B_1 u_\xi + B_2 u_\eta + Cu + F] = 0 \quad (10.5.5)$$

Examples of parabolic PDEs include the diffusion equation.

(3) *Elliptic equations:* $\Delta < 0$

For this type of equation, the characteristic equation has no real solutions. However, we can formally consider complex constants C as solutions:

$$\xi(x,y) = \bar{\eta}(x,y) = C$$

Although it cannot be interpreted as a curve in the plane, by taking (ξ, η) as new independent variables, we still have

$$A_{11} = 0, \quad A_{22} = 0$$

as well as the equation:

$$u_{\xi\eta} = -\frac{1}{2A_{12}}[B_1 u_\xi + B_2 u_\eta + Cu + F] = 0$$

Noting the constraint $\xi = \bar{\eta}$, we make the substitution:

$$\begin{cases} \xi = \alpha + \mathrm{i}\beta \\ \eta = \alpha - \mathrm{i}\beta \end{cases} \quad (\alpha, \beta \in \mathbb{R})$$

This leads to the standard elliptic form:

$$u_{\alpha\alpha} + u_{\beta\beta} = -\frac{1}{A_{12}}[(B_1 + B_2)u_\alpha + \mathrm{i}(B_1 - B_2)u_\beta + 2Cu + 2F] \tag{10.5.6}$$

Examples of elliptic PDEs include the Poisson equation.

ENDNOTE
According to the equation

$$A_{12}^2 - A_{11}A_{22} = (a_{12}^2 - a_{11}a_{22})(\xi_x\eta_y - \xi_y\eta_x)^2$$

the sign of Δ remains unchanged under variable substitution, which means the type of the equation remains the same. Furthermore, since $\Delta = a_{12}^2 - a_{11}a_{22}$ is a function of (x, y), general PDEs may exhibit different types in different regions. If the Jacobian determinant $\det J = \xi_x\eta_y - \xi_y\eta_x \equiv 1$, then Δ remains invariant under coordinate transformations.

General conic curves can be described by quadratic functions, which are classified into three types: elliptic, hyperbolic, and parabolic. For second-order partial differential equations with two independent variables (x, y), if the coefficients are constants, transforming the equation through Fourier transformation reduces it to an algebraic equation of quadratic form, also classified into elliptic, hyperbolic, and parabolic types. Although variable coefficient second-order partial differential equations cannot be explicitly transformed into algebraic equations through integral transformations, some hints can still be inferred from them.

Exercises

[1] Transform the following equation into standard form:

$$\text{(a) } au_{xx} + 2au_{xy} + au_{yy} + bu_x + cu_y + u = 0;$$

$$\text{(b) } u_{xx} - 2u_{xy} - 3u_{yy} + 2u_x + 6u_y = 0;$$

$$\text{(c) } 4y^2 u_{xx} - \mathrm{e}^{2x} u_{yy} - 4y^2 u_x = 0$$

Answer:

$$\text{(a) } \xi = y - x, \eta = x, \quad u_{\eta\eta} + \frac{c-b}{a} u_\xi + \frac{b}{a} u_\eta + \frac{1}{a} u = 0$$

$$\text{(b) } \xi = x - y, \eta = 3x + y, \quad 4u_{\xi\eta} - u_\xi + 3u_\eta = 0$$

$$\text{(c) } \xi = y^2 + \mathrm{e}^x, \eta = y^2 - \mathrm{e}^x, \quad 4(\xi + \eta) u_{\xi\eta} + u_\xi + u_\eta = 0$$

Chapter 11

Separation of Variables Methods

This method is applicable to the boundary value problems of finite systems. The basic idea is to decompose the partial differential equation into several ordinary differential equations, some of which constitute an eigenvalue problem due to the presence of homogeneous boundary conditions. Since the eigenfunctions in this chapter are all trigonometric functions, the essence of this method is the Fourier series expansion, which requires the determination of the expansion coefficients based on the initial and boundary conditions.

11.1 Homogeneous Boundary Problems

This section discusses solutions of differential equations subjected to homogeneous boundary conditions. In most cases, homogeneous boundary conditions are a prerequisite for solving differential equation initial boundary value problems. We will discuss two situations: homogeneous differential equations and inhomogeneous differential equations.

11.1.1 *Homogeneous differential equations*

Example 11.1. Solving the initial boundary value problem for a string with both ends fixed:

$$\begin{cases} u_{tt} = a^2 u_{xx} \quad (0 \leq x \leq l) \\ u(0,t) = u(l,t) = 0 \\ u(x,0) = \phi(x), \quad u_t(x,0) = \psi(x) \end{cases}$$

Solution: The wave reflects back and forth between the two endpoints, forming a stable standing wave. A standing wave is merely a collective vibration that does not propagate energy. Assuming that the solution of the PDE can be decomposed into the product of two functions with independent variables, $u(x,t) = X(x)T(t)$, substituting it into the wave equation, we have

$$XT'' = a^2 X''T \longrightarrow \frac{T''}{a^2 T} = \frac{X''}{X} \equiv -\lambda$$

Here, λ is a constant obtained through the separation of variables. This leads to two ODEs:

$$X''(x) + \lambda X(x) = 0$$

$$T''(t) + \lambda a^2 T(t) = 0$$

Based on the homogeneous boundary conditions:

$$\begin{cases} X(0)T(t) = 0 \\ X(l)T(t) = 0 \end{cases} \longrightarrow X(0) = 0, \quad X(l) = 0$$

Since the homogeneous boundary conditions must be satisfied, it follows that $\lambda \geq 0$. In this case, the solutions of the equations are

$$X_n(x) = \sin \frac{n\pi x}{l}$$

$$\lambda_n = \left(\frac{n\pi}{l}\right)^2 \quad (n = 1, 2, 3, \ldots)$$

It is observed that the separation constant λ can only take certain discrete values, known as eigenvalues of the equation. The corresponding solutions $X_n(x)$ are called eigenfunctions. Problems of this kind are collectively referred to as eigenvalue problems.

Substituting the eigenvalues λ_n into the time-related equation, we obtain

$$T_n(t) = A_n \cos\frac{n\pi at}{l} + B_n \sin\frac{n\pi at}{l} \quad (n = 1, 2, 3, \ldots)$$

Thus, the eigenmode of vibration is given by

$$\begin{aligned} u_n(x,t) &= \sin\frac{n\pi x}{l}\left[A_n \cos\frac{n\pi at}{l} + B_n \sin\frac{n\pi at}{l}\right] \\ &= N_n \sin\frac{n\pi x}{l} \sin\left(\frac{n\pi at}{l} + \phi_n\right) \end{aligned}$$

where the amplitude and initial phase reads

$$N_n = \sqrt{A_n^2 + B_n^2}, \quad \phi_n = \arctan\frac{A_n}{B_n}$$

The above equation can be interpreted as the string undergoing eigen oscillations at a frequency $\omega_n = \frac{n\pi a}{l}$ with an amplitude of $a_n = N_n \sin\frac{n\pi x}{l}$. Different eigenmodes of vibration are shown in Figure 11.1, where each mode has some points of zero displacement called nodes. The string between two nodes vibrates in phase, while on either side of a node, the string vibrates out of phase.

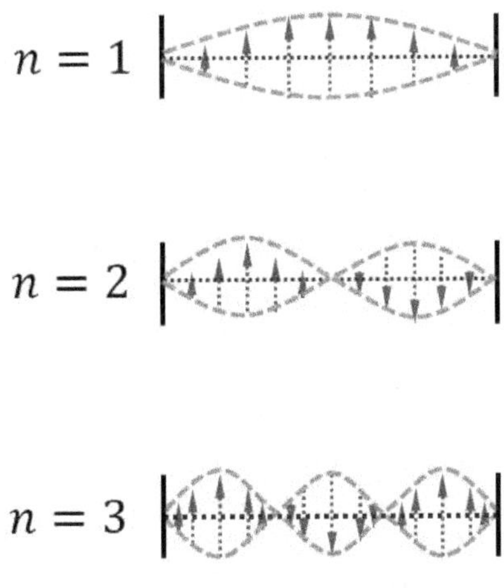

Figure 11.1.

Note that the eigen vibration function $u_n(x,t)$ alone does not satisfy the initial conditions. To obtain a solution that satisfies the initial conditions, we need to linearly superpose the eigenfunctions as follows:

$$u(x,t) = \sum_{n=1}^{\infty} \left[A_n \cos \frac{n\pi at}{l} + B_n \sin \frac{n\pi at}{l} \right] \sin \frac{n\pi x}{l}$$

By substituting the initial conditions, we can determine the coefficients of the superposition A_n and B_n as

$$u(x,0) = \sum_{n=1}^{\infty} A_n \sin \frac{n\pi x}{l} = \phi(x)$$

$$u_t(x,0) = \sum_{n=1}^{\infty} \frac{n\pi a}{l} B_n \sin \frac{n\pi x}{l} = \psi(x)$$

Actually, A_n and B_n are the coefficients of the Fourier sine series expansion of $\phi(x)$ and $\psi(x)$, respectively,

$$A_n = \frac{2}{l}\int_0^l \phi(\xi) \sin \frac{n\pi\xi}{l} d\xi, \quad B_n = \frac{2}{n\pi a}\int_0^l \psi(\xi) \sin \frac{n\pi\xi}{l} d\xi$$

Discussion:

(1) Supposing $\lambda < 0$, what would be the consequences?
(2) The frequency of vibration is inversely proportional to the length of the string, explaining the naive observation made by Pythagoras. Furthermore, the same string can have eigen frequencies that are integer multiples of a basic frequency.
(3) As the eigen mode index n increases, the number of nodes in the vibration increases, and the frequency of the string's vibration becomes higher. This has profound implications, as the energy of vibration is directly proportional to the vibration frequency, suggesting that high-frequency Fourier modes of a function would be suppressed; otherwise, the "ultraviolet catastrophe" would occur.

Example 11.2. Solve the wave equation for a free rod with both ends free:

$$\begin{cases} u_{tt} = a^2 u_{xx} & (0 \le x \le l) \\ u_x(0,t) = 0, & u_x(l,t) = 0 \\ u(x,0) = \phi(x), & u_t(x,0) = \psi(x) \end{cases}$$

Solution: Assume $u(x,t) = X(x)T(t)$, and substitute it into the equation to separate variables. First, we solve the equation for x,

$$\begin{cases} X''(x) + \lambda X(x) = 0 \\ X'(0) = X'(l) = 0 \end{cases}$$

These are the second type of homogeneous boundary conditions. The eigenfunctions and eigenvalues are

$$X_n(x) = \cos \frac{n\pi x}{l}$$

$$\lambda_n = \left(\frac{n\pi}{l}\right)^2 \quad (n = 0, 1, 2, 3, \ldots)$$

Substituting the eigenvalues λ_n into the equation for $T(t)$, we find

$$T_0(t) = A_0 + B_0 t \quad (n = 0)$$

$$T_n(t) = A_n \cos \frac{n\pi at}{l} + B_n \sin \frac{n\pi at}{l} \quad (n = 1, 2, \ldots)$$

where A_0, B_0, A_n, and B_n are undetermined constants. The eigen vibration solutions are

$$u_0(x,t) = A_0 + B_0 t(n = 0)$$

$$u_n(x,t) = \left[A_n \cos \frac{n\pi at}{l} + B_n \sin \frac{n\pi at}{l}\right] \cos \frac{n\pi x}{l} \quad (n = 1, 2, \ldots)$$

Similarly, the eigen vibrations alone cannot satisfy the initial conditions. We need to linearly superpose all the eigen modes:

$$u(x,t) = A_0 + B_0 t + \sum_{n=1}^{\infty} \left[A_n \cos \frac{n\pi at}{l} + B_n \sin \frac{n\pi at}{l}\right] \cos \frac{n\pi x}{l}$$

The constants A_0, B_0, A_n, and B_n are determined by the initial conditions:

$$A_0 + \sum_{n=1}^{\infty} A_n \cos\frac{n\pi x}{l} = \phi(x)$$

$$B_0 + \sum_{n=1}^{\infty} \frac{n\pi a}{l} B_n \cos\frac{n\pi x}{l} = \psi(x)$$

This is equivalent to expanding $\phi(x)$ and $\psi(x)$ into a Fourier cosine series with the expansion coefficients:

$$A_0 = \frac{1}{l}\int_0^l \phi(\xi)\mathrm{d}\xi, \quad A_n = \frac{2}{l}\int_0^l \phi(\xi)\cos\frac{n\pi\xi}{l}\mathrm{d}\xi$$

$$B_0 = \frac{1}{l}\int_0^l \psi(\xi)\mathrm{d}\xi, \quad B_n = \frac{2}{n\pi a}\int_0^l \psi(\xi)\cos\frac{n\pi\xi}{l}\mathrm{d}\xi$$

Example 11.3. A uniform slender rod of length l is insulated at its right end and maintained at zero temperature at its left end. The initial temperature distribution inside the rod is given by $\phi(x)$. We want to determine the temperature distribution of the rod at any given time in the absence of a heat source.

Solution: Based on the given conditions, we can write the heat equation and the boundary and initial conditions as follows:

$$\begin{cases} u_t = a^2 u_{xx} \quad (0 \le x \le l) \\ u(0,t) = u_x(l,t) = 0 \\ u(x,0) = \phi(x) \end{cases}$$

Assuming a separable solution of the form $u(x,t) = X(x)T(t)$, this leads to the following ODEs with boundary conditions:

$$\begin{cases} X''(x) + \lambda X(x) = 0 \\ X(0) = 0, \quad X'(l) = 0 \end{cases}$$

$$T' + a^2\lambda T = 0$$

We first need to determine the eigenvalues and eigenfunctions:

$$X_n(x) = \sin\left[\left(n + \frac{1}{2}\right)\frac{\pi x}{l}\right]$$

$$\lambda_n = \left[\left(n + \frac{1}{2}\right)\frac{\pi}{l}\right]^2 \quad (n = 0, 1, 2, 3\ldots)$$

Next, we find the expression for $T(t)$. $T(t)$ will be different for different eigenvalues λ_n and can be written as

$$T_n(t) = A_n \exp\left[-\left(n + \frac{1}{2}\right)^2 \frac{\pi^2 a^2 t}{l^2}\right]$$

To satisfy the initial condition, the general solution of the equation needs to be expressed as a linear combination of all the eigen solutions:

$$u(x,t) = \sum_{n=0}^{\infty} A_n \exp\left[-\left(n + \frac{1}{2}\right)^2 \frac{\pi^2 a^2 t}{l^2}\right] \sin\left[\left(n + \frac{1}{2}\right)\frac{\pi x}{l}\right]$$

The coefficients A_n can be determined using the initial condition:

$$A_n = \frac{2}{l}\int_0^l \phi(\xi)\sin\left[\left(n + \frac{1}{2}\right)\frac{\pi \xi}{l}\right] \mathrm{d}\xi$$

For example, if we take the initial condition $\phi(x) = \frac{u_0}{l}x$, we have

$$u(x,t) = \frac{2u_0}{\pi^2}\sum_{n=0}^{\infty} \frac{(-1)^n}{\left(n + \frac{1}{2}\right)^2} \exp\left[-\left(n + \frac{1}{2}\right)^2 \frac{\pi^2 a^2 t}{l^2}\right]$$
$$\times \sin\left[\left(n + \frac{1}{2}\right)\frac{\pi x}{l}\right]$$

For heat conduction problems, the coefficients in the series expansion decay exponentially with time, unlike in wave equations where they oscillate with time. The temperature decay rate of different eigen modes is not the same, which can be understood as the different rates of heat conduction for different eigen modes.

This example highlights the importance of homogeneous boundary conditions. In general, if the boundary conditions are non-homogeneous, we cannot directly solve the boundary value problem of the equation.

11.1.2 Non-homogeneous differential equation

For the non-homogeneous equation $\hat{L}u = f(x,t)$, the method of eigenfunction expansion (Fourier series) is commonly used to solve it. The basic approach is as follows:

(1) Assume that the eigenfunctions of the corresponding homogeneous equation $\hat{L}u = 0$ are $X_n(x)$. Expand the general solution u of the equation in terms of these eigenfunctions, i.e.,

$$u(x,t) = \sum_n T_n(t)X_n(x)$$

Treat the time-dependent functions $T_n(t)$ as "undetermined coefficients".

(2) Expand the non-homogeneous term on the right side of the equation as well:

$$f(x,t) = \sum_n f_n(t)X_n(x)$$

(3) Substitute the eigenfunction expansion $u(x,t)$ into the original differential equation. By utilizing the orthogonality of the eigenfunctions, separate out the ordinary differential equations that the "coefficients" $T_n(t)$ must satisfy.

(4) The initial conditions also need to be expanded using the eigenfunctions:

$$u(x,0) \equiv \phi(x) = \sum_n \phi_n X_n(x)$$

This allows us to determine the initial conditions for $T_n(t)$ and obtain their specific expressions.

By following these steps, we can solve the non-homogeneous differential equation and find the solution $u(x,t)$ in terms of the given eigenfunctions and their corresponding coefficients.

Example 11.4. Solving the boundary value problem:

$$\begin{cases} u_{tt} - a^2 u_{xx} = A\cos\dfrac{\pi x}{l}\sin\omega t \\ u_x|_{x=0} = 0, \quad u_x|_{x=l} = 0 \\ u|_{t=0} = \phi(x), \quad u_t|_{t=0} = \psi(x) \end{cases}$$

Solution 1: This is a second-type homogeneous boundary value problem, and the eigenfunctions corresponding to the homogeneous equation are

$$X_n(x) = \cos\frac{n\pi x}{l} \quad (n = 0, 1, 2, 3, \ldots)$$

Assume the solution of the equation has the following form:

$$u(x,t) = \sum_{n=0}^{\infty} T_n(t)\cos\frac{n\pi x}{l}$$

Substituting it into the non-homogeneous equation, we obtain

$$\sum_{n=0}^{\infty}\left[T_n''(t) + \frac{n^2\pi^2a^2}{l^2}T_n(t)\right]\cos\frac{n\pi x}{l} = A\cos\frac{\pi x}{l}\sin\omega t$$

By comparing the coefficients of the eigenfunctions on both sides, we get the ODEs for $T_n(t)$:

$$\begin{cases} T_1''(t) + \dfrac{\pi^2a^2}{l^2}T_1(t) = A\sin\omega t \\[2ex] T_n''(t) + \dfrac{n^2\pi^2a^2}{l^2}T_n(t) = 0 \quad (n \neq 1) \end{cases}$$

To solve for $T_n(t)$, we also need to know the initial conditions it satisfies. Therefore, we expand the initial conditions that $u(x,t)$ satisfies using the eigenfunctions as well:

$$u(x,0) = \sum_{n=0}^{\infty} T_n(0)\cos\frac{n\pi x}{l} \equiv \phi(x) = \sum_{n=0}^{\infty}\phi_n\cos\frac{n\pi x}{l}$$

$$u_t(x,0) = \sum_{n=0}^{\infty} T_n'(0)\cos\frac{n\pi x}{l} \equiv \psi(x) = \sum_{n=0}^{\infty}\psi_n\cos\frac{n\pi x}{l}$$

Hence, the initial conditions for $T_n(t)$ are

$$T_0(0) = \phi_0 = \frac{1}{l}\int_0^l \phi(\xi)\mathrm{d}\xi, \quad T_n(0) = \phi_n = \frac{2}{l}\int_0^l \phi(\xi)\cos\frac{n\pi\xi}{l}\mathrm{d}\xi$$

$$T_0'(0) = \psi_0 = \frac{1}{l}\int_0^l \psi(\xi)\mathrm{d}\xi, \quad T_n'(0) = \psi_n = \frac{2}{l}\int_0^l \psi(\xi)\cos\frac{n\pi\xi}{l}\mathrm{d}\xi$$

Solving these equations, we obtain

$$T_0(t) = \phi_0 + \psi_0 t$$

$$T_n(t) = \phi_n \cos\frac{n\pi at}{l} + \frac{l}{n\pi a}\psi_n \sin\frac{n\pi at}{l} \quad (n \neq 0, 1)$$

$$T_1(t) = \frac{Al}{\pi a}\frac{1}{\omega^2 - \pi^2 a^2/l^2}\left(\omega \sin\frac{\pi at}{l} - \frac{\pi a}{l}\sin\omega t\right) + \phi_1 \cos\frac{\pi at}{l} + \frac{l}{\pi a}\psi_1 \sin\frac{\pi at}{l}$$

The final solution to the equation is

$$u(x,t) = \frac{Al}{\pi a}\frac{1}{\omega^2 - \pi^2 a^2/l^2}\left(\omega \sin\frac{\pi at}{l} - \frac{\pi a}{l}\sin\omega t\right)\cos\frac{\pi x}{l} + \phi_0 + \psi_0 t + \sum_{n=1}^{\infty}\left(\phi_n \cos\frac{n\pi at}{l} + \frac{l}{n\pi a}\psi_n \sin\frac{n\pi at}{l}\right)\cos\frac{n\pi x}{l}$$

Solution 2: There is another simpler method to solve this problem, which involves guessing a particular solution that satisfies the equation, thereby transforming the equation into a homogeneous equation while simultaneously satisfying the homogeneous boundary conditions. In this case, based on the characteristics of the non-homogeneous term, we can try the following particular solution:

$$v(x,t) = B\cos\frac{\pi x}{l}\sin\omega t$$

Substituting it into the equation, we solve for B and obtain

$$B = \frac{A}{\left(\frac{\pi a}{l}\right)^2 - \omega^2}$$

Now, let $u(x,t) = v(x,t) + w(x,t)$. Then, $w(x,t)$ satisfies both the homogeneous differential equation and the homogeneous boundary conditions:

$$\begin{cases} w_{tt} - a^2 w_{xx} = 0 \\ w_x|_{x=0} = 0, \qquad w_x|_{x=l} = 0 \\ w|_{t=0} = \phi(x), \qquad w_t|_{t=0} = \psi(x) - B\omega\cos\frac{\pi x}{l} \end{cases}$$

We can solve this using the method described in Example 11.2. This approach appears more concise and convenient, but it relies on finding an appropriate particular solution, which requires careful comparison between the non-homogeneous term of the equation and the characteristics of the eigenfunctions.

Discussion:

(1) How to understand the physical meaning of the solution? It involves forced oscillation and resonance.
(2) Why does only the fundamental frequency $\omega_0 = \frac{\pi a}{l}$ resonate with the externally applied frequency ω?
(3) If the external driving term on the right side of the equation is $f(x,t) = A\sin\omega t$, will resonance still occur?

11.1.3 *Rectangular domain problem*

Example 11.5. To solve the boundary value problem for the two-dimensional Poisson's equation on a rectangular domain, we have the following conditions:

$$\begin{cases} \Delta_2 u = -2 \\ u|_{x=0} = 0, \quad u|_{x=a} = 0 \\ u|_{y=0} = 0, \quad u|_{y=b} = 0 \end{cases}$$

Solution: Since the non-homogeneous term of the equation is a polynomial, we can try a polynomial particular solution to transform it into a homogeneous equation. Let us take the particular solution as

$$v(x,y) = x(a-x)$$

Now, let $u(x,y) = v(x,y) + w(x,y)$, where $w(x,y)$ satisfies the following boundary value problem:

$$\begin{cases} \Delta_2 w = 0 \\ w|_{x=0} = 0, & w|_{x=a} = 0 \\ w|_{y=0} = x(x-a), & w|_{y=b} = x(x-a) \end{cases}$$

We separate variables by assuming $w(x,y) = X(x)Y(y)$. This gives us the following equations:

$$\begin{cases} X''(x) + \lambda X(x) = 0 \\ X(0) = X(a) = 0 \end{cases}, \quad Y''(y) - \lambda Y(y) = 0$$

Since the x-direction satisfies the homogeneous boundary conditions, we first solve the equation for $X(x)$:

$$\begin{cases} X_n(x) = \sin\dfrac{n\pi x}{a} \\ \lambda_n = \left(\dfrac{n\pi}{a}\right)^2 \quad (n = 1, 2, 3, \ldots) \end{cases}$$

Next, we substitute the eigenvalues λ_n into the equation for $Y(y)$ and obtain the general solution:

$$Y_n(y) = A_n \mathrm{e}^{\frac{n\pi y}{a}} + B_n \mathrm{e}^{-\frac{n\pi y}{a}}$$

The coefficients A_n and B_n are determined by the boundary conditions in the y-direction. Let's express $w(x, y)$ as a linear combination of eigenfunctions:

$$w(x, y) = \sum_{n=1}^{\infty} \left(A_n \mathrm{e}^{\frac{n\pi y}{a}} + B_n \mathrm{e}^{-\frac{n\pi y}{a}}\right) \sin\frac{n\pi x}{a}$$

We expand $w(x, y)$ according to the non-homogeneous boundary conditions in the y-direction:

$$w|_{y=0} = w|_{y=b} = x(x-a) = \sum_{n=1}^{\infty} C_n \sin\frac{n\pi x}{a}$$

$$C_n = \frac{2}{a}\int_0^a x(x-a)\sin\frac{n\pi x}{a}\mathrm{d}x = \frac{4a^2}{n^3\pi^3}[(-1)^n - 1]$$

By comparing coefficients, we obtain the following equations:

$$\begin{cases} A_n + B_n = C_n \\ A_n \mathrm{e}^{\frac{n\pi b}{a}} + B_n \mathrm{e}^{-\frac{n\pi b}{a}} = C_n \end{cases}$$

Solving for A_n and B_n, and simplifying, we have

$$w(x, y) = -\frac{8a^2}{\pi^3}\sum_{n=1}^{\infty} \frac{\cosh[(2n-1)\pi(y - b/2)/a]}{(2n-1)^3\cosh[(2n-1)\pi b/(2a)]}\sin\frac{(2n-1)\pi x}{a}$$

The solution to the equation is given by

$$u(x, y) = x(a-x) - \frac{8a^2}{\pi^3}\sum_{n=1}^{\infty} \frac{\cosh[(2n-1)\pi(y - b/2)/a]}{(2n-1)^3\cosh[(2n-1)\pi b/(2a)]} \times \sin\frac{(2n-1)\pi x}{a}$$

Discussion:

(1) The selection of particular solution must ensure that it still satisfies homogeneous boundary conditions in one direction. This is to ensure that the particular solution does not introduce any additional boundary conditions that would conflict with the given problem.
(2) Can we solve the equation for $Y(y)$ first? What would be the consequence? Just have a try.

Example 11.6. Solving for the eigenmodes of a two-dimensional rectangular membrane with dimensions $a \times b$:

$$\begin{cases} u_{tt} - a^2\Delta_2 u = 0 \\ u|_{x=0} = u|_{x=a} = 0 \\ u|_{y=0} = u|_{y=b} = 0 \\ u|_{t=0} = \phi(x,y), \quad u_x|_{t=0} = \psi(x,y) \end{cases}$$

Solution: First, let us separate variables: $u(x,y,t) = X(x)Y(y)T(t)$. This leads to the following eigenvalue problem equations:

$$\begin{cases} X'' + \lambda X = 0 \\ X(0) = 0, \quad X(a) = 0 \end{cases}, \qquad \begin{cases} Y'' + \chi Y = 0 \\ Y(0) = 0, \quad Y(b) = 0 \end{cases}$$

and

$$T'' + k^2a^2T = 0$$

where $k^2 = \lambda^2 + \chi^2$. Both sets of equations satisfy homogeneous boundary conditions. Solving for the eigenvalues respectively gives:

$$\lambda_m^2 = \frac{m^2\pi^2}{a^2}, \quad \chi_n^2 = \frac{n^2\pi^2}{b^2} \quad (m, n = 1, 2, 3, \ldots)$$

The corresponding eigenfunctions are:

$$u_{mn}(x,y) = \sin\frac{m\pi x}{a}\sin\frac{n\pi y}{b}$$

The eigenfrequencies of the 2D membrane are given by $\omega_{mn} = k_{mn}a$, with

$$k_{mn} = \sqrt{\frac{m^2\pi^2}{a^2} + \frac{n^2\pi^2}{b^2}}$$

An interesting phenomenon occurs when $a : b$ is a rational number. The equation

$$\frac{m^2}{a^2} + \frac{n^2}{b^2} = \frac{m'^2}{a^2} + \frac{n'^2}{b^2}$$

has non-zero integer solutions $\{m, n; m', n'\}$, leading to degenerate eigenvalues. The degeneracy of eigenvalues becomes a number theory problem: How many distinct ways are there to represent a number as the sum of two squares? For example, $65 = 7^2 + 4^2 = 8^2 + 1^2$.

The eigenfunctions $u_{mn}(x, y) = \sin \frac{m\pi x}{a} \sin \frac{n\pi y}{b}$ have nodal lines composed of continuous zeros. Nodal lines are typically straight segments parallel to the coordinate axes. However, with degenerate eigenvalues, eigenfunctions can be linear combinations of different degenerate modes, leading to nodal lines of different shapes. For example, for a square membrane $a = b = \pi$, eigenmodes can be a linear combination of sine functions of different frequencies,

$$u_{mn}(x, y) = \alpha \sin(mx) \sin(ny) + \beta \sin(nx) \sin(my)$$

resulting in different nodal line shapes as shown in Figure 11.2.

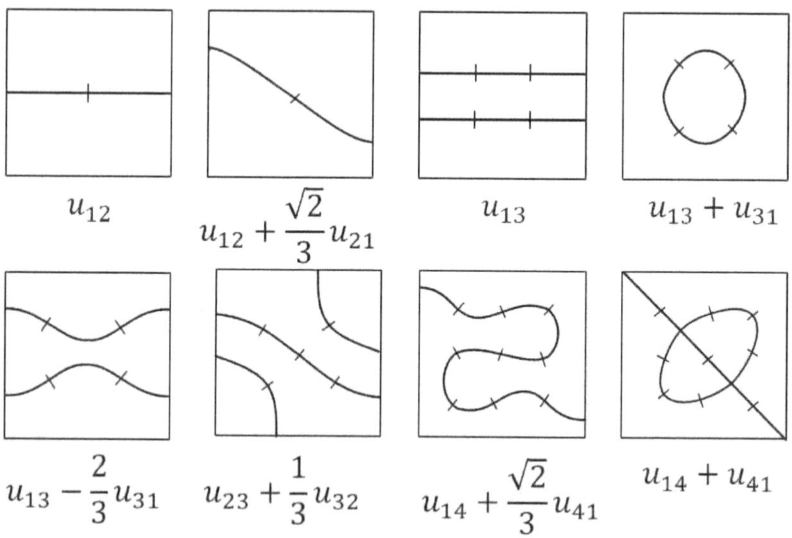

Figure 11.2.

ENDNOTE

The study of string vibrations led to the development of Fourier series theory, which is arguably the only important branch of mathematics that originated from auditory perception. Pythagoras discovered the relationship between pitch and string length: when the ratio of two string lengths is a rational number, the resulting sound is harmonious to the ear. Ancient scholars believed that the basis of pitch was the size of objects, and the inverse relationship between frequency and string length was not discovered until the 17th century by I. Bézout, a teacher of Descartes. In 1625, Mersenne discovered that the relationship between frequency, tension, cross-sectional area, and length of the string was given by

$$\nu \propto \frac{1}{l}\sqrt{\frac{T}{A}}$$

Taylor was the first to derive Mersenne's law from mathematical assumptions, and he found that the simplest shape of a vibrating string is a sine wave, given by

$$y = k \sin \frac{\pi x}{l}$$

He also concluded that the tension T in the string is proportional to the second derivative

In 1747, D'Alembert used Newton's laws of motion to derive the wave equation, obtaining D'Alembert's formula for string motion. He remarked that "the equation contains an infinity of curves", which sparked a significant controversy. Regarding the question of what types of initial displacement function $\phi(x)$ were allowed, d'Alembert believed that $\phi(x)$ must be a periodic and twice differentiable continuous function with a period of $2l$ for the equation to have a solution. Euler held a different view, arguing that $\phi(x)$ could be any curve defined on the interval $[0, l]$. D'Alembert used the method of separation of variables to find one solution to the equation as

$$y(x,t) = A \sin Nx \cos Nt$$

Daniel Bernoulli attempted to bring the debate back to practical physical problems. In 1753, he proposed that the vibrations of a string could consist of an infinite superposition of tones, given by the formula:

$$y(x,t) = c_1 \sin\frac{\pi x}{l}\cos\frac{\pi ct}{l} + c_2 \sin\frac{2\pi x}{l}\cos\frac{2\pi at}{l} + \cdots + c_n \sin\frac{n\pi x}{l}\cos\frac{n\pi at}{l} + \cdots$$

This formula asserted that the string vibration is composed of simple sine modes added together. The nth term in the series represents the nth vibration mode and has a corresponding vibration frequency. The initial displacement is expressed as an infinite sum:

$$\phi(x) = y(x,0) = c_1 \sin\frac{\pi x}{l} + c_2 \sin\frac{2\pi x}{l} + \cdots + c_n \sin\frac{n\pi x}{l} + \cdots$$

Bernoulli believed that this series could represent any initial displacement function $\phi(x)$, but he failed to provide a method for computing the coefficients c_n. Euler disagreed with Bernoulli's view, as he not only could not see any way to compute the coefficients c_n, but also believed that the series required the initial displacement to be a periodic function — although he recognized that a function defined on $[0,l]$ could be made periodic by translation. The debate among the three individuals lasted for several decades, involving many mathematicians. It was not until the early 19th century, with a deepened understanding of trigonometric series, that the controversy gradually subsided.

In the study of heat conduction theory, Fourier used the orthogonality of trigonometric functions to discover a method for determining the coefficients c_n. Fourier's application of trigonometric series achieved such great success that the series later became known as Fourier series. Fourier himself became a fortunate figure in the history of science — his name is destined to be remembered by every student of mathematics and physics.

Finally, let us speak with reverence about the human ear. The ear is a natural Fourier spectrum analyzer, perceiving not the spatiotemporal shape of sound vibrations, but rather decomposing a sound wave into its constituent frequencies, allowing us to distinguish different timbres and pitches, as well as their arbitrary combinations. People can easily identify who is speaking in the adjacent room, whether they are whispering or speaking loudly. A skilled conductor can discern whether each instrument in the orchestra is playing the correct notes. Compared to the ear, the eye seems to be lacking. Although the eye can identify colors, namely the wavelength of light, it lacks the ability to perform spectral analysis on composite light waves. The human eye cannot distinguish that white light is composed of different colors of light, and it is difficult to differentiate between a single wavelength of yellow light at 590 nm and the yellow color created by a mixture of red light at 760 nm and green light at 535 nm. Therefore, unfortunately, humans cannot appreciate symphonies created using light.

Each has its strengths and weaknesses. Due to its exceptional ability to discriminate between different frequencies, the ear can become overwhelmed when multiple frequencies of sound waves mix together, resulting in a sensation of noise. Continuous frequency sound waves can be irritating and disturbing, making it highly unpleasant to honk a car horn randomly. On the other hand, because the eye is not as sensitive to mixed wavelengths, the combination of various colors not only does not cause annoyance but can produce a more pleasant sensation of color. For example, the combination of red and blue light can produce the color purple, which cannot be achieved by a single wavelength of light. Painters create colorful oil paintings using a palette, and the colors created by television technology or textile dyeing technology are far more vivid and rich than natural light — thanks to pseudocolor.

Exercises

[1] Apply pressure to a uniform rod of length l, causing it to contract to length $l(1-2\varepsilon)$. Determine the longitudinal vibrations of the rod after removing the external forces.

[2] Consider a rectangular heat sink with dimensions $a \times b$. One side, $y = b$, is exposed to a higher temperature u_0, while the other three sides are maintained at zero degrees. Determine the steady-state temperature distribution on the cross-section.

[3] Solve the boundary value problem:

$$\begin{cases} u_{tt} - a^2 u_{xx} = bx(l-x) \\ u|_{x=0} = 0, \quad u|_{x=l} = 0 \\ u|_{t=0} = 0, \quad u_t|_{t=0} = 0 \end{cases}$$

[4] Solve the heat conduction problem:

$$\begin{cases} u_t - a^2 u_{xx} = A\sin\omega t \\ u_x|_{x=0} = 0, \quad u|_{x=l} = 0 \\ u|_{t=0} = \phi(x) \end{cases}$$

[5] Solve the transport problem:

$$\begin{cases} u_t - a^2 u_{xx} + bu_x = 0 \\ u|_{x=0} = 0, \quad u|_{x=l} = 0 \\ u|_{t=0} = \phi(x) \end{cases}$$

11.2 Non-Homogeneous Boundary Problems

In Section 11.1, the initial boundary value problems share a common characteristic, which is that the boundary conditions are homogeneous, allowing us to obtain eigenfunctions and eigenvalues. However, how should we deal with non-homogeneous boundary conditions? The approach is to first homogenize the non-homogeneous boundary conditions. The usual method is to attempt a particular solution, $v(x,t)$, which simultaneously satisfies the differential equation and the non-homogeneous boundary conditions. Then, using the linear superposition principle, we set

$$u(x,t) = v(x,t) + w(x,t)$$

By doing this, the original problem with non-homogeneous boundary conditions is transformed into a homogeneous boundary problem for the unknown function $w(x,t)$.

As $w(x,t)$ may satisfy a non-homogeneous equation, it is convenient to find a "better" particular solution, $v(x,t)$, which makes $w(x,t)$ simultaneously satisfies the homogeneous equation and the homogeneous boundary conditions. How can we find this improved particular solution? This requires observing the characteristics of the equation and boundary conditions.

Example 11.7. Solving the non-homogeneous boundary problem:

$$\begin{cases} u_{tt} - a^2 u_{xx} = 0 \\ u|_{x=0} = 0, \quad u|_{x=l} = A\sin\omega t \\ u|_{t=0} = 0, \quad u_t|_{t=0} = 0 \end{cases}$$

Solution: It is evident that $v(x,t) = \frac{Ax}{l}\sin\omega t$ is a particular solution of the equation and satisfies the non-homogeneous boundary conditions. However, the resulting $w(x,t)$ satisfies the homogeneous boundary conditions but follows a non-homogeneous equation. Although it can be solved using the method demonstrated in Section 11.1, the process can be cumbersome at times. We attempt to find a better particular solution that assure both the homogeneous boundary conditions and the homogeneous equation. For this purpose, we try the particular solution:

$$v(x,t) = X(x)\sin\omega t$$

where $X(x)$ is an undetermined function satisfying the boundary conditions $X(0) = 0$ and $X(l) = A$. Substituting $v(x,t)$ into the equation, we can solve for

$$X(x) = A\frac{\sin\frac{\omega x}{a}}{\sin\frac{\omega l}{a}}$$

Thus,

$$v(x,t) = A\frac{\sin\frac{\omega x}{a}}{\sin\frac{\omega l}{a}}\sin\omega t$$

At this point, $w(x,t)$ satisfies the initial value problem:

$$\begin{cases} w_{tt} - a^2 w_{xx} = 0 \\ w|_{x=0} = 0, \quad w|_{x=l} = 0 \\ w|_{t=0} = 0, \quad w_t|_{t=0} = -A\omega\dfrac{\sin\frac{\omega x}{a}}{\sin\frac{\omega l}{a}} \end{cases}$$

Note that the initial conditions for $w(x,t)$ have changed accordingly. We can solve this using the method demonstrated in Example 11.1. Here, we will not repeat the details. The final result is

$$u(x,t) = A\frac{\sin\frac{\omega x}{a}}{\sin\frac{\omega l}{a}}\sin\omega t + \frac{2A\omega}{al}\sum_{n=1}^{\infty}\frac{1}{\frac{\omega^2}{a^2} - \frac{n^2\pi^2}{l^2}}\sin\frac{n\pi at}{l}\sin\frac{n\pi x}{l}$$

Discussion:

(1) Selective Resonance: When the frequency of external driving matches one of the natural frequencies of the system, i.e., $\omega = \frac{n\pi a}{l}$, resonance absorption or release occurs.
(2) Comparison with non-homogeneous equations with homogeneous boundary conditions.
(3) The essence of homogeneous boundary conditions is to isolate the system or keep it in a bound state. Correspondingly, non-homogeneous boundary conditions indicate that the system is open and constantly exchanges and transfers energy with the external environment. Thus, it is fundamentally impossible to reach a stable state. For the vibration equation, there will be no fixed nodes or nodal lines, and the system is in a "forced" vibrational state. Under specific conditions, the system resonates with the external environment, and the relevant physical quantities tend to diverge.

ENDNOTE

So far, we have discussed the solutions to problems of finite-sized systems, and the obtained eigenvalues are discrete. The general solution of the equation requires superimposing all the eigenfunctions to satisfy the initial conditions. If the system is infinite, the eigenvalues may be continuous. For example, considering the variation of surface temperature with depth due to alternating day-night or seasonal changes in atmospheric temperature, a simplified one-dimensional heat conduction problem can be used to model it. In this case, the equation and boundary conditions are as follows:

$$\begin{cases} u_t - a^2 u_{xx} = 0 & (0 \leq x < \infty) \\ u(0,t) = u_0 + A \sin \Omega t, & u|_{t=0} = u_0 \end{cases}$$

Shifting the zero point by $u = v + u_0$, then $v(x,t)$ satisfies the equation and boundary conditions:

$$\begin{cases} v_t - a^2 v_{xx} = 0 & (0 \leq x < \infty) \\ v(0,t) = A \sin \Omega t, & v|_{t=0} = 0 \end{cases}$$

The heat conduction equation resembles the wave equation. Assuming the solution has the form:

$$v(x,t) \propto \mathrm{e}^{\pm \mathrm{i}(kx - \omega t)}$$

Substituting it into the equation, we have

$$\mp \mathrm{i}\omega = -a^2 k^2$$

There can be four values for k, but considering the finiteness of temperature as $x \to \infty$, only two values can be taken

$$k = (\pm 1 + \mathrm{i}) \sqrt{\frac{\omega}{2a^2}}$$

The general solution of the equation is a superposition of solutions with respect to ω:

$$v(x,t) = \sum_{\omega>0} \left[C(\omega)\mathrm{e}^{\mathrm{i}\left(\sqrt{\frac{\omega}{2a^2}}x-\omega t\right)} + D(\omega)\mathrm{e}^{-\mathrm{i}\left(\sqrt{\frac{\omega}{2a^2}}x-\omega t\right)} \right] \mathrm{e}^{-\sqrt{\frac{\omega}{2a^2}}x}$$

From the boundary conditions, we know that

$$C(\Omega) = -D(\Omega) = \frac{A}{2\mathrm{i}}$$

All other $C(\omega)$ and $D(\omega)$ are zero, and thus the solution of the equation is

$$u(x,t) = u_0 + A\mathrm{e}^{-\frac{x}{\delta}} \sin\left(\Omega t - \frac{x}{\delta}\right)$$

where $\delta = a\sqrt{\frac{2}{\Omega}}$. From the final result, the following conclusions can be drawn:

(1) Due to atmospheric influence, underground temperature decays exponentially with depth.

(2) As $\delta \propto \sqrt{\frac{1}{\Omega}}$, the deeper the frequency variation, the greater the decay gradient of underground temperature. This is similar to the penetration of electromagnetic waves into a metal surface, and δ is also referred to as the skin depth.

Exercises

[1] Solve the boundary value problem:

$$\begin{cases} u_{tt} - a^2 u_{xx} = 0 & (0 \le x \le l) \\ u|_{x=0} = \cos\dfrac{\pi a t}{l}, & u_x|_{x=l} = 0 \\ u|_{t=0} = \cos\dfrac{\pi x}{l}, & u_t|_{t=0} = \sin\dfrac{\pi x}{2l} \end{cases}$$

[2] In the rectangular domain $0 \le x \le a, 0 \le y \le b$, solve the problem:

$$\begin{cases} \Delta_2 u = -x^2 y \\ u|_{x=0} = u|_{x=a} = A \\ u|_{y=0} = u|_{y=b} = 0 \end{cases}$$

[3] Solve the boundary value problem:

$$\begin{cases} u_t - \kappa u_{xx} = 0 \\ u|_{x=0} = A\mathrm{e}^{-\alpha^2 \kappa t}, \quad u|_{x=l} = B\mathrm{e}^{-\beta^2 \kappa t} \\ u|_{t=0} = 0 \end{cases}$$

Hint: Assume a particular solution:

$$v(x,t) = f(x)\mathrm{e}^{-\alpha^2 \kappa t} + g(x)\mathrm{e}^{-\beta^2 \kappa t}$$

and substitute it into the equation to determine $f(x)$ and $g(x)$.

[4] Consider a uniform rod of length l, with one end fixed and the other end subjected to a force $F(t) = F_0 \sin \omega t$. The initial displacement and velocity are given by $\phi(x)$ and $\psi(x)$ respectively. Solve for the longitudinal vibration of the rod.
Hint: Let $u(x,t) = f(x)\mathrm{e}^{\mathrm{i}\omega t}$, and then take the imaginary part. Alternatively, the problem can be solved using Laplace transform.

[5] Solve the boundary value problem in the rectangular region $0 \le x \le a$, $0 \le y \le b$:

$$\begin{cases} \Delta_2 u = 0 \\ u|_{x=0} = Ay(b-y), \quad u|_{x=a} = 0 \\ u|_{y=0} = B \sin \dfrac{\pi x}{a}, \quad u|_{y=b} = 0 \end{cases}$$

Hint: Let $u = v + w$, where v and w satisfy the following conditions:

$$\begin{cases} \Delta_2 v = 0 \\ v|_{x=0} = 0, \quad v|_{x=a} = 0 \\ v|_{y=0} = B \sin \dfrac{\pi x}{a}, \quad v|_{y=b} = 0 \end{cases} \qquad \begin{cases} \Delta_2 w = 0 \\ w|_{x=0} = Ay(b-y), \quad w|_{x=a} = 0 \\ w|_{y=0} = 0, \quad w|_{y=b} = 0 \end{cases}$$

11.3 Periodic Boundary Problems

For circular domain problems, it is often convenient to use polar coordinates. Due to the periodicity of the quantity distribution, $u(r, \theta + 2\pi) = u(r, \theta)$, it determines the eigenvalues in the θ coordinate direction. Therefore, homogeneous boundary conditions are no longer necessary, and this constraint is called periodic boundary conditions.

11.3.1 *Homogeneous equation*

Example 11.8. Boundary value problem in a circular domain: Consider a thin circular disk of radius a with adiabatic top and bottom surfaces. The temperature distribution along the circumference is a known function $f(x, y)$. We need to find the temperature distribution inside the disk in the steady state.

$$\begin{cases} \Delta_2 u = 0 \quad (x^2 + y^2 < a^2) \\ u|_{x^2+y^2=a^2} = f(x, y) \end{cases}$$

Solution: Since the boundary shape has axial symmetry, we adopt polar coordinates:

$$\begin{cases} \dfrac{1}{\rho}\dfrac{\partial}{\partial \rho}\left(\rho \dfrac{\partial u}{\partial \rho}\right) + \dfrac{1}{\rho^2}\dfrac{\partial^2 u}{\partial \theta^2} = 0 \\ u(a, \theta) = f(\theta) \end{cases}$$

The periodic condition is $u(\rho, \theta + 2\pi) = u(\rho, \theta)$. Additionally, it needs to satisfy the physical condition that $u(0, \theta)$ must be finite. Assuming a separation of variables form, $u(\rho, \theta) = R(\rho)\Theta(\theta)$, we obtain

$$\begin{cases} \Theta'' + \lambda\Theta = 0 \\ \Theta(\theta + 2\pi) = \Theta(\theta) \end{cases}$$

The eigenvalues and eigenfunctions for the angular part can be solved as

$$\Theta_m(\theta) = a_m \cos m\theta + b_m \sin m\theta$$
$$\lambda_m = m^2 \quad (m = 0, 1, 2, \dots)$$

The radial part satisfies the Euler-type equation:

$$\rho^2 R'' + \rho R' - \lambda R = 0$$

Substituting the eigenvalue λ_m, we get

$$R_m(\rho) = \begin{cases} c_0 + d_0 \ln \rho & (m = 0) \\ c_m \rho^m + \dfrac{d_m}{\rho^m} & (m > 0) \end{cases}$$

Considering that the physical solution at $\rho = 0$ should be finite, we set $d_0 = d_m = 0$ to discard the diverging solutions. The general solution is obtained by superposition of all eigen-solutions:

$$u(\rho, \theta) = a_0 + \sum_{m=1}^{\infty} (a_m \cos m\theta + b_m \sin m\theta)\rho^m$$

Using the boundary condition $u(r)|_{\rho=a} = f(\theta)$, we determine the superposition coefficients a_m and b_m as

$$a_0 = \frac{1}{2\pi}\int_0^{2\pi} f(t)\mathrm{d}t, \quad a_m = \frac{1}{a^m\pi}\int_0^{2\pi} f(t)\cos mt\mathrm{d}t$$

$$b_m = \frac{1}{a^m\pi}\int_0^{2\pi} f(t)\sin mt\,\mathrm{d}t$$

Finally, the solution is given by

$$\begin{aligned} u(r, \theta) &= \frac{1}{\pi}\int_0^{2\pi} f(t)\left\{\frac{1}{2} + \sum_{m=1}^{\infty}\left(\frac{r}{a}\right)^m \cos\left[m(\theta - t)\right]\right\}\mathrm{d}t \\ &= \frac{1}{2\pi}\int_0^{2\pi} \frac{f(t)(a^2 - r^2)}{a^2 + r^2 - 2ar\cos(\theta - t)}\mathrm{d}t \end{aligned}$$

This exactly is the Poisson formula for circular domains addressed in Section 2.3.

Practice: Try deriving the last step of the above equation.

Example 11.9. A grounded conductor cylinder of radius a is placed in an infinite uniform electric field $\mathbf{E}_0$, with symmetry axis perpendicular to the direction of the electric field. Find the electric field distribution near the conductor cylinder.

Solution: This problem can be treated as a two-dimensional problem that satisfies the Laplace's equation: $\Delta_2 u = 0$. Since the boundary is a circle, we choose polar coordinates:

$$\begin{cases} \dfrac{\partial^2 u}{\partial \rho^2} + \dfrac{1}{\rho}\dfrac{\partial u}{\partial \rho} + \dfrac{1}{\rho^2}\dfrac{\partial^2 u}{\partial \theta^2} = 0 \quad (\rho > a) \\ u|_{\rho=a} = 0 \\ u|_{\rho\to\infty} = -E_0 \rho \cos\theta + \dfrac{q_0}{2\pi\varepsilon_0} \ln \dfrac{a}{\rho} \end{cases}$$

The boundary conditions are derived from physical analysis: At infinity, it can be considered as a uniform electric field. The term associated to q_0 accounts for the induced charges on the grounded conductor in the external electric field. It also generates an electric field distribution in space, as shown in Figure 11.3. By separating variables and using periodic conditions, we obtain the eigenfunctions in the angular direction:

$$\Phi_m(\theta) = A_m \cos m\theta + B_m \sin m\theta$$
$$\lambda_m = m^2 \quad (m = 0, 1, 2, \ldots)$$

The corresponding solution for the radial part is

$$R_m(\rho) = \begin{cases} C_0 + D_0 \ln \rho & (m = 0) \\ C_m \rho^m + \dfrac{D_m}{\rho^m} & (m > 0) \end{cases}$$

The general solution is the linear superposition of all eigen-solutions:

$$u(\rho, \theta) = C_0 + D_0 \ln \rho + \sum_{m=1}^{\infty} (A_m \cos m\theta + B_m \sin m\theta)\rho^m + \sum_{m=1}^{\infty} (C_m \cos m\theta + D_m \sin m\theta)\rho^{-m}$$

By applying the boundary condition $u|_{\rho=a} = 0$, we have

$$C_0 + D_0 \ln a = 0, \quad A_m a^m + C_m a^{-m} = 0, \quad B_m a^m + D_m a^{-m} = 0$$

With the condition as $\rho \to \infty$, there are other relationships between the coefficients:

$$C_0 = \frac{q_0}{2\pi\varepsilon_0}\ln a, \quad D_0 = -\frac{q_0}{2\pi\varepsilon_0}$$
$$A_1 = -E_0, \quad A_m = C_m = 0 \ (m \neq 1), \quad C_1 = E_0 a^2$$
$$B_m = D_m = 0$$

The final result is

$$u(\rho,\theta) = \frac{q_0}{2\pi\varepsilon_0}\ln\frac{a}{\rho} - E_0\rho\cos\theta + E_0\frac{a^2}{\rho}\cos\theta$$

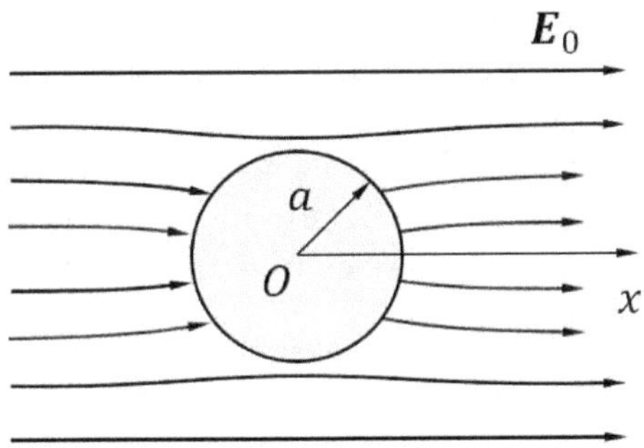

Figure 11.3.

11.3.2 *Non-homogeneous equation*

For the two-dimensional Poisson's equation $\Delta u = f(r,\theta)$, the method of particular solutions is commonly used. This involves finding a particular solution $v(x)$ for the equation and then letting $u = v + w$, transforming the problem into the homogeneous Laplace's equation $\Delta w = 0$, which can be solved.

Example 11.10. For solving the boundary value problem of the Poisson's equation on a circular domain with radius ρ_0:

$$\begin{cases} \Delta_2 u = a + b(x^2 - y^2) \\ u|_{\rho=\rho_0} = c \end{cases}$$

Solution: The non-homogeneous term of the equation is a polynomial, so we choose a polynomial as the particular solution. Since the boundary is a circle, the particular solution should be expressed in terms of polar coordinates. Therefore, we take

$$v(x,y) = \frac{a}{4}(x^2+y^2) + \frac{b}{12}(x^4-y^4) = \frac{a}{4}\rho^2 + \frac{b}{12}\rho^4 \cos 2\theta$$

The problem regarding $w(x,y)$ becomes a well-defined problem:

$$\begin{cases} \Delta_2 w = 0 \\ w|_{\rho=\rho_0} = c - \frac{a}{4}\rho_0^2 - \frac{b}{12}\rho_0^4 \cos 2\theta \equiv f(\theta) \end{cases}$$

This reduces to the homogeneous equation problem as in Example 11.8 and we can proceed step by step to solve it.

The vibrations of a circular membrane may be associated to Bessel functions, which will be discussed separately in the future. Figure 11.4 illustrates some eigenmodes of vibrations, where the nodal lines form concentric circles and radial lines. Generally, there are no degenerate eigenvalues.

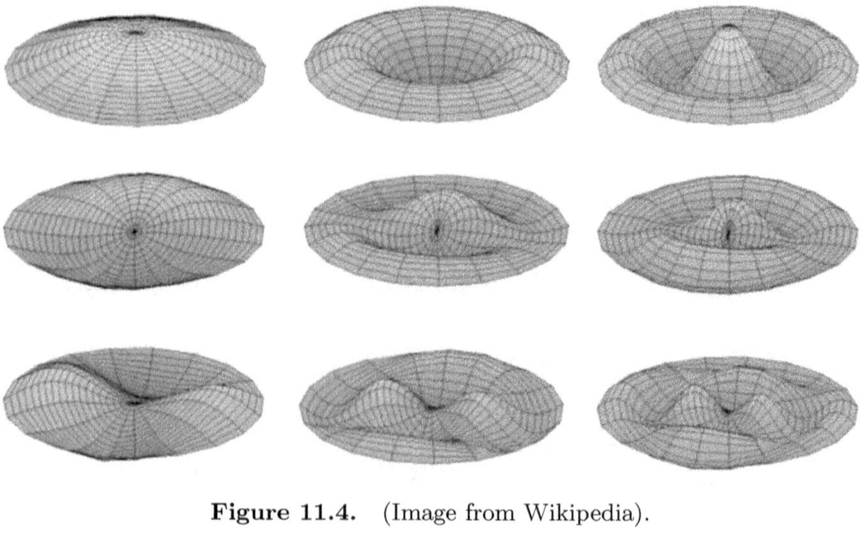

Figure 11.4. (Image from Wikipedia).

ENDNOTE

Periodic boundary conditions and homogeneous boundary conditions are similar in that they both constitute eigenvalue problems to ensure the stability of eigenmodes. The difference between the two lies in the fact that eigenmodes under homogeneous boundary conditions are non-degenerate, while eigenmodes under periodic conditions are degenerate. For the wave equation, the single non-propagating eigenmode under homogeneous boundary conditions is given by

$$\begin{aligned} u_n(x,t) &= \sin\frac{n\pi x}{l}\left[A_n \sin\frac{n\pi at}{l} + B_n \cos\frac{n\pi at}{l}\right] \\ &= N_n \sin\frac{n\pi x}{l}\sin\left(\frac{n\pi at}{l} + \phi_n\right) \end{aligned}$$

This represents a current-free standing wave state. On the other hand, under periodic boundary conditions, the doubly degenerate eigenmodes are

$$\begin{cases} u_m^{(1)} = \cos\dfrac{2m\pi x}{l}\left[A_m \cos\dfrac{m\pi at}{l} + B_m \sin\dfrac{m\pi at}{l}\right] \\ u_m^{(2)} = \sin\dfrac{2m\pi x}{l}\left[C_m \cos\dfrac{m\pi at}{l} + D_m \sin\dfrac{m\pi at}{l}\right] \end{cases}$$

These degenerate standing wave modes can be recombined into two counter-propagating traveling wave modes:

$$\begin{cases} u_m^{(-)} = N_m^{(-)} \sin\left[\dfrac{2m\pi(x-at)}{l} + \phi_m^{(-)}\right] \\ u_m^{(+)} = N_m^{(+)} \sin\left[\dfrac{2m\pi(x+at)}{l} + \phi_m^{(+)}\right] \end{cases}$$

Each mode represents a stable current-carrying state.

In the early development of quantum physics, Erwin Schrödinger used the de Broglie's matter wave concept to explain Niels Bohr's quantized model of atomic orbits. He believed that electron waves in atom appear as stable standing wave patterns with orbital circumference of $2\pi r_n = n\lambda$, as shown in Figure 11.5(a). Such standing wave states have no Amperean molecular current, resulting in the absence of atomic magnetism. However, the traveling wave mode shown in Figure 11.5(b) is also stable and carries orbital currents, exhibiting paramagnetism and Zeeman effect under an external magnetic field. The standing wave atomic model should incorporate both of these modes simultaneously.

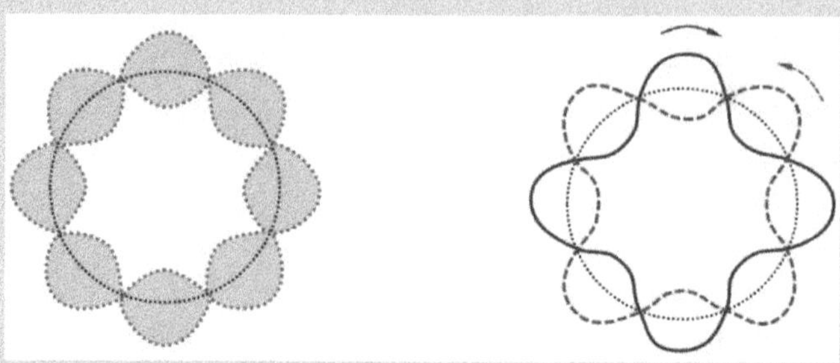

(a) (b)

Figure 11.5.

For the Poisson equation, if a suitable particular solution cannot be found, one has to resort to the method described in Section 10.1, which involves first obtaining the eigenfunctions of the corresponding homogeneous equation, i.e., the Laplace equation. Then, the solution of the equation, as well as the non-homogeneous term, are both expanded as a sum of these eigenfunctions. This results in a system of ordinary differential equations that the undetermined coefficients satisfy. For example, we need to solve the following Poisson equation with boundary conditions:

$$\begin{cases} \Delta_2 u = f(\rho, \theta) \\ u|_{\rho=\rho_0} = \phi(\theta) \end{cases}$$

The eigenfunctions of the Laplace equation are given by

$$\Phi_m(\theta) = A_m \cos m\theta + B_m \sin m\theta$$
$$\lambda_m = m^2 \quad (m = 0, 1, 2, \ldots)$$

The solution of the equation can be expressed as

$$u = \sum_{m=0}^{\infty} [A_m(\rho) \cos m\theta + B_m(\rho) \sin m\theta]$$

Assuming that the "undetermined coefficients" depend on the radial coordinate ρ, the non-homogeneous term of the equation can also be expanded as

$$f(\rho, \theta) = \sum_{m=0}^{\infty} [g_m(\rho) \cos m\theta + h_m(\rho) \sin m\theta]$$

Substituting these expansions into the original equation, we obtain the ordinary differential equations that $A_m(\rho)$ and $B_m(\rho)$ satisfy

$$\frac{\mathrm{d}^2 A_m(\rho)}{\mathrm{d}\rho^2} + \frac{1}{\rho}\frac{\mathrm{d}A_m(\rho)}{\mathrm{d}\rho} - \frac{m^2}{\rho^2} A_m(\rho) = g_m(\rho)$$
$$\frac{\mathrm{d}^2 B_m(\rho)}{\mathrm{d}\rho^2} + \frac{1}{\rho}\frac{\mathrm{d}B_m(\rho)}{\mathrm{d}\rho} - \frac{m^2}{\rho^2} B_m(\rho) = h_m(\rho)$$

The corresponding homogeneous equations are of the Euler type and their solutions are ρ^m and ρ^{-m}. Using the method described in Section 9.2, we can find a particular solution for the non-homogeneous equations. The coefficients in the general solution are determined by the boundary conditions.

Therefore, by expanding the boundary condition $u|_{\rho=\rho_0} = \phi(\theta)$ in terms of the eigenfunctions, we have

$$\phi(\theta) = \sum_{m=0}^{\infty} [a_m \cos m\theta + b_m \sin m\theta]$$

Thus, $A_m(\rho)$ and $B_m(\rho)$ satisfy the boundary conditions on the circular boundary:

$$A_m(\rho_0) = a_m, \quad B_m(\rho_0) = b_m$$

Taking into account physical considerations, $A_m(0)$ and $B_m(0)$ should be finite, which leads to a well-posed boundary value problem for the equation.

Exercises

[1] Solve the boundary value problem in the circular domain $\rho < \rho_0$:

$$\text{(a)} \begin{cases} \Delta_2 u = -xy \\ u|_{\rho=\rho_0} = 0 \end{cases}; \quad \text{(b)} \begin{cases} \Delta_2 u = -x^4 y \\ u|_{\rho=\rho_0} = 0 \end{cases}; \quad \text{(c)} \begin{cases} \Delta_2 u = -x^3 y^2 \\ u|_{\rho=\rho_0} = c \end{cases}$$

[2] Solve the boundary value problem in the circular domain $\rho_1 < \rho < \rho_2$:

$$\begin{cases} \Delta_2 u = 1 - 2\sin 2\varphi \\ u|_{\rho=\rho_1} = u|_{\rho=\rho_2} = 0 \end{cases}$$

[3] A long dielectric cylinder of radius a is placed in an infinite uniform electric field $\boldsymbol{E}_0$, with symmetry axis perpendicular to the direction of the electric field. The electrical permittivity is ε. Find the electric field distribution inside and outside of the cylinder.

[4] A hollow cylindrical conductor with radius R is cut into two semicircles with the upper part having potential V and the lower part having potential $-V$, find the potential distribution inside the cylinder.

Answer:

$$u(\rho,\theta) = \frac{4V}{\pi}\sum_{n=1}^{\infty}\frac{1}{2k+1}\left(\frac{\rho}{R}\right)^{2k+1}\sin(2k+1)\theta$$

[5] Solve the wave equation for a circular closed string and analyze the difference with the fixed boundary condition:

$$\begin{cases} u_{tt} - a^2 u_{xx} = 0 \quad (0 \le x \le l) \\ u(0,t) = u(l,t) \\ u(x,0) = \phi(x), \quad u_t(x,0) = \psi(x) \end{cases}$$

[6] Consider a segment of an annular region as shown in Figure 11.6. Solve the steady-state temperature distribution under the following boundary conditions:

$$\text{(a)} \begin{cases} u|_{\rho=\rho_1} = f_1(\theta) \\ u|_{\rho=\rho_2} = f_2(\theta) \end{cases}, \quad \begin{cases} u|_{\theta=0} = 0 \\ u|_{\theta=\alpha} = 0 \end{cases}$$

$$\text{(b)} \begin{cases} u|_{\rho=\rho_1} = 0 \\ u|_{\rho=\rho_2} = 0 \end{cases}, \quad \begin{cases} u|_{\theta=0} = g_1(\rho) \\ u|_{\theta=\alpha} = g_2(\rho) \end{cases}$$

$$\text{(c)} \begin{cases} u|_{\rho=\rho_1} = f_1(\theta) \\ u|_{\rho=\rho_2} = f_2(\theta) \end{cases}, \quad \begin{cases} u|_{\theta=0} = g_1(\rho) \\ u|_{\theta=\alpha} = g_2(\rho) \end{cases}$$

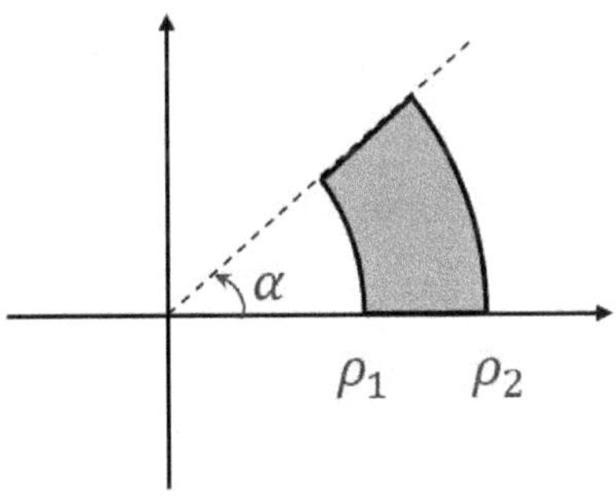

Figure 11.6.

Hint: The eigenvalues and eigenfunctions are given by

$$\text{(a) } \Theta_n(\theta) = \sin\frac{n\pi\theta}{\alpha}, \quad \lambda_n = \left(\frac{n\pi}{\alpha}\right)^2 \quad (n = 1, 2, 3\ldots)$$

(b) $R_n(\rho) = \sin\left[\frac{n\pi}{\ln\left(\frac{\rho_2}{\rho_1}\right)} \ln \frac{\rho}{\rho_1}\right], \quad \lambda_n = \left[\frac{n\pi}{\ln\left(\frac{\rho_2}{\rho_1}\right)}\right]^2$

$$(n = 1, 2, 3 \cdots)$$

(c) $u = v + w$

$$\begin{cases} v|_{\rho=\rho_1} = f_1(\theta) \\ v|_{\rho=\rho_2} = f_2(\theta) \end{cases}, \quad \begin{cases} v|_{\theta=0} = 0 \\ v|_{\theta=\alpha} = 0 \end{cases}$$

$$\begin{cases} w|_{\rho=\rho_1} = 0 \\ w|_{\rho=\rho_2} = 0 \end{cases}, \quad \begin{cases} w|_{\theta=0} = g_1(\rho) \\ w|_{\theta=\alpha} = g_2(\rho) \end{cases}$$

11.4 Joining Problems

In this section, we discuss the well-posed problem of non-uniform systems with joining conditions. Through this, we can further understand the relationship between Fourier series expansion and dynamic processes, particularly in the context of heat conduction problems, where Fourier eigenmodes play an incredible role.

Example 11.11. Consider two segments of lightweight and flexible strings with lengths l_1 and l_2, densities ρ_1 and ρ_2, respectively. The segments are perfectly connected with fixed ends. Letting $l = l_1 + l_2$ and assuming the tension in the string is T, we aim to study the eigenmodes of this system.

Solution: The wave equations and the boundary conditions are

$$\begin{cases} \frac{\partial^2 u_1}{\partial t^2} = a_1^2 \frac{\partial^2 u_1}{\partial x^2} \qquad (0 \le x \le l_1) \\ u_1(0, t) = 0 \\ u_1(x, 0) = \phi_1(x), \quad u_{1t}(x, 0) = \psi_1(x) \end{cases}$$

$$\begin{cases} \frac{\partial^2 u_2}{\partial t^2} = a_2^2 \frac{\partial^2 u_2}{\partial x^2} \qquad (l_1 \le x \le l) \\ u_2(l, t) = 0 \\ u_2(x, 0) = \phi_2(x), \quad u_{2t}(x, 0) = \psi_2(x) \end{cases}$$

Here, $a_1^2 = \frac{T}{\rho_1}$ and $a_2^2 = \frac{T}{\rho_2}$. The joining conditions at $x = l_1$ are given by

$$u_1(l_1, t) = u_2(l_1, t), \quad u_{1x}(l_1, t) = u_{2x}(l_1, t)$$

We separate the variables as $u_j(x, t) = T_j(t)X_j(x)$, yielding

$$\begin{cases} T_1''(t) + \omega^2 T_1(t) = 0 \\ X_1''(x) + \dfrac{\omega^2}{a_1^2} X_1(x) = 0 \end{cases}, \quad \begin{cases} T_2''(t) + \omega^2 T_2(t) = 0 \\ X_2''(x) + \dfrac{\omega^2}{a_1^2} X_2(x) = 0 \end{cases}$$

For the eigenmodes, the two segments of the string should have the same vibrational frequency to maintain continuity at the interface. Thus, we take a common separation constant ω^2. Since both ends are fixed, the eigenfunctions take the following form:

$$\begin{cases} X_1(x) = A_1 \sin \dfrac{\omega x}{a_1} & (0 \le x \le l_1) \\ X_2(x) = A_2 \sin \dfrac{\omega(l - x)}{a_2} & (l_1 \le x \le l) \end{cases}$$

with

$$T_1(t) \equiv T_2(t) = C \cos \omega t + D \sin \omega t$$

From the joining conditions, we have

$$A_1 \sin \frac{\omega l_1}{a_1} = A_2 \sin \frac{\omega l_2}{a_2}$$
$$\frac{A_1}{a_1} \cos \frac{\omega l_1}{a_1} = -\frac{A_2}{a_2} \cos \frac{\omega l_2}{a_2}$$

Therefore

$$a_1 \tan \frac{\omega l_1}{a_1} + a_2 \tan \frac{\omega l_2}{a_2} = 0, \quad \frac{A_1}{A_2} = \frac{\sin \frac{\omega l_2}{a_2}}{\sin \frac{\omega l_1}{a_1}}$$

This is a bi-periodic equation with infinitely many discrete eigen frequencies ω_n. Figure 11.7 depicts the distribution of eigen frequencies ω_n, which corresponds to the abscissa values of the function zeros. In this case, we take $l_1 = l_2 = \frac{l}{2}$ and $a_1 = \frac{5}{2}a_2$. Table 11.1 lists some lower eigen values ω_n. The corresponding eigen functions are given by

$$\begin{cases} X_{1n}(x) = \sin\dfrac{\omega_n l_2}{a_2}\sin\dfrac{\omega_n x}{a_1} & (0 \le x \le l_1) \\ X_{2n}(x) = \sin\dfrac{\omega_n l_1}{a_1}\sin\dfrac{\omega_n (l-x)}{a_2} & (l_1 \le x \le l) \end{cases}$$

Figure 11.8 illustrates some low-order eigen functions, showing that the number of nodes in the eigen functions continues to increase.

Table 11.1.

n	1	2	3	4	5
ω_n	3.912	9.204	14.019	17.411	22.208

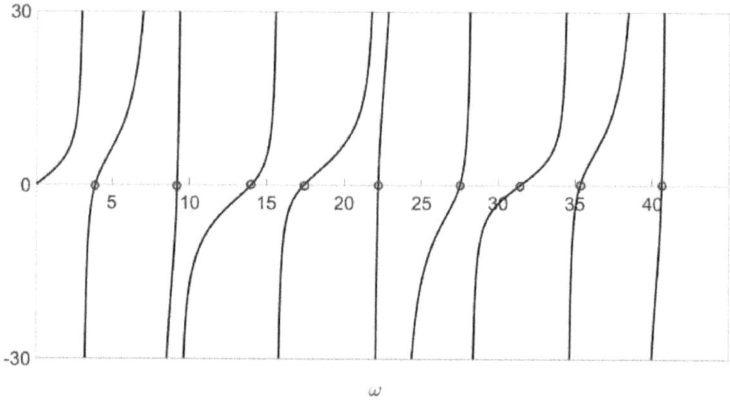

Figure 11.7.

It will be seen later that the orthogonality and completeness of the eigen functions are guaranteed by general eigenvalue theory. Orthogonality can also be directly verified, although the calculation process

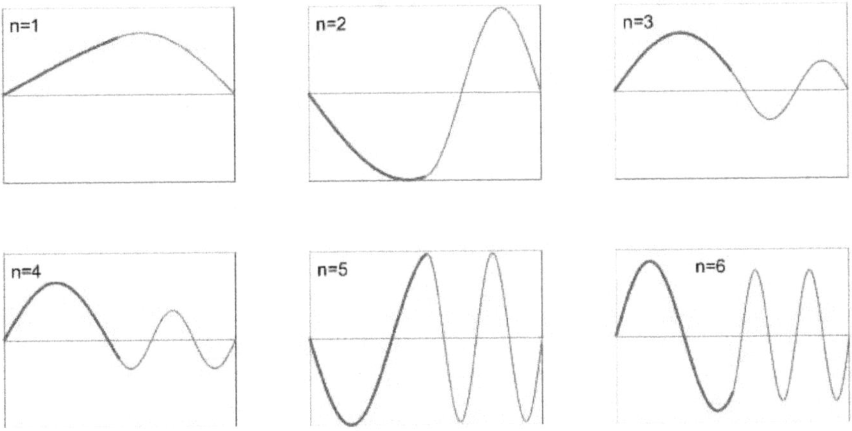

Figure 11.8.

is a bit tedious. Readers who lack patience can skip it.

$$
\begin{aligned}
&\int_0^l X_n(x)X_m(x)\mathrm{d}x \\
&= \int_0^{l_1} X_{1n}(x)X_{1m}(x)\mathrm{d}x + \int_{l_1}^{l} X_{2n}(x)X_{2m}(x)\mathrm{d}x \\
&= \frac{a_1}{\omega_n^2-\omega_m^2}\sin\frac{\omega_n l_2}{a_2}\sin\frac{\omega_m l_2}{a_2}\left[\omega_m \sin\frac{\omega_n l_1}{a_1}\cos\frac{\omega_m l_1}{a_1}\right. \\
&\quad \left. -\,\omega_n\cos\frac{\omega_n l_1}{a_1}\sin\frac{\omega_m l_1}{a_1}\right] + \frac{a_2}{\omega_n^2-\omega_m^2}\sin\frac{\omega_n l_1}{a_1}\sin\frac{\omega_m l_1}{a_1} \\
&\quad \times\left[\omega_m\sin\frac{\omega_n l_2}{a_2}\cos\frac{\omega_m l_2}{a_2} - \omega_n\cos\frac{\omega_n l_2}{a_2}\sin\frac{\omega_m l_2}{a_2}\right] \\
&= \frac{a_1}{\omega_n^2-\omega_m^2}\sin\frac{\omega_n l_2}{a_2}\sin\frac{\omega_m l_2}{a_2}\cos\frac{\omega_m l_1}{a_1}\cos\frac{\omega_n l_1}{a_1} \\
&\quad \times\left[\omega_m\tan\frac{\omega_n l_1}{a_1} - \omega_n\tan\frac{\omega_m l_1}{a_1}\right] \\
&\quad + \frac{a_2}{\omega_n^2-\omega_m^2}\sin\frac{\omega_n l_1}{a_1}\sin\frac{\omega_m l_1}{a_1}\cos\frac{\omega_m l_2}{a_2}\cos\frac{\omega_n l_2}{a_2} \\
&\quad \times\left[\omega_m\tan\frac{\omega_n l_2}{a_2} - \omega_n\tan\frac{\omega_m l_2}{a_2}\right]
\end{aligned}
$$

$$
\begin{aligned}
&= \frac{a_1}{\omega_n^2 - \omega_m^2} \sin\frac{\omega_n l_2}{a_2} \sin\frac{\omega_m l_2}{a_2} \cos\frac{\omega_m l_1}{a_1} \cos\frac{\omega_n l_1}{a_1} \\
&\quad \times \left[\omega_m \tan\frac{\omega_n l_1}{a_1} - \omega_n \tan\frac{\omega_m l_1}{a_1}\right] \\
&\quad - \frac{a_1}{\omega_n^2 - \omega_m^2} \sin\frac{\omega_n l_1}{a_1} \sin\frac{\omega_m l_1}{a_1} \cos\frac{\omega_m l_2}{a_2} \cos\frac{\omega_n l_2}{a_2} \\
&\quad \times \left[\omega_m \tan\frac{\omega_n l_1}{a_1} - \omega_n \tan\frac{\omega_m l_1}{a_1}\right] \\
&= \frac{a_1}{\omega_n^2 - \omega_m^2} \left[\omega_m \tan\frac{\omega_n l_1}{a_1} - \omega_n \tan\frac{\omega_m l_1}{a_1}\right] \\
&\quad \times \left[\sin\frac{\omega_n l_2}{a_2} \sin\frac{\omega_m l_2}{a_2} \cos\frac{\omega_m l_1}{a_1} \cos\frac{\omega_n l_1}{a_1}\right. \\
&\quad \left. - \sin\frac{\omega_n l_1}{a_1} \sin\frac{\omega_m l_1}{a_1} \cos\frac{\omega_m l_2}{a_2} \cos\frac{\omega_n l_2}{a_2}\right] \\
&= \frac{a_1}{\omega_n^2 - \omega_m^2} \left[\omega_m \tan\frac{\omega_n l_1}{a_1} - \omega_n \tan\frac{\omega_m l_1}{a_1}\right] \\
&\quad \times \left\{-\frac{1}{2}\left[\cos\frac{l_2(\omega_n + \omega_m)}{a_2} - \cos\frac{l_2(\omega_n - \omega_m)}{a_2}\right]\right. \\
&\quad \left. \times \frac{1}{2}\left[\cos\frac{l_1(\omega_n + \omega_m)}{a_1} + \cos\frac{l_1(\omega_n - \omega_m)}{a_1}\right]\right\} \\
&\quad - \left\{-\frac{1}{2}\left[\cos\frac{l_1(\omega_n + \omega_m)}{a_1} - \cos\frac{l_1(\omega_n - \omega_m)}{a_1}\right]\right. \\
&\quad \left. \times \frac{1}{2}\left[\cos\frac{l_2(\omega_n + \omega_m)}{a_2} + \cos\frac{l_2(\omega_n - \omega_m)}{a_2}\right]\right\} \\
&= \frac{a_1}{\omega_n^2 - \omega_m^2} \left[\omega_m \tan\frac{\omega_n l_1}{a_1} - \omega_n \tan\frac{\omega_m l_1}{a_1}\right] \\
&\quad \times \frac{1}{2}\left[\cos\frac{l_2(\omega_n - \omega_m)}{a_2} \cos\frac{l_1(\omega_n + \omega_m)}{a_1}\right. \\
&\quad \left. - \cos\frac{l_1(\omega_n - \omega_m)}{a_1} \cos\frac{l_2(\omega_n + \omega_m)}{a_2}\right] = 0
\end{aligned}
$$

Hence, the eigen functions $X_n(x)$ are orthogonal. The final step utilizes symmetry:

$$l_1 \longleftrightarrow l_2, \quad a_1 \longleftrightarrow a_2$$

The calculation of orthogonal modes is not listed here. The general solution of the equation is constructed by superposing the eigenfunctions, and the superposition coefficients are determined by the initial conditions, which are omitted.

ENDNOTE

In studying heterogeneous systems, it is crucial to have a unified description of physical quantities that characterize the system. For vibrational or wave systems, it is generally assumed that the eigenfrequencies should be consistent across different parts so that they can match and resonate with each other at the interfaces, leading to a stable state of resonance throughout the system. This can be easily understood in physics. For quantum systems (see Exercise [1]), due to energy conservation, the total energy of particles in different locations is the same, which is also unquestionable. With the de Broglie relation, having the same eigenenergy implies having the same frequencies at different locations, which aligns perfectly with the concept of resonance based on eigenfrequencies.

However, for systems involving heat conduction or diffusion (see Exercise [2]), considering eigenfrequencies as a unified characteristic of motion seems to lack physical basis since the system does not exhibit vibrational features. One possible explanation is that under a single eigenmode, the temperatures of different parts of the system decrease at a consistent rate, which enables the overall harmonicity of the evolution equation. Another perspective could be to interpret the temperature of eigenmodes as exponentially decaying "imaginary frequency" oscillations:

$$\tilde{\omega}_n = \frac{in\pi a}{l}$$

This suggests that the process of heat conduction or diffusion actually occurs in a synchronized manner through "imaginary frequency resonance". After all, the heat conduction equation differs from the Schrödinger equation by only an imaginary unit, i.

Resonance seems to manifest harmoniously in various aspects of the universe, and the underlying physics behind the Fourier series expansion still remains not fully understood.

Exercises

[1] Find the eigenenergy spectrum of an electron in a one-dimensional potential well ($\hbar = m = 1$):

$$-\frac{1}{2}\frac{\mathrm{d}^2}{\mathrm{d}x^2}\psi + V(x)\psi = E\psi$$

where the potential well is defined as

$$V(x) = \begin{cases} 0 & \left(0 < x < \dfrac{d}{2}\right) \\ V_0 & \left(\dfrac{d}{2} < x < d\right) \\ \infty & (x < 0, x > d) \end{cases}$$

Hint: At $x = \frac{d}{2}$, both $\psi(x)$ and $\psi'(x)$ should be continuous.

[2] Consider two rods with lengths l_1 and l_2, and thermal conductivity coefficients κ_1 and κ_2, respectively. After perfectly connecting them, one end is maintained at a constant temperature of zero, and the other end is adiabatic. Study the process of heat conduction.

Hint: The continuity conditions are given by

$$\begin{cases} u_1(l_1, t) = u_2(l_1, t) \\ \kappa_1 u_{1x}(l_1, t) = \kappa_2 u_{2x}(l_1, t) \end{cases}$$

[3] As shown in Figure 11.9, a one-dimensional circular ring with a radius of R is formed by connecting two semicircular rings of different materials, each with thermal conductivity coefficients, specific heat and mass density κ_1, c_1, ρ_1 and κ_2, c_2, ρ_2, respectively. Let T_1 and T_2 be the initial temperatures of the semicircular rings. Find the variation of temperature on the ring over time.
Hint: Let $u_j(\theta, t) = \mathrm{e}^{-k^2 a_j^2 t} v_j(\theta)$, where $a_j^2 = \frac{\kappa_j}{c_j \rho_j}$.

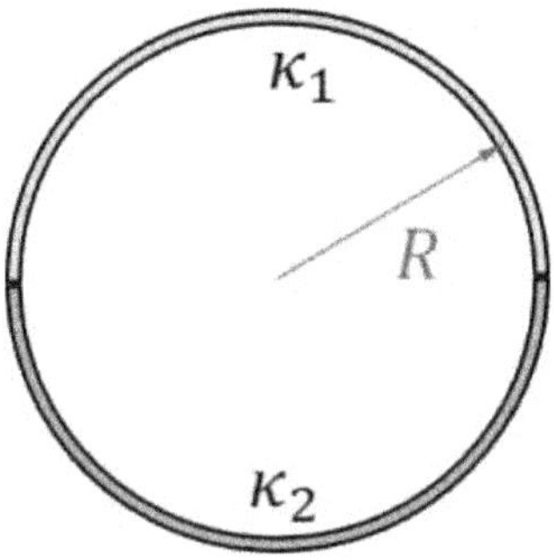

Figure 11.9.

Chapter 12

Integral Transform Methods

Chapter 11 introduced the method of solving boundary value problems for finite systems, which involves expanding the physical quantities of the system in terms of Fourier series and determining the expansion coefficients based on the given conditions. In this chapter, we will discuss infinite systems that are not constrained by boundary conditions. For such problems, Fourier integral expansion, also known as Fourier transformation, can be used. For initial value problems, the Laplace transformation method can still be employed. Both of these methods for solving differential equations fall under the category of integral transformations.

12.1 Generalized Functions

12.1.1 *The Dirac delta function*

In physics, concepts or models with singularity often arise, such as point masses or point charges. While these models have clear physical meanings and are suitable for intuitive analysis, they are not appropriate for describing equations of motion. Therefore, there is a need for a function that describes the distribution of point sources. This is where the Dirac δ function, which we have already encountered in Chapter 6, comes into play.

(1) *Formal definition*

For example, the density distribution of a one-dimensional point mass with unit mass can be represented as

$$\delta(x)=\begin{cases}0 & (x\neq 0)\\ \infty & (x=0)\end{cases},\quad \int_{-\infty}^{\infty}\delta(x)\mathrm{d}x=1$$

This is not a strictly defined function. It is neither continuous nor differentiable, and it exhibits infinities. It can hardly be called a well-defined function. A seemingly more reasonable definition is

$$\rho_{\varepsilon}(x)=\lim_{\varepsilon\to 0}\begin{cases}0 & (x\notin[-\varepsilon,\varepsilon])\\ \dfrac{1}{2\varepsilon} & (x\in[-\varepsilon,\varepsilon])\end{cases}$$

However, this definition still does not address these two flaws: the occurrence of infinity and the function's lack of continuity and differentiability.

(2) *Limit definition*

To address the second flaw, a distribution function with continuity and differentiability can be employed. One example is the normalized Gaussian function shown in Figure 12.1(a),

$$\rho_t(x)=\frac{1}{\sqrt{\pi t}}\mathrm{e}^{-\frac{x^2}{2t}}$$

As t approaches zero, the distribution range of the function becomes narrower, and the value at $x=0$ becomes larger and tends to infinity. The Dirac δ function can be defined as

$$\delta(x)\overset{\text{def}}{=}\lim_{t\to 0^+}\frac{1}{\sqrt{\pi t}}\mathrm{e}^{-\frac{x^2}{2t}}$$

An alternative option is to use the Lorenzian function shown in Figure 12.1(b),

$$\rho_a(x)=\frac{a}{\pi(a^2+x^2)}$$

As a decreases, the value of the function at $x=0$ becomes larger. By taking the limit, the δ function can be defined as

$$\delta(x)\overset{\text{def}}{=}\lim_{a\to 0}\frac{a}{\pi(a^2+x^2)}$$

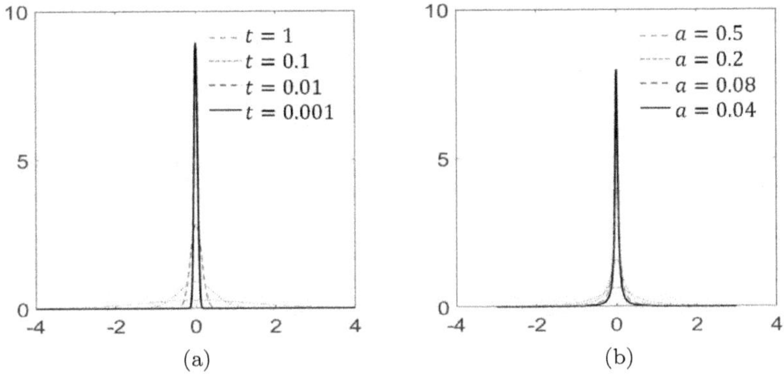

Figure 12.1.

(3) *Integral definition*

The aforementioned function definitions still contain undefined infinities. To address this flaw, note that for any smooth and continuous function $f(x)$, we have

$$\lim_{t\to 0^+}\int_{-\infty}^{\infty}\rho_t(x)f(x)\mathrm{d}x = f(0)$$

The δ function can be regarded as the following integral representation:

$$\int_{-\infty}^{\infty}\delta(x)\mathrm{d}x = 1, \quad \int_{-\infty}^{\infty}\delta(x)f(x)\mathrm{d}x = f(0) \tag{12.1.1}$$

Therefore, the δ function is a generalized function, and it only has a strict definition in the sense of integration.

Based on the definition (12.1.1), the δ function possesses the characteristics of density. The density distribution of a point charge located at $x = x_0$ can be represented as

$$\rho(x) = q\delta(x - x_0)$$

(4) *Three-dimensional Cartesian coordinates*

In a three-dimensional Cartesian coordinate system, the δ function is expressed as

$$\delta(\boldsymbol{r} - \boldsymbol{r}_0) = \delta(x - x_0)\delta(y - y_0)\delta(z - z_0)$$

The Poisson equation for the electrostatic potential of a point charge is

$$\Delta\varphi(\boldsymbol{r}) = -\frac{q}{\varepsilon_0}\delta(\boldsymbol{r}-\boldsymbol{r}_0)$$

The δ function in three-dimensional space is also often represented using cylindrical or spherical coordinate systems. For more details, please refer to the endnote.

12.1.2 *Basic properties*

[1] $\delta(x)$ is an even function:

$$\delta(-x) = \delta(x), \quad \delta'(-x) = -\delta'(x) \tag{12.1.2}$$

[2] For a continuous function $f(x)$ defined on the interval $(-\infty+\infty)$, we have

$$\int_{-\infty}^{\infty}\delta(x-x_0)f(x)\mathrm{d}x = f(x_0) \tag{12.1.3}$$

[3] If $\varphi(x)=0$ has distinct real roots x_k $(k=1,2,\ldots)$, then

$$\delta[\varphi(x)] = \sum_k \frac{\delta(x-x_k)}{|\varphi'(x_k)|} \tag{12.1.4}$$

Proof. Since x_k $(k=1,2,\ldots)$ are the distinct real roots of $\varphi(x)=0$, we have $\varphi'(x_k)\neq 0$. Therefore,

$$\delta[\varphi(x)] = \sum_k c_k\delta(x-x_k)$$

Considering the nth discrete root x_n and assuming that there are no other roots in the neighborhood $(x_n-\varepsilon, x_n+\varepsilon)$, we integrate over this interval to get

$$\int_{x_n-\varepsilon}^{x_n+\varepsilon}\delta[\varphi(x)]\mathrm{d}x = \sum_k c_k\int_{x_n-\varepsilon}^{x_n+\varepsilon}\delta(x-x_k)\mathrm{d}x = c_n$$

Let $y=\varphi(x)$, the left side of the above equation becomes

$$\int_{x_n-\varepsilon}^{x_n+\varepsilon}\delta[\varphi(x)]\mathrm{d}x = \int_{\varphi(x_n-\varepsilon)}^{\varphi(x_n+\varepsilon)}\delta(y)\frac{1}{\varphi'(x)}\mathrm{d}y = \frac{1}{|\varphi'(x_n)|}$$

The appearance of the absolute value is because $\varphi'(x)$ may be negative. In this case, $\varphi(x_n+\varepsilon) < \varphi(x_n-\varepsilon)$, and reversing the limits of integration would introduce a negative sign. Thus, we have

$$c_n = \frac{1}{|\varphi'(x_n)|}$$

[4] Fourier transform of the δ function:

$$\mathfrak{F}[\delta(x)] = \frac{1}{2\pi}\int_{-\infty}^{\infty} \delta(x)\mathrm{e}^{-ikx}\mathrm{d}k = \frac{1}{2\pi} \tag{12.1.5}$$

As we can see, the Fourier transform of the δ function is a constant. The Fourier integral representation of the δ function is given by

$$\delta(x) = \frac{1}{2\pi}\int_{-\infty}^{\infty} \mathrm{e}^{ikx}\mathrm{d}k, \quad \delta(x-x_0) = \frac{1}{2\pi}\int_{-\infty}^{\infty} \mathrm{e}^{\mathrm{i}k(x-x_0)}\mathrm{d}k$$

This is a very important relationship in physics and can be considered as another definition of the δ function.

Practice: Prove the following results:

$$\text{(a) } \delta(x^2-a^2) = \frac{1}{2|a|}[\delta(x-a)+\delta(x+a)]; \quad \text{(b) } \delta(ax) = \frac{1}{|a|}\delta(x)$$

□

Example 12.1. Find the Fourier transform of a radial function in three dimensions:

$$f(r) = \frac{\delta(r-c)}{r} \quad (c>0)$$

Solution: The three-dimensional Fourier transform is given by

$$\mathfrak{F}[f(r)] = \frac{1}{(2\pi)^3}\iiint \frac{\delta(r-c)}{r}\mathrm{e}^{-\mathrm{i}\boldsymbol{k}\cdot\boldsymbol{r}}\mathrm{d}\boldsymbol{r}$$

Using spherical coordinates with $\boldsymbol{k}$ as the polar axis and integrating over the spherical coordinates, we have

$$\begin{aligned}
\mathfrak{F}[f(r)] &= \frac{1}{(2\pi)^3}\int_0^{\infty}\mathrm{d}r\int_0^{\pi}\mathrm{d}\theta\int_0^{2\pi}\frac{\delta(r-c)}{r}\mathrm{e}^{-ikr\cos\theta}r^2\sin\theta\,\mathrm{d}\varphi \\
&= \frac{1}{(2\pi)^2}\int_0^{\infty}\mathrm{d}r\int_0^{\pi} r\delta(r-c)\mathrm{e}^{-ikr\cos\theta}\sin\theta\,\mathrm{d}\theta
\end{aligned}$$

$$= \frac{1}{(2\pi)^2} \int_0^{\infty} \delta(r-c) \frac{\mathrm{e}^{\mathrm{i}kr} - \mathrm{e}^{-\mathrm{i}kr}}{\mathrm{i}k} \mathrm{d}r$$

$$= \frac{1}{(2\pi)^2} \frac{\mathrm{e}^{\mathrm{i}kc} - \mathrm{e}^{-\mathrm{i}kc}}{\mathrm{i}k} = \frac{1}{2\pi^2} \frac{\sin kc}{k}$$

12.1.3 *Step function*

12.1.3.1 *Formal definition*

$$\mathrm{H}(x) = \begin{cases} 1 & (x \geq 0) \\ 0 & (x < 0) \end{cases}$$

Its derivative is given by

$$\mathrm{H}'(x) = \begin{cases} 0 & (x \neq 0) \\ \infty & (x = 0) \end{cases}$$

The step function is also known as the Heaviside step function. It is a generalized function, and we can give it an integral expression using the Dirac δ function:

$$\mathrm{H}(x) \stackrel{\text{def}}{=} \int_{-\infty}^{x} \delta(x) \mathrm{d}t = \begin{cases} 1 & (x \geq 0) \\ 0 & (x < 0) \end{cases} \tag{12.1.6}$$

12.1.3.2 *Fourier transform of step function*

We introduce a limiting process using a continuous function to describe the step function:

$$\mathrm{H}(x) \stackrel{\text{def}}{=} \lim_{\beta \to 0} \mathrm{H}(x, \beta) = \lim_{\beta \to 0} \begin{cases} \mathrm{e}^{-\beta x} & (x \geq 0) \\ 0 & (x < 0) \end{cases}$$

Its Fourier transform is given by

$$\mathfrak{F}[\mathrm{H}(x)] = \lim_{\beta \to 0} \mathfrak{F}[\mathrm{H}(x, \beta)] = \lim_{\beta \to 0} \frac{1}{2\pi} \int_0^{\infty} \mathrm{e}^{-\beta x} \mathrm{e}^{-\mathrm{i}kx} \mathrm{d}x$$

$$= \frac{1}{2\pi} \lim_{\beta \to 0} \left(\frac{\beta}{\beta^2 + k^2} - \mathrm{i} \frac{k}{\beta^2 + k^2} \right)$$

According to the definition of delta function, we have

$$\lim_{\beta \to 0} \frac{\beta}{\beta^2 + k^2} = \pi \delta(k)$$

We use the principal value notation:

$$\mathscr{P}\frac{1}{k} \equiv \lim_{\beta \to 0} \frac{k}{\beta^2 + k^2} = \begin{cases} 0 & (k = 0) \\ \dfrac{1}{k} & (k \neq 0) \end{cases}$$

Therefore,

$$\mathfrak{F}[\mathrm{H}(x)] = \frac{1}{2}\delta(k) - \frac{\mathrm{i}}{2\pi}\mathscr{P}\frac{1}{k} \tag{12.1.7}$$

Example 12.2. Solve the initial value problem for the ODE:

$$\begin{cases} \dfrac{\mathrm{d}^2 f(x,t)}{\mathrm{d}x^2} = \delta(x-t) \quad (x, t > 0) \\ f(0,t) = 0, \qquad\qquad f'(0,t) = 0 \end{cases}$$

Solution: Integrating both sides of the equation with respect to x, we have

$$\frac{\mathrm{d}f(x,t)}{\mathrm{d}x} = \mathrm{H}(x-t) + A(t)$$

Integrating again, we get

$$f(x,t) = (x-t)\mathrm{H}(x-t) + A(t)x + B(t)$$

Applying the boundary conditions, we find $A(t) = 0$ and $B(t) = 0$, so

$$f(x,t) = (x-t)\mathrm{H}(x-t)$$

Utilizing the step function, we can conveniently express the signum function as

$$\mathrm{sgn}(x) = 2\mathrm{H}(x) - 1$$

and the Harr function as

$$f(x) = \mathrm{H}(d - |x|)$$

ENDNOTE

Sometimes, we need to study the relationship between the δ functions represented in orthogonal curvilinear coordinate systems and Cartesian coordinate systems. Based on the definition of the δ function in n-dimensional space, we have

$$\int \delta(\boldsymbol{x}-\boldsymbol{a})f(\boldsymbol{x})\mathrm{d}^n\boldsymbol{x} \equiv f(\boldsymbol{a}), \quad \int_{-\infty}^{\infty} \delta(\boldsymbol{x})\mathrm{d}^n\boldsymbol{x} = 1$$

When performing a coordinate transformation:

$$x_j = x_j(\xi_1, \xi_2, \ldots, \xi_n) \equiv x_j(\boldsymbol{\xi})$$

$$a_j = x_j(\alpha_1 \alpha_2, \ldots, \alpha_n) \equiv a_j(\boldsymbol{\alpha})$$

The Jacobian determinant is $\det J = |\frac{\partial x_i}{\partial \xi_i}|$. Since $\mathrm{d}^n\boldsymbol{x} = \det J\, \mathrm{d}^n\boldsymbol{\xi}$, we have

$$\int \delta(\boldsymbol{x}-\boldsymbol{a})f(\boldsymbol{x})\mathrm{d}^n\boldsymbol{x}$$

$$= \int f(\boldsymbol{\xi}) \prod_{i=1}^{n} \delta[x_i(\boldsymbol{\xi}) - a_i(\boldsymbol{\alpha})] \det J\, \mathrm{d}^n\boldsymbol{\xi} = f(\boldsymbol{\alpha})$$

Therefore,

$$\det J \prod_{i=1}^{n} \delta[x_i(\boldsymbol{\xi}) - a_i(\boldsymbol{\alpha})] = \prod_{i=1}^{n} \delta(\xi_i - \alpha_i)$$

which yields

$$\delta(\boldsymbol{x}-\boldsymbol{a}) = \frac{1}{\det J}\delta(\boldsymbol{\xi}-\boldsymbol{\alpha})$$

The points where the Jacobian determinant is zero are called the singular points of the coordinate transformation. Some coordinates become negligible at these singular points, and they are referred to as ignorable coordinates. This arises because the dimensions of the curvilinear coordinates are compact and can be continuously contracted to the singular points, whereas the radial dimensions in Cartesian or spherical coordinate systems are non-compact. Suppose in the $(\xi_1, \xi_2, \ldots, \xi_n)$ coordinate system, the components $(\xi_{k+1}, \xi_{k+2}, \ldots, \xi_n)$ are ignorable coordinates at the singular point $\boldsymbol{a}$, indicating that any function does not depend on these coordinates. Therefore, they can be integrated out. Let

$$\det J_k = \int \det J \, \mathrm{d}\xi_{k+1} \cdots \mathrm{d}\xi_n$$

then we have

$$\delta(\boldsymbol{r} - \boldsymbol{a}) = \frac{1}{\det J_k} \prod_{i=1}^{k} \delta(\xi_i - \alpha_i)$$

For example, in the two-dimensional polar coordinate system, $\det J = r$, which is zero at $r = 0$, and θ is an ignorable coordinate. Thus,

$$\det J_1 = \int_0^{2\pi} \det J \, \mathrm{d}\theta = 2\pi r$$

which yields

$$\delta(\boldsymbol{r}) = \delta(x)\delta(y) = \frac{\delta(r)}{2\pi r}$$

In the three-dimensional spherical coordinate system, $\det J = r^2 \sin\theta$, which is zero at $r = 0$. Both θ and φ are ignorable coordinates. Hence,

$$\det J_1 = \int_0^{2\pi} \mathrm{d}\varphi \int_0^{\pi} r^2 \sin\theta \, \mathrm{d}\theta = 4\pi r^2$$

Therefore,

$$\delta(\boldsymbol{r}) = \frac{\delta(r)}{4\pi r^2}$$

In general, for an n-dimensional spherical coordinate system, we have

$$\delta(\boldsymbol{r}) = \delta(x_1)\delta(x_2)\cdots\delta(x_n) = \frac{\Gamma(\frac{n}{2})\delta(r)}{2\pi^{n/2}r^{n-1}}$$

Exercises

[1] Proof:

$$\text{(a) } \delta'(-x) = -\delta'(x); \quad \text{(b) } x\delta'(x) = -\delta(x)$$

[2] Proof:

$$\text{(a) } \int_{-\infty}^{\infty} \delta^{(n)}(x-a)f(x)\mathrm{d}x = (-1)^n f^{(n)}(a);$$

$$\text{(b) } \int_{-\infty}^{\infty} \delta^{(n)}(x)f(x)\mathrm{d}x = (-1)^n f^{(n)}(0)$$

Hint: Treat a as the independent variable, $\int_{-\infty}^{\infty} \delta(x-a)f(x)\mathrm{d}x = f(a)$. Differentiate the equation n times with respect to a, and then set $a = 0$.

[3] Proof:

$$\text{(a) } \delta(x) = \lim_{n\to\infty} \frac{\sin nx}{\pi x}; \quad \text{(b) } \delta(x) = \lim_{n\to\infty} \frac{\sin^2 nx}{\pi n x^2}$$

Hint: Analyze the values at $x = 0$ and $x \neq 0$ to infer

$$\lim_{n\to\infty} \int_{-\infty}^{\infty} \frac{\sin nx}{\pi x} f(x)\mathrm{d}x = \lim_{n\to\infty} \int_{-\frac{\pi}{2n}}^{\frac{\pi}{2n}} \frac{\sin nx}{\pi x} f(x)\mathrm{d}x = f(0)$$

and

$$\lim_{n\to\infty} \int_{-\infty}^{\infty} \frac{\sin nx}{\pi x}\mathrm{d}x = \lim_{n\to\infty} \int_{-\infty}^{\infty} \frac{\sin^2 nx}{\pi n x^2}\mathrm{d}x = 1$$

[4] Simplify the Dirac δ function:

$$\text{(a) } \delta[(x+2)\sin x]; \quad \text{(b) } \delta[(x-2)\ln x]$$

Answer:

$$\text{(a)}\ \frac{\delta(x+2)}{\sin 2} + \sum_{k=-\infty}^{\infty} \frac{\delta(x-k\pi)}{|k\pi+2|\pi}; \quad \text{(b)}\ \frac{\delta(x-2)}{\ln 2} + \delta(x-1)$$

[5] Compute integrals:

$$\text{(a)}\ \int_{0.5}^{\infty} \delta(\sin x) \left(\frac{2}{3}\right)^x \mathrm{d}x;$$

$$\text{(b)}\ \int_{-\infty}^{\infty} \delta(x^2-3x+2) \int_0^x \delta\left(t^2-2t+\frac{3}{4}\right) \mathrm{e}^{-t^2} \mathrm{d}t\, \mathrm{d}x;$$

$$\text{(c)}\ \int_{-\infty}^{\infty} \delta(x^2-2x-3) \int_0^x \delta\left(t^2-2xt+\frac{1}{2}\right) \mathrm{e}^{-xt^2} \mathrm{d}t\, \mathrm{d}x$$

Answer:

$$\text{(a)}\ \frac{1}{(3/2)^{\pi}-1}; \quad \text{(b)}\ \mathrm{e}^{-\frac{9}{4}} + 2\mathrm{e}^{-\frac{1}{4}};$$

$$\text{(c)}\ \frac{1}{4\sqrt{34}} \mathrm{e}^{-3(3-\sqrt{17}/2)^2} - \frac{1}{4\sqrt{2}} \mathrm{e}^{(-1+1/\sqrt{2})^2}$$

[6] Solve boundary value problems for ordinary differential equations:

$$\begin{cases} \dfrac{\mathrm{d}^2 f(x,t)}{\mathrm{d}x^2} = \delta(x-t) \quad (t>0) \\ f(a,t)=0, \quad f(b,t)=0 \end{cases}$$

[7] Solve initial value problems for ordinary differential equations:

$$\begin{cases} \dfrac{\mathrm{d}^2 f(x,t)}{\mathrm{d}x^2} + k^2 f(x,t) = \delta(x-t) \quad (x,t>0) \\ f(0,t)=0, \quad f'(0,t)=0 \end{cases}$$

Hint: Consider solutions in the regions $x<t$ and $x>t$, and the function $f(x,t)$ is continuous at $x=t$.
Answer: $f(x,t) = \frac{1}{k}\sin k(x-t)\mathrm{H}(x-t)$

12.2 Fourier Transform Method

We have previously used the Laplace transform method to transform time-dependent ordinary differential equations into ordinary algebraic equations. After obtaining the image function, we apply inverse transformation to solve initial value problems. For partial differential equations, appropriate integral transformations, including Fourier transform or Laplace transform, can be employed to convert them into ordinary differential equations or algebraic equations. After solving for the image function, inverse transformation is applied to obtain the desired solution.

12.2.1 *Infinite space problems*

For bounded domain problems, the eigenvalues are discrete, and the general solution is expressed as the superposition of discrete eigenfunctions (trigonometric functions), known as the eigenfunction expansion method. For unbounded domain problems, the eigenvalues are generally continuous, and the general solution of the equation can be represented as a Fourier integral. The coefficients are the image function of the Fourier transform,

$$\begin{cases} F(k) \equiv \mathfrak{F}[f(x)] = \dfrac{1}{2\pi}\displaystyle\int_{-\infty}^{\infty} f(x)\mathrm{e}^{-\mathrm{i}kx}\mathrm{d}x \\ f(x) \equiv \mathfrak{F}^{-1}[F(k)] = \displaystyle\int_{-\infty}^{\infty} F(k)\mathrm{e}^{\mathrm{i}kx}\mathrm{d}k \end{cases}$$

The Fourier transform method can be used to solve partial differential equations of unbounded systems. The following examples exhibit this approach.

Example 12.3. Solving the free vibrations of an infinite string:

$$\begin{cases} u_{tt} - a^2 u_{xx} = 0 \quad (-\infty < x < \infty) \\ u|_{t=0} = \varphi(x), \qquad u_t|_{t=0} = \psi(x) \end{cases}$$

Solution: Let $U(k,t) = \mathfrak{F}[u(x,t)]$. Applying the Fourier transform to both sides of the equation and the initial conditions simultaneously, then

$$\begin{cases} U_{tt}(k,t) + k^2 a^2 U(k,t) = 0 \\ U(k,t)|_{t=0} = \Phi(k), \quad U_t(k,t)|_{t=0} = \Psi(k) \end{cases}$$

where

$$\Phi(k) = \mathfrak{F}[\varphi(x)], \quad \Psi(k) = \mathfrak{F}[\psi(x)]$$

Solving this, we obtain

$$U(k,t) = A(k)\cos kat + B(k)\sin kat$$

Substituting the initial conditions, we have

$$\begin{aligned} U(k,t) &= \Phi(k)\cos kat + \frac{\Psi(k)}{ka}\sin kat \\ &= \Phi(k)\cos kat + \int_0^t \Psi(k)\cos ka\tau \,\mathrm{d}\tau \end{aligned}$$

Taking the inverse Fourier transform of $U(k)$, using $\cos kat = \frac{1}{2}(\mathrm{e}^{\mathrm{i}kat} + \mathrm{e}^{-\mathrm{i}kat})$ and the convolution theorem, we obtain the original function as

$$\begin{aligned} u(x,t) &= \frac{1}{2}[\varphi(x+at)+\varphi(x-at)] + \frac{1}{2}\int_0^t [\psi(x-a\tau)+\psi(x+a\tau)]\mathrm{d}\tau \\ &= \frac{1}{2}[\varphi(x+at)+\varphi(x-at)] + \frac{1}{2a}\int_{x-at}^{x+at} \psi(\xi)\mathrm{d}\xi \end{aligned}$$

Thus, we obtain the one-dimensional D'Alembert solution for the motion of an infinite string once again.

Example 12.4. Solving the heat conduction problem of an infinite rod:

$$\begin{cases} u_t - a^2 u_{xx} = 0 \quad (-\infty < x < \infty) \\ u|_{t=0} = \varphi(x) \end{cases}$$

Solution: Applying the Fourier transform to the equation, we have

$$\begin{cases} U_t(k,t) + k^2a^2U(k,t) = 0 \\ U(k,t)|_{t=0} = \Phi(k) \end{cases}$$

Solving for the image function, we get

$$U(k,t) = \Phi(k)\mathrm{e}^{-k^2a^2t}$$

To obtain the original function, we refer to the Appendix Table:

$$\mathfrak{F}^{-1}\left[\mathrm{e}^{-\frac{k^2}{4\sigma^2}}\right] = 2\sqrt{\pi}\sigma\mathrm{e}^{-\sigma^2x^2}$$

By making use of the convolution theorem,

$$\mathfrak{F}^{-1}[F_1(k)F_2(k)] = \frac{1}{2\pi} f_1(x) * f_2(x)$$

we get

$$u(x,t) = \frac{1}{2a\sqrt{\pi t}} \int_{-\infty}^{\infty} \varphi(\xi) \mathrm{e}^{-\frac{(x-\xi)^2}{4a^2 t}} \mathrm{d}\xi$$

The inverse Fourier transform in this example can also be directly derived from the definition.

Example 12.5. Solving the time evolution of a one-dimensional free electron wave packet in quantum mechanics:

$$\begin{cases} \mathrm{i}\hbar\partial_t \psi(x,t) = -\dfrac{\hbar^2}{2m}\partial_x^2 \psi(x,t) \\ \psi(x,t)|_{t=0} = \varphi(x) \end{cases}$$

Solution: Let $\Psi(k,t) = \mathfrak{F}[\psi(x,t)]$. Applying the Fourier transform to the equation, we have

$$\mathrm{i}\hbar\Psi_t(k,t) = \frac{\hbar^2 k^2}{2m} \Psi(k,t)$$

Thus,

$$\Psi(k,t) = \Phi(k)\mathrm{e}^{-\mathrm{i}\frac{\hbar k^2}{2m}t}$$

Using the convolution theorem for the inverse transformation, we obtain the original function as

$$\psi(x,t) = \varphi(x) * \sqrt{\frac{2\pi \mathrm{i}m}{\hbar t}} \mathrm{e}^{-\frac{\mathrm{i}m}{2\hbar t}x^2} = \sqrt{\frac{2\pi \mathrm{i}m}{\hbar t}} \int_{-\infty}^{\infty} \varphi(\xi) \mathrm{e}^{-\frac{\mathrm{i}m}{2\hbar t}(x-\xi)^2} \mathrm{d}\xi$$

Assuming the initial state is a Gaussian wave packet,

$$\varphi(x) = \frac{1}{\sqrt{2\pi}\sigma} \mathrm{e}^{-\frac{x^2}{2\sigma^2}}$$

After time t, the wave packet evolves to

$$\begin{aligned}\psi(x,t) &= \sqrt{\frac{2\pi i m}{\hbar t}}\int_{-\infty}^{\infty}\frac{1}{\sqrt{2\pi}\sigma}\mathrm{e}^{-\frac{\xi^2}{2\sigma^2}}\mathrm{e}^{-\frac{im}{2\hbar t}(x-\xi)^2}\mathrm{d}\xi \\ &= \frac{1}{\sqrt{2\pi(\sigma^2+\mathrm{i}\hbar t/m)}}\exp\left[-\frac{x^2}{2(\sigma^2+\mathrm{i}\hbar t/m)}\right]\end{aligned}$$

This corresponds to the evolution of the wave packet's width:

$$\sigma \longrightarrow \sigma\sqrt{1+\frac{\mathrm{i}\hbar t}{m\sigma^2}}$$

Figure 12.2(a) depicts the time evolution of a one-dimensional Gaussian wave packet, which appears to resemble classical particle diffusion behavior. To demonstrate the wave nature of microscopic particles, we consider the initial configuration of two isolated wave packets that spread out and eventually overlap in space after a certain time. Figure 12.2(b) shows the interference phenomenon during the spreading process of the wave packet, which is absent in classical diffusion motion.

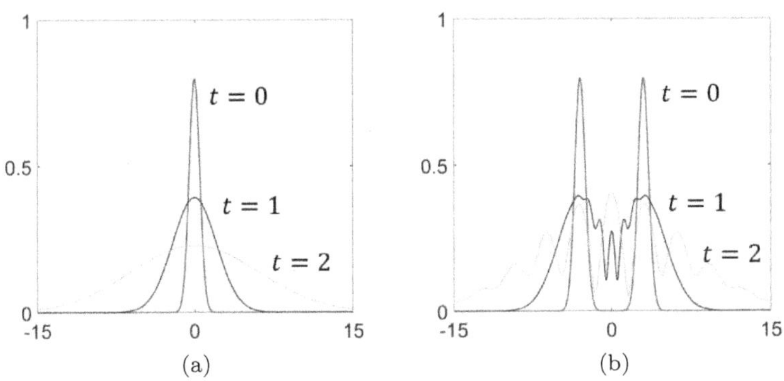

Figure 12.2.

Although both are wave equations, there are fundamental differences between the results in quantum mechanics and the D'Alembert solution for mechanical waves. One-dimensional mechanical waves adhere to the principle of Huygens, whereas quantum wave packets exhibit behavior more akin to diffusion. During the propagation of electron wave packets, dispersion occurs as $E = \frac{\hbar^2 k^2}{2m}$, where harmonics of different energies have distinct phase velocities. Consequently, the shape of the wave packet changes over time due to this dispersion effect, leading to gradual spreading of the initial wave packet.

Example 12.6. Solving the wave propagation problem in three-dimensional unbounded space:

$$\begin{cases} u_{tt} - a^2 \Delta u = 0 \\ u|_{t=0} = \varphi(\boldsymbol{r}), \quad u_t|_{t=0} = \psi(\boldsymbol{r}) \end{cases}$$

Solution: We apply a three-dimensional Fourier transform to both the equation and initial conditions:

$$\begin{cases} U_{tt}(\boldsymbol{k},t) + k^2 a^2 U(\boldsymbol{k},t) = 0 \\ U(\boldsymbol{k},t)|_{t=0} = \Phi(\boldsymbol{k}), \quad U_t(\boldsymbol{k},t)|_{t=0} = \Psi(\boldsymbol{k}) \end{cases}$$

where $k = |\boldsymbol{k}|$. The solution is given by

$$\begin{aligned} U(\boldsymbol{k},t) &= \Phi(\boldsymbol{k}) \cos kat + \frac{\Psi(\boldsymbol{k})}{ka} \sin kat \\ &= \frac{\partial}{\partial t}\left[\frac{\Phi(\boldsymbol{k})}{ka} \sin kat\right] + \frac{\Psi(\boldsymbol{k})}{ka} \sin kat \end{aligned}$$

We first perform the inverse Fourier transform on the second term on the right-hand side:

$$\begin{aligned} \mathfrak{F}^{-1}\left[\frac{\Psi(\boldsymbol{k})}{ka} \sin kat\right] &= \iiint_{-\infty}^{\infty} \frac{\Psi(\boldsymbol{k})}{ka} \sin kat \, \mathrm{e}^{\mathrm{i}\boldsymbol{k}\cdot\boldsymbol{r}} \mathrm{d}\boldsymbol{k} \\ &= \frac{1}{(2\pi)^3} \iiint_{-\infty}^{\infty} \iiint_{-\infty}^{\infty} \psi(\boldsymbol{r}') \mathrm{e}^{-\mathrm{i}\boldsymbol{k}\cdot\boldsymbol{r}'} \mathrm{d}\boldsymbol{r}' \\ &\quad \times \frac{1}{ka} \sin kat \, \mathrm{e}^{\mathrm{i}\boldsymbol{k}\cdot\boldsymbol{r}} \mathrm{d}\boldsymbol{k} \end{aligned}$$

$$= \frac{1}{(2\pi)^3 a} \iiint_{-\infty}^{\infty} \int_0^{\infty} k\mathrm{d}k \int_0^{\pi} \mathrm{d}\theta \int_0^{2\pi} \mathrm{d}\phi \, \psi(\boldsymbol{r}')$$

$$\times \mathrm{e}^{-\mathrm{i}|\boldsymbol{k}|\cdot|\boldsymbol{r}-\boldsymbol{r}'|\cos\theta} \sin kat \sin\theta \, \mathrm{d}\boldsymbol{r}'$$

$$= -\frac{1}{2a(2\pi)^2} \iiint_{-\infty}^{\infty} \int_0^{\infty} \frac{\psi(\boldsymbol{r}')}{|\boldsymbol{r}-\boldsymbol{r}'|}$$

$$\times \left[\mathrm{e}^{\mathrm{i}k(|\boldsymbol{r}-\boldsymbol{r}'|+at)} - \mathrm{e}^{-\mathrm{i}k(|\boldsymbol{r}-\boldsymbol{r}'|-at)}\right] \mathrm{d}k\mathrm{d}\boldsymbol{r}'$$

$$= -\frac{1}{4\pi a} \iiint_{-\infty}^{\infty} \frac{\psi(\boldsymbol{r}')}{|\boldsymbol{r}-\boldsymbol{r}'|} [\delta(|\boldsymbol{r}-\boldsymbol{r}'|+at)$$

$$- \delta(|\boldsymbol{r}-\boldsymbol{r}'|-at)] \mathrm{d}\boldsymbol{r}'$$

$$= \frac{1}{4\pi a} \iint_{Sr} \frac{\psi(\boldsymbol{r}')}{at} \mathrm{d}s'$$

The first term has the same form as the second term. Therefore, the solution can be written as

$$u(\boldsymbol{r},t) = \frac{1}{4\pi a} \frac{\partial}{\partial t} \oiint_{S_r} \frac{\varphi(\boldsymbol{r}')}{at} \mathrm{d}s' + \frac{1}{4\pi a} \oiint_{S_r} \frac{\psi(\boldsymbol{r}')}{at} \mathrm{d}s' \tag{12.2.1}$$

This equation is known as the Poisson formula, which indicates that the amplitude of the wave is inversely proportional to time or propagation distance, consistent with the principle of energy conservation. The surface integral region S_r represents the spherical surface composed of all points $\boldsymbol{r}'$ at a distance of at from the field point $\boldsymbol{r}$, as shown in Figure 12.3(a). The interpretation of this integral is that the motion at point $\boldsymbol{r}$ is the superposition of the motions of all the spherical points $\boldsymbol{r}'$ at a distance of at at the previous time t. The earlier or later motion at point $\boldsymbol{r}'$ does not affect the motion at point $\boldsymbol{r}$, which is in accordance with the Huygens' principle in wave optics. The gray area at the origin represents the initial wave packet. Therefore, the effective integration region in Equation (12.2.1) is the portion of the spherical surface within the gray area.

For the wave propagation problem in two-dimensional unbounded space,

$$\begin{cases} u_{tt} - a^2 \Delta_2 u = 0 \\ u|_{t=0} = \varphi(\boldsymbol{r}), \quad u_t|_{t=0} = \psi(\boldsymbol{r}) \end{cases}$$

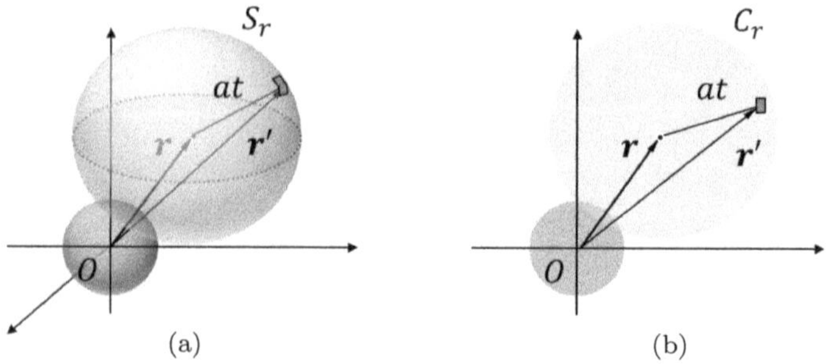

Figure 12.3.

We can similarly obtain the solution as (see Exercise [6] in Section 15.1):

$$u(\boldsymbol{r},t)=\frac{1}{2\pi a}\frac{\partial}{\partial t}\iint_{C_r}\frac{\varphi(\boldsymbol{r}')}{\sqrt{(at)^2-|\boldsymbol{r}-\boldsymbol{r}'|^2}}\mathrm{d}x'\mathrm{d}y'$$
$$+\frac{1}{2\pi a}\iint_{C_r}\frac{\psi(\boldsymbol{r}')}{\sqrt{(at)^2-|\boldsymbol{r}-\boldsymbol{r}'|^2}}\mathrm{d}x'\mathrm{d}y' \tag{12.2.2}$$

Here, C_r represents the region within the circle $\sqrt{x^2+y^2}\le at$, as shown in Figure 12.2(b). The gray area at the origin represents the initial wave packet. The effective integration region in Equation (12.2.2) is the overlapping area between the circular region C_r and the gray area.

Reflection: Consider the physical differences behind the integration regions in two-dimensional and three-dimensional cases.

12.2.2 *Half-infinite space problem*

Example 12.7. Solving the constant surface concentration diffusion problem:

$$\begin{cases} u_t-a^2u_{xx}=0 & (x\ge 0)\\ u|_{x=0}=N_0, & u|_{t=0}=0 \end{cases}$$

Solution: First, we need to homogenize the boundary conditions of this equation. Let $u(x,t)=N_0+w(x,t)$. We have

$$\begin{cases} w_t-a^2w_{xx}=0 & (x\ge 0)\\ w|_{x=0}=0, & w|_{t=0}=-N_0 \end{cases}$$

Since the equation has the first type of homogeneous boundary condition at $x=0$, we extend the equation and initial condition to an infinite heat conduction problem:

$$\begin{cases} w_t - a^2 w_{xx} = \quad (-\infty < x < \infty) \\ w|_{t=0} = \varphi(x) = \begin{cases} -N_0 & (x \geq 0) \\ N_0 & (x < 0) \end{cases} \end{cases}$$

Taking the Fourier transform, we obtain

$$W(k,t) = \Phi(k)\mathrm{e}^{-k^2a^2t}$$

Then performing the inverse transform and noting that the image function is the product of two terms, it yields

$$\begin{aligned} w(x,t) &= \frac{1}{2\pi}\varphi(x) * \frac{\sqrt{\pi}}{a\sqrt{t}}\mathrm{e}^{-\frac{x^2}{4a^2t}} = \frac{1}{2a\sqrt{\pi t}}\int_{-\infty}^{\infty}\varphi(\xi)\mathrm{e}^{-\frac{(x-\xi)^2}{4a^2t}}\,\mathrm{d}\xi \\ &= \frac{N_0}{2a\sqrt{\pi t}}\left[\int_{-\infty}^{0}\mathrm{e}^{-\frac{(x-\xi)^2}{4a^2t}}\,\mathrm{d}\xi - \int_{0}^{\infty}\mathrm{e}^{-\frac{(x-\xi)^2}{4a^2t}}\,\mathrm{d}\xi\right] \\ &= \frac{N_0}{\sqrt{\pi}}\left[-\int_{\infty}^{\frac{x}{2a\sqrt{t}}}\mathrm{e}^{-y^2}\,\mathrm{d}y + \int_{\frac{x}{2a\sqrt{t}}}^{-\infty}\mathrm{e}^{-y^2}\,\mathrm{d}y\right] \\ &= -\frac{2N_0}{\sqrt{\pi}}\int_{0}^{\frac{x}{2a\sqrt{t}}}\mathrm{e}^{-y^2}\,\mathrm{d}y \\ &= -N_0\,\mathrm{erf}\left(\frac{x}{2a\sqrt{t}}\right) \end{aligned}$$

The final result is

$$u(x,t) = N_0\left[1 - \mathrm{erf}\left(\frac{x}{2a\sqrt{t}}\right)\right] \equiv N_0\,\mathrm{erfc}\left(\frac{x}{2a\sqrt{t}}\right)$$

where the error function and the complementary error function are respectively defined as

$$\mathrm{erf}(x) \overset{\mathrm{def}}{=} \frac{2}{\sqrt{\pi}}\int_0^x \mathrm{e}^{-y^2}\,\mathrm{d}y, \quad \mathrm{erfc}(x) \overset{\mathrm{def}}{=} 1 - \mathrm{erf}(x)$$

For the half-infinite space problem, if it satisfies either the first or the second type of homogeneous boundary conditions, the problem can be extended to an infinite space problem using odd or even extension methods, and then solved using the Fourier transform method. If it has a non-homogeneous boundary condition, we can first try to homogenize it and then perform odd or even extension. However, it is also possible to directly solve it using the Fourier sine transform (for the first type of boundary conditions) or the Fourier cosine transform (for the second type of boundary conditions), as shown in the following example.

Example 12.8. Solving the Laplace equation in the half-unbounded space:

$$\begin{cases} \dfrac{\partial^2 u}{\partial x^2} + \dfrac{\partial^2 u}{\partial y^2} = 0 \quad (-\infty < x < \infty, y \geq 0) \\ u|_{y=0} = f(x) \end{cases}$$

Solution: We perform a Fourier sine transform with respect to the variable y,

$$U(x,k) = \mathfrak{F}_s[u(x,y)] = \frac{1}{\pi}\int_0^\infty u(x,y)\sin ky\,\mathrm{d}y \quad (k \geq 0)$$

Thus, it has

$$\mathfrak{F}_s\left[\frac{\partial^2}{\partial y^2}u(x,y)\right] = \frac{1}{\pi}kf(x) - k^2U(x,k)$$

The PDE is then transformed into an ODE with respect to x:

$$\frac{\mathrm{d}^2}{\mathrm{d}x^2}U(x,k) - k^2U(x,k) = -\frac{1}{\pi}kf(x)$$

Following the method in Section 9.1, we obtain the general solution to the equation as

$$U(x,k) = A\mathrm{e}^{-kx} + B\mathrm{e}^{kx} + \frac{1}{2\pi}\int_{-\infty}^{\infty} \mathrm{e}^{-k|x-x'|}f(x')\mathrm{d}x' \quad (k \geq 0)$$

Considering the condition that the temperature should remain finite as $x \to \pm\infty$, we find that $A = B = 0$. Thus, we have

$$U(x,k) = \frac{1}{2\pi}\int_{-\infty}^{\infty} \mathrm{e}^{-k|x-x'|}f(x')\mathrm{d}x'$$

Performing the inverse Fourier sine transform, we obtain the original function:

$$\begin{aligned} u(x,y) &= \frac{1}{\pi}\int_0^{\infty}\left[\int_{-\infty}^{\infty} \mathrm{e}^{-k|x-x'|} f(x')\mathrm{d}x'\right]\sin ky\,\mathrm{d}k \\ &= \frac{y}{\pi}\int_{-\infty}^{\infty} \frac{f(x')}{(x-x')^2+y^2}\mathrm{d}x' \end{aligned}$$

This is known as the upper half-plane Poisson's formula which has been discussed in Chapter 2.

Discussion:

(1) From the form of the solution, it can be observed that as $y \to 0$, $u(x,y) \to 0$. Is this contradict to the given boundary condition?
(2) Can we use the Fourier cosine transform to solve the problem? Why?

Practice: Perform a Fourier transform with respect to the x-direction in this example.

ENDNOTE
In solving the one-dimensional wave equation, we obtained the D'Alembert solution, which states that the initial wave will decompose into two independent sub-waves propagating uniformly in the positive and negative directions, while the waveform remains unchanged. The motion at any point is determined solely by the motion states of the points at a distance of *at* and the previous time t, which clearly complies with Huygens' principle. The solution to the three-dimensional wave equation also satisfies Huygens' principle, as the wave propagates without leaving any trace. We can compare the similarities between the three-dimensional wave equation and the one-dimensional wave equation in spherical coordinates by considering the radial component of a spherically symmetric wave:

$$u_{tt} - a^2\Delta u = 0 \longrightarrow \frac{\partial^2 u}{\partial r^2} + \frac{2}{r}\frac{\partial u}{\partial r} - \frac{1}{a^2}\frac{\partial^2 u}{\partial t^2} = 0$$

This equation can be written as a one-dimensional wave equation:

$$\frac{\partial^2(ru)}{\partial t^2} - a^2\frac{\partial^2(ru)}{\partial r^2} = 0$$

Its solution is

$$u(r,t) = \frac{1}{r}f_1(r-at) + \frac{1}{r}f_2(r+at)$$

It has a similar form to the D'Alembert formula for the one-dimensional wave equation, with the two terms respectively represent the diverging wave and the converging wave on the spherical surface.

In contrast, the radial component of the two-dimensional wave equation appears more like an inhomogeneous equation:

$$\frac{\partial^2(\sqrt{r}u)}{\partial t^2} - a^2\frac{\partial^2(\sqrt{r}u)}{\partial r^2} = \frac{a^2}{4r^2}\sqrt{r}u$$

This equation does not have independent propagating solutions. The term on the right-hand side associated to $u(r,t)$ indicates that the motion at any point in space will continuously affect the entire wave field, which clearly violates Huygens' principle. In fact, wave equations in even-dimensional spaces suffer from the same problem. It seems that the waves generated by a source not only propagate forward but also propagate backward, as shown in Figure 12.4(a), with "lingering sound". This indicates that Huygens' principle, based on naive intuition, does not hold in two-dimensional and even-dimensional spaces as well.

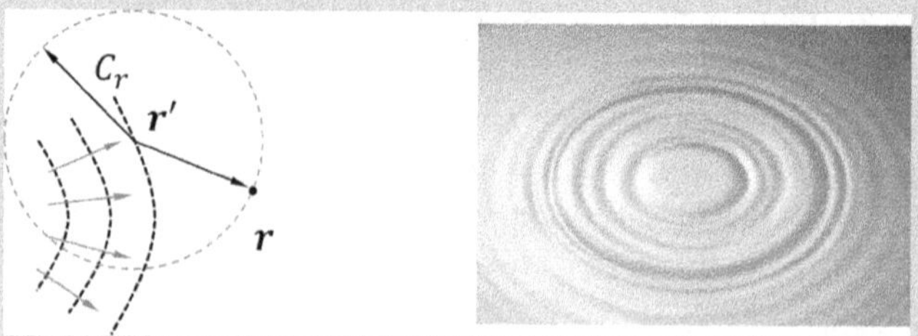

(a) (b)

Figure 12.4.

Figure 12.5 intuitively illustrates the propagation of waves in (a) two-dimensional and (b) three-dimensional spaces, respectively. Assuming an initial Gaussian wave packet at the origin, after some time, they appear similar. However, in reality, the interior of three-dimensional space has completely returned to its equilibrium position (no signal), while in two-dimensional space, there is always a displacement distribution throughout.

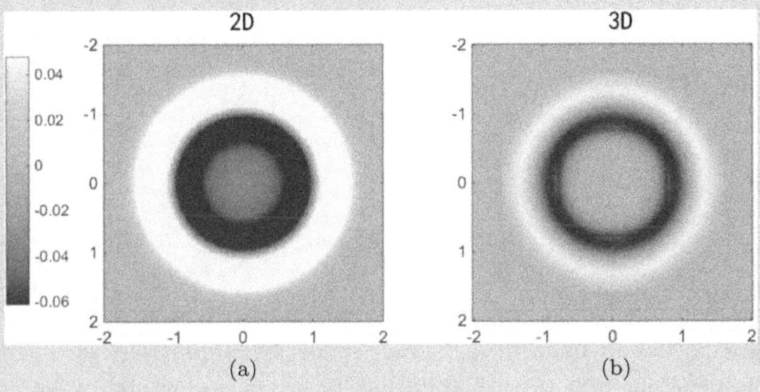

Figure 12.5.

To highlight this characteristic more clearly, Figure 12.6 depicts the radial distributions of one-dimensional, two-dimensional, and three-dimensional waves at different times along the x-axis. It can be seen that the motion characteristics of waves in three-dimensional and one-dimensional spaces are consistent, and the wave propagation complies with Huygens' principle. On the other hand, the propagation of waves in two-dimensional space resembles a diffusion process, which is why surface waves on water (Figure 12.4(b)) do not satisfy Huygens' principle.

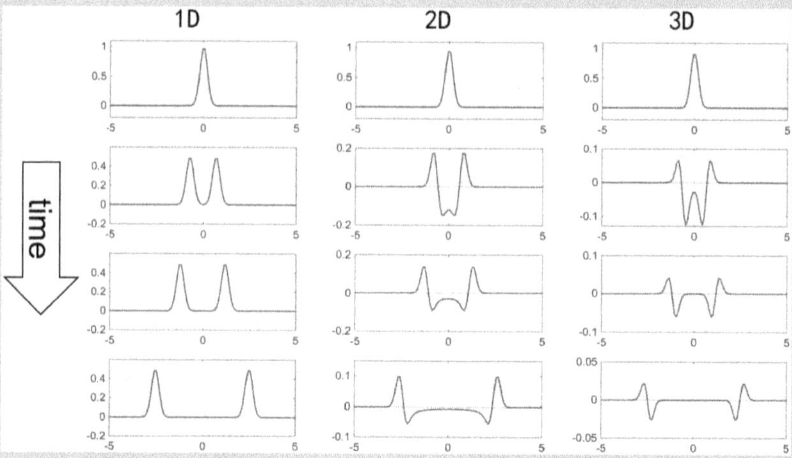

Figure 12.6.

When mechanical or electromagnetic waves propagate in one-dimensional or three-dimensional space, the signals are clear, distinguishable, and do not interfere with each other. However, in a two-dimensional world, speaking would result in a continuous buzzing sound mixed with the past and present, making it impossible to hear beautiful melodies. The faithfulness of signal propagation in space is not as certainly as people imagine. In terms of its intrinsic harmony, the three-dimensional world is truly unique in various dimensional spaces.

How fortunate we are to live in such a world!

Exercises

[1] Solve the active heat conduction problem for an infinitely long slender rod:

$$\begin{cases} u_t - a^2 u_{xx} = f(x,t) \\ u|_{t=0} = 0 \end{cases}$$

Answer:

$$u(x,t)=\int_0^t\int_{-\infty}^{\infty}\frac{f(\xi,\tau)}{2a\sqrt{\pi(t-\tau)}}\exp\left[-\frac{(x-\xi)^2}{4a^2(t-\tau)}\right]\mathrm{d}\xi\,\mathrm{d}\tau$$

[2] Solve the one-dimensional transport problem in a half-unbounded space:

$$\begin{cases}\dfrac{\partial u}{\partial t}=a^2\dfrac{\partial^2 u}{\partial x^2} & (0\le x<\infty)\\ u(0,t)=At, & u|_{t=0}=0\end{cases}$$

Answer:

$$u(x,t)=At-A\int_0^t\operatorname{erf}\left(\frac{x}{2a\sqrt{t-\tau}}\right)\mathrm{d}\tau$$

[3] Solve the vibration problem of an unbounded string:

$$\begin{cases}\dfrac{\partial^2 u}{\partial t^2}=a^2\dfrac{\partial^2 u}{\partial x^2}\\ u|_{t=0}=u_0\mathrm{e}^{-\frac{x^2}{\sigma^2}}, & u_t|_{t=0}=0\end{cases}$$

Answer:

$$u(x,t)=\frac{u_0}{2}\left[\mathrm{e}^{-\frac{(x+at)^2}{\sigma^2}}+\mathrm{e}^{-\frac{(x-at)^2}{\sigma^2}}\right]$$

[4] Solve the differential equation in one-dimensional unbounded space:

$$\text{(a)}\begin{cases}\dfrac{\partial^2 u}{\partial t\partial x}=\dfrac{\partial^2 u}{\partial x^2} & -\infty<x<\infty)\\ u|_{t=0}=\mathrm{e}^{-|x|}\end{cases}$$

$$\text{(b)}\begin{cases}t\dfrac{\partial u}{\partial x}+\dfrac{\partial u}{\partial t}=0 & (-\infty<x<\infty)\\ u|_{t=0}=f(x)\end{cases}$$

Answer: (a) $u(x,t)=\mathrm{e}^{-|x+t|}$; (b) $u(x,t)=f\left(x-\frac{1}{2}t^2\right)$

[5] Solve the three-dimensional Poisson equation in infinite space:

$$\Delta u(\boldsymbol{r})=\frac{1}{r^2}\quad (r\neq 0)$$

Answer: $u(\boldsymbol{r})=\frac{1}{2}\ln r$

12.3 Laplace Transform Method

The Fourier transformation method is only applicable for solving spatial distribution problems without boundaries. For initial value problems, the Laplace transformation method is employed for solving.

Example 12.9. Solving the surface concentration diffusion problem of a silicon wafer:

$$\begin{cases} u_t - a^2 u_{xx} = 0 & (x \geq 0) \\ u|_{x=0} = N_0, & u|_{t=0} = 0 \end{cases}$$

Solution: Let $\bar{u}(x,p) = \mathscr{L}[u(x,t)]$ be the Laplace transform of $u(x,t)$. Applying the Laplace transform to both sides of the equation, we have

$$p\bar{u} - a^2 \bar{u}_{xx} = 0, \quad \bar{u}|_{x=0} = \frac{N_0}{p}$$

Solving this, we obtain

$$\bar{u} = Ae^{\frac{\sqrt{p}}{a}x} + Be^{-\frac{\sqrt{p}}{a}x}$$

Considering the finiteness of the solution as $x \to \infty$, we have $A = 0$. Substituting this into the boundary condition, we get

$$\bar{u} = N_0 \frac{1}{p} \mathrm{e}^{-\frac{\sqrt{p}x}{a}}$$

Using the inverse Laplace transform or referring to the Laplace transform table in the appendix directly, we have

$$u(x,t) = N_0 \operatorname{erfc}\left(\frac{x}{2a\sqrt{t}}\right)$$

It is consistent with the results obtained using the Fourier transformation method.

Since the equation becomes a second-order ordinary differential equation with respect to x after the Laplace transformation, the solution process is relatively simple if the equation or initial conditions are homogeneous. Otherwise, the calculation process can be somewhat tedious, as shown in the following example.

Example 12.10. Solve the vibration problem of an infinite string:

$$\begin{cases} u_{tt} - a^2 u_{xx} = 0 & (-\infty < x < \infty) \\ u|_{t=0} = \varphi(x), & u_t|_{t=0} = \psi(x) \end{cases}$$

Solution 1: Let the Laplace transform of the function be $\bar{u}(x,p) = \mathscr{L}[u(x,t)]$. Then we have

$$p^2\bar{u}(x,p) - a^2\bar{u}_{xx}(x,p) = p\varphi(x) + \psi(x)$$

The general solution to this equation is

$$\bar{u}(x,p) = A\mathrm{e}^{\frac{px}{a}} + B\mathrm{e}^{-\frac{px}{a}} - \frac{1}{2a}\mathrm{e}^{\frac{px}{a}} \int^x \frac{\mathrm{e}^{-p\xi/a}}{p}[\psi(\xi) + p\varphi(\xi)]\mathrm{d}\xi$$
$$+ \frac{1}{2a}\mathrm{e}^{-\frac{px}{a}} \int^x \frac{\mathrm{e}^{p\xi/a}}{p}[\psi(\xi) + p\varphi(\xi)]\mathrm{d}\xi$$

Since the amplitude remains finite as $x \to \pm\infty$, we have $A = B = 0$. Thus,

$$\begin{aligned} \bar{u}(x,p) &= -\frac{1}{2a}\int_{\infty}^{x} \frac{\mathrm{e}^{-p(\xi-x)/a}}{p}[\psi(\xi) + p\varphi(\xi)]\mathrm{d}\xi \\ &\quad + \frac{1}{2a}\int_{-\infty}^{x} \frac{\mathrm{e}^{-p(x-\xi)/a}}{p}[\psi(\xi) + p\varphi(\xi)]\mathrm{d}\xi \\ &= \left[\frac{1}{2a}\int_{x}^{\infty} \frac{\mathrm{e}^{-p(\xi-x)/a}}{p}\psi(\xi)\mathrm{d}\xi + \frac{1}{2a}\int_{-\infty}^{x} \frac{\mathrm{e}^{-p(x-\xi)/a}}{p}\psi(\xi)\mathrm{d}\xi\right] \\ &\quad + \left[\frac{1}{2a}\int_{x}^{\infty} \frac{\mathrm{e}^{-p(\xi-x)/a}}{p}p\varphi(\xi)\mathrm{d}\xi + \frac{1}{2a}\int_{-\infty}^{x} \frac{\mathrm{e}^{-p(x-\xi)/a}}{p}p\varphi(\xi)\mathrm{d}\xi\right] \end{aligned}$$

The integration limits in the above equation are taken as $\pm\infty$ to ensure that the solution does not diverge at infinity. In order to invert $\bar{u}(x,p)$, we can use the delay theorem to obtain

$$\mathscr{L}^{-1}\left[\frac{\mathrm{e}^{-p(\xi-x)/a}}{p}\right] = \mathrm{H}\left(t - \frac{\xi - x}{a}\right) = \begin{cases} 1 & (\xi - x < at) \\ 0 & (\xi - x > at) \end{cases}$$

Hence

$$\mathscr{L}^{-1}\left[\int_x^{\infty}\frac{\mathrm{e}^{-p(\xi-x)/a}}{p}\psi(\xi)\mathrm{d}\xi\right]=\int_x^{x+at}\psi(\xi)\mathrm{d}\xi$$

$$\mathscr{L}^{-1}\left[\int_{-\infty}^{x}\frac{\mathrm{e}^{-p(x-\xi)/a}}{p}\psi(\xi)\mathrm{d}\xi\right]=\int_{x-at}^{x}\psi(\xi)\mathrm{d}\xi$$

and

$$\begin{aligned}
&\frac{1}{2a}\mathscr{L}^{-1}\left[\int_x^{\infty}\frac{\mathrm{e}^{-p(\xi-x)/a}}{p}p\varphi(\xi)\mathrm{d}\xi\right]\\
&=\frac{1}{2}\mathscr{L}^{-1}\left[\int_x^{\infty}\frac{\partial}{\partial x}\frac{\mathrm{e}^{-p(\xi-x)/a}}{p}\varphi(\xi)\mathrm{d}\xi\right]\\
&=\frac{1}{2}\int_x^{\infty}\frac{\partial}{\partial x}H\left(t-\frac{\xi-x}{a}\right)\varphi(\xi)\mathrm{d}\xi\\
&=\frac{1}{2}\int_x^{\infty}\delta[x-(\xi-at)]\varphi(\xi)\mathrm{d}\xi=\frac{1}{2}\varphi(x+at)\\
&\times\mathscr{L}^{-1}\left[\frac{1}{2a}\int_{-\infty}^{x}\frac{\mathrm{e}^{-p(x-\xi)/a}}{p}p\varphi(\xi)\mathrm{d}\xi\right]=\frac{1}{2}\varphi(x-at)
\end{aligned}$$

Therefore, we obtain the D'Alembert solution for the third time:

$$u(x,t)=\frac{1}{2}[\varphi(x+at)+\varphi(x-at)]+\frac{1}{2a}\int_{x-at}^{x+at}\psi(\xi)\mathrm{d}\xi$$

From this example, we can see that the Laplace transformation method can also be used to solve the wave equation for an infinite string. However, the solving process appears somewhat cumbersome. The reason is that the transformed equation is a non-homogeneous second-order ODE, and the expression for its particular solution is somewhat complicated. If both Fourier transformation and Laplace transformation are used in combination, the solution process can be more concise, as described in the following solution.

Solution 2: First, we apply the Laplace transform to the equation, $\bar{u}(x,p) = \mathscr{L}[u(x,p)]$, which transforms the PDE to an ODE,

$$p^2\bar{u}(x,p) - a^2\bar{u}_{xx}(x,p) = p\varphi(x) + \psi(x)$$

Next, we apply the Fourier transform, $\bar{U}(k,p) = \mathfrak{F}[\bar{u}(x,p)]$, which further simplifies the differential equation to an algebraic equation:

$$p^2\bar{U}(k,p) + a^2k^2\bar{U}(k,p) = p\Phi(k) + \Psi(k)$$

Solving for $\bar{U}(k,p)$, we get

$$\bar{U}(k,p) = \Phi(k)\frac{p}{p^2+a^2k^2} + \Psi(k)\frac{1}{p^2+a^2k^2}$$

Now, we apply the inverse Laplace transform,

$$\begin{aligned} U(k,t) &= \Phi(k)\cos kat + \frac{1}{ka}\Psi(k)\sin kat \\ &= \Phi(k)\cos kat + \int_0^t \Psi(k)\cos ka\tau\,\mathrm{d}\tau \end{aligned}$$

Finally, we apply the inverse Fourier transform to obtain

$$u(x,t) = \frac{1}{2}[\varphi(x+at) + \varphi(x-at)] + \frac{1}{2a}\int_{x-at}^{x+at}\psi(\xi)\mathrm{d}\xi$$

Example 12.11. Given a semi-infinite string with densities ρ_1 and ρ_2, connected at $x = 0$, and assuming the displacement distribution on the string ρ_1 at $x < 0$ is $u_1|_{t=0} = \varphi(x)$, and the string ρ_2 at $x > 0$ is in equilibrium with $u_2|_{t=0} = 0$, we are required to find the propagation of waves on the string.

Solution: Based on the given problem, we can write the differential equation and boundary conditions as follows:

$$\begin{cases} \partial_{tt}u_1 - a_1^2\partial_{xx}u_1 = 0 \quad (x < 0) \\ \partial_{tt}u_2 - a_2^2\partial_{xx}u_2 = 0 \quad (x > 0) \\ u_1|_{x=0} = u_2|_{x=0}, \quad \partial_x u_1|_{x=0} = \partial_x u_2|_{x=0} \\ u_1|_{t=0} = \varphi(x), \quad u_{1t}|_{t=0} = 0 \\ u_2|_{t=0} = 0, \quad u_{2t}|_{t=0} = 0 \end{cases}$$

where $a_1^2 = \frac{T}{\rho_1}$ and $a_2^2 = \frac{T}{\rho_2}$, with T denoting tension. Applying the Laplace transform to the wave equations for both segments of the string,

$$\bar{u}_j(x,p) = \mathscr{L}[u_j(x,t)] \quad (j = 1,2)$$

we obtain the following ODEs for the transformed functions:

$$\begin{cases} p^2\bar{u}_1 - p\varphi(x) - a_1^2\bar{u}_1'' = 0 & (x < 0) \\ p^2\bar{u}_2 - a_2^2\bar{u}_2'' = 0 & (x > 0) \end{cases}$$

The solutions are given by

$$\begin{aligned} \bar{u}_1 &= A_1 e^{\frac{p}{a_1}x} + B_1 e^{-\frac{p}{a_1}x} - \frac{1}{2a_1} e^{\frac{p}{a_1}x} \int_0^x \varphi(\xi) e^{-\frac{p}{a_1}\xi} d\xi \\ &\quad + \frac{1}{2a_1} e^{-\frac{p}{a_1}x} \int_{-\infty}^x \varphi(\xi) e^{\frac{p}{a_1}\xi} d\xi \\ \bar{u}_2 &= A_2 e^{\frac{p}{a_2}x} + B_2 e^{-\frac{p}{a_2}x} \end{aligned}$$

Since the amplitude of vibration approaches zero as $x \to \pm\infty$, hence $B_1 = A_2 = 0$. Considering the continuity conditions at $x = 0$, we have

$$A_1 + \frac{1}{2a_1} \int_{-\infty}^0 \varphi(\xi) e^{\frac{p}{a_1}\xi} d\xi = B_2$$

$$A_1 \frac{p}{a_1} - \frac{1}{2a_1} \frac{p}{a_1} \int_{-\infty}^0 \varphi(\xi) e^{\frac{p}{a_1}\xi} d\xi = -B_2 \frac{p}{a_2}$$

The solutions for $A_1(p)$ and $B_2(p)$ are given by

$$A_1(p) = \frac{1}{2a_1} \frac{a_2 - a_1}{a_1 + a_2} \int_{-\infty}^0 \varphi(\xi) e^{\frac{p}{a_1}\xi} d\xi$$

$$B_2(p) = \frac{1}{a_1} \frac{a_2}{a_1 + a_2} \int_{-\infty}^0 \varphi(\xi) e^{\frac{p}{a_1}\xi} d\xi$$

Notably,

$$\frac{1}{2a_1}\int_{-\infty}^{0}\varphi(\xi)\mathrm{e}^{\frac{p}{a_1}\xi}\mathrm{d}\xi = \frac{1}{2}\mathscr{L}[\varphi(-a_1t)]$$

which can be obtained using the delay theorem. Applying the inverse Laplace transform, the solution to the equation is given by

$$\begin{aligned} u_1(x,t) &= \frac{1}{2}\varphi(x+a_1t) + \frac{1}{2}\varphi(x-a_1t)\mathrm{H}(-x-a_1t) \\ &\quad + \frac{a_2-a_1}{2(a_1+a_2)}\varphi(-x-a_1t)\mathrm{H}(x+a_1t) \quad (x<0) \\ u_2(x,t) &= \frac{a_2}{a_1+a_2}\varphi\left[\frac{a_1}{a_2}(x-a_2t)\right]\mathrm{H}(-x+a_2t) \quad (x>0) \end{aligned}$$

Here, the first term in $u_1(x,t)$ represents the incident wave propagating in the negative direction, the second term represents the incident wave propagating towards the interface in the positive direction, and the third term represents the reflected wave. $u_2(x,t)$ represents the transmitted wave. When $a_2 < a_1$ or $\rho_1 < \rho_2$, the reflected wave undergoes a half-wave inversion, resulting in half-wave loss. When $a_2 > a_1$ or $\rho_1 > \rho_2$, there is no half-wave loss. In both cases, the reflected and transmitted waves maintain the same waveform as the incident wave, but the width of the transmitted wave is stretched or compressed by a factor of $\frac{a_1}{a_2}$. Figure 12.7 illustrates the reflection and transmission processes of waves on different interfaces.

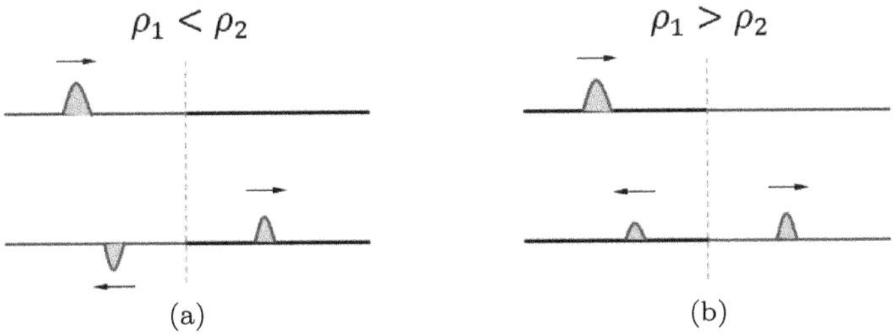

Figure 12.7.

It is worth noting that when $a_2 = a_1$, there is no reflected wave, and all waves pass through to the other medium without obstruction. This effect can be utilized to create absorbing materials.

ENDNOTE

Introducing integral transform via the Dirac delta function may shed some insights. By definition, we have

$$\delta(x) \stackrel{\text{def}}{=} \lim_{n\to\infty} \frac{\sin nx}{\pi x} = \frac{1}{2\pi}\int_{-\infty}^{\infty} \mathrm{e}^{-ikx}\mathrm{d}k$$

From this, it can be derived that

$$\begin{aligned} f(x) &= \int_{-\infty}^{\infty} f(x')\delta(x'-x)\mathrm{d}\xi \\ &= \frac{1}{2\pi}\int_{-\infty}^{\infty} f(x')\left[\int_{-\infty}^{\infty} \mathrm{e}^{-ik(x'-x)}\mathrm{d}k\right]\mathrm{d}x' \\ &= \frac{1}{2\pi}\int_{-\infty}^{\infty} \mathrm{e}^{ikx}\left[\int_{-\infty}^{\infty} f(x')\mathrm{e}^{-ikx'}\mathrm{d}x'\right]\mathrm{d}k = \int_{-\infty}^{\infty} F(k)\mathrm{e}^{ikx}\mathrm{d}k \end{aligned}$$

This equation represents the Fourier integral representation of the function $f(x)$, where $F(k)$ is called the Fourier transform function of $f(x)$:

$$F(k) \stackrel{\text{def}}{=} \frac{1}{2\pi}\int_{-\infty}^{\infty} f(x)\mathrm{e}^{-ikx}\mathrm{d}x$$

This method is distinct from extending Fourier series of periodic functions to Fourier integrals of non-periodic functions.

Analogously, we further introduce the Laplace transform via the Dirac delta function. Defining the function $f(t) \equiv f(t)H(t)$, where $H(t)$ is the Heaviside step function, we have

$$\begin{aligned} f(t) &= \int_{-\infty}^{\infty} f(t')\mathrm{H}(t')\delta(t'-t)\mathrm{d}t' \\ &= \int_{-\infty}^{\infty} f(t')\mathrm{H}(t')\mathrm{e}^{-\sigma(t'-t)}\delta(t'-t)\mathrm{d}t' \end{aligned}$$

where σ is a positive real constant. Inserting

$$\delta(t'-t) = \frac{1}{2\pi}\int_{-\infty}^{\infty} \mathrm{e}^{-\mathrm{i}(t'-t)x}\mathrm{d}x$$

we have

$$\begin{aligned} f(t) &= \frac{1}{2\pi}\int_{-\infty}^{\infty} f(t')\mathrm{H}(t')\mathrm{e}^{-\sigma(t'-t)}\int_{-\infty}^{\infty} \mathrm{e}^{-\mathrm{i}(t'-t)x}\mathrm{d}x\,\mathrm{d}t' \\ &= \frac{1}{2\pi}\int_{-\infty}^{\infty} \mathrm{e}^{(\sigma+\mathrm{i}x)t}\left[\int_{-\infty}^{\infty} f(t')\mathrm{H}(t')\mathrm{e}^{-(\sigma+\mathrm{i}x)}t'\,\mathrm{d}t'\right]\mathrm{d}x \end{aligned}$$

Letting $p = \sigma + \mathrm{i}x$, we get

$$f(t) = \frac{1}{2\pi\mathrm{i}}\int_{\sigma-\mathrm{i}\infty}^{\sigma+\mathrm{i}\infty} \mathrm{e}^{pt}F(p)\mathrm{d}p$$

where

$$F(p) \overset{\text{def}}{=} \int_{0}^{\infty} f(t)\mathrm{e}^{-pt}\mathrm{d}t$$

This is exactly the Laplace transform for function $f(t)$. In order to guarantee the integrability of the function, it requires that $\mathrm{Re}\,p > 0$.

Due to the truncation of the Laplace transform for functions with $t > 0$, it is not applicable to functions defined on $(-\infty, \infty)$. Can we generalize it? Let us take a general examination. As long as the transfer function $\rho(x', x)$ is nonsingular and satisfies $\rho(x, x) \equiv 1$, the following equation holds:

$$f(x) = \int_{-\infty}^{\infty} f(x')\rho(x', x)\delta(x'-x)\mathrm{d}x'$$

Inserting the expression of the delta function, we have

$$f(x) = \frac{1}{2\pi}\int_{-\infty}^{\infty} f(x')\rho(x', x)\int_{-\infty}^{\infty} \mathrm{e}^{-\mathrm{i}(x'-x)k}\mathrm{d}k\,\mathrm{d}x'$$

If $\rho(x',x)$ can be factorized into two parts, $\rho(x',x) = \rho_1(x')\rho_2(x)$, then the integral can also be factorized as

$$\begin{aligned} f(x) &= \frac{1}{2\pi}\int_{-\infty}^{\infty}\left[\int_{-\infty}^{\infty} f(x')\rho_1(x')\mathrm{e}^{-ikx'}\mathrm{d}x'\right]\rho_2(x)\mathrm{e}^{ikx}\mathrm{d}k \\ &= \int_{-\infty}^{\infty} F(k)\rho_2(x)\mathrm{e}^{ikx}\mathrm{d}k \end{aligned}$$

where

$$F(k) = \frac{1}{2\pi}\int_{-\infty}^{\infty} f(x)\rho_1(x)\mathrm{e}^{-ikx}\mathrm{d}x$$

This is the general form of integral transform for functions and its inverse transform, which is denoted by a symbol y:

$$F(k) = \mathscr{Y}[f(x)], \quad f(x) = y^{-1}[F(k)]$$

It may be called the y-transform for function $f(x)$.

When taking $\rho_1(x) = \rho_2(x) = 1$, the y-transform is simply the Fourier transform. When taking $\rho_1(x) = \mathrm{e}^{-\sigma x}$, $\rho_2(x) = \mathrm{e}^{\sigma x}$, it becomes the Laplace transform. If we take

$$\rho(x',x) = \mathrm{e}^{-\sigma|x'|+\sigma|x|}$$

Then the bilateral Laplace transform follows,

$$\begin{cases} F(k) \overset{\text{def}}{=} \displaystyle\int_{-\infty}^{\infty} f(x)\mathrm{e}^{-\sigma|x|-ikx}\mathrm{d}x \\ f(x) \overset{\text{def}}{=} \displaystyle\frac{1}{2\pi\mathrm{i}}\int_{-\infty}^{\infty} F(k)\mathrm{e}^{\sigma|x|+ikx}\mathrm{d}k \end{cases}$$

It extends the restriction of $x > 0$ in the Laplace transform to the real axis.

Now let us choose

$$\rho_1(x) = \mathrm{e}^{-x^2}, \quad \rho_2(x) = \mathrm{e}^{x^2}$$

Hence the transfer function $\rho(x',x) = \mathrm{e}^{-x'^2+x^2}$. Then a set of function transform and its inverse transform is defined:

$$\begin{cases} F(k) \stackrel{\text{def}}{=} \dfrac{1}{2\pi}\displaystyle\int_{-\infty}^{\infty} f(x)\mathrm{e}^{-x^2-\mathrm{i}kx}\mathrm{d}x \\ f(x) \stackrel{\text{def}}{=} \displaystyle\int_{-\infty}^{\infty} F(k)\mathrm{e}^{x^2+\mathrm{i}kx}\mathrm{d}k \end{cases}$$

It is known as Gabor transform. The essence of this transform is to damp the function $f(x)$ with a Gaussian function and then perform the Fourier transform. The advantage is that, as long as the function is non-singular, the transform is finite over $(-\infty,\infty)$, which greatly expands the applicability of the Fourier transform. The Gaussian damp factor is also called the window function. If we take a shift to the transfer function,

$$\rho_1(x,\xi) = \mathrm{e}^{-\pi(x-\xi)^2}, \quad \rho_2(x,\xi) = \mathrm{e}^{\pi(x-\xi)^2}$$

then general Gabor transform and its inverse transform are,

$$\begin{cases} G(\xi,k) \stackrel{\text{def}}{=} \displaystyle\int_{-\infty}^{\infty} f(x)\mathrm{e}^{-\pi(x-\xi)^2-\mathrm{i}kx}\mathrm{d}x \\ f(x) \stackrel{\text{def}}{=} \dfrac{1}{2\pi}\displaystyle\int_{-\infty}^{\infty} G(\xi,k)\mathrm{e}^{\pi(x-\xi)^2+\mathrm{i}kx}\mathrm{d}k \end{cases}$$

Surely, we can introduce other types of transfer functions, e.g., $\rho(x',x) = \mathrm{e}^{-x'^4+x^4}$, which damps the function more severely.

By making use of the orthogonality of the Bessel function (see Chapter 15),

$$\int_0^{\infty} J_n(kx)J_n(kx')k\mathrm{d}k = \frac{\delta(x-x')}{x'}$$

where $x, x' > 0$, insert it into the Dirac formula, then

$$\begin{aligned} f(x) &= \int_0^{\infty} f(x')\delta(x'-x)\mathrm{d}x' \\ &= \int_0^{\infty} f(x') \int_0^{\infty} J_n(kx)J_n(kx')k\,\mathrm{d}k\,x'\mathrm{d}x' \\ &= \int_0^{\infty} J_n(kx)\left[\int_0^{\infty} f(x')J_n(kx')x'\mathrm{d}x'\right] k\mathrm{d}k \end{aligned}$$

We obtain the Hankel transform and its inverse transform ($n \geq -1/2$):

$$\begin{cases} F_n(k) = \int_0^{\infty} f(x)J_n(kx)x\mathrm{d}x \\ f(x) = \int_0^{\infty} F_n(k)J_n(kx)k\mathrm{d}k \end{cases}$$

In the microscopic world, the classical concepts of point particles and trajectory motion are no longer applicable. The state of particles should not be described by the position x and momentum p. Assuming a particle of mass m is moving in an external potential $V(x)$, its state can be fully described by a certain function $\psi(x)$. Using the Dirac delta function, it can be expressed as

$$\psi(x) = \int_{-\infty}^{\infty} \psi(x')\delta(x'-x)\mathrm{d}x = \frac{1}{\sqrt{2\pi}}\int_{-\infty}^{\infty} \varphi(k)\mathrm{e}^{\mathrm{i}kx}\mathrm{d}k$$

where

$$\varphi(k) = \frac{1}{\sqrt{2\pi}}\int_{-\infty}^{\infty} \psi(x)\mathrm{e}^{-\mathrm{i}kx}\mathrm{d}x$$

Since x has dimensions of length [L], the physical quantity k has dimensions of $[\mathrm{L}^{-1}]$. Its physical meaning arises from the existence of the Planck constant $\hbar$, which has the dimensions of $[\mathrm{ML}^2\mathrm{T}^{-1}]$. Let $k = \frac{p}{\hbar}$, then the physical quantity p has the dimensions of momentum, which is simply referred to as the momentum of the particle. The above expressions are then reformulated as

$$\psi(x) = \frac{1}{\sqrt{2\pi\hbar}}\int_{-\infty}^{\infty}\varphi(p)\mathrm{e}^{\frac{\mathrm{i}px}{\hbar}}\mathrm{d}p, \quad \varphi(p) = \frac{1}{\sqrt{2\pi\hbar}}\int_{-\infty}^{\infty}\psi(x)\mathrm{e}^{-\frac{\mathrm{i}px}{\hbar}}\mathrm{d}x$$

Applying the Laplace operator to the function $\psi(x)$, we have

$$\nabla^2\psi(x) = \frac{1}{\sqrt{2\pi\hbar}}\int_{-\infty}^{\infty}\frac{(-p^2)}{\hbar^2}\varphi(p)\mathrm{e}^{\frac{\mathrm{i}px}{\hbar}}\mathrm{d}p$$

Thus

$$\begin{aligned}\left[-\frac{\hbar^2}{2m}\nabla^2 + V(x)\right]\psi(x) &= \frac{1}{\sqrt{2\pi\hbar}}\int_{-\infty}^{\infty}\left[\frac{p^2}{2m} + V(x)\right]\varphi(p)\mathrm{e}^{\frac{\mathrm{i}px}{\hbar}}\mathrm{d}p \\ &= \frac{1}{\sqrt{2\pi\hbar}}\int_{-\infty}^{\infty}H(x,p)\varphi(p)\mathrm{e}^{\frac{\mathrm{i}px}{\hbar}}\mathrm{d}p\end{aligned}$$

where $H(x,p) = \frac{p^2}{2m} + V(x)$ is the classical Hamiltonian of a particle in an external potential.

Since the energy remains an invariant physical quantity for conservative potentials, $H(x,p) \equiv E$, indicated by classical physics, the state function satisfies the following equation:

$$\left[-\frac{\hbar^2}{2m}\nabla^2 + V(x)\right]\psi(x) = \frac{1}{\sqrt{2\pi\hbar}}\int_{-\infty}^{\infty}E\varphi(p)\mathrm{e}^{\frac{\mathrm{i}px}{\hbar}}\mathrm{d}p = E\psi(x)$$

This is the stationary Schrödinger equation of quantum mechanics, where the state function $\psi(x)$ is traditionally referred to as the wave function of the particle and $\varphi(p)$ as the wave function in the momentum space.

Following the same reasoning, we can obtain the time-evolution Schrödinger equation for the state function:

$$\mathrm{i}\hbar\frac{\partial}{\partial t}\psi(x,t) = \left[-\frac{\hbar^2}{2m}\nabla^2 + V(x)\right]\psi(x,t)$$

This deduction seems the most reasonable route of introducing Schrödinger equation into quantum mechanics textbooks.

Exercises

[1] Use the Laplace transform method to solve the active heat conduction problem in an infinite long slender rod:

$$\begin{cases} u_t - a^2 u_{xx} = f(x,t) \quad (-\infty < x < \infty) \\ u|_{t=0} = 0 \end{cases}$$

[2] A uniform rod has a length of l. One end is kept at a temperature of u_0, and the other end is insulated. The rod's initial temperature is zero. Find the variation of temperature in the rod with time.

[3] A semi-infinite string is initially in equilibrium state. At $t > 0$, there is a small oscillation $A \sin \omega t$ at the $x = 0$ end of the string. Solve the motion of the string.

$$\begin{cases} \dfrac{\partial^2 u}{\partial t^2} - a^2 \dfrac{\partial^2 u}{\partial x^2} = 0 \quad (0 \leq x < \infty) \\ u|_{x=0} = A \sin \omega t \\ u|_{t=0} = 0, u_t|_{t=0} = 0 \end{cases}$$

Answer: $u(x,t) = A \sin \omega \left(t - \frac{x}{a}\right) \mathrm{H}\left(t - \frac{x}{a}\right)$

[4] Solve the following differential equation:

$$\text{(a)} \begin{cases} \dfrac{\partial^2 u}{\partial x \partial t} = 1 \quad (0 \leq x < \infty) \\ u|_{x=0} = t + 1 \\ u|_{t=0} = 1 \end{cases}$$

$$\text{(b)} \begin{cases} \dfrac{\partial u}{\partial t} + x \dfrac{\partial u}{\partial x} = x \quad (0 \leq x < \infty) \\ u|_{x=0} = 0 \\ u|_{t=0} = 0 \end{cases}$$

Answer: (a) $u(x,t) = xt + t + 1$; (b) $u(x,t) = x(1 - \mathrm{e}^{-t})$

[5] Solve the forced vibration of an unbounded string:

$$\begin{cases} u_{tt} - a^2 u_{xx} = f(x,t) \\ u|_{t=0} = \varphi(x), \quad u_t|_{t=0} = \psi(x) \end{cases}$$

Hint: Use both the Fourier transform method and the Laplace transform method simultaneously.

Answer:

$$u(x,t)=\frac{1}{2}[\varphi(x+at)+\varphi(x-at)]+\frac{1}{2a}\int_{x-at}^{x+at}\psi(\xi)\mathrm{d}\xi$$
$$+\frac{1}{2a}\int_0^t\int_{x-a(t-\tau)}^{x+a(t-\tau)}f(\xi,\tau)\mathrm{d}\xi\,\mathrm{d}\tau$$

[6] Two semi-infinite uniform rods with initial temperatures of zero and u_0 are perfectly connected at $x=0$. Solve the variations of the temperature on the rod with time.
Hint: The joining conditions at $x=0$ for the two rods are:

$$u_1|_{x=0}=u_2|_{x=0},\quad \partial_x u_1|_{x=0}=\partial_x u_2|_{x=0}$$

Answer:

$$u(x,t)=\begin{cases}\dfrac{u_0}{2}\operatorname{erfc}\left(-\dfrac{x}{2\sqrt{\kappa t}}\right) & (x<0)\\[2ex] u_0-\dfrac{u_0}{2}\operatorname{erfc}\left(\dfrac{x}{2\sqrt{\kappa t}}\right) & (x>0)\end{cases}$$

[7] Two semi-infinite slender rods with densities of $\rho_1\rho_2$ and Young's moduli of E_1E_2, respectively, are perfectly connected at $x=0$. Assuming that the rod in the region $x<0$ has a displacement distribution $u_1|_{t=0}=\varphi(x)$, and the rod in the region $x>0$ is in its equilibrium position $u_2|_{t=0}=0$. At the initial moment, the rods are at rest. Solve for the propagation of waves along the rods and investigate the conditions of half-wave loss as well as perfect transmission.
Hint: The boundary condition at $x=0$ for the two rods is

$$u_1|_{x=0}=u_2|_{x=0},\quad E_1\partial_x u_1|_{x=0}=E_2\partial_x u_2|_{x=0}$$

Answer:

$$u_1(x,t)=\frac{1}{2}\varphi(x+a_1t)+\frac{1}{2}\varphi(x-a_1t)\mathrm{H}(-x-a_1t)$$
$$+\frac{a_2E_1-a_1E_2}{2(a_1E_2+a_2E_1)}\varphi(-x-a_1t)\mathrm{H}(x+a_1t)\quad (x<0)$$
$$u_2(x,t)=\frac{a_2E_1}{a_1E_2+a_2E_1}\varphi\left[\frac{a_1}{a_2}(x-a_2t)\right]\mathrm{H}(-x+a_2t)\quad (x>0)$$

[8] Perform the Gaussian-type y-transform for the following functions:

$$\text{(a) } f(x) = x; \quad \text{(b) } f(x) = x^2; \quad \text{(c) } f(x) = \mathrm{e}^{\alpha x}; \quad \text{(d) } f(x) = \sin x$$

Hint: Utilize the Gaussian integral and differentiate with respective to k,

$$\int_{-\infty}^{\infty} \mathrm{e}^{-x^2 - ikx} \mathrm{d}x = \sqrt{\pi} \mathrm{e}^{-\frac{1}{4}k^2}$$

[9] Write out the y-transform and its inverse transform for the following transfer functions ($\sigma > 0$):

$$\text{(a) } \rho(x', x) = \operatorname{sech} \sigma x' \cosh \sigma x; \quad \text{(b) } \rho(x', x) = \frac{1 + x^2}{1 + x'^2}$$

Chapter 13

Spherical Harmonics

13.1 Legendre Equation

13.1.1 *Spherical coordinate system*

For three-dimensional spherically symmetric problems, it is more convenient to express the Laplace equation in spherical coordinates for the purpose of variable separation and handling boundary conditions. Applying the method of orthogonal curvilinear coordinates from Chapter 10, the Laplace equation takes the following form:

$$\frac{1}{r^2}\frac{\partial}{\partial r}\left(r^2\frac{\partial u}{\partial r}\right)+\frac{1}{r^2\sin\theta}\frac{\partial}{\partial\theta}\left(\sin\theta\frac{\partial u}{\partial\theta}\right)+\frac{1}{r^2\sin^2\theta}\frac{\partial^2 u}{\partial\phi^2}=0 \quad (13.1.1)$$

Firstly, we separate the variable $u(r,\theta,\phi)=R(r)Y(\theta,\phi)$. The radial part $R(r)$ satisfies the Euler-type equation:

$$\frac{\mathrm{d}}{\mathrm{d}r}\left(r^2\frac{\mathrm{d}R}{\mathrm{d}r}\right)-l(l+1)R=0$$

The reason for choosing the separation constant as $l(l+1)$ will be explained later. The solution to the equation is given by

$$R(r)=C_l r^l+D_l\frac{1}{r^{l+1}}$$

The equation satisfied by the angular part is

$$\frac{1}{\sin\theta}\frac{\partial}{\partial\theta}\left(\sin\theta\frac{\partial Y}{\partial\theta}\right)+\frac{1}{\sin^2\theta}\frac{\partial^2 Y}{\partial\phi^2}+l(l+1)Y=0$$

This equation can be further separated into $Y(\theta, \phi) = \Theta(\theta)\Phi(\phi)$, resulting in

$$\Phi'' + \lambda\Phi = 0$$

$$\frac{1}{\sin\theta}\frac{\mathrm{d}}{\mathrm{d}\theta}\left(\sin\theta\frac{\mathrm{d}\Theta}{\mathrm{d}\theta}\right) + \left[l(l+1) - \frac{\lambda}{\sin^2\theta}\right]\Theta = 0 \tag{13.1.2}$$

Here, a second separation constant λ appears. The function $\Phi(\phi)$ satisfies the periodic condition:

$$\Phi(\phi + 2\pi) = \Phi(\phi)$$

The eigenvalues are given by $\lambda = m^2 (m = 0, 1, 2, \ldots)$, and the corresponding eigenfunctions are

$$\Phi(\phi) = A\cos m\phi + B\sin m\phi$$

Regarding the equation satisfied by $\Theta(\theta)$, let $x = \cos\theta$ and express the function using the variable $y(x)$. We obtain the equation

$$(1 - x^2)\frac{\mathrm{d}^2 y}{\mathrm{d}x^2} - 2x\frac{\mathrm{d}y}{\mathrm{d}x} + \left[l(l+1) - \frac{m^2}{1 - x^2}\right]y = 0 \tag{13.1.3}$$

This equation is called the associated Legendre equation of order l. Specifically, when $m = 0$, it is referred to as the Legendre equation of order l:

$$(1 - x^2)\frac{\mathrm{d}^2 y}{\mathrm{d}x^2} - 2x\frac{\mathrm{d}y}{\mathrm{d}x} + l(l+1)y = 0 \tag{13.1.4}$$

Chapter 9 has derived the recurrence relation between the different coefficients a_k of the Legendre equation using the power series expansion method in the neighborhood of $x_0 = 0$:

$$a_{k+2} = \frac{(k-l)(k+l+1)}{(k+2)(k+1)}a_k \tag{13.1.5}$$

The two linearly independent solutions of the Legendre equation are

$$\begin{aligned} y_1(x) = 1 &+ \frac{(-l)(l+1)}{2!}x^2 + \frac{(2-l)(-l)(l+1)(l+3)}{4!}x^4 + \cdots \\ &+ \frac{(2k-2-l)(2k-4-l)\cdots(-l)(l+1)\cdots(l+2k-1)}{(2k)!}x^{2k} \\ &+ \cdots \end{aligned}$$

and

$$y_2(x) = x + \frac{(1-l)(l+2)}{3!}x^3 + \frac{(3-l)(1-l)(l+2)(l+4)}{5!}x^5 + \cdots$$
$$+ \frac{(2k-1-l)(2k-3-l)\cdots(1-l)(l+2)\cdots(l+2k)}{(2k+1)!}x^{2k+1}$$
$$+ \cdots$$

Both of their convergence radii are

$$R = \lim_{k\to\infty}\left|\frac{a_k}{a_{k+2}}\right|^{\frac{1}{2}} = \lim_{k\to\infty}\left|\frac{(k+2)(k+1)}{(k-l)(k+l+1)}\right|^{\frac{1}{2}} = 1$$

Now the problem arises: According to Gauss's criterion, it was proved in Section 3.1 that the series $y_1(x)$ and $y_2(x)$ diverge at $x = \pm 1$, corresponding to the two poles in the spherical coordinates. This divergence renders the series solutions physically meaningless.

Infinite series diverge at both endpoints $x = \pm 1$. If certain conditions are satisfied, it is possible to make the series converge at one endpoint while still diverging at the other endpoint. However, for physical problems, finite values are required throughout the entire closed interval $[-1, +1]$. What should we do in this case?

13.1.2 *Eigenvalue problem*

(1) *Legendre polynomials*

By observing the recurrence relation for the coefficients (Equation (13.1.5)), we notice that if l is a non-negative integer, the infinite power series is truncated into a finite polynomial, naturally avoiding the divergence issue at $x = \pm 1$. For example, when $l = 2n$, $y_1(x)$ is truncated to a polynomial of degree x^{2n}, while the other solution $y_2(x)$ still diverges at $x = \pm 1$. On the other hand, when $l = 2n + 1$, $y_2(x)$ is truncated to a polynomial of degree x^{2n+1}, but $y_1(x)$ diverges at $x = \pm 1$. Therefore, regardless of whether l is even or odd, the equation has a unique finite solution in the interval $x \in [-1, +1]$, known as the lth-order Legendre polynomial, denoted as $\mathrm{P}_l(x)$,

$$\mathrm{P}_l(x) = \sum_{k=0}^{[l/2]} (-1)^k \frac{(2l-2k)!}{2^l k!(l-k)!(l-2k)!} x^{l-2k} \qquad (13.1.6)$$

Below are several low-order Legendre polynomials:

$$\mathrm{P}_0(x) = 1$$

$$\mathrm{P}_1(x) = x = \cos\theta$$

$$\mathrm{P}_2(x) = \frac{1}{2}(3x^2 - 1) = \frac{1}{4}(3\cos 2\theta + 1)$$

$$\mathrm{P}_3(x) = \frac{1}{2}(5x^3 - 3x) = \frac{1}{8}(5\cos 3\theta + 3\cos\theta)$$

$$\mathrm{P}_4(x) = \frac{1}{8}(35x^4 - 30x^2 + 3) = \frac{1}{64}(35\cos 4\theta + 20\cos 2\theta + 9)$$

Their function curves are shown in Figure 13.1. When l is even, $\mathrm{P}_l(x)$ is an even function, and when l is odd, $\mathrm{P}_l(x)$ is an odd function. Here are some specific function values of $\mathrm{P}_l(x)$:

$$\mathrm{P}_l(1) = 1, \quad \mathrm{P}_l(-1) = (-1)^l$$

$$\mathrm{P}_{2l+1}(0) = 0, \quad \mathrm{P}_{2l}(0) = (-1)^l \frac{(2l)!}{(2^l l!)^2}$$

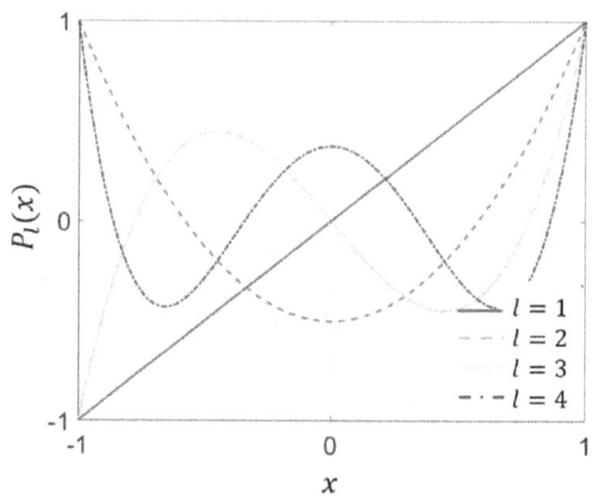

Figure 13.1.

(2) *Natural boundary conditions*

Due to the physical requirement that the solution of the equation must be finite over the closed interval $[-1, +1]$, it is determined that the constant introduced by the separation of variables, l, can only take the values of zero or positive integers. This constraint is known as the natural boundary condition of the equation, which restricts the eigenvalues of the Legendre equation to be the discrete values $l(l+1)$, where l is a non-negative integer. The corresponding eigenfunctions are given by $\mathrm{P}_l(x)$.

It will be seen later that the natural boundary condition is a common phenomenon in second-order variable coefficient ODEs with regular singularities. Similar to homogeneous boundary conditions, it determines the eigenvalue problem of the equation.

(3) *Second solution of the equation*

Given one finite solution of the equation, $\mathrm{P}_l(x)$, using the methods from Chapter 9, we can express the other linearly independent solution of the Legendre equation as

$$\begin{aligned}\mathrm{Q}_l(x) &= \mathrm{P}_l(x)\int_\alpha^x \frac{1}{[\mathrm{P}_l(s)]^2}\exp\left[-\int_0^s \frac{2t}{1-t^2}\mathrm{d}t\right]\mathrm{d}s \\ &= \mathrm{P}_l(x)\int_\alpha^x \frac{1}{(1-s^2)[\mathrm{P}_l(s)]^2}\mathrm{d}s \qquad (13.1.7)\end{aligned}$$

$\mathrm{Q}_l(x)$ is called the second kind Legendre function. We list several low-order second kind Legendre functions as follows:

$$\begin{aligned}\mathrm{Q}_0(x) &= \frac{1}{2}\ln\frac{1+x}{1-x} \\ \mathrm{Q}_1(x) &= \frac{1}{2}\mathrm{P}_1(x)\ln\frac{1+x}{1-x} - 1 \\ \mathrm{Q}_2(x) &= \frac{1}{2}\mathrm{P}_2(x)\ln\frac{1+x}{1-x} - \frac{3}{2}x \\ \mathrm{Q}_3(x) &= \frac{1}{2}\mathrm{P}_3(x)\ln\frac{1+x}{1-x} - \frac{5}{2}x^2 + \frac{2}{3} \\ \mathrm{Q}_4(x) &= \frac{1}{2}\mathrm{P}_4(x)\ln\frac{1+x}{1-x} - \frac{35}{8}x^3 + \frac{55}{24}x \\ &\cdots\end{aligned}$$

The second kind Legendre functions always exhibit logarithmic divergence at $x = \pm 1$. For physical problems involving a spherical region with the north and south poles, these divergent solutions must be discarded. If the physical boundary conditions do not include the north and south poles, there are no restrictions on the parameter l. In this case, $\mathrm{P}_l(x)$ and $\mathrm{Q}_l(x)$ need to be recombined to satisfy the homogeneous boundary conditions, thus determining the eigenvalues l. Anyway, for each discrete l, there is only one finite eigenfunction.

13.1.3 *Basic properties*

(1) *Derivative representation*

$$\mathrm{P}_l(x) = \frac{1}{2^l l!}\frac{\mathrm{d}^l}{\mathrm{d}x^l}\left(x^2-1\right)^l \tag{13.1.8}$$

This formula is known as the Rodrigues formula.

Proof. Using the binomial theorem, we have

$$(x^2-1)^l = \sum_{k=0}^{l}\frac{l!}{(l-k)!k!}(-1)^k x^{2l-2k}$$

Taking the lth derivative on both sides, the terms in the summation with powers lower than l become zero after differentiation. We only need to keep the terms where $2l-2k \geq l$ or $k \leq \frac{l}{2}$. Therefore,

$$\begin{aligned}
&\frac{1}{2^l l!}\frac{\mathrm{d}^l}{\mathrm{d}x^l}\left(x^2-1\right)^l \\
&\quad = \sum_{k=0}^{[l/2]}(-1)^k\frac{(2l-2k)(2l-2k-1)\cdots(l-2k+1)}{2^l(l-k)!k!}x^{l-2k} \\
&\quad = \sum_{k=0}^{[l/2]}(-1)^k\frac{(2l-2k)!}{2^l k!(l-k)!(l-2k)!}x^{l-2k} \equiv P_l(x)
\end{aligned}$$

□

(2) *Integral representation*

$$\mathrm{P}_l(x) = \frac{1}{2\pi\mathrm{i}2^l}\oint_L \frac{\left(z^2-1\right)^l}{(z-x)^{l+1}}\mathrm{d}z \tag{13.1.9}$$

Here, L is any closed contour in the z-plane around the point $z = x$. This formula is known as the Schlaefli integral. It can be proved using the Rodrigues formula and the Cauchy integral formula. Taking L as a closed circle centered at x with a radius of $\sqrt{|x^2-1|}$, we have

$$z - x = \sqrt{x^2-1}\mathrm{e}^{\mathrm{i}\psi}$$

Substituting this into the integral formula and simplifying, we obtain the Laplace integral formula:

$$\begin{aligned}\mathrm{P}_l(x) &= \frac{1}{\pi}\int_0^{\pi}\left[x+\mathrm{i}\sqrt{1-x^2}\cos\psi\right]^l\mathrm{d}\psi \\ \mathrm{P}_l(\cos\theta) &= \frac{1}{\pi}\int_{-\pi}^{\pi}\left[\cos\theta+\mathrm{i}\sin\theta\cos\psi\right]^l\mathrm{d}\psi\end{aligned} \tag{13.1.10}$$

(3) *Orthogonality*

$$\int_{-1}^{+1}\mathrm{P}_l(x)\mathrm{P}_k(x)\mathrm{d}x = N_l^2\delta_{kl} \tag{13.1.11}$$

where the normalization factor is given by

$$N_l^2 = \int_{-1}^{+1}\left[\mathrm{P}_l(x)\right]^2\mathrm{d}x = \frac{2}{2l+1} \tag{13.1.12}$$

Proof. First, let us prove the orthogonality of the Legendre polynomials. Without loss of generality, assume $l > k$. Using the Rodrigues formula and performing integration by parts, we have

$$\begin{aligned}&\int_{-1}^{+1}\mathrm{P}_l(x)\mathrm{P}_k(x)\mathrm{d}x \\ &= \frac{1}{2^l l! 2^k k!}\int_{-1}^{+1}\frac{\mathrm{d}^l}{\mathrm{d}x^l}\left(x^2-1\right)^l\frac{\mathrm{d}^k}{\mathrm{d}x^k}\left(x^2-1\right)^k\mathrm{d}x \\ &= \frac{1}{2^l l! 2^k k!}\int_{-1}^{+1}\frac{\mathrm{d}}{\mathrm{d}x}\left[\frac{\mathrm{d}^{l-1}}{\mathrm{d}x^{l-1}}\left(x^2-1\right)^l\right]\frac{\mathrm{d}^k}{\mathrm{d}x^k}\left(x^2-1\right)^k\mathrm{d}x \\ &= \frac{1}{2^l l! 2^k k!}\left[\frac{\mathrm{d}^{l-1}}{\mathrm{d}x^{l-1}}\left(x^2-1\right)^l\frac{\mathrm{d}^k}{\mathrm{d}x^k}\left(x^2-1\right)^k\right]_{-1}^{+1} \\ &\quad - \frac{1}{2^l l! 2^k k!}\int_{-1}^{+1}\frac{\mathrm{d}^{l-1}}{\mathrm{d}x^{l-1}}\left(x^2-1\right)^l\frac{\mathrm{d}^{k+1}}{\mathrm{d}x^{k+1}}\left(x^2-1\right)^k\mathrm{d}x\end{aligned}$$

Since $x = \pm 1$ are the lth-order zeros of $(x^2-1)^l$, after taking the $(l-1)$th derivative, they become first-order zeros. Therefore, the terms obtained from integrating the first part of the equation become zero. Hence,

$$\int_{-1}^{+1} \mathrm{P}_l(x)\mathrm{P}_k(x)\mathrm{d}x = \frac{(-1)^1}{2^l l! 2^k k!}\int_{-1}^{+1} \frac{\mathrm{d}^{l-1}}{\mathrm{d}x^{l-1}}(x^2-1)^l \frac{\mathrm{d}^{k+1}}{\mathrm{d}x^{k+1}}(x^2-1)^k \mathrm{d}x$$

Continuing with k successive integration by parts, the integrated part again becomes zero. Hence

$$\int_{-1}^{+1} \mathrm{P}_l(x)\mathrm{P}_k(x)\mathrm{d}x = \frac{(-1)^{k+1}}{2^l l! 2^k k!}\int_{-1}^{+1} \frac{\mathrm{d}^{l-k-1}}{\mathrm{d}x^{l-k-1}}(x^2-1)^l \times \frac{\mathrm{d}^{2k+1}}{\mathrm{d}x^{2k+1}}(x^2-1)^k \mathrm{d}x$$

Noting that in the integral sign we have

$$\frac{\mathrm{d}^{2k+1}}{\mathrm{d}x^{2k+1}}(x^2-1)^k = 0$$

Consequently,

$$\int_{-1}^{+1} \mathrm{P}_l(x)\mathrm{P}_k(x)\mathrm{d}x = 0$$

Next, looking at the magnitude of the eigenfunction, let us set $k = l$. After l successive integration by parts, we have

$$N_l^2 = \int_{-1}^{+1} [\mathrm{P}_l(x)]^2 \mathrm{d}x = \frac{(-1)^l}{2^{2l}(l!)^2}\int_{-1}^{+1} (x^2-1)^l \frac{\mathrm{d}^{2l}}{\mathrm{d}x^{2l}}(x^2-1)^l \mathrm{d}x$$

The highest power of $(x^2-1)^l$ within the integral sign is $2l$. Taking its $(2l)$th derivative yields $(2l)!$ Therefore,

$$N_l^2 = \frac{(-1)^l (2l)!}{2^{2l}(l!)^2}\int_{-1}^{+1} (x-1)^l (x+1)^l \mathrm{d}x$$

Performing integration by parts once more, we obtain

$$N_l^2 = \frac{(-1)^l(2l)!}{2^{2l}(l!)^2}\frac{1}{l+1}\left[(x-1)^l(x+1)^{l+1}|_{-1}^{+1} - l \times \int_{-1}^{+1}(x-1)^{l-1}(x+1)^{l+1}\mathrm{d}x\right]$$

The part that has been integrated vanishes, resulting in a decrease of the power of $(x-1)$ by one and an increase of the power of $(x+1)$ by one. Continuing with successive integration by parts, we eventually arrive at

$$\begin{aligned} N_l^2 &= \frac{(-1)^l(2l)!}{2^{2l}(l!)^2}\cdot(-1)^l\frac{l}{(l+1)}\cdot\frac{l-1}{(l+2)}\cdots\frac{1}{2l} \\ &\quad\times\int_{-1}^{+1}(x-1)^0(x+1)^{2l}\mathrm{d}x \\ &= \frac{1}{2^{2l}(2l+1)}(x+1)^{2l+1}\Big|_{-1}^{+1} = \frac{2}{2l+1} \end{aligned}$$

□

(4) *Recurrence relations*

$$\begin{aligned} &(k+1)\mathrm{P}_{k+1}(x) - (2k+1)x\mathrm{P}_k(x) + k\mathrm{P}_{k-1}(x) = 0 \\ &\mathrm{P}_k(x) = \mathrm{P}'_{k+1}(x) - 2x\mathrm{P}'_k(x) + \mathrm{P}'_{k-1}(x) \\ &(2k+1)\mathrm{P}_k(x) = \mathrm{P}'_{k+1}(x) - \mathrm{P}'_{k-1}(x) \\ &\mathrm{P}'_{k+1}(x) = x\mathrm{P}'_k(x) + (k+1)\mathrm{P}_k(x) \end{aligned} \tag{13.1.13}$$

Starting from the definition of the Legendre polynomials or using the generating function that will be introduced later, these relationships can be directly proven. Here, we will omit the detailed proof.

Example 13.1. Calculate the integral:

$$\int_{-1}^{+1} x\mathrm{P}_k(x)\mathrm{P}_l(x)\mathrm{d}x$$

Solution: Using the recurrence relations, we have

$$\begin{aligned}
&\int_{-1}^{+1} x\mathrm{P}_k(x)\mathrm{P}_l(x)\mathrm{d}x \\
&= \int_{-1}^{+1} \frac{1}{2k+1}[(k+1)\mathrm{P}_{k+1}(x)\mathrm{P}_l(x) + k\mathrm{P}_{k-1}(x)\mathrm{P}_l(x)]\mathrm{d}x \\
&= \frac{k+1}{2k+1}\cdot\frac{2}{2l+1}\delta_{k+1,l} + \frac{k}{2k+1}\cdot\frac{2}{2l+1}\delta_{k-1,l} \\
&= \begin{cases} \dfrac{2k}{(2k+1)(2k-1)} & (l = k-1) \\ \dfrac{2(k+1)}{(2k+3)(2k+1)} & (l = k+1) \\ 0 & (l \neq k \pm 1) \end{cases}
\end{aligned}$$

13.1.4 *Generalized Fourier series*

In Chapter 14, it will be explained that the Legendre equation belongs to the Sturm–Liouville type eigenvalue equation, and its eigenfunctions, the Legendre polynomials $P_l(x)$, form an orthogonal and complete set. Similar to the Fourier series expansion using trigonometric functions, a function $f(x)$ defined on the interval $[-1, +1]$ can be expressed as a linear combination of Legendre polynomials, known as the generalized Fourier series or Legendre series:

$$f(x) = \sum_{l=0}^{\infty} f_l \mathrm{P}_l(x)$$

Where the coefficients

$$f_l = \frac{2l+1}{2}\int_{-1}^{+1} f(x)\mathrm{P}_l(x)\mathrm{d}x$$

or in terms of polar coordinates:

$$f(\theta) = \sum_{l=0}^{\infty} f_l \mathrm{P}_l(\cos\theta)$$

where

$$f_l = \frac{2l+1}{2}\int_0^{\pi} f(\theta)\mathrm{P}_l(\cos\theta)\sin\theta\mathrm{d}\theta$$

Example 13.2. Expand the function $f(x) = 2x^3 + 3x + 4$ in terms of the Legendre polynomials as the basis on the interval $[-1, +1]$.

Solution: We can use the standard expansion method to perform the calculations. However, considering that $f(x)$ is a polynomial of degree no higher than three, we can directly represent it as a linear combination of $\mathrm{P}_0(x), \mathrm{P}_1(x), \mathrm{P}_2(x)$ and $\mathrm{P}_3(x)$,

$$f(x) = f_0\mathrm{P}_0(x) + f_1\mathrm{P}_1(x) + f_2\mathrm{P}_2(x) + f_3\mathrm{P}_3(x)$$

Substituting the expressions for $P_l(x)$ and comparing coefficients, we obtain

$$f_0 = 4, \quad f_1 = \frac{21}{5}, \quad f_2 = 0, \quad f_3 = \frac{4}{5}$$

Example 13.3. Solve the Laplace equation $\Delta u = 0$ inside the sphere with $r = r_0$ and satisfying the boundary condition $u|_{r=r_0} = \cos^2\theta$.

Solution: We start by considering the axisymmetric case with $m = 0$. The solution has the following form:

$$u(r, \theta) = \sum_{l=0}^{\infty} \left(A_l r^l + \frac{B_l}{r^{l+1}} \right) \mathrm{P}_l(\cos\theta)$$

Considering the finiteness of physical quantities at $r = 0$, we discard the second term of the above equation by setting $B_l = 0$. Hence,

$$u(r, \theta) = \sum_{l=0}^{\infty} A_l r^l \mathrm{P}_l(\cos\theta)$$

To determine the values of coefficients A_l, we use the boundary condition:

$$u(r_0, \theta) = \sum_{l=0}^{\infty} A_l r_0^l \mathrm{P}_l(\cos\theta) = \cos^2\theta = x^2$$

This is equivalent to expanding x^2 in terms of Legendre polynomials:

$$x^2 = \frac{1}{3}[1 + 2\mathrm{P}_2(x)] = \frac{1}{3}P_0(x) + \frac{2}{3}P_2(x)$$

The final result for is

$$u(r, \theta) = \frac{1}{3} + \frac{2r^2}{3r_0^2}\mathrm{P}_2(x)$$

Example 13.4. In a uniform static electric field $\boldsymbol{E}_0$, a homogeneous dielectric sphere with radius r_0 and relative dielectric constant ε_r is placed, and we are to determine the electric field distribution inside and outside the dielectric sphere.

Solution: Taking the center of the sphere as the origin, as shown in Figure 13.2, we establish a spherical coordinate system with the direction of the external electric field as the polar axis. According to electromagnetics, the polarized charge of a homogeneous dielectric only appears on the interface, i.e., the surface of the dielectric sphere. There is no net charge inside or outside the sphere. Let u_1 and u_2 represent the potential distribution inside and outside the sphere, respectively, both satisfying Laplace's equation:

$$\Delta u_1 = 0, \quad \Delta u_2 = 0$$

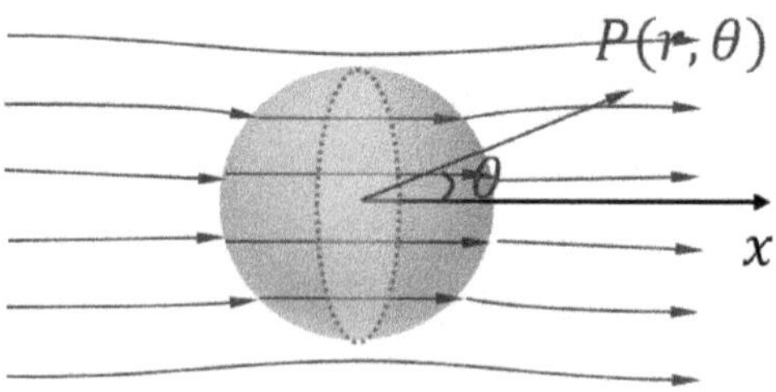

Figure 13.2.

The potential at infinity is still the potential of a uniform electric field, given by

$$u_2|_{r\to\infty} = -E_0 r\cos\theta$$

On the surface of the sphere, according to electromagnetism knowledge, the normal components of the potential and the displacement vector must be continuous, resulting in the following interface conditions:

$$\begin{cases} u_1|_{r=r_0} = u_2|_{r=r_0} \\ \varepsilon_r \dfrac{\partial}{\partial r}u_1\bigg|_{r=r_0} = \dfrac{\partial}{\partial r}u_2\bigg|_{r=r_0} \end{cases}$$

The general solution to Laplace's equation is

$$u(r,\theta) = \sum_{l=0}^{\infty}\left(A_l r^l + \frac{B_l}{r^{l+1}}\right)\mathrm{P}_l(\cos\theta)$$

Due to the finite nature of the potential at $r = 0$, the second term must be discarded inside the sphere by setting $B_l = 0$. Therefore,

$$u_1(r,\theta) = \sum_{l=0}^{\infty} A_l r^l \mathrm{P}_l(\cos\theta)$$

The potential outside the sphere is constrained by the potential at infinity:

$$u_2(r,\theta) = \sum_{l=0}^{\infty} \left(C_l r^l + \frac{D_l}{r^{l+1}} \right) \mathrm{P}_l(\cos\theta) \overset{r\to\infty}{\longrightarrow} -E_0 r\cos\theta$$

By comparing coefficients, we find $C_1 = E_0$ and $C_l = 0 (l \neq 0, 1)$. Thus, the potential outside the sphere is

$$u_2(r,\theta) = C_0 - E_0 r \mathrm{P}_1(\cos\theta) + \sum_{l=0}^{\infty} \frac{D_l}{r^{l+1}} \mathrm{P}_l(\cos\theta)$$

The coefficients A_l and D_l need to be determined by the interface condition at the sphere's surface:

$$\sum_{l=0}^{\infty} A_l r_0^l \mathrm{P}_l(\cos\theta) = C_0 - E_0 r_0 \mathrm{P}_1(\cos\theta) + \sum_{l=0}^{\infty} \frac{D_l}{r_0^{l+1}} \mathrm{P}_l(\cos\theta)$$

$$\varepsilon_r \sum_{l=0}^{\infty} l A_l r_0^{l-1} \mathrm{P}_l(\cos\theta) = -E_0 \mathrm{P}_1(\cos\theta) - \sum_{l=0}^{\infty} D_l \frac{l+1}{r_0^{l+2}} \mathrm{P}_l(\cos\theta)$$

By comparing the coefficients, we obtain

$$A_0 = C_0, \quad D_0 = 0, \quad A_1 = -\frac{3}{\varepsilon_r + 2} E_0, \quad D_1 = \frac{\varepsilon_r - 1}{\varepsilon_r + 2} r_0^3 E_0$$

$$A_l = 0, \quad D_l = 0 \ (l \geq 2)$$

The final solutions are

$$u_1(r,\theta) = A_0 - \frac{3}{\varepsilon_r + 2} E_0 r\cos\theta$$

$$u_2(r,\theta) = A_0 - E_0 r\cos\theta + \frac{\varepsilon_r - 1}{\varepsilon_r + 2} r_0^3 E_0 \frac{1}{r^2}\cos\theta$$

Discussion: The magnitude of the electric field inside the sphere is

$$E_1 = -\nabla u_1 = \frac{3}{\varepsilon_r + 2} E_0$$

which is a constant. The direction of the electric field at every point is along the positive x-axis. Therefore, the electric field inside the sphere is uniform, indicating that the dielectric sphere is uniformly polarized.

13.1.5 *Generating function*

Consider the static electric potential generated by a point charge $q = 4\pi\varepsilon_0$ at the north pole of the unit sphere in Figure 13.3. We know that the electric potential at point $M(r,\theta)$ inside the sphere is simply $\frac{1}{d}$, which can also be expressed by Legendre polynomials and their coefficients A_l and B_l:

$$\frac{q}{4\pi\varepsilon_0 d} = \frac{1}{\sqrt{1 - 2r\cos\theta + r^2}} \equiv \sum_{l=0}^{\infty} \left(A_l r^l + \frac{B_l}{r^{l+1}} \right) \mathrm{P}_l(\cos\theta)$$

To ensure that the potential at the center of the sphere is finite, it requires $B_l = 0$. To determine the coefficients A_l, we can directly set $\theta = 0$ and use the property $P_l(1) = 1$:

$$\frac{1}{1-r} = \sum_{l=0}^{\infty} A_l r^l = 1 + r + r^2 + \cdots + r^l + \cdots$$

So we have $A_l = 1$ $(l = 0, 1, 2, \ldots)$. Therefore,

$$G(r,\theta) \equiv \frac{1}{\sqrt{1 - 2r\cos\theta + r^2}} = \sum_{l=0}^{\infty} r^l \mathrm{P}_l(\cos\theta) \quad (r < 1) \qquad (13.1.14)$$

Thus, we obtain the relationship:

$$\mathrm{P}_l(\cos\theta) = \frac{1}{l!} \frac{\partial^l G(r,\theta)}{\partial r^l}\bigg|_{r=0} \qquad (13.1.15)$$

We refer to $G(r,\theta)$ as the generating function or the generating series of the Legendre polynomials $\mathrm{P}_l(\cos\theta)$.

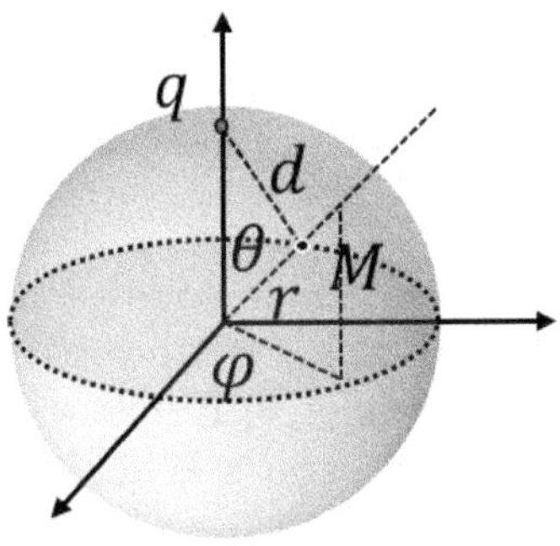

Figure 13.3.

Discussion: Based on dimensional analysis, we can derive the generating functions for the following scenarios:

(1) Generating function for inside the sphere with radius R:

$$G(r,\theta) = \frac{1}{\sqrt{R^2 - 2Rr\cos\theta + r^2}} = \sum_{l=0}^{\infty} \frac{r^l}{R^{l+1}} \mathrm{P}_l(\cos\theta) \quad (r < R)$$

(2) Generating function for outside the sphere with radius R:

$$G(r,\theta) = \frac{1}{\sqrt{R^2 - 2Rr\cos\theta + r^2}} = \sum_{l=0}^{\infty} \frac{R^l}{r^{l+1}} \mathrm{P}_l(\cos\theta) \quad (r > R)$$

Example 13.5. Prove the orthogonality of Legendre polynomials by using the generating function.

Proof. From the generating function, we have

$$\frac{1}{\sqrt{1-2xu+u^2}} \cdot \frac{1}{\sqrt{1-2xv+v^2}} = \sum_{m,n=0}^{\infty} \mathrm{P}_m(x)\mathrm{P}_n(x)u^m v^n$$

Integrating both sides with respect to x and expanding the left side in a Taylor series, we get

$$\int_{-1}^{+1} \frac{1}{\sqrt{1-2xu+u^2}} \cdot \frac{1}{\sqrt{1-2xv+v^2}} \mathrm{d}x$$

$$= \frac{1}{\sqrt{uv}} \ln \frac{1+\sqrt{uv}}{1-\sqrt{uv}} = \sum_{n=0}^{\infty} \frac{2}{2n+1} u^n v^n$$

By comparing the coefficients of $u^m v^n$ on both sides, we obtain

$$\int_{-1}^{1} \mathrm{P}_m(x)\mathrm{P}_n(x)\mathrm{d}x = \frac{2}{2n+1}\delta_{mn}$$

□

Example 13.6. Placing a grounded conducting sphere with a radius of r_0 in the electric field of a point charge $4\pi\varepsilon_0 q$, where the center of the sphere is located at a distance r_1 from the point charge ($r_1 > r_0$), we seek to solve for this electrostatic field.

Solution: We set the spherical coordinates with the center of the sphere as the origin and the point charge on the polar axis. We first start by considering the potential in the absence of the conducting sphere, which is given by

$$v(r,\theta) = \frac{q}{\sqrt{r_1^2 - 2r_1 r\cos\theta + r^2}}$$

The presence of the conducting sphere induces charge redistribution on the surface and modifies the potential distribution $w(r,\theta)$ in the outside space, which satisfies the Laplace's equation:

$$\Delta w(r,\theta) = 0$$

The total potential distribution outside the sphere can be expressed as $u = w + v$, satisfying the boundary condition:

$$u|_{r=r_0} = 0, \quad u|_{r\to\infty} = 0$$

Hence, the potential induced by the charge distribution on the conducting sphere satisfies the boundary conditions:

$$w|_{r=r_0} = -v(r_0,\theta) = -\frac{q}{\sqrt{r_1^2 - 2r_1 r_0\cos\theta + r_0^2}}, \qquad w|_{r\to\infty} = 0$$

The general solution to the equation is given by

$$w = \sum_{l=0}^{\infty}\left(A_l r^l + B_l\frac{1}{r^{l+1}}\right)\mathrm{P}_l(\cos\theta)$$

Taking account of $w \to 0$ as $r \to \infty$, we have $A_l = 0$. Thus, the solution becomes

$$w = \sum_{l=0}^{\infty} B_l\frac{1}{r^{l+1}}\mathrm{P}_l(\cos\theta)$$

Substituting the boundary conditions and using the generating function yield

$$\sum_{l=0}^{\infty} B_l \frac{1}{r_0^{l+1}} \mathrm{P}_l(\cos\theta) = -\frac{q}{\sqrt{r_1^2 - 2r_1 r_0 \cos\theta + r_0^2}}$$
$$= -q \sum_{l=0}^{\infty} \frac{r_0^l}{r_1^{l+1}} \mathrm{P}_l(\cos\theta)$$

By comparing the coefficients of both sides, we obtain $B_l = -q\frac{r_0^{2l+1}}{r_1^{l+1}}$. The final result is

$$u(r,\theta) = \frac{q}{\sqrt{r_1^2 - 2r_1 r\cos\theta + r^2}} + \sum_{l=0}^{\infty} (-q)\frac{r_0^{2l+1}}{r_1^{l+1}} \cdot \frac{1}{r^{l+1}} \mathrm{P}_l(\cos\theta)$$
$$= \frac{q}{\sqrt{r_1^2 - 2r_1 r\cos\theta + r^2}} + \frac{-q(r_0/r_1)}{\sqrt{(r_0^2/r_1)^2 - 2(r_0^2/r_1)r\cos\theta + r^2}}$$

In the last step, we utilize the generating function property for a sphere with radius $R = \frac{r_0^2}{r_1}$.

Discussion: This problem can also be solved using the method of images (see Figure 13.4), assuming a negative image charge q' at the conjugate point B of point A, where $\frac{BP}{AP} = \frac{OP}{OA} = \frac{r_0}{r_1}$. The charges q and q' together ensure that the electric potential on the sphere's surface is zero. By simple geometric calculations, it can be shown that $q' = -\frac{r_0}{r_1}q$. The electric potential in the space outside the sphere is then the superposition of the potentials produced by the point charges at A and B. Readers can try to work out the details on their own.

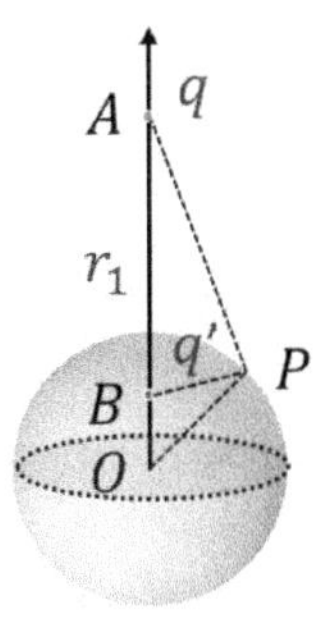

Figure 13.4.

ENDNOTE

We present an alternative way to derive the generating function for Legendre polynomials. Consider the inverse distance between two points, when $r > r'$, one has

$$\frac{1}{|\boldsymbol{r}-\boldsymbol{r}'|} = \frac{1}{\sqrt{r^2 - 2rr'\cos\theta + r'^2}} = \frac{1}{r}\frac{1}{\sqrt{1 - \frac{2r'}{r}\cos\theta + \left(\frac{r'}{r}\right)^2}}$$

Let $\varepsilon = -\frac{2r'}{r}\cos\theta + \left(\frac{r'}{r}\right)^2$, by making Tailor series expansion, we have

$$\frac{1}{|\boldsymbol{r}-\boldsymbol{r}'|} = \frac{1}{r}\frac{1}{\sqrt{1+\varepsilon}} = \frac{1}{r}\left(1 - \frac{1}{2}\varepsilon + \frac{3}{8}\varepsilon^2 - \frac{5}{16}\varepsilon^3 + \cdots\right)$$

Set $x = \cos\theta$ and $\alpha = \frac{r'}{r}$, so $\varepsilon = -2\alpha x + \alpha^2$. Therefore,

$$\frac{1}{|\boldsymbol{r}-\boldsymbol{r}'|} = \frac{1}{r}\left[1 + \alpha x + \alpha^2\left(\frac{3}{2}x^2 - \frac{1}{2}\right) + \alpha^3\left(\frac{5}{2}x^3 - \frac{3}{2}x\right) + \cdots\right]$$

Notice that on the right-hand side the terms inside the lth parentheses are exactly the Legendre polynomials $\mathrm{P}_l(x)$. Hence we deduce that

$$\frac{1}{|\boldsymbol{r}-\boldsymbol{r}'|} = \frac{1}{r}\sum_{l=0}^{\infty}\alpha^l \mathrm{P}_l(x) = \sum_{l=0}^{\infty}\frac{r'^l}{r^{l+1}}\mathrm{P}_l(x)$$

Exercises

[1] Prove that

$$\text{(a)} \int_{-1}^{1} x^m \mathrm{P}_n(x)\mathrm{d}x = 0 \ (m < n);$$

$$\text{(b)} \int_{-1}^{1} x^n \mathrm{P}_n(x)\mathrm{d}x = \frac{2n!}{(2n+1)!!}$$

Hint: Expand x^m in terms of $P_n(x)$.

[2] Calculate the integral:

$$\text{(a)} \int_{-1}^{+1} x^2 P_2(x)P_5(x)\mathrm{d}x; \quad \text{(b)} \int_{-1}^{+1} x^3 P_2(x)P_5(x)\mathrm{d}x$$

[3] Using the Legendre equation, $[(1-x^2)\mathrm{P}_l'(x)]' + l(l+1)\mathrm{P}_l(x) = 0$, prove the orthogonality of Legendre polynomials.
Hint: Multiply the equation by $\mathrm{P}_k(x)$, integrate over $[-1, +1]$, interchange k and l in the equation, and subtract the two equations.

$$[k(k+1) - l(l+1)] \int_{-1}^{1} \mathrm{P}_k(x)\mathrm{P}_l(x)\mathrm{d}x = 0$$

[4] Expand $f(x) = |x|$ in terms of generalized Fourier series using Legendre polynomials as the bases on the interval $[-1, +1]$.
Answer:

$$|x| = \frac{1}{2}\mathrm{P}_0(x) + \sum_{n=1}^{\infty} (-1)^{n+1} \frac{(4n+1)(2n-1)!!}{(2n-1)(2n+2)!!} \mathrm{P}_{2n}(x)$$

[5] Using the generating function of Legendre polynomials, prove the recurrence relation:

$$(k+1)\mathrm{P}_{k+1}(x) - (2k+1)x\mathrm{P}_k(x) + k\mathrm{P}_{k-1}(x) = 0$$

Hint: Differentiating the generating function with respect to variable r.

[6] For a hemisphere with radius r_0, where the temperature on the spherical surface is maintained at $u_0 \cos\theta$ and the base is insulated, determine the stable temperature distribution of the hemisphere.
Hint: Extend the hemisphere to a complete sphere using even symmetry.
Answer:

$$u(r,\theta) = \frac{1}{2}u_0 + u_0 \sum_{n=1}^{\infty} (-1)^{n+1} \frac{(4n+1)(2n-1)!!}{(2n-1)(2n+2)!!} \frac{r^{2n}}{r_0^{2n}} \mathrm{P}_{2n}(\cos\theta)$$

[7] For a grounded metal conducting sphere with radius r_0 placed in a uniform electric field $\boldsymbol{E}_0$, determine the distribution of the static electric field outside the sphere.

Answer:

$$u(r,\theta) = -E_0 r\cos\theta + E_0\frac{r_0^3}{r^2}\cos\theta$$

[8] Given a uniformly charged circular ring with radius a and total charge q, calculate the electric potential distribution it generates in three-dimensional space.

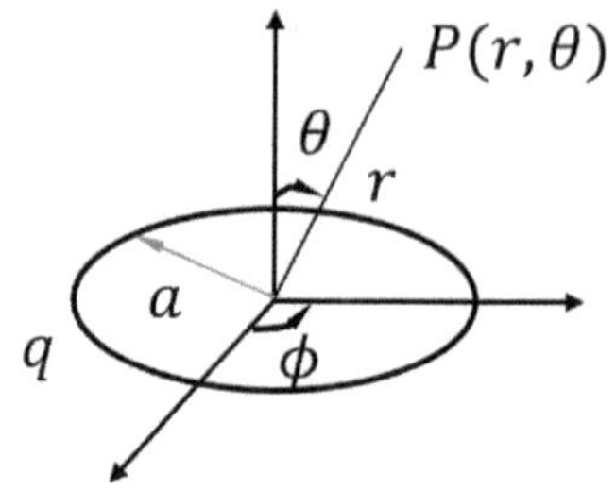

Figure 13.5.

Hint:

$$u(r,\theta) = \sum_{l=0}^{\infty} B_l \frac{a^l}{r^{l+1}} \mathrm{P}_l(\cos\theta) \quad (r > a)$$

$$u(z,0) = \frac{q}{4\pi\varepsilon_0}\frac{1}{\sqrt{z^2+a^2}} = \frac{q}{4\pi\varepsilon_0 z}\sum_{l=0}^{\infty}(-1)^l \times \frac{(2l-1)!!}{(2l)!!}\left(\frac{a}{z}\right)^{2l} \quad (z > a)$$

Answer:

$$u(r,\theta) = \frac{q}{4\pi\varepsilon_0 r}\sum_{l=0}^{\infty}(-1)^n\frac{(2l-1)!!}{(2l)!!}\left(\frac{a}{r}\right)^{2l}\mathrm{P}_{2l}(\cos\theta) \quad (r > a)$$

[9] Solve the Legendre-type inhomogeneous equation:

$$-(1-x^2)\frac{\mathrm{d}^2y}{\mathrm{d}x^2} + 2x\frac{\mathrm{d}y}{\mathrm{d}x} - y = \mathrm{e}^{-x}$$

Hint: The eigenvalue problem is

$$-(1-x^2)\frac{\mathrm{d}^2}{\mathrm{d}x^2}\mathrm{P}_l(x) + 2x\frac{\mathrm{d}}{\mathrm{d}x}\mathrm{P}_l(x) = l(l+1)\mathrm{P}_l(x) \quad (l \in \mathbb{N})$$

Expand both sides of the non-homogeneous equation in terms of eigenfunctions $P_l(x)$:

$$y(x) = \sum_{l=0}^{\infty} a_l \mathrm{P}_l(x), \quad \mathrm{e}^{-x} = \sum_{l=0}^{\infty} b_l \mathrm{P}_l(x)$$

13.2 Associated Legendre Equation

13.2.1 *Associated Legendre functions*

We continue the discussion of the associated Legendre equation of order l when $m \neq 0$:

$$(1-x^2)\frac{\mathrm{d}^2\Theta}{\mathrm{d}x^2} - 2x\frac{\mathrm{d}\Theta}{\mathrm{d}x} + \left[l(l+1) - \frac{m^2}{1-x^2}\right]\Theta = 0$$

Since $x_0 = 0$ is a regular singular point of the equation, we can still solve it using a Taylor series expansion. However, here we will use another method that is more insightful. First, we make a transformation

$$\Theta(x) = (1-x^2)^{\frac{m}{2}} y(x)$$

The equation becomes

$$(1-x^2)y'' - 2(m+1)xy' + [l(l+1) - m(m+1)]y = 0$$

On the other hand, for the Legendre equation,

$$(1-x^2)\mathrm{P}_l''(x) - 2x\mathrm{P}_l'(x) + l(l+1)\mathrm{P}_l(x) = 0$$

Taking m derivatives term by term and applying Leibniz's rule for differentiation, we have

$$(1-x^2)(\mathrm{P}_l^{[m]})'' - 2(m+1)x(\mathrm{P}_l^{[m]})' + [l(l+1) - m(m+1)]\mathrm{P}_l^{[m]} = 0$$

Here, $[m]$ denotes taking m derivatives. Comparing the two equations, we can see that the solution is given by $y(x) = \mathrm{P}_l^{[m]}(x)$. Therefore, the eigenfunction of the lth order associated Legendre equation is

$$\mathrm{P}_l^m(x) = (1-x^2)^{\frac{m}{2}}\mathrm{P}_l^{[m]}(x) \tag{13.2.1}$$

We call $\mathrm{P}_l^m(x)$ the associated Legendre function of order l, and its corresponding eigenvalue is also $l(l+1)$. Since $\mathrm{P}_l(x)$ is a polynomial of degree l, it can be differentiated at most l times. Therefore, $m \leq l$, that is, $m = 0, 1, 2, \ldots, l$.

Below are some low-order associated Legendre functions:

$$\mathrm{P}_1^1(x) = (1 - x^2)^{\frac{1}{2}} = \sin\theta$$
$$\mathrm{P}_2^1(x) = 3x(1 - x^2)^{\frac{1}{2}} = 3\sin\theta\cos\theta$$
$$\mathrm{P}_2^2(x) = 3(1 - x^2) = 3\sin^2\theta$$
$$\mathrm{P}_3^1(x) = \frac{3}{2}(5x^2 - 1)(1 - x^2)^{\frac{1}{2}} = \frac{3}{8}(\sin\theta + 5\sin 3\theta)$$
$$\mathrm{P}_3^2(x) = 15x(1 - x^2) = 15\sin^2\theta\cos\theta$$
$$\mathrm{P}_3^3(x) = 15(1 - x^2)^{\frac{3}{2}} = 15\sin^3\theta$$

13.2.2 *Basic properties*

(1) *Derivative representation (Rodrigues' formula)*

$$\mathrm{P}_l^m(x) = \frac{(1 - x^2)^{m/2}}{2^l l!} \frac{\mathrm{d}^{l+m}}{\mathrm{d}x^{l+m}}(x^2 - 1)^l \tag{13.2.2}$$

It can be shown that $P_l^m(x)$ and $P_l^{-m}(x)$ are linearly related:

$$\mathrm{P}_l^{-m}(x) = \frac{(1 - x^2)^{-m/2}}{2^l l!} \frac{\mathrm{d}^{l-m}}{\mathrm{d}x^{l-m}}(x^2 - 1)^l = (-1)^m \frac{(l - m)!}{(l + m)!} \mathrm{P}_l^m(x)$$

(2) *Integral representation (Schlafli integral)*

$$\mathrm{P}_l^m(x) = \frac{(1 - x^2)^{m/2}}{2\pi \mathrm{i}} \frac{(l + m)!}{2^l l!} \oint_C \frac{(z^2 - 1)^l}{(z - x)^{l+m+1}} \mathrm{d}z \tag{13.2.3}$$

Here, C is any closed contour in the z-plane around the point $z = x$. Similarly, we have the Laplace integral formula:

$$\mathrm{P}_l^m(\cos\theta) = \frac{\mathrm{i}^m}{2\pi} \frac{(l + m)!}{l!} \int_{-\pi}^{\pi} \mathrm{e}^{-\mathrm{i}m\psi} (\cos\theta + \mathrm{i}\sin\theta\cos\psi)^l \mathrm{d}\psi \tag{13.2.4}$$

(3) *Orthogonality*

$$\int_{-1}^{+1} \mathrm{P}_l^m(x)\mathrm{P}_k^m(x)\mathrm{d}x = (N_l^m)^2\delta_{kl}$$
$$\int_0^{\pi} \mathrm{P}_l^m(\cos\theta)\mathrm{P}_k^m(\cos\theta)\sin\theta\,\mathrm{d}\theta = (N_l^m)^2\delta_{kl} \tag{13.2.5}$$

The normalization factor is given by

$$(N_l^m)^2 = \frac{2(l+|m|)!}{(2l+1)(l-|m|)!}$$

(4) *Recurrence relations*

$$(2k+1)x\mathrm{P}_k^m(x) = (k-m+1)\mathrm{P}_{k+1}^m(x) + (k+m)\mathrm{P}_{k-1}^m(x)$$
$$(2k+1)(1-x^2)^{\frac{1}{2}}\mathrm{P}_k^m(x) = \mathrm{P}_{k+1}^{m+1}(x) - \mathrm{P}_{k-1}^{m+1}(x)$$
$$(2k+1)(1-x^2)^{\frac{1}{2}}\mathrm{P}_k^m(x) = (k+m)(k+m-1)\mathrm{P}_{k-1}^{m-1}(x) - (k-m+2)(k-m+1)\mathrm{P}_{k+1}^{m-1}(x)$$
$$(2k+1)(1-x^2)^{\frac{1}{2}}\frac{\mathrm{d}\mathrm{P}_k^m(x)}{\mathrm{d}x} = (k+m)(k+1)\mathrm{P}_{k-1}^m(x) - k(k-m+1)\mathrm{P}_{k+1}^m(x)$$

The proof of these recurrence relations can be carried out by applying the Rodrigues' formula, which is omitted here.

13.2.3 *Generalized Fourier series*

The associated Legendre equation also belongs to the Sturm–Liouville eigenvalue problem and therefore forms a complete set. The associated Legendre functions with the same m satisfy the orthogonality and completeness conditions and can serve as the basis for a generalized Fourier series expansion. For a function $f(x)$ defined on the interval $[-1,+1]$, we have

$$\begin{aligned} f(x) &= \sum_{l=0}^{\infty} f_l \mathrm{P}_l^m(x) \\ f_l &= \frac{2l+1}{2}\frac{(l-m)!}{(l+m)!}\int_{-1}^{+1} f(x)\mathrm{P}_l^m(x)\mathrm{d}x \end{aligned} \tag{13.2.6}$$

Alternatively,

$$\begin{aligned} f(\theta) &= \sum_{l=0}^{\infty} f_l \mathrm{P}_l^m(\cos\theta) \\ f_l &= \frac{2l+1}{2}\frac{(l-m)!}{(l+m)!}\int_{0}^{\pi} f(\theta)\mathrm{P}_l^m(\cos\theta)\sin\theta\mathrm{d}\theta \end{aligned} \tag{13.2.7}$$

It should be noted that the orthogonality of functions holds for the same m and different eigenvalues l. The generalized Fourier series expansion is performed using the eigenfunctions with the same m and different l.

Exercises

[1] Prove that the normalization factor for associated Legendre functions is given by

$$(N_l^m)^2 = \int_{-1}^{+1} |\mathrm{P}_l^m(x)|^2 \mathrm{d}x = \frac{2(l+|m|)!}{(2l+1)(l-|m|)!}$$

[2] Using $\mathrm{P}_l^2(x)$ as a basis, expand the function $f(x) = 1 - x^2$ into a generalized Fourier series on the interval $[-1,+1]$.
Answer: $f(x) = \frac{1}{3}\mathrm{P}_2^2(x)$

[3] Expand the function $f(x)$ using $P_l^2(x)$ as a basis in the interval $[-1,+1]$ in a generalized Fourier series:

$$f(x) = \begin{cases} u_0 & (0 < x \leq 1) \\ 0 & (-1 < x \leq 0) \end{cases}$$

[4] Prove the recurrence relation:

$$(2k+1)x\mathrm{P}_k^m(x) = (k-m+1)\mathrm{P}_{k+1}^m(x) + (k+m)\mathrm{P}_{k-1}^m(x)$$

13.3 Spherical Harmonic Functions

13.3.1 *Spherical harmonic equation*

The spherical harmonic equation is

$$\frac{1}{\sin\theta}\frac{\partial}{\partial\theta}\left(\sin\theta\frac{\partial Y}{\partial\theta}\right)+\frac{1}{\sin^2\theta}\frac{\partial^2 Y}{\partial\phi^2}+l(l+1)Y=0 \tag{13.3.1}$$

The general solution of the equation, combined with spherical coordinates (θ, ϕ), is referred to as the lth-order spherical harmonic functions:

$$\mathrm{Y}_{lm}(\theta,\phi)=\mathrm{P}_l^m(\cos\theta)\begin{Bmatrix}\sin m\phi\\ \cos m\phi\end{Bmatrix}\begin{pmatrix}m=0,1,2,\ldots,l\\ l=0,1,2,3,\ldots\end{pmatrix} \tag{13.3.2}$$

Alternatively, they can be expressed in terms of complex exponential functions:

$$\mathrm{Y}_l^m(\theta,\phi)=\mathrm{P}_l^{|m|}(\cos\theta)\mathrm{e}^{im\phi}\begin{pmatrix}m=-l,-l+1,\ldots,l\\ l=0,1,2,3,\ldots\end{pmatrix} \tag{13.3.3}$$

In classical physics, the trigonometric spherical harmonics (13.3.2) are often used, while in quantum mechanics, the complex exponential spherical harmonics (13.3.3) are more commonly employed. Spherical harmonics also exhibit orthogonality, given by the equation,

$$\begin{aligned}&\int_0^{2\pi}\int_0^{\pi}\mathrm{Y}_l^m(\theta,\phi)[\mathrm{Y}_k^n(\theta,\phi)]^*\sin\theta\mathrm{d}\theta\mathrm{d}\phi\\ &=\int_0^{\pi}\mathrm{P}_l^m(\cos\theta)\mathrm{P}_k^n(\cos\theta)\sin\theta\mathrm{d}\theta\int_0^{2\pi}\mathrm{e}^{im\phi}\mathrm{e}^{-in\phi}\mathrm{d}\phi\\ &=(N_l^m)^2\delta_{kl}\delta_{mn}\end{aligned} \tag{13.3.4}$$

The normalization factor is given by

$$(N_l^m)^2=\frac{4\pi(l+|m|)!}{(2l+1)(l-|m|)!}$$

In quantum mechanics, it is common to use orthonormalized spherical harmonics and conventionally include a factor of $(-1)^m$, known as the Condon–Shortley phase:

$$\begin{aligned} \mathrm{Y}_l^m(\theta,\phi) &\longmapsto \frac{(-1)^m}{N_l^m}\mathrm{Y}_l^m(\theta,\phi) \\ &= (-1)^m\sqrt{\frac{(2l+1)(l-|m|)!}{4\pi(l+|m|)!}}\mathrm{P}_l^{|m|}(\cos\theta)\mathrm{e}^{\mathrm{i}m\phi} \end{aligned}$$

Hence,

$$\int_0^{2\pi}\int_0^{\pi}\mathrm{Y}_l^m(\theta,\phi)[\mathrm{Y}_k^n(\theta,\phi)]^*\sin\theta\mathrm{d}\theta\mathrm{d}\phi = \delta_{lk}\delta_{mn}$$

Below are several low-order spherical harmonics:

$$\begin{aligned} \mathrm{Y}_0^0(\theta,\phi) &= \frac{1}{2}\sqrt{\frac{1}{\pi}}; \\ \mathrm{Y}_1^0(\theta,\phi) &= \frac{1}{2}\sqrt{\frac{3}{\pi}}\cos\theta = \frac{1}{2}\sqrt{\frac{3}{\pi}}\frac{z}{r}; \\ \mathrm{Y}_1^{\pm1}(\theta,\phi) &= \frac{1}{2}\sqrt{\frac{3}{2\pi}}\sin\theta\mathrm{e}^{\pm\mathrm{i}\phi} = \frac{1}{2}\sqrt{\frac{3}{2\pi}}\frac{x\pm\mathrm{i}y}{r}; \\ \mathrm{Y}_2^0(\theta,\phi) &= \frac{1}{4}\sqrt{\frac{5}{\pi}}(3\cos^2\theta-1) = \frac{1}{4}\sqrt{\frac{5}{\pi}}\frac{-x^2-y^2+2z^2}{r^2}; \\ \mathrm{Y}_2^{\pm1}(\theta,\phi) &= \mp\frac{1}{2}\sqrt{\frac{15}{2\pi}}\sin\theta\cos\theta\mathrm{e}^{\pm\mathrm{i}\phi} = \mp\frac{1}{2}\sqrt{\frac{15}{2\pi}}\frac{(x\pm\mathrm{i}y)z}{r}; \\ \mathrm{Y}_2^{\pm2}(\theta,\phi) &= \frac{1}{4}\sqrt{\frac{15}{2\pi}}\sin^2\theta\mathrm{e}^{\pm2\mathrm{i}\phi} = \frac{1}{4}\sqrt{\frac{15}{2\pi}}\frac{(x\pm\mathrm{i}y)^2}{r} \end{aligned}$$

The spherical harmonics $\mathrm{Y}_{lm}(\theta,\varphi)$ can be represented using the nodal lines on the surface of a sphere. Specifically, the real part of $\mathrm{Y}_{lm}(\theta,\varphi)$ where $\mathrm{Re}[\mathrm{Y}_{lm}(\theta,\varphi)] = 0$ forms m longitudinal lines crossing through the north and south poles on the sphere and $l-m$ latitudinal lines parallel to the equator, as shown in Figure 13.6. The spherical harmonics changes sign each time they cross a nodal line.

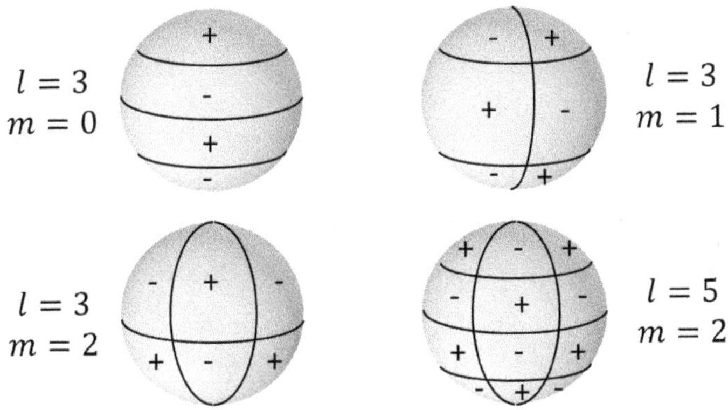

Figure 13.6.

13.3.2 *Generalized Fourier series*

A function $f(\theta,\phi)$ defined on the sphere can be expanded using spherical harmonics as a generalized Fourier series:

$$f(\theta,\phi)=\sum_{l=0}^{\infty}\sum_{m=0}^{l}C_{lm}\mathrm{Y}_{lm}(\theta,\phi)$$
$$C_{lm}=\int_0^{2\pi}\int_0^{\pi}f(\theta,\varphi)\mathrm{Y}_{lm}^*(\theta,\phi)\sin\theta\mathrm{d}\theta\mathrm{d}\phi \tag{13.3.5}$$

This equation is also known as the Laplace series. The expansion process can be divided into two steps. Let us consider the case where the spherical harmonics Y_{lm} are in trigonometric form.

(1) First, we expand $f(\theta,\phi)$ in a Fourier series with respect to ϕ:

$$f(\theta,\phi)=\sum_{m=0}^{\infty}[A_m(\theta)\cos m\phi+B_m(\theta)\sin m\phi]$$

where $A_m(\theta)$ and $B_m(\theta)$ are given by

$$A_m(\theta)=\frac{1}{\pi\delta_m}\int_0^{2\pi}f(\theta,\phi)\cos m\phi\,\mathrm{d}\phi,$$
$$B_m(\theta)=\frac{1}{\pi}\int_0^{2\pi}f(\theta,\phi)\sin m\phi\,\mathrm{d}\phi$$

Here, $\delta_0=2$ and $\delta_m=1(m\neq 0)$.

(2) Second, the expansion coefficients $A_m(\theta)$ and $B_m(\theta)$ still depend on θ. For a given m, we expand them in terms of associated Legendre polynomials:

$$A_m(\theta) = \sum_{l=m}^{\infty} A_l^m \mathrm{P}_l^m(\cos\theta), \quad B_m(\theta) = \sum_{l=m}^{\infty} B_l^m \mathrm{P}_l^m(\cos\theta)$$

Hence, we obtain the expansion of $f(\theta, \phi)$ using spherical harmonics:

$$f(\theta, \phi) = \sum_{m=0}^{\infty}\sum_{l=m}^{\infty} [A_l^m \cos m\phi + B_l^m \sin m\phi]\mathrm{P}_l^m(\cos\theta)$$

If we use the exponential form of spherical harmonics Y_l^m as the basis, then we have

$$f(\theta, \phi) = \sum_{l=0}^{\infty}\sum_{m=-l}^{l} C_l^m \mathrm{Y}_l^m(\theta, \phi) \tag{13.3.6}$$

where C_l^m are the expansion coefficients given by

$$C_l^m = \int_0^{\pi}\int_0^{2\pi} f(\theta, \phi)[\mathrm{Y}_k^n(\theta, \phi)]^* \sin\theta \mathrm{d}\theta\mathrm{d}\phi$$

Example 13.7. Using the spherical harmonics Y_{lm}, expand the given function:

$$f(\theta, \phi) = 3\sin^2\theta\cos^2\phi - 1$$

Solution: Start by expanding $f(\theta, \phi)$ in terms of ϕ using the Fourier series:

$$f(\theta, \phi) = \left(\frac{3}{2}\sin^2\theta - 1\right) + \frac{3}{2}\sin^2\theta\cos 2\phi$$

We observe that $m = 0$ and $m = 2$. We will consider these two terms separately:

Step 1: For the $m = 0$ term, expand $\frac{3}{2}\sin^2\theta - 1$ in terms of the associated Legendre polynomial $\mathrm{P}_l(\cos\theta)$:

$$\frac{3}{2}\sin^2\theta - 1 = -\frac{1}{2}(3\cos^2\theta - 1) = -\mathrm{P}_2(\cos\theta)$$

Step 2: For the $m = 2$ term, expand $\frac{3}{2}\sin^2\theta$ in terms of the associated Legendre polynomial $P_l^2(\cos\theta)$:

$$\frac{3}{2}\sin^2\theta = \frac{1}{2}\mathrm{P}_2^2(\cos\theta)$$

Putting it all together, we get the final result:

$$f(\theta,\phi) = 3\sin^2\theta\cos^2\phi - 1 = -\mathrm{P}_2(\cos\theta) + \frac{1}{2}\mathrm{P}_2^2(\cos\theta)\cos 2\phi$$

Example 13.8. Inside the spherical region with radius r_0, where there are no charges, the potential distribution on the spherical surface is given by

$$f(\theta,\phi) = u_0\sin^2\theta\cos\phi\sin\phi$$

We are asked to determine the potential distribution within this spherical region.

Solution: Due to the absence of axial symmetry, the general solution of the Laplace equation is

$$\begin{aligned} u(r,\theta,\phi) &= \sum_{m=0}^{\infty}\sum_{l=m}^{\infty} r^l[A_l^m\cos m\phi + B_l^m\sin m\phi]\mathrm{P}_l^m(\cos\theta) \\ &\quad + \sum_{m=0}^{\infty}\sum_{l=m}^{\infty}\frac{1}{r^{l+1}}[C_l^m\cos m\phi + D_l^m\sin m\phi]\mathrm{P}_l^m(\cos\theta) \end{aligned}$$

Considering the finiteness of the potential at the center, $C_l^m = D_l^m = 0$. Substituting the boundary condition,

$$\begin{aligned} u(r_0,\theta,\phi) &= \sum_{m=0}^{\infty}\sum_{l=m}^{\infty} r_0^l[A_l^m\cos m\phi + B_l^m\sin m\phi]\mathrm{P}_l^m(\cos\theta) \\ &= u_0\sin^2\theta\cos\phi\sin\phi \end{aligned}$$

Expand the right side of the equation in terms of spherical harmonics:

$$u_0\sin^2\theta\cos\phi\sin\phi = \frac{1}{6}u_0\mathrm{P}_2^2(\cos\theta)\sin 2\phi$$

By comparing coefficients, we find $A_l^m = 0$ and

$$\begin{cases} B_2^2 = \dfrac{u_0}{6r_0^2} & (l = 2, m = 2) \\ B_l^m = 0 & (l \neq 2, m \neq 2) \end{cases}$$

Hence,

$$u(r, \theta, \phi) = \frac{u_0}{6r_0^2} r^2 \mathrm{P}_2^2(\cos\theta) \sin 2\phi$$

13.3.3 *Addition formula*

By expanding the axisymmetric Legendre polynomials using spherical harmonics, we obtain the addition formula:

$$\mathrm{P}_l(\cos\gamma) = \frac{4\pi}{2l+1} \sum_{m=-l}^{l} \mathrm{Y}_{lm}(\theta_1, \phi_1) \mathrm{Y}_{lm}^*(\theta_2, \phi_2) \tag{13.3.7}$$

where γ is the angle between the unit vectors $\boldsymbol{r}_1(\theta_1, \phi_1)$ and $\boldsymbol{r}_2(\theta_2, \phi_2)$ as shown in Figure 13.7. When $\boldsymbol{r}_1$ and $\boldsymbol{r}_2$ coincide ($\gamma = 0$), we have

$$\sum_{m=-l}^{l} |\mathrm{Y}_{lm}(\theta, \phi)|^2 = \frac{2l+1}{4\pi} \tag{13.3.8}$$

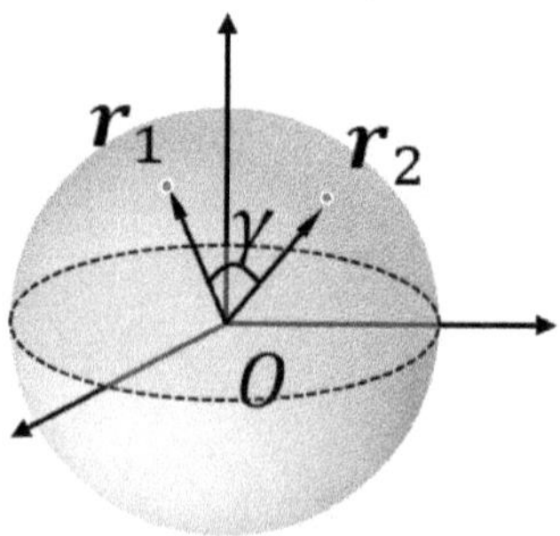

Figure 13.7.

The addition formula for spherical harmonics can be viewed as a generalization of the addition formula for circular trigonometry. In formula (13.3.7), we have by taking $l = 1$,

$$\begin{aligned} P_1(\cos\gamma) &= \frac{4\pi}{3}[Y_{10}(\theta_1,\phi_1)Y_{10}^*(\theta_2,\phi_2) + Y_{11}(\theta_1,\phi_1)Y_{11}^*(\theta_2,\phi_2) \\ &\quad + Y_{1-1}(\theta_1,\phi_1)Y_{1-1}^*(\theta_2,\phi_2)] \\ &= \cos\theta_1\cos\theta_2 + \frac{1}{2}\sin\theta_1\sin\theta_2 e^{i(\phi_1-\phi_2)} \\ &\quad + \frac{1}{2}\sin\theta_1\sin\theta_2 e^{-i(\phi_1-\phi_2)} \end{aligned}$$

Let $\phi_1 = \phi_2$, then $\gamma = \theta_1 - \theta_2$, leading to

$$\cos(\theta_1 - \theta_2) = \cos\theta_1\cos\theta_2 + \sin\theta_1\sin\theta_2$$

As $\theta_1 \to \theta_2$, formula (13.3.8) returns to the trigonometry identity:

$$\cos^2\theta + \sin^2\theta = 1$$

The addition formula is often used in dealing with gravity or Coulomb interaction problems. Considering the expansion of $\frac{1}{|\boldsymbol{r}-\boldsymbol{r}'|}$, let us take $|\boldsymbol{r}'| \equiv r'$ and $|\boldsymbol{r}| \equiv r$ with $r' < r$. We have

$$\frac{1}{|\boldsymbol{r}-\boldsymbol{r}'|} = \frac{1}{\sqrt{r^2 + r'^2 - 2rr'\cos\gamma}}$$

By using the generating function of Legendre polynomials and the addition formula of spherical harmonics, we obtain

$$\begin{aligned} \frac{1}{|\boldsymbol{r}-\boldsymbol{r}'|} &= \sum_{l=0}^{\infty} \frac{r'^l}{r^{l+1}} P_l(\cos\gamma) \\ &= \sum_{l=0}^{\infty} \frac{4\pi}{2l+1}\frac{r'l}{r^{l+1}} \sum_{m=-l}^{l} Y_{lm}(\theta,\phi)Y_{lm}^*(\theta',\phi') \end{aligned}$$

It factorizes the angular part from the radial part.

In physics, this formula is usually applied to the multipole expansions, which is very useful for approximation calculations. If $r' > r$, the expansion should be done with respect to $\frac{r}{r'}$.

ENDNOTE

Multipole expansion: Given a localized charge distribution $\rho(\boldsymbol{r}')$, the goal is to find its potential distribution in space,

$$V(r) = \int \frac{\rho(\boldsymbol{r}')}{|\boldsymbol{r}-\boldsymbol{r}'|}\mathrm{d}\boldsymbol{r}'$$

If the charge distribution lacks symmetry, exact integration can be impractical or very difficult. The multipole expansion is an effective approximation method to tackle this problem.

The multipole expansion can be expressed as a sum of terms involving progressively finer angular measurements, known as moments. The zeroth-order term is called the monopole moment, the first-order term is the dipole moment, the second-order term is the quadrupole moment, the third-order term is the octupole moment, and so on. Multipole moments often include powers (or inverse powers) of the distance from the origin and dependencies on angles. In principle, the multipole expansion can precisely describe the potential distribution. If the source or charge is localized near the origin, then the series converges.

In spherical coordinates, the multipole expansion can be represented using spherical harmonics as follows:

$$f(r,\theta,\phi) = \sum_{l=0}^{\infty}\sum_{m=-l}^{l} C_{lm}(r)\mathrm{Y}_{lm}(\theta,\phi)$$

where $C_{00}(r)$ corresponds to the monopole, $C_{1-1}(r), C_{10}(r)$, and $C_{1+1}(r)$ correspond to the dipole, and so on. Making Use of the expansion formula of $\frac{1}{|\boldsymbol{r}-\boldsymbol{r}'|}$, the potential distribution in space can be expressed as

$$\begin{aligned} V(r) &= \int \frac{\rho(\boldsymbol{r}')}{|\boldsymbol{r}-\boldsymbol{r}'|}\mathrm{d}\boldsymbol{r}' = \int \sum_{l=0}^{\infty} \frac{r'^l}{r^{l+1}}\mathrm{P}_l(\cos\gamma)\rho(\boldsymbol{r}')\mathrm{d}\boldsymbol{r}' \\ &= \int \sum_{l=0}^{\infty} \frac{4\pi}{2l+1}\frac{r'^l}{r^{l+1}} \sum_{m=-l}^{l} \mathrm{Y}_{lm}(\theta,\phi)\mathrm{Y}^*_{lm}(\theta',\phi')\rho(\boldsymbol{r}')\mathrm{d}\boldsymbol{r}' \end{aligned}$$

$$= \sum_{l=0}^{\infty} \frac{4\pi}{2l+1} \frac{1}{r^{l+1}} \sum_{m=-l}^{l} \mathrm{Y}_{lm}(\theta, \phi) \left[\int \mathrm{Y}^*_{lm}(\theta', \phi') r'^l \rho(\boldsymbol{r}') \mathrm{d}\boldsymbol{r}' \right]$$

$$= \sum_{l=0}^{\infty} \frac{4\pi}{2l+1} \sum_{m=-l}^{l} \frac{q_{lm} \mathrm{Y}_{lm}(\theta, \phi)}{r^{l+1}}$$

where the coefficients

$$q_{lm} \stackrel{\text{def}}{=} \int \mathrm{Y}^*_{lm}(\theta', \phi') r'^l \rho(\boldsymbol{r}') \mathrm{d}\boldsymbol{r}'$$

are called the $2l$th order multipole moments and they have $2l + 1$ components. In practical applications, the multipole expansion is often truncated after the first non-zero term as leading approximation.

The Laplace's equation in spherical coordinates can be written as

$$\frac{1}{r^2} \frac{\partial}{\partial r} \left(r^2 \frac{\partial u}{\partial r} \right) + \frac{1}{r^2} \hat{L}^2 u = 0$$

where the differential operator related to the angular coordinates is

$$\hat{L}^2 = \frac{1}{\sin\theta} \frac{\partial}{\partial \theta} \left(\sin\theta \frac{\partial}{\partial \theta} \right) + \frac{1}{\sin^2\theta} \frac{\partial^2}{\partial \phi^2}$$

This operator is the same as the angular momentum operator in quantum mechanics, $\hat{L}^2 = \hat{L}_x^2 + \hat{L}_y^2 + \hat{L}_z^2$, with a factor of the Planck constant $\hbar^2$. The components of the angular momentum satisfy the commutation relations:

$$[\hat{L}_i, \hat{L}_j] = \mathrm{i}\epsilon_{ijk} \hat{L}_k$$

The eigen equation for the angular momentum operator is

$$\hat{L}^2 |Y\rangle = \lambda |Y\rangle$$

It is worth noting that due to $[\hat{L}^2, \hat{L}_z] = 0$, there can be many eigenstates corresponding to the eigenvalue λ, or in other words, the eigenstates are degenerate. To distinguish them, the z-component operator of angular momentum is introduced

$$\hat{L}_z = -\mathrm{i}\frac{\partial}{\partial\phi}$$

The eigen equation is

$$\hat{L}_z|Y\rangle = m|Y\rangle$$

Since $[\hat{L}^2, \hat{L}_z]= 0$, we can take the common eigenstates of the operators $\{\hat{L}^2, \hat{L}_z\}$ and label them with two indices (λ, m): $|Y\rangle \longmapsto |Y_{\lambda m}\rangle$. We have shown that the eigenvalue λ can only take discrete values, namely,

$$\hat{L}^2|Y_{lm}\rangle = l(l+1)|Y_{lm}\rangle$$
$$\hat{L}_z|Y_{lm}\rangle = m|Y_{lm}\rangle$$

The eigenstates $|Y_{lm}\rangle$ in the coordinate representation are the spherical harmonics: $\mathrm{Y}_{lm}(\theta, \phi) = \langle \mathbf{x} \mid Y_{lm}\rangle$. The integer l is called the orbital angular momentum quantum number, and the integer m is called the azimuthal quantum number or magnetic quantum number.

Given a set of quantum numbers (l, m), we can determine a complete eigenstate of the electron cloud. Table 13.1 lists several combinations of smaller quantum numbers, and their corresponding eigen wavefunctions are represented by the electron cloud models shown in Figure 13.8, where the distance from the model surface to the origin represents $|\mathrm{Y}_{lm}(\theta, \phi)|^2$.

Table 13.1.

$l=0$	$l=1$	$l=2$	$l=3$	$l=4$
s	p	d	f	g
$m=0$	$m=0,\pm1$	$m=0,\pm1,\pm2$	$m=0,\pm1,$ $\pm2,\pm3$	$m=0,\pm1,$ $\pm2,\pm3,\pm4$

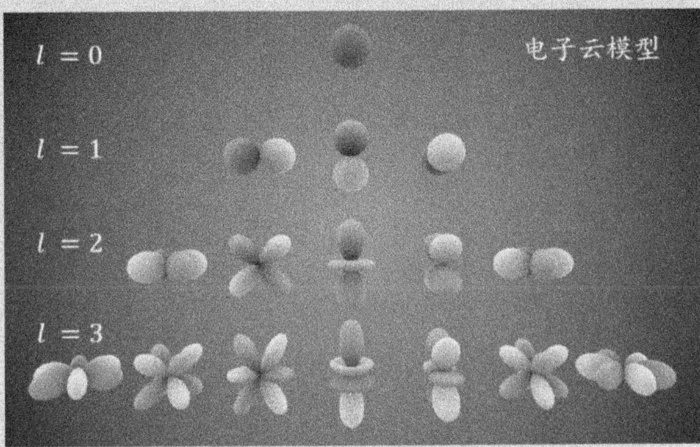

Figure 13.8. (image from Wikipedia).

Exercises

[1] Expand the following functions in terms of spherical harmonics:

$$\text{(a) } f(\theta,\phi)=\sin\theta\cos\phi; \quad \text{(b) } (1+3\cos\theta)\sin\theta\cos\phi$$

[2] Solve the equation outside a spherical region of radius r_0:

$$\begin{cases} \Delta u = 0 \quad (r > r_0) \\ \left.\dfrac{\partial u}{\partial r}\right|_{r=r_0} = u_0\left(\sin^2\theta\sin^2\phi - \dfrac{1}{3}\right) \end{cases}$$

Answer:

$$u(r,\theta,\phi)=\frac{u_0}{9}\frac{r_0^4}{r^3}\mathrm{P}_2^0(\cos\theta)+\frac{u_0}{18}\frac{r_0^4}{r^3}\mathrm{P}_2^2(\cos\theta)\cos 2\phi$$

[3] Solve the Poisson equation inside a spherical region of radius r_0:

$$\begin{cases} \Delta u = Ar\cos\theta \quad (r < r_0) \\ u|_{r=r_0} = 0 \end{cases}$$

Answer: $u(r,\theta) = Ar(r^2 - r_0^2)\mathrm{P}_1(\cos\theta)$

[4] Prove the spherical angle formula:

$$\cos\gamma = \cos\theta_1\cos\theta_2 + \sin\theta_1\sin\theta_2\cos(\phi_1 - \phi_2)$$

where γ is the angle between (θ_1, ϕ_1) and (θ_2, ϕ_2).

[5] Solve the Poisson problem inside a sphere of radius r_0:

$$\begin{cases} \Delta u = A + Br^2\sin 2\theta\cos\phi \quad (r < r_0) \\ u|_{r=r_0} = 0 \end{cases}$$

Hint: Guess a particular solution based on the non-homogeneous term, with the radial part being a polynomial of no higher than the fourth power of r. Alternatively, you can directly expand both sides of the non-homogeneous equation using the eigenfunctions.

$$u(r,\theta,\phi) = \sum_{l=0}^{\infty}\sum_{m=0}^{l}\mathrm{P}_l^m(\cos\theta)[R_{lm}(r)\sin m\phi + S_{lm}(r)\cos m\phi]$$

Answer: $u(r,\theta,\phi) = \frac{1}{6}A(r^2 - r_0^2) + \frac{1}{21}Br^2(r_0^2 - r^2)\mathrm{P}_2^1(\cos\theta)\cos\phi$

[6] Study the eigen vibrations of the two-dimensional wave equation on the surface of a sphere with radius R:

$$\partial_{tt}u - a^2\Delta_2 u = 0$$

Answer:

$$u_{lm}(\theta,\phi,t) = \mathrm{Y}_{lm}(\theta,\phi)[\cos\omega_l t, \sin\omega_l t]$$

$$\omega_l = \frac{\sqrt{l(l+1)}a}{R} \quad \begin{pmatrix} l = 0,1,2,3,4,\ldots \\ m = 0,1,2,\ldots l \end{pmatrix}$$

Since the eigen frequencies ω_l are degenerate for different m values, the eigen modes will exhibit a series of spherical distortion nodal lines similar to Figure 11.2.

[7] Given a uniformly charged circular ring with radius a and total charge q, the electric potential distribution it generates in three-dimensional space is

$$u(r,\theta) = \frac{q}{4\pi\varepsilon_0 r}\sum_{l=0}^{\infty}(-1)^l\frac{(2l-1)!!}{(2l)!!}\left(\frac{a}{r}\right)^{2l}\mathrm{P}_{2l}(\cos\theta) \quad (r > a)$$

Find the monopole moment, the dipole moment and the quadrupole moment of the electric charge.

[8] Taking the CO_2 molecule as a rod with point charges $-q$ at two endpoints and a point charge $+2q$ at the center. Find the monopole moment, the dipole moment and the quadrupole moment of the molecule.

Chapter 14

Eigenfunction Theory

Chapter 9 introduced methods for finding general solutions to differential equations, where the differential equations were considered as general equations without boundary or initial conditions. However, for physical equations, the determination of a well-posed solution requires constraints imposed by boundary values. In Chapters 11 and 13, we observed that certain differential equations, due to homogeneous boundary conditions or natural boundary conditions, lead to eigenvalues and mutually orthogonal eigenfunctions. Without these constraints, there would be no finite physical solutions. This chapter first discusses the general theory of eigenvalue problems for linear differential equations, followed by an introduction to several classical orthogonal polynomials.

14.1 Linear Spaces

14.1.1 *Metric space*

A vector space $\mathbb{V}$ is a collection of vectors satisfying addition, i.e., if $\boldsymbol{x}, \boldsymbol{y} \in \mathbb{V}$, then $\boldsymbol{x} + \boldsymbol{y} \in \mathbb{V}$. Additionally, vectors in the complex number field $\mathbb{C}$ are endowed with scalar multiplication, i.e., if $\alpha \in \mathbb{C}$, then $\alpha\boldsymbol{x} \in \mathbb{V}$. The operations of vector addition and scalar multiplication in a vector space exhibit linearity, hence it is also called a linear vector space. Each vector in the space has a unique inverse element $-\boldsymbol{x}$, satisfying $\boldsymbol{x} + (-\boldsymbol{x}) = 0$.

A norm is used to measure the length of a vector $\boldsymbol{x}$ in the linear vector space $\mathbb{V}$, denoted as $\|\boldsymbol{x}\|$, satisfying $\|\boldsymbol{x}\| \geq 0$. $\boldsymbol{x} = 0$ if and only if $\|\boldsymbol{x}\| = 0$. A vector space equipped with a defined norm is called a normed space.

A metric space extends the concept of distance from Euclidean space and is an abstract mathematical structure. It replaces the notion of distance between two points in Euclidean space with a measure of distance between two elements in a set. A metric can include vector distance, function distance, surface distance, etc., and is defined as follows:

Let $\mathbb{W}$ be a non-empty set. For any two elements $\boldsymbol{x}$ and $\boldsymbol{y}$ in $\mathbb{W}$, introduce a real number $d(\boldsymbol{x}, \boldsymbol{y})$ that satisfies the following properties:

[1] Positive definiteness: $d(\boldsymbol{x}, \boldsymbol{y}) \geq 0$, and $d(\boldsymbol{x}, \boldsymbol{y}) = 0$ if and only if $\boldsymbol{x} = \boldsymbol{y}$.
[2] Symmetry: $d(\boldsymbol{x}, \boldsymbol{y}) = d(\boldsymbol{y}, \boldsymbol{x})$.
[3] Triangle inequality: $d(\boldsymbol{x}, \boldsymbol{y}) \leq d(\boldsymbol{x}, \boldsymbol{z}) + d(\boldsymbol{z}, \boldsymbol{y})$.

Then, $d(\boldsymbol{x}, \boldsymbol{y})$ defines a metric on the set $\mathbb{W}$, which is now referred to as a metric space.

The difference between norm and metric lies in the fact that a metric is defined on any non-empty set, whereas a norm is only defined in a vector space.

14.1.2 *Completeness of space*

Once a space is equipped with a metric, it becomes possible to compare the distances between points in that space. The notion of Cauchy sequences can be expressed in a metric space as follows: Let $\{\boldsymbol{x}_1, \boldsymbol{x}_2, \ldots, \boldsymbol{x}_k, \ldots\}$ be an infinite sequence in the metric space $(\mathbb{W}, d)$. If for any positive real number $\varepsilon > 0$, there exists a positive integer $N(\varepsilon)$ such that for all $n, m > N(\varepsilon)$, the metric $d(\boldsymbol{x}_n, \boldsymbol{x}_m) < \varepsilon$, then the sequence is called a Cauchy sequence. This can be represented by the limit:

$$\lim_{n,m\to\infty} d(\boldsymbol{x}_n, \boldsymbol{x}_m) = 0$$

Any convergent sequence in a metric space is necessarily a Cauchy sequence. However, the converse is not always true. While Cauchy sequences are bounded, they may not converge to a point within that space. It can be proven that Cauchy sequences in finite-dimensional spaces always converge within that space. However, in infinite-dimensional spaces, whether Cauchy sequences converge within that space is not obvious and is closely related to the concept of completeness. A space is said to be complete if any Cauchy sequence in that metric space converges to a point within that space. In other words, if $\{\boldsymbol{x}_n\}_{n=1}^{\infty}$ is a Cauchy sequence in space $\mathbb{W}$, there exists a vector $\boldsymbol{x} \in \mathbb{W}$ such that

$$\lim_{n\to\infty} d(\boldsymbol{x}_n - \boldsymbol{x}) = 0$$

The set $\mathbb{R}$ of real numbers is complete. Any Cauchy sequence consisting of real numbers converges to a real number. The set $\mathbb{C}$ of complex numbers is also complete. However, the set $\mathbb{Q}$ of rational numbers is not complete. A counterexample can be given by considering a Cauchy sequence consisting of rational numbers:

$$x_0 = 1, \quad x_{n+1} = \frac{1}{2}\left(x_n + \frac{2}{x_n}\right)$$

This sequence converges to an irrational number, $\lim_{n\to\infty} x_n = \sqrt{2}$, which does not belong to the set of rational numbers $\mathbb{Q}$. Another familiar example is the sequence of rational numbers:

$$x_n = \left(1 + \frac{1}{n}\right)^n \quad (n \in \mathbb{N})$$

which converges to the transcendental number, $\lim_{n\to\infty} x_n = e$, also not belonging to the set $\mathbb{Q}$.

Any compact set is complete, but the converse is not true. For example, although the set of real numbers $\mathbb{R}$ is complete, it is not compact. Only by adding infinity does it become a compact closed set, equivalent to a closed circle. Finite-dimensional Euclidean spaces are complete under the usual distance definition, while the completeness of infinite-dimensional spaces is a topic that requires special exploration in Section 14.2.

14.1.3 *Inner product space*

In general, the definition of a linear vector space does not include multiplication between vectors. To address this, the concept of an inner product is introduced. The symbol $\langle a| \equiv (|a\rangle)^\dagger$ is used to represent the dual vector of a vector $|a\rangle$, defined as

$$(\alpha|a\rangle + \beta|b\rangle)^\dagger = \alpha^*\langle a| + \beta^*\langle b|$$

where the asterisk denotes complex conjugation. Dirac notation takes half of the word "bracket", representing the vector $|a\rangle$ as a "ket" and the vector $\langle a|$ as a "bra", thereby representing the vectors in the space and the conjugate vectors in the dual space.

Let $\mathbb{V}$ be an n-dimensional linear vector space, and let $|a\rangle, |b\rangle, |c\rangle \in \mathbb{V}$. The inner product $\langle a|b\rangle \in \mathbb{C}$ is defined over the complex number field and satisfies the following conditions:

[1] $\langle a|b\rangle = \langle b|a\rangle^*$;
[2] $\langle \alpha a + \beta b|c\rangle = \alpha^*\langle a|c\rangle + \beta^*\langle b|c\rangle$;
[3] $\langle a|a\rangle \geq 0$, if and only if $\langle a|a\rangle = 0$.

The inner product, sometimes called the scalar product, associates a pair of vectors with a real or complex number, establishing a mapping $\mathbb{V} \times \mathbb{V} \longmapsto \mathbb{C}$. A linear vector space equipped with an inner product is called an inner product space. When the inner product is real, it is referred to as a Euclidean space, and when the inner product is complex, it is called a unitary space. If two non-zero vectors $|a\rangle$ and $|b\rangle$ satisfy $\langle a|b\rangle = 0$, they are said to be orthogonal to each other.

The definition of an inner product distinguishes an inner product space from a general vector space. It includes three operations: vector addition, scalar-vector multiplication, and vector-vector multiplication. The inner product essentially defines a metric, making the inner product space also a metric space.

14.1.4 *Matrix representation*

Let $\{|e_1\rangle, |e_2\rangle, \ldots, |e_n\rangle\}$ be a complete set of orthonormal basis vectors in an n-dimensional vector space $\mathbb{V}$, where $\langle e_i|e_j\rangle = \delta_{ij}$. The completeness property states that any vector $|a\rangle \in \mathbb{V}$ can be represented as a linear combination of the basis vectors:

$$|a\rangle = \sum_{k=1}^{n} \alpha_k|e_k\rangle, \quad \alpha_k = \langle e_k|a\rangle$$

where α_k is called the projection or expansion coefficient of the vector $|a\rangle$ in the direction of the basis vector $|e_k\rangle$. The inner product of vectors can be expressed using the projections as

$$\langle a|b\rangle = \sum_{k=1}^{n} \alpha_k^* \left\langle \boldsymbol{e}_k \left| \sum_{j=1}^{n} \beta_j \right| \boldsymbol{e}_j \right\rangle = \sum_{j=1}^{n} \alpha_j^* \beta_j$$

The norm or magnitude of the vector $|a\rangle$ is given by

$$\|a\| = \sqrt{\langle a|a\rangle} = \left[\sum_{k=1}^{n} |\alpha_k|^2\right]^{\frac{1}{2}}$$

We can choose a set of basis vectors and represent the vector $|a\rangle$ as a column matrix,

$$|a\rangle = \alpha_1 \begin{pmatrix} 1 \\ 0 \\ \vdots \\ 0 \end{pmatrix} + \alpha_2 \begin{pmatrix} 0 \\ 1 \\ \vdots \\ 0 \end{pmatrix} + \cdots + \alpha_n \begin{pmatrix} 0 \\ 0 \\ \vdots \\ 1 \end{pmatrix} = \begin{pmatrix} \alpha_1 \\ \alpha_2 \\ \vdots \\ \alpha_n \end{pmatrix}$$

The dual vector $\langle a|$ is represented as a row matrix with complex conjugate elements,

$$\langle a| = (\alpha_1^* \quad \alpha_2^* \quad \cdots \quad \alpha_n^*)$$

14.1.5 *Hermitian operator*

In a linear vector space $\mathbb{V}$, an operator $\hat{A}$ acts on a vector $|a\rangle$, and the resulting vector $|b\rangle = \hat{A}|a\rangle$ also belongs to the same space, satisfying

$$\hat{A}(\alpha|a\rangle + \beta|b\rangle) = \alpha\hat{A}|a\rangle + \beta\hat{A}|b\rangle$$

and

$$(\hat{A}\hat{B})|a\rangle = \hat{A}(\hat{B}|a\rangle)$$

Two linear operators generally do not commute, meaning $\hat{A}\hat{B} \neq \hat{B}\hat{A}$. A linear operator $\hat{A}$ can also be represented with a basis vector set $\{|\boldsymbol{e}_1\rangle, |\boldsymbol{e}_2\rangle, \ldots, |\boldsymbol{e}_n\rangle\}$. Let

$$\hat{A}|\boldsymbol{e}_i\rangle = \sum_{k=1}^{n} a_{ik}|\boldsymbol{e}_k\rangle$$

Then we have

$$\langle \boldsymbol{e}_j|\hat{A}|\boldsymbol{e}_i\rangle = \left\langle \boldsymbol{e}_j \left| \sum_{k=1}^{n} a_{ik} \right| \boldsymbol{e}_k \right\rangle = \sum_{k=1}^{n} a_{ik}\langle \boldsymbol{e}_j|\boldsymbol{e}_k\rangle = a_{ij}$$

This is called the matrix representation of the operator $\hat{A}$ in the basis vector set $\{|\boldsymbol{e}_1\rangle, |\boldsymbol{e}_2\rangle, \ldots, |\boldsymbol{e}_n\rangle\}$,

$$[\hat{A}]_{ij} \overset{\text{def}}{=} \langle \boldsymbol{e}_j|\hat{A}|\boldsymbol{e}_i\rangle = a_{ij}$$

The operator $\hat{A}^\dagger$ is called the Hermitian conjugate or adjoint operator of $\hat{A}$, satisfying

$$\langle a|\hat{A}|b\rangle^* = \langle b|\hat{A}^\dagger|a\rangle$$

Hence $[\hat{A}^\dagger]_{ij} = a^*_{ji}$. If an operator and its Hermitian conjugate are the same, $\hat{A}^\dagger = \hat{A}$, then it is called a Hermitian operator or self-adjoint operator, and its matrix elements satisfy the conjugate transpose relation: $a_{ij} = a^*_{ji}$. If $\hat{A}$ is a Hermitian operator and $|a\rangle$ is a non-zero vector satisfying

$$\hat{A}|a\rangle = \lambda|a\rangle$$

then $|a\rangle$ is called an eigenvector of $\hat{A}$, and λ is called the eigenvalue, indicating that the vector $|a\rangle$ remains unchanged under the action of the operator $\hat{A}$.

Linear algebra theory proves the following fundamental theorems about Hermitian operators:

[1] All eigenvalues of a Hermitian operator are real, and eigenvectors corresponding to different eigenvalues are orthogonal to each other.
[2] The set of eigenvectors of a Hermitian operator is complete, meaning any vector can be represented as a linear combination of eigenvectors.
[3] The operator $\hat{A}$ has an inverse operator if and only if the determinant of its matrix $\det[\hat{A}] \neq 0$.
[4] If two Hermitian operators commute, $\hat{A}\hat{B} = \hat{B}\hat{A}$, then they share a common set of eigenvectors.

ENDNOTE

In 1845, Hermann Grassmann proposed the theory of linear vector spaces in his book titled "Die lineale Ausdehnungslehre" (*The Theory of Linear Extension*). He introduced concepts such as linear independence, dimension, basis, subspace, and projection, laying the foundation for abstract vector spaces and algebras. This development eventually evolved into a broad branch of mathematics known as linear algebra.

The concept of matrices has a long history. Scholars in ancient China used matrices to solve systems of linear equations as early as the Han Dynasty. By the 18th century, people were already calculating determinants and deriving concepts such as similarity, diagonalization, and eigenvalues from Gauss' theory of quadratic forms. In 1850, James Joseph Sylvester coined the term "matrix", and Arthur Cayley introduced the use of a single symbol to represent matrices and established the rules for matrix addition, subtraction, and multiplication. Cayley used inverse matrices to represent solutions of linear equations and noted that an inverse matrix does not exist when the determinant is zero. Charles Jordan introduced the concept of canonical forms, now known as Jordan canonical forms, for classifying matrices and extended the notion of eigenvalues and eigenvectors. Augustin-Louis Cauchy proved that all eigenvalues of a real symmetric matrix are real. In 1878, Georg Frobenius synthesized the ideas of his predecessors, providing a comprehensive exposition of matrix theory. He introduced the concepts of matrix similarity and orthogonal transformations, and established a correspondence between four 2×2 matrices and quaternions. However, it was not until the 1940s that people truly recognized the relationship between matrices and linear transformations in vector spaces.

Dual vectors are defined as

$$(\alpha|a\rangle + \beta|b\rangle)^\dagger = \alpha^*\langle a| + \beta^*\langle b|$$

This definition ensures that the length of a vector, $\langle a|a\rangle$, is non-negative. If the complex conjugate is not taken, then

$$\langle \mathrm{i}a|\mathrm{i}a\rangle = \mathrm{i}^2\langle a|a\rangle = -\langle a|a\rangle$$

which would not guarantee the positive-definiteness of the vector norm.

Intuitively, convergence implies a notion of approaching. This requires the concept of distance. The inner product of vectors defines the notion of norm, which represents the distance from the endpoint of a vector to the origin. However, norms can also be defined without an inner product. For complex or real fields, norms are commonly defined as

$$\|\boldsymbol{a}\|_p \equiv \left(\sum_{k=1}^{n} |\alpha_k|^p\right)^{1/p}$$

The usual norms defined by inner products correspond to $p = 2$. Thus, the distance between two points depends on the choice of norm. For example, in a 1000-dimensional space, consider a point or vector $\boldsymbol{b} = (0.1, 0.1, \ldots, 0.1)$. Its norms can be

$$\|\boldsymbol{b}\|_1 = 100, \quad \|\boldsymbol{b}\|_2 = 3.16, \ldots, \quad \|\boldsymbol{b}\|_{10} = 0.2$$

The choice of p value influences the norm. This may not provide a clear impression of approximation. However, the concept of distance between two vectors is relative. Although it may not make sense to say that vector $\boldsymbol{a}$ approximates vector $\boldsymbol{b}$, it is always possible to say that vector $\boldsymbol{a}$ approximates vector $\boldsymbol{b}$ more closely than vector $\boldsymbol{c}$. The natural choice for the norm definition is $p = 2$, and the distance between two vectors is defined as

$$d(\boldsymbol{a}, \boldsymbol{b}) = \|\boldsymbol{a} - \boldsymbol{b}\| = \sqrt{\sum_{k=1}^{n} |\alpha_k - \beta_k|^2}$$

The concept of compactness arises from the extension of finite sets to infinite sets. For a long time, people have recognized that compactness is essential for proving theorems in infinite-dimensional spaces. Initially, "compact" meant "sequentially compact", i.e., all sequences have convergent subsequences. A compact set is necessarily a bounded closed set, but the converse is not always true. If a subset of Euclidean space R^n is closed and bounded, it is called compact. For example, the unit interval $[0, 1]$ is compact, but the set of integers $\mathbb{Z}$ is not, and the half-open interval $[0, 1)$ is not compact either. Compact sets leverage many trivial properties of finite sets, such as the fact that any sequence in a finite set must have a convergent subsequence, as there are repeated points from the finite set in an infinitely long sequence. This conclusion holds for infinite sets as well, with the only modification being the replacement of "finite" with "compact".

Exercises

[1] Prove that the open interval $(0, 1)$ is not complete.
Hint: Consider the Cauchy sequence: $\left\{\frac{1}{k}\right\}_{k=1}^{\infty}$

[2] Prove that any convergent sequence is a Cauchy sequence.

[3] Prove that the sequence $\{x_n\}$ converges:

$$x_0 = 1, \quad x_{n+1} = \frac{1}{2}\left(x_n + \frac{2}{x_n}\right)$$

Hint: Construct the sequence:

$$y_{n+1} = \frac{x_{n+1} - \sqrt{2}}{x_{n+1} + \sqrt{2}} = \left(\frac{x_n - \sqrt{2}}{x_n + \sqrt{2}}\right)^2 = (y_n)^2$$

[4] For any pair of vectors $|a\rangle$ and $|b\rangle$ in an inner product space, prove the Schwarz inequality:

$$\langle a|a\rangle\langle b|b\rangle \geq |\langle a|b\rangle|^2$$

Hint: Let $|c\rangle = |b\rangle - \frac{\langle a|b\rangle}{\langle a|a\rangle}|a\rangle$, which gives rise to $\langle a|c\rangle = 0$.

[5] The set of all $n \times n$ matrices constitutes a n^2-dimensional linear vector space. By properly defining the inner product, prove it becomes an inner product space.

14.2 Hilbert Spaces

In Section 14.1, we briefly reviewed the fundamental concepts of finite-dimensional linear vector spaces, such as linear combination, linear independence, inner product, and subspace. These concepts can be directly extended to infinite-dimensional spaces. However, there is an important matter to consider, which is the convergence of infinite sums of vectors or the completeness of infinite-dimensional vector spaces. This non-trivial issue gives infinite-dimensional spaces deeper properties.

14.2.1 *Completeness relationship*

Let $\{|e_k\rangle\}_{k=1}^{\infty}$ be a set of orthonormal basis vectors in the linear vector space $\mathbb{V}$. For any vector $|f\rangle \in \mathbb{V}$, we define its projection onto the basis vectors as $f_k = \langle e_k|f\rangle$, where f_k is generally a complex number. Due to the norm inequality,

$$\left\| |f\rangle - \sum_{k=1} f_k|e_k\rangle \right\| \geq 0$$

we have

$$\langle f|f\rangle - \sum_{k=1}^{\infty} f_k\langle f|e_k\rangle - \sum_{k=1}^{\infty} f_k^*\langle e_k|f\rangle + \sum_{k=1}^{\infty} |f_k|^2 = \langle f|f\rangle - \sum_{k=1}^{\infty} |f_k|^2 \geq 0$$

which leads to the Bessel's inequality:

$$\langle f|f\rangle \geq \sum_{k=1}^{\infty} |f_k|^2 \tag{14.2.1}$$

Bessel's inequality states that the vector $\sum_{k=1}^{\infty} f_k|e_k\rangle$ is convergent, meaning that the norm of the vector is finite. Conversely, however, it does not imply that the infinite series $\sum_{k=1}^{\infty} f_k|e_k\rangle$ necessarily converges to a point in the space. In a linear space spanned by the basis vectors $\{|e_k\rangle\}_{k=1}^{\infty}$, if every infinite series $\sum_{k=1}^{\infty} f_k|e_k\rangle$ uniformly converges to a vector $|f\rangle$ within the space, i.e.,

$$\lim_{n\to\infty}\left(|f\rangle - \sum_{k=1}^{n} f_k|e_k\rangle\right) = 0 \tag{14.2.2}$$

then the space is said to be complete.

A complete inner product space is called a Hilbert space, denoted by $\mathcal{H}$. All complete finite or infinite-dimensional inner product spaces are Hilbert spaces. The completeness of a space implies that the Parseval identity holds:

$$\langle f|f\rangle = \sum_{k=1}^{\infty} |\langle f_k|e_k\rangle|^2 = \sum_{k=1}^{\infty} |f_k|^2 \tag{14.2.3}$$

If in a Hilbert space, the only vector orthogonal to all vectors $\{|e_k\rangle\}_{k=1}^{\infty}$ is the zero vector, then the orthogonal set of vectors $\{|e_k\rangle\}_{k=1}^{\infty}$ is called a complete set. Based on the definition of completeness, any vector $|f\rangle \in \mathcal{H}$ can be expanded using this complete set, and $f_k = \langle e_k|f\rangle$ is referred to as the generalized Fourier coefficient. Using the Parseval identity,

$$\langle f|f\rangle = \sum_{k=1}^{\infty} |f_k|^2 \equiv \sum_{k=1}^{\infty} \langle f|e_k\rangle\langle e_k|f\rangle$$

it follows that

$$\sum_{k=1}^{\infty} |e_k\rangle\langle e_k| = 1 \tag{14.2.4}$$

This equation is also known as the completeness relationship of the basis vectors. The operator $\hat{P}_k = |e_k\rangle\langle e_k|$ is called a projection operator because

$$\hat{P}_k|f\rangle = |e_k\rangle\langle e_k|f\rangle = f_k|e_k\rangle$$

which projects the vector $|f\rangle$ onto the basis $|e_k\rangle$.

Let $\{|e_k\rangle\}_{k=1}^{\infty}$ be a countable set of orthogonal vectors in the Hilbert space $\mathcal{H}$. For any vectors $|f\rangle, |g\rangle \in \mathcal{H}$, the following statements are equivalent:

[1] $\{|e_k\rangle\}_{k=1}^{\infty}$ is complete.

[2] $|f\rangle = \sum_{k=1}^{\infty} f_k|e_k\rangle = \sum_{k=1}^{\infty} |e_k\rangle\langle e_k|f\rangle$.

[3] $\sum_{k=1}^{\infty} |e_k\rangle\langle e_k| = 1$.

[4] $\langle f|f\rangle = \sum_{k=1}^{\infty} |f_k|^2$.

[5] $\langle f|g\rangle = \sum_{k=1}^{\infty}\langle f|e_k\rangle\langle e_k|g\rangle \equiv \sum_{k=1}^{\infty} f_k^* g_k$.

14.2.2 *Function space*

Continuous functions defined on an interval $[a, b]$ can be viewed as vectors, and the set of all continuous functions on this interval forms a linear vector space. However, this space is not complete. How can we construct a complete space? First, we need to define the inner product between two function vectors, which naturally relates to the integration of functions. Generally, the inner product of the function space is defined as

$$\langle g|f\rangle \overset{\text{def}}{=} \int_a^b g^*(x)f(x)\rho(x)\mathrm{d}x \tag{14.2.5}$$

where $\rho(x)$ is a positive definite real function called the weighting function, which we will discuss its significance later. In some cases, for convenience, we simply take $\rho(x) = 1$. Let $f(x) = g(x)$, then the norm of the function vector is given by

$$\langle f|f\rangle = \int_a^b |f(x)|^2\rho(x)\mathrm{d}x$$

Therefore, the prerequisite for forming an inner product space is that all functions within the interval $[a, b]$ must be square (weight) integrable.

After defining the inner product, we can discuss the completeness of the function space. Let $\mathcal{L}_\rho^2(a, b)$ denote the inner product space consisting of all square (weight) integrable functions on the interval $[a, b]$, where $\mathcal{L}$ indicates Lebesgue, who extended the conventional Riemann integral to the case of highly discontinuous functions, the superscript 2 represents the square of the function, and the subscript ρ represents the weighting function. The following fundamental theorem holds:

[1] **Riesz–Fischer theorem:** *The space $\mathcal{L}_\rho^2(a,b)$ is complete.*
According to the Cauchy criterion, if a sequence of functions $\{f_k(x)\}$ is square integrable and satisfies:

$$\lim_{m,n\to\infty}\int_a^b [f_m(x)-f_n(x)]^2\mathrm{d}x = 0$$

then there exists a square integrable function $f(x) \in \mathcal{L}_\rho^2(a,b)$ such that

$$\lim_{n\to\infty}\int_a^b [f_n(x)-f(x)]^2\mathrm{d}x = 0$$

[2] Every complete inner product space with a countable basis is isomorphic to $\mathcal{L}_\rho^2(a,b)$.
Infinite-dimensional complete inner product spaces are called Hilbert spaces. This theorem claims that the number of Hilbert spaces is limited, as it is equivalent to the space of square integrable functions. Generally, square integrability does not require functions to be continuous; it only requires them to be piecewise smooth. In fact, the space $\mathcal{L}_\rho^2(a,b)$ necessarily includes discontinuous functions with finitely many jump discontinuities.

[3] **Stone–Weierstrass Theorem:** *The sequence of monomials $\{x^k\}_{k=0}^{\infty}$ forms a complete basis for the space $\mathcal{L}_\rho^2(a,b)$.*
This implies that any square integrable function $f(x)$ on $[a,b]$ can be expressed using monomial basis functions. The space $\mathcal{L}_\rho^2(a,b)$ is a separable Hilbert space.
Differential or integral operators acting on functions correspond to linear transformations on function vectors, and the transformed functions still belong to the $\mathcal{L}_\rho^2(a,b)$ space. Among them, only certain types of differential or integral operators belong to the class of Hermitian or self-adjoint operators in the Hilbert space. They form eigenvalue problems, which will be discussed in detail in Section 14.3.

14.2.3 *Coordinate representation*

In an infinite-dimensional Hilbert space, there are two cases: countably infinite dimension and uncountably infinite dimension. The former corresponds to natural numbers, while the latter corresponds to real numbers. The components of a vector can be seen as a set of indices. For example, in a finite N-dimensional vector $|f\rangle$, the component f_k can be viewed as a function of the finite set of ordinal numbers $\{1, 2, 3, \ldots, N\}$, and it can be rewritten as $f(k)$: $f_k \longrightarrow f(k)$. For countably infinite space vectors $|f\rangle$, represented in the infinite basis $\{|\boldsymbol{e}_k\rangle\}_{k=1}^{\infty}$, the components $f(k)$ can be viewed as functions of the countably infinite set of natural numbers $\{1, 2, 3, \ldots, \infty\}$:

$$f : \mathbb{N} \longmapsto \mathbb{C}$$

Similarly, for uncountably infinite space vectors $|f\rangle$, represented in the continuous basis $\{|\boldsymbol{e}_x\rangle\}_{x\in\mathbb{R}}$, the components $f(x)$ can be viewed as functions of the continuous set of real numbers:

$$f : \mathbb{R} \longmapsto \mathbb{C}$$

The vector components $f(x)$ are precisely the functions of the vector $|f\rangle$ in the continuous basis representation.

Due to the isomorphism theorem, the following discussion is limited to the $\mathcal{L}_\rho^2(a, b)$ space. Let $\{|\boldsymbol{e}_x\rangle\}_{x\in\mathbb{R}}$ be a set of continuous coordinate basis vectors, and let $f(x)$ be viewed as the projection of the vector $|f\rangle$ onto $\{|\boldsymbol{e}_x\rangle\}_{x\in\mathbb{R}}$:

$$f(x) = \langle \boldsymbol{e}_x | f \rangle$$

Based on the inner product definition of $\mathcal{L}_\rho^2(a, b)$, we have

$$\begin{aligned}\langle g|f\rangle &= \int_a^b g^*(x) f(x) \rho(x) \mathrm{d}x = \int_a^b \langle g|\boldsymbol{e}_x\rangle\langle \boldsymbol{e}_x|f\rangle \rho(x) \mathrm{d}x \\ &= \left\langle g \left| \int_a^b \right| \boldsymbol{e}_x \right\rangle \rho(x) \langle \boldsymbol{e}_x|\mathrm{d}x|f\rangle\end{aligned}$$

This implies

$$\int_a^b |\boldsymbol{e}_x\rangle \rho(x) \langle \boldsymbol{e}_x| \mathrm{d}x = 1$$

Physicists usually omit the $\boldsymbol{e}$ and write the continuous coordinate basis vectors as $|x\rangle$. Thus, the completeness relation becomes

$$\int_a^b |x\rangle \rho(x) \langle x| \mathrm{d}x = 1$$

Expanding the vector $|f\rangle$ in terms of $|x\rangle$, we have

$$|f\rangle = \int_a^b |x\rangle\rho(x)\langle x|\mathrm{d}x|f\rangle = \int_a^b f(x)\rho(x)|x\rangle\mathrm{d}x$$

For a given $x' \in (a, b)$, we can write

$$f(x') = \langle x'|f\rangle = \int_a^b f(x)\rho(x)\langle x'|x\rangle\mathrm{d}x$$

It is evident that $\rho(x)\langle x'|x\rangle = \delta(x - x')$, which is the Dirac delta function. Taking $\rho(x) = 1$, we obtain the orthogonal completeness relation for the continuous basis:

$$\int_a^b |x\rangle\langle x|\mathrm{d}x = 1, \quad \langle x'|x\rangle = \delta(x - x') \tag{14.2.6}$$

In quantum mechanics, the state vectors in Hilbert space expressed using a continuous basis are called representations. If the continuous basis is the position basis, it is known as the coordinate representation. If the continuous basis is the momentum basis, it is known as the momentum representation.

Let $\{|f_n\rangle\}_{n=1}^{\infty}$ be a set of orthonormal basis vectors in Hilbert space, and their functions in the continuous basis representation are given by $f_n(x) = \langle x|f_n\rangle$. The Dirac δ function can be expanded using this set of basis functions:

$$\delta(x - x') = \sum_{n=1}^{\infty} c_n f_n(x)$$

The expansion coefficients are given by

$$c_n = \int_a^b \delta(x - x')f_n^*(x)\mathrm{d}x = f_n^*(x')$$

From this, we obtain an important formula:

$$\delta(x - x') = \sum_{n=1}^{\infty} f_n^*(x')f_n(x)$$

This formula can also be derived directly using the completeness relation,

$$\delta(x - x') = \langle x'|x\rangle = \left\langle x' \middle| \sum_{n=1}^{\infty} \middle| f_n \right\rangle \langle f_n|x\rangle = \sum_{n=1}^{\infty} f_n(x')f_n^*(x)$$

Example 14.1. The Hilbert space is spanned by an infinite countable orthogonal basis: $\{\varphi_n(x) = \langle x|\varphi_n\rangle = N_n \mathrm{H}_n(x)\mathrm{e}^{-\frac{x^2}{2}}\ (n \in \mathbb{N})\}$, where $\mathrm{H}_n(x)$ is the Hermite polynomials and $N_n = (2^n n!\sqrt{\pi})^{-\frac{1}{2}}$ is the normalization coefficient,

$$\begin{aligned}
&\mathrm{H}_0(x) = 1, \quad \mathrm{H}_1(x) = 2x \\
&\mathrm{H}_2(x) = 4x^2 - 2, \quad \mathrm{H}_3(x) = 8x^3 - 12x \\
&\cdots\cdots
\end{aligned}$$

Write the position operator $\hat{x}$ in matrix representation.

Solution: The matrix elements of the operator $\hat{x}$ are given by

$$x_{mn} = \langle\varphi_m|\hat{x}|\varphi_n\rangle = N_m N_n \int_{-\infty}^{\infty} \mathrm{H}_m^*(x) x \mathrm{H}_n(x)\mathrm{e}^{-x^2}\mathrm{d}x$$

By substituting the Hermite polynomials one by one and evaluating the integrals, we obtain the matrix representation of the operator $\hat{x}$:

$$\hat{x} = [x_{mn}] = \begin{bmatrix} 0 & \sqrt{2}/2 & 0 & 0 & \cdots \\ \sqrt{2}/2 & 0 & 1 & 0 & \cdots \\ 0 & 1 & 0 & \sqrt{6}/2 & \cdots \\ 0 & 0 & \sqrt{6}/2 & 0 & \cdots \\ \cdots & \cdots & \cdots & \cdots & \cdots \end{bmatrix}$$

This is an infinite-dimensional real symmetric matrix, and since $\hat{x} = \hat{x}^\dagger$, it is a Hermitian operator.

ENDNOTE

Consider a sequence of continuous functions $\{f_k(x)\}_{k=1}^{\infty}$ defined on the interval $[-1, 1]$. The functions are given by

$$f_k(x) = \begin{cases} 1 & (1/k \leq x \leq 1) \\ \dfrac{1}{2}(kx + 1) & (-1/k \leq x \leq 1/k) \\ -1 & (-1 \leq x \leq -1/k) \end{cases}$$

This sequence belongs to the function inner product space $\mathcal{L}_1^2(-1,1)$. It can be easily verified that

$$\|f_k - f_j\|^2 = \int_{-1}^{1} |f_k(x) - f_j(x)|^2 \mathrm{d}x \overset{j,k\to\infty}{\longrightarrow} 0$$

Hence, $\{f_k(x)\}_{k=1}^{\infty}$ is a Cauchy sequence. The limit of this sequence is given by

$$f(x) = \lim_{k\to\infty} f_k(x) = \begin{cases} 1 & (0 < x < 1) \\ -1 & (-1 < x < 0) \end{cases}$$

The function $f(x)$ has a discontinuity at $x = 0$ and does not belong to the continuous function space where the sequence $\{f_k(x)\}_{k=1}^{\infty}$ resides. Therefore, the continuous function space is incomplete. Completeness of the $\mathcal{L}_\rho^2(a,b)$ space requires it to contain smooth functions with discontinuities.

A linear transformation of a vector space is called a bounded linear transformation if its norm is bounded. A bounded linear transformation that maps a Hilbert space onto itself is called a bounded operator. Unfortunately, most differential operators are not bounded operators. For example, the first derivative operator $\hat{\mathrm{D}} = \frac{\mathrm{d}}{\mathrm{d}x}$ is not a bounded operator on the square-integrable function space $\mathcal{L}^2(a,b)$. To illustrate this, consider the function $f(x) = \sqrt{x-a}$. Its norm is finite:

$$\|f\|^2 = \int_a^b (x-a)\mathrm{d}x = \frac{1}{2}(b-a)^2$$

However, the norm of the derivative $\hat{D}f = \frac{1}{2\sqrt{x-a}}$ is

$$\|\hat{D}f\|^2 = \frac{1}{4}\int_a^b \frac{1}{(x-a)}\mathrm{d}x \to \infty$$

This indicates that the derivative operator $\hat{D}$ cannot act on all functions in the $\mathcal{L}^2(a,b)$ space.

Fortunately, many differential operators that frequently appear in physics can be discussed in the context of compact operators. This means that the domain of the operator (where it acts) is not the entire Hilbert space $\mathcal{L}^2(a,b)$, but rather a submanifold of it. For example, the derivative operator $\hat{D}$ can act on a linear submanifold spanned by the bases:

$$\left\{\mathrm{e}^{\frac{2n\pi x\mathrm{i}}{L}}\right\}(L = b - a, n \in \mathbb{Z})$$

This touches upon the essence of Fourier series: Any function in the $\mathcal{L}^2(a,b)$ space can be approximated arbitrarily closely by a Fourier series. Smooth functions with discontinuities are also essential elements for the completeness of the $\mathcal{L}^2(a,b)$ space. In fact, many unbounded operators in Hilbert spaces possess such properties. Section 14.3 will introduce the Sturm–Liouville operator, which is another example of an unbounded operator.

Exercises

[1] Expand the Dirac δ function using Legendre polynomials and prove:

$$\text{(a) } \delta(1-x) = \sum_{l=0}^{\infty} \frac{2l+1}{2} \mathrm{P}_l(x);$$

$$\text{(b) } \delta(1+x) = \sum_{l=0}^{\infty} (-1)^l \frac{2l+1}{2} \mathrm{P}_l(x)$$

[2] Proof of the addition theorem for spherical harmonics:

$$\mathrm{P}_l(\cos\gamma) = \frac{4\pi}{2l+1} \sum_{m=-l}^{l} \mathrm{Y}_{lm}(\theta_1, \phi_1)\mathrm{Y}^*_{lm}(\theta_2, \phi_2)$$

Hint: Rotate the polar axis of the spherical coordinate system to the direction of $\boldsymbol{r}'$ and use the identity $\delta(\boldsymbol{\xi} - \boldsymbol{\xi}') = \det J\delta(\boldsymbol{r})$.

This leads to

$$\delta(\cos\theta - \cos\theta')\delta(\phi - \phi') = \frac{1}{2\pi}\delta(\cos\gamma)$$

Expand both sides of the equation using Legendre polynomials and spherical harmonics.

[3] In the Hilbert space spanned by the countable orthonormal basis:

$$\{\varphi_n(x) = \langle x|\varphi_n\rangle = N_n \mathrm{H}_n(x)\mathrm{e}^{-\frac{x^2}{2}} \ (n = 0, 1, 2, \ldots)\}$$

write down the matrix representation of the momentum operator $\hat{p} = -\mathrm{i}\hbar\frac{\partial}{\partial x}$.

14.3 Sturm–Liouville Systems

14.3.1 *Adjoint operator*

Let $\hat{L}$ be an operator acting on any function vector $|f\rangle$ in a Hilbert space $\mathcal{H}$, and let $|g\rangle = \hat{L}|f\rangle$. If there exists an operator $\hat{L}^\dagger$ such that

$$\langle f|\hat{L}|g\rangle^* = \langle g|\hat{L}^\dagger|f\rangle$$

then $\hat{L}^\dagger$ is called the Hermitian conjugate operator of $\hat{L}$, more commonly known as the adjoint operator of $\hat{L}$.

Consider a second-order differential operator $\hat{L}$ given by

$$\hat{L}u \equiv \left[p_2(x)\frac{\mathrm{d}^2}{\mathrm{d}x^2} + p_1(x)\frac{\mathrm{d}}{\mathrm{d}x} + p_0(x)\right]u \tag{14.3.1}$$

What is the adjoint operator $\hat{L}^\dagger$ of $\hat{L}$? To determine this, we multiply both sides by a function $v(x)$ and try to interchange the position of $u(x)$ and $v(x)$,

$$\begin{aligned}
v\hat{L}u &= vp_2(x)u'' + vp_1(x)u' + vp_0(x)u \\
&= (p_2vu' + p_1vu)' - (p_2v)'u' - (p_1v)'u + p_0uv \\
&= [p_2vu' - (p_2v)'u + p_1vu]' + u[(p_2v)'' - (p_1v)' + p_0v] \\
&= u\hat{M}v + [p_2vu' - (p_2v)'u + p_1vu]'
\end{aligned}$$

where the operator $\hat{M}$ is given by

$$\hat{M} = p_2 \frac{\mathrm{d}^2}{\mathrm{d}x^2} + [2p_2' - p_1] \frac{\mathrm{d}}{\mathrm{d}x} + [p_2'' - p_1' + p_0] \tag{14.3.2}$$

Therefore,

$$v\hat{L}u - u\hat{M}v = \frac{\mathrm{d}}{\mathrm{d}x}[p_2(vu' - v'u) + (p_1 - p_2')uv]$$

Integrating both sides over the interval $[a, b]$, we obtain

$$\int_a^b \{v\hat{L}u - u\hat{M}v\}\mathrm{d}x = [p_2(vu' - v'u) + (p_1 - p_2')uv]_{x=a}^{x=b} \tag{14.3.3}$$

This equation is known as the Lagrange identity, which can be viewed as an alternative form of the Green's formula.

By representing the functions as abstract vectors $|u\rangle$ and $|v\rangle$, and $\hat{L}$ and $\hat{M}$ as Hilbert space operators defined by the inner product, the Lagrange identity becomes

$$\langle v|\hat{L}|u\rangle - \langle u|\hat{M}|v\rangle = [p_2 vu' - (p_2 v)'u + p_1 vu]_{x=a}^{x=b} \tag{14.3.4}$$

If we set the right-hand side of the equation equal to zero,

$$[p_2(vu' - v'u) + (p_1 - p_2')uv]|_a^b = 0$$

then

$$\langle v|\hat{L}|u\rangle = \langle u|\hat{M}|v\rangle$$

Therefore, for a real inner product space, $\hat{M}$ is the adjoint operator of $\hat{L}$: $\hat{M} \equiv \hat{L}^\dagger$. The equation $\hat{L}^\dagger v = 0$ is called the adjoint equation of the equation $\hat{L}u = 0$. It can be easily verified that this result holds for complex inner product spaces as well.

The adjoint operator and its boundary conditions are determined by the form of the operator and its boundary conditions.

Example 14.2. Find the adjoint operator and adjoint boundary conditions of the differential operator $\hat{L}$ with the given conditions:

$$\begin{cases} \hat{L}u = x^2u'' + u' + 2u \\ u(1) = 0, \quad u'(2) + u(2) = 0 \end{cases}$$

Solution: We perform integration by parts on the following expression,

$$\begin{aligned} \langle v|\hat{L}|u\rangle &= \int_1^2 v[x^2u'' + u' + 2u]\mathrm{d}x \\ &= [x^2vu' - (x^2v)'u + vu]_{x=1}^{x=2} + \int_1^2 u[(x^2v)'' - v' + 2v]\mathrm{d}x \end{aligned}$$

For the integral expression on the right-hand side, we define

$$\hat{L}^\dagger v = (x^2v)'' - v' + 2v = x^2v'' + (4x-1)v' + 4v$$

Let us take

$$P(x) = x^2vu' - (x^2v)'u + vu = x^2(vu' - v'u) + (-2x+1)vu$$

By setting the boundary values $P(1) = P(2) = 0$, and letting $\langle v|\hat{L}|u\rangle = \langle u|\hat{L}^\dagger|v\rangle$, we obtain the adjoint operator of $\hat{L}$:

$$\hat{L}^\dagger = x^2\frac{\mathrm{d}^2}{\mathrm{d}x^2} + (4x-1)\frac{\mathrm{d}}{\mathrm{d}x} + 4$$

Since

$$\begin{aligned} P(2) &= 4v(2)u'(2) - 4v'(2)u(2) - 3v(2)u(2) \\ &= -[7v(2) + 4v'(2)]u(2) = 0 \end{aligned}$$

The boundary condition for the adjoint operator $\hat{L}^\dagger$ at $x = 2$ is

$$P(2) = 0 \longrightarrow 7v(2) + 4v'(2) = 0$$

Similarly, the boundary condition at $x = 1$ is

$$P(1) = 0 \longrightarrow v(1) = 0$$

It can be observed that the adjoint operator $\hat{L}^\dagger$ satisfies different boundary conditions compared to the operator $\hat{L}$.

14.3.2 *Self-adjoint operators*

Self-adjoint operators are those whose adjoint operators are equal to themselves, $\hat{L} = \hat{L}^\dagger$. They are also known as Hermitian operators. In Example 14.2, the operator $\hat{L}$ is not equal to its adjoint $\hat{L}^\dagger$, so it is not a self-adjoint operator. Now, let us discuss the general form of a second-order self-adjoint differential operator.

Comparing the expressions of $\hat{L}^\dagger$ and $\hat{L}$ in equations (14.3.1) and (14.3.2), we can draw the following conclusion:

A second-order linear differential operator is self-adjoint if and only if $p_1(x) = p_2'(x)$. It can also be rephrased as

$$\hat{L}^\dagger = \hat{L} \equiv -\frac{\mathrm{d}}{\mathrm{d}x}\left[p(x)\frac{\mathrm{d}}{\mathrm{d}x}\right] + q(x) \tag{14.3.5}$$

This form of differential operator is known as the Sturm–Liouville operator. It is worth emphasizing that unlike abstract operators in finite-dimensional vector spaces, differential operators are closely related to boundary conditions. This may be the distinction between the terminologies of Hermitian operator and self-adjoint operator. As a self-adjoint operator, the Sturm–Liouville operator must satisfy the boundary condition:

$$p(x)[v(x)u'(x) - v'(x)u(x)]|_a^b = 0 \tag{14.3.6}$$

For a general second-order homogeneous differential equation:

$$p_2(x)y'' + p_1(x)y' + p_0(x)y = 0$$

Multiplying both sides of the equation by the weight function,

$$\rho(x) = \frac{1}{p_2(x)} \exp\left[\int^{(x)} \frac{p_1(t)}{p_2(t)}\mathrm{d}t\right]$$

we can transform it into the Sturm–Liouville form, where

$$p(x) = p_2(x)\rho(x), \quad q(x) = -p_0(x)\rho(x)$$

It can be seen that the appearance of the weight function $\rho(x)$ is due to the desire for the differential operator to have a self-adjoint form. Next, we will prove that the introduction of weighted inner product definitions ensures the orthogonality of eigenfunctions with different eigenvalues.

The following form of the Legendre equation is not in the Sturm–Liouville form,

$$y'' - \frac{2x}{1-x^2}y' + \frac{\lambda}{1-x^2}y = 0$$

However, if we multiply both sides by $(1-x^2)$, it can be transformed into the Sturm–Liouville form:

$$(1-x^2)y'' - 2xy + \lambda y = 0$$

This can be further written as

$$\frac{\mathrm{d}}{\mathrm{d}x}\left[(1-x^2)\frac{\mathrm{d}}{\mathrm{d}x}y\right] + \lambda y = 0$$

which has a total derivative form.

Example 14.3. Convert the differential equation into the Sturm–Liouville form:

$$x^3y'' - xy' + 2y = 0$$

Solution: Multiplying both sides by the weight function,

$$\rho(x) = \frac{1}{x^3}\exp\left[\int -\frac{x}{x^3}\mathrm{d}x\right] = \frac{1}{x^3}\mathrm{e}^{\frac{1}{x}}$$

This results in

$$\mathrm{e}^{\frac{1}{x}}y'' - \frac{1}{x^2}\mathrm{e}^{\frac{1}{x}}y' + \frac{2}{x^3}\mathrm{e}^{\frac{1}{x}}y = 0$$

Now, the equation is in the Sturm–Liouville form:

$$\frac{\mathrm{d}}{\mathrm{d}x}\left[\mathrm{e}^{\frac{1}{x}}\frac{\mathrm{d}}{\mathrm{d}x}y\right] + \frac{2}{x^3}\mathrm{e}^{\frac{1}{x}}y = 0$$

This form allows us to combine terms involving derivatives into complete derivative form.

14.3.3 *Sturm–Liouville eigenvalue equation*

A second-order ODE defined on the interval $x \in [a,b]$ in the form,

$$\hat{L}y \equiv -\frac{\mathrm{d}}{\mathrm{d}x}\left[p(x)\frac{\mathrm{d}y}{\mathrm{d}x}\right] + q(x)y = \lambda\rho(x)y \tag{14.3.7}$$

is known as the Sturm–Liouville eigenvalue equation. The constant λ is referred to as the eigenvalue. Since the weight function $\rho(x) > 0$, let

$$u(x) \longmapsto \sqrt{\rho(x)}y(x), \quad \hat{L} \longmapsto [\rho(x)]^{-\frac{1}{2}}\hat{L}[\rho(x)]^{-\frac{1}{2}}$$

This transformation allows us to rewrite Equation (14.3.7) into the standard form of eigenvalue equations in linear algebra:

$$\hat{L}u = \lambda u \tag{14.3.8}$$

Suppose y_1 and y_2 are eigenfunctions of the Sturm–Liouville operator $\hat{L}$, with corresponding eigenvalues λ_1 and λ_2,

$$\hat{L}y_i = \lambda_i\rho(x)y_i \quad (i = 1, 2)$$

Then we have

$$y_1\hat{L}y_2 - y_2\hat{L}y_1 = (\lambda_1 - \lambda_2)\rho y_1 y_2$$

Using the Lagrange identity, we obtain

$$(\lambda_1 - \lambda_2)\int_a^b \rho(x)y_1(x)y_2(x)\mathrm{d}x = p(x)[y_1(x)y_2'(x) - y_1'(x)y_2(x)]|_{x=a}^{x=b}$$

Note that the Sturm–Liouville operator is self-adjoint and the right-hand side of the equation above should be zero,

$$p(x)[y_1(x)y_2'(x) - y_1'(x)y_2(x)]_{x=a}^{x=b} = 0 \tag{14.3.9}$$

Therefore, for distinct eigenvalues $\lambda_1 \neq \lambda_2$, their corresponding eigenfunctions are (weighted) orthogonal, or their inner product is zero,

$$\int_a^b \rho(x)y_1(x)y_2(x)\mathrm{d}x = 0 \tag{14.3.10}$$

The eigenvalue problems can be implemented in three different ways based on Equation (14.3.9):

(1) If the endpoints satisfy the first type of homogeneous boundary condition (Dirichlet boundary condition) or the second type of homogeneous boundary condition (Neumann boundary condition), the condition can be satisfied.

(2) If the endpoints satisfy the third type of homogeneous boundary condition (Cauchy boundary condition), for example, at $x = a$,

$$(\alpha y_1 + \beta y_1')|_{x=a} = (\alpha y_2 + \beta y_2')|_{x=a} = 0$$

Then, $(y_1 y_2' - y_1' y_2)|_{x=a} = 0$, and the condition can be satisfied.

(3) Although not satisfying homogeneous boundary conditions, if $p(x)$ is zero at the endpoint, such as $p(a) = 0$, the condition can still be satisfied. This is known as the natural boundary condition mentioned in Chapter 13.

Applying the Sturm–Liouville-type equation with the first/second/third type of homogeneous boundary conditions or the natural boundary condition constitutes a Sturm–Liouville eigenvalue problem. What is the difference between satisfying homogeneous boundary conditions and satisfying natural boundary conditions at endpoints? Given that second-order ODEs inherently possess two linearly independent solutions, different boundary conditions will lead to distinct eigenvalue determination schemes:

(1) **Both endpoints satisfy homogeneous boundary conditions**
The two solutions will combine to form a unique eigenfunction, and the eigenvalues will be discrete. This situation arises, for example, when solving the vibration equation using the method of separation of variables in Chapter 11.

(2) **Both endpoints satisfy the natural boundary condition**
Both endpoints are first-order zeros of $p(x)$ and, at most, second-order poles of $q(x)$. As implied by the Sturm–Liouville equation, both endpoints will be regular singular points of the ODE. In this case, the two linearly independent series solutions of the ordinary differential equation will diverge at the endpoints. To obtain physically meaningful finite solutions, one of the solutions must be truncated into a polynomial form. As a result, λ can only take some discrete eigenvalues. The other solution will still diverge at the endpoints, leading to only one physically meaningful eigenfunction. This situation arises in the solution of the Legendre equation discussed in Chapter 13, as well as in many solutions of the Schrödinger equation in quantum mechanics.

(3) **Each endpoint respectively has a combination of homogeneous boundary conditions and the natural boundary condition**
In this case, one of the two linearly independent series solutions of the ordinary differential equation will converge within the region, while the other will diverge at the natural boundary where $p(x)$ is zero. Therefore, there will still be only one physically meaningful eigen solution, and the corresponding eigenvalues will be determined by the homogeneous boundary conditions. This situation will be discussed in Chapter 15, where Bessel's equation is introduced.

ENDNOTE

We have attributed the natural boundary conditions to the finiteness of the physical quantities described by the differential equation, thereby restricting the eigenvalues to a discrete set of discontinuous values. However, this is merely an apparent explanation. The essence of the eigenvalue problem lies in the singularity of the differential equation, where the solutions of the equation often exhibit singularities at the singular points. To obtain finite solutions without singularities, it is necessary to analyze the types of singularities in the equation. Frobenius studied second-order differential equations with coefficients given by rational functions and found that if the coefficient functions have regular singular points, at least one solution of the equation can be represented in the form of a Frobenius series. If the regular singular points are boundary points of the system, only when specific discrete values are chosen for the eigenvalues can the solutions of the equation eliminate the singularities, hence the terminology "natural boundary conditions" is also referred to as "singular boundary conditions". In this case, the equation has only one finite eigenfunction for each eigenvalue, while the other solution of the differential equation still exhibits singular behavior. Therefore, for each eigenvalue, the equation has only one finite eigenfunction.

If the equation has non-regular singularities, unless certain special circumstances arise, the equation has no finite solutions under any conditions and cannot be represented in the form of a Frobenius series. Unless homogeneous boundary conditions are imposed to exclude the singular points from the system, non-regular singularities generally cannot form the boundaries of an eigenvalue problem.

Although periodic boundary conditions may appear to satisfy the condition for the right side of the Lagrange identity to be zero, they often lead to degenerate eigenfunctions. Therefore, periodic boundary conditions should not be included in this category.

Exercises

[1] Find the adjoint operator and adjoint boundary conditions for the differential operator $\hat{L}$:

$$\text{(a)} \begin{cases} \hat{L}u = u'' + x^2u' + \dfrac{2}{x}u \\ u(a) = 0, u'(b) = 0 \end{cases}; \quad \text{(b)} \begin{cases} \hat{L}u = xu'' + u' + u \\ u(1) = 0, u(2) + u'(2) = 0 \end{cases}$$

Answer:

$$\text{(a) } \hat{L}^+ = \frac{d^2}{dx^2} + x^2\frac{\mathrm{d}}{\mathrm{d}x} + 2x + \frac{2}{x}; \quad v(a) = 0,\ b^2v(b) - v'(b) = 0$$

$$\text{(b) } \hat{L}^+ = x\frac{\mathrm{d}^2}{\mathrm{d}x^2} + \frac{\mathrm{d}}{\mathrm{d}x} + 1; \quad v(1) = 0, \quad v(2) + v'(2) = 0$$

[2] Convert the following equations into Sturm–Liouville form:

$$\text{(a) } x^2y'' + xy' + y = 0; \quad \text{(b) } y'' - 2xy' + 2\alpha y = 0$$
$$\text{(c) } xy'' + 2y' + (x + \lambda)y = 0; \quad \text{(d) } y'' + \cot xy + \lambda y = 0$$

Answer:

$$\text{(a) } (xy')' + \frac{1}{x}y = 0; \quad \text{(b) } (\mathrm{e}^{-x^2}y')' + 2\alpha\mathrm{e}^{-x^2}y = 0$$
$$\text{(c) } (x^2y')' + (x^2 + \lambda x)y = 0; \quad \text{(d) } (\sin xy')' + \lambda \sin xy = 0$$

[3] Convert the following equations into Sturm–Liouville form:

$$\text{(a) } x(x-1)y'' + [(1+\alpha+\beta)x - \gamma]y' + \alpha\beta y = 0$$
$$\text{(b) } xy'' + (\gamma - x)y' + \alpha y = 0$$

Answer:

$$\text{(a) } \frac{\mathrm{d}}{\mathrm{d}x}\left[x^{\gamma}(x-1)^{1+\alpha+\beta-\gamma}\frac{\mathrm{d}y}{\mathrm{d}x}\right] + \alpha\beta x^{\gamma-1}(x-1)^{\alpha+\beta-\gamma}y = 0$$
$$\text{(b) } \frac{\mathrm{d}}{\mathrm{d}x}\left(x^{\gamma}\mathrm{e}^{-x}\frac{\mathrm{d}y}{\mathrm{d}x}\right) + \alpha x^{\gamma-1}\mathrm{e}^{-x}y = 0$$

[4] For the Sturm–Liouville-type equation:

$$-\frac{\mathrm{d}}{\mathrm{d}x}\left[p(x)\frac{\mathrm{d}y}{\mathrm{d}x}\right] + q(x)y - \lambda\rho(x)y = 0$$

where $\rho(x) > 0$. If $x = a$ is a first-order zero of $p(x)$ and a pole of $q(x)$ no higher than the first order, prove that it is a regular singularity of the equation.

14.4 Eigenvalue Theory

14.4.1 *Basic properties*

For the eigenvalue equation of the Sturm–Liouville operator, assume that its coefficients $p(x), q(x)$ and $\rho(x)$ are all positive and $p(x)$ has a continuous derivative. If the equation satisfies one of the three types of homogeneous boundary conditions or the natural boundary condition, the following basic properties hold:

(1) *Discreteness of eigenvalues*

If $p(x), p'(x)$, and $q(x)$ are continuous, then there exist infinitely many discrete real eigenvalues, $\lambda_1 \leq \lambda_2 \leq \lambda_3 \leq \cdots$, and correspondingly infinitely many eigenfunctions $y_1(x), y_2(x), y_3(x), \ldots$. Here, we only prove that the eigenvalues are real. Let λ_n be an eigenvalue with its corresponding eigenfunction $y_n(x)$, then

$$\hat{L}y_n = \lambda_n y_n \longrightarrow \langle y_n \hat{L} y_n\rangle = \lambda_n \langle y_n | y_n\rangle$$

Since $\hat{L}$ is a self-adjoint operator and the norm $\|y_n\| = \langle y_n|y_n\rangle \neq 0$, we have

$$\langle y_n \hat{L} y_n\rangle = \langle y_n \hat{L} y_n\rangle^* == \lambda_n^* \langle y_n|y_n\rangle$$

Thus, we must have $\lambda_n^* = \lambda_n$. Proof completed.

(2) *Positivity of eigenvalues*

Based on the eigenvalue equation

$$-\frac{\mathrm{d}}{\mathrm{d}x}\left[p(x)\frac{\mathrm{d}y_n}{\mathrm{d}x}\right] + q(x)y_n = \lambda_n \rho(x) y_n$$

Multiplying both sides by $y_n^*(x)$ and integrating, we have

$$\begin{aligned}
&\lambda_n \int_a^b \rho(x)|y_n|^2 \mathrm{d}x \\
&= -\int_a^b y_n^*(x)\frac{\mathrm{d}}{\mathrm{d}x}\left[p(x)\frac{\mathrm{d}y_n}{\mathrm{d}x}\right]\mathrm{d}x + \int_a^b q(x)|y_n|^2\mathrm{d}x \\
&= -\left[p(x)y_n^*(x)\frac{\mathrm{d}y_n}{\mathrm{d}x}\right]_a^b + \int_a^b p(x)\left|\frac{\mathrm{d}y_n}{\mathrm{d}x}\right|^2\mathrm{d}x + \int_a^b q(x)|y_n|^2\mathrm{d}x \\
&= \int_a^b p(x)\left|\frac{\mathrm{d}y_n}{\mathrm{d}x}\right|^2\mathrm{d}x + \int_a^b q(x)|y_n|^2\mathrm{d}x
\end{aligned}$$

Due to homogeneous boundary conditions or the natural boundary condition, the integrated terms on the second line of the above equation are zero. Moreover, since $p(x), q(x)$, and $\rho(x) \geq 0$, the remaining integral terms are positive, leading to $\lambda_n \geq 0$.

(3) *Orthogonality of eigenfunctions*

The eigenfunctions $y_m(x)$ and $y_n(x)$ corresponding to different eigenvalues λ_m and λ_n are orthogonal on the interval $[a, b]$ with the weight $\rho(x)$, i.e.,

$$\langle y_m|y_n\rangle \equiv \int_a^b y_n(x)y_m^*(x)\rho(x)dx = 0 \quad (m \neq n)$$

In fact, the orthogonality of eigenfunctions is guaranteed by the self-adjointness condition of the Sturm–Liouville equation.

(4) *Non-degenerate eigenvalues*

We use a proof by contradiction to show that the eigenvalues are non-degenerate. Let's assume that there are two eigenfunctions, $y_1(x)$ and $y_2(x)$, corresponding to the eigenvalue λ_n:

$$-\frac{\mathrm{d}}{\mathrm{d}x}\left[p(x)\frac{\mathrm{d}y_1}{\mathrm{d}x}\right]+q(x)y_1=\lambda_n\rho(x)y_1$$

$$-\frac{\mathrm{d}}{\mathrm{d}x}\left[p(x)\frac{\mathrm{d}y_2}{\mathrm{d}x}\right]+q(x)y_2=\lambda_n\rho(x)y_2$$

Multiply the first equation by $y_2(x)$ and the second equation by $y_1(x)$, and then subtract the two equations:

$$y_2\frac{\mathrm{d}}{\mathrm{d}x}\left[p(x)\frac{\mathrm{d}y_1}{\mathrm{d}x}\right]-y_1\frac{\mathrm{d}}{\mathrm{d}x}\left[p(x)\frac{\mathrm{d}y_2}{\mathrm{d}x}\right]=0$$

This simplifies to

$$p(x)(y_2y_1'-y_1y_2')'+p'(x)(y_2y_1'-y_1y_2')=0$$

Integrating both sides gives

$$p(x)(y_2y_1'-y_1y_2')=C$$

Based on the homogeneous boundary conditions or natural boundary conditions, we know that $C=0$. Also, since $p(x)\neq 0$ for $x\in(a,b)$, the Wronskian determinant gives

$$W[y_1(x),y_2(x)]=y_2y_1'-y_1y_2'\equiv 0$$

This implies that $y_1(x)$ and $y_2(x)$ are linearly dependent.

(5) *Oscillation of eigenfunctions*

In general, if $y_1(x)$ and $y_2(x)$ are two linearly independent solutions of a second-order ordinary differential equation, their zeros are distinct and alternate. This means that if $y_1(x)$ has a zero between two consecutive zeros of $y_2(x)$, then $y_2(x)$ must have a zero between two consecutive zeros of $y_1(x)$, and vice versa. This proposition is known as the Sturm separation theorem.

The proof is as follows: Since $y_1(x)$ and $y_2(x)$ are linearly independent, their Wronskian determinant $W[y_1(x),y_2(x)]\neq 0$. Also, since

$y_1(x)$ and $y_2(x)$ are continuous functions, the Wronskian determinant $W[y_1(x), y_2(x)]$ maintains a constant sign. Therefore, $y_1(x)$ and $y_2(x)$ do not share any common zeros. Let us assume x_1 and x_2 are consecutive zeros of $y_2(x)$. Then, $y_2'(x_1)$ and $y_2'(x_2)$ must have opposite signs, and at the two zero points, we have

$$W[y_1(x_k), y_2(x_k)] = y_1(x_k)y_2'(x_k) \quad (k = 1, 2)$$

That is, both $y_1(x_k)$ and $y_2'(x_k)$ are non-zero. Thus, $y_1(x_1)$ and $y_1(x_2)$ must have opposite signs. Consequently, $y_1(x)$ must have a zero within the interval $[x_1, x_2]$. Furthermore, it can be deduced that $y_1(x)$ can have only one zero because if it had multiple zeros, it would contradict the assumption that there is no zero of $y_2(x)$ between two consecutive zeros of $y_1(x)$.

(6) *Completeness of eigenfunctions*

The collection of all eigenfunctions forms a complete set for the Hilbert space.

14.4.2 *Generalized Fourier series*

Due to the weighted orthogonality and completeness of the eigenfunctions of the Sturm–Liouville type equation, if a function $f(x)$ has a continuous first derivative and piecewise continuous second derivative, and satisfies the boundary conditions satisfied by the eigenfunction family, it can be expressed as a linear combination of these eigenfunctions $y_1(x), y_2(x), y_3(x), \ldots$. This expression is called the generalized Fourier series expansion:

$$f(x) = \sum_{n=1}^{\infty} f_n y_n(x) \tag{14.4.1}$$

The eigenfunction family serves as the basis for the series expansion, and the expansion coefficients are given by

$$f_n = \frac{1}{N^2} \int_a^b f(\xi) y_n^*(\xi) \rho(\xi) \mathrm{d}\xi \tag{14.4.2}$$

where the normalization factor is $N_n^2 = \int_a^b |y_n(\xi)|^2 \rho(\xi) \mathrm{d}\xi$. It is common to choose $N_n^2 = 1$, and the corresponding eigenfunctions are called orthogonally normalized eigenfunctions:

$$\int_a^b y_n(x) y_m^*(x) \rho(x) \mathrm{d}x = \delta_{mn}$$

14.4.3 *Several Eigenvalue problems*

(1) *Vibrational equation*

The equation is characterized by the coefficients:

$$p(x) = 1, \quad q(x) = 0, \quad \rho(x) = 1$$

which is

$$\frac{\mathrm{d}^2}{\mathrm{d}x^2} y + \lambda y = 0$$

The equation must satisfy one of the three types of homogeneous boundary conditions at the endpoints $x = a$ and $x = b$.

(2) *Legendre equation*

The equation is characterized by the coefficients

$$p(x) = 1 - x^2, \quad q(x) = 0, \quad \rho(x) = 1$$

which is

$$\frac{\mathrm{d}}{\mathrm{d}x}\left[(1 - x^2)\frac{\mathrm{d}y}{\mathrm{d}x}\right] + \lambda y = 0$$

The equation has natural boundary conditions at the two endpoints, namely, the two zeros of $p(x)$ at $x = \pm 1$.

(3) *Associated Legendre equation*

The equation is characterized by the coefficients:

$$p(x) = 1 - x^2, \quad q(x) = \frac{m^2}{1 - x^2}, \quad \rho(x) = 1$$

which is

$$\frac{\mathrm{d}}{\mathrm{d}x}\left[(1 - x^2)\frac{\mathrm{d}y}{\mathrm{d}x}\right] + \left(\lambda - \frac{m^2}{1 - x^2}\right) y = 0$$

The equation also has natural boundary conditions at the two endpoints $x = \pm 1$.

(4) *Bessel equation*

The equation is characterized by the coefficients

$$p(x) = x, \quad q(x) = \frac{m^2}{x}, \quad \rho(x) = x$$

which is

$$\frac{\mathrm{d}}{\mathrm{d}x}\left[x\frac{\mathrm{d}y}{\mathrm{d}x}\right] + \left(\lambda x - \frac{m^2}{x}\right)y = 0$$

The equation has a natural boundary condition at the zero of $p(x)$ at $x = 0$ and requires a homogeneous boundary condition at the endpoint $x = x_0$. It should be noted that $x = \infty$ is an irregular singular point, and without the restriction of homogeneous boundary conditions, it cannot form an eigenvalue problem.

(5) *Hermite equation*

The equation is characterized by the coefficients:

$$p(x) = \mathrm{e}^{-x^2}, \quad q(x) = 0, \quad \rho(x) = \mathrm{e}^{-x^2}$$

which is

$$\frac{\mathrm{d}}{\mathrm{d}x}\left[\mathrm{e}^{-x^2}\frac{\mathrm{d}y}{\mathrm{d}x}\right] + \lambda \mathrm{e}^{-x^2}y = 0$$

or

$$y'' - 2xy' + \lambda y = 0$$

The equation has natural boundary conditions at the two zeros of $p(x)$ at $x = \pm\infty$. This equation is related to the harmonic oscillator motion in quantum mechanics.

(6) *Laguerre equation*

The equation is characterized by the coefficients:

$$p(x) = x\mathrm{e}^{-x}, \quad q(x) = 0, \quad \rho(x) = \mathrm{e}^{-x}$$

which is

$$\frac{\mathrm{d}}{\mathrm{d}x}\left[x\mathrm{e}^{-x}\frac{\mathrm{d}y}{\mathrm{d}x}\right] + \lambda \mathrm{e}^{-x}y = 0$$

or

$$xy'' + (1 - x)y' + \lambda y = 0$$

The equation has natural boundary conditions at the two zeros of $p(x)$ at $x = 0$ and $x = \infty$. This equation is related to the motion of electrons in the hydrogen atom and the motion of electrons in a magnetic field.

The Legendre equation introduced Chapter 13 is an example of Sturm–Liouville eigenvalue problems. Chapter 15 will further discuss several other Sturm–Liouville type equations and their applications in physics.

ENDNOTE

In classical mechanics, the state of a particle is fully described by its position and momentum (q, p). Such a pair of physical quantities is called canonical conjugate variables, which satisfy the Hamiltonian equations of motion:

$$\dot{q}_j = \frac{\partial H}{\partial p_j}, \quad \dot{p}_j = -\frac{\partial H}{\partial q_j}$$

where $j = 1, 2, \ldots, n$ represents the independent coordinates in an n-dimensional space. For an N-particle system, the phase space is unfolded into a $2n\,N$-dimensional space with independent variables (q, p). A point in the phase space represents a complete state of the particle system, and the time evolution of the particle system corresponds to a trajectory in the phase space. For an isolated system, this trajectory is constrained to lie on an energy surface in the phase space, given by $H(q, p) \equiv E$.

The situation is different in quantum mechanics, where position is not suitable for describing the state of microscopic particles. We imagine that there still exists a state vector $|\psi\rangle$ that fully describes all physical properties of the particle. In this context, several questions need to be addressed:

(1) In what kind of space does the state vector $|\psi\rangle$ reside?
(2) What are the fundamental characteristics of the state vector $|\psi\rangle$?

(3) How does the state vector $|\psi\rangle$ evolve in this space over time?

In contrast to the phase space of classical mechanics, the set of all possible states of a quantum system forms a complete complex inner product space, known as a Hilbert space, as named by John von Neumann. The state vector $|\psi\rangle$ is a vector in the Hilbert space, satisfying the principle of linear superposition and the normalization condition:

$$\langle\psi|\psi\rangle \equiv 1$$

It is analogized to the motion of a classical particle on an energy surface in the phase space. All observables are considered as operators that perform linear transformations on the quantum state, while preserving the state within the original Hilbert space. This requires all observables to be self-adjoint operators.

To obtain the dynamical equation for the time evolution of the state vector $|\psi(t)\rangle$, the evolution operator $\hat{U}(t,t_0)$ is introduced:

$$|\psi(t)\rangle = \hat{U}(t,t_0)|\psi(t_0)\rangle$$

where $|\psi(t_0)\rangle$ is the state vector at time t_0. Since $\langle\psi(t)|\psi(t)\rangle = \langle\psi(t_0)|\psi(t_0)\rangle = 1$, it follows that

$$\hat{U}^\dagger(t,t_0)\hat{U}(t,t_0) = 1$$

Therefore, the evolution operator $\hat{U}(t,t_0)$ must be a unitary operator and can be expressed as $\hat{U}(t,t_0) = \mathrm{e}^{\mathrm{i}\hat{\Lambda}(t,t_0)}$, where $\hat{\Lambda}=\hat{\Lambda}^\dagger$ is a Hermitian operator.

Now, let us consider a system with time translation invariance. What form should the state vector $|\psi(t)\rangle$ have? If the system is in an energy eigenstate, since nothing has changed formally, a likely scenario is

$$|\psi(t)\rangle \sim \mathrm{e}^{-\mathrm{i}\alpha(t-t_0)}|\psi(t_0)\rangle$$

where α is a constant. In classical mechanics, for systems with time translation invariance, energy conservation is guaranteed. Therefore, the constant α should be related to the energy E of the system. Considering the principle of linear superposition for state vectors (different states may have different energies), it is reasonable to guess that the general formula for the evolution of the state is

$$|\psi(t)\rangle = \mathrm{e}^{-\frac{\mathrm{i}}{\hbar}\hat{H}(t-t_0)}|\psi(t_0)\rangle$$

where the Planck constant $\hbar$ introduced to eliminate dimensional inconsistency. This constant comes out so timely! Without it, the assumption of unitarity in the time evolution of states under time translation would not hold. If the Hamiltonian is time-dependent, the general equation for state evolution should be

$$|\psi(t)\rangle = \mathrm{e}^{-\frac{\mathrm{i}}{\hbar}\int_{t_0}^{t}\hat{H}(t')\mathrm{d}t'}|\psi(t_0)\rangle$$

If the above guess is correct, it is easy to derive the dynamical equation satisfied by the state $|\psi(t)\rangle$:

$$\mathrm{i}\hbar\frac{\mathrm{d}}{\mathrm{d}t}|\psi(t)\rangle = \hat{H}|\psi(t)\rangle$$

This is the Schrödinger equation in the Hilbert space. At this point, we still don't know the specific form of the Hamiltonian $\hat{H}$ and other mechanical operators. It is worth noting that this evolution equation is universal. It applies to systems such as central force fields with classical correspondences, as well as to spin systems without classical correspondences. Moreover, it is equally applicable to non-relativistic and relativistic cases. Since quantum states depend on time only through first-order derivatives, it implies that the energy-momentum relation in relativity should be linearized. From the later development of Dirac's relativistic quantum theory, it is indeed the case.

To obtain the evolution equation in a specific representation, let us assume $\{|\psi_k\rangle\}_{k=1}^N$ is a countable orthonormal complete basis in the Hilbert space, satisfying

$$\langle\psi_i|\psi_j\rangle = \delta_{ij}, \quad \sum_{k=1}^{N}|\psi_k\rangle\langle\psi_k| = 1$$

The state vector $|\psi(t)\rangle$ can be expressed as a linear combination of this complete basis as

$$|\psi(t)\rangle = \sum_{k=1}^{N} c_k(t)|\psi_k\rangle$$

The matrix representation of the Hamiltonian operator is

$$[H]_{ij} \stackrel{\text{def}}{=} \langle\psi_i|\hat{H}|\psi_j\rangle \equiv h_{ij}$$

The state evolution equation is

$$\begin{aligned} \mathrm{i}\hbar\frac{\mathrm{d}}{\mathrm{d}t}\sum_{i=1}^{N} c_i(t)|\psi_i\rangle &= \hat{H}\sum_{i=1}^{N} c_i(t)|\psi_i\rangle \\ &= \sum_{i=1}^{N} c_i(t)\sum_{k=1}^{N}|\psi_k\rangle\langle\psi_k|\hat{H}|\psi_i\rangle \end{aligned}$$

Then we obtain the evolution equation for the coefficients:

$$\mathrm{i}\hbar\frac{\mathrm{d}}{\mathrm{d}t}c_j(t) = \sum_{i=1}^{N} h_{ji}c_i(t)$$

Next, let us discuss the theory of representations and representation transforms. Taking the one-dimensional case as an example, the eigen equation of the position operator $\hat{x}$ is

$$\hat{x}|x\rangle = x|x\rangle$$

The eigenvalues are the positions x, and $|x\rangle$ is the corresponding eigenvector. The set $\{|x\rangle\}_{-\infty}^{\infty}$ forms an orthonormal complete set in the Hilbert space. Since the position coordinate is continuous, we have

$$\langle x'|x\rangle = \delta(x - x'), \quad \int_{-\infty}^{\infty} |x\rangle\langle x|\mathrm{d}x = 1$$

The space represented by the position basis vectors is called the coordinate representation. Quantum states in the coordinate representation are represented as

$$|\psi(t)\rangle = \int_{-\infty}^{\infty} |x\rangle\langle x|\psi(t)\rangle\mathrm{d}x = \int_{-\infty}^{\infty} \psi(x,t)|x\rangle\mathrm{d}x$$

The projection $\psi(x,t)$ is the familiar wave function. From the normalization condition $\langle\psi(t)|\psi(t)\rangle = 1$, we have

$$\int_{-\infty}^{\infty} \langle\psi(t)|x\rangle\langle x|\psi(t)\rangle\mathrm{d}x = \int_{-\infty}^{\infty} |\psi(x,t)|^2\mathrm{d}x = 1$$

The Copenhagen interpretation interprets $|\psi(x,t)|^2$ as the probability density of finding the particle at position x.

If there exists another operator $\hat{p}$ with a continuous spectrum, and its eigenvectors are represented by $|p\rangle$, satisfying

$$\hat{p}|p\rangle = p|p\rangle$$

the set of vectors $\{|p\rangle\}_{-\infty}^{\infty}$ also forms an orthogonal and complete set, i.e.,

$$\langle p'|p\rangle = \delta(p - p'), \quad \int_{-\infty}^{\infty} |p\rangle\langle p|\mathrm{d}p = 1$$

Using the properties of the Dirac δ function, we have

$$\langle x'|x\rangle = \int_{-\infty}^{\infty} \langle x'|p\rangle\langle p|x\rangle\mathrm{d}p = \delta(x - x') = \frac{1}{2\pi}\int_{-\infty}^{\infty} \mathrm{e}^{-\mathrm{i}p(x-x')}\mathrm{d}p$$

$$\langle p'|p\rangle = \int_{-\infty}^{\infty} \langle p'|x\rangle\langle x|p\rangle\mathrm{d}x = \delta(p - p') = \frac{1}{2\pi}\int_{-\infty}^{\infty} \mathrm{e}^{\mathrm{i}(p-p')x}\mathrm{d}x$$

We have

$$\langle p|x\rangle = \frac{1}{\sqrt{2\pi}}\mathrm{e}^{-\mathrm{i}px}, \quad \langle x|p\rangle = \frac{1}{\sqrt{2\pi}}\mathrm{e}^{\mathrm{i}px}$$

If p is taken as the momentum, then px has the dimension of action $[\mathrm{J}\cdot\mathrm{s}]$. We see that the Planck constant again appear timely. The phase factor should be dimensionless so it is proper to divide the Planck constant $\hbar$, i.e.,

$$\delta(x-x') = \frac{1}{2\pi\hbar}\int_{-\infty}^{\infty} \mathrm{e}^{-\frac{\mathrm{i}p(x-x')}{\hbar}}\,\mathrm{d}p$$

and

$$\langle p|x\rangle = \frac{1}{\sqrt{2\pi\hbar}}\mathrm{e}^{-\frac{\mathrm{i}px}{\hbar}}$$

The inner product $\langle p|x\rangle$ gives the relationship between the two sets of basis vectors $\{|x\rangle\}_{-\infty}^{\infty}$ and $\{|p\rangle\}_{-\infty}^{\infty}$, known as representation transform. It can be interpreted as the projection of the momentum eigenvector $|p\rangle$ onto the position representation, or equivalently, the projection of the position eigenvector $|x\rangle$ onto the momentum representation. These projections form a unitary matrix. The derivation above shows that without the presence of the Planck constant $\hbar$, there would be no continuous momentum spectrum, and hence no representation transform. The matrix elements of the representation transform have an equivalent and symmetric exponential form, which has significant implications. For example, the wave function in the momentum representation is given by

$$\begin{aligned}\varphi(p,t) &\overset{\text{def}}{=} \langle p|\psi(t)\rangle = \int_{-\infty}^{\infty} \langle p|x\rangle\langle x|\psi(t)\rangle\mathrm{d}x \\ &= \frac{1}{\sqrt{2\pi\hbar}}\int_{-\infty}^{\infty} \psi(x,t)\mathrm{e}^{-\frac{\mathrm{i}px}{\hbar}}\,\mathrm{d}x \equiv \mathfrak{F}[\psi(x,t)]\end{aligned}$$

This shows the complementary relationship between the wave function in the position representation $\psi(x,t)$ and the wave function in the momentum representation $\varphi(p,t)$ as a Fourier transform pair. According to the scaling property of the Fourier transform, the spread of the distribution in position space is inversely proportional to the spread of the distribution in momentum space, i.e.,

$$\Delta x \Delta p \sim \hbar$$

Classical particles have only one position, while quantum particles have multiple positions simultaneously. Similarly, classical particles have only one momentum, while quantum particles have multiple momenta simultaneously. The uncertainty principle, ever considered as the core of quantum mechanics, is simply a consequence of the representation transform. It is more properly referred to as the "uncertainty relation", as it has no inherent logical connection to the dynamics of quantum states. Even relativistic particles moving at speeds close to the speed of light satisfy the same relation.

From this, it can be seen that the Planck constant $\hbar$ can be regarded as an invariant of the representation transformation in quantum systems, similar to how the speed of light c is regarded as an invariant of Lorentz transformations in classical mechanics, and the charge e is regarded as an invariant of gauge transformations in electromagnetic systems. This discovery reveals that the three fundamental constants of physics are related to the invariance of specific transformations. If this judgment holds true, then the efforts to find magnetic monopoles may likely be in vain because, in existing electromagnetic theories, there is no additional transformation invariance. Perhaps it is the symmetry that excludes fundamental physical constants such as magnetic charge.

Applying the momentum operator $\hat{p}$ to a position basis vector, we have

$$\begin{aligned}\langle x'|\hat{p}|x\rangle &= \int_{-\infty}^{\infty} \langle x'|\hat{p}|p\rangle\langle p|x\rangle \mathrm{d}p \\ &= \int_{-\infty}^{\infty} \frac{1}{\sqrt{2\pi\hbar}}\mathrm{e}^{\frac{ipx'}{\hbar}}\mathrm{d}p\frac{1}{\sqrt{2\pi\hbar}}\mathrm{e}^{-\frac{ipx}{\hbar}}\mathrm{d}p\end{aligned}$$

$$= \frac{1}{2\pi\hbar}\int_{-\infty}^{\infty}\left(-\mathrm{i}\hbar\frac{\partial}{\partial x'}\right)\mathrm{e}^{-\frac{\mathrm{i}p(x-x')}{\hbar}}\mathrm{d}p$$

$$= -\mathrm{i}\hbar\frac{\partial}{\partial x'}\int_{-\infty}^{\infty}\langle x'|p\rangle\langle p|x\rangle\mathrm{d}p = -\mathrm{i}\hbar\frac{\partial}{\partial x'}\langle x'|x\rangle$$

Therefore, the momentum operator in the position representation is given by $\hat{p} = -\mathrm{i}\hbar\frac{\partial}{\partial x}$. Similarly, one can derive that the position operator $\hat{x}$ in the momentum representation is given by $\hat{x} = \mathrm{i}\hbar\frac{\partial}{\partial p}$. Since

$$\langle x'|\hat{p}\hat{x}|x\rangle = \left\langle x'\left|-\mathrm{i}\hbar\frac{\partial}{\partial x}x\right|x\right\rangle = -\mathrm{i}\hbar\langle x'|x\rangle + x\langle x'|\hat{p}|x\rangle$$

and

$$\langle x'|\hat{x}\hat{p}|x\rangle = x'\langle x'|\hat{p}|x\rangle$$

it follows that

$$\langle x'|\hat{x}\hat{p}-\hat{p}\hat{x}|x\rangle = \mathrm{i}\hbar\langle x'|x\rangle$$

which indicates that the Poisson bracket of the operators $\hat{x}$ and $\hat{p}$ has a constant eigenvalue. Therefore, they satisfy the canonical commutation relation

$$[\hat{x},\hat{p}] = \mathrm{i}\hbar$$

This relation serves as the starting point of the Heisenberg–Dirac quantum mechanics theory.

In the position representation, the Hamiltonian operator is given by

$$\hat{H}(\hat{x},\hat{p}) = \frac{\hat{p}^2}{2m} + V(\hat{x}) = -\frac{\hbar^2}{2m}\frac{\partial^2}{\partial x^2} + V(x)$$

The evolution equation for a state vector is represented as

$$\mathrm{i}\hbar\frac{\mathrm{d}}{\mathrm{d}t}|\psi(t)\rangle = \hat{H}|\psi(t)\rangle$$

Which can be written as

$$i\hbar\frac{d}{dt}\langle x|\psi(t)\rangle = \int_{-\infty}^{\infty}\langle x|\hat{H}|x'\rangle\langle x'|\psi(t)\rangle dx'$$

Leading to

$$i\hbar\frac{\partial}{\partial t}\psi(x,t) = H\psi(x,t)$$

This is the Schrödinger wave equation in the position representation.

In quantum mechanics, all physical observables are represented by Hermitian operators in the Hilbert space, $\hat{A} = \hat{A}^{\dagger}$. The measured value of a physical quantity corresponds to the average value of the corresponding operator in a quantum state, given by

$$\begin{aligned}\langle A\rangle \stackrel{\text{def}}{=} \langle\psi(t)|\hat{A}|\psi(t)\rangle &= \int_{-\infty}^{\infty}\langle\psi(t)|x\rangle\langle x|\hat{A}|x'\rangle\langle x'|\psi(t)\rangle dx dx' \\ &= \int_{-\infty}^{\infty}\psi^*(x,t)A\psi(x,t)dx\end{aligned}$$

Thus, we have derived the basic framework of quantum mechanics. Now, the remaining task is to verify the validity of the Schrödinger equation. Logically, it is required that the results obtained from it smoothly transition to classical mechanics in the limit $\hbar \to$. In classical mechanics, the evolution of a dynamical quantity with time satisfies the Hamiltonian equations of motion:

$$\frac{d}{dt}A = \frac{\partial A}{\partial q}\frac{\partial H}{\partial p} - \frac{\partial A}{\partial p}\frac{\partial H}{\partial q} \equiv \{A, H\}$$

where $\{\cdot,\cdot\}$ represents the classical Poisson bracket. The equation for the time evolution of the expectation value of a quantum mechanical quantity is

$$\frac{\mathrm{d}}{\mathrm{d}t}\langle A\rangle = \frac{\mathrm{d}}{\mathrm{d}t}\langle\psi(t)|\hat{A}|\psi(t)\rangle = \left(\frac{\mathrm{d}}{\mathrm{d}t}\langle\psi(t)|\right)\hat{A}|\psi(t)\rangle + \langle\psi(t)|\hat{A}\left(\frac{\mathrm{d}}{\mathrm{d}t}|\psi(t)\rangle\right) = \frac{1}{\mathrm{i}\hbar}\overline{[\hat{A},\hat{H}]}$$

Using the correspondence between Dirac's classical-quantum Poisson brackets:

$$\lim_{\hbar\to 0}\frac{1}{\mathrm{i}\hbar}[\hat{A},\hat{B}] \longleftrightarrow \{A,B\}$$

we obtain

$$\frac{\mathrm{d}}{\mathrm{d}t}\langle A\rangle = \frac{1}{\mathrm{i}\hbar}\overline{[\hat{A},\hat{H}]} \overset{\hbar\to 0}{\longrightarrow} \{A,H\}$$

Thus, the quantum equation of motion for a dynamical quantity and the classical equation of motion are logically consistent.

Finally, we give a coincide summary of the quantum mechanics theory:

(1) Particle's state space: Quantum mechanics uses Hilbert space to describe the state of a particle. State vectors (or wavefunctions) completely describe the particle's state, and these state vectors.
(2) There exists a action quanta, the Planck constant $\hbar$, with which the evolution of the quantum state is governed by

$$|\psi(t)\rangle = \mathrm{e}^{-\frac{\mathrm{i}}{\hbar}\hat{H}(t-t_0)}|\psi(t_0)\rangle$$

where $\hat{H}$ is the Hamiltonian describing the motion of the particle.

Exercises

[1] Prove the following differential equation constitutes an eigenvalue problem:

$$xy'' + (1 - x)y' + \lambda y = 0$$

where $y(x)$ is finite on $[0, \infty)$.

[2] Explain why the following differential equation cannot constitute an eigenvalue problem:

$$\frac{\mathrm{d}}{\mathrm{d}x}\left[x\frac{\mathrm{d}y}{\mathrm{d}x}\right] - \frac{\nu^2}{x}y - \lambda xy = 0$$

[3] Solve the non-homogeneous differential equation:

$$\begin{cases} xy'' + 2y' + 2xy = x \\ y(1) = 0, \quad y(0) \text{ is finite} \end{cases}$$

Hint: Transform $xy'' + 2y' + \lambda xy = 0$ into Sturm–Liouville eigenvalue equation form, find its eigenfunctions and eigenvalues, and expand both sides of the equation using the eigenfunctions.

Answer:

$$y(x) = \sum_{n=1}^{\infty} b_n y_n(x) = \sum_{n=1}^{\infty} \frac{\sqrt{2}(-1)^{n+1}}{n\pi(2 - n^2\pi^2)} \frac{\sqrt{2}\sin n\pi x}{x}$$

[4] The Legendre operator is

$$\hat{L} = -\frac{\mathrm{d}}{\mathrm{d}x}\left[(1 - x^2)\frac{\mathrm{d}}{\mathrm{d}x}\right]$$

Solve the non-homogeneous differential equations:

$$\text{(a) } \hat{L}y - \frac{3}{2}y = x; \quad \text{(b) } \hat{L}y - 2y = x$$

Hint: Calculate $(\hat{L} - 2)[x \ln(1 \pm x)]$ to find particular solutions of the equations.

Answer:

$$\text{(a) } y(x) = 2P_1(x); \quad \text{(b) } y(x) = aP_1(x) + bQ_1(x) + \frac{x}{6}\ln(1 - x^2)$$

[5] To understand the correspondence between quantum mechanics and classical mechanics, consider using Poisson brackets:

(a) Using the quantum operator relation $[\hat{x}, \hat{p}] = i\hbar$, prove $[\hat{x}^m, \hat{p}] = i\hbar m\hat{x}^{m-1}$.

(b) Prove: $[\hat{x}^m, \hat{p}^n] = i\hbar mn\hat{x}^{m-1}\hat{p}^{n-1} + \mathcal{O}(\hbar^2)$.

(c) Show that the quantum and classical Poisson brackets satisfy the relationship:

$$\lim_{h\to 0} \frac{1}{\mathrm{i}\hbar}[\hat{x}^m, \hat{p}^n] = mn\hat{x}^{m-1}\hat{p}^{n-1} \longrightarrow \{x^m, p^n\}$$

In particular,

$$[\hat{x}, \hat{p}] = \mathrm{i}\hbar \longrightarrow \{x, p\} = 1$$

(d) Any observable can be expanded using operators $(\hat{x}, \hat{p})$:

$$\hat{A} = \sum_{kl} a_{kl}\hat{x}^k\hat{p}^l, \quad \hat{B} = \sum_{mn} b_{mn}\hat{x}^m\hat{p}^n$$

Prove that

$$\lim_{h\to 0} \frac{1}{\mathrm{i}\hbar}[\hat{A}, \hat{B}] \longleftrightarrow \{A, B\}$$

[6] The expectation value of an operator $\hat{A}$ in a normalized quantum state $|\psi\rangle$ is given by $\bar{A} = \langle\psi|\hat{A}|\psi\rangle$, and the uncertainty in the measurement is defined as

$$\Delta A \stackrel{\text{def}}{=} \sqrt{\langle\psi|(\hat{A} - \langle\hat{A}\rangle)^2|\psi\rangle}$$

(a) Prove that for any two Hermitian operators $\hat{A}$ and $\hat{B}$,

$$|\langle\psi|\hat{A}\hat{B}|\psi\rangle|^2 \leq \langle\psi|\hat{A}^2|\psi\rangle\langle\psi|\hat{B}^2|\psi\rangle$$

(b) Prove that for the Poisson bracket,

$$|\langle\psi|[\hat{A}, \hat{B}]|\psi\rangle|^2 \leq 4\langle\psi|\hat{A}^2|\psi\rangle\langle\psi|\hat{B}^2|\psi\rangle$$

(c) Prove that

$$(\Delta A)(\Delta B) \geq \frac{1}{2}|\langle\psi|[\hat{A}, \hat{B}]|\psi\rangle|$$

(d) Take $\hat{A} = \hat{x}$ and $\hat{B} = \hat{p}$, satisfying the commutation relation $[\hat{x}, \hat{p}] = \mathrm{i}\hbar$, and prove the Heisenberg uncertainty relation:

$$(\Delta x)(\Delta p) \geq \frac{1}{2}\hbar$$

14.5 Classical Orthogonal Polynomials

We know that Legendre polynomials have many special properties, such as orthogonality, completeness, as well as Rodrigues' formula and generating functions. However, these characteristics are not exclusive to Legendre polynomials. There is a large class of polynomials that also possess such properties. Most of them are eigenfunctions of specific ordinary differential equations, some of which will be discussed in Chapter 15. In this section, we provide a rough description of their properties and types.

14.5.1 *Rodrigues' formula*

Earlier, we mentioned Stone–Weierstrass theorem, which states that the monomial sequence $\{x^k\}_{k=0}^{\infty}$ forms a complete basis for $\mathcal{L}_\rho^2(a,b)$. This theorem indicates that the monomial sequence $\{x^k\}_{k=0}^{\infty}$ can be considered as a linearly independent but not orthogonal basis. If we want to obtain orthogonal bases, it is necessary to recombine them into polynomials. The orthogonal and complete bases formed by these polynomials are called classical orthogonal polynomials. The Gram–Schmidt process is commonly used for orthogonalization, but we decide to take a different approach and use a more enlightening orthogonalization scheme for Sturm–Liouville type equations.

Consider the second-order Sturm–Liouville eigenvalue equation:

$$s(x)y'' + t(x)y' + \lambda y = 0 \tag{14.5.1}$$

It can be easily shown that the self-adjointness of the equation requires $s'(x) = t(x)$. We restrict $s(x)$ and $t(x)$ to be polynomials no higher than quadratic and linear, respectively,

$$s(x) = \alpha x^2 + \beta x + \gamma, \quad t(x) = \mu x + \nu$$

This ensures that both endpoints are regular singular points, i.e., $x = a$ and $x = b$ are first-order zeros of $s(x)$, $s(a) = s(b) = 0$. Due to the constraints of natural boundary conditions, the infinite series solution of the equation is truncated to an nth-degree polynomial, denoted as

$$y_n(x) = \sum_{k=0}^{n} a_k x^k$$

where $a_n \neq 0$. Substituting this into the differential equation (14.5.1), we obtain the coefficients of x^n, which satisfy

$$n(n-1)\alpha a_n + n\mu a_n + \lambda a_n = 0$$

Solving this equation yields the eigenvalues:

$$\lambda_n = -n(n-1)\alpha - n\mu \tag{14.5.2}$$

If the second-order differential equation is not self-adjoint, i.e., $s'(x) \neq t(x)$, then following the procedure in Section 14.3, we can transform it into a Sturm–Liouville type equation by introducing a weight function $\rho(x)$ and letting $(\rho s)' = \rho t$. We have

$$\rho(x) = \frac{1}{s(x)} \exp\left[\int^{(x)} \frac{t(x')}{s(x')} \mathrm{d}x'\right]$$

Which yields

$$\frac{\mathrm{d}}{\mathrm{d}x}[\rho(x)s(x)y'] + \lambda\rho(x)y = 0$$

Thus, the natural boundary conditions become

$$\rho(a)s(a) = \rho(b)s(b) = 0 \tag{14.5.3}$$

For certain specific weight functions $\rho(x)$, the eigenfunctions of the equation can be expressed using a generalized Rodrigues' formula:

$$y_n(x) = \frac{1}{\rho(x)} \frac{\mathrm{d}^n}{\mathrm{d}x^n}[\rho(x)s^n(x)] \tag{14.5.4}$$

Substituting $\rho(x)$ into the equation can prove this formula. By taking $s(x)$ as a constant, a linear function, and a quadratic function, we can verify one by one that $y_n(x)$ is an nth-degree polynomial.

By performing integration by parts and noting that

$$\rho(a)s(a) = \rho(b)s(b) = 0$$

it can be shown that $y_n(x)$ is orthogonal to monomials $x^k (k < n)$ with the weight $\rho(x)$:

$$\int_a^b x^k y_n(x)\rho(x)\mathrm{d}x = 0 \quad (k < n)$$

From this, it can be proven that all polynomials $y_n(x)$ are mutually orthogonal:

$$\int_a^b y_m(x)y_n(x)\rho(x)\mathrm{d}x = N_n^2\delta_{mn} \tag{14.5.5}$$

14.5.2 *Classification of orthogonal polynomials*

Based on the different forms of $s(x)$, orthogonal polynomials can be classified into three cases:

(1) *Constant* $s(x)$

Taking $s(x) = 1$ and $t(x) = -2x$, the Sturm–Liouville equation is transformed into the Hermite equation:

$$y'' - 2xy' + \lambda y = 0$$

The weight function is found to be

$$\rho(x) = \exp\left[\int^{(x)} (-2x')\mathrm{d}x'\right] = \mathrm{e}^{-x^2}$$

he natural boundary is $(-\infty, \infty)$. The eigenvalues are $\lambda_n = 2n$, and the eigenfunctions are Hermite polynomials:

$$\mathrm{H}_n(x) = \frac{(-1)^n}{\rho(x)}\frac{\mathrm{d}^n}{\mathrm{d}x^n}[\rho(x)s^n(x)] = (-1)^n\mathrm{e}^{x^2}\frac{\mathrm{d}^n}{\mathrm{d}x^n}\mathrm{e}^{-x^2}$$

Here are a few low-order Hermite polynomials, and their function curves are shown in Figure 14.1:

$$\begin{aligned}
\mathrm{H}_0(x) &= 1\\
\mathrm{H}_1(x) &= 2x\\
\mathrm{H}_2(x) &= 4x^2 - 2\\
\mathrm{H}_3(x) &= 8x^3 - 12x\\
\mathrm{H}_4(x) &= 16x^4 - 48x^2 + 12\\
\mathrm{H}_5(x) &= 32x^5 - 160x^3 + 120x\\
&\cdots\cdots
\end{aligned}$$

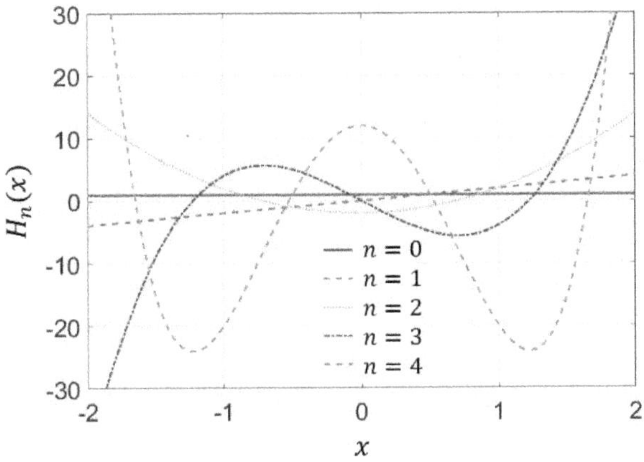

Figure 14.1.

(2) *Linear* $s(x)$

Taking $s(x) = x$ and $t(x) = 1 - x$, the Sturm–Liouville equation is transformed into the Laguerre equation:

$$p(x) = x\mathrm{e}^{-x}, \quad q(x) = 0, \quad \rho(x) = \mathrm{e}^{-x}$$

$$\frac{\mathrm{d}}{\mathrm{d}x}\left[x\mathrm{e}^{-x}\frac{\mathrm{d}y}{\mathrm{d}x}\right] + \lambda \mathrm{e}^{-x} y = 0$$

$$\longrightarrow xy'' + (1-x)y' + \lambda y = 0$$

The weight function is $\rho(x) = \mathrm{e}^{-x}$, and the natural boundary is $[0, \infty)$. The eigenvalues are $\lambda_n = n$, and the eigenfunctions are Laguerre polynomials:

$$\mathrm{L}_n(x) = \frac{1}{\rho(x)}\frac{\mathrm{d}^n}{\mathrm{d}x^n}[\rho(x)s^n(x)] = \mathrm{e}^x \frac{\mathrm{d}^n}{\mathrm{d}x^n}[x^n \mathrm{e}^{-x}]$$

Here are a few low-order Laguerre polynomials:

$$\begin{aligned}
\mathrm{L}_0(x) &= 1\\
\mathrm{L}_1(x) &= -x + 1\\
\mathrm{L}_2(x) &= x^2 - 4x + 2\\
\mathrm{L}_3(x) &= -x^3 + 9x^2 - 18x + 6\\
\mathrm{L}_4(x) &= x^4 - 16x^3 + 72x^2 - 96x + 24\\
\mathrm{L}_5(x) &= -x^5 + 25x^4 - 200x^3 + 600x^2 - 600x + 120\\
&\cdots\cdots
\end{aligned}$$

More generally, taking $s(x) = x$ and $t(x) = k + 1 - x$ ($k \in \mathbb{N}$), we obtain the associated Laguerre equation:

$$x\frac{\mathrm{d}^2 y}{\mathrm{d}x^2} + (k + 1 - x)\frac{\mathrm{d}y}{\mathrm{d}x} + \lambda y(x) = 0$$

The weight function $\rho(x) = x^k \mathrm{e}^{-x}$, and the natural boundary is $[0, \infty)$. The eigenvalues remain $\lambda_n = n$, and the eigenfunctions are associated Laguerre polynomials:

$$\mathrm{L}_n^k(x) = \frac{1}{n! x^k \mathrm{e}^{-x}} \frac{\mathrm{d}^n}{\mathrm{d}x^n}(x^k \mathrm{e}^{-x} x^n)$$

It can be shown that associated Laguerre polynomials are related to Laguerre polynomials as follows:

$$\mathrm{L}_n^k(x) = (-1)^k \frac{\mathrm{d}^k}{\mathrm{d}x^k} \mathrm{L}_{n+k}(x)$$

(3) *Quadratic* $s(x)$

Taking $s(x) = 1 - x^2$ and $t(x) = \mu - \nu - (\mu + \nu + 2)x$, the Sturm–Liouville equation is transformed into the Jacobi equation:

$$(1 - x^2)\frac{\mathrm{d}^2 y}{\mathrm{d}x^2} + [\mu - \nu - (\mu + \nu + 2)x]\frac{\mathrm{d}y}{\mathrm{d}x} + \lambda y(x) = 0$$

The weight function is

$$\begin{aligned} \rho(x) &= \frac{1}{1 - x^2} \exp\left[\int^{(x)} \left(\frac{\mu + 1}{1 + x'} - \frac{\nu + 1}{1 - x'}\right) \mathrm{d}x'\right] \\ &= (1 + x)^\mu (1 - x)^\nu \end{aligned}$$

and the natural boundary is $[-1, 1]$. The eigenvalues are $\lambda_n = n(n + \mu + \nu + 1)$, and the eigenfunctions are Jacobi polynomials:

$$\mathrm{P}_n^{\mu,\nu}(x) = \frac{(-1)^n}{2^n n!}(1 + x)^{-\mu}(1 - x)^{-\nu} \frac{\mathrm{d}^n}{\mathrm{d}x^n}[(1 + x)^{\mu+n}(1 - x)^{\nu+n}]$$

Figure 14.2 illustrates several low-order Jacobi polynomials for the case $\mu = \nu = 3$.

Jacobi polynomials can be further classified based on the values of μ and ν. For example, when $\mu = \nu = 0$, they reduce to Legendre polynomials, whose properties were discussed in detail in Chapter 13. Other classifications are listed in Table 14.1. By following the aforementioned rules, readers can easily derive the differential representations of these polynomials and their corresponding eigenvalues (omitted here).

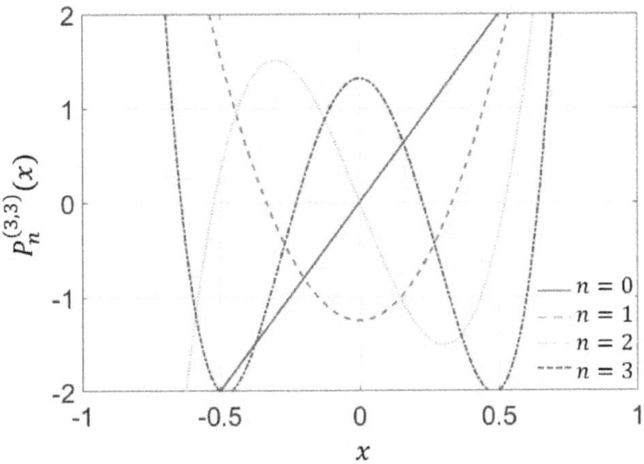

Figure 14.2.

Table 14.1.

μ	ν	$\rho(x)$	多项式
0	0	1	Legendre $\mathrm{P}_n(x)$
$\lambda - \frac{1}{2}$	$\lambda - \frac{1}{2}$	$(1-x^2)^{\lambda-\frac{1}{2}}$	Gegenbauer $\mathrm{C}_n^{\lambda}(x)$
$-\frac{1}{2}$	$-\frac{1}{2}$	$(1-x^2)^{-\frac{1}{2}}$	Chebyshev of the First Kind $\mathrm{T}_n(x)$
$\frac{1}{2}$	$\frac{1}{2}$	$(1-x^2)^{\frac{1}{2}}$	Chebyshev of the Second Kind $\mathrm{U}_n(x)$

14.5.3 *Generating functions*

One of the great advantages of the Rodrigues formula for orthogonal polynomials is that it can be used to obtain integral representations of higher-order derivatives, known as the Schläfli integral formula. By applying the Cauchy integral formula, it can be easily shown that

$$y_n(x) = \frac{1}{\rho(x)} \frac{n!}{2\pi \mathrm{i}} \oint_L \frac{\rho(z) s^n(z)}{(z-x)^{n+1}} \mathrm{d}z \qquad (14.5.6)$$

Here, the integration path L is a smooth curve that contains x, and $\rho(z)s^n(z)$ is an analytic function within the closed region $\bar{B}$ enclosed by the loop. Starting from the Schläfli formula, it can be proven that $y_n(x)$ is a solution to the corresponding second-order differential equation, thus establishing the Rodrigues formula.

Classical orthogonal polynomials can be generated by a generating function of the form:

$$G(x,t) = \sum_n a_n y_n(x) t^n \tag{14.5.7}$$

In principle, the generating function can be derived using the Schläfli integral formula:

$$G(x,t) = \frac{1}{\rho(x)} \sum_n a_n t^n \frac{n!}{2\pi \mathrm{i}} \oint_L \frac{\rho(z) s^n(z)}{(z-x)^{n+1}} \mathrm{d}z \tag{14.5.8}$$

By interchanging the order of summation and integration, first summing over n and then performing the loop integration, different orthogonal polynomials have their own techniques in the derivation process (omitted here).

Orthogonal polynomials $y_n(x)$ can be generated by taking n derivatives of $G(x,t)$ with respect to t. According to the residue theorem, we have

$$a_n y_n(x) = \frac{1}{2\pi \mathrm{i}} \oint_L \frac{G(x,t)}{t^{n+1}} \mathrm{d}t$$

The generating functions $G(x,t)$ for various orthogonal polynomials are listed in Table 14.2.

Table 14.2.

Orthogonal Polynomial	Generating Function	a_n
$\mathrm{P}_n(x)$	$\dfrac{1}{(t^2-2xt+1)^{1/2}}$	1
$\mathrm{H}_n(x)$	e^{-t^2+2xt}	$\dfrac{1}{n!}$
$\mathrm{L}_n^k(x)$	$\dfrac{1}{(1-t)^{k+1}}\mathrm{e}^{-\frac{xt}{1-t}}$	1
$\mathrm{C}_n^\lambda(x)$	$\dfrac{1}{(t^2-2xt+1)^\lambda}$	1
$\mathrm{T}_n(x)$	$\dfrac{1-t^2}{t^2-2xt+1}$	$2(n \neq 0, a_0 = 1)$
$\mathrm{U}_n(x)$	$\dfrac{1}{t^2-2xt+1}$	1

14.5.4 *Recurrence relations*

By taking derivatives of the generating function $G(x,t)$ with respect to x or t, many recurrence relations between different orders of orthogonal polynomials can be obtained. For example, for Hermite polynomials:

$$G(x,t) = \mathrm{e}^{-t^2+2xt} = \sum_{n=0}^{\infty} \frac{1}{n!} \mathrm{H}_n(x) t^n$$

By differentiating with respect to t, we obtain

$$\mathrm{H}_{n+1}(x) = 2x\mathrm{H}_n(x) - 2n\mathrm{H}_{n-1}(x)$$

By differentiating with respect to x, we obtain

$$\mathrm{H}'_n(x) = 2n\mathrm{H}_{n-1}(x)$$

Similar recurrence relations can be derived for other orthogonal polynomials.[1] The derivations are not provided here.

ENDNOTE
Gram-Schmidt orthogonalization procedure, shown in Figure 14.3, is performed as follows:

Given a set of linearly independent basis vectors $\{|a_i\rangle\}_{i=1}^N$, let $|e_1\rangle = \frac{|a_1\rangle}{\sqrt{\langle a_1|a_1\rangle}}$ such that $\langle e_1|e_1\rangle = 1$. Subtract the projection of $|a_2\rangle$ onto $|e_1\rangle$ from $|a_2\rangle$ to obtain a vector orthogonal to $|e_1\rangle$:

$$|e'_2\rangle = |a_2\rangle - \langle e_1|a_2\rangle|e_1\rangle$$

Normalize $|e_2\rangle = \frac{|e'_2\rangle}{\sqrt{\langle e'_2|e'_2\rangle}}$, resulting in $\langle e_1|e_2\rangle = 0$ and $\langle e_2|e_2\rangle = 1$. Similarly, subtract the projections of $|a_3\rangle$ onto both $|e_1\rangle$ and $|e_2\rangle$ to obtain

$$|e'_3\rangle = |a_3\rangle - \langle e_1|a_3\rangle|e_1\rangle - \langle e_2|a_3\rangle|e_2\rangle$$

---

[1]Readers are referred to Zhuxi Wang and Dunren Guo, *Introduction to Special Functions*.

Normalize $|e_3\rangle = \frac{|e_3'\rangle}{\sqrt{\langle e_3'|e_3'\rangle}}$. It can be seen that $|e_3\rangle$ is orthogonal to both $|e_1\rangle$ and $|e_2\rangle$. Repeat this process for all basis vectors to obtain an orthogonal and normalized set of basis vectors.

For example, using the linearly independent monomials $\{y_n\} = \{1, x, x^2, x^3, \ldots\}$ in the interval $x \in [-1, +1]$, we can construct orthogonal polynomials using the following procedure:

$$\psi_0 = y_0 \longrightarrow \phi_0 = \frac{\psi_0}{\sqrt{\langle\psi_0|\psi_0\rangle}} = \sqrt{\frac{1}{2}}$$

$$\psi_1 = y_1 - \phi_0\langle\phi_0|y_1\rangle \longrightarrow \phi_1 = \frac{\psi_1}{\sqrt{\langle\psi_1|\psi_1\rangle}} = \sqrt{\frac{3}{2}}x$$

$$\psi_2 = y_2 - \phi_0\langle\phi_0|y_2\rangle - \phi_1\langle\phi_1|y_2\rangle \longrightarrow \phi_2 = \frac{\psi_2}{\sqrt{\langle\psi_2|\psi_2\rangle}}$$

$$= \frac{1}{2}\sqrt{\frac{5}{2}}(3x^2 - 1)$$

$$\ldots$$

These polynomials are the well-known normalized Legendre polynomials:

$$\phi_l = \sqrt{\frac{2l+1}{2}}\mathrm{P}_l(x)$$

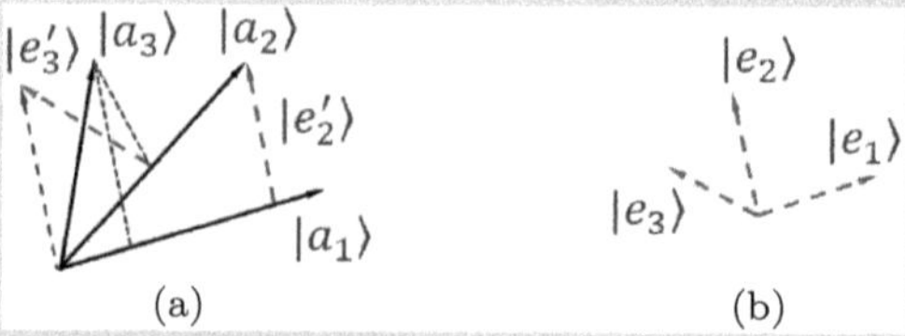

Figure 14.3.

Exercises

[1] Prove that

$$\text{(a)} \int_{-1}^{1} x^m \mathrm{H}_n(x) \mathrm{e}^{-x^2} \mathrm{d}x = 0 \ (m < n);$$

$$\text{(b)} \left(2x - \frac{\mathrm{d}}{\mathrm{d}x}\right)^n 1 = \mathrm{H}_n(x)$$

Hint: For part (b), use mathematical induction.

[2] Prove the orthogonality of polynomials expressed by the Rodrigues formula:

$$\int_a^b y_m(x) y_n(x) \rho(x) \mathrm{d}x = N_n^2 \delta_{mn}$$

[3] Assume that Hermite polynomials are solutions to the Hermite equation:

$$\mathrm{H}_n''(x) - 2x\mathrm{H}_n'(x) + 2n\mathrm{H}_n(x) = 0$$

Prove that the generating function of Hermite polynomials is $G(x,t) = \mathrm{e}^{-t^2+2xt}$, and use it to prove the orthogonality of Hermite polynomials:

$$\int_{-\infty}^{\infty} \mathrm{H}_m(x) \mathrm{H}_n(x) \mathrm{e}^{-x^2} \mathrm{d}x = 2^n \sqrt{\pi} \delta_{mn}$$

Hint:

$$\int_{-\infty}^{\infty} \mathrm{e}^{-x^2} \mathrm{e}^{-s^2+2sx} \mathrm{e}^{-t^2+2tx} \mathrm{d}x = \sqrt{\pi} \mathrm{e}^{2st} = \sqrt{\pi} \sum_{n=0}^{\infty} \frac{2^n s^n t^n}{n!}$$

[4] Prove the recurrence relation of associated Laguerre polynomials:

$$(n+1)\mathrm{L}_{n+1}^k(x) = (2n+k+1-x)\mathrm{L}_n^k(x) - (n+k)\mathrm{L}_{n-1}^k(x)$$

[5] The first kind of Chebyshev polynomials $T_n(x)$ satisfies the differential equation:

$$(1-x^2)\frac{\mathrm{d}^2 y}{\mathrm{d}x^2} - x\frac{\mathrm{d}y}{\mathrm{d}x} + \lambda y = 0$$

(a) Prove that its eigenvalues are $\lambda_n = n^2$, and the Rodrigues formula is

$$\mathrm{T}_n(x) = \frac{(-1)^n\sqrt{\pi}}{2^n\Gamma\left(n+\frac{1}{2}\right)}(1-x^2)^{\frac{1}{2}}\frac{\mathrm{d}^n}{\mathrm{d}x^n}\left[(1-x^2)^{n-\frac{1}{2}}\right]$$

(b) Let $x = \cos\theta$, prove that the Chebyshev equation can be transformed into

$$\frac{\mathrm{d}^2\mathrm{T}_n}{\mathrm{d}\theta^2} + n^2\mathrm{T}_n = 0$$
$$\longrightarrow \begin{cases}\mathrm{T}_n(x) = \cos n\theta = \cos(n \arccos x)\\ \mathrm{V}_n(x) = \sin n\theta = \sin(n \arccos x)\end{cases}$$

Note that $\mathrm{V}_n(x)$ is not a polynomial.

[6] Using the Gram-Schmidt orthogonalization method, construct the first three orthogonal polynomials with weight function $\rho(x) = \mathrm{e}^{-x}$ from the linearly independent monomials $1, x, x^2, x^3, \ldots\}$ in the range $0 \leq x < \infty$.
Answer: $F_0 = 1, F_1 = x - 1, F_2 = \frac{1}{2}(x^2 - 4x + 2)$

Chapter 15

Special Functions

15.1 Bessel Functions

15.1.1 *Cylindrical coordinate system*

Chapter 13 discussed the eigenvalue problem of the Laplace equation in spherical coordinates and introduced the Legendre functions and spherical harmonics. For physical problems with cylindrical symmetry, it is generally convenient to use cylindrical coordinate system. According to Section 10.5, the Laplace equation in cylindrical coordinates can be expressed as follows:

$$\frac{1}{\rho}\frac{\partial}{\partial \rho}\left(\rho\frac{\partial u}{\partial \rho}\right)+\frac{1}{\rho^2}\frac{\partial^2 u}{\partial \phi^2}+\frac{\partial^2 u}{\partial z^2}=0$$

Separating variables with $u(\rho,\phi,z) = R(\rho)\Phi(\phi)Z(z)$. The part related to $\Phi(\phi)$ satisfies the equation:

$$\Phi''+\lambda\Phi=0$$

Due to periodicity conditions, its eigenfunctions are

$$\begin{aligned}&\Phi(\phi)=A\cos m\phi+B\sin m\phi\\&\lambda=m^2\quad(m=0,1,2,\ldots)\end{aligned}$$

The part related to $R(\rho)$ and $Z(z)$ is

$$\frac{\rho^2}{R}\frac{\mathrm{d}^2R}{\mathrm{d}\rho^2}+\frac{\rho}{R}\frac{\mathrm{d}R}{\mathrm{d}\rho}+\rho^2\frac{Z''}{Z}=\lambda$$

Further separating variables, we obtain

$$Z'' - \mu Z = 0$$

$$\frac{\mathrm{d}^2 R}{\mathrm{d}\rho^2} + \frac{1}{\rho}\frac{\mathrm{d}R}{\mathrm{d}\rho} + \left(\mu - \frac{m^2}{\rho^2}\right) R = 0 \tag{15.1.1}$$

The solution of this equation needs to be discussed in three cases:
(1) $\mu = 0$

The solution of the equation is

$$Z(z) = C + Dz$$

$$R(\rho) = \begin{cases} E + F\ln\rho & (m = 0) \\ E\rho^m + \dfrac{F}{\rho^m} & (m = 1, 2, \ldots) \end{cases} \tag{15.1.2}$$

(2) $\mu > 0$

The solution of the $Z(z)$ equation is

$$Z(z) = C\mathrm{e}^{\sqrt{\mu}z} + D\mathrm{e}^{-\sqrt{\mu}z}$$

Making the variable substitution $x = \sqrt{\mu}\rho$, the equation satisfied by $R(\rho)$ becomes

$$\frac{\mathrm{d}^2 R}{\mathrm{d}x^2} + \frac{1}{x}\frac{\mathrm{d}R}{\mathrm{d}x} + \left(1 - \frac{m^2}{x^2}\right) R = 0 \tag{15.1.3}$$

This equation is called the mth-order Bessel equation.
(3) $\mu < 0$

Let $\mu = -\nu^2$, then

$$Z(z) = C\cos\nu z + D\sin\nu z$$

Taking $x = \nu\rho$, the radial equation becomes

$$\frac{\mathrm{d}^2 R}{\mathrm{d}x^2} + \frac{1}{x}\frac{\mathrm{d}R}{\mathrm{d}x} - \left(1 + \frac{m^2}{x^2}\right) R = 0 \tag{15.1.4}$$

This equation is called the mth-order modified Bessel equation or Bessel function of imaginary argument, which is equivalent to replacing x with $\mathrm{i}x$ in the Bessel equation.

The next task is to study the boundary value problems of the Bessel equation (15.1.3) and the modified Bessel equation (15.1.4).

15.1.2 *Three types of Bessel functions*

The Bessel equation has a regular singularity at $x = 0$. In Chapter 9, two linearly independent series solutions were obtained in the neighborhood of $x = 0$. One of them is the Bessel function of the νth order:

$$\mathrm{J}_\nu(x) = \sum_{k=0}^{\infty} (-1)^k \frac{1}{k!\Gamma(\nu+k+1)} \left(\frac{x}{2}\right)^{\nu+2k} \tag{15.1.5}$$

This infinite series is also known as the Bessel function of the first kind, and its radius of convergence is $R = \infty$. Another solution is the νth order Neumann function:

$$\mathrm{N}_\nu(x) = \frac{\mathrm{J}_\nu(x)\cos\nu\pi - \mathrm{J}_{-\nu}(x)}{\sin\nu\pi}$$

It is also known as the Bessel function of the second kind. This solution diverges at the natural boundary $x = 0$.

In certain problems involving waves scattering on a cylindrical surface, it is sometimes more convenient to use another set of linearly independent solutions. These are called Hankel functions, which are a linear combination of Bessel functions and Neumann functions:

$$\mathrm{H}_\nu^{(1)}(x) = \mathrm{J}_\nu(x) + \mathrm{i}\mathrm{N}_\nu(x)$$
$$\mathrm{H}_\nu^{(2)}(x) = \mathrm{J}_\nu(x) - \mathrm{i}\mathrm{N}_\nu(x)$$

Hankel functions are also known as the Bessel functions of the third kind. All three types of cylindrical functions exhibit oscillatory behavior and have alternating zeros. Figures 15.1(a) and 15.1(b) respectively illustrate the curves of integer-order Bessel functions and Neumann functions.

When $\nu = \frac{1}{2}$, the Bessel functions can be expressed in terms of elementary trigonometric functions:

$$\mathrm{J}_{\frac{1}{2}}(x) = \sqrt{\frac{2}{\pi x}}\sin x, \quad \mathrm{J}_{-\frac{1}{2}}(x) = \sqrt{\frac{2}{\pi x}}\cos x$$

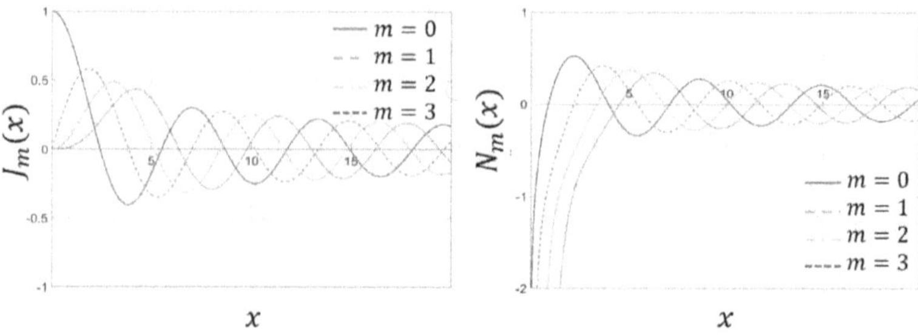

(a) (b)

Figure 15.1.

15.1.3 *Basic properties*

(1) *Limiting behavior*

As $x \to 0$,

$$\mathrm{J}_0(0) = 1, \quad \mathrm{J}_\nu(0) = 0, \quad \mathrm{J}_{-\nu}(0) \to \infty$$
$$\mathrm{N}_0(0) \to -\infty, \quad \mathrm{N}_\nu(0) \to \pm\infty \ (\nu \neq 0)$$

As $x \to \infty$,

$$\mathrm{J}_\nu(x) \sim \sqrt{\frac{2}{\pi x}} \cos\left(x - \frac{\nu\pi}{2} - \frac{\pi}{4}\right),$$
$$\mathrm{N}_\nu(x) \sim \sqrt{\frac{2}{\pi x}} \sin\left(x - \frac{\nu\pi}{2} - \frac{\pi}{4}\right)$$
$$\mathrm{H}_\nu^{(1)}(x) \sim \sqrt{\frac{2}{\pi x}} \mathrm{e}^{\mathrm{i}\left(x - \frac{\nu\pi}{2} - \frac{\pi}{4}\right)}, \quad \mathrm{H}_\nu^{(2)}(x) \sim \sqrt{\frac{2}{\pi x}} \mathrm{e}^{-\mathrm{i}\left(x - \frac{\nu\pi}{2} - \frac{\pi}{4}\right)}$$

It can be observed that the Hankel functions exhibit cylindrical wave behavior as $x \to \infty$.

(2) *Recurrence relations*

Using $Z_\nu(x)$ to represent the three types of cylindrical functions, from the series representation of Bessel functions, the general recurrence relations can be proven as follows:

$$\frac{\mathrm{d}}{\mathrm{d}x}\left[\frac{Z_\nu(x)}{x^\nu}\right] = -\frac{Z_{\nu+1}(x)}{x^\nu}$$

$$\frac{\mathrm{d}}{\mathrm{d}x}[x^\nu Z_\nu(x)] = x^\nu Z_{\nu-1}(x)$$
$$Z_{\nu-1}(x) - Z_{\nu+1}(x) = 2\frac{\mathrm{d}}{\mathrm{d}x}[Z_\nu(x)] \tag{15.1.6}$$
$$Z_{\nu+1}(x) - \frac{2\nu Z_\nu(x)}{x} + Z_{\nu-1}(x) = 0$$

15.1.4 *Eigenvalue problem*

The Bessel equation belongs to the Sturm–Liouville eigenvalue problem. It has a natural boundary condition at $\rho = 0$ and satisfies the first, second, or third type of homogeneous boundary conditions on the surface of the cylinder. For example, the first type homogeneous boundary problem is given by

$$\begin{cases} -\dfrac{\mathrm{d}}{\mathrm{d}\rho}\left[\rho\dfrac{\mathrm{d}R}{\mathrm{d}\rho}\right] + \dfrac{m^2}{\rho}R = \mu\rho R \\ R|_{\rho=\rho_0} = 0, \quad R|_{\rho=0} \text{ is finite} \end{cases}$$

Here, ρ_0 is the radius of the cylinder. A homogeneous boundary condition must be applied at finite radius of the cylinder since the infinity of the Bessel equation is a non-regular singularity. The solution to the equation is the Bessel function $\mathrm{J}_m(\sqrt{\mu}\rho)$, and the eigenvalue μ is determined by the homogeneous boundary conditions. It corresponds to the nth zero of the mth order Bessel function $\mathrm{J}_m(x)$, denoted as $x_n^{(m)}$, i.e., $\mu_n^{(m)} = \left[\frac{x_n^{(m)}}{\rho_0}\right]^2$, and the corresponding eigenfunction is $\mathrm{J}_m\left(\sqrt{\mu_n^{(m)}}\rho\right)$. Table 15.1 lists the values of the first few zeros of Bessel functions.

Table 15.1.

$x_n^{(m)}$	$n = 1$	$n = 2$	$n = 3$	$n = 4$	$n = 5$	$n = 6$	$n = 7$
$J_0(x)$	2.4048	5.5201	8.6537	11.7915	14.9309	18.0711	21.2116
$J_1(x)$	3.8317	7.0156	10.1735	13.3237	16.4706	19.6159	22.7601

The eigenfunctions satisfy the orthogonality with the weight function ρ:

$$\int_0^{\rho_0} \mathrm{J}_m\left(\sqrt{\mu_n^{(m)}}\rho\right)\mathrm{J}_m\left(\sqrt{\mu_k^{(m)}}\rho\right)\rho\,\mathrm{d}\rho = \delta_{nk}[N_n^{(m)}]^2 \tag{15.1.7}$$

Note that orthogonality refers to the orthogonality between eigenfunctions with different eigenvalues $\mu_n^{(m)}$ for the same order m. The normalization factor for Bessel functions is somewhat complicated to derive, and the result is as follows:

$$[N_n^{(m)}]^2 = \frac{1}{2}\left(\rho_0^2 - \frac{m^2}{\mu_n^{(m)}}\right)\left[\mathrm{J}_m\left(\sqrt{\mu_n^{(m)}}\rho_0\right)\right]^2 + \frac{1}{2}\rho_0^2\left[\mathrm{J}_m'\left(\sqrt{\mu_n^{(m)}}\rho_0\right)\right]^2 \quad (15.1.8)$$

The normalization factor depends on different types of boundary conditions and can be expressed as follows:

(1) First type homogeneous boundary condition

Due to $\mathrm{J}_m\left(\sqrt{\mu_n^{(m)}}\rho_0\right) = 0$, from (15.1.8) it yields

$$[N_n^{(m)}]^2 = \frac{1}{2}\rho_0^2\left[\mathrm{J}_{m+1}\left(\sqrt{\mu_n^{(m)}}\rho_0\right)\right]^2 \quad (15.1.9)$$

(2) Second type homogeneous boundary condition

Due to $\mathrm{J}_m'\left(\sqrt{\mu_n^{(m)}}\rho_0\right) = 0$, it yields

$$[N_n^{(m)}]^2 = \frac{1}{2}\left(\rho_0^2 - \frac{m^2}{\mu_n^{(m)}}\right)\left[\mathrm{J}_m\left(\sqrt{\mu_n^{(m)}}\rho_0\right)\right]^2 \quad (15.1.10)$$

(3) Third type homogeneous boundary condition

Due to $\mathrm{J}_m'(x) = -\frac{\mathrm{J}_m(x)}{\sqrt{\mu_n^{(m)}}H}$, it yields

$$[N_n^{(m)}]^2 = \frac{1}{2}\left(\rho_0^2 - \frac{m^2}{\mu_n^{(m)}} + \frac{\rho_0^2}{\mu_n^{(m)}H}\right)\left[\mathrm{J}_m\left(\sqrt{\mu_n^{(m)}}\rho_0\right)\right]^2 \quad (15.1.11)$$

15.1.5 *Generalized Fourier series*

The eigenfunctions of the Bessel equation are (weighted) orthogonal and complete, which allows us to expand functions defined on the interval $[0, \rho_0]$ into a generalized Fourier series:

$$
\begin{aligned}
f(\rho) &= \sum_{n=1}^{\infty} f_n \mathrm{J}_m\left(\sqrt{\mu_n^{(m)}}\rho\right) \\
f_n &= \frac{1}{[N_n^{(m)}]^2}\int_0^{\rho_0} f(\rho)\mathrm{J}_m\left(\sqrt{\mu_n^{(m)}}\rho\right)\rho d\rho
\end{aligned}
\tag{15.1.12}
$$

Example 15.1. Consider a homogeneous cylindrical object with radius ρ_0 and height L. The lateral surface is adiabatic, and the temperatures on the top and bottom surfaces are given by $f_1(\rho)$ and $f_2(\rho)$, respectively. We want to find the steady temperature distribution inside the cylinder.

Solution: The equation and boundary conditions are

$$
\begin{cases}
\Delta u = 0 \\
u_\rho|_{\rho=\rho_0} = 0, \quad u|_{\rho=0} \text{ is finite} \\
u|_{z=0} = f_1(\rho), \quad u|_{z=L} = f_2(\rho)
\end{cases}
$$

After separation of variables, the linearly independent solutions of the Laplace equation in cylindrical coordinates in three directions are

$$
u(\rho,\phi,z) \sim [\mathrm{J}_m(\sqrt{\mu_n}\rho), \mathrm{N}_m(\sqrt{\mu_n}\rho)](\mathrm{e}^{\sqrt{\mu_n}z}, \mathrm{e}^{-\sqrt{\mu_n}z})(\cos m\phi, \sin m\phi)
$$

For this problem, the boundary conditions have axial symmetry, so we set $m = 0$. Since the cylinder contains the cylindrical axis, the radial part can only have finite solutions due to the finiteness of physical quantities. Therefore, we discard the $\mathrm{N}_m(\sqrt{\mu_n}\rho)$ solution. Additionally, since the lateral surface of the cylinder satisfies the second type homogeneous boundary condition, we need to consider two possible cases, $\mu > 0$ and $\mu = 0$, respectively.

(1) For $\mu > 0$, from the homogeneous boundary condition:

$$
\mathrm{J}_0'(\sqrt{\mu}\rho)|_{\rho=\rho_0} = -\mathrm{J}_1(\sqrt{\mu}\rho_0) = 0
$$

we obtain the eigenvalue $\mu_n = \left(\frac{x_n^{(1)}}{\rho_0}\right)^2$, where $x_n^{(1)}$ is the nth zero of the first-order Bessel function $\mathrm{J}_1(x)$. The corresponding eigenfunction is $\mathrm{J}_0(\sqrt{\mu_n}\rho)$.

(2) For $\mu = 0$, since $m = 0$, we have

$$R(\rho) \sim (1, \ln\rho), \quad Z(z) \sim (1, z)$$

Discarding the divergent solution $\ln\rho$ at $\rho = 0$, the general solution of the equation is

$$u(\rho, z) = A_0 + B_0 z + \sum_{n=1}^{\infty} (A_n \mathrm{e}^{\sqrt{\mu_n} z} + B_n \mathrm{e}^{-\sqrt{\mu_n} z}) \mathrm{J}_0(\sqrt{\mu_n}\rho)$$

The coefficients A_0, B_0, A_n, B_n are determined by the boundary values on the top and bottom surfaces of the cylinder:

$$A_0 + \sum_{n=1}^{\infty} (A_n + B_n) \mathrm{J}_0(\sqrt{\mu_n}\rho) = f_1(\rho)$$

$$A_0 + B_0 L + \sum_{n=1}^{\infty} (A_n \mathrm{e}^{\sqrt{\mu_n} L} + B_n \mathrm{e}^{-\sqrt{\mu_n} L}) \mathrm{J}_0(\sqrt{\mu_n}\rho) = f_2(\rho)$$

The above equations correspond to expanding $f_1(\rho)$ and $f_2(\rho)$ in the generalized Fourier series with zero-order Bessel function $\mathrm{J}_0(\sqrt{\mu_n}\rho)$ as the basis. Under the first type homogeneous boundary condition, the expansion coefficients are given by

$$A_0 = \frac{2}{\rho_0^2} \int_0^{\rho_0} f_1(\rho)\rho \, \mathrm{d}\rho \equiv f_{10}$$

$$A_0 + B_0 L = \frac{2}{\rho_0^2} \int_0^{\rho_0} f_2(\rho)\rho \, \mathrm{d}\rho \equiv f_{20}$$

$$A_n + B_n = \frac{2}{\rho_0^2 [\mathrm{J}_0(x_n^{(0)})]^2} \int_0^{\rho_0} f_1(\rho) \mathrm{J}_0(\sqrt{\mu_n}\rho)\rho \, \mathrm{d}\rho \equiv f_{1n}$$

$$A_n \mathrm{e}^{\sqrt{\mu_n} L} + B_n \mathrm{e}^{-\sqrt{\mu_n} L} = \frac{2}{\rho_0^2 [\mathrm{J}_0(x_n^{(0)})]^2} \int_0^{\rho_0} f_2(\rho) \mathrm{J}_0(\sqrt{\mu_n}\rho)\rho \, \mathrm{d}\rho \equiv f_{2n}$$

The results are as follows:

$$A_0 = f_{10}, \quad A_n = \frac{f_{1n} \mathrm{e}^{-\sqrt{\mu_n} L} - f_{2n}}{\mathrm{e}^{-\sqrt{\mu_n} L} - \mathrm{e}^{\sqrt{\mu_n} L}}$$

$$B_0 = \frac{f_{20} - f_{10}}{L}, \quad B_n = \frac{f_{1n} \mathrm{e}^{\sqrt{\mu_n} L} - f_{2n}}{\mathrm{e}^{\sqrt{\mu_n} L} - \mathrm{e}^{-\sqrt{\mu_n} L}}$$

Example 15.2. A cylindrical volume with a radius of ρ_0 and height L, where the lateral and lower base surfaces maintain a temperature of u_0 and the upper base is adiabatic, assuming an initial temperature of $u_0 + f_1(\rho)f_2(z)$, seeks the temporal variation of temperature at various points within the cylindrical volume.

Solution: According to the given conditions, the formulated equation and initial conditions are as follows:

$$\begin{cases} u_t - a^2\Delta u = 0 \\ u|_{\rho=\rho_0} = u_0, \quad u|_{\rho=0} \text{ is finite} \\ u|_{z=0} = u_0, \quad u_z|_{z=L} = 0 \\ u|_{t=0} = u_0 + f_1(\rho)f_2(z) \end{cases}$$

Homogenize the boundary conditions by shifting $u = u_0 + v$, yielding

$$\begin{cases} v_t - a^2\Delta v = 0 \\ v|_{\rho=\rho_0} = 0, v|_{\rho=0} \text{ is finite} \\ v|_{z=0} = 0, \quad v_z|_{z=L} = 0 \\ v|_{t=0} = f_1(\rho)f_2(z) \end{cases}$$

Let $v = \mathrm{e}^{-k^2a^2t}\tilde{v}$, resulting in the Helmholtz equation $\Delta\tilde{v} + k^2\tilde{v} = 0$, with separation of variables leading to

$$\Phi'' + m^2\Phi = 0 \ (m = 0, 1, 2, \dots)$$

$$Z'' + \nu^2 Z = 0$$

$$\frac{\mathrm{d}^2R}{\mathrm{d}\rho^2} + \frac{1}{\rho}\frac{\mathrm{d}R}{\mathrm{d}\rho} + \left(k^2 - \nu^2 - \frac{m^2}{\rho^2}\right)R = 0$$

Let $\mu' = k^2 - \nu^2$, and the third equation remains the Bessel equation. Considering the axial symmetry due to the boundary conditions, we have $m = 0$ and constraints ensuring finiteness at the axis, leading to the general form of the solution:

$$v \propto \mathrm{e}^{-k^2a^2t}\mathrm{J}_0(\sqrt{\mu'}\rho)\sin\nu z$$

The eigenvalues in the z-direction are obtained from the homogeneous boundary conditions at the top and bottom surfaces:

$$\nu_p = \frac{(p + 1/2)\pi}{L} \quad (p = 0, 1, 2, \dots)$$

From the homogeneous boundary conditions on the lateral surface of the cylinder, we acquire the eigenvalue for the radial equation,

$\mu'_n = \left(\frac{x_n^{(0)}}{\rho_0}\right)^2$, where $x_n^{(0)}$ represents the zeros of the zeroth-order Bessel function. Thus, $k_{np} = \sqrt{\mu'_n + \nu_p^2}$. To obtain the solution of the equation, a linear superposition of all eigenfunctions is necessary:

$$v(\rho, z, t) = \sum_{n=1,p=0}^{\infty} A_{np}\mathrm{e}^{-k_{np}^2 a^2 t}\mathrm{J}_0(\sqrt{\mu'_n}\rho)\sin\nu_p z$$

The coefficients A_{np} are determined based on the orthogonality of the eigenfunctions:

$$A_{np} = \frac{2}{\rho_0^2[\mathrm{J}_1(x_n^{(0)})]^2}\int_0^{\rho_0} f_1(\rho)\mathrm{J}_0(\sqrt{\mu'_n}\rho)\rho\,\mathrm{d}\rho \times \frac{2}{L}\int_0^L f_2(z)\sin\nu_p z\mathrm{d}z$$

The complete solution reads

$$u(\rho, z, t) = u_0 + \sum_{n=1,p=0}^{\infty} A_{np}\mathrm{e}^{-k_{np}^2 a^2 t}\mathrm{J}_0(\sqrt{\mu'_n}\rho)\sin\nu_p z$$

15.1.6 *Generating function*

The generating function for Bessel functions is given by

$$F(x, z) = \mathrm{e}^{\frac{1}{2}x\left(z - \frac{1}{z}\right)} = \sum_{m=-\infty}^{\infty} \mathrm{J}_m(x)z^m \quad (0 < |z| < \infty) \tag{15.1.13}$$

This equation can be viewed as the Laurent series expansion of the function $F(x, z)$ with respect to z, as well as the Bessel series expansion with respect to x. If we let $z = \mathrm{e}^{\mathrm{i}\zeta}$, then

$$F(x, z) = \mathrm{e}^{\mathrm{i}x\sin\zeta} = \sum_{m=-\infty}^{\infty} \mathrm{J}_m(x)\mathrm{e}^{im\zeta} \tag{15.1.14}$$

This can be viewed as the complex Fourier series expansion of the function $F(x, z)$. The expansion coefficient is given by

$$\mathrm{J}_m(x) = \frac{1}{2\pi}\int_{-\pi}^{\pi} \mathrm{e}^{\mathrm{i}x\sin\zeta}\mathrm{e}^{-im\zeta}\mathrm{d}\zeta = \frac{1}{2\pi}\int_{-\pi}^{\pi} \cos(x\sin\zeta - m\zeta)\mathrm{d}\zeta \tag{15.1.15}$$

This equation is known as the integral representation of Bessel functions. If we further let $\zeta = \psi - \frac{\pi}{2}$, then

$$F(x,z) = \mathrm{e}^{-\mathrm{i}x\cos\psi} = \sum_{m=-\infty}^{\infty} (-\mathrm{i})^m \mathrm{J}_m(x)\mathrm{e}^{\mathrm{i}m\psi}$$

The expansion coefficient is given by

$$\mathrm{J}_m(x) = \frac{(-\mathrm{i})^m}{2\pi}\int_{-\pi}^{\pi} \mathrm{e}^{\mathrm{i}x\cos\psi + \mathrm{i}m\psi}\mathrm{d}\psi \tag{15.1.16}$$

Alternatively, if we let $\zeta = \theta + \frac{\pi}{2}$, then

$$F(x,z) = \mathrm{e}^{\mathrm{i}x\cos\theta} = \sum_{m=-\infty}^{\infty} \mathrm{i}^m \mathrm{J}_m(x)\mathrm{e}^{\mathrm{i}m\theta} \tag{15.1.17}$$

The meaning of this formula is to expand a plane wave in terms of cylindrical waves.

Using the relationship:

$$\sum_{m=-\infty}^{\infty} \mathrm{J}_m(x+y)z^m = \mathrm{e}^{\frac{1}{2}(x+y)\left(z-\frac{1}{z}\right)} = \mathrm{e}^{\frac{1}{2}x\left(z-\frac{1}{z}\right)} \cdot \mathrm{e}^{\frac{1}{2}y\left(z-\frac{1}{z}\right)}$$

$$= \sum_{k=-\infty}^{\infty} \mathrm{J}_k(x)z^k \sum_{n=-\infty}^{\infty} \mathrm{J}_n(y)z^n$$

we obtain the addition formula for Bessel functions:

$$\mathrm{J}_m(x+y) = \sum_{k=-\infty}^{\infty} \mathrm{J}_k(x)\mathrm{J}_{m-k}(y) \tag{15.1.18}$$

ENDNOTE

From the recursive relation of Bessel functions (15.1.6), we can obtain an infinite continued fraction expansion:

$$\frac{\mathrm{J}_{m-1}(x)}{\mathrm{J}_m(x)} = \frac{2m}{x} - \frac{1}{\frac{\mathrm{J}_m(x)}{\mathrm{J}_{m+1}(x)}} = \frac{2m}{x} - \frac{1}{\frac{2m+2}{x} - \frac{1}{\frac{\mathrm{J}_{m+1}(x)}{\mathrm{J}_{m+2}(x)}}}$$

$$= \frac{2m}{x} - \frac{1}{\frac{2m+2}{x} - \frac{1}{\frac{2m+4}{x} - \cdots}}$$

Taking $m = \frac{1}{2}$, since

$$\frac{\mathrm{J}_{-\frac{1}{2}}(x)}{\mathrm{J}_{\frac{1}{2}}(x)} = \cot x$$

we have

$$\tan x = \cfrac{1}{\frac{1}{x} - \cfrac{1}{\frac{3}{x} - \cfrac{1}{\frac{5}{x} - \cdots}}}$$

This continued fraction was discovered by Johann Lambert in 1761, and he used it to prove that π is irrational: The infinite continued fraction implies that x and $\tan x$ cannot both be rational numbers. However, when $x = \frac{\pi}{4}$, $\tan\left(\frac{\pi}{4}\right) = 1$, which is a rational number. Therefore, it follows that π must be irrational.

The addition formula for Bessel functions has a convolution form in terms of the function sequence $J_m(x)$ $(m \in \mathbb{Z})$:

$$\mathrm{J}_m(x+y) = \sum_{k=-\infty}^{\infty} \mathrm{J}_k(x) \cdot \mathrm{J}_{m-k}(y) \stackrel{\text{def}}{=} \mathrm{J}_m(x) * \mathrm{J}_m(y)$$

According to the convolution theorem of the Z-transform, we have

$$\mathcal{Z}[\mathrm{J}_m(x+y)] = \mathcal{Z}[\mathrm{J}_m(x)] \cdot \mathcal{Z}[\mathrm{J}_m(y)]$$

This equation maps the addition operation of the independent variable to the multiplication operation of functions, suggesting that the Z-transform of $\mathrm{J}_m(x)$ has exponential function characteristics. Indeed, the generating function of Bessel functions is in exponential form, and they are closely related.

Keeping this in mind, we note that Hermite polynomials also have generating functions in exponential form:

$$G(x,t) = \mathrm{e}^{-t^2+2xt} = \sum_{n=0}^{\infty} \frac{1}{n!}\mathrm{H}_n(x)t^n$$

It makes one wonder if Hermite polynomials also have an addition formula? Curious readers can try it out.

Exercises

[1] Calculate the integral:

$$\text{(a)} \int_0^a x^3 \mathrm{J}_0(x)\mathrm{d}x; \quad \text{(b)} \int_0^\infty \mathrm{e}^{-ax}\mathrm{J}_0(\sqrt{bx})\mathrm{d}x$$

Answer: (a) $a^3\mathrm{J}_1(a) - 2a^2\mathrm{J}_2(a)$; (b) $\frac{1}{a}\mathrm{e}^{-\frac{b}{4a}}$

[2] A solid cylinder of uniform density with radius ρ_0 and height L has a uniform heat flux q_0 entering the upper base and an equal heat flux leaving the lower base. The lateral surface is kept at zero temperature. Find the steady temperature distribution inside the cylinder.

Answer:

$$u(\rho,z) = q_0 \sum_{n=1}^{\infty} \frac{2\rho_0}{\kappa[x_n^{(0)}]^2 \mathrm{J}_1(x_n^{(0)})\sinh(x_n^{(0)}L/\rho_0)} \left[\cosh\left(\frac{x_n^{(0)}}{\rho_0}z\right) - \cosh\frac{x_n^{(0)}(L-z)}{\rho_0}\right] \mathrm{J}_0\left(\frac{x_n^{(0)}}{\rho_0}\rho\right)$$

[3] A circular membrane with radius ρ_0 is fixed along the boundary. The initial shape of the membrane is given by

$$u(\rho)|_{t=0} = \left(1 - \frac{\rho^2}{\rho_0^2}\right)u_0$$

Assuming the initial velocity is zero, find the solution for the vibration of the membrane.

Answer:

$$u(\rho,t)=\sum_{n=1}^{\infty}\frac{8u_0}{[x_n^{(0)}]^3\mathrm{J}_1(x_n^{(0)})}\mathrm{J}_0\left(\frac{x_n^{(0)}}{\rho_0}\rho\right)\cos\left(\frac{x_n^{(0)}}{\rho_0}at\right)$$

[4] Derive the normalization mode of Bessel functions:

$$[N_n^{(m)}]^2=\frac{1}{2}\left(\rho_0^2-\frac{m^2}{\mu_n^{(m)}}\right)\left[\mathrm{J}_m\left(\sqrt{\mu_n^{(m)}}\rho_0\right)\right]^2+\frac{1}{2}\rho_0^2\left[\mathrm{J}_m'\left(\sqrt{\mu_n^{(m)}}\rho_0\right)\right]^2$$

[5] Prove the generating function of Bessel functions:

$$F(x,z)=\mathrm{e}^{\frac{1}{2}x\left(z-\frac{1}{z}\right)}=\sum_{m=-\infty}^{\infty}\mathrm{J}_m(x)z^m\quad(0<|z|<\infty)$$

Hint: Expand $\mathrm{e}^{\frac{1}{2}x\left(z-\frac{1}{z}\right)}=\mathrm{e}^{\frac{1}{2}xz}\cdot\mathrm{e}^{-\frac{x}{2z}}$ as a Taylor series/Laurent series in the neighborhood of $z=0$, and then recombine them.

[6] Solve the wave equation in two-dimensional unbounded space:

$$\begin{cases}u_{tt}-a^2\Delta_2 u=0\\ u|_{t=0}=\varphi(\boldsymbol{r}),\quad u_t|_{t=0}=\psi(\boldsymbol{r})\end{cases}$$

Hint: Use the integral formula

$$\int_0^{\infty}\mathrm{J}_\nu(\alpha x)\sin\beta x\,\mathrm{d}x=\begin{cases}\dfrac{\sin\left(\nu\arcsin\frac{\beta}{\alpha}\right)}{\sqrt{\alpha^2-\beta^2}} & (\beta<\alpha)\\[2ex] \dfrac{\alpha^\nu\cos\frac{\nu\pi}{2}}{\sqrt{\beta^2-\alpha^2}\left(\beta+\sqrt{\beta^2-\alpha^2}\right)^\nu} & (\beta>\alpha>0)\end{cases}$$

Answer:

$$u(\boldsymbol{r},t)=\frac{1}{2\pi a}\frac{\partial}{\partial t}\iint_{C_r}\frac{\varphi(\boldsymbol{r}')}{\sqrt{(at)^2-|\boldsymbol{r}-\boldsymbol{r}'|^2}}\mathrm{d}x'\mathrm{d}y'+\frac{1}{2\pi a}\iint_{C_r}\frac{\psi(\boldsymbol{r}')}{\sqrt{(at)^2-|\boldsymbol{r}-\boldsymbol{r}'|^2}}\mathrm{d}x'\mathrm{d}y'$$

[7] Proof:

$$\text{(a)} \int_0^{\infty} \mathrm{J}_\nu(ax)\mathrm{e}^{-px}\mathrm{d}x = \frac{(\sqrt{p^2+a^2}+p)^{-\nu}}{a^\nu\sqrt{p^2+a^2}} \quad (\mathrm{Re}\, p > 0)$$

$$\text{(b)} \int_0^{\infty} x^{\nu+1}\mathrm{e}^{-b^2x^2}\mathrm{J}_\nu(ax)\mathrm{d}x = \frac{a^\nu}{(2b^2)^{\nu+1}}\mathrm{e}^{-\frac{a^2}{4b^2}}$$

$$\text{(c)} \int_0^{\infty} x^{\nu+1}\mathrm{e}^{-bx}\mathrm{J}_\nu(ax)\mathrm{d}x = \frac{2b(2a)^\nu}{\sqrt{\pi}(a^2+b^2)^{\nu+\frac{3}{2}}}\Gamma\left(\nu+\frac{3}{2}\right)$$

Hint: Substitute the series representation of Bessel functions and convert them into Gamma integrals.

[8] Proof:

$$\text{(a)} \int_0^{\infty} \mathrm{J}_0(ax)\frac{\sin bx}{x}\mathrm{d}x = \begin{cases} \dfrac{\pi}{2} & (0 \le a < b) \\ \arcsin\dfrac{b}{a} & (a > b) \end{cases}$$

$$\text{(b)} \int_0^{\infty} \mathrm{J}_0(ax)\sin bx\mathrm{d}x = \begin{cases} \dfrac{1}{\sqrt{b^2-a^2}} & (0 < a < b) \\ 0 & (a > b) \end{cases}$$

Hint: (a) Substitute the integral representation of Bessel functions and exchange the order of integration; (b) Use the Laplace transform formula for Bessel functions.

[9] For the thin plate shown in Figure 15.2, investigate the eigenmodes of vibration with fixed boundaries.

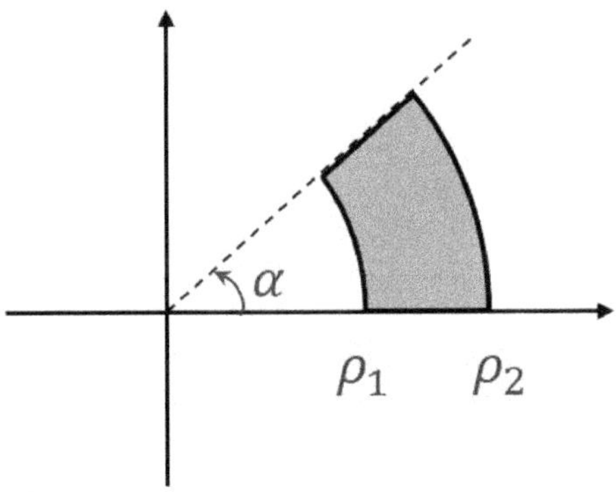

Figure 15.2.

Answer:

$$u_{np}(\rho,\theta) = [\mathrm{N}_{\lambda_n}(k_{np}\rho_1)\mathrm{J}_{\lambda_n}(k_{np}\rho) - \mathrm{J}_{\lambda_n}(k_{np}\rho_1)\mathrm{N}_{\lambda_n}(k_{np}\rho)]\sin\lambda_n\theta$$

$$\lambda_n = \frac{n\pi}{\alpha} \quad (n, p \in \mathbb{N})$$

15.2 Modified Bessel Function

In Section 15.1, we discussed the eigenvalue problem of the Bessel equation, where the homogeneous boundary conditions on the cylindrical surface determine the problem when $\mu \geq 0$. When $\mu < 0$, let $\mu = -\nu^2$ and $x = \nu\rho$. The Bessel equation for imaginary argument, also known as the modified Bessel function of the first kind, is given by

$$\frac{\mathrm{d}^2R}{\mathrm{d}x^2} + \frac{1}{x}\frac{\mathrm{d}R}{\mathrm{d}x} - \left(1 + \frac{m^2}{x^2}\right)R = 0$$

A series solution to this equation is called the mth order Bessel function of imaginary argument or the modified Bessel function of the first kind. We define $I_\nu(x) \overset{\text{def}}{=} \mathrm{i}^{-\nu}J_\nu(\mathrm{i}x)$, and it can be expressed as

$$\mathrm{I}_\nu(x) = \sum_{k=0}^{\infty}\frac{1}{k!\Gamma(\nu+k+1)}\left(\frac{x}{2}\right)^{\nu+2k} \tag{15.2.1}$$

Similar to the Neumann function, the second solution to the Bessel equation for imaginary argument can be taken as the Hankel function of imaginary argument or the modified Bessel function of the second kind:

$$\mathrm{K}_\nu(x) \overset{\text{def}}{=} \frac{\pi\mathrm{i}}{2}\mathrm{e}^{\frac{\mathrm{i}\pi\nu}{2}}\mathrm{H}_\nu^{(1)}(\mathrm{i}x) = \frac{\pi}{2}\frac{\mathrm{I}_{-\nu}(x) - \mathrm{I}_\nu(x)}{\sin\nu\pi} \tag{15.2.2}$$

When $\nu = m \in \mathbb{Z}$, we have

$$\mathrm{K}_m(x) = \lim_{v\to m}\frac{\pi}{2}\frac{\mathrm{I}_{-\nu}(x) - \mathrm{I}_\nu(x)}{\sin\nu\pi}$$

The general solution to the equation is given by

$$y(x) = C_1\mathrm{I}_\nu(x) + C_2\mathrm{K}_\nu(x)$$

Figures 15.3(a) and 15.3(b) illustrate several low-order Bessel functions of imaginary argument and Hankel functions of imaginary argument, which differ fundamentally from Bessel functions and Neumann functions in that their function curves do not oscillate. The relationship between Bessel functions and Bessel functions of imaginary argument is similar to the relationship between trigonometric functions and exponential functions. Their limiting behavior is as follows:

As $x = 0$,

$$\mathrm{I}_0(0) = 1, \quad \mathrm{I}_m(0) = 0, \quad \mathrm{K}_m(0) \to \infty$$

As $x \to \infty$,

$$\mathrm{I}_m(x) \sim \frac{1}{2\sqrt{x}}\mathrm{e}^x \to \infty, \quad \mathrm{K}_m(x) \sim \frac{\pi}{2\sqrt{x}}\mathrm{e}^{-x} \to 0$$

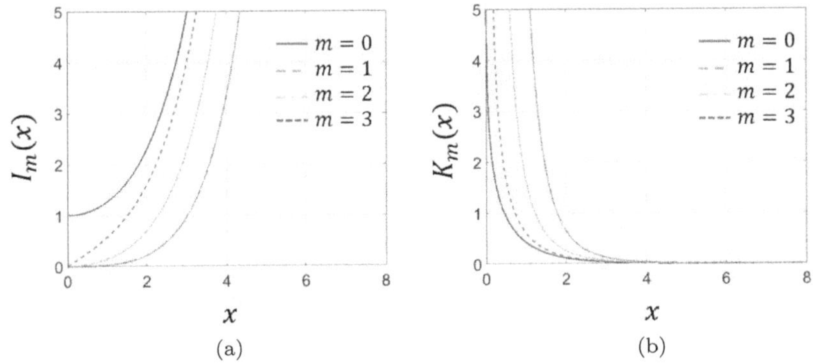

Figure 15.3.

Reflection: Is the complex variable Bessel equation belong to a Sturm–Liouville system? Does it constitute an eigenvalue problem? Why?

Bessel functions of imaginary argument, $\mathrm{I}_m(x)$, do not have any zeros other than at $x = 0$. They do not satisfy homogeneous boundary conditions and do not possess natural boundary condition as well. Since $\mathrm{K}_m(x)$ diverges at $x = 0$, it must be discarded when dealing with physical problems that involve the cylinder's axis. However, when the system does not include the axis, such as in the case of a hollow cylinder, the contribution of the Hankel function of imaginary argument should be taken account.

For problems with homogeneous boundary conditions on the top and bottom surfaces, we should consider the case where $\mu \leq 0$. In this case, the radial part satisfies the Bessel equation for imaginary argument.

Example 15.3. Consider a homogeneous cylinder with radius ρ_0 and height L. The lateral surface has a uniform heat flux q_0 entering the cylinder. The top and bottom surfaces have a constant temperature u_0. We need to solve for the steady-state temperature distribution inside the cylinder.

Solution: Based on the given problem, we can express the boundary value problem as follows:

$$\begin{cases} \Delta u = 0 \\ k u_\rho|_{\rho=\rho_0} = q_0, \quad u|_{\rho=0} \text{ is finite} \\ u|_{z=0} = u_0, \quad u|_{z=L} = u_0 \end{cases}$$

This problem is suitable for homogenizing the boundary conditions on the top and bottom surfaces by setting $u = u_0 + v$. Then

$$\begin{cases} \Delta v = 0 \\ k v_\rho|_{\rho=\rho_0} = q_0, \quad v|_{\rho=0} \text{ is finite} \\ v|_{z=0} = 0, \quad v|_{z=L} = 0 \end{cases}$$

Due to the axial symmetry of the boundary conditions in this problem, we have $m = 0$. Since the top and bottom surfaces both satisfy homogeneous boundary conditions, we only need to consider the case of $\mu \leq 0$. Let $\mu = -\nu^2$.

(1) For $\mu < 0$, we have

$$v(\rho, z) \propto [\mathrm{I}_0(\nu\rho), \mathrm{K}_0(\nu\rho)][\cos\nu z, \sin\nu z]$$

Considering the constraint that the physical quantities at the cylinder axis are finite, we discard the divergent solution $\mathrm{K}_0(\nu\rho)$. The eigenvalues are determined by the homogeneous boundary conditions on the top and bottom surfaces, and the eigenfunctions are given by

$$v(\rho, z) \propto \mathrm{I}_0(\nu_p\rho)\sin\nu_p z$$

with eigenvalues

$$\nu_p = \frac{p\pi}{L} \quad (p = 1, 2, 3, \ldots)$$

(2) For $\mu = 0$, the first type of homogeneous boundary conditions on the top and bottom surfaces lead to $v(\rho, z) = 0$, so we do not need to consider this case.

The general solution of the equation is a linear combination of the eigenfunctions,

$$v(\rho, z) = \sum_{p=1}^{\infty} A_p \mathrm{I}_0\left(\frac{p\pi}{L}\rho\right) \sin\frac{p\pi}{L}z$$

The coefficients A_p can be determined from the boundary conditions on the lateral surface of the cylinder:

$$A_p = \frac{2Lq_0}{p^2\pi^2 k} \frac{1}{\mathrm{I}_0'(p\pi\rho_0/L)}[1 - (-1)^p]$$

Therefore,

$$u(\rho, z) = u_0 + \frac{4Lq_0}{\pi^2 k} \sum_{n=0}^{\infty} \frac{1}{(2n+1)^2} \frac{1}{\mathrm{I}_0'((2n+1)\pi\rho_0/L)}$$
$$\times \mathrm{I}_0\left(\frac{(2n+1)\pi\rho}{L}\right) \sin\frac{(2n+1)\pi z}{L}$$

Reflection: Can we homogenize the boundary conditions on the lateral surface of the cylinder in this problem? Why or why not?

Example 15.4. Study the evolution of a two-dimensional isotropic quantum wave packet:

$$\begin{cases} \mathrm{i}\hbar\partial_t\psi(x, y, t) = -\dfrac{\hbar^2}{2m}\nabla^2\psi(x, y, t) \\ \psi(x, y, t)|_{t=0} = \varphi(r) \end{cases}$$

Solution: Perform a two-dimensional Fourier transform $\Psi(k_x, k_y, t) = \mathfrak{F}[\psi(x, y, t)]$ on the Schrödinger equation and obtain

$$\Psi(k_x, k_y, t) = \Phi(k)\mathrm{e}^{-\mathrm{i}\frac{\hbar k^2}{2m}t}$$

where $k^2 = k_x^2 + k_y^2$ and $\Phi(k)$ is the two-dimensional Fourier transform of $\varphi(r)$, given by

$$\Phi(k) = \mathfrak{F}[\varphi(r)] = \frac{1}{(2\pi)^2}\int_0^{\infty}\int_0^{2\pi} \mathrm{d}\theta\varphi(r)\mathrm{e}^{-\mathrm{i}kr\cos\theta} r\mathrm{d}r\mathrm{d}\theta$$
$$= \frac{1}{2\pi}\int_0^{\infty} \varphi(r)\mathrm{J}_0(kr)r\mathrm{d}r$$

Performing the inverse Fourier transform, we obtain the original function by making use of the generating function,

$$\begin{aligned}\psi(r,t) &= \int_0^\infty \int_0^{2\pi} \mathrm{d}\theta \Phi(k) \mathrm{e}^{-\mathrm{i}\frac{\hbar k^2}{2m}t} \mathrm{e}^{\mathrm{i}kr\cos\theta} k\mathrm{d}k\mathrm{d}\theta \\ &= 2\pi \int_0^\infty \Phi(k) \mathrm{J}_0(kr) \mathrm{e}^{-\mathrm{i}\frac{\hbar k^2}{2m}t} k\mathrm{d}k \\ &= \int_0^\infty \int_0^\infty \varphi(r') \mathrm{J}_0(kr') r' \mathrm{d}r' \mathrm{J}_0(kr) e^{-\mathrm{i}\frac{\hbar k^2}{2m}t} k\mathrm{d}k \\ &= \int_0^\infty \varphi(r') r' \left[\int_0^\infty \mathrm{J}_0(kr') \mathrm{J}_0(kr) \mathrm{e}^{-\mathrm{i}\frac{\hbar k^2}{2m}t} k\mathrm{d}k \right] \mathrm{d}r' \\ &= \frac{m}{\mathrm{i}\hbar t} \mathrm{e}^{\frac{imr^2}{2\hbar t}} \int_0^\infty \mathrm{e}^{\frac{imr'^2}{2\hbar t}} \mathrm{J}_0\left(\frac{mr}{\hbar t} r'\right) \varphi(r') r' \mathrm{d}r'\end{aligned}$$

In the final step we have used the integral formula,

$$\int_0^\infty \mathrm{J}_n(\alpha k) \mathrm{J}_n(\beta k) \mathrm{e}^{-\rho^2 k^2} k\mathrm{d}k = \frac{1}{2\rho^2} \mathrm{e}^{-\frac{\alpha^2+\beta^2}{4\rho^2}} \mathrm{I}_n\left(\frac{\alpha\beta}{2\rho^2}\right)$$

Assuming an initial Gaussian wave packet at the origin:

$$\varphi(r) = \frac{1}{4\pi\sigma^2} \mathrm{e}^{-\frac{r^2}{2\sigma^2}}$$

use the integral formula,

$$\int_0^\infty \mathrm{J}_n(\beta x) \mathrm{e}^{-\alpha x^2} x\mathrm{d}x = \frac{\sqrt{\pi}\beta}{8\alpha^{3/2}} \mathrm{e}^{-\frac{\beta^2}{8\alpha}} \left[\mathrm{I}_{\frac{n-1}{2}}\left(\frac{\beta^2}{8\alpha}\right) - \mathrm{I}_{\frac{n+1}{2}}\left(\frac{\beta^2}{8\alpha}\right) \right]$$

and let

$$\beta = \frac{mr}{\hbar t}, \quad \alpha = \left(\frac{1}{2\sigma^2} - \frac{im}{2\hbar t}\right), \quad \frac{\beta^2}{4\alpha} = \frac{m^2\sigma^2 r^2}{2(\hbar t - im\sigma^2)\hbar t}$$

we have

$$\begin{aligned}\psi(r,t) &= \frac{m}{4\pi\mathrm{i}\sigma^2\hbar t} \mathrm{e}^{\frac{imr^2}{2\hbar t}} \int_0^\infty \mathrm{e}^{-\left(\frac{1}{2\sigma^2} - \frac{im}{2\hbar t}\right) r'^2} \mathrm{J}_0\left(\frac{mr}{\hbar t} r'\right) r' \mathrm{d}r' \\ &= \frac{m}{4\pi\mathrm{i}\sigma^2\hbar t} \mathrm{e}^{\frac{imr^2}{2\hbar t}} \frac{\sqrt{\pi}\beta}{8\alpha^{3/2}} \mathrm{e}^{-\frac{\beta^2}{8\alpha}} \left[\mathrm{I}_{-\frac{1}{2}}\left(\frac{\beta^2}{8\alpha}\right) - \mathrm{I}_{\frac{1}{2}}\left(\frac{\beta^2}{8\alpha}\right) \right]\end{aligned}$$

$$= \frac{m}{4\pi \mathrm{i}\sigma^2 \hbar t} \mathrm{e}^{\frac{imr^2}{2\hbar t}} \frac{\sqrt{\pi}\beta}{8\alpha^{3/2}} \sqrt{\frac{16\alpha}{\pi\beta^2}} \mathrm{e}^{-\frac{\beta^2}{4\alpha}} = \frac{m}{2\mathrm{i}\sigma^2 \hbar t} \frac{1}{\alpha} \mathrm{e}^{\frac{imr^2}{2\hbar t}} \mathrm{e}^{-\frac{\beta^2}{4\alpha}}$$

$$= \frac{1}{4\pi(\sigma^2 + \mathrm{i}\hbar t/m)} \exp\left[-\frac{r^2}{2(\sigma^2 + \mathrm{i}\hbar t/m)}\right]$$

In the derivation we have used the equality:

$$\mathrm{I}_{-\frac{1}{2}}(x) - \mathrm{I}_{\frac{1}{2}}(x) = \sqrt{\frac{2}{\pi x}} \mathrm{e}^{-x}$$

Comparing with the one-dimensional wave packet diffusion in Example 12.5, we observe that the behavior of the two-dimensional wave packet is similar. The diffusion of quantum wave packets and classical particle diffusion (Brownian motion) exhibit similar characteristics, where the diffusion range is proportional to the square root of time. This similarity arises from the formal equivalence between the Schrödinger equation and the diffusion equation, both involving first-order time derivatives and differing from the motion of mechanical waves which involve second-order time derivatives.

ENDNOTE

The difference between the Bessel equation and the modified Bessel equation lies in the separation constant μ, which is determined by the homogeneous boundary conditions. Specifically,

(1) For a cylindrical body, if the lateral surface satisfies homogeneous boundary conditions, then $\mu \geq \theta$, resulting in the Bessel equation. The eigenvalues are determined by the zeros of the oscillating Bessel functions.

(2) If the top and bottom surfaces satisfy homogeneous boundary conditions, then $\mu \leq 0$, leading to the non-oscillating modified Bessel equation in the radial part. In this case, the eigenvalue problem cannot be determined solely by the radial part and must consider the homogeneous boundary conditions in the z-direction. A similar situation was encountered in Exercise [3] in Section 11.3.

(3) In particular, when both the top and bottom surfaces have second-type homogeneous boundary conditions, the case of $\mu = 0$ needs to be considered.

These rules are determined by the general Sturm–Liouville eigenvalue theory, where homogeneous and natural boundary conditions require positive eigenvalues.

Exercises

[1] Proof:

$$\text{(a) } \mathrm{e}^x = \mathrm{I}_0(x) + 2\sum_{n=1}^{\infty} \mathrm{I}_n(x)$$

$$\text{(b) } \mathrm{e}^{-x} = \mathrm{I}_0(x) + 2\sum_{n=1}^{\infty} (-1)^n \mathrm{I}_n(x)$$

$$\text{(c) } \sinh x = 2\sum_{n=1}^{\infty} \mathrm{I}_{2n-1}(x)$$

$$\text{(d) } \cosh x = \mathrm{I}_0(x) + 2\sum_{n=1}^{\infty} \mathrm{I}_{2n}(x)$$

$$\text{(e) } 1 = \mathrm{I}_0(x) + 2\sum_{n=1}^{\infty} (-1)^n \mathrm{I}_{2n}(x)$$

$$\text{(f) } \mathrm{I}_n(x) = \frac{1}{\pi}\int_0^{\pi} \mathrm{e}^{x\cos\theta} \cos n\theta \mathrm{d}\theta$$

Hint: Use the plane wave expansion of the Bessel function's generating function $F(x, z) = \mathrm{e}^{\mathrm{i}x\cos\theta}$.

[2] Consider a homogeneous cylinder with radius ρ_0 and height L. The bottom surface has a temperature of u_0, and there is a uniform heat flux q_0 entering from the top surface. The temperature distribution on the lateral surface is given by $f(z)$. Find the steady-state temperature distribution inside the cylinder.

Answer:

$$u(\rho, z) = u_0 + \frac{q_0}{\kappa} z + \sum_{p=1}^{\infty} A_p \mathrm{I}_0 \left(\frac{(p+1/2)\pi\rho}{L} \right) \sin \frac{(p+1/2)\pi z}{L}$$

[3] Consider a homogeneous cylinder with radius ρ_0 and height L. There is a uniform heat flux q_0 entering from the lateral surface of the cylinder, while the top and bottom surfaces maintain temperatures $f_2(\rho)$ and $f_1(\rho)$ respectively. Find the steady-state temperature distribution inside the cylinder.
Hint: Divide the function into two parts, where one part satisfies the homogeneous boundary conditions of the top and bottom surfaces, and the other part satisfies the homogeneous boundary conditions of the lateral surface.

[4] Proof:

$$\text{(a)} \int_0^\infty \mathrm{e}^{-\frac{ax}{2}} \sin bx \mathrm{I}_0 \left(\frac{ax}{2} \right) \mathrm{d}x = \frac{1}{\sqrt{2b}} \frac{1}{\sqrt{a^2+b^2}} \sqrt{b + \sqrt{a^2+b^2}}$$

$$\text{(b)} \int_0^\infty \mathrm{e}^{-\frac{ax}{2}} \cos bx \mathrm{I}_0 \left(\frac{ax}{2} \right) \mathrm{d}x = \frac{1}{\sqrt{2b}} \frac{1}{\sqrt{a^2+b^2}} \frac{a}{\sqrt{b + \sqrt{a^2+b^2}}}$$

Hint: Use the integral representation of the Bessel function.

15.3 Spherical Bessel Functions

The wave equation $u_{tt} - a^2 \Delta u = 0$ and the transport equation $u_t - a^2 \Delta u = 0$, after separating the time variable, both lead to the equation $\Delta v + k^2 v = 0$, which is known as the Helmholtz equation.

15.3.1 *Helmholtz equation in spherical coordinates*

$$\frac{1}{r^2} \frac{\partial}{\partial r} \left(r^2 \frac{\partial v}{\partial r} \right) + \frac{1}{r^2 \sin\theta} \frac{\partial}{\partial \theta} \left(\sin\theta \frac{\partial v}{\partial \theta} \right) + \frac{1}{r^2 \sin^2\theta} \frac{\partial^2 v}{\partial \phi^2} + k^2 v = 0$$

Its radial part satisfies the so-called spherical Bessel equation:

$$r^2\frac{\mathrm{d}^2R}{\mathrm{d}r^2}+2r\frac{\mathrm{d}R}{dr}+[k^2r^2-l(l+1)]R=0 \tag{15.3.1}$$

Making the variable substitutions

$$x=kr,\quad R(r)=\sqrt{\frac{\pi}{2x}y(x)}$$

the equation can be transformed into the Bessel equation of half-odd-integer order:

$$x^2\frac{\mathrm{d}^2y}{\mathrm{d}x^2}+x\frac{\mathrm{d}y}{\mathrm{d}x}+\left[x^2-\left(l+\frac{1}{2}\right)^2\right]y=0$$

The solutions to this equation are called the lth order spherical Bessel functions:

$$\mathrm{j}_l(x)=\sqrt{\frac{\pi}{2x}}\mathrm{J}_{l+\frac{1}{2}}(x),\quad \mathrm{j}_{-l}(x)=\sqrt{\frac{\pi}{2x}}\mathrm{J}_{-l+\frac{1}{2}}(x) \tag{15.3.2}$$

as well as the spherical Neumann functions $\mathrm{n}_l(x)$:

$$\mathrm{n}_l(x)=\sqrt{\frac{\pi}{2x}}\mathrm{N}_{l+\frac{1}{2}}(x)$$

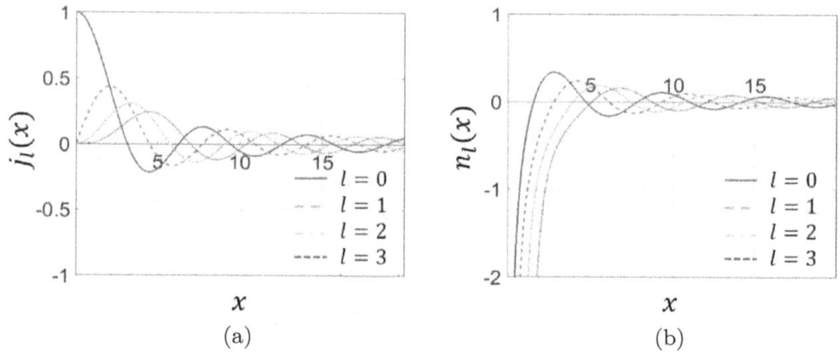

Figure 15.4.

Figures 15.4(a) and 15.4(b), respectively, depict several low-order spherical Bessel functions and spherical Neumann functions. Similar to trigonometric functions, they exhibit oscillatory behavior. In fact, low-order spherical Bessel functions can be expressed using elementary functions:

$$\mathrm{j}_0(x) = \frac{\sin x}{x}, \quad \mathrm{j}_{-1}(x) = \frac{\cos x}{x}, \quad \mathrm{j}_1(x) = \frac{\sin x - x\cos x}{x^2}, \ldots$$

$$\mathrm{n}_0(x) = -\frac{\cos x}{x}, \quad \mathrm{n}_1(x) = -\frac{\cos x + x\sin x}{x^2}, \ldots$$

When dealing with problems of waves scattering on spherical objects or central force fields, spherical Hankel functions are often used:

$$\mathrm{h}_l^{(1)}(x) = \mathrm{j}_l(x) + \mathrm{i}\mathrm{n}_l(x)$$

$$\mathrm{h}_l^{(2)}(x) = \mathrm{j}_l(x) - \mathrm{i}\mathrm{n}_l(x)$$

15.3.2 *Basic properties*

(1) *Limiting behavior*

When $x = 0$,

$$\mathrm{j}_0(0) = 1, \quad \mathrm{j}_l(0) = 0, \quad \mathrm{n}_l(x) \to \infty$$

When $x \to \infty$,

$$\mathrm{j}_l(x) \sim \frac{1}{x}\cos\left(x - \frac{l+1}{2}\pi\right), \quad \mathrm{n}_l(x) \sim \frac{1}{x}\sin\left(x - \frac{l+1}{2}\pi\right)$$

$$\mathrm{h}_l^{(1)}(x) \sim (-\mathrm{i})^{l+1}\frac{1}{x}\mathrm{e}^{\mathrm{i}x}, \quad \mathrm{h}_l^{(2)}(x) \sim \mathrm{i}^{l+1}\frac{1}{x}\mathrm{e}^{-\mathrm{i}x}$$

Spherical Hankel functions exhibit spherical wave behavior in the far field.

(2) *Recurrence relations*

$$z_{l+1}(x) = \frac{2l+1}{x}z_l(x) - z_{l-1}(x)$$

In addition to this, there exist other recurrence relations similar to Bessel functions, which are omitted here.

15.3.3 *Eigenvalue problem*

The spherical Bessel equation belongs to the Sturm–Liouville eigenvalue problem:

$$\begin{cases} -\dfrac{\mathrm{d}}{\mathrm{d}r}\left(r^2\dfrac{\mathrm{d}R}{\mathrm{d}r}\right) + l(l+1)R = k^2r^2R \\ R|_{r=r_0} = 0, \quad R|_{r=0} \text{ is finite} \end{cases}$$

The eigenvalue k is determined by the homogeneous boundary conditions on the spherical surface: $k_n = \frac{x_n^{(l)}}{r_0}$, where $x_n^{(l)}$ denotes the nth zero of the lth order spherical Bessel function, $\mathrm{j}_l(x)$. The corresponding eigenfunctions are $\mathrm{j}_l(k_nr)$, and they satisfy the orthogonal relationship with the weight function $\rho(r) = r^2$:

$$\int_0^{r_0} \mathrm{j}_l(k_mr)\mathrm{j}_l(k_nr)r^2\mathrm{d}r = [N_n]^2\delta_{mn} \tag{15.3.3}$$

The normalization factor is given by

$$[N_n]^2 = \int_0^{r_0} [\mathrm{j}_l(k_nr)]^2r^2\mathrm{d}r = \frac{\pi}{2k_n}\int_0^{r_0} [\mathrm{J}_{l+\frac{1}{2}}(k_nr)]^2r\mathrm{d}r$$

Note that orthogonality holds between eigenfunctions with the same l but different eigenvalues k_n. Table 15.2 lists the zeros of several spherical Bessel functions.

Table 15.2.

$x_n^{(m)}$	$n=1$	$n=2$	$n=3$	$n=4$	$n=5$	$n=6$	$n=7$
$\mathrm{j}_0(x)$	π	2π	3π	4π	5π	6π	7π
$\mathrm{j}_1(x)$	4.4934	7.7252	10.9041	14.0662	17.2201	20.2713	23.5194

15.3.4 *Generalized Fourier series*

A function $f(r)$ defined on the interval $[0, r_0]$ can be expanded using the orthogonal and complete eigenfunctions $\mathrm{j}_l(k_nr)$ in a generalized Fourier series:

$$f(r) = \sum_{m=1}^{\infty} f_m\mathrm{j}_l(k_mr) \tag{15.3.4}$$

The coefficients f_m are given by

$$f_m = \frac{1}{[N_m]^2} \int_0^{r_0} f(r) \mathrm{j}_l(k_m r) r^2 \mathrm{d}r$$

Example 15.5. Consider a homogeneous sphere of radius r_0, initially at a uniform temperature u_0, placed in an environment with temperature u_1. We want to solve for the temperature distribution within the sphere.

Solution: The problem can be formulated as follows:

$$\begin{cases} \dfrac{\partial u}{\partial t} - a^2 \Delta u = 0 \\ u\big|_{t=0} = u_0, \quad u\big|_{r=r_0} = u_1 \end{cases}$$

First, we homogenize the boundary condition by taking $u = v + u_1$. Since the boundary condition has spherical symmetry, we have $l = 0$ and $m = 0$. Considering that the temperature at the center of the sphere remains finite, we deduce that the radial part of the solution has the form:

$$v(r,t) = \mathrm{e}^{-k^2 a^2 t} \mathrm{j}_0(kr)$$

Here, k is determined by the homogeneous boundary condition on the spherical surface:

$$\mathrm{j}_0(k_n r_0) = \frac{\sin k r_0}{k r_0} = 0$$

Which yields

$$k_n = \frac{n\pi}{r_0}$$

Thus, the solution becomes

$$v(r,t) = \sum_{n=1}^{\infty} A_n \frac{\sin(n\pi r/r_0)}{(n\pi r/r_0)} \mathrm{e}^{-\left(\frac{n\pi a}{r_0}\right)^2 t}$$

The coefficients A_n are determined by the initial conditions. The final result for the temperature distribution is

$$u(r,t) = u_1 + 2(u_1 - u_0) \sum_{n=1}^{\infty} (-1)^n \frac{\sin(n\pi r/r_0)}{n\pi r/r_0} \mathrm{e}^{-\left(\frac{n\pi a}{r_0}\right)^2 t}$$

This result reveals that the rate of temperature decrease of an object in an environment depends on its surface area. Objects with larger surface areas or volumes exhibit slower temperature decreases.

15.3.5 *Spherical wave expansion*

When a plane wave is incident on a spherical object, the scattered waves can be represented as spherical waves. This type of problem often arises in the scattering processes of particles in quantum mechanics. Consider a plane wave propagating along the z-axis:

$$u(r,t) = \mathrm{e}^{\mathrm{i}k(z-at)} = \mathrm{e}^{\mathrm{i}k(r\cos\theta - at)}$$

Separating out the time variable, the wave equation becomes the three-dimensional Helmholtz equation: $\Delta v + k^2 v = 0$. In spherical coordinates, the finite eigenfunctions of this equation are given by

$$v(r,\theta) \sim \mathrm{j}_l(kr) P_l^m(\cos\theta) \begin{pmatrix} \cos m\phi \\ \sin m\phi \end{pmatrix}$$

Since the plane wave does not depend on ϕ, we set $m = 0$. The plane wave can be expressed as a superposition of these eigenstates:

$$\mathrm{e}^{ikr\cos\theta} = \sum_{l=0}^{\infty} A_l \mathrm{j}_l(kr) \mathrm{P}_l(\cos\theta)$$

The expansion coefficients in terms of the Legendre polynomials are given by

$$A_l \mathrm{j}_l(kr) = \frac{2l+1}{2} \int_{-1}^{1} \mathrm{e}^{ikrx} \mathrm{P}_l(x) \mathrm{d}x$$

To compute A_l, we consider the asymptotic behavior as $r \to \infty$

$$A_l \mathrm{j}_l(kr) \sim A_l \frac{1}{kr} \cos\left(kr - \frac{l+1}{2}\pi\right)$$

We perform integration by parts on the right-hand side of the equation:

$$\begin{aligned}
&\frac{2l+1}{2} \int_{-1}^{1} \mathrm{e}^{ikrx} \mathrm{P}_l(x) \mathrm{d}x \\
&= \frac{2l+1}{2\mathrm{i}kr} [\mathrm{e}^{ikrx} \mathrm{P}_l(x)]_{-1}^{1} - \frac{2l+1}{2\mathrm{i}kr} \int_{-1}^{1} \mathrm{e}^{ikrx} [\mathrm{P}_l(x)]' \mathrm{d}x \\
&= \frac{2l+1}{2\mathrm{i}kr} [\mathrm{e}^{ikr} - (-1)^l \mathrm{e}^{-ikr}] - \frac{2l+1}{2\mathrm{i}kr} \int_{-1}^{1} \mathrm{e}^{ikrx} [\mathrm{P}_l(x)]' \mathrm{d}x
\end{aligned}$$

$$= \frac{2l+1}{\mathrm{i}kr}\mathrm{e}^{\frac{\mathrm{i}(l+1)\pi}{2}}\cos\left(kr - \frac{l+1}{2}\pi\right) - \frac{2l+1}{2\mathrm{i}kr}\int_{-1}^{1}\mathrm{e}^{\mathrm{i}krx}[\mathrm{P}_l(x)]'\mathrm{d}x$$

$$\xrightarrow{r\to\infty} \frac{2l+1}{\mathrm{i}kr}\mathrm{e}^{\frac{\mathrm{i}(l+1)\pi}{2}}\cos\left(kr - \frac{l+1}{2}\pi\right) - \mathcal{O}\left(\frac{1}{r^2}\right)$$

The second term on the right-hand side, when integrated again, will be proportional to $\frac{1}{r^2}$. As $r \to \infty$, it decays faster than the leading term with $\frac{1}{r}$. Therefore,

$$A_l = -\mathrm{i}(2l+1)\mathrm{e}^{\frac{\mathrm{i}(l+1)\pi}{2}} = (2l+1)\mathrm{i}^l$$

Finally, we obtain the plane wave expansion in terms of spherical waves,

$$\mathrm{e}^{\mathrm{i}kr\cos\theta} = \sum_{l=0}^{\infty}(2l+1)\mathrm{i}^l\mathrm{j}_l(kr)\mathrm{P}_l(\cos\theta) \tag{15.3.5}$$

In the scattering processes of particles in a central potential, angular momentum is conserved. Quantum mechanics utilizes the method of partial wave expansion, which expands the scattered waves in terms of angular momentum eigenstates.

15.3.6 *Other modified Bessel equations*

Many modified Bessel equations, including the modified Bessel equation of imaginary argument and the spherical Bessel equation, can be derived through the following transformations. Let

$$u(z) = z^{\alpha}Z_{\nu}(\lambda z^{\beta})$$

then $u(z)$ satisfies the equation,

$$z^2\frac{\mathrm{d}^2u}{\mathrm{d}z^2} + (1-2\alpha)z\frac{\mathrm{d}u}{\mathrm{d}z} + (\lambda^2\beta^2z^{2\beta} + \alpha^2 - \nu^2\beta^2)u = 0$$

Alternatively, let

$$u(z) = z^{-\nu}\mathrm{e}^{\frac{\lambda x}{2}}Z_{\nu}\left(\frac{\mathrm{i}\lambda z}{2}\right)$$

then $u(z)$ satisfies the equation,

$$z\frac{d^2u}{dz^2} + [(2\nu+1) - \lambda z]\frac{du}{dz} - \left(\nu + \frac{1}{2}\right)\lambda u = 0$$

Or, let

$$u(z) = \frac{1}{\cos z} Z_\nu(z)$$

then $u(z)$ satisfies the equation:

$$z^2\frac{d^2u}{dz^2} + (z - 2z\tan z)\frac{du}{dz} - (\nu^2 + z\tan z)\lambda u = 0$$

Finally, let

$$u(z) = \frac{1}{\sin z} Z_\nu(z)$$

then $u(z)$ satisfies the equation

$$z^2\frac{d^2u}{dz^2} + (z - 2z\cot z)\frac{du}{dz} - (\nu^2 - z\cot z)\lambda u = 0$$

ENDNOTE

The rate of temperature decrease is inversely proportional to the size or surface area of an object, meaning that larger objects have more stable temperatures and are less influenced by the environment. When Sir Kelvin speculated about the age of the Earth, estimating it to be around 100 million years based on the cooling of the initial fiery state, his assertion displeased both those who believed in the biblical creation of a 6,000-year-old Earth and those who believed in evolution, considering the Earth too young. However, Kelvin was unaware that the presence of radioactive substances within the Earth acted as an additional heat source, significantly slowing down the cooling process and leading to his calculation being nearly two orders of magnitude smaller than the true age of the Earth.

This distinguished Lord, who established the absolute temperature scale in thermodynamics, also calculated that the oxygen on Earth could only sustain an industrialized society for about 400 years, a conclusion that deeply perplexed him.

The metabolic rate of animals also correlates to their size. Generally, smaller animals have faster metabolisms because they lose heat more rapidly in the same environment. In the Arctic Ocean, apart from cold-blooded fish, warm-blooded animals such as the bowhead whale tend to have large bodies. The frigid temperatures in the polar regions pose significant challenges to young animals, as their body temperatures rapidly decrease due to excessive heat loss. They need to continuously swim or occasionally dive into deeper waters, where the environmental temperature is relatively higher. The long-distance migration of bowhead whales may also be related to maintaining body temperature and rapid growth in young whales. On the other hand, the primary reason for the migration of many bird species may not only be food but also to breed in warmer zones and help their fledglings survive the vulnerable period, as their small bodies cannot withstand the cold northern winters.

During the long process of evolution, many behaviors and physiological traits of organisms have physical reasons behind them. This leads to a question: Why did dinosaurs on land evolve such enormous bodies? Can we speculate that the temperature on Earth during that era was relatively low? Or were there unusual chemical compositions in the atmosphere? Were the plants during the age of dinosaurs similarly gigantic? If these factors were not synchronized, it could be considered as evidence that climate played a role in determining the body size of warm-blooded animals.

Could this line of reasoning fall into the same pitfall as Kelvin's?

Exercises

[1] Prove the recurrence relations:

$$\text{(a) } \mathrm{j}_{k+1}(x) = \frac{2k+1}{x}\mathrm{j}_k(x) - \mathrm{j}_{k-1}(x)$$

$$\text{(b) } k\mathrm{j}_{k-1}(x) - (k+1)\mathrm{j}_{k+1}(x) = (2k+1)\mathrm{j}_k'(x)$$

[2] A homogeneous sphere with a radius of r_0, with an initial temperature distribution of $f(r)\cos\theta$, and maintaining a zero temperature at its surface, determine the temperature variation at different points within the sphere.
Answer:

$$u(r,\theta,t) = \frac{2}{r_0^3}\sum_{n=1}^{\infty}\frac{\int_0^{r_0}\mathrm{j}_1(k_n r)f(r)r^2\mathrm{d}r}{[\mathrm{j}_0(k_n r_0)]^2}\mathrm{P}_1$$
$$\times (\cos\theta)\mathrm{j}_1(k_n r)\mathrm{e}^{-a^2k_n^2 t}$$

where $k_n = \frac{x_n}{r_0}$, and x_n is the nth root of the equation $x = \tan x$.

[3] The radial velocity distribution on the surface of a sphere with a radius of r_0 is given by $v = v_0\cos\theta\cos\omega t$. Assuming that r_0 is much smaller than the wavelength of the sound wave, determine the velocity potential distribution of the steady-state vibrations emitted by this spherical surface.
Hint: Take the scattering wave solution as

$$u(r,\theta,t) = A\mathrm{h}_1^{(1)}(kr)\mathrm{P}_1(\cos\theta)\mathrm{e}^{-\mathrm{i}\omega t} \longrightarrow k = \frac{\omega}{a}$$

and use long-wave approximation to determine the parameter A
Answer:

$$u(r,\theta,t) = -\frac{v_0\omega r_0^3}{2ar}\mathrm{P}_1(\cos\theta)\sin\frac{\omega(r-at)}{a}$$

[4] A drum of radius R and height h is shown in Figure 15.5. Suppose the lateral is fixed and the upper and lower planes are free. Study the Eigen modes of vibratation.

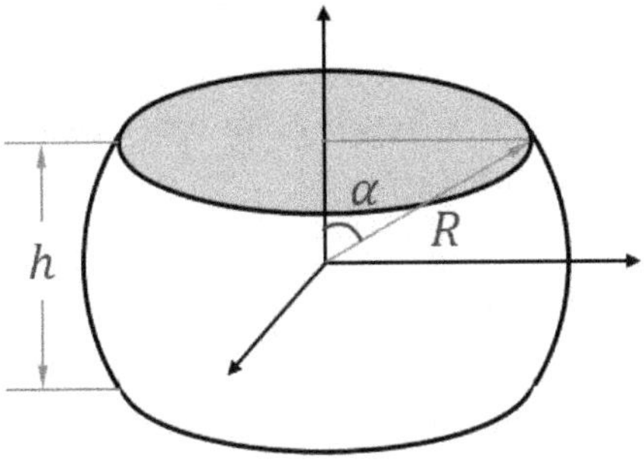

Figure 15.5.

Answer:

$$u_{mpn}(r,\theta,\phi) = j_{\lambda_{mp}}(k_{mpn}r)\{[Q^m_{\lambda_{mp}}(x)]'|_{x=\frac{h}{2R}} P^m_{\lambda_{mp}}(\cos\theta) - [P^m_{\lambda_{mp}}(x)]'|_{x=\frac{h}{2R}} Q^m_{\lambda_{mp}}(\cos\theta)\} \begin{pmatrix} \cos m\phi \\ \sin m\phi \end{pmatrix}$$

Where the eigenvalue $\lambda = \lambda_{mp}$ is determined by

$$\begin{vmatrix} P'^m_\lambda\left(\frac{h}{2R}\right) & Q'^m_\lambda\left(\frac{h}{2R}\right) \\ P'^m_\lambda\left(-\frac{h}{2R}\right) & Q'^m_\lambda\left(-\frac{h}{2R}\right) \end{vmatrix} = 0$$

The second eigenvalue is determined by the zeros of the spherical Bessel function $j_{\lambda_{mp}}(kr)$ as

$$k_{mpn} = \frac{x_n^{(\lambda_{mp})}}{R} \quad (m,n,p \in \mathbb{N})$$

15.4 Classification of Special Functions

In quantum mechanics, the behavior of differential equations at infinity is often important. For example, in the Schrödinger equation, the bound states require the probability of a particle appearing at infinity to be zero. We have seen that the form of the solution to a differential equation is determined by the coefficient function. In this section, we study a class of special second-order ODEs called Fuchsian equations and reveal the relationships among the special functions that serve as solutions to these equations.

15.4.1 *Fuchsian equations*

A Fuchsian equation is a second-order ODE with coefficient functions that are single-valued analytic functions and have all their singularities (including infinity) as regular singularities. The majority of physical equations belong to or can be transformed into Fuchsian equations, making it possible to classify them by analyzing Fuchsian equations. In the following discussion, we consider the case where the coefficient functions are rational functions (ratios of two polynomials) on the extended complex plane $\bar{\mathbb{C}}$.

Consider a second-order ODE with a complex variable:

$$\frac{\mathrm{d}^2 w}{\mathrm{d}z^2} + p(z)\frac{\mathrm{d}w}{\mathrm{d}z} + q(z)w = 0 \tag{15.4.1}$$

To determine the behavior of the equation at infinity, let $z = \frac{1}{t}$ and define $v(t) = w\left(\frac{1}{t}\right)$. Substituting into the equation, we have

$$\frac{\mathrm{d}^2 v}{\mathrm{d}t^2} + \left[\frac{2}{t} - \frac{1}{t^2}r(t)\right]\frac{\mathrm{d}v}{\mathrm{d}t} + \frac{1}{t^4}s(t)v = 0$$

where

$$r(t) = p\left(\frac{1}{t}\right), \quad s(t) = q\left(\frac{1}{t}\right)$$

If $z = \infty$ is a regular singularity of the equation, then $t = 0$ must be a regular singularity of the transformed equation. Therefore, $r(t)$ and $s(t)$ can only take the following form, respectively,

$$r(t) = a_1 t + a_2 t^2 + \cdots = \sum_{k=1}^{\infty} a_k t^k$$

$$s(t) = b_2 t^2 + b_3 t^3 + \cdots = \sum_{k=2}^{\infty} b_k t^k$$

Thus, the series forms of $p(z)$ and $q(z)$ are restricted to

$$p(z) = \frac{a_1}{z} + \frac{a_2}{z^2} + \cdots = \sum_{k=1}^{\infty} \frac{a_k}{z^k}, \quad q(z) = \frac{b_2}{z^2} + \frac{b_3}{z^3} + \cdots = \sum_{k=2}^{\infty} \frac{b_k}{z^k} \tag{15.4.2}$$

According to Section 9.2, the equation has at least one Frobenius series solution:

$$v_1(t) = t^s \left(1 + \sum_{k=1}^{\infty} c_k t^k\right)$$

where s is the larger root of the indicial equation. Hence, one solution to Equation (15.4.1) is

$$w_1(z) = z^{-s} \left(1 + \sum_{k=1}^{\infty} c_k \frac{1}{z^k}\right)$$

15.4.2 *Regular singularities*

Chapter 6 discussed fractional linear transformations that preserve conformality, meaning the pre-image and the transformed region have the same topological structure. This property allows us to change the positions of singularities of a differential equation without affecting the properties of the equation itself.

(1) *Two regular singularities*

If a Fuchsian equation has two regular singularities z_1 and z_2, a fractional linear transformation can be performed as

$$\zeta(z) = \frac{z - z_1}{z - z_2}$$

This transformation maps the two regular singularities to $z_1 \longmapsto 0$ and $z_2 \longmapsto \infty$. According to Equation (15.4.2), an equation with regular singularities at $z = 0$ and $z = \infty$ can only have the following form:

$$w'' + \frac{a_1}{z} w' + \frac{b_2}{z^2} w = 0 \tag{15.4.3}$$

This is a Euler-type equation. As discussed in Section 9.1, Euler-type equations can be transformed into ODEs with constant coefficients. Therefore, Fuchsian equations with only two regular singularities do not possess any special properties and are simply equivalent to second-order ODEs with constant coefficients.

(2) *Three regular singularities*

Suppose a Fuchsian equation has three regular singularities z_1, z_2 and z_3. We can choose a fractional linear transform as

$$\zeta(z) = \frac{(z-z_1)(z_3-z_2)}{(z-z_2)(z_3-z_1)}$$

This transformation maps the three singularities to

$$z_1 \longmapsto 0, \quad z_2 \longmapsto \infty, \quad z_3 \longmapsto 1$$

It can be shown that the characteristic indices of each regular singularity remain unchanged before and after the transformation. Therefore, we can directly assume that the equation has regular singularities at $z = 0, 1,$ and ∞. Based on the previous analysis, the most general forms of $p(z)$ and $q(z)$ are

$$p(z) = \frac{A_1}{z} + \frac{B_1}{z-1}, \quad q(z) = \frac{A_2}{z^2} + \frac{B_2}{(z-1)^2} - \frac{A_3}{z(z-1)} \tag{15.4.4}$$

Here, A_1, A_2, A_3, B_1 and B_2 are constants. This type of differential equation is called a Riemann equation.

Write down the indicial equations for each regular singularity:

$$\begin{aligned}
&\lambda^2 + (A_1 - 1)\lambda + A_2 = 0\\
&\mu^2 + (B_1 - 1)\mu + B_2 = 0\\
&\nu^2 + (1 - A_1 - B_1)\nu + (A_2 + B_2 - A_3) = 0
\end{aligned}$$

Expressing the above equations in the form $(\lambda - \lambda_1)(\lambda - \lambda_2) = 0$, we can directly solve for A_1, A_2, A_3, B_1, and B_2:

$$\begin{aligned}
&A_1 = 1 - \lambda_1 - \lambda_2, \quad A_2 = \lambda_1\lambda_2\\
&B_1 = 1 - \mu_1 - \mu_2, \quad B_2 = \mu_1\mu_2\\
&A_1 + B_1 = 1 + \nu_1 + \nu_2, \quad A_2 + B_2 - A_3 = \nu_1\nu_2
\end{aligned}$$

As a result, we obtain three sets of characteristic indices (λ_1, λ_2), (μ_1, μ_2), and (ν_1, ν_2), which satisfy the Riemann identity:

$$\lambda_1 + \lambda_2 + \mu_1 + \mu_2 + \nu_1 + \nu_2 \equiv 1 \tag{15.4.5}$$

Thus, a Fuchsian equation with three regular singularities can be transformed into a Riemann equation of the following form:

$$\begin{aligned} w'' &+ \left(\frac{1 - \lambda_1 - \lambda_2}{z} + \frac{1 - \mu_1 - \mu_2}{z - 1} \right) w' \\ &+ \left[\frac{\lambda_1 \lambda_2}{z^2} + \frac{\mu_1 \mu_2}{(z-1)^2} + \frac{\nu_1 \nu_2 - \lambda_1 \lambda_2 - \mu_1 \mu_2}{z(z-1)} \right] w = 0 \end{aligned} \tag{15.4.6}$$

The form of the equation is completely determined by the characteristic indices of the regular singularities. After an appropriate variable substitution, the Riemann equation can be transformed into a hypergeometric equation,

$$z(z-1)u'' + [(\alpha + \beta + 1)z - \gamma]u' + \alpha\beta u = 0 \tag{15.4.7}$$

The Legendre equation mentioned earlier, which has three regular singularities at $z = \pm 1$ and ∞, is a special case of the hypergeometric equation.

(3) *Five regular singularities*

If a Fuchsian equation has five regular singularities z_k $(k = 1, 2, 3, 4)$ and ∞, it can be generally expressed as follows:

$$\frac{d^2 w}{dz^2} + \left(\sum_{k=1}^{4} \frac{\alpha_k}{z - z_k} \right) \frac{dw}{dz} + \left[\sum_{k=1}^{4} \frac{\beta_k}{(z - z_k)^2} + \frac{Az^2 + Bz + C}{\prod_{k=1}^{4} (z - z_k)} \right] w = 0 \tag{15.4.8}$$

The properties of the regular singularities may change when they are merged. Some of them may remain regular, while others may no longer be regular. Differential equations categorized by the types of singularities are listed in Table 15.3.

The Mathieu equation is a second-order ordinary differential equation that arises when separating variables in the two-dimensional

Table 15.3.

Number of regular singularities	Number of non-regular singularities	Second-order differential equations
4	0	Lamé equation
3	0	Hypergeometric, Jacobian, Legendre, and Chebyshev
2	1	Mathieu equation
1	1	Confluent hypergeometric, Bessel, Laguerre, and Weber
0	1	Stokes, harmonic oscillator, and hermite

Helmholtz equation in elliptic coordinates:

$$y'' + (\lambda - 2q\cos 2x)y = 0 \tag{15.4.9}$$

and

$$y'' - (\lambda - 2q\cosh 2x)y = 0 \tag{15.4.10}$$

By making the variable substitution $t = \cos x$, the algebraic form of the Mathieu equation is obtained,

$$(1 - t^2)y'' - ty' + [\lambda + 2q(1 - 2t^2)]y = 0 \tag{15.4.11}$$

It has two regular singularities at $t = \pm 1$ and a non-regular singularity at $t = \infty$.

The Mathieu equation appears in various theoretical models, including the two-dimensional wave equation, orbital stability, seasonal population migration, electronic states in periodic potentials, and Floquet theory. For the properties of the equation and its solutions, readers can refer to literatures.[1]

15.4.3 *Hypergeometric functions*

Different types of Sturm–Liouville equations have different sets of special functions as their eigenfunctions, depending on the three

[1]Zhuxi Wang and Dunren Guo, *Introduction to Special Functions*.

classes of homogeneous boundary conditions or the natural boundary conditions. These special functions are not isolated; they have close relationships with each other, which originate from the number of regular singularities of the hypergeometric equation and their convergence. For the hypergeometric equation with three regular singularities at $z = 0, 1$, and ∞, we have obtained a series solution in the neighborhood of $z = 0$ in Section 9.2, known as the hypergeometric series:

$$w_1(z) = \mathrm{F}(\alpha, \beta, \gamma; z) \equiv \frac{\Gamma(\gamma)}{\Gamma(\alpha)\Gamma(\beta)} \sum_{k=0}^{\infty} \frac{\Gamma(\alpha+k)\Gamma(\beta+k)}{k!\Gamma(\gamma+k)} z^k \tag{15.4.12}$$

The series converges for $|z| < 1$. When $\alpha = 1, \beta = \gamma$, it reduces to the ordinary geometric series:

$$\mathrm{F}(1, \beta, \beta; z) = \sum_{k=0}^{\infty} z^k$$

This is why Equation (15.4.9) is called the hypergeometric series. To ensure convergence of the hypergeometric series at the regular singularity $z = 1$, the infinite series must be truncated to a polynomial. This requires setting the parameters $\alpha = -n$ or $\beta = -n$, which gives rise to the hypergeometric function.

The hypergeometric function possesses many remarkable properties. Here are a few simple recurrence relations:

$$(\gamma-1)\mathrm{F}(\gamma-1) - \alpha\mathrm{F}(\alpha+1) - (\gamma-\alpha-1)\mathrm{F} = 0$$

$$\gamma\mathrm{F} - \beta z\mathrm{F}(\beta+1, \gamma+1) - \gamma\mathrm{F}(\alpha-1) = 0$$

$$\alpha\mathrm{F}(\alpha+1) - \beta\mathrm{F}(\beta+1) - (\alpha-\beta)\mathrm{F} = 0$$

$$\gamma(1-z)\mathrm{F} - \gamma\mathrm{F}(\alpha-1) + (\gamma-\beta)z\mathrm{F}(\gamma+1)\frac{\mathrm{d}}{\mathrm{d}z}\mathrm{F}(\alpha, \beta, \gamma; z)$$

$$= \frac{\alpha\beta}{\gamma}\mathrm{F}(\alpha+1, \beta+1, \gamma+1; z)$$

And $\mathrm{F}(\alpha, \beta, \gamma; z) = \mathrm{F}(\beta, \alpha, \gamma; z)$. Many elementary functions can be expressed in terms of hypergeometric functions. For example,

$$(1+z)^\alpha = \mathrm{F}(-\alpha, \beta, \beta; -z)$$

$$\arcsin z = z\mathrm{F}\left(\frac{1}{2}, \frac{1}{2}, \frac{3}{2}; z^2\right)$$

$$\arctan z = z\mathrm{F}\left(\frac{1}{2}, 1, \frac{3}{2}; -z^2\right)$$

$$\ln(1+z) = z\mathrm{F}(1, 1, 2; -z)$$

The complete elliptic integrals discussed in Section 2.5 can also be expressed in terms of hypergeometric functions:

$$\mathrm{K}(k) \equiv \mathrm{F}\left(\frac{\pi}{2}, k\right) = \frac{\pi}{2}\mathrm{F}\left(\frac{1}{2}, \frac{1}{2}, 1; k^2\right)$$

$$\mathrm{E}(k) \equiv \mathrm{E}\left(\frac{\pi}{2}, k\right) = \frac{\pi}{2}\mathrm{F}\left(-\frac{1}{2}, \frac{1}{2}; 1, k\right)$$

Figure 15.6 depicts the variation of the values of the first and second kind complete elliptic integrals with the modulus k. As $k \to 0$, or $k \to 1$, we have

$$\lim_{k\to 0} \mathrm{K}(k) = \frac{\pi}{2}, \quad \lim_{k\to 0} \mathrm{E}(k) = \frac{\pi}{2}$$

$$\lim_{k\to 1} \mathrm{K}(k) \to \infty, \quad \lim_{k\to 1} \mathrm{E}(k) = 1$$

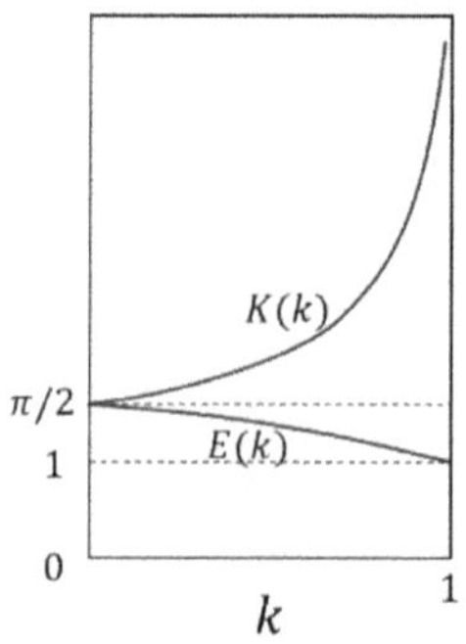

Figure 15.6.

When $\gamma \notin \mathbb{Z}$, the second solution of the hypergeometric equation is given by

$$w_2(z) = z^{1-\gamma}\mathrm{F}(\alpha - \gamma + 1, \beta - \gamma + 1, 2 - \gamma; z) \qquad (15.4.13)$$

When $\gamma \in \mathbb{Z}$, the second solution of the equation is more complex and depends on the specific relationships between α and β. However, it generally exhibits divergence at $z = 0$ or $z = 1$, which is not physically meaningful and is not discussed here.

15.4.4 *Special function classes*

In mathematical physics, many special functions are related to hypergeometric functions. Most equations with regular singular points can be transformed into hypergeometric equations through fractional linear transformations, and thus can be expressed in terms of hypergeometric functions. Here are some examples:

(1) *Jacobi functions*

They are solutions to the following Jacobi equation:

$$(1-x^2)u'' + [\beta - \alpha - (\alpha+\beta+2)x]u' + \lambda(\lambda+\alpha+\beta+1)u = 0$$

By making the variable substitution $x = 1 - z$ and defining $\alpha_1 = \lambda, \beta_1 = \lambda + \alpha + \beta + 1, \gamma_1 = 1 + \alpha$, we obtain a solution to the equation known as the first-kind Jacobi function:

$$\mathrm{P}_\lambda^{(\alpha,\beta)}(z) = \frac{\Gamma(\lambda+\alpha+1)}{\Gamma(\lambda+1)\Gamma(\alpha+1)}\mathrm{F}\left(-\lambda, \lambda+\alpha+\beta+1, 1+\alpha; \frac{1-z}{2}\right)$$

When $\lambda = n$ is a non-negative integer, the Jacobi function is truncated to a Jacobi polynomial of degree n.

(2) *Gegenbauer functions*

They are a special case of the Jacobi functions when $\alpha = \beta = \mu - \frac{1}{2}$, which are defined as

$$\mathrm{C}_\lambda^\mu(z) = \frac{\Gamma(\lambda+2\mu)}{\Gamma(\lambda+1)\Gamma(2\mu)}\mathrm{F}\left(-\lambda, \lambda+2\mu, \mu+\frac{1}{2}; \frac{1-z}{2}\right)$$

When $\lambda = n$ is an integer, the infinite series is truncated to a Gegenbauer polynomial.

(3) *Legendre functions*

They are a special case of the Jacobi functions when $\alpha = \beta = 0$:

$$\mathrm{P}_\lambda(z) = \mathrm{P}_\lambda^{(0,0)}(z) = \mathrm{C}_\lambda^{\frac{1}{2}}(z) = \mathrm{F}\left(-\lambda, \lambda+1, 1; \frac{1-z}{2}\right) \quad (|1-z| < 2)$$

When $\lambda = n$ is an integer, the infinite series is truncated to a Legendre polynomial.

(4) *Associated Legendre functions*

$$P_\lambda^\mu(z) = \frac{1}{\Gamma(1-\mu)}\left(\frac{z+1}{z-1}\right)^{\frac{\mu}{2}} F\left(-\lambda, \lambda+1, 1-\mu; \frac{1-z}{2}\right)$$
$$(|1-z|<2)$$

(5) *Chebyshev functions*

The Chebyshev equation is

$$(1-z^2)u'' - zu' + \lambda u = 0$$

The solutions to this equation are Chebyshev functions $T_\lambda(z)$, which can be expressed in terms of hypergeometric functions as

$$T_\lambda(z) = F\left(\lambda, -\lambda, \frac{1}{2}; \frac{1-z}{2}\right)$$

(6) *Confluent hypergeometric functions*

As described in Section 9.2, a hypergeometric equation can be transformed into a confluent hypergeometric equation through suitable transformations. The equation becomes

$$zu'' + (\gamma - z)u' - \alpha u = 0$$

where $z = 0$ remains a regular singular point, but the regular singular points $z = 1$ and $z = \infty$ merge into a non-regular singular point $z = \infty$. The solution to this equation is called the confluent hypergeometric function or Kummer function:

$$F(\alpha, \gamma; z) = \lim_{\beta\to 0} F\left(\alpha, \beta, \gamma; \frac{z}{\beta}\right) = \frac{\Gamma(\gamma)}{\Gamma(\alpha)}\sum_{k=0}^{\infty}\frac{\Gamma(\alpha+k)}{\Gamma(k+1)\Gamma(\gamma+k)}z^k$$

When $\alpha = -n$ is an integer, $F(\alpha, \gamma; z)$ is truncated to a polynomial.

(7) *Whittaker functions*

In the confluent hypergeometric equation, by setting

$$y = e^{\frac{z}{2}} z^{-\frac{\gamma}{2}} w(z)$$

the first-order derivative term can be eliminated, resulting in the Whittaker equation:

$$w'' + \left[-\frac{1}{4} + \left(\frac{\gamma}{2} - \alpha\right)\frac{1}{z} + \frac{\gamma}{2}\left(1 - \frac{\gamma}{2}\right)\frac{1}{z^2}\right] w = 0$$

Setting $\gamma = 1 + 2m$ and $\frac{\gamma}{2} - \alpha = k$, we arrive at

$$w'' + \left(-\frac{1}{4} + \frac{k}{z} + \frac{1/4 - m^2}{z^2}\right) w = 0$$

The solutions to this equation are called Whittaker functions:

$$\mathrm{M}_{k,m}(z) = \mathrm{e}^{-\frac{z}{2}} z^{\frac{\gamma}{2}} \mathrm{F}(\alpha, \gamma; z) = \mathrm{e}^{-\frac{z}{2}} z^{\frac{1}{2}+m} \mathrm{F}\left(\frac{1}{2} + m - k, 1 + 2m; z\right)$$

For more general forms of the Whittaker equation:

$$y'' + \left(\frac{1 - \lambda - 2\beta}{z} - 2\lambda\alpha z^{\lambda-1}\right) y' + \left[\lambda^2\left(\alpha^2 - \frac{A^2}{4}\right) z^{2\lambda-2} + \lambda(2\alpha\beta + Ak\lambda) z^{\lambda-2} + \frac{\beta(\beta + \lambda) + \lambda^2(1/4 - m^2)}{z^2}\right] y = 0$$

The solutions are expressed in terms of Whittaker functions as

$$y(z) = z^{\beta} \mathrm{e}^{\alpha\lambda z} \mathrm{M}_{k,m}(A z^{\lambda})$$

Essentially, some of the equations and their special functions that we have encountered before are special cases of the general Whittaker equation or confluent hypergeometric equation. We list some below.

• Bessel equation

Taking $\lambda = 1, \alpha = 0, \beta = -\frac{1}{2}, k = 0, A = 2\mathrm{i}$, we have

$$y'' + \frac{1}{z} y' + \left(1 - \frac{m^2}{z^2}\right) y = 0$$

The Bessel equation shares the same singularity structure as the confluent hypergeometric equation: $z = 0$ is a regular singular point, and $z = \infty$ is a non-regular singular point. The solution, Bessel function, can be expressed in terms of Kummer functions as

$$\mathrm{J}_{\nu}(z) = z^{\nu} \mathrm{e}^{-\mathrm{i}z} \mathrm{F}\left(\nu + \frac{1}{2}, 2\nu + 1; z\right)$$

• Weber equation

Taking $\lambda = 2, \alpha = 0, \beta = -\frac{1}{2}, k = \frac{n}{2} + \frac{1}{4}, m = \pm\frac{1}{4}, A = \frac{1}{2}$, we have

$$y'' + \left(n + \frac{1}{2} - \frac{z^2}{4}\right) y = 0$$

The solution to the Weber equation is the Weber function:

$$\mathrm{D}_n(z) = 2^{\frac{n}{2}+\frac{1}{4}} z^{-\frac{1}{2}} \mathrm{M}_{\frac{n}{2}+\frac{1}{4},-\frac{1}{4}}\left(\frac{z^2}{2}\right)$$

• Hermite equation

Taking $\lambda = 2$, $\alpha = \frac{1}{2}$, $\beta = -\frac{1}{2}$, $k = \frac{n}{2}$, $m = \pm\frac{1}{4}$, $A = 1$, we have

$$y'' - 2zy' + 2ny = 0$$

The solution to the Hermite equation can be truncated to the Hermite polynomials:

$$\mathrm{H}_n(z) = 2^{\frac{n}{2}} \mathrm{e}^{\frac{z^2}{2}} \mathrm{D}_n(\sqrt{2}z)$$

• Laguerre equation

Taking $\lambda = 1$, $\alpha = \frac{1}{2}$, $\beta = -\frac{1+\mu}{2}$, $k = n + \frac{1}{2}(1+\mu)$, $m = \frac{\mu}{2}$, $A = 1$, we have

$$zy'' + (\mu + 1 - z)y' + ny = 0$$

The solution to the Laguerre equation can be truncated to the Laguerre polynomials:

$$\mathrm{L}_n^{\mu}(z) = \frac{\Gamma(\mu + 1 + n)}{n!\Gamma(\mu + 1)} \mathrm{F}(-n, \mu + 1; z)$$

ENDNOTE

In the neighborhood of a singularity, the solutions of a differential equation can be represented by series expansions. However, the appropriate form of the series must be determined before the calculations, and this information can only be obtained from the differential equation itself. Although the coefficients $p(z)$ and $q(z)$ of the differential equation (15.4.1) are assumed to have single-valued analytic solutions outside certain isolated singularities, the solutions of the equation may not be single-valued throughout the entire region under consideration. Therefore, when the solutions $w_i(z)$ of the equation along a closed path enclosing the singularities are traversed, they may change to another branch of the same function, denoted as $\tilde{w}_i(z)$. It can still be expressed as a linear combination of two solutions:

$$\begin{cases} \tilde{w}_1(z) = c_{11}w_1(z) + c_{12}w_2(z) \\ \tilde{w}_2(z) = c_{21}w_1(z) + c_{22}w_2(z) \end{cases}$$

In other words, when z completes a closed loop around the singularities, the two solutions of the equation undergo a linear transformation. These transformations form a group known as the monodromy group of the differential equation. Riemann demonstrated that the properties of the solutions of a second-order differential equation depend entirely on the variation of the two linearly independent solutions when the independent variable revolves around three singularities, without needing to know the specific form of the differential equation itself. Thus, knowledge of the monodromy group can be used to derive the properties of the functions defined by the differential equation.

Inspired by Riemann's work, Fuchs further advanced the study of singularities. Instead of following Riemann's path, he directly started from differential equations with rational functions of z as coefficients, and by examining the convergence of series, he discovered that the singularities of the solutions correspond to the poles of the coefficient functions.

Special functions originate from the series solutions of ordinary differential equations with variable coefficients. The hypergeometric equation and its series solution $\mathrm{F}(\alpha, \beta, \gamma; z)$ were first studied by Euler. Generally, the series converges for $|z| < 1$, and for $z = 1$, the series converges if and only if $\alpha + \beta < \gamma$, while for $z = -1$, the series converges if and only if $\alpha+\beta<\gamma+1$. Gauss established the famous relationship:

$$\mathrm{F}(\alpha,\beta,\gamma;1)=\frac{\Gamma(\gamma)\Gamma(\gamma-\alpha-\beta)}{\Gamma(\gamma-\alpha)\Gamma(\gamma-\beta)}\quad(\alpha+\beta<\gamma)$$

Gauss realized that when the parameters (α,β,γ) take certain special values, this series encompasses almost all known elementary functions, as well as transcendental functions such as Bessel functions and Legendre functions. It really can be considered the mother of all functions.

In the separation of variables for the Laplace equation in n-dimensional hyperspherical coordinates (see Section 10.5), the angular part satisfies the following ODEs ($k=1,2,\ldots,$ $n-2$):

$$\frac{1}{\sin^{n-k-1}\theta_k}\frac{\mathrm{d}}{\mathrm{d}\theta_k}\left(\sin^{n-k-1}\theta_k\frac{\mathrm{d}}{\mathrm{d}\theta_k}\right)\Theta_k+\left(\lambda_k-\frac{\lambda_{k+1}}{\sin^2\theta_k}\right)\Theta_k=0$$

where λ_k are separation constant. Let $x_k=\cos\theta_k$, the above equation becomes

$$(1-x_k^2)\frac{\mathrm{d}^2\Theta_k}{\mathrm{d}x_k^2}-(n-k)x_k\frac{\mathrm{d}\Theta_k}{\mathrm{d}x_k}+\left(\lambda_k-\frac{\lambda_{k+1}}{1-x_k^2}\right)\Theta_k=0$$

The solutions of this equation are the Jacobi functions when $\alpha_k=\beta_k=(n-k-2)/2$:

$$\Theta_k=\mathrm{P}_{\lambda_k}^{(\alpha_k,\beta_k)}(x_k)$$

When $\lambda_k=l_k(l_k+n-k-1)$ $(l_k\in\mathbb{N})$, the Jacobi functions are truncated to Jacobi polynomials, where $0\le l_{n-1}\le\cdots\le l_1$.

For three-dimensional space, the Jacobi functions degenerate into associated Legendre functions.

Exercises

[1] Prove that the equation has an essential singularity at $z = 0$:

$$w' + \frac{1}{z^2}w = 0$$

[2] Prove that $z = \infty$ is an irregular singularity of the confluent hypergeometric equation.

[3] Prove that the Legendre equation has three regular singularities: $z = \pm 1$ and $z = \infty$.

[4] Prove that the Bessel equation has one regular singularity at $z = 0$ and one irregular singularity at $z = \infty$.

[5] Prove the following:

$$\text{(a) } \arcsin z = z\mathrm{F}\left(\frac{1}{2}, \frac{1}{2}, \frac{3}{2}; z^2\right); \quad \text{(b) } \ln(1+z) = z\mathrm{F}(1, 1, 2; -z)$$

[6] Let $0 \leq k^2 < 1$. The complete elliptic integrals of the first and second kind are respectively defined as:

$$\mathrm{K}(k) = \int_0^{\frac{\pi}{2}} \frac{1}{\sqrt{1 - k^2 \sin^2\theta}} \mathrm{d}\theta, \quad \mathrm{E}(k) = \int_0^{\frac{\pi}{2}} \sqrt{1 - k^2 \sin^2\theta}\, \mathrm{d}\theta$$

Prove the following:

$$\mathrm{K}(k) = \frac{\pi}{2}\mathrm{F}\left(\frac{1}{2}, \frac{1}{2}, 1; k^2\right), \quad E(k) = \frac{\pi}{2}\mathrm{F}\left(-\frac{1}{2}, \frac{1}{2}, 1; k^2\right)$$

Hint: Expand the integrand as a power series, integrate over the interval $\left[0, \frac{\pi}{2}\right]$, and use the Beta function integral formula:

$$\int_0^{\frac{\pi}{2}} \cos^n\theta \,\mathrm{d}\theta = \int_0^{\frac{\pi}{2}} \sin^n\theta \,\mathrm{d}\theta = \frac{\sqrt{\pi}[(n-1)/2]!}{2(n/2)!}$$

[7] Consider a second-order Fuchsian equation with three regular singularities at $z = 0, 1$ and ∞. Prove that the most general forms of $p(z)$ and $q(z)$ are

$$p(z) = \frac{A_1}{z} + \frac{B_1}{z-1}, \quad q(z) = \frac{A_2}{z^2} + \frac{B_2}{(z-1)^2} - \frac{A_3}{z(z-1)}$$

[8] Let the Fuchsian equation have three regular singularities with characteristic indices $(\lambda_1, \lambda_2), (\mu_1, \mu_2)$, and (ν_1, ν_2). Prove Riemann's identity:

$$\lambda_1 + \lambda_2 + \mu_1 + \mu_2 + \nu_1 + \nu_2 \equiv 1$$

15.5 Confluent Hypergeometric Function

In this section, we introduce another typical application of a special function known as the confluent hypergeometric equation:

$$xy'' + (\gamma - x)y' - \alpha y = 0$$

Here, $x = 0$ is a regular singular point of the equation, and $x = \infty$ is an irregular singular point. In Chapter 9, we obtained a series solution of the equation, which is known as the confluent hypergeometric function or Kummer function:

$$\mathrm{F}(\alpha, \gamma, x) = \sum_{k=0}^{\infty} \frac{\Gamma(k+\alpha)\Gamma(\gamma)}{k!\Gamma(\alpha)\Gamma(k+\gamma)} x^k \quad (|x| < \infty)$$

The convergence radius of this series is $R = \infty$, but it diverges at $x = \infty$. For bound states in quantum mechanics, it is necessary to truncate the function and express it as a polynomial. Below, we demonstrate this by solving the Schrödinger equation for the hydrogen atom.

Example 15.6. Assuming natural units $\hbar = m = 1$, solve the three-dimensional time-independent Schrödinger equation for a hydrogen-like atom:

$$-\frac{1}{2}\nabla^2\psi - \frac{Ze^2}{r}\psi = E\psi$$

Solution: We separate the variables in the spherical coordinates as $\psi(r, \theta, \varphi) = R(r)\mathrm{Y}_{lm}(\theta\varphi)$, where the angular part corresponds to the spherical harmonics $\mathrm{Y}_{lm}(\theta, \varphi)$, with eigenvalues $l(l+1)$. The radial part satisfies the equation:

$$-\frac{\mathrm{d}}{\mathrm{d}r}\left(r^2\frac{\mathrm{d}R}{\mathrm{d}r}\right) + [-2Ze^2r + l(l+1)]R = 2Er^2R$$

It can be verified that $r = 0$ is a regular singular point, while $r = \infty$ is a non-regular singularity. Since it is uncertain whether a natural boundary condition exists at infinity, expressing the solution directly as a Frobenius series may be inappropriate. Therefore, we first study the limiting behavior of the equation. By setting $u = rR(r)$ to eliminate the first-order derivative term, we have

$$\frac{\mathrm{d}^2u}{\mathrm{d}r^2} + \left[2E + \frac{2Ze^2}{r} - \frac{l(l+1)}{r^2}\right]u = 0$$

As $r \to 0$, the equation can be asymptotically represented as

$$\frac{\mathrm{d}^2 u}{\mathrm{d}r^2} - \frac{l(l+1)}{r^2}u = 0$$

Considering the finiteness of the physical solution, we take $u \sim r^{l+1}$. On the other hand, as $r \to \infty$, the equation can be asymptotically expressed as

$$\frac{\mathrm{d}^2 u}{\mathrm{d}r^2} + 2Eu = 0$$

Since we are only interested in bound states ($E < 0$), we take $u \sim \mathrm{e}^{-\sqrt{-2E}r}$. Combining the two limiting behaviors, we introduce a function $y(r)$ as $u = r^{l+1}\mathrm{e}^{-\beta r}y(r)$, where $\beta = \sqrt{-2E}$. The ODE for function $y(r)$ reads,

$$ry'' + [2(l+1) - 2\beta r]y' - 2[(l+1)\beta - 1]y = 0$$

The solution of this equation may behavior more regularly. By defining the parameters,

$$\xi = 2\beta r, \quad \gamma = 2(l+1), \quad \alpha = (l+1) - \frac{1}{\beta}$$

the equation can be simplified to

$$\xi y'' + (\gamma - \xi)y' - \alpha y = 0$$

Its standard Sturm–Liouville eigenvalue equation is

$$-\frac{\mathrm{d}}{\mathrm{d}\xi}\left(\xi^{\gamma}\mathrm{e}^{-\xi}\frac{\mathrm{d}}{\mathrm{d}\xi}y\right) = -\alpha\xi^{\gamma-1}\mathrm{e}^{-\xi}y$$

where

$$p(\xi) = \xi^{\gamma}\mathrm{e}^{-\xi}, \quad q(\xi) = 0, \quad \rho(\xi) = \xi^{\gamma-1}\mathrm{e}^{-\xi}$$

As desired, this equation satisfies natural boundary conditions at $\xi = 0$ and $\xi = \infty$, thus forming a complete eigenvalue problem. An infinite series solution of the equation is the Kummer function,

$$y(\xi) = \mathrm{F}(\alpha, \gamma, \xi) = \sum_{k=0}^{\infty} \frac{\Gamma(k+\alpha)\Gamma(\gamma)}{k!\Gamma(\alpha)\Gamma(k+\gamma)}\xi^k$$

Although the convergence radius of the Kummer series is infinite, it still diverges at $\xi = \infty$. Therefore, it is necessary to truncate the infinite series into a polynomial by setting the parameter α to either zero or a negative integer:

$$\alpha = -n_r \quad (n_r = 0, 1, 2, \ldots)$$

Let $n = n_r + l + 1$, then $\beta = \frac{1}{n}$. Thus,

$$E = -\frac{1}{2}\beta^2 = -\frac{1}{2n^2}$$

In international units, we obtain the quantized energy levels for hydrogen-like atoms:

$$E_n = -\frac{mZ^2e^4}{2\hbar^2}\frac{1}{n^2} = -\frac{Z^2e^2}{2a}\frac{1}{n^2} \quad (n = 1, 2, 3, \ldots)$$

Here, $a = \frac{\hbar^2}{me^2}$ is known as the Bohr radius. The normalized radial wave function is given by

$$R_{nl}(r) = N_{nl}\xi^l \mathrm{e}^{-\frac{\xi}{2}}\mathrm{F}(-n + l + 1, 2l + 2, \xi)$$

The normalization coefficient is

$$N_{nl} = \frac{2}{a^{3/2}n^2(2l+1)!}\sqrt{\frac{(n+l)!}{(n-l-1)!}}$$

Combining it with the angular part given by the spherical harmonics, the complete eigenwave function for hydrogen-like atoms is then

$$\psi_{nlm}(r, \theta, \varphi) = R_{nl}(r)\mathrm{Y}_l^m(\theta, \varphi)$$
$$(l = 0, 1, 2, \ldots, n-1, m = -l, -l+1, \ldots, l)$$

Here are several low-order radial wave functions $R_{nl}(r)$:

$$R_{10}(r) = \left(\frac{Z}{a}\right)^{3/2} 2\exp\left(-\frac{Z}{a}r\right)$$

$$R_{20}(r) = \left(\frac{Z}{2a}\right)^{3/2}\left(2 - \frac{Z}{a}r\right)\exp\left(-\frac{Z}{2a}r\right)$$

$$R_{21}(r) = \left(\frac{Z}{2a}\right)^{3/2}\frac{Z}{\sqrt{3}a}r\exp\left(-\frac{Z}{2a}r\right)$$

$$R_{30}(r) = \left(\frac{Z}{3a}\right)^{3/2} \left[2 - \frac{4Z}{3a}r + \frac{4}{27}\left(\frac{Z}{a}r\right)^2\right] \exp\left(-\frac{Z}{3a}r\right)$$

$$R_{31}(r) = \left(\frac{2Z}{a}\right)^{3/2} \left(\frac{2}{27\sqrt{3}} - \frac{Z}{81\sqrt{3}a}r\right) \frac{Z}{a}r \exp\left(-\frac{Z}{3a}r\right)$$

$$R_{32}(r) = \left(\frac{2Z}{a}\right)^{3/2} \frac{Z}{81\sqrt{15}} \left(\frac{Z}{a}r\right)^2 \exp\left(-\frac{Z}{3a}r\right)$$

$$\cdots$$

For each set of eigenvalues (n, l, m), the probability density of an electron in space is given by $|\psi_{nlm}(r, \theta, \varphi)|^2$. Figure 15.7 shows the distribution shape of the electron in three-dimensional space, commonly referred to as an electron cloud.

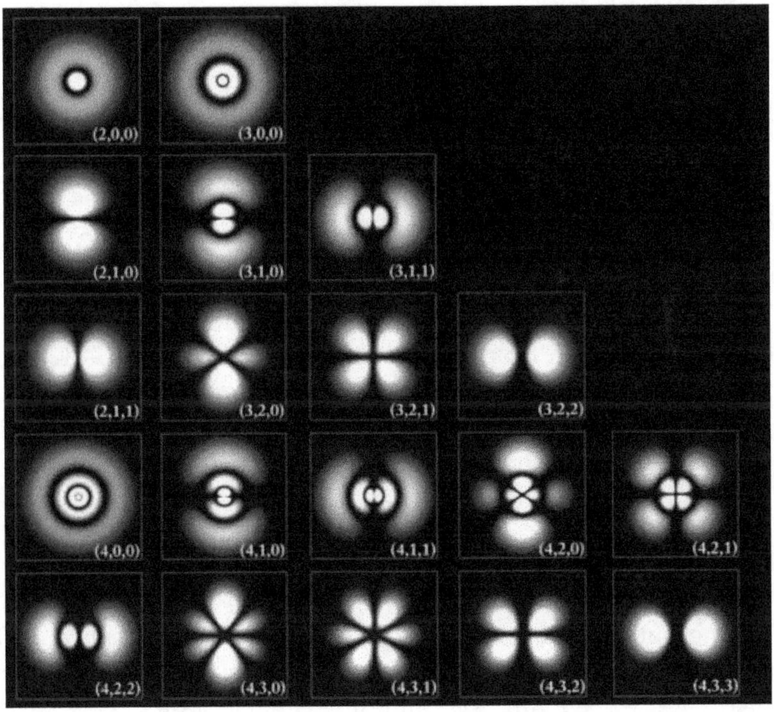

Figure 15.7. (image from Wikipedia).

ENDNOTE

If $r = \infty$ is a non-regular singularity of the equation, there usually are no natural boundary conditions. In specific cases, a function transformation can be performed, such as $u \longmapsto y$, to change the singularity at $r = \infty$ and make the function y satisfy the Sturm–Liouville eigenvalue equation and natural boundary conditions. The divergent series can then be truncated into a polynomial, and the eigenvalues determined accordingly. In quantum mechanics, it is customary to assume that the wave function $\psi(r) \to 0$ as $r \to \infty$, which is an additional physical requirement outside the Schrödinger equation. In fact, this is a natural consequence of the equation itself implying the presence of natural boundary conditions. As they say, the equation determines the physics.

The Bessel equation has a non-regular singularity at $x = \infty$, and the series solution has an infinite convergence radius, so there is no need for truncation. However, since there is a lack of natural boundary conditions at $x = \infty$, homogeneous boundary conditions must be imposed on a finite radius circumference and combined with the natural boundary at $x = 0$ to form a complete eigenvalue problem.

In quantum mechanics, the three eigenvalues of the wave function, n, l and m, are referred to as the principal quantum number, angular momentum quantum number, and magnetic quantum number, respectively. Here, $l = 0, 1, \ldots, n-1$, and $m = -l, -l+1 \ldots, l$. Since the quantized energy levels E_n of the electron only depend on n and are independent of l and m, the degeneracy (number of states) of the nth energy level is

$$w_n = \sum_{l=0}^{n-1}(2l+1) = n^2$$

Taking into account the two possible spin orientations of the electron, $s_z = \pm\frac{1}{2}\hbar$, and the Pauli exclusion principle, each energy level can accommodate $2n^2$ electrons. These $2n^2$ orbitals have the same energy and form the electron shell structure, successfully explaining atomic structure and the periodic table.

Nature has shown great generosity in revealing its mysteries to humanity. By allowing the exact solution of the motion equation for electrons in hydrogen atoms, we are able to gain profound insights into the hidden secrets of the microcosm. When Schrödinger obtained the analytical solution for the hydrogen atom system, he was greatly encouraged and his confidence in the quantum mechanical equations he discovered was strengthened. Imagining a scenario where the hydrogen atom equation had no analytical solution, the establishment of quantum mechanics would have been a long and painful process. It would have been challenging to understand the various quantum numbers that govern electron motion and transition rules, and consequently, our understanding of spectra, atomic structure, and the periodic table would not be as clear as it is today. Not to mention the comprehension of spontaneous transitions between different orbitals. Without these profound insights, our understanding of numerous microcosmic phenomena, including nuclear physics, would be superficial at best. How lucky!

Exercises

[1] Prove that the Weber–Hermite equation can be transformed into the confluent hypergeometric equation:

$$y'' + \left(\nu + \frac{1}{2} - \frac{1}{4}z^2\right) y = 0$$

Hint: Perform the variable substitution $y(z) = \mathrm{e}^{-\frac{z^2}{4}} u(z)$.

[2] Prove that the one-dimensional quantum mechanical stationary equation for harmonic oscillator motion constitutes a Sturm–Liouville eigenvalue problem:

$$-\frac{1}{2}\frac{\mathrm{d}^2}{\mathrm{d}x^2}\psi + \frac{1}{2}\omega^2 x^2 \psi = E\psi$$

Hint: $x = \pm\infty$ are non-regular singular points. Consider the limiting behavior to change their singularity. Let $\xi = \sqrt{\omega}x$, $E = \lambda\omega$, and $\psi = \mathrm{e}^{-\frac{\xi^2}{2}}u$. This can be transformed into the Hermite equation:

$$\frac{\mathrm{d}^2}{\mathrm{d}\xi^2}u - 2\xi\frac{\mathrm{d}}{\mathrm{d}\xi}u + (\lambda - 1)u = 0$$

or

$$-\frac{\mathrm{d}}{\mathrm{d}\xi}\left(\mathrm{e}^{-\xi^2}\frac{\mathrm{d}}{\mathrm{d}\xi}u\right) + \mathrm{e}^{-\xi^2}u = \lambda\mathrm{e}^{-\xi^2}u$$

Therefore, $p(\xi) = \mathrm{e}^{-\xi^2}$. The function $u(\xi)$ has natural boundary conditions at $\xi = \pm\infty$, constituting a Sturm–Liouville eigenvalue problem.

[3] Prove that the radial stationary equation for three-dimensional harmonic oscillator motion constitutes a Sturm–Liouville eigenvalue problem:

$$-\frac{1}{2}\frac{1}{r^2}\frac{\mathrm{d}}{\mathrm{d}r}\left(r^2\frac{\mathrm{d}}{\mathrm{d}r}\psi\right) + \frac{l(l+1)}{2r^2}\psi + \frac{1}{2}\omega^2r^2\psi = E\psi$$

Chapter 16

Green's Functions

16.1 Fundamental Theory

In the previous chapters, we have learned about solving various boundary value problems using the method of separation of variables or integral transform methods. In this chapter, we introduce another integral method called the Green's function method. The Green's function, also known as the point source influence function, represents the field distribution generated by a point source under certain boundary or initial conditions. For linear differential equations, knowing the Green's function allows us to use the principle of linear superposition and the given boundary conditions to calculate the field distribution generated by continuous distributed sources in integral form. Therefore, the Green's function method is commonly used to deal with non-homogeneous linear differential equation problems with external sources.

16.1.1 *Formal description*

Let the operator $\hat{L}$ act on the vector $|a\rangle$ in the linear vector space $\mathbb{V}$, resulting in a vector $|b\rangle = \hat{L}|a\rangle$. If $\hat{L}$ is non-singular, then there exists an inverse operator $\hat{L}^{-1}$ such that $|a\rangle = \hat{L}^{-1}|b\rangle$. For a non-homogeneous nth order linear ODE,

$$\hat{L}u(x) = f(x)$$

where the differential operator is

$$\hat{L} = p_n(x)\frac{\mathrm{d}^n}{\mathrm{d}x^n} + p_{n-1}(x)\frac{\mathrm{d}^{n-1}}{\mathrm{d}x^{n-1}} + \cdots + p_1(x)\frac{\mathrm{d}}{\mathrm{d}x} + p_0(x) \quad (16.1.1)$$

We can formally solve for

$$u(x) = \hat{L}^{-1} f(x)$$

provided that the linear differential operator $\hat{L}$ in the Hilbert space $\mathcal{H}$ has an inverse operator $\hat{G} \equiv \hat{L}^{-1}$. Since $\hat{L}$ is a differential operator, it can be inferred that its inverse operator $\hat{G}$ is an integral operator, known as the Green's operator. The solution to the equation can be abstractly represented as

$$|u\rangle = \hat{G}|f\rangle$$

By left-multiplying the equation by $\langle x|$ and inserting the complete relation $\hat{I} = \int \mathrm{d}x'|x'\rangle\langle x'|\rho(x')$ between $\hat{G}$ and $|f\rangle$,

$$\langle x|u\rangle = \int_\Omega \langle x|\hat{G}|x'\rangle\langle x'|f\rangle\rho(x')\mathrm{d}x'$$

we obtain the integral form of the non-homogeneous equation solution:

$$u(x) = \int_\Omega G(x, x') f(x')\rho(x')\mathrm{d}x' \quad (16.1.2)$$

where the integral kernel $G(x, x') = \langle x|\hat{G}|x'\rangle$ is called the Green's function.

Applying the differential operator $\hat{L}$ to the formal solution,

$$\hat{L}u(x) = \int_\Omega \hat{L}G(x, x') f(x')\rho(x')\mathrm{d}x' = f(x)$$

we have

$$\hat{L}G(x, x') = \frac{\delta(x - x')}{\rho(x')} \quad (16.1.3)$$

This equation is called the Green's function equation for the differential operator $\hat{L}$. When taking the weighting function $\rho(x') = 1$, the Green's function equation reduces to the familiar form:

$$\hat{L}G(x, x') = \delta(x - x') \quad (16.1.4)$$

It indicates the relationship between the Green's function and the Dirac δ function, representing the field distribution generated by a point source in space. Therefore, the Green's function can be viewed as a distribution function.

The question now is, under what conditions is the operator $\hat{L}$ reversible? We know that for finite-dimensional linear vector spaces, an operator is reversible when its matrix determinant is non-zero. Operators in infinite-dimensional Hilbert spaces do not have a matrix representation, so how can we determine their reversibility? It can be proven that the operator $\hat{L}$ is reversible when the homogeneous equation $\hat{L}|u\rangle = 0$ has no non-trivial solution, or in other words, if the equation has a non-zero solution $|u\rangle \neq 0$, then its corresponding eigenvalue is $\lambda = 0$. This proposition is stated as follows:

The sufficient and necessary condition for an operator $\hat{L}$ to be reversible in a Hilbert space is that $\lambda = 0$ is not an eigenvalue of $\hat{L}$.

Since the eigenvalue problem is related to boundary conditions, the reversibility of a differential operator is also dependent on boundary conditions. It significantly differs to the linear algebra of finite dimensional space.

16.1.2 *Green's integral formula*

The integral solution (16.1.2) for the previous non-homogeneous equation is an illustrative derivation and not the final conclusion since it does not consider the influence of boundaries. In fact, even without external sources, as long as finite boundaries exist, the system will have non-trivial field distributions. The rigorous theoretical derivation is as follows:

$$\hat{L}u(\boldsymbol{r}) = f(\boldsymbol{r})$$

Taking $\rho(\boldsymbol{r}) = 1$ and using the Green's function equation,

$$\hat{L}G(\boldsymbol{r}, \boldsymbol{r}') = \delta(\boldsymbol{r} - \boldsymbol{r}')$$

we have

$$G(\boldsymbol{r}, \boldsymbol{r}')\hat{L}u(\boldsymbol{r}) - u(\boldsymbol{r})\hat{L}G(\boldsymbol{r}, \boldsymbol{r}') = G(\boldsymbol{r}, \boldsymbol{r}')f(\boldsymbol{r}) - u(\boldsymbol{r})\delta(\boldsymbol{r} - \boldsymbol{r}')$$

Integrating both sides with respect to $\boldsymbol{r}$ yields

$$u(\boldsymbol{r}') = \int_\Omega G(\boldsymbol{r}, \boldsymbol{r}')f(\boldsymbol{r})\mathrm{d}\boldsymbol{r} - \int_\Omega [G(\boldsymbol{r}, \boldsymbol{r}')\hat{L}u(\boldsymbol{r}) - u(\boldsymbol{r})\hat{L}G(\boldsymbol{r}, \boldsymbol{r}')]\mathrm{d}\boldsymbol{r}$$

Alternatively,

$$u(\boldsymbol{r}) = \int_{\Omega'} G(\boldsymbol{r}', \boldsymbol{r}) f(\boldsymbol{r}') \mathrm{d}\boldsymbol{r}' - \int_{\Omega'} [G(\boldsymbol{r}', \boldsymbol{r}) \hat{L} u(\boldsymbol{r}') - u(\boldsymbol{r}') \hat{L} G(\boldsymbol{r}', \boldsymbol{r})] \mathrm{d}\boldsymbol{r}' \tag{16.1.5}$$

where $G(\boldsymbol{r}', \boldsymbol{r})$ is the transpose of $G(\boldsymbol{r}, \boldsymbol{r}')$. The second term on the right-hand side can be transformed into a surface integral, which is related to the field distribution $u(\boldsymbol{r}')|_{\Sigma}$ on the boundary (surface).

The adjoint equation of the Green's function is

$$\hat{L}^{\dagger} \tilde{G}(\boldsymbol{r}, \boldsymbol{r}'') = \delta(\boldsymbol{r} - \boldsymbol{r}'')$$

$\tilde{G}(\boldsymbol{r}, \boldsymbol{r}')$ is called the adjoint Green's function, and combined with Equation (16.1.4), we have

$$\begin{aligned} &\tilde{G}^*(\boldsymbol{r}, \boldsymbol{r}'') \hat{L} G(\boldsymbol{r}, \boldsymbol{r}') - G(\boldsymbol{r}, \boldsymbol{r}') [\hat{L}^{\dagger} \tilde{G}(\boldsymbol{r}, \boldsymbol{r}'')]^* \\ &= \tilde{G}^*(\boldsymbol{r}, \boldsymbol{r}'') \delta(\boldsymbol{r} - \boldsymbol{r}') - G(\boldsymbol{r}, \boldsymbol{r}') \delta(\boldsymbol{r} - \boldsymbol{r}'') \end{aligned}$$

Integrating both sides with respect to $\boldsymbol{r}$ and using the definition of the adjoint operator, we find that the left-hand side is zero. Therefore,

$$\tilde{G}^*(\boldsymbol{r}, \boldsymbol{r}') = G(\boldsymbol{r}', \boldsymbol{r}) \tag{16.1.6}$$

It reveals that the Green's function and the adjoint Green's function satisfy the transpose complex conjugate relation.

If the operator $\hat{L}$ is self-adjoint, $\hat{L}^{\dagger} = \hat{L}$, then their Green's functions satisfy the same boundary conditions, we have

$$G(\boldsymbol{r}, \boldsymbol{r}') = G^*(\boldsymbol{r}', \boldsymbol{r})$$

In particular, when the coefficients of the differential operator $\hat{L}$ are real, the Green's function is also real, leading to

$$G(\boldsymbol{r}, \boldsymbol{r}') = G(\boldsymbol{r}', \boldsymbol{r}) \tag{16.1.7}$$

This leads to the Green's integral formula:

$$u(\boldsymbol{r}) = \int_{\Omega'} G(\boldsymbol{r}, \boldsymbol{r}') f(\boldsymbol{r}') \mathrm{d}\boldsymbol{r}' - \int_{\Omega'} [G(\boldsymbol{r}, \boldsymbol{r}') \hat{L} u(\boldsymbol{r}') - u(\boldsymbol{r}') \hat{L} G(\boldsymbol{r}, \boldsymbol{r}')] \mathrm{d}\boldsymbol{r}' \tag{16.1.8}$$

When there are no external sources, $f(\boldsymbol{r}') = 0$, the field distribution in space is determined solely by the values on the boundary. Before further discussing equation (16.1.8), let us examine an example of a Green's function for a non-self-adjoint operator.

Example 16.1. Find the Green's function and the adjoint Green's function of operator $\hat{L}$,

$$\begin{cases} \hat{L}u = u'' - 2u' - 3u \\ u(0) = 0, \quad u'(1) = 0 \end{cases}$$

Solution: According to the definition of the adjoint operator:

$$v\hat{L}u - u\hat{L}^{\dagger}v = 0$$

Following the method in Section 14.3, integration by parts yields the adjoint equation and the adjoint boundary conditions:

$$\begin{cases} \hat{L}^{\dagger}v = v'' + 2v' - 3v \\ v(0) = 0, \quad v'(1) + 2v(1) = 0 \end{cases}$$

Thus, $\hat{L}$ is not a self-adjoint operator. The Green's function equation $\hat{L}G = \delta(x - x')$ has two linearly independent solutions for $x < x'$ and $x > x'$: e^{-x} and e^{3x}, respectively. According to the boundary condition requirements, the Green's function can be expressed as

$$\begin{cases} G(x, x') = A(\mathrm{e}^{-x} - \mathrm{e}^{3x}) & (x < x') \\ G(x, x') = B\left(\mathrm{e}^{-x} + \dfrac{1}{3}\mathrm{e}^{3x-4}\right) & (x > x') \end{cases}$$

From the continuity of the Green's function at $x = x'$ and the jump conditions for derivatives (by integrating the Green's function equation once), we have

$$\begin{cases} A(\mathrm{e}^{-x'} - \mathrm{e}^{3x'}) = B\left(\mathrm{e}^{-x'} + \dfrac{1}{3}\mathrm{e}^{3x'-4}\right) \\ B(-\mathrm{e}^{-x'} + \mathrm{e}^{3x'-4}) - A(-\mathrm{e}^{-x'} - 3\mathrm{e}^{3x'}) = 1 \end{cases}$$

Solving for the Green's function gives

$$G(x,x') = \begin{cases} \dfrac{1}{4(1+3\mathrm{e}^4)}(\mathrm{e}^{-x} - \mathrm{e}^{3x})(\mathrm{e}^{x'} + 3\mathrm{e}^{4-3x'}) & (x < x') \\ \dfrac{1}{4(1+3\mathrm{e}^4)}(\mathrm{e}^{-3x'} - \mathrm{e}^{x'})(\mathrm{e}^{3x} + 3\mathrm{e}^{4-x}) & (x > x') \end{cases}$$

Similarly, solving the adjoint Green's function equation $\hat{L}^\dagger\tilde{G} = \delta(x-x')$ gives the adjoint Green's function:

$$\tilde{G}(x,x') = \begin{cases} \dfrac{1}{4(1+3\mathrm{e}^4)}(\mathrm{e}^{-3x} - \mathrm{e}^{x})(\mathrm{e}^{3x'} + 3\mathrm{e}^{4-x'}) & (x < x') \\ \dfrac{1}{4(1+3\mathrm{e}^4)}(\mathrm{e}^{-x'} - \mathrm{e}^{3x'})(\mathrm{e}^{x} + 3\mathrm{e}^{4-3x}) & (x > x') \end{cases}$$

The result verifies that the Green's function and its adjoint satisfy

$$\tilde{G}(x,x') = G(x',x)$$

However, the reciprocity relation for Green's functions does not hold,

$$G(x',x) \neq G(x,x')$$

Thus, the Green's integral formula (6.1.8) is only applicable to self-adjoint operators $\hat{L}$. For non-self-adjoint operators, they should be transformed into self-adjoint operators, with attention to the change in the weight function $\rho(x)$, and the corresponding modification of the type of boundary conditions.

16.1.3 *Sturm–Liouville equation*

(1) *One-dimensional equation*

For the non-homogeneous equation of a self-adjoint Sturm–Liouville operator, given by

$$\hat{L}u \equiv -\frac{\mathrm{d}}{\mathrm{d}x}\left[p(x)\frac{\mathrm{d}}{\mathrm{d}x}\right]u(x) + q(x)u(x) = f(x) \tag{16.1.9}$$

The corresponding Green's function equation is $\hat{L}G(x,x') = \delta(x-x')$. According to the Green's integral formula (16.1.8), we have

$$u(x) = \int_a^b G(x,x')f(x')\mathrm{d}x' - \int_a^b [G(x,x')\hat{L}u(x') - u(x')\hat{L}G(x,x')]\mathrm{d}x'$$

By performing integration by parts on the second term on the right-hand side, we obtain

$$u(x) = \int_a^b G(x,x')f(x')\mathrm{d}x' - p(x')[G(x,x')u'(x') - u(x')G'(x,x')]|_{x'=a}^{x'=b} \tag{16.1.10}$$

It can be seen that the distribution of $u(x)$ is related to the values at both endpoints.

(2) *Three-dimensional equation*

For the non-homogeneous equation of a three-dimensional Sturm–Liouville operator, given by

$$\hat{L}u \equiv -\nabla \cdot [p(\boldsymbol{r})\nabla]u(\boldsymbol{r}) + q(\boldsymbol{r})u(\boldsymbol{r}) = f(\boldsymbol{r}) \tag{16.1.11}$$

its Green's function equation is $\hat{L}G(\boldsymbol{r},\boldsymbol{r}') = \delta(\boldsymbol{r}-\boldsymbol{r}')$. Applying the second Green's formula, we have

$$u(\boldsymbol{r}) = \iiint_\Omega G(\boldsymbol{r},\boldsymbol{r}')f(\boldsymbol{r}')\mathrm{d}V' - \oiint_\Sigma p(\boldsymbol{r}')\left[G(\boldsymbol{r},\boldsymbol{r}')\frac{\partial u(\boldsymbol{r}')}{\partial n'} - u(\boldsymbol{r}')\frac{\partial G(\boldsymbol{r},\boldsymbol{r}')}{\partial n'}\right]\mathrm{d}S' \tag{16.1.12}$$

The second term arises from boundary effects, where Σ represents the boundary of Ω.

Now, can we use the integral formulas (16.1.10) and (16.1.12) to solve the boundary value problems of the equation? It should be noted that the integrals in these formulas require knowledge of the physical quantities and their derivative values on the boundary Σ. However, the three types of boundary conditions do not simultaneously provide the values of u and $\frac{\partial u}{\partial n}$ on the boundary Σ. Therefore, further processing is needed for the second-term boundary integral of the equation. Since the self-adjoint Green's function equation requires it to satisfy homogeneous boundary conditions, we have

- For the first type of boundary value problem, where $u|_\Sigma = \varphi(\boldsymbol{r})$, we can take $G|_\Sigma = 0$, leading to

$$u(\boldsymbol{r}) = \iiint_\Omega G(\boldsymbol{r},\boldsymbol{r}')f(\boldsymbol{r}')\mathrm{d}V' + \oiint_\Sigma \varphi(\boldsymbol{r}')p(\boldsymbol{r}')\frac{\partial G(\boldsymbol{r},\boldsymbol{r}')}{\partial n'}\mathrm{d}S' \tag{16.1.13}$$

- For the third type of boundary value problem, where $\left[\alpha\frac{\partial u}{\partial n} + \beta u\right]_\Sigma = \varphi(\boldsymbol{r})$, we can set $\left[\alpha\frac{\partial G}{\partial n} + \beta G\right]_\Sigma = 0$, resulting in

$$u(\boldsymbol{r}) = \iiint_\Omega G(\boldsymbol{r},\boldsymbol{r}')f(\boldsymbol{r}')\mathrm{d}V' - \frac{1}{\alpha}\oiint_\Sigma G(\boldsymbol{r},\boldsymbol{r}')p(\boldsymbol{r}')\varphi(\boldsymbol{r}')\mathrm{d}S' \tag{16.1.14}$$

- For the second type of boundary value problem, it is possible that the potential equation does not have a Green's function! This is because the Green's function equation is given by

$$-\nabla\cdot[p(\boldsymbol{r})\nabla]G(\boldsymbol{r},\boldsymbol{r}') + q(\boldsymbol{r})G(\boldsymbol{r},\boldsymbol{r}') = \delta(\boldsymbol{r}-\boldsymbol{r}')$$

  By integrating both sides, we have

$$-\oiint_\Sigma p(\boldsymbol{r}')\frac{\partial G(\boldsymbol{r},\boldsymbol{r}')}{\partial n}\mathrm{d}S + \iiint_\Omega q(\boldsymbol{r})G(\boldsymbol{r},\boldsymbol{r}')\mathrm{d}V = 1$$

  If we require the Green's function to satisfy the homogeneous boundary condition of the second type, i.e., $\frac{\partial G}{\partial n}|_\Sigma = 0$, then it must hold that

$$\iiint_\Omega q(\boldsymbol{r})G(\boldsymbol{r},\boldsymbol{r}')\mathrm{d}V \equiv 1$$

  However, this is not always satisfied. For example, for the Poisson equation with $q(\boldsymbol{r}) = 0$, there is an obvious contradiction. From a physical point of view, in the case of heat conduction, when there is a point heat source in the system, it is impossible to reach a steady state if the heat flows out is prohibited by adiabatic conditions imposed on the boundaries. Therefore, there is no Green's function that satisfies the homogeneous boundary conditions of the second type. However, extended Green's functions can be introduced, and a brief discussion on this is provided in the endnote.

Till now, we can in principle use Green's function to solve the boundary value problems of differential equations. Here is an example:

Example 16.2. Solve the equation by the Green's Function method:

$$\begin{cases} (x^2u')' - \dfrac{3}{4}u = \sqrt{x} \\ u(0) = 0, \quad u(1) = A \end{cases}$$

Solution: We need to find the Green's function for the differential equation with the boundary conditions:

$$\begin{cases} [x^2G'(x,x')]' - \dfrac{3}{4}G(x,x') = -\delta(x-x') \\ G(0) = 0, \quad G(1) = 0 \end{cases}$$

When $x \neq x'$, this equation is a self-adjoint Euler-type equation. By solving it with homogeneous boundary conditions, we obtain

$$G(x,x') = \begin{cases} Cx^{\frac{1}{2}} & (x < x') \\ D(x^{\frac{1}{2}} - x^{-\frac{3}{2}}) & (x > x') \end{cases}$$

To determine the constants C and D, the continuity condition at $x = x'$ gives

$$G(x,x') = \begin{cases} \dfrac{1}{2}x^{\frac{1}{2}}\left(x'^{\frac{1}{2}} - x'^{-\frac{3}{2}}\right) & (x < x') \\ \dfrac{1}{2}x'^{\frac{1}{2}}\left(x^{\frac{1}{2}} - x^{-\frac{3}{2}}\right) & (x > x') \end{cases}$$

This result confirms that $G(x,x') = G(x',x)$. Employing the Green's integral formula, we can solve the differential equation as

$$\begin{aligned} u(x) &= \int_0^1 G(x,x')f(x')\mathrm{d}x' - p(x')[G(x,x')u'(x') \\ &\quad - u(x')G'(x,x')]|_{x'=0}^{x'=1} \\ &= \int_0^x \frac{1}{2}x'\left(x^{\frac{1}{2}} - x^{-\frac{3}{2}}\right)\mathrm{d}x' + \int_x^1 \frac{1}{2}x^{\frac{1}{2}}\left(x' - \frac{1}{x'}\right)\mathrm{d}x' \\ &\quad + x'^2\left[\frac{A}{2}x^{\frac{1}{2}}\left(\frac{1}{2}x'^{-\frac{1}{2}} + \frac{3}{2}x'^{-\frac{5}{2}}\right)\right]\Bigg|_{x'=1} = \frac{1}{2}\sqrt{x}\ln x + A\sqrt{x} \end{aligned}$$

ENDNOTE

George Green was the son of a baker in Nottingham, England. He had little formal education and taught himself mathematics and physics while working in his father's factory. In 1828, at the age of 35, Green published his work "An Essay on the Application of Mathematical Analysis to the Theories of Electricity and Magnetism", in which he derived the famous Green's theorem relating area and volume integrals and established the method of using Green's functions to solve differential equations. This work is considered the beginning of mathematical physics in England. Green did not receive academic recognition during his lifetime and there are no known photographs of him. In the 20th century, Green's windmill, the mill where he lived, was restored and converted into the Nottingham Mathematical Centre for visitors.

The potential $v(\boldsymbol{r}, \boldsymbol{r}')$ generated by a unit point charge is determined by the equation $\Delta v(\boldsymbol{r}, \boldsymbol{r}') = \delta(\boldsymbol{r} - \boldsymbol{r}')$. According to the superposition principle, the potential produced by a continuous charge distribution in space satisfies the Poisson's equation:

$$\Delta u(\boldsymbol{r}) = \rho(\boldsymbol{r})$$

$$\longrightarrow \Delta u(\boldsymbol{r}) = \int_{\Omega} \rho(\boldsymbol{r}')\delta(\boldsymbol{r} - \boldsymbol{r}')\mathrm{d}\boldsymbol{r}' = \Delta \int_{\Omega} \rho(\boldsymbol{r}')v(\boldsymbol{r}, \boldsymbol{r}')\mathrm{d}\boldsymbol{r}'$$

Formally, we have

$$u(\boldsymbol{r}) \sim \int_{\Omega} \rho(\boldsymbol{r}')v(\boldsymbol{r}, \boldsymbol{r}')\mathrm{d}\boldsymbol{r}'$$

This means that the potential generated by a continuous charge distribution can be obtained by superposing the potentials produced by point charges. However, this derivation is valid only for infinite systems and does not consider boundary effects. For finite systems, even in the absence of charge, there can be an electric field distribution depending on the potential distribution on the boundaries. Therefore, the correct $v(\boldsymbol{r}, \boldsymbol{r}')$ must satisfy certain boundary conditions, and the complete expression is given by Green's function integral formula. The field distribution generated by a point source satisfying homogeneous boundary conditions is called the Green's function, denoted by $G(\boldsymbol{r}, \boldsymbol{r}')$.

For Green's function of a point charge, Gauss's theorem implies that the electric flux through a closed surface enclosing the point charge is non-zero. This implies that $\frac{\partial G(\boldsymbol{r},\boldsymbol{r}')}{\partial n}$ cannot be zero everywhere on the closed surface, and thus, it is not possible to impose the second type of homogeneous boundary condition. Specifically, for the Helmholtz equation:

$$\Delta u + k^2 u = 0$$

under the second type of homogeneous boundary condition, there exist zero eigenvalues for one-dimensional, two-dimensional, and three-dimensional cases. Therefore, the operator $\hat{L} = \Delta + k^2$ is non-invertible, and the Green's function does not exist.

The relationship between the normal derivative and the Laplacian operator can be established using the divergence theorem. For the second type of boundary value problem:

$$\begin{cases} \Delta u = f(\boldsymbol{r}) & (\boldsymbol{r} \in \Omega) \\ \dfrac{\partial u}{\partial n} = g(\boldsymbol{r}) & (\boldsymbol{r} \in \Sigma) \end{cases}$$

Integrating the first equation over the volume Ω, we have

$$\int_\Omega f(\boldsymbol{r}) \mathrm{d}\boldsymbol{r} = \int_\Omega \nabla \cdot (\nabla u) \mathrm{d}\boldsymbol{r} = \int_\Sigma n \cdot \nabla u \mathrm{d}\mathbf{S} = \int_\Sigma \frac{\partial u}{\partial n} \mathrm{d}\mathbf{S}$$

This means that the boundary condition is connected to the differential equation and it is not possible to impose arbitrary values of $\frac{\partial u}{\partial n}$ on the boundary, especially in the case of Green's function, which requires homogeneous boundary values. To satisfy this condition, a special solvable condition needs to be imposed by subtracting the average value of a non-homogeneous term in the differential equation:

$$\Delta u = f(\boldsymbol{r}) - \bar{f}, \quad \bar{f} = \frac{1}{V_\Omega}\int_\Omega f(\boldsymbol{r})\mathrm{d}\boldsymbol{r}$$

which gives rise to

$$\int_\Sigma \frac{\partial u}{\partial n}\mathrm{d}\mathbf{S} = 0$$

This ensures the homogeneous boundary value condition for Green's function. Thus, Green's function equation for the second type of boundary value problem needs to be modified to

$$\begin{cases} \Delta G(\boldsymbol{r}, \boldsymbol{r}') = \delta(\boldsymbol{r} - \boldsymbol{r}') - \dfrac{1}{V_\Omega} & (\boldsymbol{r} \in \Omega) \\ \dfrac{\partial G(\boldsymbol{r}, \boldsymbol{r}')}{\partial n} = 0 & (\boldsymbol{r} \in \Sigma) \end{cases}$$

The corresponding Green's integral formula for the second type of boundary value problem is

$$u(\boldsymbol{r}) = \iiint_\Omega G(\boldsymbol{r}, \boldsymbol{r}')f(\boldsymbol{r}')\mathrm{d}V' - \oiint_\Sigma G(\boldsymbol{r}, \boldsymbol{r}')\frac{\partial u}{\partial n}\mathrm{d}S' + \bar{u}$$

with

$$\bar{u} = \frac{1}{V_\Omega}\int_\Omega u(\boldsymbol{r})\mathrm{d}\boldsymbol{r}$$

Similar treatments can be applied to other types of self-adjoint differential equations, but they are not discussed in detail here.

Exercises

[1] Find the Green's function for the following equation:

$$\text{(a)}\quad \begin{cases} \dfrac{\mathrm{d}^2}{\mathrm{d}x^2} G(x, x') = -\delta(x - x') \\ G(x, x')|_{x=0} = G(x, x')|_{x=1} = 0 \end{cases};$$

$$\text{(b)}\quad \begin{cases} \dfrac{\mathrm{d}^2}{\mathrm{d}x^2} G(x, x') + 4G(x, x') = -\delta(x - x') \\ G(x, x')|_{x=0} = \left.\dfrac{\mathrm{d}}{\mathrm{d}x} G(x, x')\right|_{x=1} = 0 \end{cases}$$

Answer:

$$\text{(a)}\ G(x, x') = \begin{cases} x(x' - 1) & (x < x') \\ x'(x - 1) & (x > x') \end{cases}$$

$$\text{(b)}\ G(x, x') = \begin{cases} -\dfrac{1}{2\cos 2} \sin 2x \cos(2x' - 2) & (x < x') \\ -\dfrac{1}{2\cos 2} \sin 2x' \cos(2x - 2) & (x > x') \end{cases}$$

[2] Convert the equations into self-adjoint form, find the Green's function, and write the integral solution:

$$\text{(a)}\ \begin{cases} u'' - 2u' + u = \mathrm{e}^x \\ u(0) = 0, \quad u(1) = 0 \end{cases} \qquad \text{(b)}\ \begin{cases} x^2 u'' - xu' + u = f(x) \\ u(0) = 0, \quad u(1) = 0 \end{cases}$$

$$\text{(c)}\ \begin{cases} xu'' + 2u' = f(x) \\ u'(0) = 0, \quad u(1) = A \end{cases} \qquad \text{(d)}\ \begin{cases} x^2 u'' + 3xu' = x^2 \\ u(1) = 1, \quad u(2) = 2 \end{cases}$$

[3] Verify by substitution that $u = \left(\frac{1-x}{1+x}\right)^{\pm h/2}$ is a solution to the self-adjoint equation:

$$[(1 - x^2)u']' - \frac{h^2}{1 - x^2} u = 0$$

and find the Green's function for u is finite at $x = \pm 1$.

[4] Find the Green's function for the infinite plate with a heat source $q(x,y)$ distributed within a finite range and κ as the thermal conductivity,

$$\nabla u = -\frac{1}{\kappa}q(x,y)$$

Answer:

$$u(x,y) = -\frac{1}{4\pi\kappa}\iint_{-\infty}^{\infty} q(\xi,\eta)\ln[(x-\xi)^2+(y-\eta)^2]\mathrm{d}\xi\mathrm{d}\eta$$

16.2 Potential Equations

According to the reasoning in Section 16.1, if the Green's function is known, the boundary value problem of the inhomogeneous equation can be solved. Solving the Green's function is equivalent to solving the boundary value problem with homogeneous boundary conditions, which is simpler than the usual non-homogeneous boundary value problems. This section introduces several basic methods for finding the Green's function of the potential equation.

16.2.1 *Unbounded region*

The solution of the Green's function equation in an unbounded region is called the fundamental solution, denoted by G_0. Due to the unbounded nature in infinite space, it can be solved using the Fourier transform method.

Example 16.3. Find the fundamental solution of the three-dimensional Poisson equation.

Solution: The three-dimensional Green's function equation in infinite space is given by

$$\Delta G_0(\boldsymbol{r},\boldsymbol{r}') = -\delta(\boldsymbol{r}-\boldsymbol{r}')$$

Taking the three-dimensional Fourier transform of the equation, we get

$$G_0(\boldsymbol{k}) = \frac{1}{(2\pi)^3}\frac{\mathrm{e}^{-\mathrm{i}\boldsymbol{k}\cdot\boldsymbol{r}'}}{\boldsymbol{k}^2}$$

where $\boldsymbol{k}^2 = k_x^2 + k_y^2 + k_z^2$. In order to performing the inverse Fourier transform, take $\boldsymbol{r} - \boldsymbol{r}'$ as the polar axis in spherical coordinates. Integrating over $\boldsymbol{k}$, we obtain

$$\begin{aligned} G_0(\boldsymbol{r}, \boldsymbol{r}') &= \frac{1}{(2\pi)^3} \iiint \frac{\mathrm{e}^{\mathrm{i}\boldsymbol{k}\cdot(\boldsymbol{r}-\boldsymbol{r}')}}{k^2} \mathrm{d}\boldsymbol{k} \\ &= \frac{1}{(2\pi)^3} \int_0^\infty \mathrm{d}k \int_0^\pi \mathrm{d}\theta \int_0^{2\pi} \frac{\mathrm{e}^{-\mathrm{i}k|\boldsymbol{r}-\boldsymbol{r}'|\cos\theta}}{k^2} k^2 \sin\theta \mathrm{d}\varphi \\ &= \frac{1}{(2\pi)^2} \int_0^\infty \frac{\mathrm{e}^{\mathrm{i}k|\boldsymbol{r}-\boldsymbol{r}'|} - \mathrm{e}^{-\mathrm{i}k|\boldsymbol{r}-\boldsymbol{r}'|}}{\mathrm{i}k|\boldsymbol{r}-\boldsymbol{r}'|} \mathrm{d}k \\ &= \frac{1}{2\pi^2} \int_0^\infty \frac{\sin k|\boldsymbol{r}-\boldsymbol{r}'|}{k|\boldsymbol{r}-\boldsymbol{r}'|} \mathrm{d}k = \frac{1}{4\pi|\boldsymbol{r}-\boldsymbol{r}'|} \end{aligned}$$

This is the potential distribution of a unit point charge in three-dimensional unbounded space in physics. Table 16.1 lists the fundamental solutions of several common operators, and Appendix A.3 provides more Green's functions for fundamental equations.

Table 16.1.

	Laplace operator Δ	Helmholtz operator $\Delta + k^2$	Modified Helmholtz operator $\Delta - k^2$
1D	$-\frac{1}{2}\lvert x_1 - x_2\rvert$	$\frac{\mathrm{i}}{2k}\mathrm{e}^{\mathrm{i}k\lvert x_1-x_2\rvert}$	$\frac{\mathrm{i}}{2k}\mathrm{e}^{-k\lvert x_1-x_2\rvert}$
2D	$-\frac{1}{2\pi}\lvert\rho_1 - \rho_2\rvert$	$\frac{\mathrm{i}}{4}H_0^{(1)}(k\lvert\rho_1-\rho_2\rvert)$	$\frac{1}{2\pi}K_0^{(1)}(k\lvert\rho_1-\rho_2\rvert)$
3D	$\frac{1}{4\pi}\frac{1}{\lvert\boldsymbol{r}_1-\boldsymbol{r}_2\rvert}$	$\frac{1}{4\pi}\frac{1}{\lvert\boldsymbol{r}_1-\boldsymbol{r}_2\rvert}\mathrm{e}^{\mathrm{i}k\lvert\boldsymbol{r}_1-\boldsymbol{r}_2\rvert}$	$\frac{1}{4\pi}\frac{1}{\lvert\boldsymbol{r}_1-\boldsymbol{r}_2\rvert}\mathrm{e}^{-k\lvert\boldsymbol{r}_1-\boldsymbol{r}_2\rvert}$

16.2.2 *Bounded region*

The Green's function G can be decomposed into two parts, $G = G_0 + G_1$, where G_0 is the fundamental solution. Then G_1 satisfies the corresponding homogeneous equation. For the first type of boundary problem, we have

$$\begin{aligned} &\hat{L}G_1 = 0 \\ &G_1|_\Sigma = (G - G_0)|_\Sigma = -G_0|_\Sigma \end{aligned}$$

Example 16.4. Find the Green's function inside a hollow conductor sphere of radius R,

$$\begin{cases} \Delta G(\boldsymbol{r}, \boldsymbol{r}') = -\delta(\boldsymbol{r} - \boldsymbol{r}') \\ G|_{r=R} = 0 \end{cases}$$

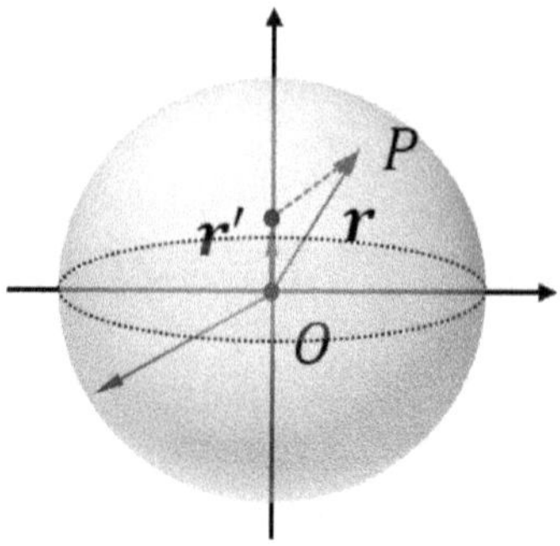

Figure 16.1.

Solution: As shown in Figure 16.1, according to Example 16.3, the fundamental solution in infinite space is given by

$$G_0(\boldsymbol{r}, \boldsymbol{r}') = \frac{1}{4\pi|\boldsymbol{r} - \boldsymbol{r}'|}$$

Therefore,

$$\begin{cases} \Delta G_1(\boldsymbol{r}, \boldsymbol{r}') = 0 \\ G_1|_{r=R} = -G_0|_{r=R} = -\dfrac{1}{4\pi\sqrt{r'^2 - 2r'R\cos\theta + R^2}} \end{cases}$$

The solution to the Laplace equation inside the sphere is given by

$$G_1(\boldsymbol{r}, \boldsymbol{r}') = \sum_{l=0}^{\infty} A_l r^l P_l(\cos\theta)$$

Using the generating function of Legendre polynomials, we can obtain, on the surface of the sphere where $r = R$,

$$\sum_{l=0}^{\infty} A_l R^l P_l(\cos\theta) = -\frac{1}{4\pi}\frac{1}{\sqrt{R^2 - 2r'R\cos\theta + r'^2}}$$

$$= -\frac{1}{4\pi}\sum_{l=0}^{\infty}\frac{1}{R^{l+1}}r'^l P_l(\cos\theta)$$

The coefficient A_l is given by

$$A_l = -\frac{1}{4\pi}\frac{r'^l}{R^{2l+1}}$$

Therefore,

$$\begin{aligned}
G_1(\boldsymbol{r}, \boldsymbol{r}') &= -\frac{1}{4\pi}\sum_{l=0}^{\infty}\frac{r'^l}{R^{2l+1}}r^l P_l(\cos\theta) \\
&= -\frac{1}{4\pi}\frac{R}{r'}\sum_{l=0}^{\infty}\frac{1}{(R^2/r')^{l+1}}r^l P_l(\cos\theta) \\
&= -\frac{1}{4\pi}\frac{R/r'}{\sqrt{\tilde{r}^2 - 2\tilde{r}r + r^2}}
\end{aligned}$$

It is equivalent to the potential generated in unbounded space by a point charge $\tilde{q} = -\frac{R}{r'}$ located on the polar axis at $\tilde{r} = \frac{R^2}{r'}$. This hypothetical charge is called the image charge. The Green's function for this problem is the sum of the potential distributions generated by the point charge and its image charge,

$$G(\boldsymbol{r}, \boldsymbol{r}') = G_0(\boldsymbol{r}, \boldsymbol{r}') + G_1(\boldsymbol{r}, \boldsymbol{r}') = \frac{1}{4\pi|\boldsymbol{r} - \boldsymbol{r}'|} - \frac{1}{4\pi}\frac{R/r'}{\sqrt{\tilde{r}^2 - 2\tilde{r}r\cos\theta + r^2}}$$

For the Poisson equation of the first type of boundary value problem, such as finding the Green's function for the static field equation, sometimes a more concise method called the electric image method can be used.

Example 16.5. For a grounded conductor sphere of radius R with a point charge $-\varepsilon_0$ placed at point Q inside the sphere, we need to find the Green's function inside the sphere:

$$\begin{cases}\Delta G = -\delta(\boldsymbol{r} - \boldsymbol{r}') \\ G|_{r=R} = 0\end{cases}$$

Solution: As shown in Figure 16.2, let us assume that the conjugate point of Q is Q', located at $r_1 = \frac{R^2}{r'}$, with a charge $Q' = -\frac{R}{r'}Q$.

Based on geometry, we know that these two charges ensure that the potential on the surface of the sphere is zero. Therefore, the

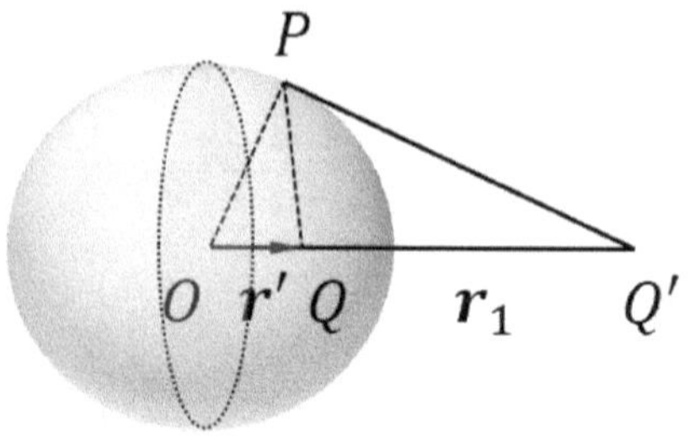

Figure 16.2.

Green's function inside the sphere is the sum of the potential distributions generated by these two point charges. This conclusion is the same as Example 16.4, but the process is more concise.

Reflection: Can the electric image method be used to find the Green's function for the first type of boundary problem of the Helmholtz equation?

16.2.3 *Method of eigenfunction expansion*

The method of eigenfunction expansion can be used to find the Green's function for finite-sized systems, analogous to the separation of variables method. Consider the Green's function equation for the Sturm–Liouville operator:

$$\begin{cases} \hat{L}G - \lambda G = \delta(\boldsymbol{r} - \boldsymbol{r}') \\ G(\boldsymbol{r})|_\Sigma = 0 \end{cases}$$

The corresponding eigenvalue problem is

$$\begin{cases} \hat{L}\psi_n(\boldsymbol{r}) = \lambda_n \psi_n(\boldsymbol{r}) \\ \psi_n(\boldsymbol{r})|_\Sigma = 0 \end{cases}$$

where λ_n is the eigenvalue and ψ_n is its eigenfunction. The Green's function G can be expanded in terms of the eigenfunctions ψ_n as follows:

$$G(\boldsymbol{r}, \boldsymbol{r}') = \sum_n C_n(\boldsymbol{r}')\psi_n(\boldsymbol{r})$$

Substituting this expansion into the equation and multiplying both sides by $\psi_m^*(\boldsymbol{r})$ and integrating, using the orthogonality of the eigenfunctions, we obtain the expansion coefficients as

$$C_n(\boldsymbol{r}') = \frac{1}{\lambda_n - \lambda}\psi_n^*(\boldsymbol{r}')$$

The Green's function is given by

$$G(\boldsymbol{r}, \boldsymbol{r}') = \sum_n \frac{1}{\lambda_n - \lambda}\psi_n^*(\boldsymbol{r}')\psi_n(\boldsymbol{r})$$

It is evident that $G(\boldsymbol{r}, \boldsymbol{r}') = G^*(\boldsymbol{r}', \boldsymbol{r})$. Taking $\lambda = 0$, we have

$$G(\boldsymbol{r}, \boldsymbol{r}') = \sum_n \frac{1}{\lambda_n}\psi_n^*(\boldsymbol{r}')\psi_n(\boldsymbol{r})$$

Example 16.6. Find the Green's function for the Poisson equation in a rectangular region:

$$\begin{cases} \Delta G(\boldsymbol{r}, \boldsymbol{r}') = -\delta(\boldsymbol{r} - \boldsymbol{r}') \\ G(\boldsymbol{r}, \boldsymbol{r}')|_{x=0} = G(\boldsymbol{r}, \boldsymbol{r}')|_{x=a} = 0 \\ G(\boldsymbol{r}, \boldsymbol{r}')|_{y=0} = G(\boldsymbol{r}, \boldsymbol{r}')|_{y=b} = 0 \end{cases}$$

Solution: This is a special case for $\lambda = 0$. The corresponding eigenvalue problem is

$$\begin{cases} -\Delta\psi(\boldsymbol{r}) = \lambda\psi(\boldsymbol{r}) \\ \psi(\boldsymbol{r})|_{x=0} = \psi(\boldsymbol{r})|_{x=a} = 0 \\ \psi(\boldsymbol{r})|_{x=0} = \psi(\boldsymbol{r})|_{x=a} = 0 \end{cases}$$

The eigenfunctions and eigenvalues are

$$\begin{cases} \psi_{mn}(x, y) = \dfrac{2}{\sqrt{ab}}\sin\mu_m x \sin\nu_n y \\ \lambda_{mn} = \mu_m^2 + \nu_n^2, \quad \mu_m = \dfrac{m\pi}{a}, \quad \nu_n = \dfrac{n\pi}{b} \end{cases}$$

Taking $\lambda = 0$, we obtain the Green's function:

$$G(\boldsymbol{r}, \boldsymbol{r}') = \frac{4}{ab}\sum_{m,n} \frac{\sin\mu_m x' \sin\nu_n y'}{\mu_m^2 + \nu_n^2}\sin\mu_m x \sin\nu_n y$$

ENDNOTE

For the eigenvalue equation of a non-self-adjoint operator, let u_n and v_n be the eigenfunctions of $\hat{L}$ and $\hat{L}^\dagger$ with eigenvalues λ_n and μ_n, respectively,

$$\hat{L}u_n = \lambda_n u_n, \quad \hat{L}^\dagger v_n = \mu_n v_n$$

If these eigenfunctions form a complete set, there exists a v_n that is not orthogonal to u_n, which leads to

$$\lambda_n \langle v_n | u_n \rangle = \langle v_n | \hat{L} | u_n \rangle = \langle u_n | \hat{L}^\dagger | v_n \rangle = \mu_n \langle u_n | v_n \rangle$$

This implies $\mu_n = \lambda_n$, indicating that $\hat{L}$ and $\hat{L}^\dagger$ have a common spectrum of eigenvalues:

$$\hat{L}u_n = \lambda_n u_n, \quad \hat{L}^\dagger v_m = \lambda_m v_m$$

Hence,

$$\langle v_m | \hat{L} | u_n \rangle - \langle u_n | \hat{L}^\dagger | v_m \rangle = (\lambda_n - \lambda_m) \langle v_n | u_m \rangle = 0$$

This shows that the eigenfunctions with different eigenvalues are biorthogonal:

$$\langle v_n | u_m \rangle = 0 (m \neq n)$$

by normalizing the eigenfunctions with the same eigenvalue, $\langle v_n | u_n \rangle = 1$.

For the non-homogeneous equation $\hat{L}u = f$, expand it using the eigenfunctions:

$$u = \sum_n a_n u_n, \quad f = \sum_n b_n u_n$$

Using the biorthogonality of u_n and v_n, we have $b_n = \langle f | v_n \rangle$, so

$$\hat{L}u = \hat{L} \sum_n a_n u_n = \sum_n a_n \lambda_n u_n = \sum_n b_n u_n$$

leading to

$$a_n = \frac{b_n}{\lambda_n} = \frac{1}{\lambda_n}\langle v_n|f\rangle$$

The solution to the equation is then

$$u(x) = \sum_n \int_a^b \frac{u_n(x)v_n^*(\xi)}{\lambda_n} f(\xi)\mathrm{d}\xi$$

By comparison, we find the Green's function is

$$G(\boldsymbol{r}, \boldsymbol{r}') = \sum_n \frac{1}{\lambda_n} u_n(\boldsymbol{r}) v_n^*(\boldsymbol{r}')$$

Using the adjoint relationship $\tilde{G}^*(\boldsymbol{r}, \boldsymbol{r}') = G(\boldsymbol{r}', \boldsymbol{r})$, the adjoint Green's function is then

$$\tilde{G}(\boldsymbol{r}, \boldsymbol{r}') = \sum_n \frac{1}{\lambda_n} v_n(\boldsymbol{r}) u_n^*(\boldsymbol{r}')$$

If $\hat{L}$ is a self-adjoint operator, we have $u_n = v_n$ leading to the reciprocity relation of the Green's function:

$$G^*(\boldsymbol{r}, \boldsymbol{r}') = G(\boldsymbol{r}', \boldsymbol{r})$$

Exercises

[1] Find the fundamental solution of the two-dimensional Poisson's equation using the Fourier transform method:

$$\Delta G_0(\boldsymbol{r}, \boldsymbol{r}') = -\delta(\boldsymbol{r} - \boldsymbol{r}')$$

Answer:

$$G_0(\boldsymbol{r}, \boldsymbol{r}') = \frac{1}{2\pi} \ln |\boldsymbol{r} - \boldsymbol{r}'|$$

[2] Using the Fourier transform method, find the fundamental solution of the three-dimensional Helmholtz equation:

$$\Delta G_0(\boldsymbol{r}, \boldsymbol{r}') + k^2 G_0(\boldsymbol{r}, \boldsymbol{r}') = -\delta(\boldsymbol{r} - \boldsymbol{r}')$$

Answer:

$$G_0(\boldsymbol{r},\boldsymbol{r}') = \frac{\mathrm{e}^{-\mathrm{i}k|\boldsymbol{r}-\boldsymbol{r}'|}}{4\pi|\boldsymbol{r}-\boldsymbol{r}'|}$$

[3] Use the method of images to find the Green's function of the Poisson's equation with the first type of boundary condition inside a circular domain.

[4] Use the method of images to find the Green's function of the Poisson's equation with the first type of boundary condition inside the upper half-sphere of radius a (see Figure 16.3).

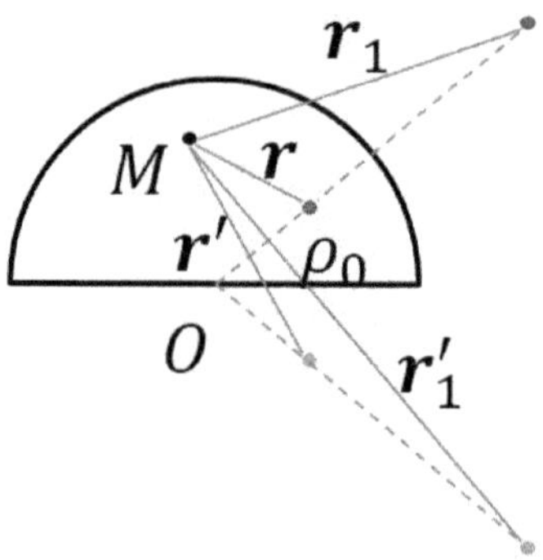

Figure 16.3.

[5] Using the method of eigenfunction expansion, find the Green's function for the following problem:

$$\begin{cases} \dfrac{\mathrm{d}^2}{\mathrm{d}x^2}G(x,x') + 4G(x,x') = -\delta(x-x') \\ G(x,x')|_{x=0} = \dfrac{\mathrm{d}}{\mathrm{d}x}G(x,x')|_{x=1} = 0 \end{cases}$$

Answer:

$$G(x,x') = \sum_{n=0}^{\infty} \frac{1}{\lambda_n - 4} \sin\left(n+\frac{1}{2}\right)\pi x' \sin\left(n+\frac{1}{2}\right)\pi x,$$

$$\lambda_n = \left(n+\frac{1}{2}\right)^2 \pi^2$$

[6] Given a second-order differential equation and homogeneous boundary conditions,

$$\begin{cases} \hat{L}u = u'' + u' \\ u(0) = 0, \quad u'(1) = 0 \end{cases}$$

(a) Write down the adjoint equation and the adjoint boundary conditions.
(b) Solve the eigenvalue equation and prove the biorthogonality of the eigenfunctions:

$$\hat{L}u = \lambda u, \quad \hat{L}^{\dagger}v = \lambda v$$

16.3 Some Applications

After obtaining the Green's function, the integral form solution to boundary value problems can be solved using the Green's integral formula.

Example 16.7. By using the obtained Green's function, solve the first boundary value problem of the Laplace equation inside a sphere:

$$\begin{cases} \Delta u = 0 \\ u|_{r=a} = f(\theta, \varphi) \end{cases}$$

Solution: As shown in Figure 16.4, according to the Green's integral formula,

$$u(r) = \iiint_{\Omega} G(\boldsymbol{r}, \boldsymbol{r}')\rho(\boldsymbol{r}')\mathrm{d}V' + \oiint_{\Sigma} f(\boldsymbol{r}')\frac{\partial G(\boldsymbol{r}, \boldsymbol{r}')}{\partial n'}\mathrm{d}S'$$

where the external source $\rho(\boldsymbol{r}') = 0$. The Green's function is given by

$$G(\boldsymbol{r}, \boldsymbol{r}') = -\frac{1}{4\pi}\frac{1}{|\boldsymbol{r} - \boldsymbol{r}'|} + \frac{a}{r'}\frac{1}{4\pi}\frac{1}{|\boldsymbol{r} - \boldsymbol{r}_1|}$$

Using the relationship:

$$\cos\gamma = \cos\theta\cos\theta' + \sin\theta\sin\theta'\cos(\varphi - \varphi')$$

we can obtain

$$\frac{\partial}{\partial n'}\frac{1}{|\boldsymbol{r}-\boldsymbol{r}'|} = \frac{\partial}{\partial r'}\frac{1}{(r^2-2rr'\cos\gamma+r'^2)^{1/2}}$$

$$= -\frac{r'-r\cos\gamma}{(r^2-2rr'\cos\gamma+r'^2)^{3/2}}$$

Eliminating the $\cos\gamma$ term in the numerator, we have

$$\left.\frac{\partial}{\partial n'}\frac{1}{|\boldsymbol{r}-\boldsymbol{r}'|}\right|_{\Sigma} = \left.\frac{r^2-|\boldsymbol{r}-\boldsymbol{r}'|-a^2}{2a|\boldsymbol{r}-\boldsymbol{r}'|^3}\right|_{\Sigma}$$

By taking the boundary condition $r'\to a$, we can obtain

$$\left.\frac{\partial}{\partial n'}\left[\frac{a}{r'}\frac{1}{|\boldsymbol{r}-\boldsymbol{r}_1|}\right]\right|_{\Sigma} = -\frac{a}{r'^2}\frac{1}{|\boldsymbol{r}-\boldsymbol{r}_1|} + \left.\frac{a}{r'}\frac{\partial}{\partial r_1}\frac{1}{|\boldsymbol{r}-\boldsymbol{r}_1|}\frac{\partial r_1}{\partial r'}\right|_{\boldsymbol{r}_1\to\boldsymbol{r}'}$$

$$= \left.\frac{a^2-r^2-|\boldsymbol{r}-\boldsymbol{r}'|}{2a|\boldsymbol{r}-\boldsymbol{r}'|^3}\right|_{\Sigma}$$

Hence,

$$\left.\frac{\partial G}{\partial n'}\right|_{\Sigma} = \left.\frac{1}{4\pi a}\frac{a^2-r^2}{|\boldsymbol{r}-\boldsymbol{r}'|^3}\right|_{\Sigma}$$

Substituting this into the integral formula, we get

$$u(r,\theta,\varphi) = \int_0^{\pi}\mathrm{d}\theta'\int_0^{2\pi}\frac{1}{4\pi a}\frac{a^2-r^2}{|\boldsymbol{r}-\boldsymbol{r}'|^3}f(\theta',\varphi')a^2\sin\theta'\mathrm{d}\varphi'$$

$$= \frac{a}{4\pi}\int_0^{\pi}\int_0^{2\pi}f(\theta',\varphi')\frac{a^2-r^2}{(a^2-2ar\cos\gamma+r^2)^{3/2}}\sin\theta'\mathrm{d}\theta'\mathrm{d}\varphi'$$

This equation is also known as the spherical Poisson's formula.

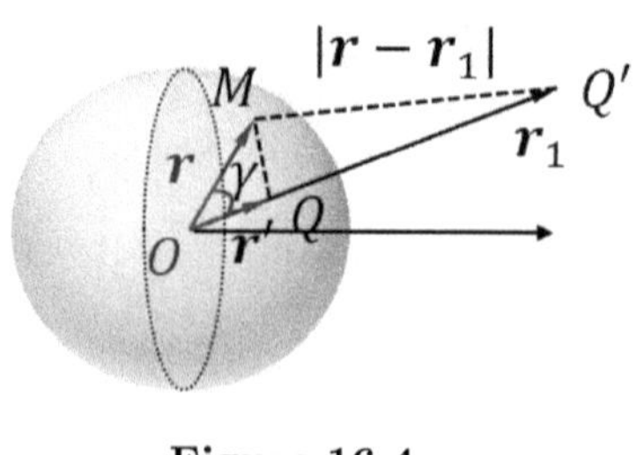

Figure 16.4.

Exercises

[1] Solve the first boundary value problem of the Laplace equation inside a circle:

$$\begin{cases} \Delta_2 u = 0 \quad (\rho < a) \\ u|_{\rho=a} = f(\varphi) \end{cases}$$

[2] Solve the first boundary value problem of the Laplace equation inside a two-dimensional half-plane:

$$\begin{cases} \Delta_2 u = 0 \quad (y \geq 0) \\ u|_{y=0} = f(x) \end{cases}$$

Answer:

$$u(x,y) = \frac{1}{\pi}\int_{-\infty}^{\infty} \frac{yf(x')}{(x-x')^2 + y^2}\mathrm{d}x'$$

[3] Solve the first boundary value problem of the Laplace equation inside a three-dimensional half-space:

$$\begin{cases} \Delta u = 0 \quad (z > 0) \\ u|_{z=0} = f(x,y) \end{cases}$$

Answer:

$$u(x,y,z) = \frac{z}{2\pi}\iint_{-\infty}^{\infty} \frac{f(x',y')}{[(x-x')^2 + (y-y')^2 + z^2]^{3/2}}\mathrm{d}x'\mathrm{d}y'$$

[4] Using the Green's function method to solve the equation:

$$\begin{cases} \dfrac{\mathrm{d}^2}{\mathrm{d}x^2}u(x) + k^2 u(x) = f(x) \\ u|_{x=a} = 0, \quad u|_{x=b} = A \end{cases}$$

Answer:

$$G(x,x') = \begin{cases} \dfrac{\sin k(x'-b)\sin k(x-a)}{k\sin k(b-a)} & (x < x') \\ \dfrac{\sin k(x'-a)\sin k(x-b)}{k\sin k(b-a)} & (x > x') \end{cases}$$

$$u(x) = \int_a^b G(x,x')f(x')\mathrm{d}x' + A\frac{\sin k(x-a)}{\sin k(b-a)}$$

16.4 Development Equation

16.4.1 *Time-dependent Green's function*

Earlier, we discussed the Green's function method for the potential equation. Similarly, for time-dependent problems such as wave or transport equations, Green's functions can also be used for solutions. It can be proven that Green's function satisfies spatial symmetry and time-reversal symmetry:

$$G(\boldsymbol{r},\boldsymbol{r}';t,t') = G(\boldsymbol{r}',\boldsymbol{r};-t',-t) \tag{16.4.1}$$

Now let us introduce several methods for computing time-dependent Green's functions.

(1) *Unbounded systems*

For the time-dependent Green's function of unbounded problems, the Fourier transform method can be employed for solution.

Example 16.8. Consider the Green's function of the free particle Schrödinger equation:

$$\begin{cases} \mathrm{i}\dfrac{\partial}{\partial t}\psi(\boldsymbol{r},t) = -\frac{1}{2}\nabla^2\psi(\boldsymbol{r},t) \\ \psi(\boldsymbol{r},t)|_{t=0} = \psi(\boldsymbol{r},0) \end{cases}$$

By performing a Fourier transform with respect to spatial coordinates, $\Psi(\boldsymbol{k},t) = \mathfrak{F}[\psi(\boldsymbol{r},t)]$, we obtain

$$\Psi(\boldsymbol{k},t) = \Psi(\boldsymbol{k},0)\mathrm{e}^{-\frac{\mathrm{i}k^2t}{2}}$$

Applying the inverse Fourier transform, we have

$$\begin{aligned} \psi(\boldsymbol{r},t) &= \int_{-\infty}^{\infty} \Psi(\boldsymbol{k},0)\mathrm{e}^{-\frac{\mathrm{i}k^2t}{2}}\mathrm{e}^{\mathrm{i}\boldsymbol{k}\cdot\boldsymbol{r}}\mathrm{d}k \\ &= \frac{1}{(2\pi)^3}\int_{-\infty}^{\infty}\int_{-\infty}^{\infty} \psi(\boldsymbol{r}',0)\mathrm{e}^{-\mathrm{i}\boldsymbol{k}\cdot\boldsymbol{r}'}\mathrm{d}\boldsymbol{r}'\mathrm{e}^{-\frac{\mathrm{i}k^2t}{2}}\mathrm{e}^{\mathrm{i}\boldsymbol{k}\cdot\boldsymbol{r}}\mathrm{d}\boldsymbol{k} \\ &= \int_{-\infty}^{\infty} \psi(\boldsymbol{r}',0)G(\boldsymbol{r},\boldsymbol{r}';t)\mathrm{d}\boldsymbol{r}' \end{aligned}$$

Considering the initial wave packet $\psi(\boldsymbol{r},0)$ as an external source, the Green's function is given by

$$G(\boldsymbol{r},\boldsymbol{r}';t)=\frac{1}{(2\pi)^3}\int_{-\infty}^{\infty}\mathrm{e}^{\mathrm{i}\boldsymbol{k}\cdot(\boldsymbol{r}-\boldsymbol{r}')-\frac{\mathrm{i}k^2t}{2}}\,\mathrm{d}\boldsymbol{k}$$

In quantum mechanics, this Green's function is also known as the propagator for a free particle. It represents the probability amplitude for a particle at point $\boldsymbol{r}'$ at time $t=0$ to propagate to point $\boldsymbol{r}$ after time t. For one-dimensional systems, we have

$$G(x,x';t)=\sqrt{\frac{2\pi\mathrm{i}m}{\hbar t}}\mathrm{e}^{-\frac{\mathrm{i}m}{2\hbar t}(x-x')^2}$$

(2) *Bounded systems*

For the time-dependent Green's function of bounded problems, the method of eigenfunction expansion can be used for solution.

Example 16.9. Consider the Green's function for the wave equation in a bounded region:

$$\partial_{tt}G-a^2\Delta G=\delta(\boldsymbol{r}-\boldsymbol{r}')\delta(t-t')$$

Solution: Let $u_n(\boldsymbol{r})$ be the eigenfunctions of the Helmholtz equation $\Delta u(\boldsymbol{r})+\lambda u(\boldsymbol{r})=0$ under homogeneous boundary conditions, with corresponding eigenvalues λ_n. Expanding the Green's function in terms of $u_n(\boldsymbol{r})$ and assuming the expansion coefficients depend on time, we have

$$G(\boldsymbol{r},\boldsymbol{r}';t-t')=\sum_{n=1}^{\infty}C_n(\boldsymbol{r}';t)u_n(\boldsymbol{r})$$

Using the identities:

$$\Delta u_n(\boldsymbol{r})+\lambda_n u_n(\boldsymbol{r})=0,\delta(\boldsymbol{r}-\boldsymbol{r}')=\sum_{n=1}^{\infty}u_n^*(\boldsymbol{r}')u_n(\boldsymbol{r})$$

Substituting them into the wave equation, we obtain

$$\sum_{n=1}^{\infty}\left[\frac{\partial^2}{\partial t^2}C_n(\boldsymbol{r}';t)+a^2\lambda_n C_n(\boldsymbol{r}';t)\right]u_n(\boldsymbol{r})=\sum_{n=1}^{\infty}[u_n^*(\boldsymbol{r}')\delta(t-t')]u_n(\boldsymbol{r})$$

Comparing the coefficients on both sides, we have

$$\frac{\partial^2}{\partial t^2}C_n(\boldsymbol{r}';t) + a^2\lambda_n C_n(\boldsymbol{r}';t) = u_n^*(\boldsymbol{r}')\delta(t-t')$$

Assuming that when $t < t'$,

$$C_n(\boldsymbol{r}';t) = 0, \quad \frac{\mathrm{d}}{\mathrm{d}t}C_n(\boldsymbol{r}';t) = 0$$

Solving the equation to get

$$C_n(\boldsymbol{r}';t) = u_n^*(\boldsymbol{r}')\frac{\sin\omega_n(t-t')}{\omega_n}H(t-t'), \quad \omega_n = a\sqrt{\lambda_n}$$

Thus, the Green's function is given by

$$G(\boldsymbol{r},\boldsymbol{r}';t-t') = \sum_{n=1}^{\infty} u_n^*(\boldsymbol{r}')u_n(\boldsymbol{r})\frac{\sin\omega_n(t-t')}{\omega_n}H(t-t')$$

The same approach can be used to obtain the Green's function for the heat conduction equation:

$$\partial_t G - a^2\Delta G = \delta(\boldsymbol{r}-\boldsymbol{r}')\delta(t-t')$$

$$G(\boldsymbol{r},\boldsymbol{r}';t-t') = \sum_{n=1}^{\infty} u_n^*(\boldsymbol{r}')u_n(\boldsymbol{r})\mathrm{e}^{-\lambda_n(t-t')}H(t-t')$$

(3) *Laplace transform method*

For time-dependent Green's functions, another method is to use the Laplace transform to convert them into potential equations in the image space. The solution can then be obtained following the procedure in Section 15.2, and finally, the inverse transform is applied to obtain the original function. Here, we will omit the detailed derivation.

16.4.2 *Time-dependent Green's integral formula*

We can also derive the integral formulas for the development equations. Here, we present the results, and readers can refer to relevant literature for detailed derivations.

(1) *Transport equation (parabolic equation)*

$$\begin{cases} u_t - a^2 \Delta u = f(\boldsymbol{r}, t) \\ \left(\alpha \dfrac{\partial u}{\partial n} + \beta u \right) \bigg|_{\Sigma} = \theta(M, t) \\ u|_{t=0} = \phi(\boldsymbol{r}) \end{cases} \tag{16.4.2}$$

The integral form solution is given by

$$\begin{aligned} u(\boldsymbol{r}, t) = & \iiint_{\Omega} \int_0^t G(\boldsymbol{r}, \boldsymbol{r}'; t - t') f(\boldsymbol{r}', t') \mathrm{d}V' \mathrm{d}t' \\ & + a^2 \iint_{\Sigma} \int_0^t \left[G \frac{\partial u}{\partial n'} - u \frac{\partial G}{\partial n'} \right] \mathrm{d}S' \mathrm{d}t' + \iiint_{\Omega'} [uG]|_{t'=0} \mathrm{d}V' \end{aligned} \tag{16.4.3}$$

(2) *Wave equation* (*hyperbolic equation*)

$$\begin{cases} u_{tt} - a^2 \Delta u = f(\boldsymbol{r}, t) \\ \left(\alpha \dfrac{\partial u}{\partial n} + \beta u \right) \bigg|_{\Sigma} = \theta(M, t) \\ u|_{t=0} = \phi(\boldsymbol{r}), \quad u_t|_{t=0} = \psi(\boldsymbol{r}) \end{cases} \tag{16.4.4}$$

The Green's function satisfies the following initial value problem:

$$\begin{cases} G_{tt} - a^2 \Delta G = \delta(\boldsymbol{r} - \boldsymbol{r}') \delta(t - t') \\ \left(\alpha \dfrac{\partial G}{\partial n} + \beta G \right) \bigg|_{\Sigma} = 0 \\ G|_{t=0} = 0, \quad G_t|_{t=0} = 0 \end{cases} \tag{16.4.5}$$

The integral form solution is given by

$$\begin{aligned} u(\boldsymbol{r}, t) = & \iiint_{\Omega'} \int_0^t G(\boldsymbol{r}, \boldsymbol{r}'; t - t') f(\boldsymbol{r}', t') \mathrm{d}V' \mathrm{d}t' \\ & + a^2 \iint_{\Sigma} \int_0^t \left[G \frac{\partial u}{\partial n'} - u \frac{\partial G}{\partial n'} \right] \mathrm{d}S' \mathrm{d}t' \\ & + \iiint_{\Omega'} [G u_{t'} - u G_{t'}]|_{t'=0} \mathrm{d}V' \end{aligned} \tag{16.4.6}$$

ENDNOTE

In quantum mechanics, the Schrödinger equation is usually considered to be source-free, meaning it is always a homogeneous equation. This ensures the conservation of particle number, and it seems improper to introduce Green's functions with external sources. We can approach it differently by defining $u(\boldsymbol{r},t) = \psi(\boldsymbol{r},t) - \psi(\boldsymbol{r},0)$. Then $u(\boldsymbol{r},t)$ satisfies the inhomogeneous equation:

$$\begin{cases} \mathrm{i}\dfrac{\partial}{\partial t}u(\boldsymbol{r},t) + \dfrac{1}{2}\nabla^2 u(\boldsymbol{r},t) = -\dfrac{1}{2}\nabla^2\psi(\boldsymbol{r},0) \\ u(\boldsymbol{r},t)|_{t=0} = 0 \end{cases}$$

We can consider $f(\boldsymbol{r}) \equiv -\frac{1}{2}\nabla^2\psi(\boldsymbol{r},0)$ as the external source of the field $u(\boldsymbol{r},t)$. Surely, $u(\boldsymbol{r},t)$ cannot be interpreted as the wave function anymore. However, we can define its Green's function equation as

$$\mathrm{i}\frac{\partial}{\partial t}G(\boldsymbol{r},\boldsymbol{r}';t-t') + \frac{1}{2}\nabla^2 G(\boldsymbol{r},\boldsymbol{r}';t-t') = \delta(\boldsymbol{r}-\boldsymbol{r}')\delta(t-t')$$

Taking the Fourier transform $\bar{G}(\boldsymbol{k};t,t') = \mathfrak{F}[G(\boldsymbol{r},\boldsymbol{r}';t,t')]$, we have

$$\mathrm{i}\frac{\mathrm{d}}{\mathrm{d}t}\bar{G}(\boldsymbol{k};t-t') - \frac{1}{2}k^2\bar{G}(\boldsymbol{k};t,t') = \delta(t-t')$$

Solving this equation, we obtain

$$\bar{G}(\boldsymbol{k};t-t') = -\mathrm{i}\mathrm{e}^{-\frac{\mathrm{i}}{2}\boldsymbol{k}^2(t-t')}$$

Thus, the Green's function is

$$G(\boldsymbol{r},\boldsymbol{r}';t-t') = \frac{1}{(2\pi)^3}\int_{-\infty}^{\infty}\mathrm{e}^{\mathrm{i}\boldsymbol{k}\cdot(\boldsymbol{r}-\boldsymbol{r}')-\frac{\mathrm{i}}{2}\boldsymbol{k}^2(t-t')}\mathrm{d}\boldsymbol{k}$$

Using this, we can derive the integral form of $u(\boldsymbol{r},t)$ and then solve for the wave function $\psi(\boldsymbol{r},t)$.

Exercises

[1] Using the method of Green's function to solve the initial value problem:

$$\begin{cases} \dfrac{\mathrm{d}^2 y(t)}{\mathrm{d}t^2} + y(t) = f(t) \\ y(0) = y'(0) = 0 \end{cases}$$

Answer:

$$G(t,t') = \begin{cases} 0 & (t < t') \\ \sin(t-t') & (t > t') \end{cases}$$

$$y(t) = \int_0^{\infty} G(t,t')f(t')\mathrm{d}t' = \int_0^t f(t')\sin(t-t')\mathrm{d}t'$$

[2] Using the method of Green's function to solve

$$\begin{cases} u_t - a^2 u_{xx} = A\sin\omega t \\ u_x|_{x=0} = 0, \quad u_x|_{x=l} = 0 \\ u|_{t=0} = 0 \end{cases}$$

Answer: $u = \frac{A}{\omega}(1-\cos\omega t)$

[3] Using the method of Green's function to solve

$$\begin{cases} u_{tt} - a^2 u_{xx} = A\cos\dfrac{x}{l}\sin\omega t \\ u_x|_{x=0} = 0, \quad u_x|_{x=l} = 0 \\ u|_{t=0} = 0, \quad u_t|_{t=0} = 0 \end{cases}$$

[4] Find the Green's function for the three-dimensional wave equation in unbounded space.

Answer:

$$G(\boldsymbol{r},\boldsymbol{r}';t-t') = \frac{1}{4\pi a}\frac{1}{|\boldsymbol{r}-\boldsymbol{r}'|}\delta(|\boldsymbol{r}-\boldsymbol{r}'| - a(t-t'))$$

[5] Find n-dimensional Green's function for the unbounded space diffusion equation.

Hint: Perform both Fourier and Laplace transforms simultaneously, integrate over the n-dimensional hyperspherical coordinates (see Section 10.5), and utilize the integral formula involving Bessel functions.

$$\int_0^\infty \mathrm{e}^{-a^2x^2} J_\nu(bx)x^{\nu+1}\mathrm{d}x = \frac{b^\nu}{(2a^2)^{\nu+1}}\mathrm{e}^{-\frac{b^2}{4a^2}}$$

Answer:

$$G(\boldsymbol{r}, \boldsymbol{r}'; t-t') = \frac{\mathrm{e}^{-\frac{r^2}{4t}}}{[4\pi a(t-t')]^{n/2}}\mathrm{H}(t-t')$$

16.5 Perturbation Method

From the previous discussion, it can be seen that solving boundary value problems exactly using Green's functions can be quite cumbersome. The true utility of Green's functions in physics lies in providing an effective method for approximating problems that do not have exact solutions, and obtaining the desired results. This is known as the perturbation expansion method. It makes Green's functions a specialized computational technique that finds widespread applications in quantum field theory and condensed matter physics. This section provides a brief introduction to its basic ideas.

16.5.1 *Scattering problem*

In quantum mechanics, the elastic scattering process of a particle in a central potential is described by the Schrödinger equation:

$$-\frac{\hbar^2}{2\mu}\Delta\psi + V(\boldsymbol{r})\psi = E\psi \quad (E > 0)$$

Taking $E = \frac{\hbar^2\boldsymbol{k}^2}{2\mu}$, the equation can be written as

$$(\Delta + \boldsymbol{k}^2)\psi = \frac{2\mu}{\hbar^2}V(\boldsymbol{r})\psi \tag{16.5.1}$$

Assuming the incident wave is a plane wave $\mathrm{e}^{\mathrm{i}\boldsymbol{k}_0\cdot\boldsymbol{r}}$, the wave function has the following asymptotic form as $\boldsymbol{r} \to \infty$,

$$\psi(\boldsymbol{r}) \overset{r\to\infty}{\longrightarrow} \mathrm{e}^{\mathrm{i}\boldsymbol{k}_0\cdot\boldsymbol{r}} + f(\theta, \varphi)\frac{\mathrm{e}^{\mathrm{i}kr}}{r}$$

The second term represents the scattered spherical wave $\psi_{sc}(\boldsymbol{r})$ with wavevector $\boldsymbol{k}$, which is effective for short-range potentials $V(\boldsymbol{r})$. $f_k(\theta, \varphi)$ represents the scattering wave amplitude. Treating the right-hand side of Equation (16.5.1) as a non-homogeneous term, the general solution can be expressed using the Green's function as

$$\psi(\boldsymbol{r}) = \mathrm{e}^{\mathrm{i}\boldsymbol{k}_0 \cdot \boldsymbol{r}} + \frac{2\mu}{\hbar^2} \int V(\boldsymbol{r}')\psi(\boldsymbol{r}')G_0(\boldsymbol{r}, \boldsymbol{r}')\mathrm{d}\boldsymbol{r}' \tag{16.5.2}$$

The Green's function for the three-dimensional unbounded space Helmholtz operator $\Delta + \boldsymbol{k}^2$ is given by

$$G_0(\boldsymbol{r}, \boldsymbol{r}') = -\frac{\mathrm{e}^{\mathrm{i}k|\boldsymbol{r}-\boldsymbol{r}'|}}{4\pi|\boldsymbol{r} - \boldsymbol{r}'|} \tag{16.5.3}$$

The solution of the equation is expressed as

$$\psi(\boldsymbol{r}) = \mathrm{e}^{\mathrm{i}\boldsymbol{k}_0 \cdot \boldsymbol{r}} - \frac{\mu}{2\pi\hbar^2} \int V(\boldsymbol{r}')\psi(\boldsymbol{r}')\frac{\mathrm{e}^{\mathrm{i}k|\boldsymbol{r}-\boldsymbol{r}'|}}{|\boldsymbol{r} - \boldsymbol{r}'|}\mathrm{d}\boldsymbol{r}' \tag{16.5.4}$$

This equation is known as the Lippmann–Schwinger equation. For elastic scattering, $|\boldsymbol{k}_0| = |k|$, and the first term $\mathrm{e}^{\mathrm{i}\boldsymbol{k}_0 \cdot \boldsymbol{r}}$ satisfies the corresponding homogeneous equation, reflecting the behavior of scattered particles at infinity.

When the scattering potential is weak or a short-range potential, $\mathrm{e}^{\mathrm{i}\boldsymbol{k}_0 \cdot \boldsymbol{r}}$ dominates the wave function and can be used to replace the wave function $\psi(\boldsymbol{r}')$ inside the integral sign, resulting in the Born approximation formula:

$$\psi(\boldsymbol{r}) = \mathrm{e}^{\mathrm{i}\boldsymbol{k}_0 \cdot \boldsymbol{r}} - \frac{\mu}{2\pi\hbar^2} \int V(\boldsymbol{r}')\mathrm{e}^{\mathrm{i}\boldsymbol{k}_0 \cdot \boldsymbol{r}}\frac{\mathrm{e}^{\mathrm{i}k|\boldsymbol{r}-\boldsymbol{r}'|}}{|\boldsymbol{r} - \boldsymbol{r}'|}\mathrm{d}\boldsymbol{r}' \tag{16.5.5}$$

The Lippmann–Schwinger equation has a nested structure for the wave function $\psi(\boldsymbol{r})$, and it can be self-iterated as follows:

$$\psi(\boldsymbol{r}) = \mathrm{e}^{\mathrm{i}\boldsymbol{k}_0 \cdot \boldsymbol{r}} - \frac{\mu}{2\pi\hbar^2} \int \left[\mathrm{e}^{\mathrm{i}\boldsymbol{k}_0 \cdot \boldsymbol{r}'} - \frac{\mu}{2\pi\hbar^2} \int V(\boldsymbol{r}'')\psi(\boldsymbol{r}'')\frac{\mathrm{e}^{\mathrm{i}k|\boldsymbol{r}-\boldsymbol{r}''|}}{|\boldsymbol{r} - \boldsymbol{r}''|}\mathrm{d}\boldsymbol{r}''\right] \times \frac{\mathrm{e}^{\mathrm{i}k|\boldsymbol{r}-\boldsymbol{r}'|}}{|\boldsymbol{r} - \boldsymbol{r}'|}V(\boldsymbol{r}')\mathrm{d}\boldsymbol{r}' \tag{16.5.6}$$

The Born formula can be viewed as a first-order approximation. By replacing $\psi(\boldsymbol{r}'')$ inside the integral sign with $\mathrm{e}^{\mathrm{i}\boldsymbol{k}_0\cdot\boldsymbol{r}''}$, the second-order approximation can be obtained

$$\begin{aligned}\psi(\boldsymbol{r}) = \mathrm{e}^{\mathrm{i}\boldsymbol{k}_0\cdot\boldsymbol{r}} &- \frac{\mu}{2\pi\hbar^2}\int V(\boldsymbol{r}')\mathrm{e}^{\mathrm{i}\boldsymbol{k}_0\cdot\boldsymbol{r}}\frac{\mathrm{e}^{\mathrm{i}k|\boldsymbol{r}-\boldsymbol{r}'|}}{|\boldsymbol{r}-\boldsymbol{r}'|}\mathrm{d}\boldsymbol{r}' \\ &+ \left(\frac{\mu}{2\pi\hbar^2}\right)^2\int\int V(\boldsymbol{r}'')\mathrm{e}^{\mathrm{i}\boldsymbol{k}_0\cdot\boldsymbol{r}''}\frac{\mathrm{e}^{\mathrm{i}k|\boldsymbol{r}-\boldsymbol{r}''|}}{|\boldsymbol{r}-\boldsymbol{r}''|}\mathrm{d}\boldsymbol{r}''\frac{\mathrm{e}^{\mathrm{i}k|\boldsymbol{r}-\boldsymbol{r}'|}}{|\boldsymbol{r}-\boldsymbol{r}'|}\mathrm{d}\boldsymbol{r}'\end{aligned} \tag{16.5.7}$$

By repeating this self-iteration process, higher-order approximation solutions can be obtained. However, it is important to ensure that the higher-order correction terms decrease progressively to ensure the convergence of the series.

Example 16.10. Solve the one-dimensional stationary Schrödinger equation for bound states:

$$-\frac{\hbar^2}{2\mu}\frac{\mathrm{d}^2\psi}{\mathrm{d}x^2} + V(x)\psi = E\psi \quad (E < 0)$$

Solution: Let $E = -\hbar^2\kappa^2/2\mu$, the differential equation becomes

$$\hat{L}\psi \equiv \left(\frac{\mathrm{d}^2}{\mathrm{d}x^2} - \kappa^2\right)\psi = \frac{2\mu}{\hbar^2}V(x)\psi$$

The formal solution of the equation is given by

$$\psi(x) = \psi_0(x) + \int_{-\infty}^{\infty}\frac{2\mu}{\hbar^2}G_0(x,x')V(x')\psi(x')\mathrm{d}x'$$

Since $\hat{L}\psi_0 = 0$, under the condition of a bound state, we have $\psi_0 = A\mathrm{e}^{\kappa x} + B\mathrm{e}^{-\kappa x} \to 0$ as $x \to \pm\infty$. It can also be shown that the Green's function of the operator $\hat{L}$ is

$$G_0(x,x') = -\frac{1}{2\kappa}\mathrm{e}^{-\kappa|x-x'|}$$

Therefore,

$$\psi(x) = -\frac{\mu}{\hbar^2\kappa}\int_{-\infty}^{\infty}\mathrm{e}^{-\kappa|x-x'|}V(x')\psi(x')\mathrm{d}x'$$

If we consider an external potential $V(x) = -V_0\delta(x-a)$, then

$$\psi(x) = -\frac{\mu}{\hbar^2\kappa}\int_{-\infty}^{\infty} \mathrm{e}^{-\kappa|x-x'|} V_0\delta(x'-a)\psi(x')\mathrm{d}x' = \frac{\mu V_0}{\hbar^2\kappa}\mathrm{e}^{-\kappa|x-a|}\psi(a)$$

The equation holds when $x = a$, and we obtain

$$\frac{\mu V_0}{\hbar^2\kappa} = 1 \longrightarrow E = -\frac{\mu V_0}{2\hbar^2}$$

This indicates that the equation has a unique bound state.

16.5.2 *Perturbation theory*

For a general inhomogeneous equation with n independent variables, $\boldsymbol{x} = (x_1, x_2, \ldots, x_n)$,

$$\hat{L}u(\boldsymbol{x}) + \alpha\hat{V}(\boldsymbol{x})u(\boldsymbol{x}) = f(\boldsymbol{x})$$

where α is a constant and $V(\boldsymbol{x})$ can be seen as an external potential, moving it to the right-hand side along with $f(\boldsymbol{x})$ as a non-homogeneous source term. Let $G_0(\boldsymbol{x}, \boldsymbol{x}')$ be the Green's function of the operator $\hat{L}$ under certain boundary conditions:

$$\hat{L}G_0(\boldsymbol{x}, \boldsymbol{x}') = \delta(\boldsymbol{x} - \boldsymbol{x}')$$

We can write the formal solution as

$$u(\boldsymbol{x}) = h(\boldsymbol{x}) + \int_D \mathrm{d}^n\boldsymbol{x}' G_0(\boldsymbol{x}, \boldsymbol{x}')[f(\boldsymbol{x}) - \alpha V(\boldsymbol{x}')u(\boldsymbol{x}')]$$

where $h(\boldsymbol{x})$ is a solution corresponding to the homogeneous equation of the operator $\hat{L}$: $\hat{L}h(\boldsymbol{x}) = 0$, satisfying the corresponding non-homogeneous boundary conditions, and D is the integration domain of the operator $\hat{L}$. Combining the integral term with $h(\boldsymbol{x})$ and denoting it as $F(\boldsymbol{x})$, we obtain the formal solution of the equation:

$$u(\boldsymbol{x}) = F(\boldsymbol{x}) - \alpha\int_D G_0(\boldsymbol{x}, \boldsymbol{x}')V(\boldsymbol{x}')u(\boldsymbol{x}')\mathrm{d}^n\boldsymbol{x}'$$

Such integral equations are generally difficult to solve analytically. However, if α is sufficiently small, we can use perturbation methods to expand the terms inside the integral using a series.

By repeatedly substituting u into the integrand, we obtain a series of integrals:

$$u(\boldsymbol{x}) = F(\boldsymbol{x}) + \sum_{m=1}^{N-1} (-\alpha)^m \int_D \mathrm{d}^n\boldsymbol{x}' K^m(\boldsymbol{x}, \boldsymbol{x}') F(\boldsymbol{x}') + (-\alpha)^N \int_D \mathrm{d}^n\boldsymbol{x}' K^N(\boldsymbol{x}, \boldsymbol{x}') u(\boldsymbol{x}')$$

where

$$K(\boldsymbol{x}, \boldsymbol{x}') = V(\boldsymbol{x}) G_0(\boldsymbol{x}, \boldsymbol{x}')$$

$$K^m(\boldsymbol{x}, \boldsymbol{x}') = \int_D \mathrm{d}^n\boldsymbol{y} K^{m-1}(\boldsymbol{x}, \boldsymbol{y}) K(\boldsymbol{y}, \boldsymbol{x}) \quad (m \geq 2)$$

Taking $N \to \infty$, we obtain the Neumann series:

$$u(\boldsymbol{x}) = F(\boldsymbol{x}) + \sum_{m=1}^{\infty} (-\alpha)^m \int_D \mathrm{d}^n\boldsymbol{x}' K^m(\boldsymbol{x}, \boldsymbol{x}') F(\boldsymbol{x}')$$

In a more concise operator form,

$$|u\rangle = |F\rangle + \sum_{m=1}^{\infty} (-\alpha)^m \mathbf{K}^m |F\rangle$$

For this series to have meaning, it must converge, which requires

$$|\alpha| \left[\int_D \mathrm{d}^n\boldsymbol{x}' \int_D \mathrm{d}^n\boldsymbol{x} |K^m(\boldsymbol{x}, \boldsymbol{x}')|^2 \right]^{1/2} < 1$$

When α is small enough, the series may converge, and this constitutes a method of using perturbation expansion to approximate the solutions of equations.

ENDNOTE

In physics, Feynman introduced the Feynman diagram technique based on the Neumann series. Since most second-order partial differential equations are homogeneous, we have $f(\boldsymbol{x}) = 0$. In this case, $\hat{L}$ and $\hat{V}(\boldsymbol{x})$ can be regarded as the operators for free particles and the interaction potential, respectively. The solution to $\hat{L}u = 0$ is called the free particle solution, denoted as $u_f(\boldsymbol{x})$. Thus, we have

$$u(\boldsymbol{x}) = u_f(\boldsymbol{x}) - \alpha \int_{R^n} \mathrm{d}^n \boldsymbol{x}' G_0(\boldsymbol{x}, \boldsymbol{x}') V(\boldsymbol{x}') u(\boldsymbol{x}')$$

Here, we denote the Green's function of the operator $\hat{L} + \alpha \hat{V}$ as $G(\boldsymbol{x}, \boldsymbol{x}')$, and we use R^n to indicate that there are no boundary constraints on u, allowing us to use the singular part of the Green's function. By assuming $\hat{G} = \hat{G}_0 + \hat{A}$, where $\hat{L}\hat{G}_0 = \hat{I}$ and $\hat{A}$ is an undetermined operator, we have

$$\hat{L}\hat{G} = \hat{L}\hat{G}_0 + \hat{L}A = \hat{I} + \hat{L}\hat{A}$$

On the other hand, $(\hat{L} + \alpha \hat{V})\hat{G} = \hat{I}$, therefore,

$$\hat{L}\hat{A} = -\alpha \hat{V}\hat{G}$$

yielding

$$\hat{A} = -\alpha \hat{L}^{-1}\hat{V}\hat{G} = -\alpha \hat{G}_0 \hat{V}\hat{G}$$

Consequently,

$$\hat{G} = \hat{G}_0 - \alpha \hat{G}_0 \hat{V} \hat{G}$$

Iteratively substitute $\hat{G}$ into itself, yielding

$$\begin{aligned}\hat{G} = \hat{G}_0 - \alpha \hat{G}_0 \hat{V} \hat{G}_0 + \alpha^2 \hat{G}_0 \hat{V} \hat{G}_0 \hat{V} \hat{G}_0 \\ - \alpha^3 \hat{G}_0 \hat{V} \hat{G}_0 \hat{V} \hat{G}_0 \hat{V} \hat{G}_0 + \cdots = \frac{\hat{G}_0}{\hat{I} - \alpha \hat{G}_0 \hat{V}}\end{aligned}$$

This formula is known as the Dyson equation in quantum field theory, and in integral form, it can be written as

$$G(\boldsymbol{x}, \boldsymbol{y}) = G_0(\boldsymbol{x}, \boldsymbol{y}) - \alpha \int_{R^n} \mathrm{d}^n \boldsymbol{x}' G_0(\boldsymbol{x}, \boldsymbol{x}') V(\boldsymbol{x}') G(\boldsymbol{x}', \boldsymbol{y})$$

Expanding the Green's function perturbatively, we obtain

$$G(\boldsymbol{x},\boldsymbol{y}) = G_0(\boldsymbol{x},\boldsymbol{y}) + \sum_{m=1}^{\infty}(-\alpha)^m \int_{R^n} \mathrm{d}^n\boldsymbol{x}' G_0(\boldsymbol{x},\boldsymbol{x}')K^m(\boldsymbol{x}',\boldsymbol{y})$$

Feynman's idea is to interpret $G(\boldsymbol{x},\boldsymbol{y})$ as the propagator of interacting particles from $\boldsymbol{x}$ to $\boldsymbol{y}$, represented by a thick line, while $G_0(\boldsymbol{x},\boldsymbol{y})$ represents the propagator of free particles, represented by a thin line. The perturbative expansion of $G(\boldsymbol{x},\boldsymbol{y})$ implies that it can be graphically represented by the propagation of a free particle from $\boldsymbol{x}$ to $\boldsymbol{x}'_1$, interaction at the vertex $-\alpha V(\boldsymbol{x}'_1)$, propagation from $\boldsymbol{x}'_1$ to $\boldsymbol{x}'_2$ with another interaction $-\alpha V(\boldsymbol{x}'_2)$, and so on, until reaching the point $\boldsymbol{y}$. All possible scenarios are superimposed, as shown in Figure 16.5.

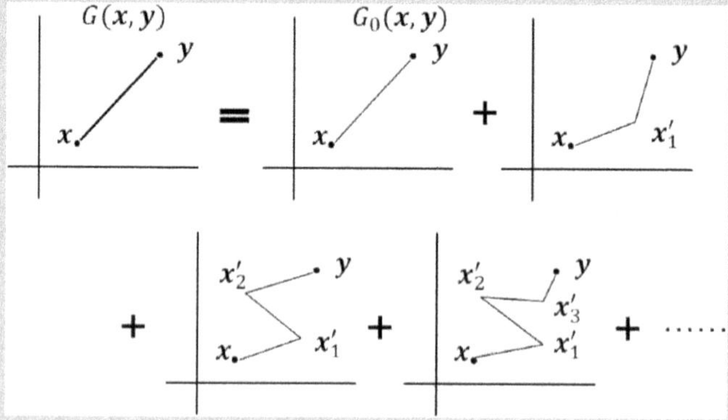

Figure 16.5.

The Feynman diagram technique has been widely applied in relativistic quantum field theory, where $n = 4$ represents four-dimensional spacetime. In quantum electrodynamics, $\alpha = \frac{e^2}{\hbar c} \approx \frac{1}{137}$ is known as the fine-structure constant, which is a good perturbative parameter. This has led to astonishingly accurate results in calculations within quantum electrodynamics. Feynman diagrams have also been introduced to non-relativistic condensed matter physics, becoming an important tool for perturbative calculations in strongly correlated systems.

Exercises

[1] Prove that the integral solution for the damped harmonic oscillator equation,

$$\begin{cases} \ddot{x} + 2\gamma\dot{x} + \omega^2 x = 0 \\ x(0) = x_0, \quad \dot{x}(0) = 0 \end{cases}$$

can be expressed in the following form:

$$x(t) = x_0 - \frac{\omega^2}{2\gamma}\int_0^t [1 - \mathrm{e}^{-2\gamma(t-t')}]x(t')\mathrm{d}t'$$

Alternatively,

$$x(t) = x_0 \cos\omega t + \frac{2\gamma x_0}{\omega}\sin\omega t - 2\gamma\int_0^t \cos\omega(t-t')x(t')\mathrm{d}t'$$

Hint: Treat $2\gamma\dot{x}$ or $\omega^2 x$ as the non-homogeneous term.

[2] For the scattering state ($E > 0$) of the one-dimensional time-independent Schrödinger equation:

(a) To prove the integral solution:

$$\psi(x) = \mathrm{e}^{\mathrm{i}kx} - \frac{\mathrm{i}\mu}{\hbar^2 k}\int_{-\infty}^{\infty} \mathrm{e}^{\mathrm{i}k|x-x'|}V(x')\psi(x')\mathrm{d}x'$$

(b) Dividing $(-\infty, \infty)$ into three regions: $R_1 = (-\infty, -a)$, $R_2 = (-a, +a)$ and $R_3 = (+a, \infty)$, with wave functions in each region denoted as $\psi_i(x)$ ($i = 1, 2, 3$). The external potential is zero in R_1 and R_3. Prove the following:

$$\psi_1(x) = \mathrm{e}^{\mathrm{i}kx} - \frac{\mathrm{i}\mu}{\hbar^2 k}\mathrm{e}^{-\mathrm{i}kx}\int_{-a}^{a} \mathrm{e}^{\mathrm{i}kx'}V(x')\psi_2(x')\mathrm{d}x'$$

$$\psi_2(x) = \mathrm{e}^{\mathrm{i}kx} - \frac{\mathrm{i}\mu}{\hbar^2 k}\int_{-a}^{a} \mathrm{e}^{\mathrm{i}k|x-x'|}V(x')\psi_2(x')\mathrm{d}x'$$

$$\psi_3(x) = \mathrm{e}^{\mathrm{i}kx} - \frac{\mathrm{i}\mu}{\hbar^2 k}\mathrm{e}^{\mathrm{i}kx}\int_{-a}^{a} \mathrm{e}^{-\mathrm{i}kx'}V(x')\psi_2(x')\mathrm{d}x'$$

[3] For a short-range scattering potential $V(\boldsymbol{r}')$, when $r \to \infty$, $|\boldsymbol{r} - \boldsymbol{r}'| \approx r\left(1 - \frac{\boldsymbol{r}'}{r^2}\boldsymbol{r}\right)$, prove the following:

(a) In the Born approximation, the scattering wave function is given by

$$\psi_{\text{sc}}(\boldsymbol{r}) \approx -\frac{\mu}{2\pi\hbar^2 r}\mathrm{e}^{ikr}\int V(\boldsymbol{r}')\mathrm{e}^{-\mathrm{i}\boldsymbol{q}\cdot\boldsymbol{r}'}\mathrm{d}\boldsymbol{r}' \equiv -\frac{4\pi^2\mu}{\hbar^2 r}\mathrm{e}^{ikr}\tilde{V}(\boldsymbol{q})$$

where $\boldsymbol{q} = \boldsymbol{k} - \boldsymbol{k}_0$ represents the momentum transfer of the scattered particle.

(b) The scattering differential cross-section is

$$\mathrm{d}\sigma = |f(\theta,\varphi)|^2\mathrm{d}\boldsymbol{q} = \left|\frac{4\pi^2\mu}{\hbar^2}\tilde{V}(\boldsymbol{q})\right|^2 \mathrm{d}\boldsymbol{q}$$

Chapter 17

Calculus of Variations

17.1 Functionals and Extremals

17.1.1 *Steepest descent problem*

Johann Bernoulli once posed the following question: Under the influence of gravity, a stationary particle slides without friction from point A to point B, as shown in Figure 17.1. Along which path does it take the shortest time?

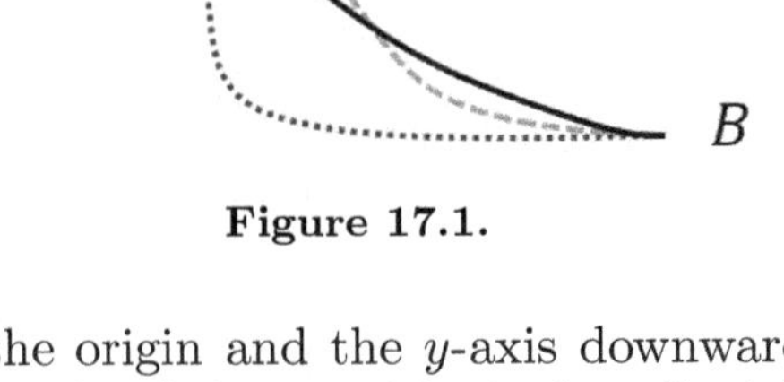

Figure 17.1.

Taking A as the origin and the y-axis downward, let us assume that the sliding path of the particle is described by a continuous smooth function $y(x)$. According to the conservation of mechanical energy, the velocity of the particle is given by

$$v = \frac{\mathrm{d}s}{\mathrm{d}t} = \sqrt{2gy}$$

where ds represents the differential element of the curve. The time required to slide from A to B is given by

$$T = \int_A^B \frac{\mathrm{d}s}{v} = \int_A^B \frac{\sqrt{1+y'^2}}{\sqrt{2gy}} \mathrm{d}x \equiv J[y(x)]$$

where J is known as the functional of $y(x)$, which represents the class of admissible functions and is referred to as the variable function of the functional $J[y(x)]$.

Difference between a functional and an ordinary function: The value of an ordinary function depends on the independent variable x, whereas the value of a functional not only depends on the independent variable x within the interval from A to B, but also on one or more functions $y(x)$. Additionally, it generally involves the derivative $y'(x)$. In fact, the domain of definition for a functional is not a region in coordinate space but a set composed of admissible functions. Functionals typically appear in integral form and are generally expressed as

$$J[y(x)] \equiv \int_a^b F(x, y, y')\mathrm{d}x \tag{17.1.1}$$

In physics, the function $F(x, y, y')$ is often referred to as the Lagrangian.

The rephrased Bernoulli's question is: What kind of function curve $y(x)$ should be chosen to minimize $J[y(x)]$?

17.1.2 *Functional variation*

Functional variation is a method for finding extrema of functionals, typically by transforming the problem into solving differential equations. Let us consider a continuous and smooth function curve $y(x)$. It is slightly deformed as

$$y(x) \longmapsto \tilde{y}(x) = y(x) + \varepsilon\eta(x)$$

where ε is a very small constant and $\eta(x)$ is a continuous and differentiable arbitrary function. Thus, we have

$$\delta y(x) = \tilde{y}(x) - y(x) = \varepsilon\eta(x)$$

where δy is the variation of the function $y(x)$.

What about the variation $\delta y'$ of the derivative function $y'(x)$? Since

$$y' = \lim_{\Delta x \to 0} \frac{y(x + \Delta x) - y(x)}{\Delta x}$$

according to the definition, the derivative of the deformed function is

$$\begin{aligned}
\tilde{y}' &= \lim_{\Delta x \to 0} \frac{\tilde{y}(x + \Delta x) - \tilde{y}(x)}{\Delta x} \\
&= \lim_{\Delta x \to 0} \frac{[y(x + \Delta x) + \varepsilon \eta(x + \Delta x)] - [y(x) + \varepsilon \eta(x)]}{\Delta x}
\end{aligned}$$

Thus,

$$\delta y' = \tilde{y}'(x) - y'(x) = \lim_{\Delta x \to 0} \frac{\varepsilon \eta(x + \Delta x) - \varepsilon \eta(x)}{\Delta x} = \varepsilon \eta'(x) \equiv (\delta y)'$$

It is observed that the variation of a function's derivative is equal to the derivative of the function's variation, indicating that taking variations of functions and differentiating functions can be interchanged.

If the Lagrangian $F(x, y, y')$ is twice differentiable with respect to x, y and y', and $y(x)$ has a continuous second derivative, then when $y(x)$ takes the variation $\delta y = \varepsilon \eta(x)$, the corresponding change in the functional $J[y(x)]$ is

$$\begin{aligned}
\delta J &= J[y(x) + \delta y] - J[y(x)] = \int_a^b [F(x, \tilde{y}, \tilde{y}') - F(x, y, y')] \mathrm{d}x \\
&= \int_a^b \left[\frac{\partial F}{\partial y} \varepsilon \eta + \frac{\partial F}{\partial y'} \varepsilon \eta' + O(\varepsilon^2) \right] \mathrm{d}x
\end{aligned}$$

where $O(\varepsilon^2)$ represents higher-order infinitesimals of ε. This yields the first-order variation of the functional:

$$\delta J = \int_a^b \left[\frac{\partial F}{\partial y} \delta y + \frac{\partial F}{\partial y'} \delta y' \right] \mathrm{d}x \tag{17.1.2}$$

This equation indicates that both the function y and its derivative y' should be regarded as independent variation quantities.

17.1.3 *Functional extrema*

Let us discuss the conditions under which a function $y(x)$ can make the functional $J[y(x)]$ attain an extremum. Suppose a function $y(x)$ yields an extremum for the functional $J[y(x)]$. When $y(x)$ undergoes a small variation $\delta y = \varepsilon\eta(x)$, the functional $J[y(x) + \varepsilon\eta(x)]$ can be treated as an ordinary function of the parameter ε, denoted as $\Phi(\varepsilon) \equiv J[y(x) + \varepsilon\eta(x)]$. The extremum problem of the functional is then transformed into an extremum problem of the ordinary function:

$$\left.\frac{\mathrm{d}\Phi(\varepsilon)}{\mathrm{d}\varepsilon}\right|_{\varepsilon=0} = 0$$

which can be expressed as

$$\left.\frac{\mathrm{d}}{\mathrm{d}\varepsilon}\int_a^b F(x, y + \varepsilon\eta, y' + \varepsilon\eta')\mathrm{d}x\right|_{\varepsilon=0} = \int_a^b \left(\frac{\partial F}{\partial y}\eta + \frac{\partial F}{\partial y'}\eta'\right)\mathrm{d}x = 0$$

Comparing this equation with Equation (17.1.2), it implies that the variation of the functional $\delta J = 0$. By performing integration by parts on the second term of the equation, we have

$$\int_a^b \frac{\partial F}{\partial y'}\delta y'\mathrm{d}x = \int_a^b \frac{\partial F}{\partial y'}\frac{\mathrm{d}}{\mathrm{d}x}(\delta y)\mathrm{d}x = \left.\frac{\partial F}{\partial y'}\delta y\right|_a^b - \int_a^b \frac{\mathrm{d}}{\mathrm{d}x}\left(\frac{\partial F}{\partial y'}\right)\delta y\mathrm{d}x$$

obtaining the necessary condition for the functional to attain an extremum:

$$\delta J = \left[\frac{\partial F}{\partial y'}\delta y\right]\Bigg|_a^b - \int_a^b \left[\frac{\partial F}{\partial y} - \frac{\mathrm{d}}{\mathrm{d}x}\left(\frac{\partial F}{\partial y'}\right)\right]\delta y\mathrm{d}x = 0 \qquad (17.1.3)$$

ENDNOTE

The study of the catenary curve, also known as the hanging chain, dates back to the observations made by Robert Hooke. He noticed that arches constructed with small stones exhibited a similar shape to hanging chains. Galileo initially believed that the hanging chain resembled a parabola. However, at the age of 17, Huygens challenged this notion. Although he could not provide the correct equation for the hanging chain, he believed that the shape of a rope under horizontal forces was a parabola.

The problem of the hanging chain was eventually solved by Jacob Bernoulli, Huygens, and Leibniz independently. As shown in Figure 17.2, the tangential force at point P is denoted as F_1, and the horizontal force is denoted as F_0. When these forces are independent of the position of P, the hanging chain is in equilibrium. The gravitational force acting on the segment OP of the rope is proportional to the arc length OP, denoted as s. Jacob Bernoulli proved that the curve satisfies the equation:

$$\frac{\mathrm{d}y}{\mathrm{d}x} = \frac{W}{F_0} = \frac{s}{a}$$

where a is a constant. After some transformations, the above equation can be rewritten as

$$\frac{\mathrm{d}y}{\mathrm{d}x} = \frac{\sqrt{y^2 - a^2}}{a}$$

The solution to this equation is given by

$$y = a\cosh\left(\frac{x}{a}\right) - a$$

Therefore, the hanging chain is a hyperbolic function, not a parabola.

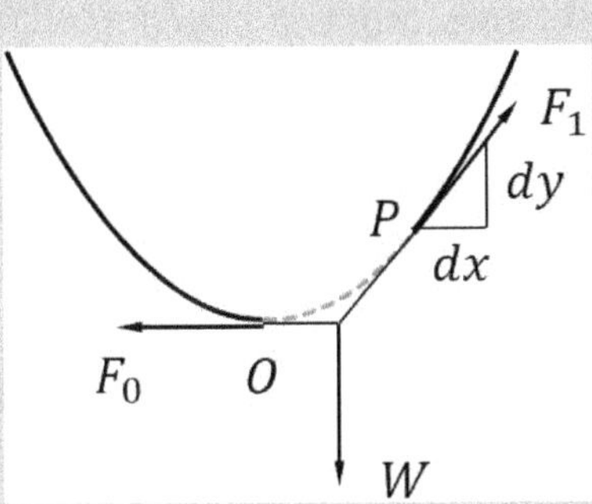

Figure 17.2.

Elastic curves refer to the shapes formed when a thin elastic plate is compressed and bent. Jacob Bernoulli demonstrated that these curves satisfy the differential equation:

$$\mathrm{d}s = \frac{\mathrm{d}x}{\sqrt{1 - x^4}}$$

To geometrically interpret this integral, he introduced the lemniscate curve and proved that its arc length corresponds exactly to the expression given by this formula. This became the starting point for the study of lemniscates, which eventually led to the development of elliptic functions theory.

In the world of music, there is a famous Bach family, whose members possess musical talents to such an extent that in some parts of Germany, the term "Bach" is synonymous with "musician". In the kingdom of mathematics, there is a Bernoulli family, which produced eight mathematicians within three generations and nurtured numerous outstanding descendants in many areas. Regarding the brachistochrone problem, Johann Bernoulli once arrogantly challenged mathematicians worldwide. Both Leibniz and Newton independently solved the problem, with the answer of the cycloid, a curve traced by a fixed point on a rolling circle along a straight line. Newton sent his solution anonymously to Bernoulli, who quickly recognized the "lion hidden behind the claws". Among all the solutions, Jacob Bernoulli's approach had the most profound significance. He imposed specific conditions on various possible paths and introduced the concept of "variable curve", which was a crucial step in the development of the calculus of variations.

In conventional extremum theory, the existence of solutions is guaranteed by the Weierstrass theorem on compact sets. However, in certain cases of the calculus of variations, it is not always possible to assume in advance that an extremum exists. The particular difficulty of the calculus of variations is that a problem can be meaningful but may lack a solution because it is generally impossible to define the region of admissible functions as a compact set, allowing for the application of the Bolzano–Weierstrass theorem. For example, consider the following problem: Find the shortest line with continuous curvature between two points on the x-axis, with the requirement that it is perpendicular to the x-axis at the endpoints. This problem has no solution! A curve is always longer than the straight line connecting the two points, and yet the curve can approach the straight line infinitely closely, so the curve's length has a lower bound but no extremum.

This fact helps us understand that functional extremum problems can have necessary conditions but not sufficient conditions.

Exercises

[1] Find the second-order variation expression for the functional.

[2] Given the equation $\frac{\mathrm{d}y}{\mathrm{d}x} = \frac{s}{a}$, where s is the arc length of the curve, prove

$$\frac{\mathrm{d}y}{\mathrm{d}x} = \frac{\sqrt{y^2 - a^2}}{a}$$

Hint: Use the relation:

$$\frac{\mathrm{d}^2 y}{\mathrm{d}x^2} = \frac{1}{2}\frac{\mathrm{d}}{\mathrm{d}y}\left(\frac{\mathrm{d}y}{\mathrm{d}x}\right)^2$$

17.2 Euler–Lagrange Equation

The variational method provides a new perspective for reexamining the fundamental principles of physics. This method is not only applicable to describing the dynamics of particle systems but also suitable for describing the motion of continuous fields. First, we state the fundamental lemma of the variational method.

Let $y(x)$ be a continuous function of x. If the relation

$$\int_a^b y(x)\eta(x)\mathrm{d}x = 0$$

holds for any boundary-fixed continuous function $\eta(x)$ with second-order derivatives, then it must be true that $y(x) = 0$.

17.2.1 *Fixed endpoints*

If the endpoints a and b are fixed, then the variation δy is always zero,

$$\delta y|_{x=a} = \delta y|_{x=b} = 0$$

From the necessary condition for a functional to have an extremum, as given in Equation (17.1.3), we have

$$\delta J = \int_a^b \left[\frac{\partial F}{\partial y} - \frac{\mathrm{d}}{\mathrm{d}x}\left(\frac{\partial F}{\partial y'}\right)\right] \delta y \mathrm{d}x = 0$$

According to the fundamental lemma of the variational method, the functional to reach extremum requires that the Lagrangian $F(x, y, y')$ satisfies

$$\frac{\partial F}{\partial y} - \frac{\mathrm{d}}{\mathrm{d}x}\left(\frac{\partial F}{\partial y'}\right) = 0 \tag{17.2.1}$$

This is known as the Euler–Lagrange equation.

Example 17.1. Solving the brachistochrone problem which minimizes the descent time:

$$T = \int_A^B \frac{\sqrt{1 + (y')^2}}{\sqrt{2gy}} \mathrm{d}x$$

Solution: Consider the Lagrangian

$$F(x, y, y') = \frac{\sqrt{1 + (y')^2}}{\sqrt{2gy}}$$

Since F does not explicitly contain the independent variable x, the Euler–Lagrange equation can be simplified as

$$\frac{\partial F}{\partial y} - \frac{\mathrm{d}}{\mathrm{d}x}\left(\frac{\partial F}{\partial y'}\right) = 0 \longrightarrow y'\frac{\partial F}{\partial y'} - F = c$$

Substituting the function F into this equation, we have

$$y'\frac{\partial}{\partial y'}\sqrt{\frac{1+(y')^2}{y}}-\sqrt{\frac{1+(y')^2}{y}}=c$$

leading to

$$\frac{(y')^2}{\sqrt{y[1+(y')^2]}}-\sqrt{\frac{1+(y')^2}{y}}=c$$

Simplifying this expression, we obtain

$$\frac{1}{y[1+(y')^2]}=c^2 \overset{c^2=1/c_1}{\longrightarrow} \frac{\sqrt{y}\mathrm{d}y}{\sqrt{c_1-y}}=\mathrm{d}x$$

To solve this equation, we introduce a parameter representation by

$$y=c_1\sin^2\left(\frac{\theta}{2}\right) \longrightarrow \mathrm{d}y=\frac{c_1}{2}\sin\theta\mathrm{d}\theta$$

Substituting this into the equation, we have

$$\mathrm{d}x=c_1\sin^2\left(\frac{\theta}{2}\right)\mathrm{d}\theta=\frac{c_1}{2}(1-\cos\theta)\mathrm{d}\theta$$

Integrating yields

$$\begin{cases} x=\dfrac{c_1}{2}(\theta-\sin\theta)+c_2 \\ y=\dfrac{c_1}{2}(1-\cos\theta) \end{cases}$$

The constants c_1 and c_2 are determined based on the endpoint positions.

This is a parametric equation of a cycloid, as shown in Figure 17.3(a). It represents the trajectory traced by a fixed point on a rolling circle. The cycloid is also known as a tautochrone or a pendulum curve because the time taken to reach the lowest point is the same regardless of the starting point (Figure 17.3(b)). In other words, the period of oscillation of a small ball moving along the cycloid is independent of the amplitude, which is quite impressive.

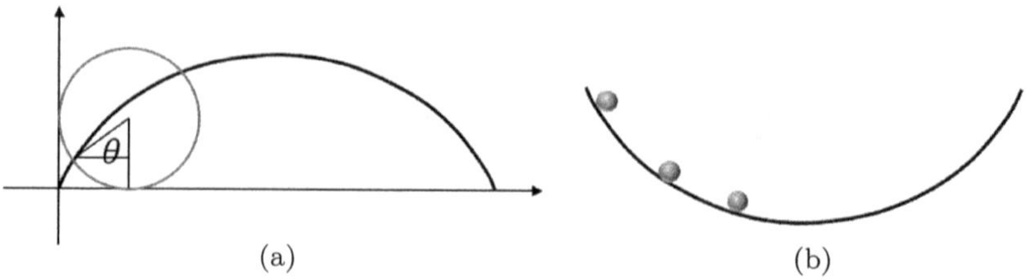

Figure 17.3.

Practice: Prove that if F does not explicitly contain x, the Euler–Lagrange equation reduces to

$$y'\frac{\partial F}{\partial y'} - F = c$$

Reflection: How to prove that the cycloid is a tautochrone?

Example 17.2. Find the extremal curve, given the functional:

$$\begin{cases} J[y(x)] = \int_0^{\frac{\pi}{2}} [(y')^2 - y^2]\mathrm{d}x \\ y(0) = 0, \quad y\left(\frac{\pi}{2}\right) = 1 \end{cases}$$

Solution: We solve the Euler–Lagrange equation,

$$\frac{\partial F}{\partial y} - \frac{\mathrm{d}}{\mathrm{d}x}\left(\frac{\partial F}{\partial y'}\right) = 0 \longrightarrow y'' + y = 0$$

The solution is

$$y = A\cos x + B\sin x$$

By substituting the values of the endpoints, we find

$$y = \sin x$$

17.2.2 *Movable endpoints*

We next consider the case that the endpoints are movable for a functional $\int_a^b F(x, y, y')\mathrm{d}x$. According to (17.1.3) for the functional to have an extremum, except satisfying the Euler–Lagrange equation, additional constraints are required:

(1) If both endpoints are free in the vertical direction, meaning $\delta y|_a$ and $\delta y|_b$ can take any values, then it must satisfy

$$\left.\frac{\partial F}{\partial y'}\right|_{x=a} = \left.\frac{\partial F}{\partial y'}\right|_{x=b} = 0 \tag{17.2.2}$$

(2) If only the endpoint A is fixed and the endpoint B is free in the vertical direction, as shown in Figure 17.4(a), then in addition to $\delta y|_{x=a} = 0$ at A, the following constraint is needed at B:

$$\left.\frac{\partial F}{\partial y'}\right|_{x=b} = 0 \tag{17.2.3}$$

(3) If the endpoint A is fixed and the endpoint B is movable on a given function curve $y = f(x)$, as shown in Figure 17.4(b), then the transversality condition must be satisfied:

$$\left.\left\{F + [f'(x) - y'(x)]\frac{\partial F}{\partial y'}\right\}\right|_{x=b} = 0 \tag{17.2.4}$$

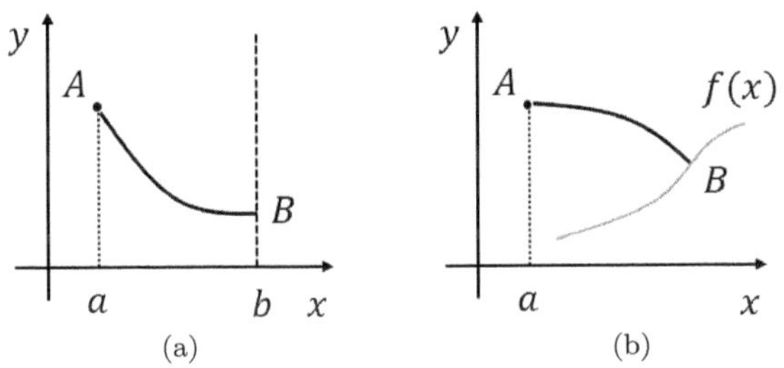

Figure 17.4.

Example 17.3. Find the equation of the shortest curve on the $x-y$ plane from point $A(0, 1)$ to the line $y = 2 - x$.

Solution: This is a variable endpoint problem. We start by calculating the differential arc length $\mathrm{d}s$:

$$\mathrm{d}s = \sqrt{(\mathrm{d}x)^2 + (\mathrm{d}y)^2} = \sqrt{1 + (y')^2}\mathrm{d}x$$

The problem is to find the extremal of the following functional:

$$\begin{cases} J[y] = \int_0^b \sqrt{1+(y')^2}\mathrm{d}x \\ y(x)|_{x=0} = 1, \quad y(x)|_{x=b} = 2-b \end{cases}$$

where b is undetermined. Since $F = \sqrt{1+(y')^2}$ does not explicitly depend on y, the Euler–Lagrange equation becomes

$$\frac{\mathrm{d}}{\mathrm{d}x}F_{y'} = \frac{\mathrm{d}}{\mathrm{d}x}\frac{y'}{\sqrt{1+(y')^2}} = 0$$

Solving this equation to yield

$$y = cx + d$$

Applying the endpoint value $A(0,1)$, we get $d = 1$. Then, using the transversality condition:

$$\left\{F + [f'(x) - y'(x)]\frac{\partial F}{\partial y'}\right\}\Bigg|_{x=b} = 0 \longrightarrow y'(x)|_{x=b} = 1$$

One finds $c = 1$ and $b = \frac{1}{2}$. Hence the curve equation is

$$y = x + 1$$

This is the equation of a perpendicular line from point A to the given straight line. This is the reason for the endpoint constrained on a curve being called the “transverse intersection condition”.

17.2.3 *Multivariable functions*

For a functional of a two-variable function:

$$J[y(x_1, x_2)] = \iint_\Omega F(x_1, x_2, y, y_{x_1}, y_{x_2})\mathrm{d}x_1\mathrm{d}x_2 \tag{17.2.5}$$

where $y(x_1, x_2)$ is a function of (x_1, x_2), we can derive the Euler–Lagrange equation that the function F must satisfy

$$\frac{\partial F}{\partial y} - \frac{\partial}{\partial y}\left(\frac{\partial F}{\partial y_{x_1}}\right) - \frac{\partial}{\partial y}\left(\frac{\partial F}{\partial y_{x_2}}\right) = 0 \tag{17.2.6}$$

For a general n-variable function F:

$$J[y(x_1, x_2)] = \iiint F(x_1, x_2, \ldots, x_n; \\ \times y, y_{x_1}, y_{x_2}, \ldots, y_{x_n}) \mathrm{d}x_1 \mathrm{d}x_2 \cdots \mathrm{d}x_n \quad (17.2.7)$$

with fixed boundaries (surfaces), the Euler–Lagrange equation becomes

$$\frac{\partial F}{\partial y} - \sum_{j=1}^{n} \frac{\partial}{\partial x_j} \left(\frac{\partial F}{\partial y_{x_j}} \right) = 0 \quad (17.2.8)$$

On the other hand, for a functional of multivariable functions with respect to a single-variable:

$$J[y_1(x), y_2(x), \ldots, y_n(x)] \\ = \int_A^B J[x, y_1(x), \ldots, y_n(x), y_1'(x), \ldots, y_n'(x)] \mathrm{d}x \quad (17.2.9)$$

it can be proven that the Euler–Lagrange equation is

$$\frac{\partial F}{\partial y_i} - \frac{\mathrm{d}}{\mathrm{d}x} \left(\frac{\partial F}{\partial y_i'} \right) = 0 \quad (i = 1, 2, 3, \ldots, n) \quad (17.2.10)$$

This is a system of n differential equations.

ENDNOTE

Lagrange introduced generalized coordinates q_i $(i = 1, 2, 3, \ldots, n)$, in which the Euler–Lagrange equation becomes

$$\frac{\partial F}{\partial q_i} - \frac{\mathrm{d}}{\mathrm{d}t} \left(\frac{\partial F}{\partial q_i'} \right) = 0 \quad (i = 1, 2, 3, \ldots, n)$$

These generalized coordinates can be regarded as coordinates in configuration space and do not necessarily have geometric or physical meanings. $q_i(t)$ represents a trajectory in configuration space, and the equations of motion in configuration space remain unchanged under any coordinate transformation.

Although the Euler–Lagrange equation is equivalent to Newton's second law, it exhibits several clear advantages:

(1) Any coordinate can be included in the Lagrangian formalism.
(2) It is convenient for dealing with constrained systems.
(3) It can describe the motion of fields in space.
(4) From this, the principle of least action, a more fundamental principle is discovered.

The transverse intersection condition, where the endpoint A is fixed and the endpoint B is constrained to a function curve $y = f(x)$, can be understood as follows: Assuming that there is a small variation Δx in the horizontal direction at the endpoint B, the variation of the functional is given by

$$\delta J = \frac{\partial F}{\partial y'}\delta y\bigg|_a^b - \int_a^b \left[\frac{\partial F}{\partial y} - \frac{\mathrm{d}}{\mathrm{d}x}\left(\frac{\partial F}{\partial y'}\right)\right]\delta y \mathrm{d}x + F(b)\Delta x$$

Therefore, in addition to satisfying the Euler–Lagrange equation, the extremum condition for the functional at the endpoint B requires

$$\left[\frac{\partial F}{\partial y'}\delta y + F(b)\Delta x\right]\bigg|_{x=b} = 0$$

Based on the analysis in Figure 17.5, it can be derived that

$$\Delta y = y(x + \Delta x) + \delta y(x + \Delta x) - y(x) \approx y'\Delta x + \delta y$$

Since $\Delta y = f'(x)\Delta x$ as $\Delta x \to 0$, we have $\delta y = [f'(x) - y']\Delta x$. Substituting this into the constraint condition, we obtain the condition that must be satisfied at the endpoint B:

$$\left\{F + [f'(x) - y'(x)]\frac{\partial F}{\partial y'}\right\}\bigg|_{x=b} = 0$$

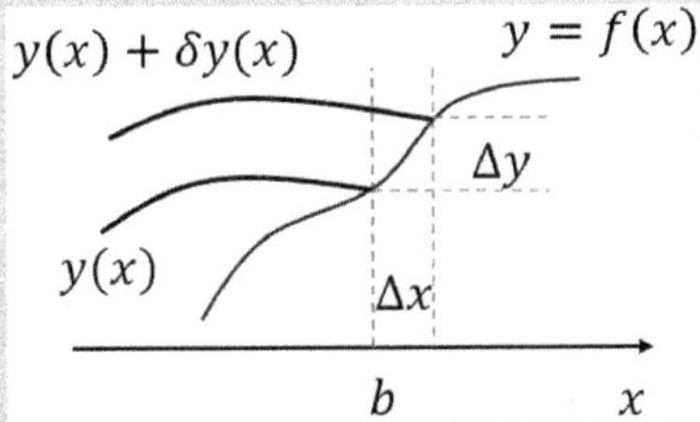

Figure 17.5.

Reflection: If the endpoint A is also constrained to another function curve $y = g(x)$, what would be the transverse intersection condition?

Exercises

[1] Solve the Euler–Lagrange equations for extremizing the functionals:

$$\text{(a)} \int_{x_1}^{x_2} x\sqrt{1-(y')^2}\mathrm{d}x; \quad \text{(b)} \int_{x_1}^{x_2} \frac{\sqrt{1+(y')^2}}{\sqrt{y}}\mathrm{d}x$$

Hint: In (b), consider $\mathrm{d}s = \sqrt{\mathrm{d}x^2 + \mathrm{d}y^2}$ and treat y as the independent variable.

[2] Prove that the Euler–Lagrange equation can be equivalently written as

$$\frac{\partial F}{\partial x} - \frac{\mathrm{d}}{\mathrm{d}x}\left(F - y'\frac{\partial F}{\partial y'}\right) = 0$$

and prove that if F does not explicitly contain x, the Euler–Lagrange equation reduces to

$$y'\frac{\partial F}{\partial y'} - F = c$$

[3] Consider a cylindrical soap film stretched between two coaxial circular rings with radii r_1 and r_2. Find the curve equation of the thin film.

Hint: The surface tension of the film minimizes the free energy, which corresponds to minimizing the surface area of the film.
Answer:

$$y = c_1 \cosh\left(\frac{x - c_2}{c_1}\right)$$

[4] Prove the Euler–Lagrange equation for an n-variable function as

$$\frac{\partial F}{\partial y} - \sum_{j=1}^{n} \frac{\partial}{\partial x_j}\left(\frac{\partial F}{\partial y_{x_j}}\right) = 0$$

[5] If the functional contains higher-order derivatives of $y(x)$:

$$J[x, y(x), y'(x), y''(x) \cdots y^{(n)}(x)]$$

with $y(x) = y'(x) = \cdots = y^{(n-1)}(x) = 0$ at both endpoints, prove that its Euler–Lagrange equation is

$$\frac{\partial F}{\partial y} - \frac{\mathrm{d}}{\mathrm{d}x}\left(\frac{\partial F}{\partial y'}\right) + \frac{\mathrm{d}^2}{\mathrm{d}x^2}\left(\frac{\partial F}{\partial y''}\right) - \cdots + (-1)^n \frac{\mathrm{d}^n}{\mathrm{d}x^n}\left(\frac{\partial F}{\partial y^{(n)}}\right) = 0$$

[6] Assuming the functional depends on multiple dependent variables:

$$J[x, y_1(x), y_2(x), \ldots, y_n(x), y_1'(x), y_2'(x), \ldots, y_n'(x)]$$

prove that its Euler–Lagrange equation is

$$\frac{\partial F}{\partial y_i} - \frac{\mathrm{d}}{\mathrm{d}x}\left(\frac{\partial F}{\partial y_i'}\right) = 0 \quad (i = 1, 2, 3, \ldots, n)$$

[7] Consider the Lagrangian of the pion field, given by

$$L(\boldsymbol{r}, t) = \frac{1}{2}\left[\dot{\phi}^2 - |\nabla\phi|^2 - \mu^2\phi^2\right]$$

where μ is the mass of the pion and ϕ is its wave function. Write down the equation of motion satisfied by ϕ.
Answer: $\nabla^2\phi - \partial_{tt}\phi = \mu^2\phi$

[8] Find the shortest line equation in the $x - y$ plane from point $A(2, 4)$ to the parabola $y = 2 - x^2$.

[9] Find the extremal curve for the functional:

$$J[y] = \int_0^{x_1} \frac{\sqrt{1 + (y')^2}}{y - 1} \mathrm{d}x$$

where one end has $y(0) = 1$ and the other end $y(x_1)$ can move on the circle $(x - 9)^2 + y^2 = 9$.
Answer: $(x - 4)^2 + (y - 1)^2 = 16$

17.3 Constrained Systems

17.3.1 *Isoperimetric problems*

In some extremal problems of functionals, the dependent variable function $y(x)$ may be subject to additional constraints, such as fixed length of the path or restricted motion of a particle on a surface. These problems are known as problems of conditional extremum for functionals. If these constraints appear in integral form, such as

$$\int_a^b G(x, y, y')\mathrm{d}x = l$$

they are collectively referred to as isoperimetric problems, which arise from finding the curve equation that encloses the maximum area for a given length of line segment.

Similar to the conditional extremum problems for ordinary functions, the Lagrange multiplier method can be employed to handle them, i.e.,

$$\delta \int_a^b [F(x, y, y') + \lambda G(x, y, y')]\mathrm{d}x = 0 \tag{17.3.1}$$

The corresponding Euler–Lagrange equation is

$$\frac{\partial F}{\partial y} + \lambda \frac{\partial G}{\partial y} - \frac{\mathrm{d}}{\mathrm{d}x}\left(\frac{\partial F}{\partial y'}\right) - \lambda \frac{\mathrm{d}}{\mathrm{d}x}\left(\frac{\partial G}{\partial y'}\right) = 0$$

Example 17.4. Seeking the extremum of $J[y(x)] = \int_0^1 [y'(x)]^2 \mathrm{d}x$, where y is normalized as

$$\begin{cases} \int_0^1 y^2(x)\mathrm{d}x = 1 \\ y(0) = 0, \quad y(1) = 0 \end{cases}$$

Solution: The Euler–Lagrange equation under the given constraints is

$$y'' - \lambda y = 0 \longrightarrow y = c_1 e^{\sqrt{\lambda}x} + c_2 e^{-\sqrt{\lambda}x}$$

Due to the homogeneous boundary conditions, λ can only take certain discrete values, i.e.,

$$y_n = c_n \sin n\pi x, \quad \lambda_n = -n^2\pi^2$$

Further, the normalization condition determines $c_n = \pm\sqrt{2}$. Thus, the extremum of the functional $J[y(x)]$ is

$$\int_0^1 2n^2\pi^2 \cos^2 n\pi x \mathrm{d}x = n^2\pi^2$$

When $n = 1$, the function $y_1 = \pm\sqrt{2}\sin \pi x$ achieves the minimum extremum of the functional $J_{\min} = \pi^2$.

Example 17.5. Hanging cable problem: Consider a flexible cable of length l, with both ends fixed at points $A(x_1, y_1)$ and $B(x_2, y_2)$. Find the equation of the cable under the influence of gravity.

Solution: Assuming the cable has a uniform density of 1, the stable state of the cable requires its overall center of mass to be at its lowest point. Thus, we seek to minimize the functional

$$J[y] = \int_{x_1}^{x_2} y(x)\sqrt{1 + (y')^2}\mathrm{d}x$$

subject to the constraint

$$\int_{x_1}^{x_2} \sqrt{1 + (y')^2}\mathrm{d}x = l$$

Let $\tilde{F} = y\sqrt{1 + (y')^2} + \lambda\sqrt{1 + (y')^2}$. As $\tilde{F}$ does not explicitly depend on x, the Euler–Lagrange equation reduces to

$$y'\frac{\partial \tilde{F}}{\partial y'} - \tilde{F} = c$$

which gives rise to

$$y + \lambda = c\sqrt{1 + (y')^2} \longrightarrow y'^2 = \left(\frac{y + \lambda}{c}\right)^2 - 1$$

Finally, we obtain

$$y + \lambda = \frac{1}{c_1}\cosh(c_1 x + c_2)$$

This is the equation of a hyperbolic function curve, where the parameters c_1, c_2, and the Lagrange multiplier λ can be determined based on the coordinates of A and B, as well as the length of the cable.

17.3.2 *Finite constraints*

For a given functional under boundary conditions:

$$J = \int_a^b F(x, y, z, y'(x), z'(x))\mathrm{d}x \tag{17.3.2}$$

If the functions $y(x)$ and $z(x)$ are subject to additional constraints $G(x, y, z) = 0$, where the variables are not independent, how do we determine the conditions for the extremum of the functional?

Since $G(x, y, z) = 0$ represents a surface in three-dimensional space, geometrically, this problem is equivalent to finding the spatial function curves $y(x)$ and $z(x)$ on the given surface that minimize the functional J. A natural approach is to solve the equation $G(x, y, z) = 0$ for one of the functions, reducing it to a functional problem with a single independent function. In fact, as long as $\frac{\partial}{\partial z}G \neq 0$ on the extremal curve, we can solve for $z = g(x, y)$, treating z' as a function of x, y and y', and utilizing the relationship:

$$G_x + y'G_y + z'G_z = 0, \quad z' = \frac{\partial}{\partial x}g + y'\frac{\partial}{\partial y}g$$

to eliminate z'. This yields

$$F(x, y, z, y', z') = F\left(x, y, g(x, y), y', \frac{\partial}{\partial x}g + y'\frac{\partial}{\partial y}g\right)$$

Then, based on the Euler–Lagrange equation satisfied by F,

$$\frac{\mathrm{d}}{\mathrm{d}x}\left(F_{y'} + F_{z'}\frac{\partial g}{\partial y}\right) - \left[F_y + F_z\frac{\partial g}{\partial y} + F_{z'}\left(\frac{\partial^2 g}{\partial x \partial y} + y'\frac{\partial^2 g}{\partial^2 y}\right)\right] = 0$$

we can simplify to

$$\left(\frac{\mathrm{d}}{\mathrm{d}x}F_{y'} - F_y\right) + \left(\frac{\mathrm{d}}{\mathrm{d}x}F_{z'} - F_z\right)\frac{\partial g}{\partial y} = 0$$

Moreover, since

$$G_y + G_z \frac{\partial g}{\partial y} = 0$$

we must have

$$\left(\frac{\mathrm{d}}{\mathrm{d}x}F_{y'} - F_y\right) : \left(\frac{\mathrm{d}}{\mathrm{d}x}F_{z'} - F_z\right) = G_y : G_z \tag{17.3.3}$$

Hence, there exists an arbitrary common function factor $\lambda(x)$ such that

$$\begin{cases} \dfrac{\mathrm{d}}{\mathrm{d}x}F_{y'} - F_y - \lambda(x)G_y = 0 \\ \dfrac{\mathrm{d}}{\mathrm{d}x}F_{z'} - F_z - \lambda(x)G_z = 0 \end{cases} \tag{17.3.4}$$

In many cases, the specific form of $\lambda(x)$ is not required, and it is sometimes referred to as an undetermined multiplier in the Lagrange method.

Example 17.6. An object of mass m sliding down a smooth spherical surface with radius R. Find the position where the object leaves the surface.

Solution: As shown in Figure 17.6, the motion of the object satisfies the constraint equation: $G(r, \theta) = r - R = 0$. When the object slides down from the highest point to the critical point θ_c, it will leave the surface and fall freely. The Lagrangian for this problem is given by

$$L = T - V = \frac{1}{2}m(\dot{r}^2 + r^2\dot{\theta}^2) - mgr\cos\theta$$

Since θ is a single-valued function of time t, we choose the Lagrange multiplier λ to be a function of θ. The Euler–Lagrange equations for this problem are

$$\begin{cases} m\ddot{r} - mr\dot{\theta}^2 + mg\cos\theta = \lambda(\theta) \\ mr^2\ddot{\theta} + 2mr\dot{r}\dot{\theta} - mgr\sin\theta = 0 \end{cases}$$

Note that the physical interpretation of $\lambda(\theta)$ is the normal pressure exerted by the object on the spherical surface. Since the object is sliding on the surface with $\dot{r} = \ddot{r} = 0$, we have

$$\begin{cases} -mR\dot{\theta}^2 + mg\cos\theta = \lambda(\theta) \\ mR^2\ddot{\theta} - mgR\sin\theta = 0 \end{cases}$$

Taking the derivative of the first equation with respect to time and substituting the value of $\ddot{\theta}$ from the second equation, we obtain

$$\frac{\mathrm{d}\lambda(\theta)}{\mathrm{d}\theta} = -3mg\sin\theta \longrightarrow \lambda(\theta) = mg(3\cos\theta - 2)$$

When $\lambda(\theta) \geq 0$, the object remains on the spherical surface. The critical value for the object to leave the surface is

$$\theta_c = \arccos\left(\frac{2}{3}\right)$$

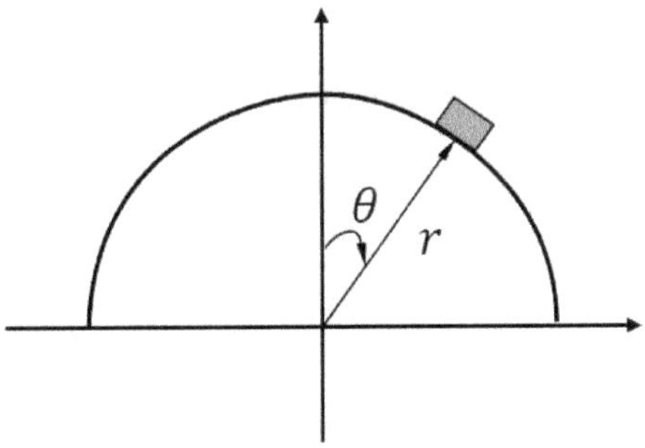

Figure 17.6.

ENDNOTE

The Lagrange multiplier method is used to find the extrema of a multivariable function subject to one or more constraints. Taking a two-variable function as an example, suppose we want to find the extrema of the function $f(x, y)$ subject to the constraint $g(x, y) = 0$. From Figure 17.7, it can be observed that when the contour lines of $f(x, y)$ are tangent to or intersect the curve $g(x, y) = 0$ at only one point, the function $f(x, y)$ achieves an extremum. At the point of tangency P, the contour line and the curve $g(x, y) = 0$ have the same normal direction, i.e., $\nabla f(x,y) = -\lambda \nabla g(x,y)$, where λ is a constant. Alternatively, let $F = f + \lambda g$, then $\nabla F(x,y) = 0$, which implies

$$\frac{\partial}{\partial x}F(x,y) = \frac{\partial}{\partial y}F(x,y) = 0$$

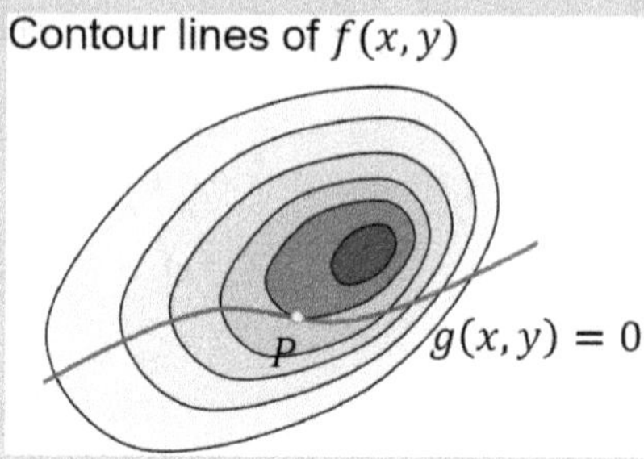

Figure 17.7.

If the system with n independent variables has M constraints, we define

$$F = f + \sum_{i=1}^{M} \lambda_i g_i$$

where λ_i $(i = 1, 2, \ldots, M)$ are M Lagrange multipliers. They satisfy the conditions:

$$\frac{\partial F}{\partial x_k} = 0 \; (k = 1, 2, \ldots, n)$$

Exercises

[1] Given the constraint $y' = u - y$, find the function that minimizes the functional:

$$J = \frac{1}{2}\int_0^{x_1} [y^2(x) + u^2(x)]\mathrm{d}x$$

with $y(0) = 0$ and $y(x_1)$ arbitrary.

Answer:

$$\begin{cases} y(x) = c_1 e^{\sqrt{2}x} + c_2 e^{-\sqrt{2}x} \\ u = c_1(1+\sqrt{2})e^{\sqrt{2}x} + c_2(1-\sqrt{2})e^{-\sqrt{2}x} \end{cases}$$

$$c_1 = \frac{y_0}{1+(1+\sqrt{2})^2 e^{2\sqrt{2}x_1}}, \quad c_2 = \frac{y_0(1+\sqrt{2})^2 e^{2\sqrt{2}x_1}}{1+(1+\sqrt{2})^2 e^{2\sqrt{2}x_1}}$$

[2] Given a line segment of length l fixed at points A and B, rotating it around the x-axis, find the equation of the curve that minimizes the surface area.

[3] Given a smooth curve $y(x)$ between two points x_1 and x_2 on the x-axis, rotating it around the x-axis, find the curve that maximizes the volume of the resulting cylinder, under the condition of a given surface area.
Answer: A circle.

[4] The classic isoperimetric problem: the largest area that can be enclosed by a line segment of length l.
Hint: This problem can also be understood as finding the equation of a curve that minimizes the perimeter given a fixed area; or taking the line element as the independent variable, with the area formula as the dependent variable.

$$A = 2\int_0^{\frac{L}{2}} y(s)\sqrt{1-[y'(s)]^2}\mathrm{d}s, \quad y(0)=0, \quad y\left(\frac{L}{2}\right)=0$$

[5] Find the equation of the short geodesic on the cone: $x^2+y^2=z^2$
Hint: Use cylindrical coordinates.
Answer:

$$r\cos\frac{\theta+c}{\sqrt{2}} = K, \quad z = r$$

[6] Prove that the shortest distance between two points on a sphere of radius R is the arc of a great circle passing through the two points.
Hint: Take point A as the north pole of the sphere and write the expression for the spherical arc.

17.4 Mathematical Principles of Physics

17.4.1 *Fermat's principle*

The actual path of light propagation between points A and B, compared to other possible neighboring paths, is the one for which the optical path length is minimized. This can be expressed as the following variational principle:

$$\delta \int_A^B n(x, y, y')\mathrm{d}s = 0 \tag{17.4.1}$$

Here, $n(x, y, y')$ represents the refractive index distribution of the medium. This rule of light propagation is known as Fermat's principle. Figure 17.8 illustrates the branching phenomenon of light rays when propagating through a soap film, and all branches follow the path of minimum optical path length.

Figure 17.8. (image from Network).

Example 17.7. Determining the propagation path of light in a uniform medium with a refractive index of n_0.

Solution: The functional variation of the optical path length is given by

$$\delta J = \delta \int_A^B n(x, y)\mathrm{d}s = \delta \int_A^B n_0\sqrt{1 + (y')^2}\mathrm{d}x = 0$$

Applying the Euler–Lagrange equation, we obtain

$$\frac{\mathrm{d}}{\mathrm{d}x}\left(\frac{y'}{\sqrt{1 + (y')^2}}\right) = 0 \longrightarrow y = cx + d$$

Therefore, light propagates in a straight line in a uniform medium.

Example 17.8. Deriving Snell's law of refraction using Fermat's principle.

Solution: Let the refractive index distribution be

$$n(x) = \begin{cases} n_1 & (y \geq 0) \\ n_2 & (y < 0) \end{cases}$$

As shown in Figure 17.9, the optical path from point A to point B is given by

$$J = \int_A^B n(y)\sqrt{1+(y')^2}\mathrm{d}x$$

Since the Lagrangian does not explicitly depend on the independent variable x, we have

$$y'\frac{\partial F}{\partial y'} - F = c \longrightarrow n(y) = c\sqrt{1+(y')^2}$$

Considering that y' represents the slope of the curve, and $y' = \tan\phi$, we find that $n\cos\phi \equiv c$, indicating that the angle ϕ remains constant in a uniform medium, and light propagates in a straight line. At the interface of two uniform media, we have

$$n_1\cos\phi_1 = n_2\cos\phi_2$$

This yields Snell's law of refraction:

$$n_1\sin\theta_1 = n_2\sin\theta_2$$

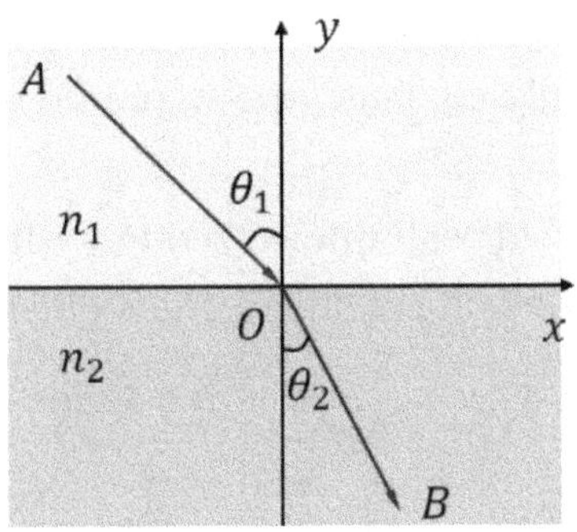

Figure 17.9.

17.4.2 The principle of least action

In Newtonian mechanics, the action of a conservative mechanical system is defined as

$$S \overset{\text{def}}{=} \int_{t_A}^{t_B} L(q_j, \dot{q}_j, t)\mathrm{d}t \tag{17.4.2}$$

Here, the Lagrangian of a particle is defined as

$$L \overset{\text{def}}{=} T(q_j, \dot{q}_j, t) - V(q_j, \dot{q}_j, t)$$

where T and V represent the kinetic and potential energies of the particle, respectively, and q_j ($j = 1, 2, \ldots, n$) are the generalized coordinates. The principle of least action states that among all possible configurations between two given points, the actual path followed by the particle minimizes the action S. Mathematically, this is expressed as

$$\delta S = 0 \tag{17.4.3}$$

The principle of least action is also known as Hamilton's principle.

Based on the previous arguments, we can directly derive the Euler–Lagrange equations of motion for a particle from the principle of least action:

$$\frac{\partial L}{\partial q_j} - \frac{\mathrm{d}}{\mathrm{d}t}\frac{\partial L}{\partial \dot{q}_j} = 0 \ (j = 1, 2, \ldots, n) \tag{17.4.4}$$

From these equations, Newton's second law can be derived:

$$m\ddot{q}_j = -\partial_j V$$

thus establishing a fully equivalent system of conservative mechanics called analytical mechanics.

Example 17.9. Consider a uniform string with length l and density ρ, under tension τ. Derive its equation of motion using Hamilton's principle.

Solution: The kinetic energy of the string's vibration is given by

$$T = \frac{1}{2}\int_0^l \rho(u_t)^2 \mathrm{d}x$$

The elastic potential energy is proportional to the square of the relative elongation of the string,

$$V = \frac{1}{2}\int_0^l \tau(u_x)^2 \mathrm{d}x$$

The Lagrangian is defined as $L = T - V$. According to Hamilton's principle, the action should be extremized:

$$S = \int_{t_0}^t L\mathrm{d}t = \frac{1}{2}\int_{t_0}^t \int_0^l [\rho(u_t)^2 - \tau(u_x)^2]\mathrm{d}x\mathrm{d}t$$

Using the Euler–Lagrange equations for a function of two variables and noting that the Lagrangian does not explicitly depend on u, we obtain the familiar equation of motion for a string:

$$u_{tt} - a^2 u_{xx} = 0, \quad a^2 = \frac{\tau}{\rho}$$

Credited to Euler and Maupertuis, the action can also be expressed in another form as $S = \int_A^B p\mathrm{d}q$, where q representing the path of the object's motion and p being the generalized momentum along the path. The actual path of an object in a conservative force field is the one that extremizes this action, sometimes referred to as Maupertuis' principle. This form of the action avoids explicitly involving time and describes the object's motion trajectory in phase space. Maupertuis' principle requires fixed endpoint positions and energy conservation along each trajectory but does not impose constraints on the time at the endpoints.

Example 17.10. Find the trajectory of an object moving from point A to point B under the influence of gravity.

Solution: According to the principle of conservation of mechanical energy,

$$\frac{1}{2}mv^2 + mgy = \frac{1}{2}mv_A^2 + mgy_A$$

Therefore, the action reads

$$\begin{aligned} S &= \int_A^B p\mathrm{d}q = m\int_A^B v\mathrm{d}s \\ &= m\int_A^B \sqrt{v_A^2 + 2g(y_A - y)} \cdot \sqrt{1 + \left(\frac{\mathrm{d}x}{\mathrm{d}y}\right)^2}\mathrm{d}y \end{aligned}$$

Denote $\frac{\mathrm{d}x}{\mathrm{d}y} = x'$, the Lagrangian is

$$F = \sqrt{v_A^2 + 2g(y_A - y)} \cdot \sqrt{1 + (x')^2}$$

Since it does not explicitly depend on $x(y)$, we obtain the Euler–Lagrange equation from the principle of least action:

$$\frac{\mathrm{d}}{\mathrm{d}y}\left[\frac{\partial F}{\partial x'}\right] = 0 \longrightarrow \frac{\sqrt{v_A^2 + 2g(y_A - y)}}{\sqrt{1 + (x')^2}} \frac{\mathrm{d}x}{\mathrm{d}y} = c_1$$
$$\longrightarrow x' = \frac{c_1}{\sqrt{v_A^2 + 2g(y_A - y) - c_1^2}}$$

Here c_1 is an integration constant. Integrating over y, we have

$$x = -\frac{c_1}{g}\sqrt{v_A^2 + 2g(y_A - y) - c_1^2} + c_2$$
$$\longrightarrow 2g(y_A - y) = \left(\frac{g}{c_1}\right)^2 (x - c_2)^2 + c_1^2 - v_A^2$$

where c_2 is another integration constant. As expected, this is the equation of a parabola. Assuming the object undergoes projectile motion with an initial position at $(0, 0)$ and initial velocity $(v_0, 0)$ in the x-direction, we have

$$y = -\frac{g}{2v_0^2}x^2$$

Reflection:

(1) Is the action defined by Euler–Maupertuis the same physical quantity as the action defined by the Lagrangian $L = T - V$?
(2) Does the action have a relation to angular momentum, considering its dimensional nature?
(3) For a planet moving in a central force field along a closed orbit, what is the action $S = \oint p \cdot \mathrm{d}q$?

17.4.3 *Symmetry and conservation laws*

In a mechanical system with a conservative potential V, the kinetic energy of a particle is given by $T = \sum_j \frac{1}{2} m\dot{q}_j^2$, and the Lagrangian is $L = T - V$. In all possible transformations between different configurations of the system, if the system possesses a certain symmetry, there exists a corresponding physical quantity that is conserved. This is known as Noether's theorem, which reveals the profound connection between symmetry and conservation laws. For a particle system, Noether's theorem is exhibited as follows:

(1) If the system exhibits time translation invariance, meaning the Lagrangian L does not explicitly depend on time t, $\frac{\partial L}{\partial t} = 0$, then

$$\frac{\mathrm{d}L}{\mathrm{d}t} = \frac{\partial L}{\partial t} + \sum_j \left(\dot{q}_j \frac{\partial L}{\partial q_j} + \ddot{q}_j \frac{\partial L}{\partial \dot{q}_j} \right)$$

$$\longrightarrow \frac{\mathrm{d}L}{\mathrm{d}t} = \sum_j \left(\dot{q}_j \frac{\mathrm{d}}{\mathrm{d}t} \frac{\partial L}{\partial \dot{q}_j} + \ddot{q}_j \frac{\partial L}{\partial \dot{q}_j} \right) = \frac{\mathrm{d}}{\mathrm{d}t} \left(\sum_j \dot{q}_j \frac{\partial L}{\partial \dot{q}_j} \right)$$

$$\longrightarrow \sum_j \dot{q}_j \frac{\partial L}{\partial \dot{q}_j} - L = C$$

Substituting the expression for the Lagrangian, we have

$$H \stackrel{\text{def}}{=} \sum_j \dot{q}_j \frac{\partial L}{\partial \dot{q}_j} - L = T + V = C \tag{17.4.5}$$

This is the conservation law of mechanical energy, where the conserved quantity H is called the Hamiltonian.

(2) In a conservative force field $V_{\text{ext}}(\boldsymbol{r})$, where the interaction potential between particles depends only on relative coordinates, $V_{\text{int}}(\boldsymbol{r}_i - \boldsymbol{r}_j)$, the Lagrangian is given by

$$L = \sum_j \frac{1}{2} m\dot{r}_j^2 - \sum_j V_{\text{ext}}(\boldsymbol{r}_i) - \sum_{i<j} V_{\text{int}}(\boldsymbol{r}_i - \boldsymbol{r}_j)$$

Assuming that the particle system exhibits spatial translational invariance along the x_j-direction, $\frac{\partial V_{\text{ext}}(\boldsymbol{r}_j)}{\partial x_j} = 0$, and due to

$$\frac{\partial V_{\text{int}}(\boldsymbol{r}_i - \boldsymbol{r}_j)}{\partial x_i} = -\frac{\partial V_{\text{int}}(\boldsymbol{r}_i - \boldsymbol{r}_j)}{\partial x_j}$$

we find that

$$\sum_{i<j} \frac{\partial V_{\text{int}}(\boldsymbol{r}_i - \boldsymbol{r}_j)}{\partial x_j} = 0$$

Applying the Euler–Lagrange equation, we have

$$\frac{\mathrm{d}}{\mathrm{d}t} \sum_j \frac{\partial L}{\partial \dot{x}_j} = 0 \longrightarrow P_x \equiv \sum_j p_{jx} = \sum_j \frac{\partial L}{\partial \dot{x}_j} = C \qquad (17.4.6)$$

This is the conservation law of momentum.

(3) In the Lagrangian formalism, generalized coordinates are often used. For example, in the case of rotational motion of a rigid body, if we choose the angle θ_j as the generalized coordinate, the kinetic energy of the rigid body is given by $T = \sum_j \frac{1}{2} I \dot{\theta}_j^2$, where I is the moment of inertia. If the system exhibits rotational symmetry, such that the Lagrangian L does not explicitly depend on θ_j, then

$$\frac{\partial L}{\partial \dot{\theta}_j} = I \dot{\theta}_j = C \qquad (17.4.7)$$

This is the conservation law of angular momentum. Similarly, for a many-particle system with rotational symmetry, one can derive the conservation law of angular momentum. The details are omitted here.

17.4.4 *Analytical mechanics*

At this point, we can provide a brief overview of the basic framework of analytical mechanics. Analytical mechanics is also known as algebraic mechanics, while Newtonian mechanics is referred to as geometric mechanics. Corresponding to the generalized coordinates q_j, we can introduce the corresponding generalized momenta:

$$p_j = \frac{\partial L}{\partial \dot{q}_j} \qquad (17.4.8)$$

If the Lagrangian L is independent of time, the Hamiltonian is defined as

$$H = \sum_j p_j \dot{q}_j - L \qquad (17.4.9)$$

At first glance, the Hamiltonian appears to depend on three quantities: p_j, q_j, and $\dot{q}_j$. However, by analyzing the expression:

$$\begin{aligned}\mathrm{d}H &= \sum_j \dot{q}_j \mathrm{d}p_j + \sum_j p_j \mathrm{d}\dot{q}_j - \sum_j \left(\frac{\partial L}{\partial q_j}\mathrm{d}q_j + \frac{\partial L}{\partial \dot{q}_j}\mathrm{d}\dot{q}_j\right)\\ &= \sum_j \dot{q}_j \mathrm{d}p_j - \sum_j \frac{\partial L}{\partial q_j}\mathrm{d}q_j\end{aligned}$$

we find that there are only two independent variables (q_j, p_j). By comparing this with

$$\mathrm{d}H = \sum_j \left(\frac{\partial H}{\partial q_j}\mathrm{d}q_j + \frac{\partial H}{\partial p_j}\mathrm{d}p_j\right)$$

we can see that the Hamiltonian only involves the independent variables p_j and q_j, and

$$\frac{\partial H}{\partial q_j} = -\frac{\partial L}{\partial q_j}, \quad \frac{\partial H}{\partial p_j} = \dot{q}_j$$

Using the Euler–Lagrange equations, we obtain the Hamilton's canonical equations:

$$\dot{q}_j = \frac{\partial H}{\partial p_j}, \quad \dot{p}_j = -\frac{\partial H}{\partial q_j} \tag{17.4.10}$$

Furthermore, the Poisson bracket is defined as

$$\{g, h\} \overset{\text{def}}{=} \sum_j \left(\frac{\partial g}{\partial q_j}\frac{\partial h}{\partial p_j} - \frac{\partial g}{\partial p_j}\frac{\partial h}{\partial q_j}\right) \tag{17.4.11}$$

Based on the total derivative,

$$\frac{\mathrm{d}}{\mathrm{d}t}g(p,q) = \sum_j \left(\frac{\partial g}{\partial q_j}\dot{q}_j + \frac{\partial g}{\partial p_j}\dot{p}_j\right) = \sum_j \left(\frac{\partial g}{\partial q_j}\frac{\partial H}{\partial p_j} - \frac{\partial g}{\partial p_j}\frac{\partial H}{\partial q_j}\right)$$

we obtain the Hamilton's equations of motion satisfied by the mechanical quantity $g(p, q)$:

$$\frac{\mathrm{d}}{\mathrm{d}t}g(p,q) = \{g(p,q), H(p,q)\} \tag{17.4.12}$$

The Hamilton's canonical equations can be expressed in a symmetric form using Poisson brackets:

$$\dot{q}_j = \{q_j, H\}, \quad \dot{p}_j = \{p_j, H\} \tag{17.4.13}$$

If we take $g = q_j$ and $h = p_j$, then

$$\{q_j, p_j\} = 1 \tag{17.4.14}$$

This pair of physical quantities for which the Poisson bracket is equal to 1 is commonly referred to as canonically conjugate variables.

ENDNOTE

Newton's second law is the differential form of the equation of motion, which describes the instantaneous relationship between the force acting on an object and its acceleration at each point in space-time. It can be used to derive the paths of planetary motion under the influence of conservative forces. Newton's second law can also be applied to non-conservative forces. After 150 years of Newton, people began to approach the problem from a different perspective: Given the initial and final positions, what possible paths can an object take under the influence of conservative forces? Rather than focusing on the motion of the object at each point in space-time, the emphasis shifted to the selection of paths under certain constraints. If the object's motion follows a unique path, as observed in the case of planets, what is the criterion for choosing this path? The principle of least action in classical mechanics states that the object will move along a path where the action is minimized. Thus, when people started looking at the same problem from different viewpoint, they finally discovered the hidden gem of action buried deep within nature!

Lagrange successfully described dynamics using the principle of least action, which suggested that this principle could be applied to other branches of physics. Hamilton also approached the problem from the perspective of the principle of least action, attempting to derive a mathematical framework similar to mechanics for optics. However, his views on this principle differed substantially from those of Maupertuis, Euler, and Lagrange. He introduced the action integral:

$$S = \int_{q_1,t_1}^{q_2,t_2} (T - V)\mathrm{d}t$$

Although $L = T - V$ is known as the Lagrangian, it was Poisson who first introduced it. Hamilton extended the principles of Euler and Lagrange, allowing for the comparison of all paths for an object without energy constraints, as long as the initial and final positions and times are the same. This means that energy is not conserved along different paths, whereas in the Euler–Lagrange principle, the conservation of energy is assumed a priori, leading to a different time required to compare paths compared to the actual path.

The Hamiltonian principle asserts that the real motion is the one that stabilizes the action, even for non-conservative forces. The potential energy V can be a function of time or even velocity. The action integral is written as

$$S = \int_{t_1}^{t_2} L(q_i, \dot{q}_i, t)\mathrm{d}t$$

where it is assumed that all paths q_i have the same values at the initial and final moments. From the principle of least action, we derive the system of equations:

$$\frac{\partial L}{\partial q_i} - \frac{\mathrm{d}}{\mathrm{d}t}\frac{\partial L}{\partial \dot{q}_i} = 0 \ (i = 1, 2, \ldots, n)$$

These are still known as the Euler–Lagrange equations of motion.

Astute readers may notice that concepts such as time, velocity, and acceleration, which describe the state of motion, have quietly disappeared from the scene. Some may wonder how an object “sniffs out” the path of least action and whether there could be more than one possible path. This seemingly naive idea is not unfounded, but it does not appear in the deterministic realm of classical mechanics. However, it does exist in the microscopic world, leading us into a realm of wonder: microscopic particles always explore all possible paths simultaneously.

The initial origins of quantum mechanics are deeply connected to action. When Niels Bohr considered the electronic structure of the hydrogen atom, he employed angular momentum quantization. Arnold Sommerfeld generalized this idea to action quantization: $S = \oint p\mathrm{d}q = n\hbar$ (where $n \in \mathbb{N}$). Although this may seem like a side note in the development of quantum mechanics, it is of profound significance. It indicates an inherent logical connection between the extremization of classical action and the quantization of atomic structure. Later, Richard Feynman's path integral formulation also utilized the concept of action. The contribution of the action along a certain path for a microscopic particle introduces a phase factor, and the probability amplitude from point A to point B is obtained by summing the phase factors over all paths:

$$K(B,A) = \sum_{\text{all paths}} \mathcal{A}\mathrm{e}^{\mathrm{i}S\frac{[q(t)]}{\hbar}} \longrightarrow \int \mathrm{e}^{\frac{\mathrm{i}S[q(t)]}{\hbar}} \mathcal{D}[q(t)]$$

Noether's theorem applies not only to particle dynamics but also more generally to electromagnetism, classical field theory, and quantum field theory, where there are infinitely many degrees of freedom. The action for a continuous physical system is the integral of the Lagrangian density over time and space, and the principle of least action governs the entire behavior of the system. The continuous symmetry of the Lagrangian density is described by Lie groups, and the corresponding conserved quantities are the generators of the Lie groups. For example, if the Lagrangian density remains invariant under continuous rotations, the angular momentum of the system is conserved.

In addition to translational and rotational invariance in space-time, there are symmetries related to internal degrees of freedom. For electromagnetic systems, the Lagrangian density remains invariant under continuous global phase transformations. Noether's theorem reveals that this symmetry corresponds to the conservation of electric charge. Such phase transformations are generally referred to as gauge transformations. When global gauge transformations are further localized, and internal degrees of freedom are combined with space-time degrees of freedom, the non-Abelian gauge field theories of Yang-Mills are induced, opening the door to the mysteries of fundamental interactions in the natural world. The laws of electromagnetic interaction, as well as strong and weak interactions, are manifested in the invariance under local gauge transformations, forming the unified standard theory of particle physics. The principle of least action, which initially emerged as a deduction from Newton's theory of conservative forces, has finally metamorphosed into the decisive cornerstone for describing all natural phenomena.

The principle of symmetry is destined to become the guiding principle of physics.

Exercises

[1] Using Fermat's principle, determine the path of light propagation in a non-uniform medium:

$$\text{(a) } n_0(x+1); \quad \text{(b) } \frac{n_0}{y}; \quad \text{(c) } n_0\sqrt{y}; \quad \text{(d) } \frac{n_0}{r}$$

Answer:

$$\text{(a) } a(x+1) = \cosh(ay+b); \quad \text{(b) } (x-a)^2 + y^2 = c^2;$$
$$\text{(c) } (x-a)^2 = 4K^2(y-c^2); \quad \text{(d)} be^{\alpha\theta}$$

[2] For a particle moving in an external potential $V(\boldsymbol{r})$, the Lagrangian is given by $L = \frac{1}{2}m\dot{r}^2 - V(\boldsymbol{r})$. Determine the Euler–Lagrange equation for the action $S[L(\boldsymbol{r})]$ to take extremal values.

[3] For a one-dimensional harmonic oscillator with Lagrangian

$$L = \frac{1}{2}m(\dot{x}^2 - \omega^2 x^2)$$

where $T = t_b - t_a$, prove that the action is given by

$$S_{cl} = \frac{m\omega}{2\sin\omega T}(x_b^2 + x_a^2)\cos\omega T - 2x_b x_a$$

[4] Derive the equations of motion for the double pendulum shown in Figure 17.10.
Hint: Write down the Lagrangian

$$L = mr^2\left[\dot{\theta}^2 + \frac{1}{2}(\dot{\theta} + \dot{\alpha})^2 + \dot{\theta}(\dot{\theta} + \dot{\alpha})\cos\alpha\right]$$
$$- mgr[2\cos\theta + \cos(\theta + \alpha)]$$

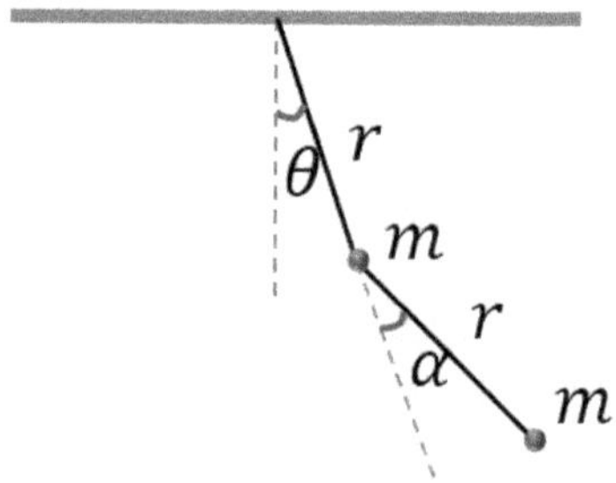

Figure 17.10.

[5] In spherical coordinates, write down the Lagrangian equations of motion for a particle constrained to equipotential surface, and identify:
(a) Centrifugal force; (b) Coriolis force.

[6] Derive the planar motion equation for a unit mass in a central force field $V(r)$, and determine the form of the central force field for the particle trajectory to be $r = a\sin\theta$.
Answer:

$$r^2\dot{\theta} = K, \quad \ddot{r} - r\dot{\theta}^2 + \frac{\mathrm{d}V}{\mathrm{d}r} = 0; \quad V(r) = -\frac{K^2a^2}{2r^4}$$

[7] The energy density of a static electric field in a charge-free space is given by

$$E = \frac{1}{2}\varepsilon_0\boldsymbol{E}^2 = \frac{1}{2}\varepsilon_0(\nabla\varphi)^2$$

Assuming a fixed boundary condition is imposed on the electric potential φ and considering the total energy to be minimized, derive the equation satisfied by φ.

Answer: $\Delta\varphi = 0$

[8] For a charge q with mass m in a static electric field $\boldsymbol{E}$ and a static magnetic field $\boldsymbol{B}$, the Lagrangian is given by

$$L = \frac{1}{2}m\boldsymbol{v}^2 - q\varphi + q\boldsymbol{A}\cdot\boldsymbol{v}$$

Write down the equations of motion for the charge.

Answer: $m\ddot{\boldsymbol{r}} = q\boldsymbol{E} + q\boldsymbol{v}\times\boldsymbol{B}$

[9] The Lagrangian density for the electromagnetic field is

$$L = \frac{1}{2}\left(\varepsilon_0\boldsymbol{E}^2 - \frac{1}{\mu_0}\boldsymbol{B}^2\right) - \rho\varphi + \boldsymbol{J}\cdot\boldsymbol{A}$$

where ρ is the free charge density and $\boldsymbol{J}$ is the current density. Derive the Maxwell's equations.

Hint: Consider x, y, z, t as independent variables, $S = \int L\mathrm{d}\boldsymbol{r}\mathrm{d}t$. Treat φ and $\boldsymbol{A}$ as independent functions. According to the definition,

$$\boldsymbol{E} = -\nabla\varphi - \frac{\partial}{\partial t}\boldsymbol{A}, \quad \boldsymbol{B} = \nabla\times\boldsymbol{A}$$

17.5 Boundary Value Problems

Basic Idea: Connect the boundary value problem of a differential equation with the extremum problem of a certain functional. The Euler–Lagrange equation satisfied by the function is the desired differential equation. Therefore, solving the initial value problem of a differential equation is equivalent to finding a function that satisfies the extremum condition of the functional.

17.5.1 *Eigenvalue problems*

(1) *Eigenvalue equation*

$$\begin{cases} \Delta u + \lambda u = 0 \\ u|_{\Sigma} = 0 \end{cases}$$

Multiply both sides of the equation by δu and then integrate,

$$\iiint_{\Omega}(\Delta u\delta u + \lambda u\delta u)\mathrm{d}\tau = 0$$

Using Green's formula, the second-order derivative terms in the functional can be transformed into first-order derivative terms:

$$\iint_{\Sigma}\frac{\partial u}{\partial n}\delta u\mathrm{d}\sigma - \frac{1}{2}\iiint_{\Omega}[\delta(\nabla u)^2 - \lambda\delta u^2]\mathrm{d}\tau = 0$$

Since the boundary is fixed, $\delta u = 0$, so $\iint_{\Sigma}\delta u\frac{\partial u}{\partial n}\mathrm{d}\sigma = 0$. Therefore, the functional is taken as

$$J[u] = \iiint_{\Omega}[(\nabla u)^2 - \lambda u^2]\mathrm{d}\tau$$

Its Euler–Lagrange equation is

$$\Delta u + \lambda u = 0$$

Hence, the original eigenvalue problem is transformed into finding the extremum of the functional $J[u]$. Considering the normalization condition,

$$\iiint_{\Omega}u^2\mathrm{d}\tau = 1$$

the extremum problem of the functional $J[u]$ is actually the extremum problem of the functional

$$J_1[u] = \iiint_{\Omega}(\nabla u)^2\mathrm{d}\tau$$

under the normalization constraint. The Lagrange multiplier λ is the smallest eigenvalue.

Similarly, the second or third kind of homogeneous boundary value problems of differential equations correspond to extremum problems of functional as well.

(2) *Sturm–Liouville eigenvalue equation*

$$\begin{cases} -\dfrac{\mathrm{d}}{\mathrm{d}x}\left[p(x)\dfrac{\mathrm{d}y}{\mathrm{d}x}\right] + q(x)y(x) = \lambda\rho(x)y(x) \\ y(a) = 0, \quad y(b) = 0 \end{cases}$$

It can be proven that it is equivalent to finding the extremum of the functional under the normalization constraint $\int_a^b \rho(x) y^2(x)\mathrm{d}x = 1$,

$$J[y] = \int_a^b [p(x)(y')^2 + q(x)y^2]\mathrm{d}x$$

Thus, the eigenvalue problem of the Sturm–Liouville type equation is equivalently transformed into the extremum problem of the functional.

17.5.2 *Inhomogeneous equations*

(1) *Poisson's equation*

$$\begin{cases} \Delta u = -\rho(\boldsymbol{r}) \\ u(\boldsymbol{r})|_\Sigma = f(\Sigma) \end{cases}$$

Poisson's equation is equivalent to the extremum problem of the functional $J[u]$ under the boundary conditions:

$$\delta J[u] = \iiint_\Omega [\Delta u + \rho(\boldsymbol{r})]\delta u \mathrm{d}V$$

By using Green's formula and the variation on the boundary $\delta u|_\Sigma = 0$, we obtain the functional:

$$J[u] = \iiint_\Omega \left[-\frac{1}{2}(\nabla u)^2 + \rho(\boldsymbol{r})u\right] \mathrm{d}V$$

Its Euler–Lagrange equation (in multiple variables) is the aforementioned Poisson's equation.

(2) *Sturm–Liouville equation*

$$\begin{cases} -\dfrac{\mathrm{d}}{\mathrm{d}x}\left[p(x)\dfrac{\mathrm{d}y}{\mathrm{d}x}\right] + q(x)y(x) = f(x) \\ y(x_0) = y_0, \quad y(x_1) = y_1 \end{cases}$$

This equation implies that the following functional $J[y]$ takes extremum,

$$\delta J[y(x)] = \int_{x_0}^{x_1} \left\{ -\frac{\mathrm{d}}{\mathrm{d}x}\left[p(x)\frac{\mathrm{d}y}{\mathrm{d}x}\right] + q(x)y(x) - f(x) \right\} \delta y(x)\mathrm{d}x$$

Since

$$\int_{x_0}^{x_1} q(x)y(x)\delta y(x)\mathrm{d}x = \frac{1}{2}\delta \int_{x_0}^{x_1} q(x)y^2(x)\mathrm{d}x$$

$$\int_{x_0}^{x_1} f(x)\delta y(x)\mathrm{d}x = \delta \int_{x_0}^{x_1} f(x)y(x)\mathrm{d}x$$

and

$$\int_{x_0}^{x_1} -\frac{\mathrm{d}}{\mathrm{d}x}\left[p(x)\frac{\mathrm{d}y}{\mathrm{d}x}\right] \delta y(x)\mathrm{d}x = \delta \int_{x_0}^{x_1} \frac{1}{2}p(x)\left(\frac{\mathrm{d}y}{\mathrm{d}x}\right)^2 \mathrm{d}x$$

the differential equation is a necessary condition for the extremum of the functional $J[y]$,

$$J[y(x)] = \int_{x_0}^{x_1} \left\{ \frac{1}{2}\left[p(x)\left(\frac{\mathrm{d}y}{\mathrm{d}x}\right)^2 - q(x)y^2(x)\right] + f(x)y(x) \right\} \mathrm{d}x$$

where $\delta y(x)|_{x_0} = \delta y(x)|_{x_1} = 0$.

Example 17.11. Derivation of the extremum problem of the functional and the Lagrangian for conservative force fields based on Newton's second law.

Solution: Let the potential function of the conservative force field be $V(q_j)$. Newton's second law states

$$m\frac{\mathrm{d}^2}{\mathrm{d}t^2}q_j(t) = -\frac{\partial}{\partial q_j}V(q_j)\ (j = 1, 2, 3)$$

It implies that the following functional takes extremum:

$$\delta S[q_j(t)] = -\int_{t_0}^{t_1} \sum_j \left[m\frac{\mathrm{d}^2}{\mathrm{d}t^2}q_j(t) + \frac{\partial}{\partial q_j}V(q_j)\right] \delta q_j, \mathrm{d}t$$

Performing integration by parts once and utilizing $\delta q_j|_{t_0} = \delta q_j|_{t_1} = 0$, we have

$$\delta S[q_j(t)] = -\sum_j m\dot{q}_j \delta q_j|_{t_0}^{t_1} + \int_{t_0}^{t_1} \left[\sum_j m\dot{q}_j \delta\dot{q}_j - \delta V(q_j)\right] \mathrm{d}t$$

$$= \delta \int_{t_0}^{t_1} \left[\sum_j \frac{1}{2} m\dot{q}_j^2 - V(q_j)\right] \mathrm{d}t$$

Here, the variation formula

$$\delta V(q_j) = \sum_j \frac{\partial}{\partial q_j} V(q_j)\delta q_j$$

is applied. Therefore, the Lagrangian of the system is

$$L = \sum_j \frac{1}{2} m\dot{q}_j^2 - V(q_j) = T - V$$

The functional $S[q_j(t)]$ represents the action in classical mechanics.

Exercises

[1] Express the eigenvalue problem as an extremum problem of a functional ($\beta \neq 0$):

$$\begin{cases} \Delta u + \lambda u = 0 \\ \left(\alpha u + \beta \dfrac{\partial u}{\partial n}\right)_\Sigma = 0 \end{cases}$$

Answer:

$$J[u] = \iiint_\Omega (\nabla u)^2 \mathrm{d}\tau + \beta \iint_\Sigma u^2 \mathrm{d}\sigma$$

[2] Prove that the following functional takes extremum, which is equivalent to the Poisson's equation,

$$J[u] = \iiint_\Omega \left[-\frac{1}{2}(\nabla u)^2 + \rho(\boldsymbol{r})u\right] \mathrm{d}V$$

17.6 Rayleigh–Ritz Approximation

By transforming the initial value problem of a differential equation into a variational problem, the variational method provides a new approach to approximating solutions of differential equations. One of these methods is called the Rayleigh–Ritz approximation. For Sturm–Liouville-type equations with three types of homogeneous boundary conditions, the eigenfunctions exhibit oscillatory behavior, with the ground state having no nodes. We can utilize this characteristic to seek approximate solutions. The following example illustrates this.

Example 17.12. Study the following initial value problem for a differential equation:

$$\begin{cases} -\dfrac{\mathrm{d}^2 y}{\mathrm{d}x^2} + ky = 1 \ (k > 0) \\ y(0) = y(1) = 0 \end{cases}$$

Solution: The solution to this differential equation problem is equivalent to finding the extremum of the functional $J[y]$ under the corresponding boundary conditions:

$$J[y] = \frac{1}{2}\int_0^1 [(y')^2 + ky^2 - 2y]\mathrm{d}x$$

Due to the homogeneous boundary conditions, the solution of the equation should possess symmetry: $y\left(x - \frac{1}{2}\right) = y\left(x + \frac{1}{2}\right)$. Therefore, the approximate solution can be chosen in the form of a series:

$$y = \alpha_1 x(1-x) + \alpha_2 x^2(1-x)^2 + \alpha_3 x^3(1-x)^3 + \cdots$$

If we consider a second-order approximate solution,

$$y = \alpha_1 x(1-x) + \alpha_2 x^2(1-x)^2$$

then we have

$$J[y] = \frac{1}{2}\int_0^1 \left[\left(\frac{\mathrm{d}y}{\mathrm{d}x}\right)^2 + ky^2 - 2y\right]\mathrm{d}x$$

$$= \left(\frac{1}{6} + \frac{k}{60}\right)\alpha_1^2 + \left(\frac{1}{15} + \frac{k}{140}\right)\alpha_1\alpha_2$$

$$+ \left(\frac{1}{105} + \frac{k}{1260}\right)\alpha_2^2 - \left(\frac{\alpha_1}{6} + \frac{\alpha_2}{30}\right)$$

The conditions for extremum are given by

$$\frac{\partial J[y]}{\partial \alpha_1} = \frac{\partial J[y]}{\partial \alpha_2} = 0$$

Solving these equations yields

$$\alpha_1 = \frac{14(k+36)}{k^2 + 112k + 1008}, \quad \alpha_2 = -\frac{42k}{k^2 + 112k + 1008}$$

The exact solution to this problem is

$$y(x) = \frac{1}{k}\left[1 - \frac{\cosh \sqrt{k}(x - 1/2)}{\cosh(\sqrt{k}/2)}\right]$$

Figure 17.11 shows the solid line representing the exact solution, while the circles represent the results of the Rayleigh–Ritz approximation (taking $k = 2$). It can be seen that the approximation is quite accurate.

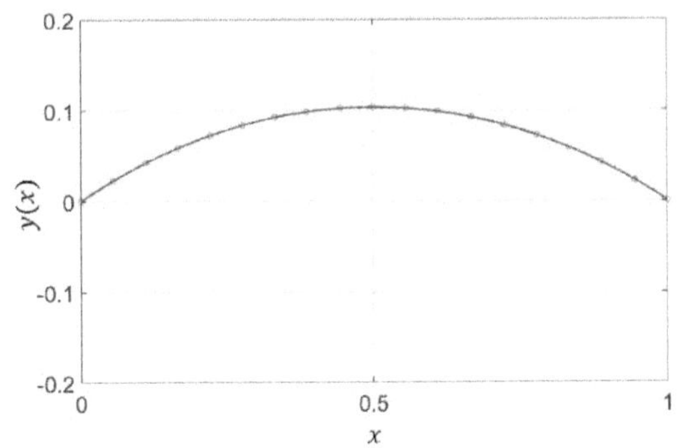

Figure 17.11.

Example 17.13. Finding the minimum eigenvalue of the equation:

$$\begin{cases} \dfrac{1}{x}\dfrac{\mathrm{d}}{\mathrm{d}x}\left(x\dfrac{\mathrm{d}y}{\mathrm{d}x}\right) + \lambda y = 0 \\ y(0)\,\text{is finite}, \quad y(1) = 0 \end{cases}$$

Solution: According to the discussion in the previous section, this eigenvalue problem is equivalent to finding the extremum of the functional:

$$J[y] = \int_0^1 x(y')^2 \mathrm{d}x$$

subject to the corresponding boundary conditions and the constraint,

$$J_1[y] = \int_0^1 xy^2 \mathrm{d}x = 1$$

Since the solution of the equation should possess even symmetry about $x = 0$, we assume it to be in the following series form:

$$y = \alpha_1(1 - x^2) + \alpha_2(1 - x^2)^2 + \alpha_3(1 - x^2)^3 + \cdots$$

Considering a second-order approximation,

$$J[y] = \int_0^1 x(y')^2 \mathrm{d}x = \alpha_1^2 + \frac{4}{3}\alpha_1\alpha_2 + \frac{2}{3}\alpha_2^2$$

The constraint is given by

$$J_1[y] = \int_0^1 xy^2 \mathrm{d}x = \frac{1}{6}\alpha_1^2 + \frac{1}{4}\alpha_1\alpha_2 + \frac{1}{10}\alpha_2^2 = 1$$

To obtain the extremum of the functional:

$$\frac{\partial(J[y] - \lambda J_1[y])}{\partial\alpha_1} = 2\alpha_1 + \frac{4}{3}\alpha_2 - \lambda\left(\frac{1}{3}\alpha_1 + \frac{1}{4}\alpha_2\right) = 0$$

$$\frac{\partial(J[y] - \lambda J_1[y])}{\partial\alpha_2} = \frac{4}{3}\alpha_1 + \frac{4}{3}\alpha_2 - \lambda\left(\frac{1}{4}\alpha_1 + \frac{1}{5}\alpha_2\right) = 0$$

The sufficient and necessary condition for the existence of non-zero solutions is

$$\begin{vmatrix} 2 - \frac{\lambda}{3} & \frac{4}{3} - \frac{\lambda}{4} \\ \frac{4}{3} - \frac{\lambda}{4} & \frac{4}{3} - \frac{\lambda}{5} \end{vmatrix} = 0 \longrightarrow 3\lambda^2 - 128\lambda + 640 = 0$$

Thus, we obtain

$$\lambda_1 = \frac{64}{3} - \frac{8}{3}\sqrt{34} \approx 5.7841\cdots$$

Solving for α_1 and α_2, we get

$$\alpha_1 \approx 1.650\ldots, \quad \alpha_2 \approx 1.053\cdots$$

The analytical solution to this equation is given by

$$R(\rho) = J_0(\sqrt{\lambda_n}\rho)$$

The minimum eigenvalue is $\lambda_1 = 2.4048^2 = 5.7831$, which is very close to the approximate result.

Note that the determinant equation has another solution λ_2, but it is not the second eigenvalue because the symmetry of the second eigenfunction is different from the ground state eigenfunction. If you want to find the second eigenvalue, you need to redefine the form of the variational function $y_1(x)$ and assume it to be orthogonal (weighted) to the lowest eigenfunction, i.e.,

$$\int_a^b y_1(x)y(x)\rho(x)\mathrm{d}x = 0$$

Please refer to relevant literature for detailed steps; they will not be elaborated here.

Example 17.14. Estimating the ground state energy of the harmonic oscillator Schrödinger equation.

Solution: The Schrödinger equation for the harmonic oscillator is given by

$$-\frac{\hbar^2}{2m}\frac{\mathrm{d}^2}{\mathrm{d}x^2}\psi + \frac{1}{2}kx^2\psi = Ey$$

Considering that $\psi \to 0$ as $x \to \pm\infty$, we choose the following trial wave function:

$$\psi = \mathrm{e}^{-\alpha x^2}, \quad \int_{-\infty}^{\infty} |\psi|^2 \mathrm{d}x = 1$$

The corresponding energy is defined as

$$E \equiv \lambda = \frac{\int_{-\infty}^{\infty} \psi^* \left(-\frac{\hbar^2}{2m}\frac{\mathrm{d}^2}{\mathrm{d}x^2} + \frac{1}{2}kx^2\right)\psi \mathrm{d}x}{\int_{-\infty}^{\infty} |\psi|^2 \mathrm{d}x} = \frac{\hbar^2\alpha}{2m} + \frac{k}{8\alpha}$$

We find the extremum of λ with respect to α:

$$\frac{\mathrm{d}\lambda}{\mathrm{d}\alpha} = 0 \longrightarrow \alpha = \frac{\sqrt{km}}{2\hbar}$$

Finally, we obtain the minimum value of the energy,

$$E_{\min} = \lambda = \frac{\hbar}{2}\sqrt{\frac{k}{m}} = \frac{1}{2}\hbar\omega$$

The result obtained from the variation calculation precisely matches the exact solution.

ENDNOTE

The Rayleigh-Ritz method is closely related to the variational principle in quantum mechanics. If H is the Hamiltonian of a system and E_0 is its ground state energy, then for any function $\psi(\boldsymbol{r})$, we have

$$\frac{\int \psi^*(\boldsymbol{r}) H \psi(\boldsymbol{r}) \mathrm{d}\boldsymbol{r}}{\int \psi^*(\boldsymbol{r})\psi(\boldsymbol{r}) \mathrm{d}\boldsymbol{r}} \geq E_0$$

The proof is straightforward. Assuming $\{\phi_n(\boldsymbol{r})\}_{n=1}^{\infty}$ is a complete set of eigenfunctions of H, we can expand the function $\psi(\boldsymbol{r})$ as

$$\psi(\boldsymbol{r}) = \sum_n a_n \phi_n(\boldsymbol{r})$$

Then we have

$$\frac{\int \psi^*(\boldsymbol{r})H\psi(\boldsymbol{r})\mathrm{d}\boldsymbol{r}}{\int \psi^*(\boldsymbol{r})\psi(\boldsymbol{r})\mathrm{d}\boldsymbol{r}} = \frac{\sum_n |a_n|^2 E_n}{\sum_n |a_n|^2} \geq E_0$$

Therefore, the Rayleigh-Ritz approximation method requires selecting a good variational function as the approximate ground state wave function. The selection of the trial variational function is mainly based on the following considerations:

(a) It must satisfy the boundary conditions.
(b) It should closely approximate the exact solution.
(c) The calculation of eigenvalues should be as simple as possible.

Exercises

[1] Use the Rayleigh–Ritz method to find the approximate solution for the differential equation:

$$\begin{cases} -y'' + ky = 1 \\ y(0) = y'(1) = 0 \end{cases}$$

Hint: Take the trial solutions as

$$\text{(a) } y = c_1 x(2-x) + c_2 x^2 (2-x)^2$$

$$\text{(b) } y = c_1 \sin\frac{\pi x}{2} + c_2 \sin\frac{3\pi x}{2}$$

[2] Use the Rayleigh-Ritz approximation method to find the minimum eigenvalue for the differential equation:

$$\begin{cases} y'' + \lambda y = 0 \\ y(-1) = y(1) = 0 \end{cases}$$

Hint: Take the trial functions as

$$\text{(a) } y = c_1(1-x^2) + c_2 x(1-x^2)$$
$$\text{(b) } y = c_1(1-x^2) + c_2 x^2(1-x^2)$$

[3] The Hamiltonian for a microscopic particle is given by

$$H = -\frac{\hbar^2}{2m}\frac{\mathrm{d}^2}{\mathrm{d}x^2} + V(x)$$

where the external potential well is given by

$$V(x) = -\mu\frac{\hbar^2\alpha^2}{m}\operatorname{sech}^2\alpha x$$

with $\mu, \alpha > 0$. Using the trial wave function $y = A\operatorname{sech}\beta x$, prove that

(a) When $\mu = 1$, there exists an exact solution with an average energy $\langle E\rangle$ equal to half the maximum depth of the potential well.
(b) When $\mu = 6$, the ground state binding energy $\langle E\rangle \geq \frac{10\hbar^2\alpha^2}{3m}$.
Hint: For $u, v \geq 0$, $\operatorname{sech} u \operatorname{sech} v \geq \sec h(u+v)$

[4] Find the extremal conditions satisfying the following eigenvalue equation for the functional $J[y]$:

$$\begin{cases} (1+x)\dfrac{\mathrm{d}^2y}{\mathrm{d}x^2} + (2+x)\dfrac{\mathrm{d}y}{\mathrm{d}x} + \lambda y = 0 \\ y(0) = 0, \quad y'(2) = 0 \end{cases}$$

(a) Using the trial function $y(x) = x\mathrm{e}^{-\frac{x}{2}}$, derive the approximate minimum eigenvalue λ_0.
(b) For what value of γ can you obtain a better estimation of λ_0 by using the trial function: $y(x) = x\mathrm{e}^{-\frac{x}{2}} + \beta\sin\gamma x$?

Hint: This differential equation is not self-adjoint, so it needs to be transformed into a self-adjoint equation.

Answer:

(a) $J[y] = \dfrac{\int_0^2 (1+x)\mathrm{e}^x(y')^2\mathrm{d}x}{\int_0^2 \mathrm{e}^x(y')^2\mathrm{d}x}, \quad \lambda_0 = \dfrac{3}{8}$; (b) $\gamma = \dfrac{\pi}{2}\left(n + \dfrac{1}{2}\right)$

Appendix

A.1 Table of Fourier Transform Functions

	Original Functions	Image Functions
1	$e^{-a^2x^2}$	$\frac{1}{2a\sqrt{\pi}}e^{-\frac{k^2}{4a^2}}$
2	$e^{-a\|x\|}$	$\frac{1}{\pi a}\frac{a^2}{a^2+k^2}$
3	$e^{-ax}H(x)$	$\frac{1}{2\pi}\frac{a-ik}{a^2+k^2}$
4	$e^{-a\|x\|}\text{sgn}x$	$-\frac{i}{\pi}\frac{k}{a^2+k^2}$
5	$\begin{cases}1\,(\|x\|<L)\\0\,(\|x\|>L)\end{cases}$	$\frac{1}{\pi}\frac{\sin kL}{k}$
6	$\begin{cases}1-\frac{\|x\|}{L}\,(\|x\|<L)\\0\quad(\|x\|>L)\end{cases}$	$\frac{L}{2\pi}\left[\frac{\sin(kL/2)}{kL/2}\right]^2$

(*Continued*)

	Original Functions	Image Functions
7	$\begin{cases} \frac{x}{L}\,(\lvert x\rvert < L) \\ 0\,(\lvert x\rvert > L) \end{cases}$	$\frac{\mathrm{i}}{\pi k}\left[\cos kL - \frac{\sin kL}{kL}\right]$
8	$\begin{cases} \frac{\lvert x\rvert}{L}\,(\lvert x\rvert < L) \\ 0\ (\lvert x\rvert > L) \end{cases}$	$\frac{L}{\pi}\left[\frac{\sin kL}{kL} - 2\left(\frac{\sin(kL/2)}{kL}\right)^2\right]$
9	$\mathrm{e}^{\pm \mathrm{i}a^2x^2}$	$\frac{1\pm\mathrm{i}}{2a\sqrt{2\pi}}\mathrm{e}^{\mp\frac{\mathrm{i}k^2}{4a^2}}$
10	$\frac{\sin k_0 x}{x}\,(k_0 > 0)$	$\begin{cases} \frac{1}{2}\,(\lvert k\rvert < k_0) \\ 0\ (\lvert k\rvert > k_0) \end{cases}$
11	$\frac{1}{\lvert x\rvert}\,(x \neq 0)$	$\frac{1}{\lvert k\rvert}\ (k \neq 0)$
12	$\frac{1}{\lvert x\rvert^a}\,(0 < \mathrm{Re}\,a < 1)$	$\frac{\sin(a\pi/2)}{\pi}\frac{\Gamma(1-a)}{\lvert k\rvert^{1-a}}$
13	$\frac{\mathrm{sech}\,ax}{\mathrm{sech}\,\pi x}\,(-\pi < a < \pi)$	$\frac{1}{2\pi}\frac{\sin a}{\cosh k + \cos a}$
14	$\frac{\cosh ax}{\cosh \pi x}\,(-\pi < a < \pi)$	$\frac{1}{\pi}\frac{\cos(a/2)\cos(k/2)}{\cosh k + \cos a}$
15	$\mathrm{H}(x)$	$\frac{1}{2\pi}\left[\pi\delta(k) - \frac{\mathrm{i}}{k}\right]$
16	$\mathrm{e}^{\mathrm{i}k_0x}$	$\delta(k-k_0)$
17	$\sum_{n=-\infty}^{\infty}\delta(x - nx_0)$	$\sum_{n=-\infty}^{\infty}\frac{1}{x_0}\delta\left(k - n\frac{2\pi}{x_0}\right)$

(*Continued*)

	Original Functions	Image Functions
18	$\cos(a\sin k_0x+bx)$	$\frac{1}{2}\sum_{n=-\infty}^{\infty}\mathrm{J}_n(a)\big[\delta(k-b-nk_0)$ $+\,\delta(k+b+nk_0)\big]$
19	$\cos(a\cos k_0x+bx)$	$\frac{1}{2}\sum_{n=-\infty}^{\infty}\mathrm{J}_n(a)\big[\mathrm{i}^n\delta(k-b-nk_0)$ $+\,(-\mathrm{i})^n\delta(k+b+nk_0)\big]$
20	$\sin(a\sin k_0x+bx)$	$\frac{\mathrm{i}}{2}\sum_{n=-\infty}^{\infty}\mathrm{J}_n(a)\big[-\delta(k-b-nk_0)$ $+\,\delta(k+b+nk_0)\big]$
21	$\sin(a\cos k_0x+bx)$	$\frac{\mathrm{i}}{2}\sum_{n=-\infty}^{\infty}\mathrm{J}_n(a)\big[-\mathrm{i}^n\delta(k-b-nk_0)$ $+\,(-\mathrm{i})^n\delta(k+b+nk_0)\big]$
22	$\mathrm{e}^{-a\cos k_0x}$	$\sum_{n=-\infty}^{\infty}(-1)^n\mathrm{J}_n(a)\delta(k-nk_0)$
23	$\mathrm{e}^{-a\sin k_0x}$	$\sum_{n=-\infty}^{\infty}\mathrm{i}^n\mathrm{J}_n(a)\delta(k-nk_0)$

(*Continued*)

A.2 Table of Laplace Transform Functions

	Original Functions	Image Functions
1	$t^a\ (a > -1)$	$\dfrac{\Gamma(a+1)}{p^{a+1}}$
2	$e^{-\lambda t}$	$\dfrac{1}{p+\lambda}$
3	$\sin(\omega t + \alpha)$	$\dfrac{\omega\cos\alpha + p\sin\alpha}{p^2+\omega^2}$
4	$\cos(\omega t + \alpha)$	$\dfrac{p\cos\alpha - \omega\sin\alpha}{p^2+\omega^2}$
5	$t^n \sin\omega t$	$n!\dfrac{\mathrm{Im}(p+i\omega)^{n+1}}{(p^2+\omega^2)^{n+1}}$
6	$t^n \cos\omega t$	$n!\dfrac{\mathrm{Re}(p+i\omega)^{n+1}}{(p^2+\omega^2)^{n+1}}$
7	$\sinh\omega t$	$\dfrac{\omega}{p^2-\omega^2}$
8	$\cosh\omega t$	$\dfrac{p}{p^2-\omega^2}$
9	$\dfrac{e^{bt}-e^{at}}{t}$	$\ln\dfrac{p-a}{p-b}$
10	$\dfrac{1}{\sqrt{\pi t}}$	$\dfrac{1}{\sqrt{p}}$
11	$\dfrac{1}{\sqrt{\pi t}}e^{-\frac{a^2}{4t}}$	$\dfrac{e^{-a\sqrt{p}}}{\sqrt{p}}$
12	$\dfrac{1}{\sqrt{\pi t}}e^{-2a\sqrt{t}}$	$\dfrac{1}{\sqrt{p}}e^{-\frac{a^2}{p}}\operatorname{erfc}\left(\dfrac{a}{\sqrt{p}}\right)$

(*Continued*)

	Original Functions	Image Functions
13	$e^{-a^2t^2}$	$\frac{\sqrt{\pi}}{2}e^{\frac{p^2}{4a^2}}\operatorname{erfc}\left(\frac{p}{2a}\right)$
14	$\frac{1}{\sqrt{\pi a}}\sin(2\sqrt{at})$	$\frac{1}{p\sqrt{p}}e^{-\frac{a}{p}}$
15	$\frac{1}{\sqrt{\pi a}}\cos(2\sqrt{at})$	$\frac{1}{\sqrt{p}}e^{-\frac{a}{p}}$
16	$\operatorname{erf}(\sqrt{at})$	$\frac{\sqrt{a}}{p\sqrt{p+a}}$
17	$\operatorname{erfc}\left(\frac{a}{2\sqrt{t}}\right)$	$\frac{1}{p}e^{-a\sqrt{p}}$
18	$\frac{1}{1\pm t}$	$\mp e^{\pm p}\operatorname{Ei}(\mp p)$
19	$\frac{1}{1+t^2}$	$\operatorname{Ci}(p)\sin p - \operatorname{Si}(p)\cos p$
20	$\frac{1}{\sqrt{1+t}}$	$\sqrt{\frac{\pi}{p}}e^{p}\operatorname{erfc}(\sqrt{p})$
21	$J_\nu(at)$	$\frac{(\sqrt{p^2+a^2}+p)^{-\nu}}{a^\nu\sqrt{p^2+a^2}}$
22	$\frac{J_\nu(at)}{t}$	$\frac{1}{\nu a^\nu}(\sqrt{p^2+a^2}+p)^{-\nu}$
23	$I_\nu(at)$	$\frac{(\sqrt{p^2-a^2}+p)^{-\nu}}{a^\nu\sqrt{p^2-a^2}}$
24	$\lambda^\nu e^{-\lambda t}I_\nu(\lambda t)$	$\frac{[\sqrt{p^2+2\lambda p}-(p+\lambda)]^\nu}{\sqrt{p^2+2\lambda p}}$

(*Continued*)

	Original Functions	Image Functions
25	$t^{\nu}\mathrm{J}_{\nu}(t)\ \left(\nu > -\dfrac{1}{2}\right)$	$\dfrac{2^{\nu}\Gamma\left(\nu+\dfrac{1}{2}\right)}{\sqrt{\pi}(p^2+1)^{\nu+1/2}}$
26	$t^{\nu/2}\mathrm{J}_{\nu}(2\sqrt{t})$	$\dfrac{1}{p^{\nu+1}}\mathrm{e}^{-1/p}$
27	$\mathrm{J}_0(a\sqrt{t^2-\tau^2})\mathrm{H}(t-\tau)$	$\dfrac{1}{\sqrt{p^2+a^2}}\mathrm{e}^{-\tau\sqrt{p^2+a^2}}$
28	$\dfrac{\mathrm{J}_1(a\sqrt{t^2-\tau^2})}{\sqrt{t^2-\tau^2}}\mathrm{H}(t-\tau)$	$\dfrac{\mathrm{e}^{-\tau p}-\mathrm{e}^{-\tau\sqrt{p^2+a^2}}}{a\tau}$
29	$\dfrac{1}{\sqrt{\pi t}}\sin\left(\dfrac{1}{2t}\right)$	$\dfrac{1}{\sqrt{p}}\mathrm{e}^{-\sqrt{p}}\sin\sqrt{p}$
30	$\dfrac{1}{\sqrt{\pi t}}\cos\left(\dfrac{1}{2t}\right)$	$\dfrac{1}{\sqrt{p}}\mathrm{e}^{-\sqrt{p}}\cos\sqrt{p}$

(1) The error function is defined by

$$\mathrm{erf}(x) = \frac{2}{\sqrt{\pi}}\int_0^x \mathrm{e}^{-t^2}\mathrm{d}t, \quad \mathrm{erfc}\,(x) = 1 - \mathrm{erf}\,(x)$$

(2) $\mathrm{Si}(x)$ and $\mathrm{Ci}(x)$ are respectively known as sine and cosine integral functions, which are defined by

$$\mathrm{Si}(x) = \int_0^x \frac{\sin t}{t}\mathrm{d}t, \quad \mathrm{Ci}(x) = -\int_x^{\infty}\frac{\cos t}{t}\mathrm{d}t$$

(3) $\mathrm{Ei}(x)$ is known as exponential function, with $x \neq 0$, and is defined by

$$\mathrm{Ei}(x) = -\int_{-x}^{\infty}\frac{\mathrm{e}^{-t}}{t}\mathrm{d}t = \int_{-\infty}^{x}\frac{\mathrm{e}^{t}}{t}\mathrm{d}t$$

A.3 Table of Z-Transform Functions

	Original Sequences	Image Functions
1	$\delta(k)$	1
2	$\delta(k-m)\ (m \geq 0)$	z^{-m}
3	$\mathrm{H}(k-m)$	$\frac{z}{z-1} \cdot z^{-m}$
4	k	$\frac{z}{(z-1)^2}$
5	k^2	$\frac{z(z+1)}{(z-1)^3}$
6	k^3	$\frac{z(z^2+4z+1)}{(z-1)^4}$
7	a^k	$\frac{z}{z-a}$
8	ka^k	$\frac{az}{(z-a)^2}$
9	$k^2 a^k$	$\frac{az(z+a)}{(z-a)^3}$
10	$k^3 a^k$	$\frac{az(z^2+4az+a^2)}{(z-a)^4}$
11	$\frac{k(k-1)\cdots(k-m+1)}{m!}$	$\frac{z}{(z-a)^{m+1}}$
12	$\sin \beta k$	$\frac{z \sin \beta}{z^2 - 2z\cos\beta + 1}$
13	$\cos \beta k$	$\frac{z(z-\cos\beta)}{z^2 - 2z\cos\beta + 1}$

(*Continued*)

	Original Sequences	Image Functions
14	$a^k \sin\beta k$	$\dfrac{az\sin\beta}{z^2-2az\cos\beta+a^2}$
15	$a^k \cos\beta k$	$\dfrac{z(z-a\cos\beta)}{z^2-2az\cos\beta+a^2}$
16	$\dfrac{1}{k}a^k\ (k>0)$	$\ln\dfrac{z}{z-a}$
17	$\dfrac{1}{k!}a^k$	$e^{\frac{a}{z}}$
18	$\dfrac{1}{k+1}$	$z\ln\dfrac{z}{z+1}$
19	$\dfrac{1}{2k+1}$	$\sqrt{z}\arctan\sqrt{\frac{1}{z}}$
20	$\dfrac{1}{(2k)!}$	$\cosh\sqrt{\frac{1}{z}}$

A.4 Table of Green's Functions

Differential Operators	Green's Functions
∂_t^{n+1}	$\dfrac{t^n}{n!}\mathrm{H}(t)$
$\partial_t+\gamma$	$e^{-\gamma t}\mathrm{H}(t)$
$(\partial_t+\gamma)^2$	$te^{-\gamma t}\mathrm{H}(t)$
$\partial_t^2+2\gamma\partial_t+\omega_0^2$	$\dfrac{\sin\omega t}{\omega}e^{-\gamma t}\mathrm{H}(t),\ \omega=\sqrt{\omega_0^2-\gamma^2}$

(Continued)

Differential Operators	Green's Functions
$\Delta_{2D} = \partial_x^2 + \partial_y^2$	$\frac{1}{2\pi} \ln \rho$
$\Delta_{3D} = \partial_x^2 + \partial_y^2 + \partial_z^2$	$-\frac{1}{2\pi r}$
$\Delta_{3D} + k^2$	$-\frac{e^{-ikr}}{2\pi r}$
$\Delta_{nD} - k^2$	$-(2\pi)^{-n/2} \left(\frac{k}{r}\right)^{n/2-1} K_{n/2-1}(kr)$
$\partial_t - k\partial_x^2$	$\left(\frac{1}{4\pi kt}\right)^{1/2} e^{-\frac{x^2}{4kt}} H(t)$
$\partial_t - k\Delta_{2D}$	$\frac{1}{4\pi kt} e^{-\frac{\rho^2}{4kt}} H(t)$
$\partial_t - k\Delta_{3D}$	$\left(\frac{1}{4\pi kt}\right)^{3/2} e^{-\frac{r^2}{4kt}} H(t)$
$\partial_t^2 - c^2\partial_x^2$	$\frac{1}{2c} H\left(t - \left\|\frac{x}{c}\right\|\right)$
$\partial_t^2 - c^2\Delta_{2D}$	$\frac{1}{2\pi c\sqrt{c^2t^2 - \rho^2}} H\left(t - \frac{\rho}{c}\right)$
$\frac{1}{c^2}\partial_t^2 - \Delta_{3D}$	$\frac{\delta(t - r/c)}{4\pi r}$
$\frac{1}{c^2}\partial_t^2 - \partial_x^2 + \mu^2$	$\frac{1}{2}\Big[(1 - \sin \mu ct)\left[\delta(ct - x) + \delta(ct + x)\right] + \mu H(ct - \|x\|) J_0(\mu\sqrt{c^2t^2 - x^2})\Big]$

(*Continued*)

Differential Operators	Green's Functions
$\frac{1}{c^2}\partial_t^2 - \Delta_{2D} + \mu^2$	$\frac{1}{4\pi}\left[(1+\cos\mu ct)\frac{\delta(ct-\rho)}{\rho} + \mu^2 \mathrm{H}(ct-\rho)\sin c(\mu\sqrt{c^2t^2-\rho^2})\right]$
$\frac{1}{c^2}\partial_t^2 - \Delta_{3D} + \mu^2$	$\frac{1}{4\pi}\left[\frac{\delta(t-r/c)}{r} + \mu c\mathrm{H}(ct-r)\frac{\mathrm{J}_1(\mu u)}{u}\right]$
$\partial_t^2 + 2\gamma\partial_t - c^2\partial_x^2$	$\frac{1}{2}\mathrm{e}^{-\gamma t}\left[\delta(ct-x)+\delta(ct+x)+H(ct-\lvert x\rvert)\left(\frac{\gamma}{c}\mathrm{I}_0\left(\frac{\gamma u}{c}\right)+\frac{\gamma t}{u}\mathrm{I}_1\left(\frac{\gamma u}{c}\right)\right)\right]$
$\partial_t^2 + 2\gamma\partial_t - c^2\Delta_{2D}$	$\frac{\mathrm{e}^{-\gamma t}}{4\pi}\left[\left(1+\mathrm{e}^{-\gamma t}+3\gamma t\right)\frac{\delta(ct-\rho)}{\rho} + \mathrm{H}(ct-\rho)\left(\frac{\gamma\sinh\frac{\gamma u}{c}}{cu}+\frac{3\gamma t\cosh\frac{\gamma u}{c}}{u^2}-\frac{3ct\sinh\frac{\gamma u}{c}}{u^3}\right)\right]$
$\partial_t^2 + 2\gamma\partial_t - c^2\Delta_{3D}$	$\frac{\mathrm{e}^{-\gamma t}}{20\pi}\left[\left(8-3\mathrm{e}^{-\gamma t}+2\gamma t+4\gamma^2t^2\right)\frac{\delta(ct-r)}{r^2} + \frac{\gamma^2}{c}\mathrm{H}(ct-r)\left(\frac{\gamma}{cu}\mathrm{I}_1\left(\frac{\gamma u}{c}\right)+\frac{4t}{u^2}\mathrm{I}_2\left(\frac{\gamma u}{c}\right)\right)\right]$

where

$$u = \begin{cases} \sqrt{c^2t^2 - x^2} & (\text{for 1D}) \\ \sqrt{c^2t^2-\rho^2} & (\text{for 2D}) \\ \sqrt{c^2t^2-r^2} & (\text{for 3D}) \end{cases}$$

Name Index

Subject Index

R

S

T

U

V

W

Z

www.ingramcontent.com/pod-product-compliance
Lightning Source LLC
LaVergne TN
LVHW020753231125
825833LV00002B/121
* 9 7 8 9 8 1 9 8 1 3 5 5 1 *